TEXT-BOOK OF ALGEBRA

WITH EXERCISES

FOR

SECONDARY SCHOOLS AND COLLEGES

BY

GEORGE EGBERT FISHER, M.A., Ph.D.

AND

ISAAC J. SCHWATT, Ph.D.

ASSISTANT PROFESSORS OF MATHEMATICS IN THE
UNIVERSITY OF PENNSYLVANIA

PART I

PHILADELPHIA
FISHER AND SCHWATT
1898

Norwood Press
J. S. Cushing & Co. — Berwick & Smith
Norwood Mass. U.S.A.

PREFACE.

In preparing this text-book the authors have assumed that mental discipline is of the first importance to every student of mathematics. They have therefore endeavored to present the elements of Algebra in a clear and logical form. Yet the needs of beginners in the subject have been kept constantly in mind.

While not minimizing the importance of facility and accuracy in performing algebraic operations, special attention has been paid to making clear to the beginner the reason for every step taken. Each principle has been first illustrated by particular examples, thus preparing the mind of the student to grasp the meaning of a formal statement of the principle and its proof. Rules and suggestions for performing the different operations have been given after these operations have been illustrated by examples.

The authors have endeavored to avoid apparent conciseness at the expense of clearness and accuracy. The full treatment of each topic, the great amount of illustrative matter given to make clear the formal statements and proofs of principles, the numerous examples which have been worked in the text, and the great number of exercises have made the book larger than the ordinary text-book covering the same topics. But the subject matter of the book has been printed in two sizes of type. The matter in smaller type consists of the formal proofs of principles, of the more difficult portions of each topic treated, and of examples whose solutions depend upon principles printed in the smaller type.

The matter given in the larger type is logically complete (except for the proofs of principles), and can be taken up as a first course in the subject. Certain portions of this matter can also be omitted by the beginner at the option of the teacher. Much of Chapter II., which is prepared especially for the teachers, can be omitted on first reading.

The attention of teachers is especially invited to the following features of the book:

The introductory chapter and the development in Chapter II. of the fundamental operations with algebraic numbers. The concrete illustrations of these operations.

The use of type-forms in multiplication and division (Chapter VI.), and in factoring (Chapter VIII.).

The application of factoring to the solution of equations (Chapters VIII. and XX.). By the early introduction of this method it has been possible to give problems which lead to quadratic equations before the formal treatment of that topic.

The solutions of equations based upon equivalent equations and equivalent systems of equations (Chapter IV., etc.). This method is of extreme importance, even to the beginner. The ordinary way of treating equations is illogical, leads to many serious errors, and is therefore also pedagogically wrong.

The treatment of irrational equations (Chapter XXII.).

The special suggestions given in the first chapter on problems (Chapter V.), and applied subsequently, to assist the student in acquiring facility in translating the verbal language of the problem into the symbolic language of the equation.

The discussion of general problems (Chapter XI.), and the interpretation of positive, negative, zero, indeterminate, and infinite solutions of problems (Chapter XII.).

The more than usually full treatment of inequalities and their applications (Chapter XVI.).

The outline of a projected discussion of irrational numbers (Chapter XVII.).

The brief introduction to imaginary and complex numbers (Chapter XIX.).

The great number of graded examples and problems. This unusual number of exercises has been given in order that the teacher from year to year may have variety with different classes.

Errors in the text and in the exercises may have been overlooked. Any suggestions from teachers and students with respect either to errors in the text and exercises, or to the mode of presenting the subject, will be highly appreciated.

The authors take pleasure in acknowledging their indebtedness to Professor William Hoover, of the University of Ohio; to Professor T. F. Leighton, of the Hyde Park High School, Chicago; to Miss Clara J. Hendley, of the Girls' High School, Philadelphia; to Professor George Q. Sheppard, of the Hill School, Pottstown, for critical reading of the manuscript and for helpful suggestions; and especially to Dr. George Bruce Halsted, Professor of Mathematics in the University of Texas, for many valuable and critical suggestions.

The authors also desire to thank their colleagues in the mathematical department of the University of Pennsylvania, and those teachers in secondary schools and colleges who have been kind enough to read the manuscript in whole or in part.

Also Messrs. J. S. Cushing & Co., of the Norwood Press, for the typographical excellence of the book.

G. E. F.

I. J. S.

University of Pennsylvania,
Philadelphia, August, 1898.

TABLE OF CONTENTS.

CHAPTER I.

INTRODUCTION.

CHAPTER II.

THE FOUR FUNDAMENTAL OPERATIONS WITH ALGEBRAIC NUMBERS.

CHAPTER III.

The Fundamental Operations with Integral Algebraic Expressions.

CHAPTER IV.

Integral Algebraic Equations.

CHAPTER V.

Problems.

CHAPTER VI.

Type-Forms.

CHAPTER VII.

PARENTHESES.

CHAPTER VIII.

FACTORS AND MULTIPLES OF INTEGRAL ALGEBRAIC EXPRESSIONS.

CHAPTER IX.

FRACTIONS.

CHAPTER X.

Fractional Equations in One Unknown Number.

CHAPTER XI.

Literal Equations in One Unknown Number.

CHAPTER XII.

Interpretation of the Solutions of Problems.

CHAPTER XIII.

Simultaneous Linear Equations.

CHAPTER XIV.

Problems which lead to Simultaneous Linear Equations.

CHAPTER XV.

Evolution.

CHAPTER XVI.

Inequalities.

CHAPTER XVII.

Irrational Numbers.

CHAPTER XVIII.

Surds.

CHAPTER XIX.

IMAGINARY AND COMPLEX NUMBERS.

CHAPTER XX.

QUADRATIC EQUATIONS.

CHAPTER XXI.

EQUATIONS OF HIGHER DEGREE THAN THE SECOND

CHAPTER XXII.

IRRATIONAL EQUATIONS.

CHAPTER XXIII.

SIMULTANEOUS QUADRATIC AND HIGHER EQUATIONS.

CHAPTER XXIV.

Ratio, Proportion, and Variation.

CHAPTER XXV.

Doctrine of Exponents.

CHAPTER XXVI.

Progressions.

CHAPTER XXVII.

The Binomial Theorem for Positive Integral Exponents.

CHAPTER I.

INTRODUCTION.

Algebra, like Arithmetic, treats of number. But the meaning of number, and the mode of representing it, are extended in passing from ordinary Arithmetic to Algebra.

§ 1. GENERAL NUMBER.

1. In ordinary Arithmetic all numbers have particular values and are represented by definite symbols, the Arabic numerals, 1, 2, 3, etc. The symbol 7, for instance, stands for a group of *seven* units.

In Algebra, however, such symbols as a, b, x, y, are used to represent numbers which may have *any values whatever,* or numbers whose values are, as yet, *unknown.*

Just as we speak of 10 miles, of 95 dollars, etc., in Arithmetic; so in Algebra we speak of a miles, meaning *any number* of miles or *an unknown number* of miles; of x dollars, meaning *any number* or *an unknown number* of dollars, etc.

For the sake of brevity, we shall say *the number a,* or simply a, meaning thereby *the number denoted by the symbol a.*

2. The symbols of Arithmetic, 1, 2, 3, etc., are retained in Algebra with their exact arithmetical meanings. The numbers represented by letters are, for the sake of distinction, called *literal* or *general numbers.* Other symbols than letters might be used to represent general numbers, but letters are more convenient to write and to pronounce.

3. The operations of Addition, Subtraction, Multiplication, and Division are denoted by the same symbols in Algebra as in Arithmetic.

4. The symbol of **Addition,** $+$, read **plus,** is placed between two numbers to indicate that the number on its right is to be added to the number on its left.

E.g., just as $5+3$, read *five plus three,* means that 3 is to be added to 5; so $a+b$, read *a plus b,* means that b is to be added to a. Just as $5+3+2$, read *five plus three plus two,* means that 3 is to be added to 5 and then 2 is to be added to that result; so $x+y+z$, read *x plus y plus z,* means that y is to be added to x and then z is to be added to that result.

As in Arithmetic, so in Algebra, the result of adding one number to another is called the **Sum.**

5. The **Symbol of Subtraction,** $-$, read **minus,** is placed between two numbers to indicate that the number on its right is to be subtracted from the number on its left.

E.g., just as $5-3$, read *five minus three,* means that 3 is to be subtracted from 5; so $a-b$, read *a minus b,* means that b is to be subtracted from a. Just as $18-5-6$, read *eighteen minus five minus six,* means that 5 is to be subtracted from 18 and then 6 is to be subtracted from that result; so $a-b-c$, read *a minus b minus c,* means that b is to be subtracted from a and then c is to be subtracted from that result.

The result of subtracting one number from another is called the **Remainder.**

6. The **Symbol of Multiplication,** $\times$, read **multiplied by,** or **times,** means that the number on its left is to be multiplied by the number on its right.

E.g., just as 5×3, read *five multiplied by three,* or *three times five,* means that 5 is to be multiplied by 3; so $a\times b$, read *a multiplied by b,* or *b times a,* means that a is to be multiplied by b. Just as $5\times 3\times 7$, read *five multiplied by three multiplied by seven,* means that 5 is to be multiplied by 3 and then that result is to be multiplied by 7; so $x\times y\times z$, read *x multiplied by y multiplied by z,* means that x is to be multiplied by y and then that result is to be multiplied by z.

The result of multiplying one number by another is called the **Product.**

A dot ($\cdot$) is frequently used, instead of the symbol $\times$, to denote multiplication. Thus, the product of a by b is indicated by $a \times b$, or by $a \cdot b$.

The symbol of multiplication between two literal numbers, or one literal number and an Arabic numeral, is frequently omitted.

E.g., the product $x \times y \times z$, or $x \cdot y \cdot z$, is frequently written, xyz; and the product $6 \times u \times v$, or $6 \cdot u \cdot v$, is frequently written, $6\,uv$.

But the symbol of multiplication between two numerals cannot be omitted without changing the meaning.

E.g., if in the indicated multiplication, 3×6, or $3 \cdot 6$, the symbol, $\times$, or $\cdot$, were omitted, we should have 36, not 18. It is better not to use the dot between numerals to denote multiplication, since it might be mistaken for the decimal point.

In like manner, in reading a product, the words *multiplied by*, or *times*, are frequently omitted. Thus, in reading $6\,ab$, instead of saying *six multiplied by a multiplied by b*, we frequently say *six a b*.

7. The Symbol of Division, $\div$, read **divided by**, is placed between two numbers to indicate that the number on its left is to be divided by the number on its right.

E.g., just as $10 \div 5$, read *ten divided by five*, means that 10 is to be divided by 5; so $a \div b$, read *a divided by b*, means that a is to be divided by b. Just as $10 \div 5 \div 3$, read *ten divided by five divided by three*, means that 10 is to be divided by 5 and then that result is to be divided by 3; so $x \div y \div z$, read *x divided by y divided by z*, means that x is to be divided by y and then that result is to be divided by z.

The result of dividing one number by another is called the **Quotient**.

8. The use of letters to represent general numbers may be illustrated by a few simple examples.

Ex. 1. If a boy has 3 books and is given 2 more, he has $3 + 2$ books. If he has a books and is given 5 more, he has $a + 5$ books. If he has 7 books and is given b more, he has $7 + b$ books. If he has m books and is given n more, he has $m + n$ books.

Ex. 2. If a boy has 5 oranges and gives away 2, he has left $5 - 2$ oranges. If he has p oranges and gives away 7, he has left $p - 7$ oranges. If he has 11 oranges and gives away q, he has left $11 - q$ oranges. If he has u oranges and gives away v, he has left $u - v$ oranges.

Ex. 3. If a man has 100 dollars, pays out 20 dollars, and then receives 30 dollars, he has finally $100 - 20 + 30$ dollars. If he has a dollars, pays out b dollars, and then receives c dollars, he has finally $a - b + c$ dollars.

Ex. 4. If a man buys 5 city lots at 120 dollars each, he pays 120×5 dollars for the lots. If he buys a lots at 150 dollars each, he pays $150 \times a$, or $150a$, dollars for the lots. If he buys 6 lots at b dollars each, he pays $b \times 6$, or $b6$ dollars, for the lots. If he buys u lots at v dollars each, he pays $v \times u$, or vu, dollars for the lots.

Ex. 5. If a train runs 60 miles in 2 hours, it runs $60 \div 2$ miles in 1 hour. If it runs a miles in 5 hours, it runs $a \div 5$ miles in 1 hour. If it runs 150 miles in b hours, it runs $150 \div b$ miles in 1 hour. If it runs p miles in q hours, it runs $p \div q$ miles in 1 hour.

Ex. 6. If a train runs 100 miles in 4 hours, how many miles will it run in 5 hours at the same rate? It will run $100 \div 4$ miles in 1 hour, and therefore in 5 hours it will run $100 \div 4 \times 5$ miles. If it runs m miles in h hours, how many miles will it run in k hours at the same rate? It will run $m \div h$ miles in 1 hour, and therefore in k hours it will run $m \div h \times k$ miles.

Ex. 7. If a pupil buys 2 note books at 10 cents each and 3 note books at 12 cents each, how much does he pay for all? He pays $10 \times 2 + 12 \times 3$ cents. If he buys a note books at m cents each and b note books at n cents each, how much does he pay for all? He pays $m \times a + n \times b$, or $ma + nb$, cents.

Ex. 8. If A earns 18 dollars in 6 days and B earns 12 dollars in 5 days, how much more does A earn than B in 1 day?

A earns $18 \div 6$ dollars in 1 day and B earns $12 \div 5$ dollars in 1 day; therefore A earns $18 \div 6 - 12 \div 5$ dollars more than B in 1 day.

If A earns a dollars in m days and B earns b dollars in n days, how much more does A earn than B in 1 day? A earns $a \div m$ dollars in 1 day and B earns $b \div n$ dollars in 1 day; therefore A earns $a \div m - b \div n$ dollars more than B in 1 day.

Ex. 9. If, in a number of *two* digits, the digit in the *units'* place is 3 and the digit in the *tens'* place is 5, the number is $10 \times 5 + 3$.

If the digit in the units' place is a and the digit in the tens' place is b, the number is $10 \times b + a$, or $10b + a$.

9. Observe that in the preceding examples the reasoning is the same whether the numbers are represented by letters or by Arabic numerals.

The results of these operations are *numbers* in all cases, whether letters or numerals, or both, are involved.

Thus, the result of adding b to a, $a + b$, is a number, just as $5 + 3$, or 8, is a number.

Likewise, $a + b - c$, $ab - cd$, $3a - 5b$, $a \div b + a \div d$, etc., are numbers, *expressed by means of the signs and symbols of Algebra.*

EXERCISES I.

1. What number exceeds 7 by 3? What number exceeds 5 by a? What number exceeds x by 4? What number exceeds m by n?

2. If 5 pupils join a class containing 20 pupils, how many pupils are then in the class? If 7 pupils join a class containing m pupils, how many pupils are then in the class? If n pupils join a class containing p pupils, how many pupils are then in the class?

3. A boy made three copy books; for the first he used a sheets of paper, for the second b sheets, for the third c sheets. How many sheets did he use altogether?

4. The width of a room is a feet, and the length is b feet more than the width. What is the length of the room?

5. A man is now n years old. How old will he be in 20 years? How old in m years?

6. What number is less than 15 by 8? Less than a by 9? Less than 11 by b? Less than m by n?

7. A has $1000, and B has $200 less; how many dollars has B? If A has m dollars and B has n dollars less, how many dollars has B?

8. A text-book consisting of two parts contains 560 pages; if the first part contains 300 pages, how many pages does the second part contain? If both parts contain n pages, and the first part contains 250 pages, how many pages are there in the second part? If both parts contain k pages and the second part contains q pages, how many pages are there in the first part?

9. A number N is divided into two unequal parts, the greater of which is 6; what is the less? If the less is a, what is the greater?

10. What number added to 16 gives 25? What number added to m gives n?

11. What number subtracted from 8 gives 5? What number subtracted from p gives q?

12. If the minuend is 10 and the remainder is 6, what is the subtrahend? If the minuend is a and the remainder is b, what is the subtrahend?

13. If the subtrahend is 8 and the remainder is 7, what is the minuend? If the subtrahend is x and the remainder is y, what is the minuend?

14. What is the integer next less than 8? Next greater than 8? If a is an integer, what is the next less integer? The next greater?

15. A man is n years old; how old was he 5 years ago? How old was he m years ago? How long must he live to be 90 years old? How long to be p years old?

16. The middle of three consecutive integers is 7; what are the first and third? If the middle integer is a, what are the first and third?

17. What are the two *even* numbers nearest to 6, one greater and the other less than 6?

18. If m is an *even* number, what are the two nearest *even* numbers, one greater and the other less than m? The two nearest *odd* numbers, one greater and the other less than m?

19. What are the two *even* numbers nearest to 9, one greater and the other less than 9?

20. If m is an odd number, what are the two nearest *even* numbers, one greater and the other less than m? The two nearest *odd* numbers, one greater and the other less than m?

21. If 1 pound of tea costs 75 cents, how much do 3 pounds cost? If 1 pound costs 75 cents, how much do n pounds cost? If 1 pound costs a cents, how much do b pounds cost?

22. If 3 men can do a piece of work in 8 hours, in how many hours can 1 man do the work? If a men can do a piece of work in 9 hours, in how many hours can 1 man do the work? If b men can do a piece of work in h hours, in how many hours can 1 man do the work?

23. 10×2, 10×3, etc., are *particular* multiples of 10; express *any* multiple of 10.

24. Write a number containing a units and b tens.

25. Write a number containing a units, b tens, and c hundreds.

26. An edition of a book consists of a copies, each containing b pages; on each page are c lines, and each line contains d letters. How many letters are there in the whole edition?

27. The speed of sound is 1100 feet per second. What is the distance of a cloud, if the thunder is heard 3 seconds after the flash of lightning? What is the distance of a cloud, if the thunder is heard b seconds after the flash?

28. If A rides a wheel 4 hours at the rate of 10 miles an hour, and B rides 3 hours at the rate of 14 miles an hour, how many miles do they both ride? How many more miles does B ride than A?

29. If A rides a wheel h hours at the rate of r miles an hour, and B rides k hours at the rate of s miles an hour, how many miles do they both ride? How many more miles does B ride than A?

30. If $100 is divided equally among 5 men, how many dollars does each man receive? If $100 is divided equally among a men, how many dollars does each man receive? If d dollars is divided equally among a men, how many dollars does each man receive?

31. By what number must 20 be multiplied to give 40? By what number must 20 be multiplied to give a? By what number must a be multiplied to give 20? By what number must n be multiplied to give b?

32. By what number must 50 be divided to give 10? By what number must 50 be divided to give a? By what number must a be divided to give 50? By what number must n be divided to give b?

33. How many revolutions does a wheel 21 feet in circumference make in passing a distance of 35 yards? How many revolutions does a wheel c feet in circumference make in passing a distance of d yards?

34. If it costs $30 to feed 5 horses 3 days, how much does it cost to feed 6 horses 4 days? If it costs $70 to feed 4 horses a days, how much does it cost to feed b horses 3 days? If it costs a dollars to feed b horses c days, how much does it cost to feed d horses e days?

35. A house costs a dollars, and rents for b dollars a month. What per cent does the investment pay?

36. In how many days can a typewriter write a pages, working c hours a day, at the rate of d pages an hour?

37. If a men are paid b dollars, how many dollars would c men receive at the same rate?

38. If c men can do a piece of work in a days, how many men can do the same work in b days?

39. If m men, working h hours a day, can dig a ditch a yards long, b feet wide, and c feet deep, in d days, in how many days will n men, working k hours a day, dig a ditch e yards long, f feet wide, and g feet deep?

40. If 22 yards of cloth cost $33, and 7 yards are sold for $14, what is the gain on each yard sold?

41. If d yards of cloth cost c dollars, and b yards are sold for a dollars, what is the gain on each yard sold?

10. Parentheses, ( ), and **Brackets,** [ ], are used to indicate that whatever is placed within them is to be treated as a whole.

E.g., $10 - (2 + 5)$ means that the result of adding 5 to 2, or 7, is to be subtracted from 10; that is,

$$10 - (2 + 5) = 10 - 7 = 3.$$

But $10 - 2 + 5$ means that 2 is to be subtracted from 10 and 5 is then to be added to that result; that is,

$$10 - 2 + 5 = 8 + 5 = 13.$$

In like manner, $[27 - (3 + 2) \times 5] \div 2$ means that the result of multiplying the sum $3 + 2$ by 5 is first to be subtracted from 27, and the remainder is then to be divided by 2; that is,

$$[27 - (3 + 2) \times 5] \div 2 = [27 - 25] \div 2 = 2 \div 2 = 1.$$

Likewise, the result of adding $a + b$ to c is $c + (a + b)$. The result of multiplying $x + y$ by z is $(x + y)z$. The result of dividing $m + n$ by p is $(m + n) \div p$.

EXERCISES II.

Find the values of the following indicated operations:

1. $18+(7-3)$.
2. $12-(8-4)$.
3. $(12+7)-(3+8)$.
4. $(25-11)-(18-7)$.
5. $(5+7)2$.
6. $(11-4)3$.
7. $12+(4-3)2$.
8. $17-(5-2)3$.
9. $(7+8)\div 5$.
10. $(12-6)\div 2$.
11. $18+(7-3)\div 4$.
12. $25-(15-7)\div 2$.
13. $11-[(5+3)-4]$.
14. $12+[8-(4+3)]$.
15. $17-[(3+5)-(2+4)]$.
16. $[7+(11-2)-(8-6)]\times 2$.
17. $[24-(8-5)-(7-4)]\div 3$.

18. What is the result of subtracting $a-b$ from $c+d$? Of multiplying $a-b$ by $c+d$? Of dividing $a-b$ by $c+d$?

19. What is the result of subtracting from 8 the product of $x+y$ by 3? Of subtracting from a the product of $x-y$ by z?

20. What is the result of subtracting from x a number 5 greater than b?

21. What is the result of multiplying $a-5$ by $b+3$? Of adding to a the result of dividing b by $c+d$?

22. One-third of a man's property is $a+100$ dollars. What is his entire property?

23. The length of a rectangular field is a rods, and its width is b rods less; what is the area of the field?

24. The older of two brothers is 20 years old; if he were 5 years younger, he would be three times as old as his younger brother. How old is his younger brother?

25. The older of two brothers is n years old; if he were a years younger, he would be b times as old as his younger brother. How old is the younger brother?

26. The younger of two brothers is a years old. How old is the older brother, if n years ago he was m times as old as his younger brother?

11. An **Algebraic Expression** is a number expressed by means of the signs and symbols of Algebra.

E.g., $a + b - c$, $ab - cd$, etc., are algebraic expressions.

12. The **Symbol of Equality**, $=$, read *is equal to*, *has the value*, etc., is placed between two numbers or expressions to indicate that they have the same or equal values.

E.g., $3 + 2 = 5$, read *three plus two is equal to five.*

An **Equation** is a statement that two numbers or expressions are equal.

E.g., $7 \times 9 = 63$, $4 \times 7 + 3 = 31$, etc.

The *first*, or *left-hand member*, or *side*, of an equation is the expression on the *left* of the symbol $=$; the *second*, or *right-hand member*, or *side*, is the expression on the *right* of the symbol $=$.

13. The **Symbol of Inequality**, $>$, read *is greater than*, is used to indicate that the number or expression on its left is greater than that on its right. *E.g.*, $7 > 5$.

The **Symbol of Inequality**, $<$, read *is less than*, is used to indicate that the number or expression on its left is less than that on its right. *E.g.*, $3 < 4 + 2$.

Axioms.

14. An **Axiom** is a truth so simple that it cannot be made to depend upon a truth still simpler.

Algebra makes frequent use of the following mathematical axioms.

(i.) *Every number is equal to itself.* *E.g.*, $7 = 7$, $a = a$.

(ii.) *The whole is equal to the sum of all its parts.*

E.g., $7 = 3 + 4$, $5 = 1 + 1 + 1 + 1 + 1$.

(iii.) *If two numbers be equal, either can replace the other in any algebraic expression in which it occurs.*

E.g., If $a + b = c$, and $b = d$, then $a + d = c$, replacing b by d.

(iv.) *Two numbers which are each equal to a third number are equal to each other.*

E.g., if $a = b$, and $c = b$, then $a = c$.

(v.) *The whole is greater than any of its parts; and, conversely, any part is less than the whole.*

E.g., $3 + 2 > 2$ and $2 < 3 + 2$.

15. *Literal numbers*, as has been stated, are used to represent numbers which may have *any values whatever*, or numbers whose values are, as yet, *unknown*. But it is frequently necessary to assign particular values to such numbers.

Substitution is the process of replacing a literal number in an algebraic expression by a particular value. See axiom (iii.).

Ex. 1. If in $a + b$, $a = 3$ (read *a has the value* 3) and $b = 5$, then

$$a + b = 3 + 5 = 8, \text{ or } a + b = 8.$$

Notice that the last step involved an application of axiom (iv.). For we have $a + b = 3 + 5$, and $3 + 5 = 8$; therefore, since $a + b$ and 8 are each equal to $3 + 5$, they are, by axiom (iv.), equal to each other. That is,

$$a + b = 8.$$

The application of this axiom will not, in subsequent work, be specifically pointed out, unless for some special reason.

Ex. 2. If in $a - (b + c)$, $a = 11$, $b = 2$, and $c = 3$, we have

$$a - (b + c) = 11 - (2 + 3) = 11 - 5 = 6.$$

Ex. 3. If, in the last example, $a = 3$, $b = 2$, and $c = 5$, we have

$$a - (b + c) = 3 - (2 + 5) = 3 - 7.$$

This result is, *as yet*, meaningless, since we are not prepared to subtract a greater number from a less.

If, in an example, a particular value be assigned to a literal number, the same value must be substituted for the number wherever it occurs in the same example.

Ex. 4. If, in $a + b - 2a + 3b - c$, we let $a = 6$, $b = 11\frac{1}{2}$, $c = \frac{5}{8}$, we have

$$\begin{aligned} a + b - 2a + 3b - c &= 6 + 11\tfrac{1}{2} - 2 \times 6 + 3 \times 11\tfrac{1}{2} - \tfrac{5}{8} \\ &= 6 + \tfrac{23}{2} - 12 + \tfrac{69}{2} - \tfrac{5}{8} \\ &= 38\tfrac{1}{8}. \end{aligned}$$

Observe that in the work of the last example, the expression $a + b - 2a + 3b - c$ is to be understood on the left of the symbol, =, in the second and third lines.

Ex. 5. If, in the last example, $a = 3$, $b = 1$, and $c = 1$, we have

$$a + b - 2a + 3b - c = 3 + 1 - 6 + 3 - 1 = 4 - 6 + 3 - 1.$$

We cannot further reduce $4 - 6 + 3 - 1$, since we are unable, *as yet*, to subtract 6 from 4.

EXERCISES III.

What are the values of the following expressions when $a = 6$, $b = 4$, $c = 2$:

1. $a + b$. **2.** $a - b$. **3.** ab. **4.** $a \div b$.

5. $b \div a$. **6.** $a + b + c$. **7.** $a + b - c$. **8.** $a - b + c$.

9. $a - b - c$. **10.** abc. **11.** $a \times b \div c$. **12.** $a \div b \times c$.

13. $a + (b - c)$. **14.** $a - (b + c)$. **15.** $a - (b - c)$. **16.** $(a - b)c$.

17. $(a - b) \div c$. **18.** $c \div (a - b)$. **19.** $[a + (b - c)]a$.

20. $[a - (b - c)] \div b$. **21.** $(a + b)(c + 1)$. **22.** $(a - b)(c - 1)$.

23. $[3 + (a - b)](c - 1)$. **24.** $(7 - a)(6 - b)(5 - c)$.

25. $(12 - a) \div (7 - b)$. **26.** $[15 - (7 - a)] \times [(25 - b) - (16 - c)]$.

27. $[8 - (b - c)] \div [(15 - a) - (6 - b)]$.

28. A man owes one creditor 150 dollars and another 135 dollars; his yearly income is 1000 dollars and his expenses are 905 dollars. In how many years can he pay his debts?

29. A man owes one creditor a dollars and another b dollars; his yearly income is m dollars and his expenses are n dollars. In how many years can he pay his debts?

30. What is the result of Ex. 29, when $a = 700$, $b = 200$, $m = 1500$, and $n = 1200$? When $a = 200$, $b = 300$, $m = 1750$, and $n = 1700$?

31. How much is a increased by b less than m multiplied by c?

32. What is the result of Ex. 31 when $a = 3$, $b = 2$, $m = 5$, and $c = 4$? When $a = 7$, $b = 11$, $m = 8$, and $c = 9$?

33. If x, y, and z are three numbers in decreasing order of magnitude, find the product of the smallest by a times the difference between the other two.

34. What is the result of Ex. 33, when $x = 15$, $y = 7$, $z = 4$, and $a = 3$? When $x = 25$, $y = 14$, $z = 9$, and $a = 2$?

16. Some of the advantages of using literal numbers are shown by the following examples:

Ex. **1.** The two equations

$$\frac{2}{7} + \frac{3}{7} = \frac{2+3}{7} = \frac{5}{7} \text{ and } \frac{5}{11} + \frac{4}{11} = \frac{5+4}{11} = \frac{9}{11}$$

are particular examples of the following arithmetical principle:

The sum of two fractions which have a common denominator is a fraction whose denominator is that common denominator, and whose numerator is the sum of the two given numerators; or,

$$\frac{\textit{1st num.}}{\textit{com. den.}} + \frac{\textit{2d num.}}{\textit{com. den.}} = \frac{\textit{1st num.} + \textit{2d num.}}{\textit{com. den.}}.$$

This principle can be stated still more concisely if the terms of the fractions, which may be *any numbers whatever,* are represented by three symbols, say a, b, c.

We then have

$$\frac{a}{c} + \frac{b}{c} = \frac{a+b}{c}.$$

This equation states by means of signs and symbols all that is contained in the verbal statement of the principle. It is thus a *symbolic statement of a general principle,* and includes all particular cases that result from assigning particular values to a, b, c.

Ex. 2. The equations

$$\frac{2}{7}+\frac{4}{9}=\frac{2\times 9}{7\times 9}+\frac{4\times 7}{7\times 9}=\frac{2\times 9+4\times 7}{7\times 9},$$

and

$$\frac{3}{6}+\frac{1}{5}=\frac{3\times 5}{6\times 5}+\frac{1\times 6}{6\times 5}=\frac{3\times 5+1\times 6}{6\times 5}$$

are particular examples of the general principle:

The sum of two fractions which do not have a common denominator is a fraction whose denominator is the product of the two denominators, and whose numerator is the sum of the product of the numerator of each fraction multiplied by the denominator of the other; or,

$$\frac{\textit{1st num.}}{\textit{1st den.}}+\frac{\textit{2d num.}}{\textit{2d den.}}=\frac{\textit{1st num.}\times\textit{2d den.}}{\textit{1st den.}\times\textit{2d den.}}+\frac{\textit{2d num.}\times\textit{1st den.}}{\textit{1st den.}\times\textit{2d den.}}$$
$$=\frac{\textit{1st num.}\times\textit{2d den.}+\textit{2d num.}\times\textit{1st den.}}{\textit{1st den.}\times\textit{2d den.}}.$$

This principle can be stated still more concisely if the terms of the fractions are represented by symbols for general numbers, say a, b, c, d.

We then have

$$\frac{a}{b}+\frac{c}{d}=\frac{a\times d}{b\times d}+\frac{c\times b}{b\times d}=\frac{a\times d+c\times b}{b\times d}.$$

This equation, like that of Ex. 1, is perfectly general, and holds for all particular values that may be assigned to the letters a, b, c, d.

Ex. 3. A, traveling 4 miles an hour, is 10 miles in advance of B, who is traveling 6 miles an hour in the same direction. After how many hours will B overtake A?

B must gain 10 miles to overtake A. In 1 hour B gains $6-4$, or 2 miles; in order to gain 10 miles, B must evidently travel $10\div 2$, or 5 hours.

If other data were given, it would be necessary to repeat the above reasoning in full in order to obtain the result.

But this example may be made general by introducing general numbers instead of numerals.

A, traveling a miles an hour, is m miles in advance of B, who is traveling b miles an hour in the same direction. After how many hours will B overtake A?

B must gain m miles to overtake A. In 1 hour B gains $b - a$ miles; in order to gain the m miles, B must evidently travel $m \div (b - a)$ hours.

This result is general.

If any particular data be given, the result can be obtained by substituting in this general result the particular values assigned to m, b, and a.

17. Notice the following advantages secured by introducing general numbers:

(i.) *General laws and relations can be expressed with great brevity, and yet include all that the most general verbal statements can express.*

(ii.) *Such symbolic statements mass under the eye the various operations involved, and thus enable the eye to assist the understanding and memory.*

See Exx. 1 and 2, which give the principles for adding fractions with a common denominator and with different denominators.

(iii.) *Such statements also show how the final result involves each given number.*

Thus, there is nothing in the result, $\frac{46}{63}$, of adding $\frac{2}{7}$ and $\frac{4}{9}$, to indicate in what way $\frac{46}{63}$ is obtained from the terms of the given fractions. But the result,

$$\frac{a \times d + c \times b}{b \times d},$$

of adding the two fractions $\frac{a}{b}$ and $\frac{c}{d}$, shows in what way the terms of these fractions are involved.

The result, 5, of Ex. 3 does not show what operations have been performed upon the given numbers 4, 10, and 6; while it is evident from the general result, $m \div (b - a)$, in what way the given numbers, a, b, and m, are involved.

EXERCISES IV.

Express *in algebraic language* (*i.e.*, by means of the signs and symbols of Algebra) the following principles of Arithmetic:

1. If a, b, and c are any three numbers, their sum diminished by any one of them is equal to the sum of the other two.

2. If a, b, c, and d are any four numbers, their sum diminished by the sum of any two of them is equal to the sum of the other two.

3. Verify the results of Exx. 1 and 2, when $a = 11$, $b = 13$, $c = 9$, $d = 4$. When $a = 5\frac{1}{2}$, $b = 2\frac{3}{4}$, $c = 7\frac{1}{4}$, $d = 3\frac{1}{2}$.

If z is the result of subtracting y from x, express in algebraic language the following principles of subtraction:

4. The minuend is equal to the subtrahend plus the remainder.

5. The subtrahend is equal to the minuend diminished by the remainder.

6. Verify the results of Exx. 4 and 5, when $x = 12$, $y = 9$, $z = 3$. When $x = 17$, $y = 3\frac{1}{2}$, $z = 13\frac{1}{2}$.

If z is the result of multiplying x by y, express in algebraic language the following principles of multiplication:

7. The multiplicand is equal to the product divided by the multiplier.

8. The multiplier is equal to the product divided by the multiplicand.

9. Verify the results of Exx. 7 and 8, when $x = 5$, $y = 15$, $z = 75$. When $x = 7\frac{1}{2}$, $y = 2$, $z = 15$.

If a is exactly divisible by b, and q is the quotient, express in algebraic language the following principles of division:

10. The dividend is equal to the divisor multiplied by the quotient.

11. The divisor is equal to the dividend divided by the quotient.

12. Verify the results of Exx. 10 and 11, when $a = 18$, $b = 2$, and $q = 9$. When $a = 75$, $b = 15$, and $q = 5$.

If a is not exactly divisible by b, and q is the quotient and r the remainder, express in algebraic language the following principles of division:

13. The dividend is equal to the divisor multiplied by the quotient, plus the remainder.

14. The divisor is equal to the dividend minus the remainder divided by the quotient.

15. Verify the results of Exx. 13 and 14, when $a = 17$, $b = 2$, $q = 8$, $r = 1$. When $a = 108$, $b = 11$, $q = 9$, $r = 9$.

If $\frac{a}{b}$ is any fraction, and m is any integer, express in algebraic language the following principles of fractions:

16. If the numerator of a fraction is multiplied by any integer, the value of the fraction is multiplied by that integer.

17. If the denominator of a fraction is multiplied by any integer, the value of the fraction is divided by that integer.

18. If both numerator and denominator of a fraction are multiplied by any number, the value of the fraction is not changed.

19. If both numerator and denominator of a fraction are divided by any number, the value of the fraction is not changed.

If $\frac{a}{b}$ and $\frac{c}{d}$ are any two fractions, state in algebraic language the following principles:

20. The product of two fractions is a fraction whose numerator is the product of the numerators of the given fractions, and whose denominator is the product of the denominators.

21. The quotient of dividing the first fraction by the second is a fraction whose numerator is the product of the numerator of the first fraction by the denominator of the second, and whose denominator is the product of the denominator of the first fraction by the numerator of the second.

Express in algebraic language the following geometrical principles:

22. The area of any triangle is equal to one-half the product of its base by its altitude. What is the area, when the base is 6 and the altitude is 5?

23. The area of a rectangle is equal to the product of its base by its altitude. What is the area, when the base is 9 and the altitude is 4?

18. General numbers are most frequently represented by the *italicized letters* of the English alphabet. But letters of other alphabets are sometimes employed, and there is often an advantage in using the same letter with some distinguishing marks to represent different numbers in the same discussion.

We add a list of the more common symbols for future reference.

Greek letters: α, β, γ, δ, etc., read *alpha, beta, gamma, delta,* etc.;

with *prime marks:* a', a'', a''', $a^{(n)}$, read *a prime, a two prime, a three prime, a n prime;*

with *subscripts:* a_1, a_2, a_3, etc., read *a sub-one, a sub-two, a sub-three,* etc., or simply *a one, a two, a three,* etc.

§ 2. POSITIVE AND NEGATIVE NUMBERS, OR ALGEBRAIC NUMBERS.

1. A still greater extension of the idea of number in passing from Arithmetic to Algebra is arrived at by the following considerations:

In ordinary Arithmetic we subtract a number from an equal

or a greater number, never a greater number from a less. We are familiar with such operations as the following:

$$\begin{rcases}\text{Min.} - \text{Subt.} = \text{Rem.}\\ 8 - 5 = 3 \\ 7 - 5 = 2 \\ 6 - 5 = 1 \\ 5 - 5 = 0 \end{rcases} \quad \text{(i.)}$$

2. In the equations (i.) of Art. 1, the subtrahend remains the same, while the minuend and the remainder decrease. When the minuend is equal to the subtrahend, the remainder is zero.

Zero *is therefore the result of subtracting any number from an equal number.* That is, $0 = n - n$, of which particular cases are $0 = 2 - 2$, $0 = 7 - 7$, $0 = 2\frac{1}{2} - 2\frac{1}{2}$.

3. If the minuend in equations (i.), Art. 1, be still further diminished, the subtrahend remaining the same, we have the indicated operations:

$$\begin{rcases}\text{Min.} - \text{Subt.}\\ 4 - 5 \\ 3 - 5 \\ 2 - 5 \end{rcases} \quad \text{(ii.)}$$

Such operations have not occurred in ordinary Arithmetic, and cannot be carried out in terms of arithmetical numbers. For, from an arithmetical point of view, we cannot subtract from a number more units than are contained in that number. In general, the indicated operation $a - b$ can, as yet, be performed only when a is greater than b. But if a and b are to have any values whatever, the case in which a is less than b, that is, in which *the minuend is less than the subtrahend*, must be included in the operation of subtraction.

4. Now observe that, *as the minuend in equations* (i.), *Art.* 1, *decreases by* 1, 2, *or more units* (*the subtrahend remaining the same*) *the remainder decreases by an equal number of units.*

When the minuend is equal to the subtrahend, the remainder is 0. If then, as in the indicated operations (ii.), Art. 3, the minuend becomes less than the subtrahend by 1, 2, or more units, the remainder must decrease by an equal number of units, and therefore become less than 0 by 1, 2, or more units.

The operation of subtracting a greater number from a less is therefore possible only when numbers less than zero are introduced.

We then have from (i.), Art. 1, and (ii.), Art. 3:

Min. − Subt. = Rem.

$8 - 5 = 3$
$7 - 5 = 2$
$6 - 5 = 1$
$5 - 5 = 0$ (iii.)
$4 - 5 =$ *a number one unit less than* 0
$3 - 5 =$ *a number two units less than* 0
$2 - 5 =$ *a number three units less than* 0

5. *Numbers less than zero* are called **Negative Numbers.** *Numbers greater than zero* are, for the sake of distinction, called **Positive Numbers.**

Positive and negative numbers are called **Algebraic** or **Relative Numbers.**

6. The **Absolute Value** of a number is the number of units contained in it without regard to their *quality* (*i.e., whether positive or negative*).

A *positive* number may be indicated by placing a small sign, $^{+}$, to the left and a little above its absolute value; as $^{+}5$, $^{+}10$, $^{+}16$; read *positive* 5, *positive* 10, *positive* 16.

A *negative* number may be indicated by placing a small sign, $^{-}$, to the left and a little above its absolute value; as, $^{-}5$, $^{-}10$, $^{-}16$, read *negative* 5, *negative* 10, *negative* 16.

We must, as yet, carefully distinguish these symbols of *quality*, $^{+}$ and $^{-}$, from the (larger) symbols of *operation*, $+$ and $-$.

7. Equations (iii.), Art. 4, can now be written as follows:

Min. − Subt. = Rem.		Min. − Subt. = Rem.	
pos. 8 − *pos.* 5 = *pos.* 3		$^+8 - {}^+5 = {}^+3$	
pos. 7 − *pos.* 5 = *pos.* 2		$^+7 - {}^+5 = {}^+2$	
pos. 6 − *pos.* 5 = *pos.* 1		$^+6 - {}^+5 = {}^+1$	
pos. 5 − *pos.* 5 = 0	or	$^+5 - {}^+5 = 0$	(iv.)
pos. 4 − *pos.* 5 = *neg.* 1		$^+4 - {}^+5 = {}^-1$	
pos. 3 − *pos.* 5 = *neg.* 2		$^+3 - {}^+5 = {}^-2$	
pos. 2 − *pos.* 5 = *neg.* 3		$^+2 - {}^+5 = {}^-3$	

8. Thus negative numbers arise in Algebra through the extension of the operation of subtraction to the case in which the minuend is less than the subtrahend.

A negative remainder does not mean that more units have been taken from the minuend than were contained in it; *such a remainder indicates that the subtrahend is greater than the minuend by as many units as are contained in the remainder.*

Thus, in $^+10 - {}^+15 = {}^-5$ and $^+87 - {}^+92 = {}^-5$, the remainder, $^-5$, indicates that the subtrahend is, in each case, 5 units greater than the minuend.

9. The results of the preceding articles, restated briefly, are:

A positive number is a number greater than zero, by as many units as are contained in its absolute value.

E.g., $^+2$ is *two* units greater than 0, $^+7\frac{1}{2}$ is *seven and one-half* units greater than 0.

A negative number is a number less than zero by as many units as are contained in its absolute value.

E.g., $^-3$ is *three* units less than 0; $^-15\frac{2}{3}$ is *fifteen and two-thirds* units less than 0.

Zero is the result of subtracting a number from an equal number.

E.g., $0 = {}^+7 - {}^+7 = {}^-5 - {}^-5 = {}^+n - {}^+n = {}^-n - {}^-n$,

wherein n denotes any absolute number.

Since zero can be neither greater nor less than itself, it is neither a positive nor a negative number. It stands by itself, as the number from which positive and negative numbers are counted.

10. Positive and negative numbers are thus completely defined and are represented by definite symbols.

The numbers of ordinary Arithmetic are the absolute values of the positive and negative numbers of Algebra.

Since letters are to represent numbers which may have any values whatever, or values which are as yet unknown, they can represent either *positive* or *negative* numbers. Thus, in one case a may have the value $^{+}2$, in another case the value $^{-}7$; in the first case the absolute value of a is 2, in the second case the absolute value of a is 7.

If, however, a have a sign of quality, $^{+}$ or $^{-}$, then a can have only absolute values. Thus, while a (*without any sign of quality*) can represent any number whatever, *positive* or *negative*, ^{+}a can represent only *any positive number*, and ^{-}a only *any negative number*.

Likewise, $^{+}(a+2)$ represents a *positive* number whose *absolute value* is $a+2$; and $^{-}(a+2)$ represents a *negative* number whose absolute value is $a+2$. In both numbers, a represents *any absolute value.*

EXERCISES V.

1. What is the absolute value of $^{+}8$? Of $^{+}17$? Of $^{-}11$? Of $^{-}21$? Of $^{+}17\frac{1}{2}$? Of $^{-}2\frac{3}{8}$? Of ^{+}a? Of ^{-}b? Of $^{-}(x+3)$? Of $^{+}(2+y)$? Of $^{-}(a+b)$? Of $^{+}(a-b)$?

What is the absolute value of x,

2. When $x = {^{-}7}$? **3.** When $x = {^{+}21}$? **4.** When $x = {^{-}2\frac{1}{2}}$?

For what values of x do the following expressions reduce to 0:

5. $x - {^{+}3}$. **6.** $x - {^{+}51}$. **7.** $x - {^{-}7}$. **8.** $x - {^{-}18}$.

9. $x - {^{+}a}$. **10.** $x - {^{+}(a+b)}$. **11.** $x - {^{-}(a+b+c)}$.

For what absolute values of x do the following expressions reduce to 0:

12. $27 - x$. **13.** $14 - 2x$. **14.** $7 - 3x$.

What are the results of the following indicated operations:

15. $^+4 - ^+3$. **16.** $^+17 - ^+2$. **17.** $^+10 - ^+11$.

18. $^+19 - ^+25$. **19.** $^+17 - ^+81$. **20.** $^+32 - ^+49$.

21. $^+100 - ^+200$. **22.** $^+199 - ^+350$.

23. What is the value of $a - ^+7$, when $a = ^+10$? When $a = ^+2$? When $a = ^+4$?

24. What is the value of $^+a - ^+11$, when $a = 15$? When $a = 10$? When $a = 1$?

What is the value of $a - b$,

25. When $a = ^+5$, $b = ^+2$? **26.** When $a = ^+5$, $b = ^+7$?

What is the value of $^+a - ^+b$,

27. When $a = 103$, $b = 205$? **28.** When $a = 2$, $b = 1$?

29. When $a = 27$, $b = 93$? **30.** When $a = 10$, $b = 83$?

What values of a make the first members of the following equations identical with the second members:

31. $a - ^+7 = ^+2$. **32.** $a - ^+7 = ^-2$. **33.** $a - ^+7 = ^-5$.

34. $^+15 - a = ^+12$. **35.** $^+15 - a = ^+1$. **36.** $^+15 - a = ^-20$.

What is the value of $^+a - ^+b$,

37. When a is 3 greater than b? **38.** When a is 5 greater than b?

39. When a is equal to b? **40.** When a is 1 less than b?

41. When a is 5 less than b? **42.** When a is 17 less than b?

What are the results of the following indicated operations:

43. $^+(a + 2) - ^+2$. **44.** $^+(a + 5) - ^+5$. **45.** $^+(a + 7) - ^+a$.

46. $^+(a + b) - ^+b$. **47.** $^+(a + b + c) - ^+(a + b)$.

48. $^+n - ^+(n + 3)$. **49.** $^+16 - ^+(n + 16)$. **50.** $^+10 - ^+(a + 10)$.

51. $^+a - ^+(a + b)$. **52.** $^+(m + n) - ^+(m + n + p)$.

53. Give three sets of positive values of a and b which will make $a - b = ^+2$.

54. Give three sets of positive values of m and n which will make $m - n = ^-5$.

Positive and Negative Numbers are Opposite Numbers.

11. If the minuend in equations (iv.), Art. 7, be still further diminished, the subtrahend remaining the same, we have (repeating the latter part of that table):

$$\begin{rcases}\text{Min.} - \text{Subt.} = \text{Rem.} \\ {}^{+}3 - {}^{+}5 = {}^{-}2 \\ {}^{+}2 - {}^{+}5 = {}^{-}3 \\ {}^{+}1 - {}^{+}5 = {}^{-}4 \\ 0 - {}^{+}5 = {}^{-}5\end{rcases} \qquad \text{(v.)}$$

The student is familiar with the principle of subtraction in Arithmetic that *the remainder added to the subtrahend is equal to the minuend.* This principle, like all principles of arithmetical operations, is retained in Algebra. Consequently table (v.) gives:

$$\begin{rcases}\text{Subt.} + \text{Rem.} = \text{Min.} \\ {}^{+}5 + {}^{-}2 = {}^{+}3 \\ {}^{+}5 + {}^{-}3 = {}^{+}2 \\ {}^{+}5 + {}^{-}4 = {}^{+}1 \\ {}^{+}5 + {}^{-}5 = 0\end{rcases} \qquad \text{(vi.)}$$

12. The last one of equations (vi.), Art. 11,

$$ {}^{+}5 + {}^{-}5 = 0, $$

furnishes an important relation between positive and negative numbers:

The sum of a positive and a negative number having the same absolute value is equal to zero; i.e., *they cancel each other when united by addition.*

E.g., ${}^{+}1 + {}^{-}1 = 0$, ${}^{+}3 + {}^{-}3 = 0$, ${}^{-}17\frac{1}{2} + {}^{+}17\frac{1}{2} = 0$.

In general, ${}^{+}n + {}^{-}n = 0$.

For that reason, positive and negative numbers in their relation to each other are called *opposite* numbers. When their absolute values are equal, they are called *equal and opposite* numbers.

13. In Art. 9, *zero* was defined as the difference between two equal numbers. From the preceding article it follows that 0 is likewise *the sum of two equal and opposite numbers.*

14. Although negative numbers arise through the extension of the operation of subtraction, it is necessary to treat them as numbers apart from this particular operation.

As in Arithmetic, so in Algebra, any integer is an aggregate of like units. Just as

$$4 = 1 + 1 + 1 + 1,$$

so $$^{+}4 = {}^{+}1 + {}^{+}1 + {}^{+}1 + {}^{+}1,$$

and $$^{-}4 = {}^{-}1 + {}^{-}1 + {}^{-}1 + {}^{-}1.$$

In like manner, any fraction is an aggregate of like fractional units. Just as

$$\tfrac{2}{3} = \tfrac{1}{3} + \tfrac{1}{3},$$

so $$^{+}(\tfrac{2}{3}) = {}^{+}(\tfrac{1}{3}) + {}^{+}(\tfrac{1}{3}),$$

and $$^{-}(\tfrac{2}{3}) = {}^{-}(\tfrac{1}{3}) + {}^{-}(\tfrac{1}{3}).$$

EXERCISES VI.

Express the following numbers as sums of like units:

1. $^{+}7$. **2.** $^{-}3$. **3.** $^{+}6$. **4.** $^{-}9$.

5. ^{+}a. **6.** ^{-}b. **7.** $^{+}(a + b)$. **8.** $^{-}(x + y)$.

Express the following numbers as sums of like fractional units:

9. $^{+}(\tfrac{3}{4})$. **10.** $^{-}(\tfrac{5}{6})$. **11.** $^{+}(\tfrac{3}{8})$. **12.** $^{-}(\tfrac{5}{11})$.

15. Any *quantities* which in their relation to each other are *opposite*, may be represented in algebra by *positive* and *negative* numbers.

Thus *credits* and *debits*, in their relation to each other, are opposite quantities, and may therefore be represented by positive and negative numbers.

Ex. 1. 100 *dollars credit* and 100 *dollars debit* cancel each other. That is, 100 *dollars credit united with* 100 *dollars debit is equal to neither credit nor debit;* or,

100 *dollars credit* + 100 *dollars debit* = *neither credit nor debit.*

If *credits* be taken *positively* and *debits negatively,* then 100 *dollars credit* may be represented by $^{+}100$, and 100 *dollars debit* by $^{-}100$. Their united effect, as stated above, may then be represented algebraically thus:

$$^{+}100 + {}^{-}100 = 0.$$

The result, 0, means *neither credit nor debit.*

Ex. 2. 100 *dollars credit* and 80 *dollars debit* is equivalent to 20 *dollars credit;* or,

100 *dollars credit* + 80 *dollars debit* = 20 *dollars credit.*

This equation may be stated algebraically thus:

$$^{+}100 + {}^{-}80 = {}^{+}20.$$

The result, $^{+}20$, means 20 *dollars credit.*

Ex. 3. 100 *dollars credit* and 120 *dollars debit* is equivalent to 20 *dollars debit;* or, stated algebraically,

$$^{+}100 + {}^{-}120 = {}^{-}20.$$

The result, $^{-}20$, means 20 *dollars debit.*

Gain and *loss* are also opposite quantities, and may be represented by positive and negative numbers.

Ex. 4. 250 *dollars gain* and 250 *dollars loss* cancel each other. If *gain* be taken *positively* and *loss negatively,* we have, in algebraic language,

$$^{+}250 + {}^{-}250 = 0.$$

The result, 0, means *neither gain nor loss.*

Ex. 5. 250 *dollars gain* and 200 *dollars loss* is equivalent to 50 *dollars net gain;* or,

$$^{+}250 + {}^{-}200 = {}^{+}50.$$

The result, $^{+}50$, means 50 *dollars net gain.*

Ex. 6. 250 *dollars gain* and 300 *dollars loss* is equivalent to 50 *dollars net loss;* or,

$$^{+}250 + {}^{-}300 = {}^{-}50.$$

The result, $^{-}50$, means 50 *dollars net loss.*

Similarly for *opposite temperatures.*

Ex. 7. If a body is first heated so as to cause its temperature to rise 10° and is then cooled down 10°, its temperature is the same as it was originally.

If *rises* in temperature be taken *positively* and *falls* in temperature *negatively,* their united effect may be represented algebraically thus:

$$^{+}10 + {}^{-}10 = 0.$$

The result, 0, means that there is finally no change in the temperature of the body.

Ex. 8. If the body is first heated 10° and then cooled down 8°, its final temperature is 2° above its original temperature; or, stated algebraically,

$$^{+}10 + {}^{-}8 = {}^{+}2.$$

The final result, $^{+}2$, means a *rise* of 2° in temperature.

Ex. 9. If the body is first heated 10° and then cooled down 12°, its final temperature is 2° below its original temperature; or,

$$^{+}10 + {}^{-}12 = {}^{-}2.$$

The result, $^{-}2$, means a *fall* of 2° in temperature.

Similar reasoning applies to *opposite directions.*

Ex. 10. A man walks 10 miles *due north,* and turning, walks 10 miles *due south.* How far is he from his starting point?

If distances *north* be taken *positively* and distances *south negatively,* we have

$$^{+}10 + {}^{-}10 = 0.$$

The result, 0, means that the man has returned to his starting point.

Ex. 11. If he walks 10 miles *due north,* and turning, walks 6 miles *due south,* he is still 4 miles *north* of his starting point; or,

$$^{+}10 + {}^{-}6 = {}^{+}4.$$

The result, $^{+}4$, means that he is still 4 miles *north* of his starting point.

Ex. 12. If he walks 10 miles due *north,* and turning, walks 14 miles due *south,* he is then 4 miles *south* of his starting point; or,

$$^{+}10 + {}^{-}14 = {}^{-}4.$$

The result, $^{-}4$, means that he is now 4 miles *south* of his starting point.

16. It is evidently immaterial which of two opposite quantities is taken positively and which negatively, in any particular problem. Thus, we might have called *gain* a *negative* quantity, and *loss* a *positive* quantity; where we had *positive* results before we should now have *negative,* and *vice versa.* So we could call distances *south positive* and distances *north negative.* We have only to interpret results differently.

Thus, a man *gains* 100 dollars and *loses* 50 dollars. What is his net gain?

Calling *gain positive* as before, and *loss negative,* we have:

$$^{+}100 + {}^{-}50 = {}^{+}50.$$

The *positive* result denotes *gain.*

Now calling *gain negative,* and *loss positive,* we have:

$$^{-}100 + {}^{+}50 = {}^{-}50.$$

The *negative* result now denotes *gain,* as the *positive* result before denoted *gain.*

That is, whichever of two opposite quantities be taken positively, the other being taken negatively, the meaning of the results is always the same.

EXERCISES VII.

State algebraically in two ways each of the following relations (by Art. 16):

1. 100 dollars gain and 20 dollars loss is equivalent to 80 dollars net gain.

2. 100 dollars loss and 300 dollars gain is equivalent to 200 dollars net gain.

3. 250 dollars gain and 250 dollars loss is equivalent to neither gain nor loss.

4. A rise of 25° in temperature followed by a fall of 10° is equivalent to a rise of 15°.

5. A rise of 15° in temperature followed by a fall of 22° is equivalent to a fall of 7°.

6. If a man ascends from the foot of a ladder 20 steps, and then descends 7 steps, he is 13 steps up.

7. If a man ascends from the foot of a ladder 10 steps, and then descends 10 steps, he is at the foot of the ladder.

8. If a man ascends from some step of a ladder 5 steps, and then descends 8 steps, he is 3 steps below the step from which he started.

9. If a man walks 150 feet to the right and then 50 feet to the left, he is 100 feet to the right of his original position.

10. If a man walks 150 feet to the right, then 50 feet to the left, and then 75 feet to the right, he is finally 175 feet to the right of his original position.

11. If a bucket is lowered to the bottom of a well 60 feet deep, and is then raised 50 feet, it is still 10 feet below the surface.

12. If a balloon ascends 2500 feet, and then falls 1500 feet, it is still 1000 feet above the surface of the earth.

13. If the minute hand of a clock is moved 5 minutes forward and then 5 minutes backward, the time indicated is unchanged.

14. If the minute hand of a clock is moved 20 minutes forward and then 12 minutes backward, the indicated time is 8 minutes later than at first.

15. If the minute hand of a clock is moved forward 6 minutes and then backward 16 minutes, the indicated time is 10 minutes earlier than at first.

16. Two men, A and B, run a race. The first minute A runs 5 feet more than B, the second minute A runs 8 feet less than B; in the two minutes A runs 3 feet less than B.

Double Series of Algebraic Numbers.

17. In forming table (iv.), Art. 7, we inferred that each remainder was less than the remainder next above it, or, what is the same, greater than the remainder next below it.

We therefore have, from the column of remainders in this table, the following relations:

$^+3$ is greater than $^+2$		$^-3$ is less than $^-2$
$^+2$ " " " $^+1$		$^-2$ " " " $^-1$
$^+1$ " " " 0	(*a*)	$^-1$ " " " 0
0 " " " $^-1$		0 " " " $^+1$
$^-1$ " " " $^-2$		$^+1$ " " " $^+2$
$^-2$ " " " $^-3$		$^+2$ " " " $^+3$ (*b*)

These statements may be expressed concisely, by using the two symbols of inequality:

$$^+3 > {}^+2,\ {}^+2 > {}^+1,\ {}^+1 > 0,\ 0 > {}^-1,\ {}^-1 > {}^-2,\ {}^-2 > {}^-3;$$

and $^-3 < {}^-2,\ {}^-2 < {}^-1,\ {}^-1 < 0,\ 0 < {}^+1,\ {}^+1 < {}^+2,\ {}^+2 < {}^+3.$

It is more convenient to combine these several statements into one continued statement, thus:

$$\left.\begin{array}{l} ^+3 > {}^+2 > {}^+1 > 0 > {}^-1 > {}^-2 > {}^-3, \\ \text{or,}\quad ^-3 < {}^-2 < {}^-1 < 0 < {}^+1 < {}^+2 < {}^+3. \end{array}\right\} \quad (c)$$

Evidently table (iii.), Art. 4, can be continued without limit, upward by increasing the minuend, downward by decreasing the minuend, the subtrahend remaining always the same. Then, since the remainder and the minuend increase, or decrease, by an equal number of units, tables (*a*) and (*b*) above can be continued upward and downward without limit. The relations (*c*) may therefore be continued toward the right and toward the left without limit.

18. The **Sign of Continuation**, ⋯, read *and so on*, or *and so on to*, is used to indicate that a succession of numbers continues without end in the way shown by those written down, as 1, 2, 3, ⋯, read, *one, two, three, and so on;* or that the succession continues as far as a certain number which is written after the sign ⋯, as 1, 2, 3, ⋯, 10, read *one, two, three as far as*, or *to*, 10. In the first illustration the series continues without end; in the second it stops at 10.

We may now write the series of algebraic numbers as follows:

$$\cdots\ {}^-4,\ {}^-3,\ {}^-2,\ {}^-1,\ 0,\ {}^+1,\ {}^+2,\ {}^+3,\ {}^+4,\ \cdots$$

In this series the numbers increase from left to right, and decrease from right to left; or a number is greater than any number on its left and less than any number on its right.

Consequently we have in Algebra the series of positive increasing numbers

$${}^+1,\ {}^+2,\ {}^+3,\ {}^+4,\ \cdots \text{ without end,}$$

and the series of negative decreasing numbers

$${}^-1,\ {}^-2,\ {}^-3,\ {}^-4,\ \cdots \text{ without end.}$$

The number 0 might be written at the beginning of both series.

Relations between Positive and Negative Numbers and Zero.

19. From the results of the preceding article, we obtain the following general relations:

(i.) *Of two positive numbers, that number is the greater which has the greater absolute value; and that number is the less which has the less absolute value.*

For example, $^+5 > {}^+3$, or $^+3 < {}^+5$, since $^+5$ is five units greater than 0, and $^+3$ is only three units greater than 0.

Likewise, $^+8 > {}^+4$, $^+23 < {}^+107$.

(ii.) *Of two negative numbers, that number is the greater which has the less absolute value; and that number is the less which has the greater absolute value.*

For example, $^-3 > {}^-5$, or $^-5 < {}^-3$, since $^-5$ is five units less than 0, and $^-3$ is only three units less than 0.

Likewise $^-17 > {}^-20$, $^-5 < {}^-2$.

(iii.) *Any positive number is greater than any negative number.*

For example, $^+7 > {}^-8$, since $^+7$ is seven units greater than 0, and $^-8$ is eight units less than 0.

Likewise $^+1 > {}^-1000$, $^+2 > {}^-3$.

(iv.) *Any negative number is less than any positive number.*

For example, $^-2 < {}^+7$, since $^-2$ is two units less than 0, and $^+7$ is seven units greater than 0.

Likewise $^-1000 < {}^+1$, $^-27 < {}^+5$.

EXERCISES VIII.

How many units is each of the following numbers greater or less than 0:

1. $^+10$. **2.** $^-3$. **3.** $^+8$. **4.** $^-19$. **5.** $^+17$.

6. $^+(\frac{3}{4})$. **7.** $^-(2\frac{5}{8})$. **8.** ^+x. **9.** ^-y. **10.** $^+(a+b)$.

Which number is greater,

11. $^+5$ or $^+2$? **12.** $^+3$ or $^-5$? **13.** $^-12$ or $^-5$?

14. 0 or $^+2$? **15.** 0 or $^-3$? **16.** $^-7$ or $^-11$?

17. $^-5$ or $^+4$? **18.** $^-2$ or $^+3$? **19.** $^-7$ or $^-5$?

20. $^+(5+a)$ or ^+a? **21.** $^+(5+a)$ or ^-a?

22. $^-(5+a)$ or ^+a? **23.** $^-(5+a)$ or ^-a?

24. $^+(6+a)$ or $^+(2+a)$? **25.** $^+(6+a)$ or $^-(2+a)$?

26. $^-(6+a)$ or $^+(2+a)$? **27.** $^+(a+1)$ or $^+(a-1)$?

28. $^-(a+1)$ or $^-(a-1)$? **29.** $^-(a+1)$ or $^+(a-1)$?

30. $^+(a+1)$ or $^-(a-1)$? **31.** $^+(1+a)$ or $^+(1+2a)$?

32. $^{-}(1+a)$ or $^{-}(1+2a)$? **33.** $^{+}(1+a)$ or $^{-}(1+2a)$?

34. $^{-}(1+a)$ or $^{+}(1+2a)$? **35.** $^{+}(a+b)$ or ^{+}a?

36. $^{-}(a+b)$ or ^{-}a? **37.** $^{-}(a+b)$ or ^{+}a?

20. We have defined negative numbers as numbers less than zero; that is, as the result of enlarging our conception of the operation of subtraction. We afterward find, as we have seen, that they often have a meaning when applied to practical problems. Yet even if they had not, we should be justified in introducing them in order to make our principles general. Sometimes, indeed, negative results have no meaning, and indicate an impossibility.

E.g., a men are at work on a building, and b men quit work. How many are still working? Evidently $a-b$. If $b > a$, the negative result has no meaning, and indicates that we have stated an impossibility.

A man has a dollars and pays out b dollars. How many dollars has he left? Evidently $a-b$.

When $b > a$, the negative result has no meaning if the money be regarded as actually handled. But in dealing with book accounts, it is quite possible that the debits shall exceed the credits; a state which would, as we have seen, be indicated by a negative result (if credits be taken positively and debits negatively). So, too, when applied to opposition in direction, etc., negative results are as intelligible as positive results.

In fact, there is no more objection to the use of negative numbers than to the use of fractions, for each kind of number may indicate an impossible state. For instance, there are a men in a company, which is divided into b equal groups. How many men are there in each group? Evidently $a \div b$. If a be not exactly divisible by b, the result is as impossible as taking b men from a men, when $b > a$.

In Arithmetic we proceed to prove all the laws of fractions, without inquiring whether they can be applied in all cases. So, in Algebra, we shall proceed to operate with and upon negative numbers without inquiring whether or not they will always have a meaning in particular problems.

CHAPTER II.

THE FOUR FUNDAMENTAL OPERATIONS WITH ALGEBRAIC NUMBER.

If, in the four indicated operations,

$$a + b,\ a - b,\ a \times b,\ a \div b,$$

particular numerical values, *positive* or *negative*, be substituted for a and b, say $^{+}5$ for a and $^{-}2$ for b, we have:

$$a+b={}^{+}5+{}^{-}2;\ a-b={}^{+}5-{}^{-}2;\ a\times b={}^{+}5\times{}^{-}2;\ a\div b={}^{+}5\div{}^{-}2.$$

The indicated operations in the second members of these equations can be performed, and the methods of carrying them out will now be given.

§ 1. ADDITION OF ALGEBRAIC NUMBERS.

1. When the numbers to be added represent only *absolute* values, addition in Algebra differs in no respect from addition in ordinary Arithmetic. But, on account of the introduction of algebraic numbers, the meaning of addition, and of the other fundamental operations, must be enlarged.

The sign of addition has already been used in connection with algebraic numbers (Ch. I., § 2, Arts. 11 and 12), and the definition of addition of algebraic numbers must be based upon that use; that is, upon the properties of positive and negative numbers.

2. Addition *of one algebraic number to another is the process of uniting it with the other into one aggregate.*

As in Arithmetic, the one number is said to be **added** to the other, and the result of the addition is called the **Sum**.

Addition of Numbers with Like Signs.

3. Consider first particular examples.

Ex. 1. Add $^+3$ to $^+4$.

The three positive units, $^+3$, when united by addition with the four positive units, $^+4$, give an aggregate of *four plus three*, or *seven*, positive units. That is,

$$^+4 + {}^+3 = {}^+(4+3) = {}^+7.$$

Ex. 2. Add $^-3$ to $^-4$.

The three negative units, $^-3$, when united by addition with the four negative units, $^-4$, give an aggregate of *four plus three*, or *seven*, negative units. That is,

$$^-4 + {}^-3 = {}^-(4+3) = {}^-7.$$

In the above examples the representation has been limited to the addition of integers with like signs. But the nature of the process, according to the definition of addition, remains the same, if one or both of the numbers be a fraction.

Ex. 3. $^+2 + {}^+(\frac{3}{4}) = {}^+(2 + \frac{3}{4}) = {}^+2\frac{3}{4}$.

Ex. 4. $^-2 + {}^-(\frac{3}{4}) = {}^-(2 + \frac{3}{4}) = {}^-2\frac{3}{4}$.

4. The examples of Art. 3 illustrate the following principle of addition:

To add one algebraic number to another, with like sign (i.e., *both numbers positive or both negative*), *add arithmetically the absolute value of the one number to the absolute value of the other, and prefix to the sum the common sign of quality.* Or, stated symbolically,

$$^+a + {}^+b = {}^+(a+b) \qquad \text{(i.).}$$

$$^-a + {}^-b = {}^-(a+b) \qquad \text{(ii.).}$$

Observe that $a+b$ corresponds to $4+3=7$, in the particular examples of Art. 3, and means the arithmetical sum of the absolute values of ^+a and ^+b (or of ^-a and ^-b), just as 7 is the sum of 3 and 4, the absolute values of $^+3$ and $^+4$ (or of $^-3$ and $^-4$).

The proof of the principle is as follows :

In (i.), the positive units and parts of positive units represented by ${}^{+}b$, when united by addition with the positive units and parts of positive units represented by ${}^{+}a$, give an aggregate of positive units and parts of positive units represented by ${}^{+}(a+b)$. In like manner (ii.) can be proved.

EXERCISES I.

Find the results of the following indicated additions:

1. ${}^{+}11 + {}^{+}5$. **2.** ${}^{+}20 + {}^{+}2$. **3.** ${}^{+}105 + {}^{+}3$.

4. ${}^{-}7 + {}^{-}12$. **5.** ${}^{-}9 + {}^{-}5$. **6.** ${}^{-}2\frac{2}{3} + {}^{-}3\frac{1}{3}$.

Add

7. ${}^{+}17$ to ${}^{+}5$. **8.** ${}^{-}6$ to ${}^{-}27$. **9.** ${}^{-}4$ to ${}^{-}1$.

10. ${}^{+}127$ to ${}^{+}108$. **11.** ${}^{+}2\frac{3}{4}$ to ${}^{+}7\frac{1}{2}$. **12.** ${}^{-}11\frac{5}{6}$ to ${}^{-}3\frac{3}{8}$.

13. ${}^{+}13$ to ${}^{+}a$. **14.** ${}^{-}103$ to ${}^{-}b$. **15.** ${}^{+}c$ to ${}^{+}104$.

16. ${}^{-}x$ to ${}^{-}7$. **17.** ${}^{+}b$ to ${}^{+}q$. **18.** ${}^{-}m$ to ${}^{-}n$.

What is the value of $m + n$,

19. When $m = {}^{+}5$, $n = {}^{+}3$? **20.** When $m = {}^{-}11$, $n = {}^{-}14$?

21. When $m = {}^{+}3\frac{1}{2}$, $n = {}^{+}2\frac{5}{6}$? **22.** When $m = {}^{-}7\frac{7}{8}$, $n = {}^{-}3\frac{3}{4}$?

What is the value of ${}^{+}a + {}^{+}b$,

23. When $a = 5$, $b = 3$? **24.** When $a = 17$, $b = 29$?

What is the value of ${}^{-}a + {}^{-}b$,

25. When $a = 4$, $b = 5$? **26.** When $a = 11$, $b = 18$?

What is the value of ${}^{+}(a+b)$,

27. When $a = 5$, $b = 6$? **28.** When $a = 4$, $b = 105$?

29. When $a = 2\frac{1}{2}$, $b = 3\frac{3}{4}$? **30.** When $a = 1$, $b = 999$?

What is the value of ${}^{-}(a+b)$,

31. When $a = 11$, $b = 14$? **32.** When $a = 16$, $b = 17$?

33. When $a = 4\frac{1}{5}$, $b = \frac{3}{5}$? **34.** When $a = 1\frac{1}{7}$, $b = 98\frac{6}{7}$?

35. A man owes A 50 dollars and B 20 dollars. Express the number of dollars that he owes, taking debts positively. Express the number, taking debts negatively.

36. A contributes 300 dollars to an undertaking, and B contributes 250 dollars. What number expresses their joint contribution, if contributions be taken positively? What number, if contributions be taken negatively?

37. A thermometer rises 10° in the forenoon and 15° in the afternoon. What number expresses the rise in temperature, if falling temperature be taken negatively?

5. Just as any two positive numbers can be united by addition into one positive number; so, conversely, any positive number can be expressed as the sum of two positive numbers which together contain as many positive units and parts of positive units as the given number contains.

E.g., $^{+}7 = {}^{+}1 + {}^{+}6 = {}^{+}2 + {}^{+}5 = {}^{+}3 + {}^{+}4 = {}^{+}1\frac{3}{4} + {}^{+}5\frac{1}{4}$.

A similar statement is true of negative numbers.

E.g., $^{-}7 = {}^{-}1 + {}^{-}6 = {}^{-}2 + {}^{-}5 = {}^{-}3 + {}^{-}4 = {}^{-}2\frac{1}{2} + {}^{-}4\frac{1}{2}$.

EXERCISES II.

Express as the sum of two positive numbers, in three ways:

1. $^{+}18$. **2.** $^{+}27$. **3.** $^{+}108$. **4.** $^{+}8\frac{1}{2}$.

Express as the sum of two negative numbers, in three ways:

5. $^{-}9$. **6.** $^{-}15$. **7.** $^{-}95$. **8.** $^{-}7\frac{2}{3}$.

9. Express $^{+}(p+q)$ as the indicated sum of two positive numbers.

10. Express $^{-}(x+y)$ as the indicated sum of two negative numbers.

11. Express $^{+}(3+m)$ as the indicated sum of two positive numbers.

12. Express $^{-}(n+5)$ as the indicated sum of two negative numbers.

13. Express $^{+}(a+b+c)$ as the sum of two positive numbers.

Addition of Numbers with Unlike Signs.

6. Consider first particular examples.

Ex. **1.** Add $^{-}2$ to $^{+}5$.

The two negative units, $^{-}2$, when united by addition with the five positive units, $^{+}5$, *cancel two of the five positive units.* There remain then *five minus two,* or *three,* positive units. That is,

$$^{+}5 + {}^{-}2 = {}^{+}(5 - 2) = {}^{+}3.$$

Ex. **2.** Add $^{+}2$ to $^{-}5$.

The two positive units, $^{+}2$, when united by addition with the five negative units, $^{-}5$, *cancel two of the five negative units.* There remain then *five minus two,* or *three,* negative units.

That is, $$^{-}5 + {}^{+}2 = {}^{-}(5 - 2) = {}^{-}3.$$

Observe that in Ex. 1 the sum, $^{+}3$, is of the same quality as the number, $^{+}5$, which has the greater absolute value; and that in Ex. 2 the sum, $^{-}3$, is of the same quality as the number, $^{-}5$, which has the greater absolute value.

In both examples, *the absolute value of the sum is obtained by subtracting the less absolute value,* 2, *from the greater,* 5.

In the above cases the representation has been limited to the addition of integers with unlike signs. But the nature of the process, according to the definition of addition, remains the same, if one or both of the numbers be a fraction.

Ex. **3.** $$^{+}2 + {}^{-}(\tfrac{3}{4}) = {}^{+}(2 - \tfrac{3}{4}) = {}^{+}1\tfrac{1}{4}.$$

Ex. **4.** $$^{-}2 + {}^{+}(\tfrac{3}{4}) = {}^{-}(2 - \tfrac{3}{4}) = {}^{-}1\tfrac{1}{4}.$$

7. The examples of Art. 6 illustrate the following principle of addition:

To add one algebraic number to another, with unlike sign, subtract arithmetically the less absolute value from the greater, and prefix to the remainder the sign of quality of the number which has the greater absolute value. Or, stated symbolically,

$$^{+}a + {}^{-}b = {}^{+}(a - b), \text{ when } a > b \quad \text{(iii.)},$$

$$^{+}a + {}^{-}b = {}^{-}(b - a), \text{ when } a < b \quad \text{(iv.)}.$$

Since the quality of the results of Exx. 1 and 2, Art. 6, depends upon the quality of the number having the greater absolute value, it is necessary in adding ^{-}b to ^{+}a to state which number has the greater absolute value.

In (iii.), $a - b$ corresponds to $5 - 2$ of Ex. 1, Art. 6, and is the remainder of subtracting *arithmetically* the less absolute value from the greater; and $^{+}(a - b)$ corresponds to $^{+}(5 - 2)$, $= {}^{+}3$.

In (iv.), $b - a$ corresponds to $5 - 2$ of Ex. 2, Art. 6, and is the remainder of subtracting *arithmetically* the less absolute value from the greater; and $^{-}(b - a)$ corresponds to $^{-}(5 - 2)$, $= {}^{-}3$.

The proof of the principle is as follows:

In (iii.), the negative units and parts of negative units represented by ^{-}b, when united by addition with the positive units and parts of positive units represented by ^{+}a, cancel an equal number of positive units and parts of positive units. There then remain positive units and parts of positive units represented by $^{+}(a - b)$.

In (iv.), when the negative units and parts of negative units represented by ^{-}b are united by addition with the positive units and parts of positive units represented by ^{+}a, from the former are cancelled an equal number of negative units and parts of negative units. There then remain negative units and parts of negative units represented by $^{-}(b - a)$.

EXERCISES III.

Find the results of the following indicated additions:

1. $^{-}16 + {}^{+}7$. **2.** $^{+}16 + {}^{-}7$. **3.** $^{+}118 + {}^{-}5$.

4. $^{-}111 + {}^{+}17$. **5.** $^{-}3\frac{3}{8} + {}^{+}1\frac{1}{4}$. **6.** $^{+}5\frac{1}{2} + {}^{-}2\frac{3}{4}$.

Add

7. $^{-}20$ to $^{+}5$. **8.** $^{+}20$ to $^{-}5$. **9.** $^{-}11$ to $^{+}4$.

10. $^{+}18$ to $^{-}6$. **11.** $^{-}2\frac{3}{4}$ to $^{+}1\frac{5}{8}$. **12.** $^{-}(\frac{1}{2})$ to $^{+}(\frac{5}{6})$.

13. $^{+}11$ to ^{-}a, when $a > 11$. **14.** $^{+}11$ to ^{-}a, when $a < 11$.

15. $^{-}17$ to ^{+}x, when $x > 17$. **16.** $^{-}17$ to ^{+}x, when $x < 17$.

17. ^{+}m to $^{-}4$, when $m > 4$. **18.** ^{+}m to $^{-}4$, when $m < 4$.

19. ^{-}n to $^{+}3$, when $n > 3$. **20.** ^{-}n to $^{+}3$, when $n < 3$.

What is the value of $p + q$,

21. When $p = {}^{+}71$, $q = {}^{-}53$? **22.** When $p = {}^{+}25$, $q = {}^{-}34$?

23. When $p = {}^{-}18$, $q = {}^{+}39$? **24.** When $p = {}^{-}105\frac{3}{4}$, $q = {}^{+}56\frac{1}{2}$?

What is the value of ${}^{+}a + {}^{-}b$,

25. When $a = 6$, $b = 5$? **26.** When $a = 17$, $b = 23$?

27. When $a = 14\frac{1}{3}$, $b = 19\frac{2}{3}$? **28.** When $a = \frac{1}{4}$, $b = \frac{7}{8}$?

What is the value of ${}^{+}(a - b)$,

29. When $a = 19$, $b = 4$? **30.** When $a = 101$, $b = 2$?

31. When $a = 5\frac{1}{2}$, $b = 2\frac{3}{4}$? **32.** When $a = \frac{1}{10}$, $b = \frac{1}{15}$?

33. Has ${}^{+}(7 - 9)$ any meaning? Give reason.

34. Has ${}^{+}(a - b)$ any meaning, when $a < b$?

What is the value of ${}^{-}(a - b)$,

35. When $a = 23$, $b = 16$? **36.** When $a = 40$, $b = 1$?

37. When $a = 3\frac{5}{6}$, $b = 1\frac{7}{8}$? **38.** When $a = \frac{1}{15}$, $b = \frac{1}{20}$?

39. Has ${}^{-}(7 - 9)$ any meaning? Give reason.

40. Has ${}^{-}(a - b)$ any meaning, when $b > a$?

41. If a man receives 1000 dollars and pays out 200 dollars, what number expresses his net receipts, taking receipts positively? What number, taking receipts negatively?

42. If a man ascends 20 steps on a ladder, and descends 8 steps, how many steps is he from the foot of the ladder, taking steps upward positively? How many, taking steps upward negatively?

43. If a thermometer rises 25° and then falls 20°, how many degrees does it register above its original temperature, taking rises in temperature positively? How many, taking falls in temperature positively?

44. If a balloon ascends 500 feet and then falls 300 feet, how many feet is it above the starting point, if ascent be taken positively? How many, if descent be taken positively?

8. Just as any two numbers with unlike signs can be united by addition into one number, so, conversely, any positive or negative number can be expressed as the sum of two numbers with unlike signs.

E.g., ${}^{+}7 = {}^{+}9 + {}^{-}2 = {}^{-}11 + {}^{+}18 = {}^{-}1 + {}^{+}8 = {}^{+}12\frac{1}{2} + {}^{-}5\frac{1}{2}$.

EXERCISES IV.

Express as the sum of a positive and a negative number, in three ways:

1. $^{+}17$. 2. $^{+}18$. 3. $^{+}29$.

4. $^{-}12$. 5. $^{-}24$. 6. $^{-}15\frac{3}{4}$.

7. Express $^{+}(p - q)$ as the indicated sum of a positive and a negative number.

8. Express $^{-}(p - q)$ as the indicated sum of a positive and a negative number.

9. Express $^{+}(3 - m)$ as the indicated sum of a positive and a negative number.

10. Express $^{-}(n - 5)$ as the indicated sum of a positive and a negative number.

Addition of Three or More Numbers.

9. To unite three or more algebraic numbers by addition, add the second to the first, to that sum add the third, again to that sum the fourth, and so on, until all the numbers have been united by addition.

Ex. 1. $^{+}2 + {}^{+}3 + {}^{+}7 = {}^{+}5 + {}^{+}7 = {}^{+}12$.

Ex. 2. $^{-}3 + {}^{-}5 + {}^{-}2 = {}^{-}8 + {}^{-}2 = {}^{-}10$.

Ex. 3. $^{+}11 + {}^{-}8 + {}^{+}2 = {}^{+}3 + {}^{+}2 = {}^{+}5$.

EXERCISES V.

Find the results of the following indicated additions:

1. $^{+}7 + {}^{-}5 + {}^{+}8$. 2. $^{-}8 + {}^{+}11 + {}^{-}3$. 3. $^{+}17 + {}^{-}23 + {}^{+}4$.

4. $^{-}25 + {}^{+}18 + {}^{-}3$. 5. $^{+}5 + {}^{-}6 + {}^{-}7 + {}^{+}9$. 6. $^{-}81 + {}^{+}70 + {}^{-}180 + {}^{+}12$.

Add

7. $^{-}2$ to $^{+}8 + {}^{-}4$. 8. $^{+}18$ to $^{-}2 + {}^{+}105$. 9. $^{-}5$ to $^{-}11 + {}^{+}15$.

Find the results of the following indicated additions, first uniting the numbers within the parentheses:

10. $^{+}7 + ({}^{+}8 + {}^{-}3)$. 11. $^{+}11 + ({}^{-}12 + {}^{+}2)$.

12. $({}^{+}2 + {}^{-}3) + ({}^{-}11 + {}^{+}12)$. 13. $({}^{+}5 + {}^{-}8) + ({}^{-}12 + {}^{+}3)$.

14. $({}^{-}1 + {}^{-}2) + ({}^{+}5 + {}^{-}6) + ({}^{+}13 + {}^{-}10 + {}^{+}12)$.

Add

15. $^-3 + {}^+12$ to $^-1 + {}^-13$.

16. $^-16 + {}^+13$ to $^+15 + {}^-108$.

17. $^-120 + {}^+133$ to $^+12 + {}^-100$.

18. $^-1\frac{3}{4} + {}^+7\frac{5}{6}$ to $^-6\frac{1}{6} + {}^+(\frac{1}{12})$.

19. Add $^+5$ to $^+a + {}^-b$, when $a > b$.

20. Add $^-5$ to $^+a + {}^-b$, when $a < b$.

21. Add $^+7$ to $^-a + {}^+b$, when $a < b$.

22. Add $^-7$ to $^-a + {}^+b$, when $a > b$.

What is the value of $^+a + {}^-b + {}^+c$,

23. When $a = 1, b = 2, c = 3$?

24. When $a = 11, b = 12, c = 18$?

What is the value of $a + b + c + d$,

25. When $a = {}^-3,\ b = {}^+5,\ c = {}^-4,\ d = {}^-7$?

26. When $a = {}^+10,\ b = {}^+12,\ c = {}^-13,\ d = {}^-14$?

What is the value of $a + b$,

27. When $a = {}^+2 + {}^-3,\ b = {}^-8 + {}^+7$?

28. When $a = {}^-5 + {}^+3,\ b = {}^+11 + {}^-4$?

What is the value of $a + b$, wherein $a = m + n$ and $b = p + q$,

29. When $m = {}^-1,\ n = {}^+2,\ p = {}^-3,\ q = {}^+4$?

30. When $m = {}^+8,\ n = {}^-3,\ p = {}^-5,\ q = {}^-9$?

31. When $m = {}^-11,\ n = {}^-13,\ p = {}^+5,\ q = {}^+6$?

32. When $m = {}^+2,\ n = {}^+3,\ p = {}^-4,\ q = {}^-5$?

33. The temperature on Sunday was 45° above zero. On Monday it fell 5°, on Tuesday 3° more, on Wednesday it rose 10°, on Thursday it fell 2°, and on Friday it rose 6°. How many degrees did it register on Friday above or below the temperature on Sunday, taking rises in temperature positively? How many, taking falls in temperature positively?

34. A man walks 15 miles due east one day, 12 miles due west the next day, and 11 miles due east the third day. How many miles is he from the starting point, if distances west be taken positively? How many, if distances east be taken positively?

The Associative and Commutative Laws for Addition.

10. In the preceding articles the process of addition has been carried out from left to right from number to number.

E.g., $$\begin{aligned} {}^{+}7 + {}^{-}3 + {}^{+}5 + {}^{-}8 &= {}^{+}4 + {}^{+}5 + {}^{-}8 \\ &= {}^{+}9 + {}^{-}8 \\ &= {}^{+}1. \end{aligned}$$

But the result is the same if two or more successive numbers be *associated* in performing the additions.

E.g., $$\begin{aligned} {}^{+}7 + {}^{-}3 + ({}^{+}5 + {}^{-}8) &= {}^{+}7 + {}^{-}3 + {}^{-}3 \\ &= {}^{+}4 + {}^{-}3 \\ &= {}^{+}1, \text{ as above.} \\ {}^{+}7 + ({}^{-}3 + {}^{+}5 + {}^{-}8) &= {}^{+}7 + ({}^{+}2 + {}^{-}8) \\ &= {}^{+}7 + {}^{-}6 \\ &= {}^{+}1, \text{ as above.} \\ {}^{+}7 + ({}^{-}3 + {}^{+}5) + {}^{-}8 &= {}^{+}7 + {}^{+}2 + {}^{-}8 \\ &= {}^{+}9 + {}^{-}8 \\ &= {}^{+}1, \text{ as above.} \end{aligned}$$

The parentheses in the above illustrations indicate that the numbers within them are to be added first, and the resulting sums then added, the additions in each case being performed from left to right.

The preceding examples illustrate the following principle:

The Associative Law. — *The sum of three or more numbers is the same in whatever way successive numbers are grouped or associated in the process of adding.* Or, stated symbolically,

$$a + b + c = a + (b + c);$$
$$a+b+c+d = a+b+(c+d) = a+(b+c+d) = a+(b+c)+d.$$

11. In an indicated addition, the number on the right of the sign + is to be added to the number on its left.

E.g., In $${}^{+}5 + {}^{-}3, = {}^{+}2,$$
${}^{-}3$ is added to ${}^{+}5$; while in
$${}^{-}3 + {}^{+}5, = {}^{+}2,$$

$^{+}5$ is added to $^{-}3$. But the result is the same, whichever of the two numbers, $^{+}5$ and $^{-}3$, be added to the other. That is,

$$^{+}5 + {}^{-}3 = {}^{-}3 + {}^{+}5.$$

In like manner, by performing the indicated additions from left to right, we can verify that

$$\begin{aligned} ^{+}4 + {}^{-}7 + {}^{-}3 + {}^{+}2 &= {}^{-}7 + {}^{+}4 + {}^{+}2 + {}^{-}3 \\ &= {}^{+}2 + {}^{-}3 + {}^{-}7 + {}^{+}4 \\ &= \text{etc.,} \end{aligned}$$

in which the numbers to be added are *commuted*, that is, the order in which they are added is changed.

The above examples illustrate the following principle:

The Commutative Law. — *The sum of two or more numbers is the same in whatever order they may be added.* Or, stated symbolically,

$$a + b = b + a$$

$$\begin{aligned} a + b + c + d &= b + a + d + c \\ &= d + c + b + a \\ &= \text{etc.} \end{aligned}$$

12. The proof of the principles enunciated in Arts. 10 and 11 is as follows:

The total number of units and parts of units, positive and negative, in the given numbers is the same in whatever way they may be grouped or arranged; a given number of positive units will always cancel an equal number of negative units, and *vice versa;* and a given number of parts of positive units will always cancel an equal number of like parts of negative units, and *vice versa.* Therefore the final result will be the same, whatever order or way of associating the units and parts of units may be used.

13. Since $a + b = b + a$, the two numbers a and b, when united by addition, are given the common name **Summand.**

14. The Associative and Commutative Laws may be applied simultaneously.

E.g., $^{-}2 + {}^{+}4 + {}^{-}3 + {}^{+}1 = ({}^{-}2 + {}^{-}3) + ({}^{+}4 + {}^{+}1) = {}^{-}5 + {}^{+}5 = 0.$

In general, $a + b + c = a + (c + b) = c + (a + b)$, etc.

15. In adding three or more numbers, some of which are positive and some negative, the Commutative and Associative Laws enable us to employ the following method:

Add all the numbers of one sign, then all the numbers of the opposite sign, and add the two resulting sums.

E.g.,

$$^-8 + {}^+3 + {}^-5 + {}^+7 + {}^+3 = {}^-8 + {}^-5 + {}^+3 + {}^+7 + {}^+3 = {}^-13 + {}^+13 = 0.$$

EXERCISES VI.

Find, in three different ways, by applying the Commutative Law, the values of:

1. $^+18 + {}^-4 + {}^+2$. **2.** $^+12 + {}^-13 + {}^+1$. **3.** $^-20 + {}^-3 + {}^+17$.

4. $^-38 + {}^+27 + {}^+5 + {}^-18$. **5.** $^+72 + {}^-18 + {}^-4 + {}^+9$.

Find, in the most convenient way, the values of:

6. $^+99 + {}^-15 + {}^+1$. **7.** $^-998 + {}^+500 + {}^-2$.

8. $^+333\frac{1}{3} + {}^-125 + {}^+66\frac{2}{3}$.

Find, in the most convenient way, the value of $a + b + c + d$,

9. When $a = {}^-5$, $b = {}^+100$, $c = {}^-95$, $d = {}^+4$.

10. When $a = {}^-763$, $b = {}^+1000$, $c = {}^-237$, $d = {}^-3$.

Find the values of Exx. 4 and 5, in the following ways:

11. Associating the third and fourth summands.

12. Associating the first and third summands.

13. Associating the second and fourth summands.

Find the values of the following expressions by the method of Art. 15:

14. $^+3 + {}^-4 + {}^-6 + {}^+9 + {}^+2$. **15.** $^-5 + {}^+7 + {}^+19 + {}^-15 + {}^-22$.

16. $^-13 + {}^+5 + {}^-15 + {}^+8 + {}^-4$. **17.** $^-(\frac{2}{3}) + {}^+(\frac{5}{6}) + {}^-(\frac{7}{8}) + {}^-(\frac{5}{9}) + {}^+(\frac{1}{5})$.

16. In ordinary Arithmetic to add a number to any number increases the latter.

E.g., $7 + 4 = 11$, and $11 > 7$.

But such is not always the case in adding one algebraic number to another.

E.g., $^+7 + {}^+4 = {}^+11$ and $^+7 + {}^-4 = {}^+3$.

In the first case $^+4$ is added to $^+7$, and the result, $^+11$, is greater than $^+7$; in the second case $^-4$ is added to $^+7$, and the result, $^+3$, is less than $^+7$.

Also, $\quad ^-7 + ^+4 = ^-3$ and $^-7 + ^-4 = ^-11$.

In the first case $^+4$ is added to $^-7$, and the result, $^-3$, is greater than $^-7$; in the second case $^-4$ is added to $^-7$, and the result, $^-11$, is less than $^-7$.

The preceding examples illustrate the following principle:

To add a positive number to any number increases the latter; while to add a negative number to any number decreases the latter.

For, $\quad ^+a + ^+b = ^+(a+b)$,

and $\quad ^+(a+b) > ^+a$, by Ch. I., § 2, Art. 19 (i.);

also, if $a > b$, $\quad ^-a + ^+b = ^-(a-b)$,

and $\quad ^-(a-b) > ^-a$, by Ch. I., § 2, Art. 19 (ii.);

finally, if $a < b$, $\quad ^-a + ^+b = ^+(b-a)$,

and $\quad ^+(b-a) > ^-a$, by Ch. I., § 2, Art. 19 (iii.).

Likewise, if $a > b$, $\quad ^+a + ^-b = ^+(a-b)$,

and $\quad ^+(a-b) < ^+a$, by Ch. I., § 2, Art. 19 (i.);

also, if $a < b$, $\quad ^+a + ^-b = ^-(b-a)$,

and $\quad ^-(b-a) < ^+a$, by Ch. I., § 2, Art. 19 (iv.);

finally, $\quad a + ^-b = ^-(a+b)$,

and $\quad ^-(a+b) < ^-a$, by Ch. I., § 2, Art. 19 (ii.).

17. The results of Art. 16 may also be stated thus:

(i.) *The sum of two algebraic numbers is greater than either of them when both are positive.*

E.g., $\quad ^+7 + ^+4 = ^+11$, and $^+11 > ^+7$, $^+11 > ^+4$.

(ii.) *The sum of two algebraic numbers is less than either of them when both are negative.*

E.g., $\quad ^-7 + ^-4 = ^-11$, and $^-11 < ^-7$, $^-11 < ^-4$.

(iii.) *When one of two algebraic numbers is positive and the other negative, their sum is less than the positive number but greater than the negative number.*

E.g., $\quad ^+7 + ^-4 = ^+3$, and $^+3 < ^+7$, $^+3 > ^-4$.

Property of Zero in Addition.

18. We have $$^{+}3 + {}^{+}2 + {}^{-}2 = {}^{+}3.$$

But $$^{+}3 + {}^{+}2 + {}^{-}2 = {}^{+}3 + ({}^{+}2 + {}^{-}2), \text{ by Assoc. Law,}$$
$$= {}^{+}3 + 0, \text{ since } {}^{+}2 + {}^{-}2 = 0.$$

Therefore, by Axiom (iv.), $^{+}3 + 0 = {}^{+}3$.

In general, $$N + {}^{+}a + {}^{-}a = N.$$

But $$N + {}^{+}a + {}^{-}a = N + ({}^{+}a + {}^{-}a) = N + 0.$$

Therefore, by Axiom (iv.), $N + 0 = N$. (i.)

From (i.), by the Commutative Law,

$$0 + N = N. \qquad \text{(ii.)}$$

In particular, $$0 + 0 = 0. \qquad \text{(iii.)}$$

19. The following principle will be useful in subsequent work:

If the same number or equal numbers be added to equal numbers, the sums will be equal.

If $a = b$, and $A = B$, then $a + A = b + B$.

For, $a + A = a + A$, by Axiom (i.). (1)

Since $b = a$, and $B = A$, we can, by Axiom (iii.), substitute b for a, and B for A, in the second member of (1). We thus obtain

$$a + A = b + B.$$

§ 2. SUBTRACTION OF ALGEBRAIC NUMBERS.

1. **Subtraction** *is the inverse of addition.* In addition two numbers are given, and it is required to find their sum. In subtraction the sum and one of the numbers are given, and it is required to find the other number.

As in ordinary Arithmetic, the given sum is called the **Minuend,** the given number the **Subtrahend,** and the required number the **Remainder.**

Ex. 1. Subtract $^{+}3$ from $^{+}7 + {}^{+}3$.

We have $({}^{+}7 + {}^{+}3) - {}^{+}3 = {}^{+}7$, by the definition of subtraction.

Ex. 2. Subtract $^+7$ from $(^+7 + {}^+3)$.

We have $(^+7 + {}^+3) - {}^+7 = {}^+3$, by the definition of subtraction.

Ex. 3. Subtract $^-2$ from $^+9 + {}^-2$.

We have $(^+9 + {}^-2) - {}^-2 = {}^+9$, by the definition of subtraction.

Ex. 4. Subtract $^-2\frac{1}{8}$ from $^-4\frac{1}{4} + {}^-2\frac{1}{8}$.

We have $(^-4\frac{1}{4} + {}^-2\frac{1}{8}) - {}^-2\frac{1}{8} = {}^-4\frac{1}{4}$, by the definition of subtraction.

That is, *if from the sum of two numbers either of the numbers be subtracted, the remainder is the other number.*

In general, if the given sum be $a + b$, we have, by the definition of subtraction,

$$(a + b) - b = a, \qquad (1)$$

and

$$(a + b) - a = b. \qquad (2)$$

2. It follows directly from (1) and (2), Art. 1, that:

(i.) *The minuend is equal to the sum of the subtrahend and the remainder.*

(ii.) *The subtrahend is equal to the minuend minus the remainder.*

EXERCISES VII.

Find the results of the following indicated subtractions:

1. $(^+2+{}^+9)-{}^+2$.
2. $(^+2+{}^+9)-{}^+9$.
3. $(^+4+{}^-5)-{}^+4$.
4. $(^+4+{}^-5)-{}^-5$.
5. $(^-7+{}^-11)-{}^-7$.
6. $(^-7+{}^-11)-{}^-11$.
7. $(^+a+{}^+1)-{}^+a$.
8. $(^+a+{}^+1)-{}^+1$.
9. $(^-x+{}^-7)-{}^-x$.
10. $(^-x+{}^-7)-{}^-7$.
11. $(^-m+{}^+n)-{}^-m$.
12. $(^-m+{}^+n)-{}^+n$.
13. $(^-2+{}^-3+{}^+8)-(^-2+{}^-3)$.
14. $(^-2+{}^-3+{}^+8)-{}^+8$.

Find the result of subtracting

15. $^+5$ from $^-7+{}^+5$.
16. $^-19$ from $^+11+{}^-19$.
17. ^+x from $^+3+{}^+x$.
18. $^+7$ from $^-a+{}^+7$.
19. ^-m from $^-3+{}^-m$.
20. $^-11$ from $^-x+{}^-11$.
21. $^+4+{}^-5$ from $^+4+{}^-5+{}^+6$.
22. $^-a+{}^+b$ from $^+c+{}^-a+{}^+b$.

3. The minuend is, as a rule, a single number, and does not appear as a sum of two numbers, one of which is the given subtrahend.

We must, therefore, derive from the definition of subtraction a principle which will enable us to subtract any one number from any other.

Ex. **1.** Subtract $^{+}5$ from $^{+}7$.

In $^{+}7 - {}^{+}5$, the minuend, $^{+}7$, is to be expressed as the sum of two numbers, *one of which is* $^{+}5$. But since $^{-}5 + {}^{+}5 = 0$, and $^{+}7 = {}^{+}7 + 0$, we have

$$\begin{aligned} ^{+}7 &= {}^{+}7 + ({}^{-}5 + {}^{+}5) \\ &= ({}^{+}7 + {}^{-}5) + {}^{+}5, \text{ by Assoc. Law.} \end{aligned}$$

That is, $^{+}7$ is the sum of the given number, $^{+}5$, and the number $^{+}7 + {}^{-}5$. Therefore, by the definition of subtraction, $^{+}7 + {}^{-}5$ is the required remainder; or

$$\begin{aligned} ^{+}7 - {}^{+}5 &= [({}^{+}7 + {}^{-}5) + {}^{+}5] - {}^{+}5 \\ &= {}^{+}7 + {}^{-}5 = {}^{+}2. \end{aligned}$$

That is, *to subtract* $^{+}5$ *from* $^{+}7$ *is equivalent to adding* $^{-}5$ *to* $^{+}7$.

Ex. **2.** Subtract $^{-}5$ from $^{+}7$.

In $^{+}7 - {}^{-}5$, the minuend, $^{+}7$, is to be expressed as the sum of two numbers, *one of which is* $^{-}5$. But since $^{+}5 + {}^{-}5 = 0$, and $^{+}7 = {}^{+}7 + 0$, we have

$$\begin{aligned} ^{+}7 &= {}^{+}7 + ({}^{+}5 + {}^{-}5) \\ &= ({}^{+}7 + {}^{+}5) + {}^{-}5, \text{ by Assoc. Law.} \end{aligned}$$

That is, $^{+}7$ is the sum of the given number, $^{-}5$, and the number $^{+}7 + {}^{+}5$. Therefore, by the definition of subtraction, $^{+}7 + {}^{+}5$ is the required remainder; or

$$\begin{aligned} ^{+}7 - {}^{-}5 &= [({}^{+}7 + {}^{+}5) + {}^{-}5] - {}^{-}5. \\ &= {}^{+}7 + {}^{+}5 = {}^{+}12. \end{aligned}$$

That is, *to subtract* $^{-}5$ *from* $^{+}7$ *is equivalent to adding* $^{+}5$ *to* $^{+}7$.

The preceding examples illustrate the following principle of subtraction:

To subtract one number from another number, reverse the sign of quality of the former (the subtrahend), from + *to* −, *or from* − *to* +, *and add.*

Or, stated symbolically,

$$\mathbf{N} - {}^{+}b = \mathbf{N} + {}^{-}b,$$

$$\mathbf{N} - {}^{-}b = \mathbf{N} + {}^{+}b.$$

E.g.,

$$^{+}2 - {}^{+}3 = {}^{+}2 + {}^{-}3, \;= {}^{-}1.$$

$$^{+}2 - {}^{-}3 = {}^{+}2 + {}^{+}3, \;= {}^{+}5.$$

$$^{-}2 - {}^{+}3 = {}^{-}2 + {}^{-}3, \;= {}^{-}5.$$

$$^{-}2 - {}^{-}3 = {}^{-}2 + {}^{+}3, \;= {}^{+}1.$$

4. The proof of the principle enunciated in Art. 3 is as follows:

Let N denote any number *positive or negative.* Then in

$$N - {}^{+}b,$$

N is to be expressed as the sum of two numbers, *one of which is* ^{+}b.

But since $\quad {}^{-}b + {}^{+}b = 0$ and $N = N + 0$, we have

$$N = N + ({}^{-}b + {}^{+}b)$$
$$= (N + {}^{-}b) + {}^{+}b,$$

by the Associative Law.

That is, N is the sum of the given number, ^{+}b, and the number $N + {}^{-}b$. Therefore, by the definition of subtraction, $N + {}^{-}b$ is the required remainder; or

$$\mathbf{N} - {}^{+}b = \mathbf{N} + {}^{-}b. \qquad (1)$$

That is, *to subtract* ^{+}b *from* N *(any number) is equivalent to adding* ^{-}b *to* N.

In

$$N - {}^{-}b,$$

N is to be expressed as the sum of two numbers, *one of which is* ^{-}b.

But since $\quad {}^{+}b + {}^{-}b = 0$ and $N = N + 0$, we have

$$N = N + ({}^{+}b + {}^{-}b)$$
$$= (N + {}^{+}b) + {}^{-}b,$$

by the Associative Law.

That is, N is the sum of the given number, ^{-}b, and the number $N + {}^{+}b$. Therefore by the definition of subtraction, $N + {}^{+}b$ is the required remainder; or

$$\mathbf{N} - {}^{-}b = \mathbf{N} + {}^{+}b. \qquad (2)$$

That is, *to subtract* ^{-}b *from* N *(any number) is equivalent to adding* ^{+}b *to* N.

EXERCISES VIII.

Find the results of the following indicated subtractions:

1. $^{+}18 - {}^{+}5$. **2.** $^{-}28 - {}^{+}17$. **3.** $^{+}105 - {}^{+}14$.

4. $^{-}204 - {}^{+}203$. **5.** $^{+}7 - {}^{+}8$. **6.** $^{+}11 - {}^{+}25$.

7. $^{+}20 - {}^{+}30$. 8. $^{+}41 - {}^{-}50$. 9. $^{+}5 - {}^{-}18$.

10. $^{-}30 - {}^{-}35$. 11. $^{-}65 - {}^{-}100$. 12. $^{-}108 - {}^{-}110$.

13. $^{-}3 - {}^{-}2$. 14. $^{-}108 - {}^{-}59$.

15. $^{-}85 - {}^{-}79$. 16. $^{+}44 - {}^{-}40$.

Subtract

17. $^{-}5$ from $^{-}8$. 18. $^{+}11$ from $^{-}6$. 19. $^{+}19$ from $^{-}27$.

20. $^{-}16$ from $^{+}11$. 21. $^{-}33$ from $^{-}52$. 22. $^{-}1$ from $^{+}10$.

23. $^{+}3$ from ^{+}a, when $a > 3$. 24. $^{+}3$ from ^{+}a, when $a < 3$.

25. $^{-}17$ from ^{-}x, when $x > 17$. 26. $^{-}17$ from ^{-}x, when $x < 17$.

27. $^{-}11$ from ^{+}x. 28. $^{+}14$ from ^{-}y.

29. ^{+}m from $^{+}3$, when $m > 3$. 30. ^{+}m from $^{+}3$, when $m < 3$.

31. ^{-}n from $^{-}17$, when $n > 17$. 32. ^{-}n from $^{-}17$, when $n < 17$.

33. ^{-}u from $^{+}9$. 34. ^{+}v from $^{-}11$.

What is the value of $m - n$,

35. When $m = {}^{+}4$, $n = {}^{+}3$? 36. When $m = {}^{+}5$, $n = {}^{+}6$?

37. When $m = {}^{-}7$, $n = {}^{+}8$? 38. When $m = {}^{-}11$, $n = {}^{+}3$?

39. When $m = {}^{-}4$, $n = {}^{-}9$? 40. When $m = {}^{-}8$, $n = {}^{-}2$?

What is the value of $^{+}a - {}^{+}b$,

41. When $a = 7$, $b = 4$? 42. When $a = 3$, $b = 5$?

What is the value of $^{+}a - {}^{-}b$,

43. When $a = 1$, $b = 2$? 44. When $a = 7$, $b = 8$?

What is the value of $^{-}a - {}^{+}b$,

45. When $a = 7$, $b = 11$? 46. When $a = 3$, $b = 8$?

What is the value of $^{-}a - {}^{-}b$,

47. When $a = 2$, $b = 5$? 48. When $a = 8$, $b = 4$?

5. The following illustrations may help the student to understand the two relations:

$$N - {}^{+}a = N + {}^{-}a \text{ and } N - {}^{-}a = N + {}^{+}a.$$

Ex. **1.** A man's net profits last year were 1200 dollars. This year his income is 150 dollars less, and his expenditures are the same. What are his net profits for this year?

This year's net profits are equal to last year's net profits minus 150 *dollars income.*

If net profits and income be taken positively, and expenditures negatively, the last statement, expressed algebraically, is

$$^{+}1200 - {}^{+}150 = {}^{+}1200 + {}^{-}150.$$

That is, *to take away* 150 *dollars income is equivalent to adding* 150 *dollars expenditures.*

Ex. 2. A man's net profits in business last year were 1200 dollars. This year his income is the same and his expenditures are 150 dollars less. What are his net profits for this year?

This year's net profits are equal to last year's net profits minus 150 *dollars expenditures.*

The algebraic statement of this relation is

$$^{+}1200 - {}^{-}150 = {}^{+}1200 + {}^{+}150.$$

That is, *to take away* 150 *dollars expenditures is equivalent to adding* 150 *dollars profits.*

6. The following examples illustrate the meaning of results in the subtraction of algebraic numbers.

Ex. 1. Two men, A and B, starting from the same point, P, walk at different rates in the same direction, A 8 miles to the point Q, B 11 miles to the point R. How far is B then from A?

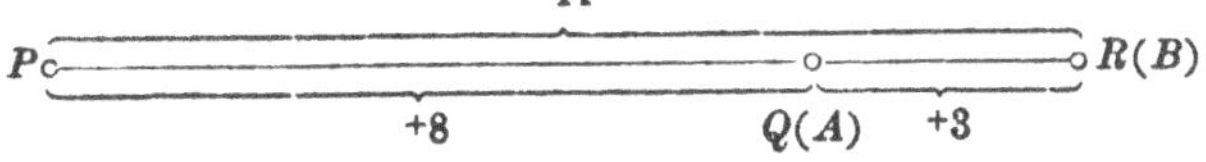

FIG. 1.

As we have seen in Chapter I., distances in one and the same direction may be represented by numbers of the same sign. Let distances toward the right be taken positively, as in Fig. 1, and consequently distances toward the left negatively.

The distance of B from A is then represented by QR, and

$$\begin{aligned} QR &= PR - PQ \\ &= {}^{+}11 - {}^{+}8 \\ &= {}^{+}3. \end{aligned}$$

The *positive* result, $^{+}3$, shows that B is 3 miles to the *right* of A.

In general, however far either may walk, the distance of B from A will always be obtained by subtracting A's distance from the starting point from B's distance from the same point.

Ex. 2. If A walks 15 miles and B walks 11 miles, both to the right, their distances from P are still both positive, as in Fig. 2.

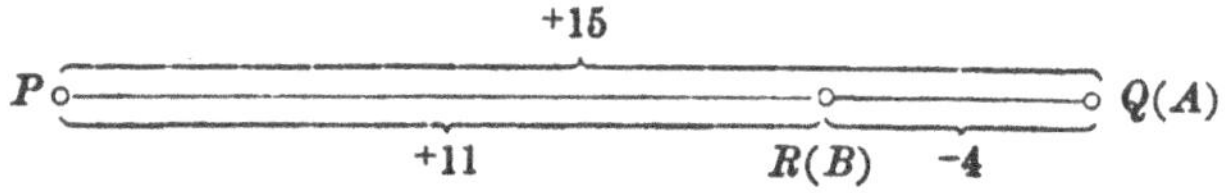

FIG. 2.

We then have

$$\begin{aligned} QR &= PR - PQ \\ &= {}^{+}11 - {}^{+}15 \\ &= {}^{-}4. \end{aligned}$$

The *negative* result, $^{-}4$, shows that B is now 4 miles to the *left* of A.

Ex. 3. If A walks 8 miles to the right and B walks 11 miles to the left, A's distance from P is positive and B's distance from P is negative, as in Fig. 3.

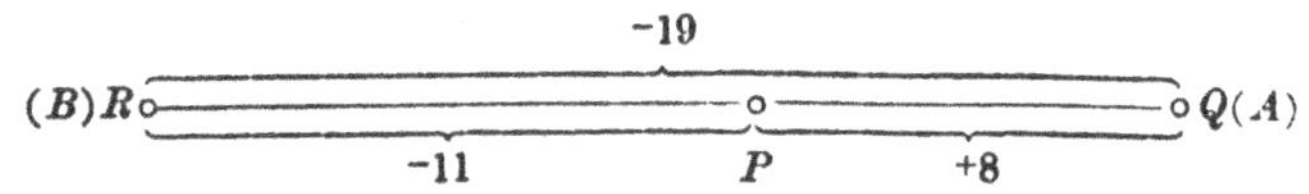

FIG. 3.

We then have

$$\begin{aligned} QR &= PR - PQ \\ &= {}^{-}11 - {}^{+}8 \\ &= {}^{-}19. \end{aligned}$$

The *negative* result, $^{-}19$, shows that B is now 19 miles to the *left* of A.

Ex. 4. If A walks 8 miles to the left and B walks 11 miles to the right, A's distance from P is now negative and B's distance from P is now positive, as in Fig. 4.

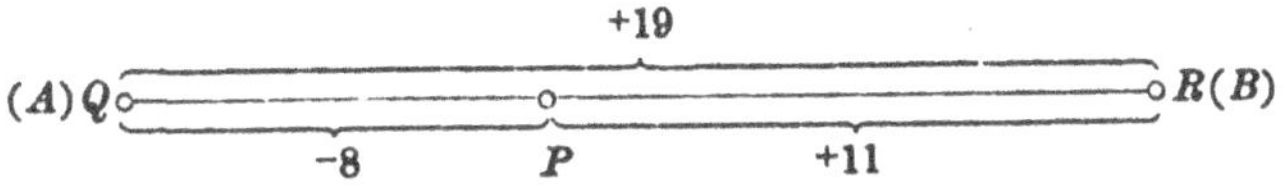

FIG. 4.

We then have
$$\begin{aligned} QR &= PR - PQ \\ &= {}^{+}11 - {}^{-}8 \\ &= {}^{+}19. \end{aligned}$$

The *positive* result, ${}^{+}19$, shows that B is now 19 miles to the *right* of A.

Ex. 5. Finally, if A walks 8 miles to the left and B 11 miles to the left, their distances from P are both negative, as in Fig. 5.

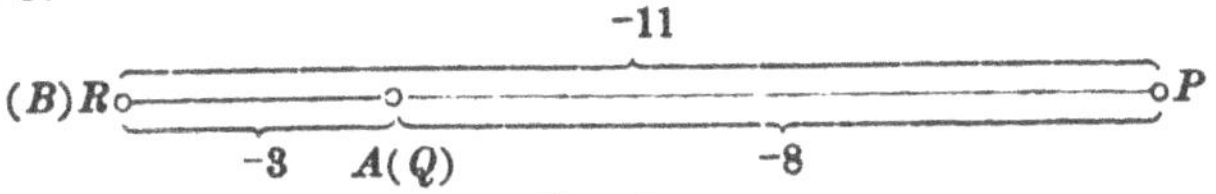

FIG. 5.

We then have
$$\begin{aligned} QR &= PR - PQ \\ &= {}^{-}11 - {}^{-}8 \\ &= {}^{-}3. \end{aligned}$$

The *negative* result, ${}^{-}3$, shows that B is 3 miles to the *left* of A.

Thus, however far the men may walk, or in whichever direction (along one and the same line), a uniform method of obtaining the distance of B from A leads to intelligible results. They therefore give intelligible meanings to all the principles of subtraction.

EXERCISES IX.

1. A's assets are 3100 dollars and B's are 2700 dollars. What number expresses the excess of A's assets over B's, if assets be taken positively? What number, if assets be taken negatively?

2. A's assets are a dollars and B's are b dollars. What number expresses the excess of A's assets over B's, if assets be taken positively? What number, if assets be taken negatively?

What are the meanings of the results of Ex. 2,

3. When $a=3500$, $b=2750$? **4.** When $a=2000$, $b=2000$?

5. When $a=2600$, $b=3000$?

6. A's assets are 2000 dollars, and B owes 500 dollars. What number expresses the excess of A's fortune over B's, if assets be taken positively? What number, if debts be taken positively?

7. The temperature within a room is 65° above zero, and out of doors it is 3° below zero. What number expresses the excess of temperature within the room over that out of doors, if temperature above zero be taken positively? What number, if temperature below zero be taken positively?

8. The temperature in Chicago on a certain day was $a°$ and in Philadelphia $b°$. What number expresses the excess of temperature in Chicago over that in Philadelphia?

What is the meaning of the result of Ex. 8, taking temperature above zero positively,

9. When $a={}^{+}90$, $b={}^{+}68$? **10.** When $a={}^{+}65$, $b={}^{+}98$?

11. When $a={}^{-}12$, $b={}^{-}4$? **12.** When $a={}^{-}5$, $b={}^{-}8$?

13. When $a={}^{+}5$, $b=\ 2$? **14.** When $a={}^{-}b$, $b={}^{+}3$?

15. Two trains, on a north-and-south track, start at the same time from the same station. After a certain time the first is a miles and the second is b miles from the starting point. How far is the first train then from the second?

What is the meaning of the result of Ex. 15, taking distances north positively,

16. When $a={}^{+}50$, $b={}^{+}32$? **17.** When $a={}^{+}75$, $b={}^{+}81$?

18. When $a={}^{+}40$, $b={}^{-}20$? **19.** When $a={}^{-}32$, $b={}^{+}58$?

20. When $a={}^{-}63$, $b={}^{-}55$?

21. The city A is in latitude $a°$ and the city B is in latitude $b°$. What number expresses the difference in latitude between A and B?

What is the meaning of the result of Ex. 21, taking latitude south positively,

22. When $a = {}^{+}22$, $b = {}^{+}17$?
23. When $a = {}^{+}28$, $b = {}^{+}21$?
24. When $a = {}^{+}44$, $b = {}^{-}4$?
25. When $a = {}^{-}23$, $b = {}^{+}11$?
26. When $a = {}^{-}14$, $b = {}^{-}16$?

Successive Additions and Subtractions.

7. Successive subtractions are carried out by applying the principle of subtraction at each step of the process.

E.g.,
$$
\begin{aligned}
{}^{+}7 - {}^{-}2 - {}^{+}4 - {}^{-}8 &= {}^{+}9 - {}^{+}4 - {}^{-}8 \\
&= {}^{+}5 - {}^{-}8 \\
&= {}^{+}13.
\end{aligned}
$$

In like manner successive additions and subtractions are performed.

E.g.,
$$
\begin{aligned}
{}^{+}9 + {}^{-}5 - {}^{+}2 + {}^{+}4 &= {}^{+}4 - {}^{+}2 + {}^{+}4 \\
&= {}^{+}2 + {}^{+}4 \\
&= {}^{+}6.
\end{aligned}
$$

8. The following definition is based upon the principle that every operation of subtraction is equivalent to an operation of addition.

An **Algebraic Sum** is an expression which consists of a chain of indicated additions and subtractions.

E.g., $a - b$, $x + y - z$, etc., are algebraic sums.

EXERCISES X.

Find the results of the following indicated operations:

1. ${}^{-}2 + {}^{+}3 - {}^{+}4$.
2. ${}^{+}11 - {}^{+}13 + {}^{-}12$.
3. ${}^{-}7 - {}^{-}11 + {}^{-}2$.
4. ${}^{+}5 - {}^{-}3 + {}^{-}15 - {}^{+}27$.
5. ${}^{-}11 - {}^{-}12 + {}^{-}15 - {}^{+}29$.

Add

6. ${}^{+}5$ to ${}^{-}8 - {}^{-}3$.
7. ${}^{-}2$ to ${}^{+}17 - {}^{-}4$.
8. ${}^{-}5$ to ${}^{-}12 + {}^{-}3$.

Subtract

9. $^-8$ from $^+15 - ^-12$.

10. $^+7$ from $^-8 - ^+15$.

11. $^-9$ from $^+8 + ^-4$.

Find the results of the following indicated operations, first uniting the numbers within the parentheses:

12. $(^-2 - ^-5) + (^-6 + ^+3)$.

13. $(^+8 - ^-2) + (^-11 - ^+2)$.

14. $(^+18 + ^-2) - (^+8 - ^-2)$.

15. $(^+11 - ^-8) - (^-2 - ^+3)$.

Add

16. $^-2 + ^-3$ to $^+4 - ^-6$.

17. $^+6 - ^-5$ to $^-15 - ^-2$.

Subtract

18. $^-2 + ^-3$ from $^-12 - ^+18$.

19. $^+4 - ^-6$ from $^-19 + ^+17$.

What is the value of $a - b$,

20. When $a = ^-2 + ^-3$, $b = ^+11 - ^+4$?

21. When $a = ^+5 - ^-4$, $b = ^-18 - ^+7$?

What is the value of $^+a - ^+b - ^-c + ^-d$,

22. When $a = 1$, $b = 2$, $c = 4$, $d = 7$?

23. When $a = 19$, $b = 14$, $c = 9$, $d = 4$?

What is the value of $a + b - c - d + e$,

24. When $a = ^+5$, $b = ^-6$, $c = ^+7$, $d = ^-8$, $e = ^+9$?

25. When $a = ^-4$, $b = ^-7$, $c = ^-5$, $d = ^+8$, $e = ^-9$?

What is the value of $a - b$, wherein $a = m + n$ and $b = p - q$,

26. When $m = ^-1$, $n = ^+2$, $p = ^-3$, $q = ^+4$?

27. When $m = ^+5$, $n = ^-6$, $p = ^-11$, $q = ^-12$?

28. A's assets are 1500 dollars and his liabilities are 100 dollars; B's assets are 1200 dollars and his liabilities are 130 dollars. What number expresses the excess of A's fortune over B's, if assets be taken positively? What number, if liabilities be taken positively?

29. In the morning the thermometer registered 52° above zero; by noon it had risen 6°, by 3 o'clock it had risen 4° more, and by 8 o'clock it had fallen 5°. What number expresses the temperature at 8 o'clock, if rises in temperature be taken positively? What number, if falls in temperature be taken positively?

9. It is frequently desirable to speak of a number as an *additive* or *a subtractive number;* that is, as a number to be added or subtracted. In that case we prefix the corresponding sign of operation to the number, and understand that it is to be added to or subtracted from any number, including 0, that may be placed on its left. Thus,

$+{}^{+}2$ means *add pos.* 2 and $+{}^{-}2$ means *add neg.* 2;

$-{}^{+}2$ means *subtract pos.* 2 and $-{}^{-}2$ means *subtract neg.* 2.

In general,

$+{}^{+}b$ means *add pos. b* and $+{}^{-}b$ means *add neg. b.*

$-{}^{+}b$ means *subtract pos. b* and $-{}^{-}b$ means *subtract neg. b.*

Since to subtract any number is equivalent to adding its opposite, we have

$$-{}^{+}b = +{}^{-}b \text{ and } -{}^{-}b = +{}^{+}b,$$

in accordance with the meanings given above.

10. If an *additive* or *subtractive* number stand first in a chain of additions and subtractions, or first within parentheses, it may be regarded as added to or subtracted from 0. Thus,

$$N + (+{}^{+}2 - {}^{+}3) = N + (0 + {}^{+}2 - {}^{+}3),$$

$$N + (-{}^{+}2 - {}^{+}3) = N + (0 - {}^{+}2 - {}^{+}3).$$

The sign of operation of an *additive number* in such a case may be omitted.

E.g., $$N + (+{}^{-}3) = N + {}^{-}3.$$

The Associative and Commutative Laws for Subtraction.

11. Since every operation of subtraction is equivalent to an operation of addition, it follows that the Associative and Commutative Laws which were proved for addition hold also for subtraction, and for successive additions and subtractions.

Ex. 1. $+{}^{+}8 - {}^{+}3 = +{}^{+}8 + {}^{-}3$, since $-{}^{+}3 = +{}^{-}3$

$= +{}^{-}3 + {}^{+}8$, by Comm. Law

$= -{}^{+}3 + {}^{+}8$, since $+{}^{-}3 = -{}^{+}3$.

Observe that *in changing the order of the operations the sign of operation, + or −, must be transferred with each number.* In a chain of additions this precaution was unnecessary, since in that case all the signs of operation were the same, namely, +.

Ex. 2. ${}^{+}9 - {}^{-}5 + {}^{-}2 - {}^{+}4 = {}^{+}14 + {}^{-}2 - {}^{+}4 = {}^{+}12 - {}^{+}4 = {}^{+}8$,

the successive operations being performed from left to right.

Or, in a different order,

${}^{+}9 + {}^{-}2 - {}^{-}5 - {}^{+}4 = {}^{+}7 - {}^{-}5 - {}^{+}4 = {}^{+}12 - {}^{+}4 = {}^{+}8$, as above.

Or, in a still different order,

${}^{+}9 - {}^{+}4 - {}^{-}5 + {}^{-}2 = {}^{+}5 - {}^{-}5 + {}^{-}2 = {}^{+}10 + {}^{-}2 = {}^{+}8$, as above.

As an illustration of the Associative Law, let the second and third numbers of Ex. 2, as first written, be associated. As in applying the Commutative Law, the numbers must be taken *with their signs of operation.*

Thus, $-{}^{-}5 + {}^{-}2 = +{}^{+}5 + {}^{-}2 = +{}^{+}3$.

Consequently, in this way of associating the numbers, we have

${}^{+}9 - {}^{-}5 + {}^{-}2 - {}^{+}4 = {}^{+}9 + {}^{+}3 - {}^{+}4 = {}^{+}12 - {}^{+}4 = {}^{+}8$, as above.

Or, associating the third and fourth numbers, we have

$$+{}^{-}2 - {}^{+}4 = +{}^{-}2 + {}^{-}4 = +{}^{-}6.$$

Consequently,

${}^{+}9 - {}^{-}5 + {}^{-}2 - {}^{+}4 = {}^{+}9 - {}^{-}5 + {}^{-}6 = {}^{+}14 + {}^{-}6 = {}^{+}8$, as above.

The method of applying the Associative Law can, however, be simplified by a proper use of parentheses. We, therefore, in the next article, consider the simpler cases of the use of parentheses.

EXERCISES XI.

Find in three different ways, by applying the Commutative Law, the values of:

1. ${}^{+}8 + {}^{-}3 - {}^{+}4$.
2. ${}^{-}17 - {}^{+}12 + {}^{-}5$.
3. ${}^{+}28 - {}^{-}14 + {}^{-}2$.
4. ${}^{+}4 - {}^{+}2 + {}^{-}1$.
5. ${}^{+}4 - {}^{-}3 - {}^{+}4 + {}^{+}3 - {}^{+}5 + {}^{-}7 - {}^{-}8 - {}^{+}11$.

6. $^-31 - ^-17 + ^-36 + ^+46 - ^-11 - ^+19 + ^-49 + ^+11.$

7. $^-45 + ^+31 - ^-15 - ^+12 + ^-5 - ^-9 + ^-8 + ^+4.$

Find, in the most convenient way, the values of:

8. $^+103 - ^-12 - ^+3.$

9. $^-799 - ^-11 + ^-1.$

10. $^-388 + ^-804 - ^+12 - ^-4.$

Find, in the most convenient way, the value of $a - b + c - d$,

11. When $a = ^+188$, $b = ^-12$, $c = ^-104$, $d = ^-4$.

12. When $a = ^+507$, $b = ^-95$, $c = ^-7$, $d = ^-5$.

Find the values of Exx. 6 and 7 in the following ways:

13. Associating the third and fourth summands.

14. Associating the second, fifth, and eighth summands.

15. Associating the first, third, seventh, and eighth summands.

16. Associating the first and third, and the fifth and seventh summands.

Removal of Parentheses.

12. Parentheses have already been used to inclose an expression which is to be treated as a single number. Thus,

$$^+7 - (^+5 - ^+2)$$

means that the result of the operation $^+5 - ^+2$ is to be subtracted from $^+7$, or

$$^+7 - (^+5 - ^+2) = ^+7 - ^+3 = ^+4.$$

But if the parentheses be omitted from the above expression, we have

$$^+7 - ^+5 - ^+2 = ^+2 - ^+2 = 0, \text{ not } ^+4, \text{ as above.}$$

It will now be shown how parentheses may be removed without changing the value of the expression.

13. We have $^+9 + (^+5 + ^+6) = ^+9 + ^+5 + ^+6$,

since to add the sum $^+5 + ^+6$ is equivalent to adding successively the single numbers of that sum.

Again, $^+9 + (^+5 - ^+6) = ^+9 + (^+5 + ^-6)$, since $- ^+6 = + ^-6$,

$= ^+9 + ^+5 + ^-6$, removing parentheses,

$= ^+9 + ^+5 - ^+6$, since $+ ^-6 = - ^+6$.

The preceding examples illustrate the following principle:

When the sign of addition, +, precedes parentheses, they may be removed; and the signs of operation, + and −, within them be left unchanged ; that is,

$$N+(a+b)=N+a+b,$$
$$N+(a-b)=N+a-b,$$
$$N+(a+b-c)=N+a+b-c, \text{ etc.}$$

For, $\quad N+({}^{+}a+{}^{+}b)=N+{}^{+}a+{}^{+}b,$

since to add the sum ${}^{+}a+{}^{+}b$ is equivalent to adding successively the single numbers of that sum.

$$\begin{aligned} N+({}^{+}a-{}^{+}b) &= N+({}^{+}a+{}^{-}b), \text{ since } -{}^{+}b=+{}^{-}b, \\ &= N+{}^{+}a+{}^{-}b, \text{ removing parentheses,} \\ &= N+{}^{+}a-{}^{+}b, \text{ since } +{}^{-}b=-{}^{+}b. \end{aligned}$$

Evidently the preceding proofs do not depend upon the signs of quality of the numbers within the parentheses, nor upon how many numbers are inclosed.

14. We have $\;{}^{+}9-({}^{+}5+{}^{+}6)={}^{+}9-{}^{+}5-{}^{+}6,$

since to subtract the sum ${}^{+}5+{}^{+}6$ is equivalent to subtracting successively the single numbers of that sum.

$$\begin{aligned} \text{Again, } {}^{+}9-({}^{+}5-{}^{+}6) &= {}^{+}9-({}^{+}5+{}^{-}6), \text{ since } -{}^{+}6=+{}^{-}6, \\ &= {}^{+}9-{}^{+}5-{}^{-}6, \text{ removing parentheses,} \\ &= {}^{+}9-{}^{+}5+{}^{+}6, \text{ since } -{}^{-}6=+{}^{+}6. \end{aligned}$$

The preceding examples illustrate the following principle:

When the sign of subtraction, −, precedes parentheses, they may be removed, if the signs of operation within them be reversed from + to −, and from − to +; that is,

$$N-(+a+b)=N-a-b,$$
$$N-(+a-b)=N-a+b,$$
$$N-(a+b-c)=N-a-b+c, \text{ etc.}$$

For $\quad N-({}^{+}a+{}^{+}b)=N-{}^{+}a-{}^{+}b,$

since to subtract the sum ${}^{+}a+{}^{+}b$ is equivalent to subtracting successively the single numbers of that sum.

$$N-({}^{+}a-{}^{+}b)=N-({}^{+}a+{}^{-}b),\text{ since }-{}^{+}b=+{}^{-}b,$$
$$=N-{}^{+}a-{}^{-}b,\text{ removing parentheses,}$$
$$=N-{}^{+}a+{}^{+}b,\text{ since }-{}^{-}b=+{}^{+}b.$$

Evidently the preceding proofs do not depend upon the signs of quality of the numbers within the parentheses, nor upon how many numbers are inclosed.

EXERCISES XII.

Find the values of the following expressions, first removing parentheses:

1. ${}^{+}12+({}^{+}4+{}^{-}6)$.
2. ${}^{-}15+({}^{-}6-{}^{+}2)$.
3. ${}^{-}22+({}^{+}11+{}^{-}2)$.
4. ${}^{-}12+({}^{-}4-{}^{-}6)$.
5. ${}^{+}28+({}^{-}5-{}^{+}6)$.
6. ${}^{+}18+({}^{-}2+{}^{-}3-{}^{-}5)$.
7. ${}^{+}8+({}^{-}2-{}^{+}3)+({}^{+}5-{}^{-}6)$.
8. ${}^{+}11-({}^{+}12+{}^{-}5)$.
9. ${}^{-}13-({}^{-}11+{}^{-}3)$.
10. ${}^{+}15-({}^{-}6-{}^{+}2)$.
11. ${}^{-}17-({}^{-}3-{}^{-}5)$.
12. ${}^{-}21-({}^{-}4+{}^{-}5-{}^{-}6)$.
13. ${}^{+}14-({}^{+}5-{}^{+}6-{}^{-}7)$.
14. ${}^{+}6-({}^{+}8-{}^{+}2)+({}^{-}12-{}^{+}8)-({}^{-}2+{}^{-}9)$.
15. $m+(n-p)$, when $m={}^{-}4$, $n={}^{-}6$, $p={}^{+}5$.
16. $x-(y-z)$, when $x={}^{+}3$, $y={}^{-}4$, $z={}^{+}5$.
17. $a-(b+c)+(d-e)$, when $a={}^{-}5$, $b={}^{-}6$, $c={}^{+}7$, $d={}^{-}8$, $e={}^{-}9$.

Insertion of Parentheses.

15. The insertion of parentheses is the converse of the process of removing them.

(i.) *An expression may be inclosed within parentheses preceded by the sign of addition, if the signs of operation, + and –, preceding the numbers inclosed within the parentheses remain unchanged.*

E.g.,
$$+{}^{+}7-{}^{+}5+{}^{-}3-{}^{-}4=+{}^{+}7+(-{}^{+}5+{}^{-}3-{}^{-}4)$$
$$=+{}^{+}7-{}^{+}5+(+{}^{-}3-{}^{-}4)$$
$$=+{}^{+}7-{}^{+}5+{}^{-}3+(-{}^{-}4).$$

(ii.) *An expression may be inclosed within parentheses preceded by the sign of subtraction, if the signs of operation preceding the numbers inclosed within the parentheses be reversed, from* $+$ *to* $-$ *and from* $-$ *to* $+$.

E.g., $+{}^{+}7 - {}^{+}5 + {}^{-}3 - {}^{-}4 = + {}^{+}7 - (+ {}^{+}5 - {}^{-}3 + {}^{-}4)$
$= + {}^{+}7 - {}^{+}5 - (- {}^{-}3 + {}^{-}4)$
$= + {}^{+}7 - {}^{+}5 + {}^{-}3 - (+ {}^{-}4)$.

The insertion of parentheses is a direct application of the Associative Law.

EXERCISES XIII.

Insert parentheses in the expression ${}^{+}8 - {}^{-}5 + {}^{-}3 - {}^{+}7$,

1. To inclose the last three numbers preceded by the sign $+$; preceded by the sign $-$.

2. To inclose the last two numbers preceded by the sign $+$; preceded by the sign $-$.

3. To inclose the first and third numbers preceded by the sign $+$; preceded by the sign $-$.

Insert parentheses in the expression $- {}^{+}15 - {}^{+}46 + {}^{+}29 + {}^{-}37 - {}^{-}42$,

4. To inclose the first and third numbers preceded by the sign $+$; preceded by the sign $-$.

5. To inclose the second, third, and fifth numbers preceded by the sign $+$; preceded by the sign $-$.

6. To inclose the first, fourth, and fifth numbers preceded by the sign $+$, and the second and third preceded by the sign $-$; the first, fourth, and fifth numbers preceded by the sign $-$, and the second and third preceded by the sign $+$.

16. In ordinary Arithmetic, to subtract a number from any number decreases the latter.

E.g., $7 - 4 = 3$, and $3 < 7$.

But such is not always the case in subtracting one algebraic number from another.

E.g., ${}^{+}7 - {}^{+}4 = {}^{+}3$, and ${}^{+}7 - {}^{-}4 = {}^{+}11$.

In the first case ${}^{+}4$ is subtracted from ${}^{+}7$, and the remainder, ${}^{+}3$, is less than ${}^{+}7$; in the second case ${}^{-}4$ is subtracted from ${}^{+}7$, and the remainder, ${}^{+}11$, is greater than ${}^{+}7$.

Also, $^-7 - {}^+4 = {}^-11$, and $^-7 - {}^-4 = {}^-3$.

In the first cases $^+4$ is subtracted from $^-7$, and the remainder, $^-11$, is less than $^-7$; in the second case $^-4$ is subtracted from $^-7$, and the remainder, $^-3$, is greater than $^-7$.

The preceding examples illustrate the following principle:

To subtract a positive number from any number decreases the latter; while to subtract a negative number from any number increases the latter.

For, if $a > b$, $^+a - {}^+b = {}^+(a - b)$,

and $^+(a - b) < {}^+a$, by Ch. I., § 2, Art. 19 (i.);

also, if $a < b$, $^+a - {}^+b = {}^-(b - a)$,

and $^-(b - a) < {}^+a$, by Ch. I., § 2, Art. 19 (iv.);

finally, $^-a - {}^+b = {}^-(a + b)$,

and $^-(a + b) < {}^-a$, by Ch. I., § 2, Art. 19 (ii.).

Likewise, $^+a - {}^-b = {}^+(a + b)$,

and $^+(a + b) > {}^+a$, by Ch. I., § 2, Art. 19 (i.);

also, if $a > b$, $^-a - {}^-b = {}^-(a - b)$,

and $^-(a - b) > {}^-a$, by Ch. I., § 2, Art. 19 (ii.);

finally, if $a < b$, $^-a - {}^-b = {}^+(b - a)$,

and $^+(b - a) > {}^-a$, by Ch. I., § 2, Art. 19 (iii.).

17. The results of Art. 16 may also be stated thus:

(i.) *The remainder of subtracting one algebraic number from another is less than the minuend when the subtrahend is positive.*

E.g., $^+7 - {}^+4 = {}^+3$, and $^+3 < {}^+7$.

(ii.) *The remainder of subtracting one algebraic number from another is greater than the minuend when the subtrahend is negative.*

E.g., $^+2 - {}^-3 = {}^+5$, and $^+5 > {}^+2$.

18. The result of subtracting b from a is $a - b$, wherein a is the minuend and b is the subtrahend. Therefore, by Art. 2 (i.),

$$b + (a - b) = a.$$

If $a - b$ be *positive*, then $b + (a - b)$, $= a$, is *greater* than b; or $a > b$, if $a - b$ be *positive*.

If $a - b$ be *negative*, then $b + (a - b)$, $= a$, is *less* than b; or $a < b$, if $a - b$ be *negative*.

The preceding results may be stated thus:

(i.) *If the remainder be positive, the minuend is greater than the subtrahend.*

(ii.) *If the remainder be negative, the minuend is less than the subtrahend.*

Property of Zero in Subtraction.

19. From § 1, Art. 18, we have

$$N + 0 = N.$$

If, therefore, from N, which is the sum of N and 0, be subtracted either N or 0, the remainder is 0 or N, by the definition of subtraction.

That is, $$N - N = 0, \quad (1)$$

and $$N - 0 = N. \quad (2)$$

In particular, $$0 - 0 = 0. \quad (3)$$

Observe that (1) agrees with the definition of 0 previously given (Ch. I., § 2, Art. 2).

20. From Ch. I., § 2, Art. 12, we have

$$^{+}N + {}^{-}N = 0. \quad (4)$$

Therefore, $$0 - {}^{+}N = {}^{-}N, \quad (5)$$

and $$0 - {}^{-}N = {}^{+}N. \quad (6)$$

That is, any negative number may be regarded as the remainder of subtracting from 0 a positive number having the same absolute value; and any positive number may be regarded as the remainder of subtracting from 0 a negative number having the same absolute value.

21. The following principle will be useful in subsequent work.

If the same number or equal numbers be subtracted from equal numbers, the remainders will be equal.

If $a = b$ and $A = B$, then $a - A = b - B$.

For $$a - A = a - A, \text{ by Axiom (i.)} \quad (1)$$

Since $b = a$ and $B = A$, we can, by Axiom (iii.), substitute b for a, and B for A, in the second member of (1). We thus obtain

$$a - A = b - B.$$

§ 3. MULTIPLICATION OF ALGEBRAIC NUMBERS.

1. In ordinary Arithmetic, multiplication by an integer is defined as an abbreviated addition. Thus, to multiply 4 by 3, the number 4 is used three times as a summand; or

$$4 \times 3 = 4 + 4 + 4.$$

Now the number 3 stands for an aggregate of three units; or

$$3 = 1 + 1 + 1.$$

We thus see that, just as 3 is obtained by taking the unit, 1, three times as a summand, so the value of 4×3 is obtained by taking 4 three times as a summand.

2. We are thus naturally led to the following definition of multiplication:

To multiply one number by a second number is to find a third number which is obtained from the first just as the second is obtained from the positive unit.

3. The above definition is an extension of the meaning of arithmetical multiplication when the multiplier is an integer, and gives an intelligible meaning to arithmetical multiplication when the multiplier is a fraction.

Thus, $\frac{2}{3}$ is obtained from the unit, 1, by taking one-third of the latter twice as a summand; or

$$\tfrac{2}{3} = \tfrac{1}{3} + \tfrac{1}{3}.$$

In like manner, to multiply 5 by $\frac{2}{3}$, we take one-third of 5 twice as a summand; or

$$5 \times \tfrac{2}{3} = \tfrac{5}{3} + \tfrac{5}{3} = \tfrac{10}{3}.$$

4. The extended definition of multiplication given in Art. 2 is applicable to the multiplication of algebraic numbers.

As in Arithmetic, the number multiplied is called the **Multiplicand**, the number that multiplies the **Multiplier**, and the result the **Product**.

5. There are two cases to be considered in the multiplication of algebraic numbers.

(i.) **The Multiplier Positive.** — Consider first particular examples.

Ex. 1. Multiply $^+4$ by $^+3$.

By the definition of multiplication, the product,

$$^+4 \times {^+3},$$

is obtained from $^+4$ just as $^+3$ is obtained from the positive unit. But $^+3$ is obtained from the positive unit by taking the latter *three times as a summand;* or

$$^+3 = {^+1} + {^+1} + {^+1}.$$

Consequently the required product is obtained by taking $^+4$ *three times as a summand;* or

$$\begin{aligned} ^+4 \times {^+3} &= {^+4} + {^+4} + {^+4} \\ &= {^+(4 + 4 + 4)} \\ &= {^+(4 \times 3)} \\ &= {^+12}. \end{aligned}$$

Ex. 2. Multiply $^-4$ by $^+3$.

By the definition of multiplication, the product,

$$^-4 \times {^+3},$$

is obtained from $^-4$ just as $^+3$ is obtained from the positive unit. Consequently, the required product is obtained by taking $^-4$ *three times as a summand;* or

$$\begin{aligned} ^-4 \times {^+3} &= {^-4} + {^-4} + {^-4} \\ &= {^-(4 + 4 + 4)} \\ &= {^-(4 \times 3)} \\ &= {^-12}. \end{aligned}$$

(ii.) **The Multiplier Negative.** — Consider first particular examples.

Ex. 3. Multiply $^+4$ by $^-3$.

By the definition of multiplication, the product,

$$^+4 \times {^-3},$$

is obtained from $^+4$ just as $^-3$ is obtained from the positive unit. But

$$^-3 = {^-1} + {^-1} + {^-1} = -{^+1} - {^+1} - {^+1};$$

that is, $^{-}3$ is obtained *by subtracting the positive unit, $^{+}1$, three times in succession from* 0. Consequently, the required product is obtained *by subtracting the multiplicand, $^{+}4$, three times in succession from* 0; or

$$
\begin{aligned}
{}^{+}4 \times {}^{-}3 &= - {}^{+}4 - {}^{+}4 - {}^{+}4 \\
&= + {}^{-}4 + {}^{-}4 + {}^{-}4 \\
&= {}^{-}(4 \times 3).
\end{aligned}
$$

Ex. 4. Multiply $^{-}4$ by $^{-}3$.

By the definition of multiplication, the product,

$$^{-}4 \times {}^{-}3,$$

is obtained from $^{-}4$ just as $^{-}3$ is obtained from the positive unit. Consequently, the required product is obtained *by subtracting the multiplicand, $^{-}4$, three times in succession from* 0; or

$$
\begin{aligned}
{}^{-}4 \times {}^{-}3 &= - {}^{-}4 - {}^{-}4 - {}^{-}4 \\
&= + {}^{+}4 + {}^{+}4 + {}^{+}4 \\
&= {}^{+}(4 \times 3).
\end{aligned}
$$

6. In Art. 5 the examples were limited to the multiplication of integers having the same or opposite signs.

But the essential part of the results therein obtained is the sign of the product.

Since this sign depends only upon the signs of the multiplicand and multiplier, and not upon their absolute values, the sign of the product in each example would have been the same as above, if the multiplicand and multiplier, either or both, had been fractions.

Those examples illustrate the following **Rule of Signs for Multiplication**:

The product of two numbers having like signs is positive; and the product of two numbers having unlike signs is negative. Or, stated symbolically,

$$
\begin{aligned}
{}^{+}a \times {}^{+}b &= {}^{+}(ab), \\
{}^{-}a \times {}^{-}b &= {}^{+}(ab). \\
{}^{-}a \times {}^{+}b &= {}^{-}(ab), \\
{}^{+}a \times {}^{-}b &= {}^{-}(ab),
\end{aligned}
$$

7. The proof of the principle enunciated in Art. 6 is as follows:

The product $$^+a \times {}^+b,$$
wherein a and b *are, as yet, limited to integral values*, is obtained from ^+a just as ^+b is obtained from the positive unit. But ^+b is obtained from $^+1$ by taking the latter *b times as a summand;* or,

$$^+b = {}^+1 + {}^+1 + {}^+1 + {}^+1 + \cdots b \text{ summands}.$$

Therefore the required product is obtained by taking ^+a *as a summand b times;* or,

$$\begin{aligned} ^+a \times {}^+b &= {}^+a + {}^+a + {}^+a + \cdots b \text{ summands} \\ &= {}^+(a + a + a \cdots b \text{ summands}) \\ &= {}^+(ab). \end{aligned}$$

Consequently, $^+a \times {}^+b = {}^+(ab)$.

In like manner, the product $^-a \times {}^+b$,
wherein a and b *are, as yet, limited to integral values*, is obtained from ^-a just as ^+b is obtained from the positive unit. Therefore, the required product is obtained by taking ^-a *as a summand b times;* or,

$$\begin{aligned} ^-a \times {}^+b &= {}^-a + {}^-a + {}^-a + \cdots b \text{ summands} \\ &= {}^-(a + a + a + \cdots b \text{ summands}) = {}^-(ab). \end{aligned}$$

Consequently, $^-a \times {}^+b = {}^-(ab)$.

The product $$^+a \times {}^-b,$$
wherein a and b *are, as yet, limited to integral values*, is obtained from ^+a just as ^-b is obtained from the positive unit.

But
$$\begin{aligned} ^-b &= {}^-1 + {}^-1 + {}^-1 + \cdots b \text{ summands} \\ &= - {}^+1 - {}^+1 - {}^+1 - \cdots b \text{ summands}; \end{aligned}$$

that is, ^-b is obtained *by subtracting the positive unit*, $^+1$, *b times in succession from* 0. Consequently the required product is obtained *by subtracting the multiplicand*, ^+a, *b times in succession from* 0; or,

$$\begin{aligned} ^+a \times {}^-b &= - {}^+a - {}^+a - {}^+a - \cdots b \text{ summands} \\ &= + {}^-a + {}^-a + {}^-a + \cdots b \text{ summands} \\ &= {}^-(a \times b). \end{aligned}$$

In like manner, the product $^-a \times {}^-b$
is obtained from ^-a just as ^-b is obtained from the positive unit. Consequently the required product is obtained *by subtracting the multiplicand*, ^-a, *b times in succession from* 0; or,

$$\begin{aligned} ^-a \times {}^-b &= - {}^-a - {}^-a - {}^-a - \cdots b \text{ summands} \\ &= + {}^+a + {}^+a + {}^+a + \cdots b \text{ summands} \\ &= {}^+(ab). \end{aligned}$$

Since the essential part of the above proof is the sign of the product, the results hold when a and b have fractional values.

EXERCISES XIV.

Find the values of the following indicated multiplications:

1. $^+2 \times {}^+3$. 2. $^+5 \times {}^+7$. 3. $^+9 \times {}^+7$. 4. $^+8 \times {}^+5$.
5. $^+6 \times {}^+2$. 6. $^-3 \times {}^+7$. 7. $^-11 \times {}^+9$. 8. $^-8 \times {}^+5$.
9. $^-17 \times {}^+3$. 10. $^-2 \times {}^+3$. 11. $^+5 \times {}^-6$. 12. $^+7 \times {}^-4$.
13. $^+15 \times {}^-2$. 14. $^+21 \times {}^-5$. 15. $^+33 \times {}^-3$. 16. $^-5 \times {}^-11$.
17. $^-7 \times {}^-9$. 18. $^-18 \times {}^-3$. 19. $^-22 \times {}^-6$. 20. $^-41 \times {}^-2$.

Find the values of:

21. $({}^+15 \times {}^-4) + ({}^-12 \times {}^+1)$. 22. $({}^+16 \times {}^-3) - ({}^+5 \times {}^+7)$.
23. $({}^-1 \times {}^+11) - ({}^-22 \times {}^+2)$.
24. $({}^-16 \times {}^+2) + ({}^+17 \times {}^+5) - ({}^-16 \times {}^+7)$.
25. $({}^+2 \times {}^+13) - ({}^+2 \times {}^-5) - ({}^-7 \times {}^+4)$.
26. $({}^+12 \times {}^-3) + ({}^+15 \times {}^-4)$. 27. $({}^+18 \times {}^-4) - ({}^+15 \times {}^-6)$.
28. $({}^-52 \times {}^-2) - ({}^+12 \times {}^-3)$.
29. $({}^-4 \times {}^-7) - ({}^+11 \times {}^-3) + ({}^+15 \times {}^-2)$.
30. $({}^-7 \times {}^-5) + ({}^-11 \times {}^-4) - ({}^+14 \times {}^+3)$.

Multiply:

31. $^+5$ by $^+4$. 32. $^+8$ by $^+3$. 33. $^-17$ by $^+4$. 34. $^-11$ by $^+5$.
35. $^+12$ by $^-4$. 36. $^+18$ by $^-5$. 37. $^-7$ by $^-8$. 38. $^-11$ by $^-3$.

What is the value of $m \times n$,

39. When $m = {}^+5$, $n = {}^+7$? 40. When $m = {}^-4$, $n = {}^+3$?
41. When $m = {}^+4$, $n = {}^-5$? 42. When $m = {}^-11$, $n = {}^-3$?

What is the value of $^+a \times {}^+b$,

43. When $a = 19$, $b = 3$? 44. When $a = 4$, $b = 15$?

What is the value of $^-a \times {}^+b$,

45. When $a = 13$, $b = 2$? 46. When $a = 9$, $b = 6$?

What is the value of $^+x \times {}^-y$,

47. When $x = 3$, $y = 8$? 48. When $x = 15$, $y = 4$?

What is the value of $^-x \times {}^-y$,

49. When $x = 7$, $y = 10$? 50. When $x = 11$, $y = 8$?

What is the value of $^{+}(xy)$,

51. When $x=3$, $y=7$?

52. When $x=4$, $y=9$?

What is the value of $^{-}(xy)$,

53. When $x=5$, $y=10$?

54. When $x=6$, $y=11$?

What is the value of $(a+b)\times c$,

55. When $a={}^{+}5$, $b={}^{+}3$, $c={}^{+}4$?

56. When $a={}^{+}7$, $b={}^{-}10$, $c={}^{+}8$?

What is the value of $(a-b)\times c$,

57. When $a={}^{+}4$, $b={}^{+}3$, $c={}^{-}5$?

58. When $a={}^{-}8$, $b={}^{-}2$, $c={}^{-}7$?

What is the value of $(a-b)(c+d)$,

59. When $a={}^{-}2$, $b={}^{+}4$, $c={}^{-}5$, $d={}^{+}6$?

60. When $a={}^{+}5$, $b={}^{-}8$, $c={}^{-}9$, $d={}^{+}10$?

What is the value of $(a+b)(c-d)$,

61. When $a={}^{-}18$, $b={}^{-}5$, $c={}^{+}4$, $d={}^{+}7$?

62. When $a={}^{+}1$, $b={}^{-}3$, $c={}^{-}5$, $d={}^{+}2$?

Concrete Illustrations.

8. At 12 M. a train passes the station A, on an east and west track, running at the rate of 20 miles an hour.

How far is the train from the station A two hours later?

Call time *after* 12 M. *positive*, and consequently time before 12 M. *negative*. Also, rate of the train *toward the east* positive, and hence *rate toward the west negative.*

(i.) If the train is running *east*, both time and rate are positive. We thus have

$$^{+}20\times{}^{+}2={}^{+}40.$$

From the conditions of the problem we know that the train is *east* of A. Consequently, the *positive* result, $^{+}40$, shows that distances measured east from A *must* be taken *positively*, and hence distances *west negatively.*

Thus, having made definite assumptions in regard to *time* and *rate*, we are not at liberty to take distances east positively or negatively, as we please.

To emphasize the last statement, call time after 12 M. *negative*, and consequently time *before* 12 M. *positive*, leaving the other assumptions unchanged. We then have

$$^{+}20 \times {}^{-}2 = {}^{-}40.$$

But, as before, the train is *east* of A. Consequently, the *negative* result, $^{-}40$, shows that distances *east* of A *must* now be *negative*.

Returning now to the original statement,

$$^{+}20 \times {}^{+}2 = {}^{+}40,$$

we conclude that the train is 40 miles *east* of A, at B say (Fig. 6).

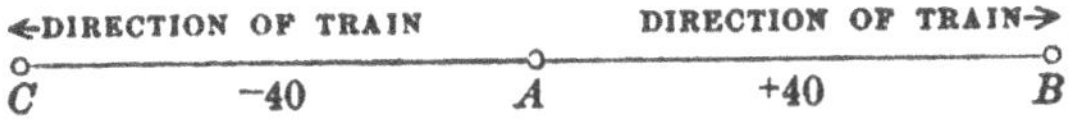

FIG. 6.

(ii.) But if the train is running *west*, *time* is *positive*, and *rate* is *negative*. We therefore have

$$^{-}20 \times {}^{+}2 = {}^{-}40.$$

Distances east having already been determined as positive, and hence distances west as negative, the *negative* result, $^{-}40$, shows that the train is now 40 miles *west* of A, at C say (Fig. 6).

(iii.) How far was the train from A two hours before 12 M., if the train is running east?

Time in this case is *negative*, and *rate* is *positive*. Consequently, $$^{+}20 \times {}^{-}2 = {}^{-}40.$$

The *negative* result, $^{-}40$, shows that the train was still 40 miles *west* of A, at C say (Fig. 7).

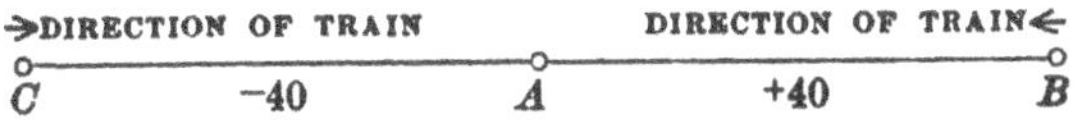

FIG. 7.

(iv.) How far was the train from A two hours before 12 M., if the train is running west? Both *time* and *rate* are *negative*. Therefore $$^{-}20 \times {}^{-}2 = {}^{+}40.$$

The *positive* result, $^{+}40$, shows that the train was 40 miles east of A, at B say (Fig. 7).

Continued Products.

9. The results of the preceding articles may be applied to determine the value of a chain of indicated multiplications, *i.e.*, of a *continued product*.

E.g.,
$$^{+}a \times {}^{+}b \times {}^{+}c = {}^{+}(ab) \times {}^{+}c = {}^{+}(abc),$$
$$^{+}a \times {}^{+}b \times {}^{-}c = {}^{+}(ab) \times {}^{-}c = {}^{-}(abc),$$
$$^{+}a \times {}^{-}b \times {}^{-}c = {}^{-}(ab) \times {}^{-}c = {}^{+}(abc),$$
$$^{-}a \times {}^{-}b \times {}^{-}c = {}^{+}(ab) \times {}^{-}c = {}^{-}(abc).$$

These equations illustrate a more general rule of signs than that given in Art. 6:

A continued product which contains no negative number, or an even number of negative numbers, is positive; one that contains an odd number of negative numbers is negative.

In practice the sign of a required product may first be determined by inspection, and that sign prefixed to the product of the absolute values of the numbers in the continued product.

E.g., the sign of the product
$$(^{+}2) \times (^{-}3) \times (^{-}7) \times (^{+}4) \times (^{-}5)$$
is *negative*, since it contains *three* negative numbers; the product of the absolute values is 840. Consequently,
$$(^{+}2) \times (^{-}3) \times (^{-}7) \times (^{+}4) \times (^{-}5) = {}^{-}840.$$

EXERCISES XV.

Find the values of the following continued products:

1. $^{-}2 \times {}^{+}4 \times {}^{-}3.$ 2. $^{-}5 \times {}^{-}6 \times {}^{+}7.$ 3. $^{+}12 \times {}^{-}2 \times {}^{-}5 \times {}^{+}4.$
4. $^{-}7 \times {}^{-}11 \times {}^{+}2 \times {}^{-}3.$ 5. $^{-}15 \times {}^{+}2 \times {}^{+}4 \times {}^{-}5 \times {}^{-}6.$

What is the value of abc,

6. When $a = {}^{-}3, b = {}^{+}5, c = {}^{-}7$? 7. When $a = {}^{-}4, b = {}^{+}1, c = {}^{+}2$?

What is the value of $abcd$,

8. When $a = {}^{-}2, b = {}^{+}3, c = {}^{-}4, d = {}^{+}5$?
9. When $a = {}^{-}7, b = {}^{+}2, c = {}^{-}3, d = {}^{-}5$?

The Commutative Law for Multiplication.

10. In an indicated multiplication, the number which follows the symbol of multiplication is the multiplier. Thus, in

$$^{-}4 \times {}^{+}3 = {}^{-}4 + {}^{-}4 + {}^{-}4, = {}^{-}12,$$

the multiplier is $^{+}3$; while in

$$^{+}3 \times {}^{-}4, = - {}^{+}3 - {}^{+}3 - {}^{+}3 - {}^{+}3, = {}^{-}12,$$

the multiplier is $^{-}4$. But the result is the same, whichever of the two numbers, $^{+}3$ or $^{-}4$, is used as the multiplier.

The above example illustrates the following principle:

The Commutative Law. — *The product of two numbers is the same, if either be taken as the multiplier and the other as the multiplicand*; or, stated symbolically,

$$a \times b = b \times a.$$

11. It follows from the rule of signs, Art. 6, that the signs of the multiplier and multiplicand may be interchanged without affecting the sign of the product.

E.g., $^{+}3 \times {}^{-}4 = {}^{-}(3 \times 4) = {}^{-}12$, and $^{-}3 \times {}^{+}4 = {}^{-}(3 \times 4) = {}^{-}12$.

This principle is called the *Commutative Law of Signs.*

We have, therefore, to prove the Commutative Law only for the multiplication of absolute numbers.

Take first a particular example: $4 \times 3 = 3 \times 4$.

Consider the following arrangement of units in rows and columns:

$$\begin{matrix} 1 & 1 & 1 & 1 \\ 1 & 1 & 1 & 1 \\ 1 & 1 & 1 & 1 \end{matrix}$$

The total number of units in this arrangement is obtained either by multiplying the number in each row, 4, by the number of rows, 3, giving 4×3; or by multiplying the number of units in each column, 3, by the number of columns, 4, giving 3×4.

Consequently, $\qquad 4 \times 3 = 3 \times 4.$

In general, consider the following arrangement of units in rows and columns:

$$b \text{ rows} \left\{ \overbrace{\begin{array}{cccccccc} 1 & 1 & 1 & 1 & . & . & . & . \\ 1 & 1 & 1 & 1 & . & . & . & . \\ 1 & 1 & 1 & 1 & . & . & . & . \\ . & . & . & . & . & . & . & . \\ . & . & . & . & . & . & . & . \end{array}}^{a \text{ units in each row}} \right.$$

The total number of units in this arrangement is obtained either by multiplying the number in each row, a, by the number of rows, b, giving $a \times b$; or by multiplying the number in each column, b, by the number of columns, a, giving $b \times a$.

Consequently, $$a \times b = b \times a,$$

wherein a and b are *absolute integers.*

Consider next the case in which a and b, either or both, denote absolute fractions.

$$3 \times \tfrac{2}{5} = \tfrac{3}{5} + \tfrac{3}{5}, \text{ by the definition of multiplication,}$$

$$= \frac{3+3}{5} = \frac{3 \times 2}{5}$$

$$= \frac{2 \times 3}{5}, \text{ since } 3 \times 2 = 2 \times 3$$

$$= \frac{2+2+2}{5} = \frac{2}{5} + \frac{2}{5} + \frac{2}{5} = \frac{2}{5} \times 3.$$

Also, $$\frac{4}{5} \times \frac{2}{3} = \frac{4}{5 \times 3} + \frac{4}{5 \times 3}, \text{ by the definition of multiplication,}$$

$$= \frac{4+4}{5 \times 3} = \frac{4 \times 2}{5 \times 3}$$

$$= \frac{2 \times 4}{3 \times 5}, \text{ since } 4 \times 2 = 2 \times 4 \text{ and } 5 \times 3 = 3 \times 5,$$

$$= \frac{2+2+2+2}{3 \times 5} = \frac{2}{3 \times 5} + \frac{2}{3 \times 5} + \frac{2}{3 \times 5} + \frac{2}{3 \times 5} = \frac{2}{3} \times \frac{4}{5}.$$

Similar reasoning can be applied to the product of any two absolute fractions.

Consequently, the Commutative Law for multiplication,

$$a \times b = b \times a,$$

holds for all values of a and b, positive or negative, integral or fractional.

EXERCISES XVI.

1. Express the sum, $^{+}4+{}^{+}4+{}^{+}4$, as a sum of summands each equal to $^{+}3$.

2. Express the sum, $^{-}5+{}^{-}5+{}^{-}5+{}^{-}5$, as a sum of summands each equal to $^{-}4$.

3. Express the sum, $^{+}5+{}^{+}5+\cdots$ 11 summands, as a sum of 5 summands each equal to $^{+}11$.

4. Express the sum, $^{-}7+{}^{-}7+\cdots$ 21 summands, as a sum of 7 summands each equal to $^{-}21$.

5. Express the sum, $^{+}(\frac{2}{3})+{}^{+}(\frac{2}{3})+{}^{+}(\frac{2}{3})+{}^{+}(\frac{2}{3})+{}^{+}(\frac{2}{3})$, as a sum of 2 summands each equal to $^{+}(\frac{5}{3})$.

6. Express the sum, $^{-}(\frac{3}{4})+{}^{-}(\frac{3}{4})+{}^{-}(\frac{3}{4})+{}^{-}(\frac{3}{4})+{}^{-}(\frac{3}{4})+{}^{-}(\frac{3}{4})$, as a sum of 3 summands each equal to $^{-}(\frac{6}{4})$.

7. Express the sum, $^{+}9+{}^{+}9+{}^{+}9+\cdots a$ summands, as a sum of 9 summands each equal to ^{+}a.

8. Express the sum, $^{-}5+{}^{-}5+\cdots x$ summands, as a sum of 5 summands each equal to ^{-}x.

The Associative Law for Multiplication.

12. In finding the value of a continued product in Art. 9, the indicated operations were performed successively from left to right.

E.g., $$({}^{+}4\times{}^{+}3)\times{}^{-}2={}^{+}12\times{}^{-}2={}^{-}24.$$

But the same result is obtained if $^{+}3$ be first multiplied by $^{-}2$ and then $^{+}4$ be multiplied by the product.

E.g., $${}^{+}4\times({}^{+}3\times{}^{-}2)={}^{+}4\times{}^{-}6={}^{-}24.$$

The parentheses in the above illustrations indicate that the numbers within them are to be multiplied first, the multiplications in each case being performed from left to right.

The above examples illustrate the following principle:

The Associative Law. — *The product of three numbers is the same in whichever way two successive numbers are grouped or associated in the process of multiplying;* or stated symbolically,

$$(ab)c = a(bc).$$

13. For the reason stated in Art. 11, it is sufficient to prove this law for absolute numbers.

Take first a particular example:

$$4 \times 3 \times 2 = 4 \times (3 \times 2).$$

Consider the following arrangement of 3's in rows and columns:

$$\begin{array}{cc} 3 & 3 \\ 3 & 3 \\ 3 & 3 \\ 3 & 3 \end{array}$$

Each row contains 3×2 units; and, since there are 4 rows, the total number of units is

$$(3 \times 2) \times 4, \text{ or } 4 \times (3 \times 2), \text{ by Art. 10.}$$

But each column contains 3×4, or 4×3 units; and, since there are 2 columns, the total number of units is $(4 \times 3) \times 2$.

Therefore $\qquad (4 \times 3) \times 2 = 4 \times (3 \times 2).$

The parentheses on the left of the sign of equality may be omitted, since they indicate only the ordinary way of performing the multiplications successively from left to right.

In general, consider the following arrangement of b's in rows and columns:

$$\begin{array}{c} \overbrace{}^{c \text{ columns of } b\text{'s}} \\ a \text{ rows of } b\text{'s} \left\{ \begin{array}{cccccccc} b & b & b & . & . & . & . & . \\ b & b & b & . & . & . & . & . \\ b & b & b & . & . & . & . & . \\ . & . & . & . & . & . & . & . \\ . & . & . & . & . & . & . & . \end{array} \right. \end{array}$$

Each row contains $b \times c$ units; and, since there are a rows, the total number of units is

$$(b \times c) \times a, \text{ or } a \times (b \times c), \text{ by Art. 10.}$$

But each column contains $b \times a$, or $a \times b$ units; and, since there are c columns, the total number of units is

$$(a \times b) \times c.$$

Therefore $$(a \times b) \times c = a \times (b \times c).$$

In the above representation, the values of a, b, and c are limited to absolute integers. It can be shown, however, as in Art. 11, that the law holds also for absolute fractional values of a, b, c.

E.g., $\frac{2}{3} \times \frac{4}{5} \times \frac{6}{7} = \frac{2 \times 4}{3 \times 5} \times \frac{6}{7} = \frac{2 \times 4 \times 6}{3 \times 5 \times 7} = \frac{2 \times (4 \times 6)}{3 \times (5 \times 7)} = \frac{2}{5} \times \left(\frac{4}{5} \times \frac{6}{7}\right).$

We conclude, therefore, that the Associative Law for multiplication holds for all values of a, b, c, *positive or negative, integral or fractional.*

EXERCISES XVII.

Find, in the most convenient way, the values of:

1. $^{-}17 \times {}^{+}5 \times {}^{-}2.$
2. $^{+}38 \times {}^{-}2\frac{1}{2} \times {}^{-}4.$
3. $^{-}139 \times {}^{-}3 \times {}^{+}33\frac{1}{3}.$
4. $^{+}228 \times {}^{+}225 \times {}^{-}4.$
5. $^{-}139 \times {}^{-}8 \times {}^{+}12\frac{1}{2}.$
6. $^{-}17 \times {}^{-}16\frac{2}{3} \times {}^{-}3.$

Find, in the most convenient way, the value of abc:

7. When $a = {}^{-}4$, $b = {}^{+}33\frac{1}{3}$, $c = {}^{-}9$.
8. When $a = {}^{+}19$, $b = {}^{-}66\frac{2}{3}$, $c = {}^{-}3$.

14. The Associative and Commutative Laws may be extended as follows:

The value of a product of three or more numbers remains the same if, in performing the indicated multiplications, the order of the numbers be changed, or if two or more numbers be associated in any way.

E.g., $$^{+}2 \times {}^{-}3 \times {}^{-}4 = {}^{+}2 \times {}^{-}4 \times {}^{-}3 = {}^{-}4 \times {}^{+}2 \times {}^{-}3 = \text{etc.}$$

$$^{+}2 \times {}^{-}3 \times {}^{-}4 \times {}^{+}5 = {}^{+}2 \times ({}^{-}3 \times {}^{-}4) \times {}^{+}5 = {}^{+}2 \times {}^{-}3 \times ({}^{-}4 \times {}^{+}5) = \text{etc.}$$

In general,

$$abc = acb = bca = bac = cab = cba.$$

$$abcd = a(bcd) = a(bc)d = ab(cd).$$

The two laws may be applied simultaneously.

E.g.,

$$^{+}2 \times {}^{-}3 \times {}^{-}4 \times {}^{+}5 = {}^{+}2 \times ({}^{+}5 \times {}^{-}3) \times {}^{-}4 = {}^{-}3 \times ({}^{-}4 \times {}^{+}2 \times {}^{+}5) = \text{etc.}$$

In general, $$abcd = a(cb)d = c(adb) = \text{etc.}$$

The proof is as follows:

(i.) *The Commutative Law for the product of three numbers.*

We have $abc = a(bc)$, by the Associative Law,

$= a(cb)$, since $bc = cb$,

$= acb$.

Again, $abc = bac$, since $ab = ba$,

$= b(ac)$, by the Associative Law,

$= b(ca)$, since $ac = ca$,

$= bca$.

In like manner the identity of the other products can be shown.

(ii.) *The Associative Law for the product of four numbers.*

Consider the following arrangement of bc's in rows and columns:

$$a \text{ rows} \left\{ \begin{array}{ccccccc} \multicolumn{7}{c}{\overbrace{\hphantom{bc\quad bc\quad bc\quad \cdot\quad \cdot\quad \cdot\quad \cdot}}^{d \text{ columns}}} \\ bc & bc & bc & \cdot & \cdot & \cdot & \cdot \\ bc & bc & bc & \cdot & \cdot & \cdot & \cdot \\ bc & bc & bc & \cdot & \cdot & \cdot & \cdot \\ \cdot & \cdot & \cdot & \cdot & \cdot & \cdot & \cdot \\ \cdot & \cdot & \cdot & \cdot & \cdot & \cdot & \cdot \end{array} \right.$$

Each row contains $bc \times d$, or bcd, units; and, since there are a rows, the total number of units is $(bcd) \times a$, or $a \times (bcd)$, by the Commutative Law for the product of two factors.

But each column contains $bc \times a$, or abc, units; and, since there are d columns, the total number of units is $(abc) \times d$, or $abcd$.

Therefore, $abcd = a(bcd)$.

Similarly, the identity of the other products can be shown.

In like manner the Commutative and the Associative Laws can be extended to a continued product of any number of numbers.

E.g., $abcd = acbd =$ etc.

$abcde = a(bcd)e = a(bcde) =$ etc.

15. Since the multiplier and the multiplicand can be interchanged without affecting the value of the product, they are both given the common name **Factor**. Thus, a and b are the factors of the product $a \times b$.

For a similar reason, each number in a continued product is called a factor of the product.

Thus a, b, c, and d are factors of $abcd$.

EXERCISES XVIII.

Find in three different ways, by the Commutative Law, the values of:

1. $^-6 \times {}^+7 \times {}^-3$. 2. $^-8 \times {}^+5 \times {}^+2$. 3. $^-4 \times {}^-5 \times {}^-2$.

Find in the most convenient way the values of:

4. $^+5 \times {}^+17 \times {}^-2$. 5. $^+8 \times {}^-5 \times {}^-12\frac{1}{2}$. 6. $^-125 \times {}^-17 \times {}^+8$.

7. $^-333\frac{1}{3} \times {}^+29 \times {}^-3$. 8. $^-37\frac{1}{2} \times {}^-10 \times {}^-4$. 9. $^-225 \times {}^-37 \times {}^+4$.

Find, in the most convenient way, the value of $abcd$,

10. When $a = {}^-37\frac{1}{2}$, $b = {}^+5$, $c = {}^-3$, $d = {}^+8$.

11. When $a = {}^-12\frac{1}{2}$, $b = {}^-16$, $c = {}^+33\frac{1}{3}$, $d = {}^-6$.

Find the value of the continued product

$$^-7 \times {}^+2 \times {}^+3 \times {}^-4 \times {}^+10:$$

12. Associating the second and third factors.
13. Associating the third and fourth factors.
14. Associating the last three factors.
15. Associating the last four factors.

Find, in the most convenient way, the values of:

16. $^-7 \times {}^+25 \times {}^-4 \times {}^-9$. 17. $^+9 \times {}^+13 \times {}^-12\frac{1}{2} \times {}^+8$.

18. $abcd$, when $a = {}^+7$, $b = {}^-33\frac{1}{3}$, $c = {}^-3$, $d = {}^-4$.

16. Since $(ab) \times c = (ac)b = a(bc)$,
the product ab is multiplied by c, if either a or b be multiplied by c.

In like manner $(abc)d = (ad)bc = a(bd)c = ab(cd)$.

Hence follows the principle:

A product of two or more factors is multiplied by a number if any one of the factors be multiplied by that number.

EXERCISES XIX.

Multiply:

1. $^-75 \times {}^-22\frac{1}{2}$ by $^-4$. 2. $^-125 \times {}^-13$ by $^+8$. 3. $^+71\frac{3}{7} \times {}^-8$ by $^-7$.

4. $^+5 \times {}^-33\frac{1}{3} \times {}^-6$ by $^-3$. 5. $^-7 \times {}^+250 \times {}^-6$ by $^-4$.

17. The following principle will be useful in subsequent work:

If equal numbers be multiplied by the same number or by equal numbers, the products will be equal.

If $a = b$, and $A = B$, then $aA = bB$.

For, $\qquad aA = aA$, by Axiom (i.), $\qquad$ (1)

Since $b = a$ and $B = A$, we can, by Axiom (iii.), substitute b for a, and B for A, in the second member of (1). We thus obtain

$$aA = bB.$$

§ 4. DIVISION OF ALGEBRAIC NUMBERS.

1. Division *is the inverse of multiplication.* In multiplication two factors are given, and it is required to find their product. In division the product of two factors and one of them are given, and it is required to find the other factor. As in ordinary Arithmetic, the given product is called the **Dividend,** the given factor the **Divisor,** and the required factor the **Quotient.**

E.g., Since $\qquad {}^-28 = {}^-4 \times {}^+7,$

therefore, $\qquad {}^-28 \div {}^+7 = {}^-4$, and ${}^-28 \div {}^-4 = {}^+7$.

2. From the definition of division we infer the following principle:

If the product of two factors be divided by either of the factors, the quotient is the other factor.

In general, if the given product be $a \times b$, we have, by the definition of division,

$$(a \times b) \div b = a, \text{ and } (a \times b) \div a = b.$$

EXERCISES XX.

Find the values of the following indicated divisions:

1. $({}^-27 \times {}^+3) \div {}^+3$.
2. $({}^-27 \times {}^+3) \div {}^-27$.
3. $({}^-15 \times {}^-8) \div {}^-8$.
4. $({}^-15 \times {}^-8) \div {}^-15$.
5. $({}^+2\tfrac{2}{3} \times {}^+5\tfrac{3}{8}) \div {}^+2\tfrac{2}{3}$.
6. $({}^+2\tfrac{2}{3} \times {}^-5\tfrac{3}{8}) \div {}^-5\tfrac{3}{8}$.

3. The dividend is, as a rule, a single number and does not appear as the product of two factors, one of which is the divisor.

Since the absolute value of the product of two factors is equal to the product of their absolute values, it follows, from the definition of division, that the absolute value of the quotient is equal to the quotient of the absolute values of the dividend and the divisor.

By the definition of division, the equations of § 3, Art. 6, may be written

$$^{+}(ab) \div {}^{+}a = {}^{+}b;\ ^{-}(ab) \div {}^{-}a = {}^{+}b;$$
$$^{-}(ab) \div {}^{+}a = {}^{-}b;\ ^{+}(ab) \div {}^{-}a = {}^{-}b.$$

From these equations, we derive the following **Rule of Signs for Division:**

Like signs of dividend and divisor give a positive quotient; unlike signs of dividend and divisor give a negative quotient.

E.g.,
$$^{+}8 \div {}^{+}2 = {}^{+}4;\ ^{-}8 \div {}^{-}2 = {}^{+}4;$$
$$^{-}8 \div {}^{+}2 = {}^{-}4;\ ^{+}8 \div {}^{-}2 = {}^{-}4.$$

EXERCISES XXI.

Find the values of the following indicated divisions:

1. $^{+}27 \div {}^{+}9$. **2.** $^{-}50 \div {}^{+}3$. **3.** $^{+}81 \div {}^{-}9$. **4.** $^{-}33 \div {}^{+}11$.

5. $^{-}34 \div {}^{+}17$. **6.** $^{+}52 \div {}^{-}4$. **7.** $^{+}75 \div {}^{+}15$. **8.** $^{-}105 \div {}^{-}7$.

Find the values of:

9. $(^{-}16 \div {}^{+}2) + (^{+}18 \div {}^{-}3)$. **10.** $(^{-}24 \div {}^{-}8) - (^{-}36 \div {}^{+}6)$.

11. $(^{+}12 \div {}^{+}4) - (^{-}28 \div {}^{-}7)$.

12. $(^{-}15 \div {}^{-}5) - (^{+}100 \div {}^{-}25) + (^{-}200 \div {}^{+}8)$.

13. $(^{+}6 \div {}^{-}2) + (^{-}72 \div {}^{-}4) + (^{+}9 \div {}^{-}3)$.

Divide:

14. $^{+}18$ by $^{-}6$. **15.** $^{-}24$ by $^{+}3$. **16.** $^{-}20$ by $^{-}3$. **17.** $^{+}72$ by $^{+}8$.

What is the value of $m \div n$,

18. When $m = {}^{+}8$, $n = {}^{+}4$? **19.** When $m = {}^{-}36$, $n = {}^{+}9$?

20. When $m = {}^{+}72$, $n = {}^{-}4$? **21.** When $m = {}^{-}75$, $n = {}^{-}15$?

What is the value of $^+a \div {}^-b$,

22. When $a = 8$, $b = 4$?
23. When $a = 36$, $b = 9$?

What is the value of $^-a \div {}^-b$,

24. When $a = 22$, $b = 11$?
25. When $a = 16$, $b = 8$?

What is the value of $(a + b) \div c$,

26. When $a = {}^-17$, $b = {}^+2$, $c = {}^+5$?
27. When $a = {}^+39$, $b = {}^-4$, $c = {}^-7$?

What is the value of $(a - b) \div (c + d)$,

28. When $a = {}^+18$, $b = {}^-2$, $c = {}^+3$, $d = {}^+2$?
29. When $a = {}^-23$, $b = {}^+5$, $c = {}^-4$, $d = {}^-3$?

4. In a chain of indicated divisions, the operations are to be performed successively from left to right.

E.g., $$^-16 \div {}^+4 \div {}^-2 = {}^-4 \div {}^-2 = {}^+2;$$

$$^+210 \div {}^-3 \div {}^-2 \div {}^+5 = {}^-70 \div {}^-2 \div {}^+5 = {}^+35 \div {}^+5 = {}^+7.$$

Likewise, in a chain of indicated multiplications and divisions, the operations are to be performed successively from left to right.

E.g., $$^-375 \times {}^+3 \div {}^-5 \times {}^+2 \div {}^-9 = {}^-1125 \div {}^-5 \times {}^+2 \div {}^-9$$

$$= {}^+225 \times {}^+2 \div {}^-9 = {}^+450 \div {}^-9 = {}^-50.$$

EXERCISES XXII.

Find the values of:

1. $^+210 \div {}^-5 \div {}^-7 \div {}^+2$.
2. $^-420 \div {}^+7 \div {}^+2 \div {}^-3$.
3. $^+375 \div {}^-5 \times {}^+2 \div {}^-3$.
4. $^+15 \times {}^-6 \div {}^-2 \div {}^+5 \times {}^+4$.
5. $^-280 \div {}^-4 \times {}^+2 \times {}^-3 \div {}^-42$.

What is the value of $a \div b \times c$,

6. When $a = {}^-32$, $b = {}^+4$, $c = {}^-5$?
7. When $a = {}^+28$, $b = {}^-7$, $c = {}^+2$?

What is the value of $a \div b \times c \div d$,

8. When $a = {}^+125$, $b = {}^-5$, $c = {}^+4$, $d = {}^-10$?
9. When $a = {}^-49$, $b = {}^-7$, $c = {}^+18$, $d = {}^-2$?

5. From the definition of division, we have

$$\textit{Quotient} \times \textit{Divisor} = \textit{Dividend}.$$

Since the quotient of a divided by b is $a \div b$, we have

$$(a \div b) \times b = a, \text{ or } a \div b \times b = a. \qquad (1)$$

From (1) and the equation which defined division, namely,

$$a \times b \div b = a \qquad (2)$$

we derive the following principle:

If a given number be first divided and then multiplied by one and the same number, or be first multiplied and then divided by one and the same number, the result is the given number. That is, these successive operations are equivalent to multiplying (or dividing) by $^{+}1$; or, stated symbolically,

$$N \times b \div b = N \div b \times b = N \times {}^{+}1 = N \div {}^{+}1 = N.$$

That is, $\times\, b \div b = \div\, b \times b = \times\, {}^{+}1 = \div\, {}^{+}1,$ (3)

whatever number be placed on the left of the two indicated operations.

E.g., $^{-}11 \times {}^{+}3 \div {}^{+}3 = {}^{-}11, \quad {}^{+}11 \div {}^{-}5 \times {}^{-}5 = {}^{+}11.$

The Commutative Law for Division.

6. In a chain of divisions, or of multiplications and divisions, the successive operations are to be performed, as has been stated, in order from left to right.

E.g.,
$$^{-}14 \div {}^{+}2 \times {}^{-}7 = {}^{-}7 \times {}^{-}7 = {}^{+}49.$$
$$^{+}8 \div {}^{-}4 \div {}^{+}2 = {}^{-}2 \div {}^{+}2 = {}^{-}1.$$

But, if the operations in the above examples be performed in a different order, the symbol of operation, $\times$ or $\div$, being carried with its proper constituent, we have

$$^{-}14 \times {}^{-}7 \div {}^{+}2 = {}^{+}98 \div {}^{+}2 = {}^{+}49, \text{ as above.}$$
$$^{+}8 \div {}^{+}2 \div {}^{-}4 = {}^{+}4 \div {}^{-}4 = {}^{-}1, \text{ as above.}$$

The above examples illustrate **The Commutative Law**:

(i.) *To multiply any number by a second number and then to divide the product by a third number, gives the same result as first to divide the given number by the third number and then to mul-*

tiply the resulting quotient by the second number; and vice versa; or, stated symbolically,

$$N \times b \div c = N \div c \times b, \text{ or } \times b \div c = \div c \times b.$$

(ii.) *If a given number be divided successively by two numbers, the result is the same whichever of the two divisions is first performed;* or, stated symbolically,

$$N \div b \div c = N \div c \div b, \text{ or } \div b \div c = \div c \div b.$$

These principles are proved as follows:

(i.) If in $N \times b \div c$,

N be replaced by $N \div c \times c$, $= N$, by (3), Art. 5, we have

$$N \times b \div c = N \div c \times c \times b \div c$$
$$= N \div c \times b \times c \div c, \text{ since } \times c \times b = \times b \times c,$$
$$= N \div c \times b, \text{ since } \times c \div c = \times {}^{+}1, \text{ by (3), Art. 5.}$$

(ii.) If in $N \div b \div c$,

N be replaced by $N \div c \times c$, $= N$, by (3), Art. 5, we have

$$N \div b \div c = N \div c \times c \div b \div c$$
$$= N \div c \div b \times c \div c, \text{ since } \times c \div b = \div b \times c, \text{ by (i.),}$$
$$= N \div c \div b, \text{ since } \times c \div c = \times {}^{+}1, \text{ by (3), Art. 5.}$$

In like manner, it can be shown that the Commutative Law holds for any number of successive multiplications and divisions.

EXERCISES XXIII.

Find, in two different ways, by the Commutative Law, the values of:

1. ${}^{-}25 \times {}^{-}12 \div {}^{+}5$. 2. ${}^{+}100 \times {}^{-}4 \div {}^{+}25$. 3. ${}^{-}1000 \div {}^{-}125 \times {}^{+}12$.

Find, in the most convenient way, the values of:

4. ${}^{-}25 \times {}^{-}12 \div {}^{+}5$. 5. ${}^{+}100 \times {}^{-}7 \div {}^{-}25$. 6. ${}^{-}1000 \times {}^{-}11 \div {}^{+}125$.

7. ${}^{+}33\frac{1}{3} \div {}^{-}20 \times {}^{+}3$. 8. ${}^{-}30 \div {}^{-}9 \times {}^{-}12$. 9. ${}^{-}10 \div {}^{+}17 \times {}^{-}34$.

Find, in the most convenient way, the value of $a \div b \div c \times d$,

10. When $a = {}^{+}170$, $b = {}^{-}3$, $c = {}^{+}17$, $d = {}^{-}6$.

11. When $a = {}^{-}125$, $b = {}^{-}7$, $c = {}^{+}25$, $d = {}^{-}14$.

The Associative Law for Division.

7. By Art. 4, we have

$$^{+}8 \div {}^{-}4 \div {}^{-}2 = {}^{-}2 \div {}^{-}2 = {}^{+}1.$$

But if the operation $^{-}4 \div {}^{-}2$ be performed first, we have

$$^{+}8 \div ({}^{-}4 \div {}^{-}2) = {}^{+}8 \div {}^{+}2 = {}^{+}4, \text{ and not } {}^{+}1, \text{ as above.}$$

It would thus appear that the Associative Law does not hold for division. But principles very much like those used in grouping or associating successive additions and subtractions by means of parentheses can be employed for grouping or associating successive multiplications and divisions. These principles stated formally are:

(i.) *A chain of multiplications and divisions may be inclosed within parentheses preceded by the symbol of multiplication, if the symbols of operation,* $\times$ *and* $\div$, *preceding the numbers inclosed within the parentheses be left unchanged*; or, stated symbolically,

$$N \times a \div b = N \times (a \div b).$$

(ii.) *A chain of multiplications and divisions may be inclosed within parentheses preceded by the symbol of division, if the symbols of operation,* $\times$ *and* $\div$, *preceding the numbers inclosed within the parentheses be reversed from* $\times$ *to* $\div$ *and from* $\div$ *to* $\times$; or, stated symbolically,

$$N \div a \div b = N \div (a \times b), \text{ and } N \div a \times b = N \div (a \div b).$$

E.g., $\quad {}^{-}32 \times {}^{+}4 \div {}^{-}2 = {}^{-}128 \div {}^{-}2 = {}^{+}64,$

and $\quad {}^{-}32 \times ({}^{+}4 \div {}^{-}2) = {}^{-}32 \times {}^{-}2 = {}^{+}64;$

$\quad {}^{+}32 \div {}^{-}4 \times {}^{+}2 = {}^{-}8 \times {}^{+}2 = {}^{-}16,$

and $\quad {}^{+}32 \div ({}^{-}4 \div {}^{+}2) = {}^{+}32 \div {}^{-}2 = {}^{-}16;$

$\quad {}^{-}32 \div {}^{-}4 \div {}^{-}2 = {}^{+}8 \div {}^{-}2 = {}^{-}4,$

and $\quad {}^{-}32 \div ({}^{-}4 \times {}^{-}2) = {}^{-}32 \div {}^{+}8 = {}^{-}4;$

$\quad {}^{+}64 \div {}^{-}8 \times {}^{-}4 \div {}^{+}2 = {}^{-}8 \times {}^{-}4 \div {}^{+}2 = {}^{+}32 \div {}^{+}2 = {}^{+}16,$

and $\quad {}^{+}64 \div ({}^{-}8 \div {}^{-}4 \times {}^{+}2) = {}^{+}64 \div ({}^{+}2 \times {}^{+}2) = {}^{+}64 \div {}^{+}4 = {}^{+}16.$

The proof is as follows :

If in $$N \div a \div b,$$

N be replaced by $N \div (a \times b) \times (a \times b), = N$, by (3), Art. 5,

we have
$$N \div a \div b = N \div (a \times b) \times (a \times b) \div a \div b$$
$$= N \div (a \times b) \times b \times a \div a \div b,$$
since $\times a \times b = \times b \times a$,
$$= N \div (a \times b) \times b \div b, \text{ since } \times a \div a = \times {}^{+}1,$$
$$= N \div (a \times b), \text{ since } \times b \div b = \times {}^{+}1. \qquad (1)$$

Again, if in $$N \times a \div b,$$

N be replaced by $N \times (a \div b) \div (a \div b), = N$,

we have
$$N \times a \div b = N \times (a \div b) \div (a \div b) \times a \div b$$
$$= N \times (a \div b) \div (a \div b) \div b \times a,$$
since $\times a \div b = \div b \times a$,
$$= N \times (a \div b) \div [(a \div b) \times b] \times a, \text{ by (1)},$$
$$= N \times (a \div b) \div a \times a, \text{ since } (a \div b) \times b = a$$
$$= N \times (a \div b), \text{ since } \div a \times a = \div {}^{+}1. \qquad (2)$$

Finally, if in $$N \div a \times b,$$

N be replaced by $N \div (a \div b) \times (a \div b), = N$,

we have
$$N \div a \times b = N \div (a \div b) \times (a \div b) \div a \times b$$
$$= N \div (a \div b) \times a \div b \div a \times b,$$
since $\times (a \div b) = \times a \div b$, by (2),
$$= N \div (a \div b) \div b \times a \div a \times b,$$
since $\times a \div b = \div b \times a$,
$$= N \div (a \div b) \div b \times b, \text{ since } \times a \div a = \times {}^{+}1,$$
$$= N \div (a \div b), \text{ since } \div b \times b = \times {}^{+}1. \qquad (3)$$

In like manner these principles can be extended to include any number of successive multiplications and divisions.

8. If a number preceded by the symbol $\times$ or $\div$ stand first in a chain of multiplications and divisions, or first within parentheses, it is to be regarded as multiplying or dividing ${}^{+}1$.

Thus, $$N \times (\times {}^{+}2 \div {}^{+}3) = N \times ({}^{+}1 \times {}^{+}2 \div {}^{+}3),$$

and $$N \times (\div {}^{+}2 \div {}^{+}3) = N \times ({}^{+}1 \div {}^{+}2 \div {}^{+}3).$$

The symbol of operation $\times$ in such a case may be omitted.

For $${}^{+}1 \times N = N.$$

Thus, $^+30 \div {}^+5 \div {}^-3 = {}^+30 \div ({}^+5 \times {}^-3) = {}^+30 \div {}^-15 = {}^-2$,

and $^+30 \div {}^+5 \div {}^-3 = {}^+30 \times (\div {}^+5 \div {}^-3) = {}^+30 \times ({}^+1 \div {}^+5 \div {}^-3)$

$= {}^+30 \times {}^-(\frac{1}{15}) = {}^-2$, as above.

EXERCISES XXIV.

Find the values of the following expressions, inclosing the last two numbers within parentheses preceded by the symbol $\times$:

1. $^+30 \times {}^-12 \div {}^+3$.
2. $^-125 \times {}^+4 \div {}^-2$.
3. $^+35 \div {}^+5 \div {}^-3$.
4. $^+140 \div {}^-15 \div {}^-2\frac{1}{2}$.
5. $^-1000 \div {}^-250 \times {}^+2$.
6. $^-1000 \div {}^+64 \times {}^+8$.
7. Find the values of the expressions in Exx. 1–6, inclosing the last two numbers within parentheses preceded by the symbol $\div$.

9. Since $(ab) \div c = a \div c \times b = b \div c \times a$,
the product ab is divided by c, if either a or b be divided by c.

In like manner,

$$(abc) \div d = a \div d \times (bc) = b \div d \times (ac) = c \div d \times (ab).$$

Hence follows the principle:

A product of two or more factors is divided by a number, if any one of the factors be divided by that number.

10. An **even number** is one whose absolute value is exactly divisible by 2.

E.g., 2, 4, 6, etc.

Since, by the preceding article,

$$2n \div 2 = 2 \div 2 \times n = 1 \times n = n,$$

$2n$ is always an *even* number when n is an integer.

11. The converse of the Associative Law evidently enables us to remove parentheses which contain a succession of multiplications and divisions.

(i.) *When the symbol $\times$ precedes parentheses, they may be removed and the symbols of operation, $\times$ and $\div$, within them be left unchanged.*

E.g.,

${}^{+}120 \times ({}^{-}15 \div {}^{+}3 \times {}^{-}2) = {}^{+}120 \times ({}^{-}5 \times {}^{-}2) = {}^{+}120 \times {}^{+}10 = {}^{+}1200$;

and

${}^{+}120 \times {}^{-}15 \div {}^{+}3 \times {}^{-}2 = {}^{-}1800 \div {}^{+}3 \times {}^{-}2 = {}^{-}600 \times {}^{-}2 = {}^{+}1200$,

as above.

(ii.) *When the symbol ÷ precedes parentheses, they may be removed, if the symbols of operation,* × *and* ÷, *within them be reversed from* × *to* ÷ *and from* ÷ *to* ×.

E.g.,

${}^{+}120 \div ({}^{-}15 \div {}^{+}3 \times {}^{-}2) = {}^{+}120 \div ({}^{-}5 \times {}^{-}2) = {}^{+}120 \div {}^{+}10 = {}^{+}12$;

and ${}^{+}120 \div {}^{-}15 \times {}^{+}3 \div {}^{-}2 = {}^{-}8 \times {}^{+}3 \div {}^{-}2 = {}^{-}24 \div {}^{-}2 = {}^{+}12$,

as above.

Observe that (i.) and (ii.) cannot be applied if the parentheses contain additions and subtractions, besides multiplications and divisions.

E.g., ${}^{-}64 \div ({}^{+}8 \times {}^{+}2 - {}^{+}4 \div {}^{+}2) = {}^{-}64 \div ({}^{+}16 - {}^{+}2) = {}^{-}4\frac{4}{7}$;

but ${}^{-}64 \div {}^{+}8 \div {}^{+}2 - {}^{+}4 \times {}^{+}2 = {}^{-}8 \div {}^{+}2 - {}^{+}8 = {}^{-}12$, not ${}^{-}4\frac{4}{7}$.

EXERCISES XXV.

Find the values of the following expressions, first removing the parentheses:

1. ${}^{+}25 \times ({}^{+}12 \div {}^{-}4)$. 2. ${}^{-}20 \times ({}^{-}5 \div {}^{+}2)$. 3. ${}^{+}100 \div ({}^{+}25 \times {}^{-}2)$.

4. ${}^{-}600 \div ({}^{-}200 \div {}^{-}25 \times {}^{+}3 \div {}^{-}4)$. 5. ${}^{+}300 \div ({}^{-}150 \div {}^{+}6 \times {}^{+}8 \div {}^{-}4)$.

12. In a succession of additions, subtractions, multiplications, and divisions, the multiplications and divisions are first to be performed, and then the additions and subtractions.

E.g., ${}^{+}2 \times {}^{-}3 + {}^{-}4 \times {}^{+}5 = {}^{-}6 + {}^{-}20 = {}^{-}26$.

When a different order of performing the operations is proposed, the required order must be indicated by the insertion of parentheses.

E.g., ${}^{+}2 \times ({}^{-}3 + {}^{-}4) \times {}^{+}5 = {}^{+}2 \times {}^{-}7 \times {}^{+}5 = {}^{-}70$,

not ${}^{-}26$, as before.

Likewise,

$$ {}^{+}12 \div {}^{-}4 + {}^{-}2 \times {}^{-}18 - {}^{-}3 \times {}^{+}5 = {}^{-}3 + {}^{+}36 - {}^{-}15 = {}^{+}48; $$

$$ {}^{+}12 \div ({}^{-}4 + {}^{-}2) \times ({}^{-}18 - {}^{-}3) \times {}^{+}5 = {}^{+}12 \div {}^{-}6 \times {}^{-}15 \times {}^{+}5 = {}^{+}150; $$

and

$$ ({}^{+}12 \div {}^{-}4 + {}^{-}2) \times {}^{-}18 - {}^{-}3 \times {}^{+}5 = {}^{-}5 \times {}^{-}18 - {}^{-}15 $$
$$ = {}^{+}90 - {}^{-}15 = {}^{+}105. $$

EXERCISES XXVI.

Find the values of the following expressions:

1. ${}^{+}180 \div {}^{-}36 + {}^{-}4 \times {}^{-}2.$
2. ${}^{-}25 \times {}^{+}4 - {}^{+}36 \div {}^{-}12.$
3. ${}^{+}48 \times {}^{-}2 - {}^{-}96 \div {}^{-}24.$
4. ${}^{-}7 + {}^{+}15 \div {}^{+}3 \times {}^{-}2 - {}^{-}28 \div {}^{+}4.$
5. ${}^{+}11 - {}^{-}4 \times {}^{+}5 \div {}^{+}2 + {}^{+}33 \div {}^{-}3 \times {}^{+}2.$
6. ${}^{-}1 - {}^{-}2 \times {}^{+}3 + {}^{+}4 \times {}^{-}5 \div {}^{+}2.$

13. The following principle will be useful in subsequent work:

If equal numbers be divided by equal numbers, or by the same number, except 0, the quotients will be equal.

If $a = b$ and $A = B$, then $a \div A = b \div B$.

For, $\qquad a \div A = a \div A$, by Axiom (i.). $\qquad$ (1)

But, since $b = a$ and $B = A$, we can substitute b for a, and B for A, in the second member of (1). We thus obtain

$$ a \div A = b \div B. $$

§ 5. ONE SET OF SIGNS FOR QUALITY AND OPERATION.

1. In conformity with the usage of most text-books of Algebra we shall in subsequent work use the one set of signs, + and −, to denote both *quality* and *operation*. For the sake of brevity the sign + is usually omitted when it denotes *quality;* the sign − is never omitted.

Thus, instead of ${}^{+}2$, we shall write $+2$, or 2;
instead of ${}^{-}2$, we shall write -2.

2. We have used the double set of signs hitherto in order to emphasize the difference between *quality* and *operation*. It should be kept clearly in mind that the same distinction still exists.

We now have

$N + {}^{+}2 = N + (+2) = N + 2$, omitting the sign of *quality*, +;

$N + {}^{-}2 = N + (-2)$, wherein + denotes *operation*, and − denotes *quality*.

$N - {}^{+}2 = N - (+2) = N - 2$, omitting the sign of *quality*, +;

$N - {}^{-}2 = N - (-2)$, wherein the first sign, −, denotes *operation*, the second sign, −, denotes *quality*.

3. In the chain of operations

$$(+2) + (-5) - (+2) - (-11)$$

the signs within the parentheses denote *quality*, those without denote *operation*. That expression reduces to

$$(+2) - (+5) - (+2) + (+11),$$

or $2 - 5 - 2 + 11$, dropping the sign of *quality*, +.

In the latter expression all the signs denote *operation*, and the numbers are all *positive*.

4. A number with the sign +, or the sign −, standing by itself, apart from any chain of additions and subtractions, admits of two meanings: It may be regarded either as a positive or a negative number (*i.e.*, as so many units greater or less than zero); or as indicating that the number is to be added to or subtracted from any number, including 0, that may be placed on its left.

E.g., −3 may be regarded as *negative* 3; *i.e.*, as a number *three units less than* 0; or as indicating that 3 is to be subtracted from any number that may be placed on its left, including 0.

5. The second meaning given to a number with the sign +, or the sign −, is preferable when the number stands first in a chain of additions and subtractions.

E.g., in $$-5 + 3 - 2$$

all the signs are then signs of operation.

Likewise, in $+3=+1+1+1$, and $-3=-1-1-1$, the signs $+$ and $-$ everywhere denote operation.

6. For reference in subsequent work, we give a summary of the more important results of this chapter with the one set of signs, $+$ and $-$.

(I.) From Ch. I., § 2, Art. 12,

$^{+}n+{}^{-}n=0$ becomes $+n+(-n)=0$, or $n+(-n)=0$.

(II.) From Ch. II., § 1.

(1) Art. 3, Ex. 1: $^{+}4+{}^{+}3={}^{+}(4+3)={}^{+}7$

becomes $+4+(+3)=+(4+3)=+7$, or $4+3=7$.

(2) Art. 3, Ex. 2: $^{-}4+{}^{-}3={}^{-}(4+3)={}^{-}7$

becomes $-4+(-3)=-(4+3)=-7$.

(3) Art. 6, Ex. 1: $^{+}5+{}^{-}2={}^{+}(5-2)={}^{+}3$

becomes $+5+(-2)=+(5-2)=+3$,

or $5+(-2)=5-2=3$.

(4) Art. 6, Ex. 2: $^{-}5+{}^{+}2={}^{-}(5-2)={}^{-}3$

becomes $-5+(+2)=-(5-2)=-3$,

or $-5+2=-(5-2)=-3$.

In general:

(5) Art. 4: $^{+}a+{}^{+}b={}^{+}(a+b)$

becomes $+a+(+b)=+(a+b)$, or $a+b=a+b$.

(6) Art. 4: $^{-}a+{}^{-}b={}^{-}(a+b)$

becomes $-a+(-b)=-(a+b)$.

(7) Art. 7: $^{+}a+{}^{-}b={}^{+}(a-b)$, when $a>b$,

becomes $+a+(-b)=+(a-b)$,

or $a+(-b)=a-b$.

(8) Art. 7: $^{+}a+{}^{-}b={}^{-}(b-a)$, when $a<b$,

becomes $+a+(-b)=-(b-a)$,

or $a+(-b)=-(b-a)$.

(III.) From Ch. II., § 2.

(1) Art. 3, Ex. 1: $^{+}7 - {}^{+}5 = {}^{+}7 + {}^{-}5 = {}^{+}2$

becomes $+7 - (+5) = +7 + (-5) = +2,$

or $7 - 5 = 7 + (-5) = 2.$

(2) Art. 3, Ex. 2: $^{+}7 - {}^{-}5 = {}^{+}7 + {}^{+}5 = {}^{+}12$

becomes $+7 - (-5) = +7 + (+5) = +12,$

or $7 - (-5) = 7 + 5 = 12.$

In general:

(3) Art. 3: $N - {}^{+}b = N + {}^{-}b$

becomes $N - (+b) = N + (-b),$

or $N - b = N + (-b).$

(4) Art. 3: $N - {}^{-}b = N + {}^{+}b$

becomes $N - (-b) = N + (+b),$

or $N - (-b) = N + b.$

(IV.) From Ch. II., § 3.

(1) Art. 5 (i.), Ex. 1: $^{+}4 \times {}^{+}3 = {}^{+}(4 \times 3) = {}^{+}12$

becomes $+4 \times (+3) = +(4 \times 3) = +12,$

or $4 \times 3 = 12.$

(2) Art. 5 (i.), Ex. 2: $^{-}4 \times {}^{+}3 = {}^{-}(4 \times 3) = {}^{-}12$

becomes $-4 \times (+3) = -(4 \times 3) = -12,$

or $-4 \times 3 = -12.$

(3) Art. 5 (ii.), Ex. 3: $^{+}4 \times {}^{-}3 = {}^{-}(4 \times 3) = {}^{-}12$

becomes $+4 \times (-3) = -(4 \times 3) = -12,$

or $4 \times (-3) = -12.$

(4) Art. 5 (ii.), Ex. 4: $^{-}4 \times {}^{-}3 = {}^{+}(4 \times 3) = {}^{+}12$

becomes $(-4)(-3) = +(4 \times 3) = +12,$

or $(-4) \times (-3) = 12.$

In general:

(5) Art. 6: $^{+}a \times {}^{+}b = {}^{+}(ab)$

becomes $(+a) \times (+b) = +(ab),$

or $a \times b = ab.$

(6) Art. 6: $^{-}a \times {}^{+}b = {}^{-}(ab)$

becomes $(-a)(+b) = -(ab)$,

or $(-a)b = -ab$.

(7) Art. 6: $^{+}a \times {}^{-}b = {}^{-}(ab)$

becomes $(+a)(-b) = -(ab)$,

or $a(-b) = -ab$.

(8) Art. 6: $^{-}a \times {}^{-}b = {}^{+}(ab)$

becomes $(-a)(-b) = +(ab)$,

or $(-a)(-b) = ab$.

(V.) From Ch. III., § 4.

(1) Art. 3: $^{+}(ab) \div {}^{+}a = {}^{+}b$

becomes $+(ab) \div (+a) = +b$,

or $ab \div a = b$.

(2) Art. 3: $^{-}(ab) \div {}^{-}a = {}^{+}b$

becomes $-(ab) \div (-a) = +b$,

or $-ab \div (-a) = b$.

(3) Art. 3: $^{-}(ab) \div {}^{+}a = {}^{-}b$

becomes $-(ab) \div (+a) = -b$,

or $-ab \div a = -b$.

(4) Art. 3: $^{+}(ab) \div {}^{-}a = {}^{-}b$

becomes $+(ab) \div (-a) = -b$,

or $ab \div (-a) = -b$.

EXERCISES XXVII.

Find the values of the expressions in Exx. 1–19, first changing them into equivalent expressions in which there is only the one set of signs, + and − :

1. $^{+}2 + {}^{+}3$.
2. $^{+}11 + {}^{+}4$.
3. $^{+}14 + {}^{-}9$.
4. $^{+}17 + {}^{-}3$.
5. $^{-}8 + {}^{+}5$.
6. $^{-}7 + {}^{+}6$.
7. $^{-}4 + {}^{-}5$.
8. $^{-}7 + {}^{-}2$.
9. $^{+}1 - {}^{+}2$.
10. $^{+}8 - {}^{+}4$.
11. $^{+}5 - {}^{-}2$.
12. $^{+}14 - {}^{-}7$.
13. $^{-}8 - {}^{+}3$.
14. $^{-}15 - {}^{+}5$.
15. $^{-}2 - {}^{-}5$.
16. $^{-}13 - {}^{-}10$.
17. $^{+}a + {}^{-}b - {}^{+}c$.
18. $^{+}x - {}^{-}y - {}^{+}z$.
19. $^{+}u + {}^{+}v - {}^{-}w$.

Find the values of the expressions in Exx. 20–27, first changing them into equivalent expressions in which there is only one set of signs, + and −, and then removing the parentheses:

20. ${}^{+}5+({}^{+}4+{}^{+}3)$. **21.** ${}^{-}9+({}^{+}11+{}^{-}5)$.

22. ${}^{+}7+({}^{-}5+{}^{+}3)$. **23.** ${}^{+}17-({}^{+}5+{}^{-}8)$.

24. ${}^{-}8-({}^{-}12+{}^{+}4)$. **25.** ${}^{+}11-({}^{-}7-{}^{-}4)$.

26. ${}^{+}5-({}^{-}3+{}^{-}4)+({}^{-}2-{}^{+}8)$. **27.** ${}^{-}7-({}^{-}5-{}^{-}8)+({}^{+}6-{}^{-}7)$.

Find the values of the expressions in Exx. 28–52, first changing them into equivalent expressions in which there is only one set of signs, + and −:

28. ${}^{+}3\times{}^{+}4$. **29.** ${}^{+}7\times{}^{+}9$. **30.** ${}^{-}9\times{}^{+}11$. **31.** ${}^{-}17\times{}^{+}3$.

32. ${}^{+}18\times{}^{-}4$. **33.** ${}^{+}19\times{}^{-}5$. **34.** ${}^{-}4\times{}^{-}7$. **35.** ${}^{-}5\times{}^{-}12$.

36. ${}^{+}18\div{}^{+}2$. **37.** ${}^{+}15\div{}^{+}5$. **38.** ${}^{-}18\div{}^{+}3$. **39.** ${}^{-}28\div{}^{+}7$.

40. ${}^{+}35\div{}^{-}5$. **41.** ${}^{+}105\div{}^{-}15$. **42.** ${}^{-}96\div{}^{-}6$. **43.** ${}^{-}27\div{}^{-}9$.

44. ${}^{+}2\times{}^{-}3\times{}^{+}5$. **45.** ${}^{-}11\times{}^{+}2\times{}^{-}4$. **46.** ${}^{+}17\times{}^{-}4\times{}^{-}3$.

47. ${}^{+}18\div{}^{-}3\times{}^{-}5$. **48.** ${}^{-}27\times{}^{-}3\div{}^{+}9$. **49.** ${}^{+}42\div{}^{-}6\div{}^{+}2$.

50. ${}^{+}18\div({}^{+}9\div{}^{-}3)$. **51.** ${}^{-}36\div({}^{-}3\times{}^{+}6)$. **52.** ${}^{+}15\times({}^{-}18\div{}^{+}9)$.

Which of the signs + and − denote operation, and which quality, in Exx. 53–59:

53. $+2+(+3)$. **54.** $+7-(+8)$. **55.** $-11+(-4)$.

56. $-17-(+5)$. **57.** $+25-(-4)$. **58.** $-38-(-5)$.

59. $-17-(-2)+(-4)$.

Find the values of the expressions in Exx. 60–79, first changing them into equivalent expressions in which the signs + and − denote only operation:

60. $+5+(+8)$. **61.** $7+(+9)$. **62.** $3+(-4)$.

63. $5+(-2)$. **64.** $15+(-3)$. **65.** $-18+(-19)$.

66. $-3+(-5)$. **67.** $-11-(-18)$. **68.** $-15-(-32)$.

69. $7+(-3)-(-5)$. **70.** $-12+(-11)-(-19)$.

71. $-3+(-2)-(+15)$.

72. $+3+(+5)-(+8)-(-12)+(-15)$.

73. $-4-(-19)+(-3)+(+7)+(-9)$.

74. $-2-(-3)-(-1)+(-9)+(+12)+(-19)-(-37)$.

75. $2+[+3-(-5)]$.

76. $8-[-3-(-2)]$.

77. $-17-[+2-(-3)]$.

78. $25-[+17-(+2)+(-8)-(-19)]$.

79. $-38-[-29-(-11)-(+15)-(-12)]$.

What are the values of $a+b-c+d-e$ and $a-(b-c+d-e)$,

80. When $a=3$, $b=-5$, $c=-8$, $d=-9$, $e=7$?

81. When $a=-9$, $b=6$, $c=-9$, $d=-11$, $e=-12$?

Find the values of the expressions in Exx. 82–91, first changing those in the brackets, [ ], into equivalent expressions in which the signs + and − denote only operation:

82. $5[-3+(-2)]$.

83. $(-7)[4-(-3)]$.

84. $[-11-(-2)]\div(-13)$.

85. $9[23-(-2)+(-5)-(-9)]$.

86. $[58+(-12)-(-4)]\div 5$.

87. $(-11)[15-(+2)-(+8)+(-11)-(-12)]$.

88. $-15\div[-3+(-2)]$.

89. $+28\div[-3-(-10)]$.

90. $[-15-(-3)]\div(+4)$.

91. $[+33-(-5)]\div(-19)$.

What is the value of $a(b-c+d)$,

92. When $a=5$, $b=-9$, $c=-11$, $d=8$?

93. When $a=-2$, $b=11$, $c=-12$, $d=-9$?

94. When $a=-13$, $b=-2$, $c=3$, $d=-3$?

95. When $a=22$, $b=-11$, $c=-8$, $d=-4$?

What is the value of $a\div(b+c-d+e)$,

96. When $a=-5$, $b=-3$, $c=4$, $d=-5$, $e=-7$?

97. When $a=12$, $b=-2$, $c=-9$, $d=13$, $e=28$?

98. When $a=-12$, $b=-10$, $c=-12$, $d=15$, $e=-23$?

Change the expressions in Exx. 99–104 into equivalent expressions, in which there are both sets of signs $^+$ and $^-$ and +

and $-$, and in each one of which the third and fourth numbers are preceded by the sign of addition:

99. $4-7+11-8.$ **100.** $-18-15+2-7.$

101. $41-11+15-7.$ **102.** $125+4-7-11.$

103. $81-9+4-2.$ **104.** $-15+23-14+5.$

Find the results of the following indicated operations:

105. $4-7.$ **106.** $-3+11.$ **107.** $+18-22.$

108. $+1-55.$ **109.** $2-4+6.$ **110.** $5-8-2.$

111. $1-17+2.$ **112.** $1-4+5-6+8-11.$

113. $-2-3+17-25-18+1.$ **114.** $-3\times 4.$

115. $-7\times 11.$ **116.** $15\times(-8).$ **117.** $7\times(-9).$

118. $(-4)\times(-5).$ **119.** $(-9)\times(-17).$ **120.** $15\times(-12).$

121. $(-9)\times(-8).$ **122.** $18\div(-2).$ **123.** $36\div(-4).$

124. $(-15)\div 5.$ **125.** $(-48)\div 12.$ **126.** $96\div(-6).$

127. $(-144)\div(-12).$ **128.** $(-200)\div(-25).$

129. $(-1000)\div 8.$ **130.** $45\div(-5)\times 3.$

131. $(-96)\times 12\div 4.$ **132.** $33\times 12\div(-3).$

133. $(-2)\times(-3)+(-4)\times 5-7\times(-3).$

134. $5\times(-8)-16\div(-4)+2\times(-5).$

§ 6. POSITIVE INTEGRAL POWERS.

1. A continued product of equal factors is called a **Power** of that factor.

Thus, 2×2 is called the *second power* of 2, or 2 *raised to the second power*; aaa is called the *third power* of a, or a *raised to the third power*.

In general, $aaa \cdots$ to n factors is called the nth power of a, or a *raised to the nth power*.

The second power of a is often called the *square* of a, or a *squared*; and the third power of a the *cube* of a, or a *cubed*.

2. The notation for powers is abbreviated as follows:

a^2 is written instead of aa; a^3 instead of aaa;

a^n instead of $aaa \cdots$ to n factors.

3. The **Base** of a power is the number which is repeated as a factor.

E.g., a is the base of a^2, a^3, $\cdots$, a^n.

The **Exponent** of a power is the number which indicates how many times the base is used as a factor, and is written to the right and a little above the base.

E.g., the exponent of a^2 is 2, of a^3 is 3, of a^n is n.

The exponent 1 is usually omitted. Thus, $a^1 = a$.

An exponent must not be confused with a subscript. Thus, a^3 stands for the product aaa; while a_3 is a notation for a single number.

4. The definition of a *power* given above requires the exponent to be a *positive integer*. In a subsequent chapter this definition will be extended to include powers with negative and fractional exponents.

Notice that the words *positive integral* refer to the exponent and not to the value of the power, which may be negative or fractional.

E.g., $(-2)^3 = (-2)(-2)(-2) = -8$; $(\frac{2}{3})^2 = \frac{2}{3} \times \frac{2}{3} = \frac{4}{9}$.

5. The base of a power must be inclosed within parentheses to prevent ambiguity:

(i.) *When the base is a negative number.* Thus,

$(-5)^2 = (-5)(-5) = 25$; while $-5^2 = -(5 \times 5) = -25$.

(ii.) *When the base is a product or a quotient.* Thus,

$$(2 \times 5)^3 = (2 \times 5)(2 \times 5)(2 \times 5) = 1000;$$

while $$2 \times 5^3 = 2(5 \times 5 \times 5) = 250.$$

Likewise $\left(\dfrac{2}{3}\right)^2 = \dfrac{2}{3} \times \dfrac{2}{3} = \dfrac{4}{9}$, while $\dfrac{2^2}{3} = \dfrac{2 \times 2}{3} = \dfrac{4}{3}$.

(iii.) *When the base is a sum.* Thus,

$$(2+3)^2 = (2+3)(2+3) = 5 \times 5 = 25;$$

while $$2 + 3^2 = 2 + 3 \times 3 = 2 + 9 = 11.$$

(iv.) *When the base is itself a power.* Thus,

$$(2^3)^2 = 2^3 \times 2^3 = (2 \times 2 \times 2)(2 \times 2 \times 2) = 64,$$

while $2^{3^2} = 2^{3 \times 3} = 2^9 = 2 \times 2 \times 2 \times 2 \times 2 \times 2 \times 2 \times 2 \times 2 = 512.$

EXERCISES XXVIII.

In the following powers which numbers are the bases? Which numbers are the exponents?

1. 2^3.
2. 3^2.
3. $(-11)^4$.
4. $(2a)^3$.
5. $(-6x)^7$.
6. $(xy)^2$.
7. $(-5mn)^{10}$.
8. 5^p.
9. $(2^5)^3$.
10. $[(-7)^2]^4$.
11. $[(-x)^m]^3$.
12. $[(ab)^x]^y$.
13. $(x+y)^2$.
14. $(a-b)^3$.
15. $(a+b+c)^4$.

Express the following powers in the abbreviated notation:

16. $2 \times 2 \times 2$.
17. $(-3)(-3)(-3)$.
18. $5 \times 5 \times 5 \times 5$.
19. $aaaaaa$.
20. $(-x)(-x)(-x)(-x)(-x)(-x)(-x)$.
21. $(xy)(xy)(xy)$.
22. $(-3a)(-3a)(-3a)(-3a)$.
23. $b \cdot b \cdot b \cdots$ to 8 factors.
24. $(-z)(-z)(-z) \cdots$ to 17 factors.
25. $5 \times 5 \times 5 \cdots$ to m factors.
26. $(-ab)(-ab)(-ab) \cdots$ to n factors.
27. $3^2 \times 3^2 \times 3^2 \times 3^2 \times 3^2$.
28. $x^n \cdot x^n \cdot x^n \cdots$ to m factors.

Express the following powers as continued products:

29. 2^5.
30. $(-3)^7$.
31. $(5x)^4$.
32. $(-6z)^6$.
33. 4^m.
34. x^n.
35. $(ab)^p$.
36. $(-7x)^x$.
37. $(2^2)^3$.
38. $(-3^4)^2$.
39. $(a^4)^2$.
40. $[(-xy)^2]^3$.
41. $(-2)^5$.
42. -2^5.
43. $(-3)^6$.
44. -3^6.
45. $(2a)^3$.
46. $2a^3$.
47. $(-5b)^4$.
48. $-5b^4$.

Express the following powers in the abbreviated notation:

49. $(a+x)(a+x)(a+x)(a+x)$.
50. $(a+x)(a+x)(a+x) \cdots$ to 8 factors.
51. $(aaa-b)(aaa-b)(aaa-b) \cdots$ to 17 factors.

Express in algebraic notation:

52. The sum of the squares of a and b.
53. The square of the sum of a and b.
54. The sum of the cubes of x, y, and z.
55. The cube of the sum of x, y, and z.

Properties of Positive Integral Powers.

6. (i.) *All (even and odd) powers of positive bases are positive.*

E.g., $2^3 = 2 \times 2 \times 2 = 8.$ $3^4 = 3 \times 3 \times 3 \times 3 = 81.$

In general, $(+a)^n = (+a)(+a)(+a) \cdots n$ factors.
$= +a^n$, *n even or odd.*

(ii.) *Even powers of negative bases are positive; odd powers of negative bases are negative.*

Notice that the words *even* and *odd* refer to the exponents. Also, that, for all integral values of n, $2n$ is *even* (Ch. II., § 4, Art. 10), and hence that $2n + 1$ or $2n - 1$ is *odd.*

E.g., $(-2)^4 = (-2)(-2)(-2)(-2) = 16;$
$(-5)^3 = (-5)(-5)(-5) = -125.$

In general,

$$(-a)^{2n} = (-a)(-a)(-a) \cdots 2n \text{ factors},$$
$$= +(aaa \cdots 2n \text{ factors}), \text{ by Ch. II., § 3, Art. 9},$$
$$= +a^{2n}.$$

And,

$$(-a)^{2n+1} = (-a)(-a)(-a) \cdots (2n+1) \text{ factors},$$
$$= -[aaa \cdots (2n+1) \text{ factors}], \text{ by Ch. II., § 3, Art. 9},$$
$$= -a^{2n+1}.$$

(iii.) In particular,

$(+1)^n = +1^n = 1$, wherein n is *even* or *odd;*
$(-1)^{2n} = +1^{2n} = 1;$ and $(-1)^{2n+1} = -1^{2n+1} = -1.$

7. Any positive integral power of 0 is 0; that is, $0^n = 0$.

For $0^n = 0 \cdot 0 \cdot 0 \cdots n$ factors,
$= 0.$

EXERCISES XXIX.

Find the values of the following powers:

1. 2^8. 2. 3^2. 3. $(-3)^5$. 4. -3^5. 5. $(-2)^6$.
6. -2^6. 7. $(-5)^4$. 8. -5^4. 9. $(-x)^4$. 10. $(-x)^7$.

Express as powers of 2:

11. 8. **12.** 64. **13.** 512. **14.** 4096.

Express as powers of -3:

15. 9. **16.** 81. **17.** -243. **18.** 729.

Express as powers of -5:

19. 25. **20.** 625. **21.** -3125. **22.** 15,625.

Express as products of powers of 2 and 3:

23. 24. **24.** 144. **25.** 1536. **26.** 2916.

Determine, by inspection, the signs of the following powers:

27. $(-5)^{17}$. **28.** $(-7)^{76}$. **29.** 8^{13}. **30.** 4^7.

31. a^7. **32.** $(-b)^{11}$. **33.** $-c^{15}$. **34.** $(-3x)^{18}$.

35. 2^n. **36.** $(-3)^n$, when n is even. **37.** $(-3)^{2p}$.

38. $(-7)^{n-1}$, when n is even. **39.** $(-4)^{2p+1}$.

40. $(-6)^{n-1}$, when $n+1$ is even.

Find the values of the following expressions:

41. 2^2+3^2. **42.** $(2+3)^2$. **43.** (3^3-5^3). **44.** $(3-5)^3$.

45. 2^4-3^4. **46.** $(2-3)^4$. **47.** 4^5+3^4. **48.** 6^3-2^4.

49. $3^5-(-7)^3$. **50.** $2^7+(-3)^3$. **51.** 2×3^2. **52.** $(2\times 3)^2$.

53. 5×4^3. **54.** $(5\times 4)^3$. **55.** $2(-7)^3$. **56.** $[2(-7)]^3$.

57. $2\times 3^4-(3\times 2)^4$. **58.** $(5\times 6)^2-6\times 5^2$.

59. $3(-4)^3-[2(-5)]^2$. **60.** $3^4-(-3)^4$. **61.** $5^3+(-5)^3$.

62. $2^7-(-2)^7$. **63.** $29^{24}-(-29)^{24}$.

64. $(-37)^{13}+37^{13}$. **65.** $[9(-4)]^{11}+36^{11}$.

Find the value of a^n,

66. When $a=2$, $n=3$. **67.** $a=-2$, $n=4$.

68. When $a=-3$, $n=5$.

Find the values of $(a+b)^n$ and $(a-b)^n$,

69. When $a=3$, $b=2$, $n=4$. **70.** When $a=5$, $b=-2$, $n=3$.

71. When $a=-6$, $b=4$, $n=5$.

When $a = 1$, $b = -3$, $c = 2$, find the values of:

72. a^c. **73.** b^a. **74.** ab^c. **75.** $(ab)^c$. **76.** $(b^a)^c$.

77. b^{ac}. **78.** $a^2 + b^2 + c^2$. **79.** $(ab)^3 + (bc)^3 + (ac)^3$.

It is proved in Geometry that the area of a square is equal to the square of its side. Find the area of a square,

80. When its side is 15 inches. **81.** When its side is a feet.

It is proved in Geometry that the volume of a cube is equal to the cube of its edge. Find the volume of a cube,

82. When its edge is 8 inches. **83.** When its edge is x feet.

Find the values of the following sums when $n = 5$, $a = 2$, $b = -3$:

84. $1^2 + 2^2 + 3^2 + \cdots + n^2$. **85.** $(1 + 2 + 3 + \cdots + n)^2$.

86. $1^3 + 2^3 + 3^3 + \cdots + n^3$. **87.** $(1 + 2 + 3 + \cdots + n)^3$.

88. $a + a^2 + a^3 + \cdots + a^n$. **89.** $b + b^2 + b^3 + \cdots + b^n$.

CHAPTER III.

THE FUNDAMENTAL OPERATIONS WITH INTEGRAL ALGEBRAIC EXPRESSIONS.

§ 1. DEFINITIONS.

1. An **Integral Algebraic Expression** is an expression which involves only additions, subtractions, multiplications, and positive integral powers of *literal* numbers; that is, in which the *literal* numbers are connected only by one or more of the symbols of operation, $+$, $-$, $\times$, but not by the symbol $\div$.

E.g., $1 + x + x^2$, $5a^2b + \frac{2}{3}cd^2$, etc., are integral algebraic expressions.

But $x \div y$, $a \div b - c \div d$, etc., are *not* integral algebraic expressions.

Notice that the word *integral* refers only to the *literal* parts of the expression. At the same time, the letters are not *limited to integral numerical values, but, as always, may have any values whatever.*

E.g., $a + b$ is *algebraically* integral; but a may have the value $\frac{1}{2}$, and b the value $\frac{3}{4}$, in which case

$$a + b = \tfrac{1}{2} + \tfrac{3}{4} = 1\tfrac{1}{4}.$$

2. Coefficients.—In a product, any factor, or product of factors, is called the **Coefficient** of the product of the remaining factors.

E.g., in $3abc$, 3 is the coefficient of abc, $3a$ of bc, $3ab$ of c, abc of 3, ab of $3c$, etc.

In $(a + b)cd$, $a + b$ is the coefficient of cd, $(a + b)c$ of d, cd of $(a + b)$, etc.

A **Numerical Coefficient** is a coefficient expressed in figures.

E.g., in $-3ab$, -3 is the numerical coefficient of ab.

A **Literal Coefficient** is a coefficient expressed in letters, or in letters and figures.

E.g., in $3\,ab$, a is the literal coefficient of $3\,b$, and $3\,a$ of b.

The coefficients $+1$ and -1 are usually omitted.

E.g., $1 \times a$ is usually written a, and $-1 \times a$ is usually written $-a$.

3. The sign $+$ or the sign $-$, preceding a product, is to be regarded as the sign of its numerical coefficient.

Thus, $+3\,a$ means the product of *positive* 3 by a; $-5\,x$ means the product of *negative* 5 by x. In particular, $+a$ means the product of positive 1 by a, and $-a$ means the product of *negative* 1 by a, unless the contrary is stated.

EXERCISES I.

What is the coefficient of x in

1. $2\,x$? 2. $-3\,x$? 3. $5\,ax$? 4. $-7\,bx$? 5. $(a+b)\,x$?

In the expression $12\,xy\,(a+b)$, what is the coefficient of

6. $a+b$? 7. $x(a+b)$? 8. $12\,y$? 9. $12\,x(a+b)$?

10. If the sum, $a+a+a+a$, be represented as a product, what is the coefficient of a?

11. If the algebraic sum, $-b-b-b-b-b$, be represented as a product, what is the coefficient of $-b$? Of b?

12. If the sum, $7+7+7+\cdots$ to a summands, be represented as a product, what is the coefficient of a?

13. If the sum, $a+a+a+\cdots$ to 8 summands, be represented as a product, what is the coefficient of a?

14. If the sum, $x+x+x+\cdots$ to n summands, be represented as a product, what is the coefficient of x? Of n?

15. If the algebraic sum, $-y-y-y-\cdots$ to a summands, be represented as a product, what is the coefficient of $-y$? Of y?

16. If the sum, $2\,ax+2\,ax+2\,ax+\cdots$ to y summands, be represented as a product, what is the coefficient of $2\,ax$? Of $2\,a$? Of ay?

17. In the sum, $(a+b)+(a+b)+(a+b)+\cdots$ to 8 summands, what is the coefficient of $(a+b)$?

4. In an expression such as

$$4a-3b,$$

the sign $-$ means *operation*, i.e., *subtraction*.

But, since $\quad 4a-3b=4a+(-3b),$

the given expression is the result of adding $4a$ and $-3b$.

In like manner, the expression

$$2x-3y+4z-5u=2x+(-3y)+4z+(-5u),$$

is the result of adding $2x$, $-3y$, $4z$, and $-5u$. Upon these considerations are based the following definitions:

The **Terms** of an algebraic expression are the *additive* and *subtractive* parts of that expression.

E.g., the terms of $4a-3b$ are $4a$ and $-3b$; the terms of $2x-3y+4z-5u$ are $2x$, $-3y$, $4z$, and $-5u$.

The **Sign of a Term** is the sign of quality, $+$ or $-$, of its numerical coefficient.

E.g., the sign of the term $2x$ is $+$, of $-3y$ is $-$, etc.

A **Positive Term** is one whose sign is $+$; as $2x$ and $4z$.

A **Negative Term** is one whose sign is $-$; as $-3y$ and $-5u$.

5. Like or **Similar Terms** are terms which do not differ, or which differ only in their numerical coefficients.

E.g., in the expression

$$+3a+6ab-5a+7ab,$$

$+3a$ and $-5a$ are like terms; so are $+6ab$ and $+7ab$.

Unlike or **Dissimilar Terms** are terms which are not like.

E.g., in the above expression $+3a$ and $+7ab$ are unlike terms.

6. A **Monomial** is an expression of one term.

E.g., $\quad a,\ 3a^2,\ -7bc.$

A **Binomial** is an expression of two terms.

E.g., $\quad a+b,\ -2a^2+3abc,\ 7a^2b-11b^2c.$

A Trinomial is an expression of three terms.

E.g., $a + b - c,\ -3a^2 + 7b^3 - 5c^4.$

A Multinomial* is an expression of two or more terms, including, therefore, binomials and trinomials as particular cases.

E.g., $a + b^2,\ a^2 + b - c^3,\ ab + bc - cd - ef.$

§ 2. ADDITION AND SUBTRACTION.

1. Like Terms can be united by addition and subtraction into a single *like* term.

Just as $2 = 1 + 1$, so $2a = a + a$;

just as $5 = 1 + 1 + 1 + 1 + 1$, so $5a = a + a + a + a + a$.

Therefore, just as $2 + 5 = 7$, so $2a + 5a = (2 + 5)a = 7a$.

That is, a is used in $2a + 5a$ just as the unit is used in $2 + 5$.

Again, just as

$$3 = 1 + 1 + 1, \text{ so } 3xy = xy + xy + xy;$$

just as

$$-4 = -1 - 1 - 1 - 1, \text{ so } -4xy = -xy - xy - xy - xy.$$

Therefore, just as $3 - (-4) = 3 + 4 = 7$,

so $$3xy - (-4xy) = [3 - (-4)]xy = 7xy.$$

That is, xy is used in $3xy - (-4xy)$ just as the unit is used in $3 - (-4)$.

Consequently the addition and subtraction of *like* terms reduces to the addition and subtraction of their numerical coefficients. That is, *to add or subtract like terms, add or subtract their numerical coefficients and annex to that result their common literal part.*

Ex. 1. Add $-7ab$ to $4ab$.

We have $4ab + (-7ab) = [4 + (-7)]ab = -3ab$.

Ex. 2. Add $-3(x + y)$ to $-2(x + y)$.

We have

$$-2(x + y) + [-3(x + y)] = [-2 + (-3)](x + y) = -5(x + y).$$

---

* The word **Polynomial** is frequently used instead of **Multinomial**.

Ex. 3. Find the sum of $3a$, $-5a$, $8a$, $-4a$.

By the Commutative Law for addition, the terms may be added in any order.

We have

$$3a + 8a + (-5a) + (-4a) = [3 + 8 + (-5) + (-4)]a = 2a.$$

Ex. 4. Subtract $-5x^2y$ from $-7x^2y$.

We have

$$-7x^2y - (-5x^2y) = -7x^2y + 5x^2y = (-7 + 5)x^2y = -2x^2y.$$

The student should accustom himself to find mentally the algebraic sum of the numerical coefficients, writing only the result. In addition, it is better to find the sum of all the positive coefficients, then the sum of all the negative coefficients, or *vice versa*, and unite the resulting sums.

EXERCISES II.

Add

1. $2a$ to $3a$.
2. $5x$ to $7x$.
3. $4b$ to $-9b$.
4. $6u$ to $-u$.
5. $-7y$ to $3y$.
6. $-5m$ to $8m$.
7. $-6c$ to $-5c$.
8. $-9v$ to $-4v$.
9. $\frac{15}{2}a^2$ to $\frac{7}{8}a^2$.
10. $-4x^3$ to $\frac{1}{3}x^3$.
11. $\frac{7}{4}ab$ to $-\frac{9}{5}ab$.
12. $-5mn^p$ to $-\frac{2}{3}mn^p$.
13. $a + b$ to $-3(a + b)$.
14. $-7(x^2 + y^2)$ to $2(x^2 + y^2)$.

Subtract

15. $7x$ from $4x$.
16. $2a$ from $5a$.
17. $3b$ from $-5b$.
18. $10m$ from $-m$.
19. $-9y$ from $2y$.
20. $-3z$ from $11z$.
21. $-4c$ from $-3c$.
22. $-5u$ from $-9u$.
23. $-16x^2$ from $-5x^2$.
24. $-3ab$ from $7ab$.
25. ax from $8ax$.
26. $\frac{3}{2}m^3$ from $-\frac{11}{8}m^3$.
27. $-\frac{3}{4}x^2y$ from $-\frac{1}{2}x^2y$.
28. $-\frac{7}{5}cy^2$ from $\frac{2}{3}cy^2$.
29. $\frac{1}{2}ax^5$ from $\frac{9}{4}ax^5$.
30. $-(a^2 + b^2)$ from $-(a^2 + b^2)$.
31. $-7x^{n-1}(y - z)$ from $2x^{n-1}(y - z)$.

Find the sum of

32. $a,\ 2a,\ -3a.$

33. $-5x,\ 2x,\ 7x.$

34. $-ab,\ -3ab,\ -7ab.$

35. $3x^n,\ -4x^n,\ -9x^n.$

36. $-9ax,\ ax,\ -10ax.$

37. $-z^3,\ -9z^3,\ -5z^3.$

38. $2a^2b^2,\ -a^2b^2,\ -5a^2b^2,\ 7a^2b^2.$

39. $-5ax^m,\ 7ax^m,\ -9ax^m,\ 3ax^m.$

40. $-abc,\ \frac{3}{2}abc,\ -\frac{5}{8}abc,\ 4abc,\ 9abc,\ -\frac{17}{6}abc.$

41. $(x^n+y^n),\ -3(x^n+y^n),\ 4(x^n+y^n),\ -7(x^n+y^n),\ 15(x^n+y^n).$

Simplify the following expressions:

42. $a+2a+3a.$

43. $5x-2x+4x.$

44. $-7m+4m-5m.$

45. $8n-2n-3n.$

46. $-9b-2b-3b.$

47. $3y-8y-7y.$

48. $-x^{2n}+7x^{2n}+2x^{2n}-5x^{2n}.$

49. $a^2b-2ba^2-3a^2b+4ba^2.$

50. $\frac{1}{2}xy-\frac{3}{4}xy+\frac{5}{6}xy-xy-\frac{7}{2}xy+\frac{3}{2}xy.$

51. $-x^n+\frac{1}{2}x^n-\frac{1}{3}x^n+\frac{1}{4}x^n-\frac{1}{5}x^n+\frac{1}{6}x^n.$

52. $2(a^2+1)-3(a^2+1)-5(a^2+1)+7(a^2+1).$

53. $-(a+b-c)+2(a+b-c)+11(a+b-c)-7(a+b-c).$

Verify the results of Exx. 50–53, by making the following substitutions in the results, and also in the given expressions before they are simplified:

54. $x=60,\ y=1,\ a=3,\ b=-4,\ c=5,\ n=2.$

55. $x=-30,\ y=2,\ a=-\frac{1}{2},\ b=\frac{1}{4},\ c=1,\ n=3.$

Simplify the following expressions:

56. $3x+(-5x).$

57. $-10a+(-12a).$

58. $12x^3-(-7x^3).$

59. $-7ab-(+6ab).$

60. $-3x^my^n+(+2x^my^n)-(-x^my^n).$

61. $a-(2a-a).$

62. $x-(5x-9x).$

63. $m+[2m-(3m-4m)].$

64. $2a-[-4a-(-6a)].$

65. $6y-[5y-4y-(-3y+2y)]-y.$

66. $a-2a-(3a-4a)-(5a-6a).$

67. $x-[x-2x-(x-3x)-(x-4x)].$

Verify the results of Exx. 61–67 by making the following substitutions in the results, and by making the same substitutions in the given expressions and reducing:

68. $a=3,\ m=-5,\ x=4,\ y=-7.$

69. $a=-(\frac{3}{4}),\ m=\frac{2}{3},\ x=-9,\ y=8.$

2. Unlike Terms are added and subtracted by writing them in succession, each preceded by the sign + if it is to be added, by the sign − if it is to be subtracted.

Ex. 1. Add $3b$ to $2a$. We have $2a + 3b$.

Ex. 2. Add $-3x^2$ to $2y^2$. We have

$$2y^2 + (-3x^2) = 2y^2 - 3x^2.$$

Ex. 3. Subtract $-11m$ from $2n$. We have

$$2n - (-11m) = 2n + 11m.$$

In thus adding and subtracting unlike terms, the student should accustom himself to write at once the final form of the result, as $2y^2 - 3x^2$ in Ex. 2, and $2n + 11m$ in Ex. 3. Such steps as changing $+(-3x^2)$ into $-3x^2$, and $-(-11m)$ into $11m$ should be performed mentally.

Ex. 4. The sum of $2a$, $-3b$, $-5c$, and $4d$ is

$$2a - 3b - 5c + 4d.$$

Ex. 5. The result of subtracting $-3b$ from $2a$, and $-5c$ from that remainder, and adding $-4d$ to the last remainder is $2a + 3b + 5c - 4d$.

EXERCISES III.

Add

1. a to 1.
2. -3 to $2x$.
3. $-4b$ to 5.
4. $-x$ to $-y$.
5. ax to by.
6. $-x^3$ to y^3.
7. x^2 to $-y^2$.
8. $3ab$ to $-2bc$.
9. $-7z$ to $-3x$.
10. a^2 to a.
11. a^2 to $2a$.
12. $-3m^n$ to $2n^m$.
13. $-2ab$ to a^2.
14. $-xy^2$ to $-x^2y$.
15. $2x^py^q$ to $-3x^pz^q$.

Subtract

16. x from 1.
17. $-a$ from 2.
18. $-3b$ from 5.
19. 7 from $-y^2$.
20. $-2m$ from $-3n$.
21. $7x^m$ from $-2y^n$.
22. a^2 from a.
23. $-x^3$ from $3x$.
24. $-bx$ from $-cy$.

Add

25. $1, -x, x^2$.
26. $-3, 2x, -3y$.
27. $-ab, -ac, -ad$.
28. $3, x^2, -2x^3, -7x^4, 3x^5$.
29. $x^n, -3x^{n-1}y, 3x^{n-2}y^2, -y^3$.

30. Subtract $-3x^2$ from the sum of 2 and $-4x$.

31. Subtract x^2y from the result of subtracting $-2xy$ from $3xy^2$.

32. Add b^2 to the result of subtracting $-2ab$ from a^2.

33. Add $-5x^2$ to the result of subtracting $-2x$ from 0.

34. Subtract ab from the sum of ax and by.

Find the values of the results of Exx. 30–34,

35. When $a = 1$, $b = 2$, $x = 4$, $y = -5$.

36. When $a = -\frac{2}{3}$, $b = \frac{7}{5}$, $x = -\frac{1}{2}$, $y = \frac{2}{3}$.

3. A multinomial consisting of two or more sets of like terms can be simplified by uniting like terms.

Ex. 1. $2a - 3b - 5a + 4b$
$= 2a - 5a - 3b + 4b$, by the Commutative Law, (1)
$= -3a + b$, by the Associative Law.

Ex. 2. $3x^2 - 2xy + 5y^2 - 5x^2 + 7xy + 3y^2$
$= 3x^2 - 5x^2 - 2xy + 7xy + 5y^2 + 3y^2$ (1)
$= -2x^2 + 5xy + 8y^2$.

Ex. 3. $3(x^2 + 1) - 2(y - 1) + 4(x^2 + 1) + 7(y - 1)$
$= 3(x^2 + 1) + 4(x^2 + 1) - 2(y - 1) + 7(y - 1)$ (1)
$= 7(x^2 + 1) + 5(y - 1)$.

The student should not rewrite the expression, as in lines (1) above, but should collect and unite mentally the numerical coefficients of like terms.

EXERCISES IV.

Simplify the following expressions by uniting like terms:

1. $a + 1 + a - 1$.
2. $2x + 5 + 3x - 7$.
3. $5a - 6b + 4a - 2b$.
4. $3x^2 - 4y^2 + 2x^2 - 6y^2$.
5. $-5mn + 3n - 2nm - 6n$.
6. $4x^2y - 2xy^2 - 5yx^2 + 3y^2x$.
7. $a + b - 3a + c - 4b + 6a - 5c - 8a - 3c + 11b$.
8. $5a^3 - 7ax^3 - 2ax^3 - a^3x - ax^3 - 5a^3 + 9a^3x$.

9. $8a^5bc^2 - 2a^5b^2c - 7a^5b^2c + a^5bc^2 + 6a^5b^2c - 3a^5bc^2$.

10. $6m^4 - cm^4 + 3 + 9cm^4 - 8m^4 + 5m^4 - m^4c + 11$.

11. $7ac + 6ac^2 - 2n^2p - 3pn^2 - 6c^2a - 10 + 4n^2p + 10 - 2ac$.

12. $3n^2x^3 + 8n^2x^2 - 6n^3x^2 - 9n^2x^2 + 2n^2x^3 - n^3x^3 - 5n^3x^2 + 4n^3x^3$.

13. $10a^nb - 6a^4 - 4ba^n + 2 - 5a^nb - 3 - a^4$.

14. $9ab^4 - bx - 13ab^4 - a^4b + 3bx - 2ab^4 + 10a^4b - 2bx - ab^4$.

15. $3(a+m) - 4(a+m) - 2(a+m) + 8(a+m)$.

16. $7(x-n)^3 - (x-n)^2 - 4(x-n) - 7(x-n)^2 + 4(x-n) - 5(x-n)^3$.

17. $(a-1)n^2 - 8(a-1) + 4a - 1 - 2(a-1) - 3(a-1)n^2 - 5a + 7$.

18. $6(x-y) + 6x - y - y - 2(x-y) - 2x + y - 5(x-y)$.

19. $(a+z)^3 - 2(a+z)^3 + 2(a+z^3) + 7(a+z)^3 - 5(a+z^3)$.

Simplify the following expressions, first removing parentheses:

20. $a + 1 - (2 - 3a)$.

21. $5x - (-2y + 3x)$.

22. $x - 2y - (4x + 3y)$.

23. $2m + 3n - (5m - 4n) - (-3m + 7n)$.

24. $x^2 + y^2 - (x^2 - y^2) + (y^2 - x^2)$.

25. $3ab - [4ab + 2b^2 - (2ab + b^2)]$.

26. $1 - [a^3 - 2 - (-2a^3 - 3)]$.

27. $x^2 - y^2 + [-3x^2 - 2y^2 - (2x^2 - 3y^2)]$.

28. $2a - [3b - (-2a + 5b) - 7a]$.

29. $2xy + 5yz - (2xy - 3yz) - [2xy - (3xy - 2yz) + 5yz]$.

Find the values of the expressions in Exx. 20–29.

30. When $a=1$, $b=-2$, $x=3$, $y=-5$, $z=10$, $m=4$, $n=-7$.

31. When $a=-3$, $b=5$, $x=6$, $y=-7$, $z=8$, $m=-1$, $n=-2$.

If $x = 2a - 3b + 4c$, $y = -3a + 2b - 7c$, $z = 9a - 7b + 6c$, find the values of

32. $x + y + z$. 33. $x - y + z$. 34. $x + y - z$. 35. $x - y - z$.

Simplify

36. $a + (a+1) + (a+2) + (a+3)$.

37. $x + (x+2) + (x+4) + \cdots + (x+10)$.

38. $2m + (2m - 1) + (2m - 3) + (2m - 5)$.

39. $2a + (2a - 1) + (2a - 3) + \cdots + (2a - 9)$.

40. $a+b+(a+b+c)+(a+b+2c)+(a+b+3c)+(a+b+4c)$.

41. $a+b+(a+b-c)+(a+b-3c)+(a+b-5c)+(a+b-7c)$.

42. Find the sum of 5 terms, the first term being ab, and each succeeding term being 1 greater than the preceding term.

43. Find the sum of 7 terms, the first term being x^2, and each succeeding term being 1 less than the preceding term.

44. Find the sum of 6 terms, the first term being $m + n$, and each succeeding term being p less than the preceding term.

Addition and Subtraction of Multinomials.

4. Ex. 1. Add $-2a + 3b$ to $3a - 5b$.

We have
$$(3a - 5b) + (-2a + 3b) = 3a - 5b - 2a + 3b,$$
$$= a - 2b.$$

Ex. 2. Subtract $-2a + 3b$ from $3a - 5b$.

We have
$$(3a - 5b) - (-2a + 3b) = 3a - 5b + 2a - 3b,$$
$$= 5a - 8b.$$

In adding multinomials, it is often convenient to write one underneath the other, placing like terms in the same column.

Ex. 3. $(3a - 5b) + (-2a + 3b)$ may be written

$$\begin{array}{r} 3a - 5b \\ -2a + 3b \\ \hline a - 2b \end{array}$$

It is evidently immaterial whether the addition is performed from left to right, or from right to left, since there is no carrying as in arithmetical addition.

Ex. 4. Find the sum of $-7a^2 + 5ab - 3b^2$,

$-2a^2 - 3ab + 9b^2$, and $11a^2 - 2ab - 7b^2$.

We have

$$\begin{array}{rrr} -7a^2 & +5ab & -3b^2 \\ -2a^2 & -3ab & +9b^2 \\ 11a^2 & -2ab & -7b^2 \\ \hline 2a^2 & +0 & -b^2, \end{array} = 2a^2 - b^2$$

Ex. 5. Find the sum of $-4x^2+3y^2-8z^2$, $2x^2-3z^2$, and $2y^2+5z^2$.

We have

$$\begin{array}{r} -4x^2+3y^2-8z^2 \\ 2x^2 \qquad\quad -3z^2 \\ 2y^2+5z^2 \\ \hline -2x^2+5y^2-6z^2 \end{array}$$

If a similar arrangement be made in subtracting multinomials, change mentally the signs of the terms of the subtrahend, and proceed as in addition.

Ex. 6. Subtract $-2a+3b$ from $3a-5b$.

Changing mentally the signs of the terms of the subtrahend, and adding, we have

$$\begin{array}{r} 3a-5b \\ -2a+3b \\ \hline 5a-8b \end{array}$$

Ex. 7. Subtract $2x^2-3z^2$ from $-4x^2+3y^2$, and from the result subtract $2y^2+5z^2$.

When several multinomials are to be subtracted in succession, the work is simplified by writing them with the signs of the terms already changed.

Changing the signs of the terms of the multinomials to be subtracted, and adding, we have

$$\begin{array}{r} -4x^2+3y^2 \qquad\quad \\ -2x^2 \qquad\quad +3z^2 \\ -2y^2-5z^2 \\ \hline -6x^2+\ y^2-2z^2 \end{array}$$

EXERCISES V.

Add

1. $a+4$ to $a-4$.
2. $2a-3$ to $3a-2$.
3. $7a-4b$ to $-3a+2b$.
4. $-x+y$ to $x-y$.
5. $6x-9y$ to $-3x-y$.
6. $8ax+2by$ to $-7ax+3by$.
7. $-3xy+10yz$ to $-xy-12yz$.
8. $2x^2-xy$ to $-x^2+y^2$.
9. x^2+x to $-x^2-x^3$.

10. $\frac{1}{2}m^2n - \frac{3}{8}mn^2$ to $\frac{3}{2}m^2n + \frac{5}{8}mn^2$.

11. $-x^m + 3y^n$ to $2x^m - 2y^n$.

12. $x^2 + x + 1$ to $x^2 - x + 1$.

13. $2a^2 - 3ab - b^2$ to $-a^2 + 5ab + 2b^2$.

14. $4x^3 - 5x + 1$ to $-3x^3 + 4x^2 - 2$.

15. $a^3 - \frac{5}{8}a^2 + \frac{1}{2}a$ to $-\frac{1}{2}a^3 + 2a^2 - \frac{1}{4}a$.

16. $a^3 - 3a^2b + 3ab^2 - b^3$ to $a^3 + 3a^2b + 3ab^2 + b^3$.

17. $x^n + 3x^{n-1} - 2x^{n-2} + x^{n-3} - 1$ to $x^n - 3x^{n-1} + 4x^{n-2} - x^{n-3} + 2$.

Subtract

18. $a - 1$ from $a + 1$.

19. $3a + 2$ from $4a - 5$.

20. $a - 2b$ from 0.

21. $8a - 3b$ from $7a - 2b$.

22. $x + y$ from $x - y$.

23. $x - y$ from $x + y$.

24. $-x^2 + xy$ from $2x^2 - 2xy$.

25. $\frac{1}{2}ax - \frac{3}{4}by$ from $-\frac{3}{8}ax - \frac{1}{2}by$.

26. $-3x^m - 4y^n$ from $2x^m - y^n$.

27. $-x^2y - xy^2$ from $x^3 + y^3$.

28. $a^3b + ab^3$ from $a^4 + b^4$.

29. $a^3 - 2a^2 - a - 2$ from $a - 5$.

30. $-x^4 + 7x^3 + 3x^2 + 3x - 9$ from 0.

31. $-5a^2 - 2a + 3$ from $-3a^2 + 5a - 1$.

32. $a^3b - 3ab^3 + a^2b^2$ from $2a^3b - 5ab^3 - a^2b^2$.

33. $x^3 - 3x^2y + 3xy^2 - y^3$ from $x^3 + 3x^2y + 3xy^2 + y^3$.

34. $x^3 - y^3 + z^3 - 3xyz$ from $x^3 + x^2 - y^3 + y^2 + z^3 - z^2$.

35. $2x^4 - 3x^3 - 7x^2 + 3x + 1$ from $2 + 4x - 6x^2 - 2x^3 + 3x^4$.

36. $x^4 - x^3 + x^2 - x + 1$ from $x^4 + x^2 + 1$.

37. $-\frac{1}{2}x^5 - \frac{3}{8}x^3 + \frac{3}{5}x^2 - 1$ from $x^5 + \frac{1}{2}x^4 - x^3 - \frac{3}{8}$.

38. $ax^{3n} - 3bx^{2n} + 4cx^n - 1$ from $x^{4n} - 2bx^{2n} + 3cx^n$.

39. $-3(x+y)^2 + 4(x+y) - 7$ from $-(x+y)^2 - 7(x+y) + 3$.

40. $7(a^2 + b^2) - 3(a + b)^2 - (a + b) - 2$ from $9(a^2 + b^2) - 5(a + b) + 3$.

Find the sum of

41. $7a - 9b - c$, $5a - 3b - 2c$, $2a + 3b - 5c$.

42. $3x^2 - 5x + 1$, $7x^2 + 2x - 3$, $-x^2 - 2x - 3$.

43. $x^2 - ax + a^2,\ 2x^2 + 3ax - 4a^2,\ x^2 + ax + 2a^2.$

44. $3a^2 - 4ab + b^2,\ a^2 - 2ab - 2b^2,\ 2a^2 - 3ab + 4b^2.$

45. $4ab - x,\ 3x^2 - 2ab,\ 2ax + ab.$

46. $a^2 - 2ab + 2b^3,\ 2a^3 - 3ab + b^3,\ a^2 + 5ab - b^3.$

47. $2x^2y^3 + 4x^3y^2,\ -5x^2y^3 + 2x^2y^3 - 3x^3y^2,\ 4x^3y^2 - 5x^2y^3 - 6x^2y^2.$

48. $3a - 2b + 5c,\ a + b - c,\ -2a + 5b - 3c,\ -2a + b - c.$

49. $x^3 + 2x^2 - 3x + 1,\ 2x^3 - 3x^2 + 4x - 2,\ 5x^3 + 4x^2 + 5,$
$6x^3 - 5x^2 - 4x - 3.$

50. $5a^3 - 3a^2 + 2a,\ a^3 - a^2,\ a^2 - a + 1,\ a^3 - 2a^2 - a - 2.$

51. $2a + b - (c + d),\ a + (b - c) - d,\ a + b - (c - d).$

52. $a + 2b + c,\ 2a - (b - c) - d,\ 3a + b - (2c + d).$

53. $3(a + b) - 4(a + b)^2 + 5(a + b)^3,\ (a + b)^2 - 2(a + b)^3,$
$-(a + b)^3 + 2(a + b)^2 - (a + b).$

54. $7(x^2 + y^2) - 3(x^2 - y^2) + 2xy,\ 2(x^2 - y^2) - 4xy,$
$3(x^2 + y^2) - (x^2 - y^2).$

55. Subtract $x^2 + x + 1$ from the sum of $2x^2 - 2x + 1$ and $3x^2 + x + 4$.

56. Subtract the sum of $a^2 + ab + b^2$ and ab from $2a^2 + 3ab + 2b^2$.

57. Subtract the sum of a^2 and $b^2 + c^2$ from the sum of b^2 and $a^2 - c^2$.

58. Add $xy - 2y^2$ to the result of subtracting $y^2 - 3xy$ from 5.

59. How much does $m^2 + n^2$ exceed $m^2 - n^2$?

60. How much does $1 - x^2$ exceed $2 - 3x^2$?

61. What expression must be added to $2a - 3b + 4c$ to give $4a + 2b - 2c$?

62. What expression must be added to $x^2 + 3xy - y^2$ to give $x^2 + y^2$?

63. What expression must be added to $xy + xz + yz$ to give $x^2 + y^2 + z^2$?

64. What expression must be subtracted from $a^2 + ab + b^2$ to give $a^2 - 2ab + b^2$?

65. What expression must be subtracted from $x^2 - 2xy + y^2$ to give $x^2 + 2xy + y^2$?

66. What expression must be added to $x^2 + x + 1$ to give 0?

67. Given the four expressions,

$$5a^2 - 3ab + b^2 - 3ac + 2bc + c^2,$$
$$2a^2 + 5ab - 3b^2 + 2ac - 4bc + 3c^2,$$
$$4a^2 - 7ab + 5b^2 - 4ac - 5bc + c^2,$$
$$2a^2 + 9ab - 8b^2 + 3ac + 3bc + 2c^2.$$

From the sum of the first two subtract the sum of the last two.

68. Given the four expressions,

$$6a^2 - 3ab + 2b^2 - 5ac + 6bc - 2c^2,$$
$$a^2 - ab + b^2 + ac - bc - c^2,$$
$$2a^2 + 3ab - 2b^2 - 3ac - 4bc - 5c^2,$$
$$3a^2 - 6ab + 3b^2 - 4ac + 10bc + 4c^2.$$

Subtract the sum of the last three from the first.

§ 3. MULTIPLICATION.

Principles of Powers.

1. Products of Powers.

Ex. 1. $2^3 \times 2^2 = (2 \times 2 \times 2)(2 \times 2) = 2 \times 2 \times 2 \times 2 \times 2 = 2^{3+2} = 2^5$.

Ex. 2. $a^4a^5 = (aaaa)(aaaaa) = aaaaaaaaa = a^{4+5} = a^9$.

Ex. 3. $xx^2x^3 = (x)(xx)(xxx) = xxxxxx = x^{1+2+3} = x^6$.

The preceding examples illustrate the following principle:

(i.) *The product of two or more powers of one and the same base is equal to a power of that base whose exponent is the sum of the exponents of the given powers;* or, stated symbolically,

$$a^ma^n = a^{m+n}; \quad a^ma^na^p = a^{m+n+p}; \text{ etc.}$$

For, $a^ma^n = (aaa \cdots \text{to } m \text{ factors})(aaa \cdots \text{to } n \text{ factors})$
$= aaa \cdots \text{to } (m+n) \text{ factors}$
$= a^{m+n}$.

$$a^ma^na^p = (a^ma^n)a^p = a^{m+n}a^p = a^{m+n+p}.$$

The law can be easily extended to the case of any number of powers.

(ii.) *The converse of the principle is evidently true.*

$$a^{m+n} = a^m a^n; \quad a^{m+n+p} = a^m a^n a^p.$$

E.g., $a^7 = a^6 a = a^5 a^2 = a^4 a^3$
$= a^4 a^2 a = a^3 a^3 a =$ etc.

EXERCISES VI.

Express each of the following products as a single power:

1. 2×2^2.
2. $5^2 \times 5^3$.
3. $(-7)^3(-7)^5$.
4. $(-2)^3 2^4$.
5. $(-3)^4 3^5$.
6. $5^4(-5)^6$.
7. $a^2 a^3$.
8. $b^6 b^7$.
9. $(-x)^6(-x)^5$.
10. $(-x)^3 x^4$.
11. $a^2 a^3 a^5$.
12. $(-x)(-x)^2(-x)^3$.
13. $(ab)^2(ab)^3$.
14. $(-xy)^3(-xy)^4(-xy)^6$.
15. $(-mn)^3(-mn)(-mn)^5$.
16. $a^3 a^n$.
17. $a^6 a^p a^q$.
18. $(-x)^7(-x)^r(-x)^s$.
19. $a^{m-1}a^{m+2}$.
20. $x^{m+n}x^{m-3n}$.
21. $y^{m+5}y^{n-6}y^{m-n+p}$.
22. $(x+y)(x+y)^2$.
23. $(a^2+b^2)^3(a^2+b^2)^5$.
24. $(2x+1)^7(2x+1)^3(2x+1)^5$.
25. $(a+b+c)^5(a+b+c)^6(a+b+c)^7$.

2. Powers of Powers.

Ex. 1. $(2^2)^3 = 2^2 \times 2^2 \times 2^2 = 2^{2+2+2} = 2^{2\times 3} = 2^6$.

Ex. 2. $(a^4)^5 = a^4 a^4 a^4 a^4 a^4 = a^{4+4+4+4+4} = a^{4\times 5} = a^{20}$.

Ex. 3. $[(a^4)^5]^2 = (a^{4\times 5})^2$, since $(a^4)^5 = a^{4\times 5}$, by Ex. 2,
$= a^{4\times 5}a^{4\times 5} = a^{4\times 5+4\times 5} = a^{4\times 5\times 2}$.

The preceding examples illustrate the following principle:

(i.) *A power of a power of a given base is equal to a power of that base whose exponent is the product of the given exponents;* or, stated symbolically,

$$(a^m)^n = a^{mn}; \quad [(a^m)^n]^p = a^{mnp}; \quad \text{etc.}$$

For, $(a^m)^n = a^m a^m a^m \cdots$ to n factors
$= a^{m+m+m+\cdots \text{ to } n \text{ summands}}$
$= a^{mn}$.

Likewise, $[(a^m)^n]^p = (a^{mn})^p = a^{mnp}$; and so on.

(ii.) *The converse of the principle is evidently true:*

$$a^{mn} = (a^m)^n;\ \ a^{mnp} = (a^{mn})^p = [(a^m)^n]^p;\ \text{etc.}$$

E.g., $$a^6 = (a^2)^3;\ a^8 = [(a^2)^2]^2.$$

(iii.) *In a power of a power the exponents are commutative:*

$$(a^m)^n = (a^n)^m;\ \ [(a^n)^m]^p = [(a^m)^n]^p = \text{etc.}$$

For, $$(a^m)^n = a^{mn} = a^{nm} = (a^n)^m.$$

In like manner,

$$[(a^n)^m]^p = [(a^m)^p]^n = [(a^n)^p]^m = \text{etc.}$$

E.g., $(3^2)^3 = (3^3)^2$; $[(x^2)^5]^7 = [(x^5)^2]^7 = [(x^7)^5]^2 = \text{etc.}$

EXERCISES VII.

Find the values of the following powers:

1. $(3^2)^3$. 2. 3^{2^3}. 3. 3^{3^2}. 4. $(4^3)^2$.
5. 4^{3^2}. 6. 4^{2^3}. 7. $[(-2)^3]^4$. 8. $(-2^3)^5$.

Simplify the following powers:

9. $(11^4)^5$. 10. $[(-18)^5]^6$. 11. $[(2^3)^2]^4$. 12. $[(3^4)^5]^6$.
13. $(a^3)^4$. 14. $(x^2)^7$. 15. $(-x^4)^3$. 16. $[(-x)^4]^3$.
17. $[(ab)^2]^5$. 18. $[(x^2)^5]^7$. 19. $[(-n^2)^5]^3$. 20. $(x^3)^m$.
21. $(x^n)^5$. 22. $(x^n)^{2n}$. 23. $(x^3)^{2n}$.
24. $[(x+y)^3]^2$. 25. $[(a^2+1)^4]^3$. 26. $[-(x^3-y^2)]^3$.

Express the following powers as powers of 2:

27. $[(2^3)^2]^4$. 28. $(2^{3^2})^4$. 29. $(2^3)^{2^4}$. 30. $2^{3^{2^4}}$.
31. 4^3. 32. 8^5. 33. 16^7. 34. 32^{25}.

Express the following powers as powers of 3^2:

35. 3^8. 36. 3^{24}. 37. 3^{80}. 38. 3^{2a}.
39. 9^4. 40. 81^5. 41. 27^6. 42. 243^2.

Express the following powers as powers of 5^3:

43. 5^{15}. 44. 5^{81}. 45. 125^6. 46. 25^9.

Simplify the following powers:

47. $(a^2a^3)^4$. 48. $(x^2x^5)^6$. 49. $[(-x)^2x^7]^3$. 50. $[(-c)^3c^4]^{5n}$.
51. $[(a+b)^2(a+b)^3]^5$. 52. $[(1-x)^4(1-x)^5]^6$.
53. $[(x^3-y^2)^4(x^3-y^2)]^n$.

Write the squares and the cubes of:

54. a^2. **55.** $-a^3$. **56.** $(x^2x^5)^3$. **57.** $[(-y)^3y^4]^5$.

58. $x+y$. **59.** $(a-b)^2$. **60.** $-(a+b-c)^3$.

Write

61. The fourth power of a. **62.** The ath power of 4.

Write the sum of *ten* terms, the first term being x,

63. When each term is the square of the preceding term.

64. When each term is the cube of the preceding term.

65. When each term is the nth power of the preceding term.

Simplify

66. $3(a^3)^4+2(a^4)^3-4(a^2)^6$. **67.** $3(a^5)^4-2(a^4)^5-5(a^{10})^2+7[(a^2)^5]^2$.

3. Like and Unlike Powers. — Two powers are said to be *like* or *unlike* powers, according as their exponents are equal or unequal, whether or not their bases are equal. Thus,

a^2, b^2 are *like* powers; a^2, a^3, a^4 are *unlike* powers.

4. Products of Like Powers.

Ex. 1.

$$\begin{aligned}2^3 \times 5^3 &= 2 \times 2 \times 2 \times 5 \times 5 \times 5\\ &= (2 \times 5)(2 \times 5)(2 \times 5), \text{ by the Commutative Law,}\\ &= (2 \times 5)^3, \text{ by the definition of a power.}\end{aligned}$$

$$\begin{aligned}a^4b^4c^4 &= (aaaa)(bbbb)(cccc),\\ &= (abc)(abc)(abc)(abc), \text{ by the Commutative Law,}\\ &= (abc)^4, \text{ by the definition of a power.}\end{aligned}$$

The preceding examples illustrate the following principle:

(i.) *The product of like powers of two or more given bases is the like power of the product of the bases;* or, stated symbolically,

$$\boldsymbol{a^mb^m = (ab)^m;\ a^mb^mc^m = (abc)^m}, \text{ etc.}$$

For, $a^mb^m = (aaa \cdots \text{ to } m \text{ factors})(bbb \cdots \text{ to } m \text{ factors})$

$= (ab)(ab)(ab) \cdots$ to m factors, by the Commutative Law,

$= (ab)^m$, by the definition of a power.

In like manner the theorem can be extended to the product of any number of like powers.

(ii.) *The converse of the principle is evidently true:*

$$(ab)^m = a^m b^m;\ (abc)^m = a^m b^m c^m, \text{ etc.}$$

E.g.,

$$(xy)^3 = x^3y^3;\ (xyz)^5 = x^5y^5z^5;$$

$$(2a^2b)^3 = 2^3(a^2)^3b^3 = 8a^6b^3;$$

$$(3ax^2y^3)^2 = 3^2a^2(x^2)^2(y^3)^2 = 9a^2x^4y^6.$$

EXERCISES VIII.

Express the following products of powers as powers of products:

1. $7^3 \times 5^3$. **2.** $(8)^4 \times (-3)^4$. **3.** a^3b^3.

4. $(-x)^7(-y)^7$. **5.** $(-x)^3y^3$. **6.** $a^4b^4c^4$.

7. $(-a)^5b^5(-c)^5$. **8.** $(-x)^3y^3z^3$. **9.** $a^2(b+c)^2$.

10. $(x^2+y^2)^4(x+y)^4$. **11.** $c^3(b^3+c^3)^3$. **12.** a^2b^4.

13. $a^6b^3c^9$. **14.** $x^{12}y^{15}z^{18}$.

Express the following powers of products as products of powers, reducing powers of any numerical factors:

15. $(ab)^2$. **16.** $(xy)^3$. **17.** $(-2u)^5$. **18.** $(-2v)^4$.

19. $(-2xy)^4$. **20.** $(-3ab)^5$. **21.** $(abc)^3$. **22.** $(x^2y)^4$.

23. $(-x^2y^3)^5$. **24.** $(a^2b^3c)^4$. **25.** $(-3x^2y)^4$. **26.** $(2ab^3c^4)^7$.

27. $(xyz)^m$. **28.** $(m^2n^3)^p$. **29.** $(2x^2y^4)^n$. **30.** $(2u^3v^m)^n$.

Write

31. The square of twice a. **32.** Twice the square of a.

33. The cube of three times b. **34.** Three times the cube of b.

35. Four times the square of the difference between x and y.

36. The square of four times the difference between x and y.

Given two numbers, a and b, write in algebraic language:

37. The square of the first number, plus twice the product of the two numbers, plus the square of the second number.

38. The cube of the first number, plus three times the product of the square of the first by the second, plus three times the product of the first by the square of the second, plus the cube of the second.

39. Write in algebraic language the verbal statements in Exx. 37 and 38, when the given numbers are $2a$ and $-3b$.

5. The following principle will be useful in subsequent work: *Like powers of the same numbers, or of equal numbers, are equal.*

If $a = b$, then $a^m = b^m$.

For, $a^m = a^m$, by Axiom (i.). (1)

Since $b = a$, we can, by Axiom (iii.), substitute b for a in the second member of (1). We thus obtain

$$a^m = b^m.$$

Degree. Homogeneous Expressions.

6. The **Degree of an Integral Term** is its form as indicated by the sum of the exponents of its *literal* factors.

E.g., $3ab$ is of the *second* degree, as indicated by the sum of the exponents of a and b. The degree of $\frac{5}{8}x^2y$ is the *third.*

The **Degree of a Multinomial** is the degree of its term of highest degree.

E.g., the degree of $a + b^2 + c^3$ is the degree of c^3; *i.e.*, the *third.*

The degree of $x^2y + xy^3 - x^2y^3z$ is the degree of x^2y^3z; *i.e.*, the *sixth.*

7. It is often desirable to speak of the degree of a term, or of an expression, in regard to one or more of its literal factors.

E.g., the term ax^2y^3 is of the *fifth* degree in x and y, of the *first* degree in a, of the *second* degree in x, of the *third* degree in y, etc.

The expression $ax^2 + 2bxy + cy^2$ is of the *second* degree in x, in y, and in x and y.

8. A **Homogeneous Expression** in one or more letters is an expression all of whose terms are of the same degree in these letters.

E.g., $a^2 + 2ab + b^2$ is homogeneous in a and b;

$x^3 + 3x^2y + 3xy^2 + y^3$ is homogeneous in x and y;

$ax^2 + 2xy + by^2$ is homogeneous in x and y.

9. If the terms of a multinomial be arranged so that the exponents of some one letter increase, or decrease, from term to term, the multinomial is said to be arranged to *ascending*, or *descending*, powers of that letter. The letter is called the *letter of arrangement.*

E.g., The multinomial $a^3 + 3a^2b + 3ab^2 + b^3$ is arranged to *descending* powers of a, which is then the letter of arrangement; or to *ascending* powers of b, which is then the letter of arrangement.

The multinomial $1 + 2x + 3x^2 + 4x^3$ is arranged to *ascending* powers of x, the letter of arrangement.

EXERCISES IX.

What is the degree of $2a^3b^2x^4y^5$

1. In a? **2.** In b? **3.** In x? **4.** In y?

5. In a and x? **6.** In b and y?

7. In b, x, and y? **8.** In a, b, x, and y?

What is the degree of the expression

$$a^3x^4 - 6a^2b^2x^3y + 5abx^2y^3$$

9. In x? **10.** In y? **11.** In a? **12.** In b?

13. Arrange $2x - 3x^5 + 7 - 2x^4 + 3x^2$ to ascending powers of x; to descending powers of x.

14. Arrange $3y - 7xy^3 + 5x^3y^2 + 4x^2y^4$ to ascending powers of x; to ascending powers of y.

15. Arrange $29a^2b^4 + 4b^6 - 30a^3b^3 + 25a^4b^2 - 12ab^5$ to descending powers of a; to descending powers of b.

10. Multiplication of Monomials by Monomials.—The product of two or more monomials which contain either numerical coefficients or common literal factors can be simplified.

Ex. 1.
$$3a \times 5b = 3 \times 5 \times a \times b$$
$$= 15ab.$$

Ex. 2.
$$2x \times (-4y^2) = 2(-4)xy^2$$
$$= -8xy^2.$$

Ex. 3. $-5\,xy \times 7\,yz \times 3\,xyz = -5 \times 7 \times 3 \times xx \times yyy \times zz$
$= -105\,x^2y^3z^2.$

Ex. 4. $\frac{2}{3}\,a^3 \times 6\,ab^2 \times 11\,b^5 = \frac{2}{3} \times 6 \times 11 \times a^3ab^2b^5 = 44\,a^4b^7.$

Ex. 5. $3\,(a+b) \times \frac{4}{5}\,(a+b)^2 = 3 \times \frac{4}{5}\,(a+b)\,(a+b)^2$
$= \frac{12}{5}\,(a+b)^3.$

Ex. 6. $3\,a^mb^2 \times 5\,a^3b^n = 3 \times 5\,a^ma^3b^2b^n = 15\,a^{m+3}b^{2+n}.$

Ex. 7. $5\,a^{n-1}x^{n+2} \times 4\,a^{n+1}x^{n-1} = 5 \times 4\,a^{n-1}a^{n+1}x^{n+2}x^{n-1}$
$= 20\,a^{2n}x^{2n+1}.$

The preceding examples illustrate the following method:

The product of two or more monomials is obtained by multiplying the product of their numerical coefficients by the product of their literal factors.

The student should accustom himself to write at once the final result, performing mentally the intermediate steps.

EXERCISES X.

Multiply

1. $3\,a$ by 4.
2. -5 by $2\,a$.
3. 7 by $-5\,x^2$.
4. $2\,a$ by $3\,a$.
5. $2\frac{1}{2}\,x$ by $-5\,x^3$.
6. $-3\,a^2$ by $-4\,a$.
7. $-2\,ab$ by $5\,ab$.
8. $3\,a^2b$ by $-7\,ab^2$.
9. $4\,b^2c$ by $-3\,b^3c^2$.
10. $\frac{2}{5}\,a^2m$ by $-5\,am^2$.
11. $7\,a^2n^5$ by $3\,b^2n^6$.
12. $9\,a^4b^3$ by $-2\,a^7b^4$.
13. $4\,ab$ by $-5\,xy$.
14. $-3\frac{1}{2}\,abc$ by $-2\frac{2}{3}\,xy$.
15. $5\,ac$ by $3\,ax$.
16. $-5\,a^2bc^3$ by $-7\,a^3b^4c^5$.
17. $12\,m^2np$ by $-7\,m^8n^9p^6$.
18. $3\,abc$ by ab^2c^3.
19. $-6\,abc$ by $-2\,bc$.
20. $-3\,x^2yz$ by xy^2.
21. $-5\,u^2vw$ by $-2\,xy^2u$.
22. a^2bxy by $ab^2x^2y^2$.
23. $a^4b^3c^2x^3y^2z^7$ by $-7\,ab^6c^3x^5y^4z^2$.
24. $2\,(a+b)^3$ by $-3\,(a+b)^2$.
25. $7\frac{1}{2}\,a^2(x-y)^3$ by $6\,a^3(x-y)^2$.
26. $-7\,x^2y^n$ by $-2\,x^my^2$.
27. $3\,a^{m-1}b^{n+1}$ by $-12\,a^5b^3$.
28. $12\,a^4b^m$ by $-\frac{3}{4}\,ab^n$.
29. $\frac{5}{12}\,m\,(2\,a-1)^2$ by $8\,m^2(2\,a-1)^n$.
30. $5\,n^px^{n-1}$ by $-8\,p^nx^{n+1}$.
31. $2\,a^{n+2}bx^2$ by $a^nc^3x^{p-2}$.

Simplify the following continued products:

32. $3ab \times 5bc \times 6ac$. **33.** $-7x^2y \times (-2y^2z) \times 3xz^3$.

34. $-5axy \times 7abx^2z \times 2bx^2yz$.

35. $(2ax^3)^4 \times (5ab^2xy)^2 \times (-2a^2x^2y^3)^3$.

36. $(1-x)^3 \times 3ab \times 4(1-x)^5 \times (-2a^2c)$.

37. $2a^2(a-b)^3 \times (-3ab)(a-b)^2 \times 5b^2(a-b)^4$.

38. $a^nb^2 \times (-3ab^m) \times 2a^pb^q$. **39.** $x^2y^{n+1} \times 5x^my^{3n} \times (-7x^{5m}y^{2n-1})$.

The Distributive Law for Multiplication.

11. If the indicated operation within the parentheses in the product, $4(2+3)$, be first performed, in accordance with the meaning of parentheses, we have

$$4(2+3) = 4 \times 5 = 20.$$

But if each term within the parentheses be multiplied by 4 and the resulting products be then added, we have

$$4 \times 2 + 4 \times 3 = 8 + 12 = 20, \text{ as above.}$$

Therefore $\quad 4(2+3) = 4 \times 2 + 4 \times 3.$

Likewise, $\quad -4(3-9) = -4(-6) = 24;$

and $\quad -4 \times 3 - (-4)9 = -12 + 36 = 24$, as above.

Therefore $\quad -4(3-9) = -4 \times 3 - (-4)9.$

The above examples illustrate the following principle:

The Distributive Law. — *The product of a multinomial by a monomial is obtained by multiplying each term of the multinomial by the monomial and adding algebraically the resulting products.* That is,

$$\boldsymbol{a(b+c-d) = ab + ac - ad.}$$

(i.) For, let a be limited to positive integral values.

Then

$$\begin{aligned} a(b+c-d) &= (b+c-d) + (b+c-d) + (b+c-d) \\ &\qquad + \cdots \text{ to } a \text{ summands,} \\ &= (b + b + \cdots \text{ to } a \text{ summands}) + (c + c + \cdots \text{ to } a \text{ summands}) \\ &\qquad - (d + d + \cdots \text{ to } a \text{ summands}), \\ &= ab + ac - ad. \end{aligned}$$

(ii.) Let a be limited to negative integral values, and be denoted by $-x$, so that x is an absolute number.

Then

$$a(b+c-d) = -x(b+c-d), \text{ replacing } a \text{ by } -x.$$
$$= -(b+c-d)-(b+c-d)- \cdots \text{ to } x \text{ summands,}$$
$$= -b-b- \cdots \text{ to } x \text{ summands } -c-c- \cdots \text{ to } x \text{ summands}$$
$$+d+d+ \cdots \text{ to } x \text{ summands,}$$
$$= +(-b-b \cdots \text{ to } x \text{ summands}) + (-c-c- \cdots \text{ to } x \text{ summands})$$
$$-(-d-d- \cdots \text{ to } x \text{ summands}),$$
$$= +(-x)b + (-x)c - (-x)d$$
$$= ab + ac - ad, \text{ replacing } -x \text{ by } a.$$

In (i.) and (ii.) a was limited to integral numerical values. Similar reasoning can, however, be applied when a has fractional numerical values.

Thus,
$$\tfrac{2}{3}(4+\tfrac{5}{7}) = \frac{4+\frac{5}{7}}{3} + \frac{4+\frac{5}{7}}{3}$$
$$= \tfrac{4}{3} + \tfrac{4}{3} + \frac{\frac{5}{7}}{3} + \frac{\frac{5}{7}}{3}$$
$$= \tfrac{2}{3} \times 4 + \tfrac{2}{3} \times \tfrac{5}{7}.$$

Multiplication of a Multinomial by a Monomial.

12. The multiplication of a multinomial by a monomial is a direct application of the Distributive Law.

Ex. 1. Multiply $(x - y)$ by 3.

We have $$3(x-y) = 3x - 3y.$$

Ex. 2. Multiply $3x - 2y - 7z$ by $-4x$.

We have
$$-4x(3x-2y-7z) = (-4x)(3x) - (-4x)(2y) - (-4x)(7z) \quad (1)$$
$$= -12x^2 + 8xy + 28xz.$$

In thus multiplying a multinomial by a monomial the student should accustom himself to write at once each term of the product in its final form. Such steps as changing $(-4x)(3x)$ into $-12x^2$, $-(-4x)(2y)$ into $8xy$, and $-(-4x)(7z)$ into $28xz$, should be performed mentally.

Ex. 3. $-2ab(3a^2 + 4ab - 7b^2) = -6a^3b - 8a^2b^2 + 14ab^3.$

Ex. 4. $4x^2y(xy - 3xz + 2yz) = 4x^3y^2 - 12x^3yz + 8x^2y^2z.$

EXERCISES XI.

Multiply

1. $a+1$ by 3. 2. $2a-5$ by -4. 3. $2a-3b$ by $-2a$.

4. $7x-8y$ by $3x$. 5. $5a^2-3ab$ by $2a^2b$.

6. ax^2-2by^2 by $3abxy$. 7. $2a^3b^5+3a^5b^3$ by $-4a^4b^4$.

Simplify the following expressions:

8. $2a-3(a-1)$. 9. $3x-2(3x-2)$. 10. $a^2-a(a-1)$.

11. $5a+2a(a-1)-3a(a+1)$.

12. $1-[5(a-b)+6(a+b)]$.

13. $5x-3(x-2y)-7[5x-3(x-3y)]$.

14. $a+a(1+a^2)-a[1-a(1-a)]$.

Multiply $4a-3b+c$ by

15. $2a$. 16. $-3b$. 17. $4a^mb^n$. 18. $\frac{5}{6}a^2b^2c$.

Multiply $a^3b-3ab^2c+4bc^2$ by

19. $2ab$. 20. $-3ac$. 21. a^2bc^{m-1}. 22. $-\frac{2}{3}ab^3c^3$.

Multiply x^3-2x^2+6x-1 by

23. -2. 24. $3x$. 25. $-5x^2$. 26. $\frac{2}{3}x^3$.

Multiply $2ax^3-3bx^2+abx-1$ by

27. ax. 28. a^2x^3. 29. $-2abx$. 30. $\frac{2}{3}a^{m-1}b^{3-n}x^{4p}$.

31. Multiply $(a+b)^2-3x(a+b)+7x^2$ by $4(a+b)$.

32. Multiply $(x^2+1)^4-5a(x^2+1)^2+3ab$ by $-2a^2b^2(x^2+1)^3$.

Simplify the result of substituting $a+b-c$ for x, and $a-b+c$ for y, in the following expressions:

33. $-3x$. 34. $2y$. 35. $2x+3y$.

36. $5bx-7ay$. 37. $3a^2bx-14ab^2y$. 38. $7abx+2bcy$.

Find the values of the results of Exx. 33–38,

39. When $a=-2,\ b=3,\ c=-4$.

40. When $a=5,\ b=-7,\ c=-5\frac{1}{2}$.

Multiply $3a^3-5abc+ba^2c^2$ by

41. a^n. 42. b^{n+1}. 43. $-a^{n+1}b^{n-1}$. 44. $\frac{2}{3}a^2b^{n+3}c^{p-4}$.

Multiply $5x^n - 3x^{n-3}y^2 + 4x^{n-5}y^4 + y^{n-4}$ by

45. x^3. **46.** $-5x^2y$. **47.** $3x^ny^4$. **48.** $-6\frac{4}{5}x^ny^m$.

49. Multiply

$3(a+b)^n(x-1) - 2(a+b)^{n-5}(x-1)^4 - 4(a+b)(x-1)^n$ by $2(a+b)^4(x-1)^5$.

Write the squares, the cubes, and the *n*th powers of:

50. a^{m+1}. **51.** x^{m-2}. **52.** $2x^{m+n}y$. **53.** $-3a^{m+n-1}y^7$.

54. $a^{5m-2}b^{3n-4}$. **55.** $2x^{5m-3n-p}y^{m+n-p-4}$. **56.** an^2.

13. *The Distributive Law holds when the multiplier is a multinomial;* that is,

$$(a+b)(c+d-e) = ac + ad - ae + bc + bd - be, \text{ etc.}$$

E.g.,
$$\begin{aligned}(2+3)(7-5) &= (2+3)7 - (2+3)5 \\ &= 2\times 7 + 3\times 7 - 2\times 5 - 3\times 5 \\ &= 2\times 7 - 2\times 5 + 3\times 7 - 3\times 5.\end{aligned}$$

For, $(a+b)(c+d-e) = (a+b)c + (a+b)d - (a+b)e$, by Art. 11,

$$\begin{aligned}&= ac + bc + ad + bd - ae - be \\ &= ac + ad - ae + bc + bd - be.\end{aligned}$$

Similarly for any number of terms in either multiplier or multiplicand.

Multiplication of Multinomials by Multinomials.

14. From the preceding article is derived the following principle for multiplying a multinomial by a multinomial:

Multiply each term of the multiplicand by each term of the multiplier, and add algebraically the resulting products.

Ex. 1. Multiply $-3a + 2b$ by $2a - 3b$.

We have

$$\begin{aligned}(2a-3b)\times(-3a+2b) &= 2a\times(-3a) + 2a\times 2b - 3b(-3a) \\ &\quad - 3b\times 2b \qquad (1) \\ &= -6a^2 + 4ab + 9ab - 6b^2 \qquad (2) \\ &= -6a^2 + 13ab - 6b^2.\end{aligned}$$

The step indicated by the line (1) should be performed mentally, and line (2) be at once written.

The work may be better arranged as follows: *Write the multiplier under the multiplicand, the first partial product,* i.e., *the product of the multiplicand by the first term of the multiplier, under the multiplier, the second partial product under the first, and so on, placing like terms of the different partial products in the same column.*

Arranging the work of Ex. 1 as suggested, we have

$$\begin{array}{rrr} -3a & +\ 2b & \\ 2a & -\ 3b & \\ \hline -6a^2 & +\ 4ab & \\ & +\ 9ab & -6b^2 \\ \hline -6a^2 & +13ab & -6b^2 \end{array}$$

It is customary to multiply from left to right, instead of from right to left as in Arithmetic.

Ex. 2. Multiply $-2x+3xy+y^2$ by $3x^2-5xy-2y^2$.

We have

$$\begin{array}{rrrrrrr} -2x & +3xy & +y^2 & & & & \\ 3x^2 & -5xy & -2y^2 & & & & \\ \hline -6x^3 & +9x^3y & +\ 3x^2y^2 & & & & \\ & & -15x^2y^2 & +10x^2y & -\ 5xy^3 & & \\ & & & & -\ 6xy^3 & +4xy^2 & -2y^4 \\ \hline -6x^3 & +9x^3y & -12x^2y^2 & +10x^2y & -11xy^3 & +4xy^2 & -2y^4 \end{array}$$

Ex. 3. Multiply $a^3+x^3-4a^2x-2ax^2$ by $x-3a$.

Arranging the multiplicand to descending powers of x, we have

$$\begin{array}{rrrrr} x^3 & -2ax^2 & -4a^2x & +a^3 & \\ x & -3a & & & \\ \hline x^4 & -2ax^3 & -4a^2x^2 & +\ \ a^3x & \\ & -3ax^3 & +6a^2x^2 & +12a^3x & -3a^4 \\ \hline x^4 & -5ax^3 & +2a^2x^2 & +13a^3x & -3a^4 \end{array}$$

In the work of Ex. 3 the literal parts of the first three terms of the second partial product could have been omitted, it being understood that the numerals remaining are the coefficients of the literal parts just above in the first partial product. Thus,

$$\begin{array}{l|l|l|l}
x^3 - 2\,ax^2 - 4\,a^2x + a^3 & & & \\
x - 3\,a & & & \\
\hline
x^4 - 2 & ax^3 - 4 & a^2x^2 + \;\;1 & a^3x \\
\;\;\;\;\; - 3 & \;\;\;\; + 6 & \;\;\;\; + 12 & \;\;\;\;\;\; - 3\,a^4 \\
\hline
\multicolumn{4}{l}{x^4 - 5\,ax^3 + 2\,a^2x^2 + 13\,a^3x - 3\,a^4}
\end{array}$$

Each vertical line is drawn to indicate that the literal part on its right belongs to all the coefficients in the same column on its left.

A similar arrangement may be made when the multiplier contains any number of terms.

Ex. 4. Multiply $a^3 - 3\,a^2b + 3\,ab^2 - b^3$ by $a^2 - 2\,ab + b^2$.
We have

$$\begin{array}{l|l|l|l|l}
a^3 - 3\,a^2b + 3\,ab^2 - b^3 & & & & \\
a^2 - 2\,ab + b^2 & & & & \\
\hline
a^5 - 3 & a^4b + \;\;3 & a^3b^2 - \;\;1 & a^2b^3 & \\
\;\;\;\;\; - 2 & \;\;\;\; + \;\;6 & \;\;\;\; - \;\;6 & \;\;\;\; + 2 & ab^4 \\
 & \;\;\;\; + \;\;1 & \;\;\;\; - \;\;3 & \;\;\;\; + 3 & \;\;\; - b^5 \\
\hline
\multicolumn{5}{l}{a^5 - 5\,a^4b + 10\,a^3b^2 - 10\,a^2b^3 + 5\,ab^4 - b^5}
\end{array}$$

Observe that in the last two examples the multiplicand and multiplier, and also the product, are *homogeneous*.

Ex. 5. Multiply $2\,x^{m+1} - 5\,x^m + 7\,x^{m-1} - 9\,x^{m-2}$ by $x^{2m} - x^{2m-1}$.
We have

$$\begin{array}{l}
2\,x^{m+1} - 5\,x^m + 7\,x^{m-1} - 9\,x^{m-2} \\
\;\;\;\; x^{2m} - x^{2m-1} \\
\hline
2\,x^{3m+1} - 5\,x^{3m} + \;\;7\,x^{3m-1} - \;\;9\,x^{3m-2} \\
\qquad\quad\;\; - 2\,x^{3m} + \;\;5\,x^{3m-1} - \;\;7\,x^{3m-2} + 9\,x^{3m-3} \\
\hline
2\,x^{3m+1} - 7\,x^{3m} + 12\,x^{3m-1} - 16\,x^{3m-2} + 9\,x^{3m-3}
\end{array}$$

EXERCISES XII.

Multiply

1. $x + 3$ by $x + 7$.
2. $x + 5$ by $x - 2$.
3. $2\,a - 7$ by $3\,a + 4$.
4. $\frac{1}{2}\,a - 5$ by $\frac{1}{3}\,a - 8$.

5. $-3ab+7$ by $2ab-5$.

6. $5xy-4$ by $-6xy+9$.

7. $x+3$ by $y+4$.

8. $-5a+7$ by $2b-3$.

9. $a+b$ by $2a-3b$.

10. $x-\frac{2}{3}a$ by $2x+\frac{3}{5}a$.

11. $ax-by$ by $ax+2by$.

12. $-3ab+ac$ by $7ab-5ac$.

13. $5x^2-x$ by $1-2x$.

14. $5a^3b^2-2a^2b^3$ by a^2-b^2.

15. $-17a^2x^7+12a^5x^4$ by $a^3x-3a^2x^2$.

16. $2a^mx^{n-1}-3a^{m-1}x^n$ by $5a^2x^n-2a^mx^2$.

17. $\frac{2}{3}a^{m+1}b^{n-1}-\frac{1}{5}a^{n-2}b^{m+2}$ by $-\frac{1}{2}a^nb^{m+1}+a^{m+1}b^n$.

18. x^2-3x+1 by $x-4$.

19. $4a^2-6a+9$ by $2a+3$.

20. $a^2x^3+5ax+7x^2$ by $3ax-5$.

21. x^2+x+1 by $x-1$.

22. a^2-ab+b^2 by $a+b$.

23. $8x^3+12x^2y+18xy^2+27y^3$ by $2x-3y$.

24. $2a^2-a$ by a^2-a+1.

25. $1+x+x^2+x^3+x^4$ by $1-x$.

26. $1-2a+4a^2-8a^3$ by $1+2a$.

27. k^2-k+1 by k^2+k+1.

28. $4a-4a^2+2a^3$ by $2+2a+a^2$.

29. $2b^2-3b+4$ by b^2-2b-3.

30. a^2b+2ab^2-1 by $2a^2-ab+1$.

31. $2x^2+3xy+4y^2$ by $3x^2-4xy+y^2$.

32. $-1+3x^3-5x^6$ by $4x-x^4+2x^7$.

33. x^3-2x^2+3x-4 by $4x^3+3x^2+2x+1$.

34. $x^4+2x^3+x^2-4x-11$ by x^2-2x+3.

35. $x^2-xy+y^2+x+y+1$ by $x+y-1$.

36. $2a^2b-ca^2-ab^2-abc+b^3$ by $2a^2b-a^2c+ab^2+abc-b^3$.

37. $3\frac{1}{3}a^4b+2\frac{3}{10}a^2b^3-4\frac{1}{2}a^3b^2$ by $2\frac{1}{5}a^3-2\frac{1}{2}a^2b+ab^2$.

38. $7\frac{2}{5}n^5l+3\frac{3}{10}n^4l^2-1\frac{1}{5}n^2l^4$ by $\frac{1}{4}n^2l^2-1\frac{1}{6}n^3l+2\frac{1}{3}n^4$.

39. $3x^m-2x^{m-1}+4x^{m-2}$ by $2x^{m-1}+4x^{m-2}-5x^{m-3}$.

40. $3a^{p-3}+a^{p-2}-2a^{p-1}-4a^p$ by $2a^{p-3}+3a^{p-4}$.

41. $4a^4x^{3n+3}-\frac{1}{5}a^2x^{2n+1}+10x^{n-1}$ by $\frac{3}{20}a^2x^{2n-1}+7\frac{1}{2}x^{n-3}$.

42. $4a^{m+1}b^2+a^{m-2}b-2a^{m+4}b^3$ by $3a^mb^2-a^{m+3}b^3-5a^{m+6}b^4$.

43. $3a^{n-2}x^2-a^{n-3}x^3+a^n$ by $a^2x^{n-1}-3x^{n+1}-2ax^n$.

44. $4x^{2n-1}-2x^{3n}-4x^{n-2}+\frac{1}{2}x^{4n+1}$ by $\frac{1}{4}x^{4n+1}+2x^{2n-1}+2x^{n-2}+x^{3n}$.

45. $5a^{n+3r}b^{r-1} - 2a^{n-r}b^{r+1} + 3a^{n+5r}b^{r-2} + a^{n+r}b^r$
by $a^{n+r}b^r + 4a^{n-5r}b^{r+3} - 2a^{n-3r}b^{r+2}$.

46. $2x(a^2 + b^2)^3 - 3x^2(a^2 + b^2)^2 + 4x^3(a^2 + b^2)$
by $x(a^2 + b^2)^2 - 2(a^2 + b^2)$.

47. $2(x + y)^3 - 5(x + y)^2(x - y) - 7(x + y)(x - y)^2 + 9(x - y)^3$
by $3(x + y)^2(x - y) - 5(x + y)(x - y)^2$.

48. $(x + y)^{n+2} + 3(x + y)^{n+1} - 5(x + y)^n$
by $6(x + y)^{n+1} + 4(x + y)^n - 2(x + y)^{n-1}$.

49. $x^4(x^2 + 2)^{n-3} + 2x^3(x^2 + 2)^{2n-1} + 4(x^2 + 2)^{3n+1}$
by $x^7(x^2 + 2)^{n-5} - 4x^3(x^2 + 2)^{3n-1} + 8x(x^2 + 2)^{5n+1}$.

Perform the following indicated operations:

50. $(x - 2)(x + 3)(x - 4)$.

51. $(x - 3)(x - 5)(x - 7)$.

52. $(2x - 3y)(4x + y)(x + 5y)$.

53. $(xy - 2z)(3z - 4xy)(z + 5xy)$.

54. $(2m^2 + 3m - 2)(m - 1)(2m + 3)$.

55. $(x^2 + 4x - 1)(x^2 - 2x + 1)(x + 2)$.

56. $(a^2 - a + 1)(a^2 + a + 1)(a^4 - a^2 + 1)$.

57. $(a^m + b^m)(a^n + b^n)(a^p - b^p)$.

58. $(1 + x^n + x^m)(3 - 2x^n + x^m)(5x^{n-1} - 3x^{m-1})$.

15. *The converse of the Distributive Law evidently holds;* that is,

$$ab + ac - ad = a(b + c - d), \text{ etc.}$$

E.g.,
$$3x + 6x - 7x = (3 + 6 - 7)x,$$
$$ax + bx = (a + b)x,$$
$$2ay - 3by = (2a - 3b)y.$$

16. If the coefficients of the multiplicand and multiplier, arranged to a common letter of arrangement, be literal, it is frequently desirable to unite the terms of the product which are *like in this letter of arrangement.*

Ex. 1. Multiply $x+a$ by $x+b$.

We have

$$\begin{array}{l} x+a \\ x+b \\ \hline x^2+ax \\ \qquad bx+ab \\ \hline x^2+ax+bx+ab = x^2+(a+b)x+ab, \text{ by Art. 15.} \end{array}$$

Such steps as uniting $ax+bx$ into $(a+b)x$ should be performed mentally, and the result be at once written.

EXERCISES XIII.

Arrange the values of the following products to descending powers of x, uniting like terms in x:

1. $(x^2+ax+b)(x+a)$.
2. $(x^3-ax^2+bx-c)(x-b)$.
3. $(px^3+qx^2+rx+s)(ax^2+bx+c)$.

17. It is frequently required to find a product of two or more factors, which are arranged to powers of a common letter of arrangement, only as far as a given power of the letter of arrangement.

Ex. 1. Find the value of the product

$$(1+2x+5x^2-7x^3+3x^4)(2-3x+4x^2-6x^3)$$

as far as, and including, the term of the *third* degree in x.

We have

$$\begin{array}{rrrr} 1+2x+ & 5x^2- & 7x^3+3x^4 \\ 2-3x+ & 4x^2- & 6x^3 \\ \hline 2+4x+ & 10x^2- & 14x^3 \\ -3x & -6x^2 & -15x^3 \\ & +4x^2 & +8x^3 \\ & & -6x^3 \\ \hline 2+x+ & 8x^2- & 27x^3 \end{array}$$

Notice that each partial product is carried only to the term of the required degree.

EXERCISES XIV.

Find the value of each of the following products as far as, and including, the term of the *fourth* degree in x:

1. $(3 - x + 5x^2 - 7x^3)(2 + x - 3x^2)$.
2. $(7 - 2x^2 + 3x^4)(1 - x - x^2)$.
3. $(a + bx + cx^2 + dx^3 + ex^4)(3 + 2x - x^2)$.
4. $(a + bx^2 + cx^4 + dx^6)(1 - 4x + 3x^2 + 5x^3)$.
5. $(a + bx + cx^2 + dx^3 + ex^4)(a + bx + cx^2 + dx^3 + ex^4)$.
6. $(2 - 3x + 5x^2 - 7x^3)(5 + 2x - 7x^2 - x^3)(x - x^2 - x^3)$.

Find the following products as far as, and including, the term of the *sixth* degree:

7. $(1 - x + x^2 - x^3 + \cdots)(1 + x + x^2 + x^3 + \cdots)$.
8. $(1 + \frac{1}{2}x + \frac{1}{3}x^2 + \frac{1}{4}x^3 + \cdots)(1 - \frac{1}{3}x + \frac{1}{5}x^2 - \frac{1}{7}x^3 + \cdots)$.

Zero in Multiplication.

18. Since $a \cdot 0 = a(b - b)$, by definition of 0,

$= ab - ab$, by the Distributive Law,

$= 0$, by definition of 0,

we have $\boldsymbol{a \cdot 0 = 0}$ and $\boldsymbol{0 \cdot a = 0}$.

In like manner, $a \cdot 0 \cdot b = 0 \cdot b = 0$; and so on.

That is, *a product is* 0 *if one of its factors be* 0. In particular,

$$0 \cdot 0 = 0(a - a) = 0 - 0 = 0.$$

19. The words *is not equal to, does not have the same value as*, etc., are frequently denoted by the symbol, $\neq$.

E.g., $7 \neq 2$, read *seven is not equal to* 2;

$P \neq Q$, read *P is not equal to Q*.

20. It follows, conversely, from Art. 18:

If a product be 0, *one or more of its factors is* 0. That is, if

$$P \times Q = 0,$$

then either $P = 0$ and $Q \neq 0$;

or $Q = 0$ and $P \neq 0$; or $P = 0$ and $Q = 0$.

EXERCISES XV.

1. What is the value of $2(a-b)$, when $b=a$?

2. What is the value of $(a+b)(c-d)$, when $c=d$?

3. What is the value of $(b+c)(a+b-c)$, when $c=a+b$?

4. What is the value of $(x-2)(x^3-7x^2+4x-3)$, when $x=2$?

5. What is the value of $(x^2-4)(x^3-6x^2+11x-6)$, when $x=2$? When $x=-2$?

6. What is the value of $(x+3)(x^2-9)(x^4-7x^3+2x-9)$, when $x=3$? When $x=-3$?

If $P \times Q \times R = 0$, what can we infer,

7. When $P \neq 0$? 8. When $Q \neq 0$? 9. When $P \neq 0$ and $R \neq 0$?

For what values of x does each of the following expressions reduce to 0:

10. $x(x-2)$.

11. $(x-1)(x-2)$.

12. $(x-4)(x+7)$.

13. $(x+9)(x+11)$.

14. $(x-6)(x+8)(x^2-25)$.

15. $x(x-a)(x-b)(x-c)$.

§ 4. DIVISION.

1. One power is said to be *higher* or *lower* than another according as its exponent is *greater* or *less* than the exponent of the other.

E.g., a^4 is a higher power than b^2 or a^3, but is a lower power than a^6 or b^7.

2. Quotient of Powers of One and the Same Base.

Ex. 1. $4^5 \div 4^2 = (4 \times 4 \times 4 \times 4 \times 4) \div (4 \times 4)$,
$= (4 \times 4 \times 4) \times (4 \times 4) \div (4 \times 4)$, by Assoc. Law,
$= 4 \times 4 \times 4$, since $(4 \times 4) \div (4 \times 4) = 1$,
$= 4^3 = 4^{5-2}$.

Ex. 2. $a^7 \div a^3 = (aaaaaaa) \div (aaa)$
$= (aaaa) \times (aaa) \div (aaa)$, by Assoc. Law,
$= aaaa$, since $(aaa) \div (aaa) = 1$,
$= a^4 = a^{7-3}$.

The preceding examples illustrate the following principle:

(i.) *The quotient of a higher power of a given base by a lower power of the same base, is equal to a power of that base whose exponent is the exponent of the dividend minus the exponent of the divisor;* or, stated symbolically,

$$a^m \div a^n = a^{m-n}, \text{ when } m > n.$$

For $a^m \div a^n = (aaa \cdots \text{to } m \text{ factors}) \div (aaa \cdots \text{to } n \text{ factors})$

$= [aaa \cdots \text{to } (m-n) \text{ factors}] \times (aaa \cdots \text{to } n \text{ factors}) \div (aaa \cdots \text{to } n \text{ factors})$

$= aaa \cdots \text{to } (m-n) \text{ factors, since } (aaa \cdots \text{to } n \text{ factors}) \div (aaa \text{ to } n \text{ factors}) = 1$

$= a^{m-n}.$

(ii.) $$a^m \div a^m = 1, \text{ when } m = n.$$

E.g., $$a^2 \div a^2 = 1.$$

(iii.) *The converse of the principle is evidently true.*

E.g., $$a^5 = a^8 \div a^3 = a^{21} \div a^{16} = \text{etc.}$$

EXERCISES XVI.

Express each of the following quotients as a single power:

1. $2^3 \div 2.$
2. $x^5 \div x^2.$
3. $(-5)^7 \div (-5)^4.$
4. $(-6)^5 \div 6^3.$
5. $a^7 \div a^3.$
6. $b^9 \div b^6.$
7. $(-x)^7 \div (-x)^4.$
8. $(-a)^9 \div a^4.$
9. $(ab)^5 \div (ab)^2.$
10. $(-xy)^{11} \div (-xy)^7.$
11. $a^n \div a^3.$
12. $3^n \div 3^m.$
13. $a^{n+1} \div a.$
14. $x^{n+7} \div x^n.$
15. $b^{x+8} \div b^{x+1}.$
16. $a^n \div a^{n-1}.$
17. $a^{n+1} \div a^{n-1}.$
18. $a^{2n} \div a^{n-1}.$
19. $a^{n+x} \div a^{n-1}.$
20. $a^{2-x} \div a^{n-x}.$
21. $(a+b)^5 \div (a+b)^2.$
22. $(1+x^2)^9 \div (1+x^2)^8.$
23. $(x-y)^{n+3} \div (x-y)^{n+2}.$
24. $(xy-1)^{2n-4} \div (xy-1)^{n-3}.$

Division of Monomials by Monomials.

3. The quotient of two monomials which contain either numerical coefficients or common literal factors can be simplified.

Ex. 1. $$12a \div 4 = 12 \div 4 \times a,$$ by the Commutative Law, $$= 3a.$$

Ex. 2. $$-27x^7 \div 3x^2 = (-27 \div 3) \times (x^7 \div x^2),$$ by the Commutative Law, $$= -9x^5.$$

Ex. 3. $$15a^3b^2 \div (-5ab^2) = [15 \div (-5)] \times (a^3 \div a) \times (b^2 \div b^2)$$ $$= -3a^2.$$

Ex. 4. $$-5x^9y^5z^3 \div 7x^3y^4z$$ $$= (-5 \div 7) \times (x^9 \div x^3) \times (y^5 \div y^4) \times (z^3 \div z) = -\tfrac{5}{7}x^6yz^2.$$

Ex. 5. $$-10(a+b)^7 \div [-5(a+b)^3]$$ $$= [-10 \div (-5)] \times [(a+b)^7 \div (a+b)^3] = 2(a+b)^4.$$

The preceding examples illustrate the following method:

The quotient of one monomial divided by another is the quotient of their numerical coefficients multiplied by the quotient of their literal factors.

EXERCISES XVII.

Divide

1. $6a$ by 3.
2. $12x$ by $-x$.
3. $-15m$ by $3m$.
4. $5x^4$ by $2x$.
5. $9x^3$ by $-3x^2$.
6. $-11a^7$ by $-5a^2$.
7. $4ab$ by $-2a$.
8. $6abc$ by $-3ac$.
9. $\frac{1}{2}a^3b$ by $3a^2b$.
10. $6x^8y$ by $5x^4$.
11. $-15a^5b^7$ by $-3ab^5$.
12. $a^5b^2x^4$ by abx^2.
13. $-24x^2yz^3$ by $-8xyz^2$.
14. $7a^7b^{10}c^{13}$ by $-5a^4b^5c^6$.
15. $\frac{5}{6}m^6n^7p^8$ by $-\frac{2}{3}m^2n^4p^6$.
16. $15(a+b)$ by $3(a+b)$.
17. $25x^2(x+1)^3$ by $-5x(x+1)^2$.
18. $21x^5(x^2+2x-1)^5$ by $7x^4(x^2+2x-1)^3$.
19. $6a^{10}(x+y)^7(x+2)^6$ by $3a^5(x+y)^3(x+2)^4$.
20. $10a^{2n}b^5$ by $-5a^nb^3$.
21. $-27x^{n+1}y^{3m}$ by $-9xy^{2m}$.
22. $-9a^{5m+1}b^{3n-1}$ by $3a^{2m+1}b^{2n-1}$.
23. $a^{n+5}b^{10}c^{n-3}$ by $-7a^{n+3}b^5c^{n-3}$.
24. $a^{n+1}b^x$ by a^nb^{x-1}.
25. $a^{n-1}b^{n-2}$ by $a^{n-3}b^{n-4}$
26. $x^{2n-1}y^{3m+2}$ by $x^{n+1}y^{2m-3}$.
27. $a^{x+1}(x-1)^7$ by $a^{x-1}(x-1)^4$.

Simplify

28. $a^3x^5 \div (-ax^3) \times 2axy.$ **29.** $35x^2y^3z \times 2xz^3 \div (7x^2y^2z^2).$

30. $a^{2n-1}b^{m+1}c^{m+n} \div a^nb^mc^m \div a^{n-2}bc^{n-3}.$

31. $6x^{m+1}y^{n-1} \div (-x^{m-1}y^{m-n}) \times (3x^2y^2z^2).$

The Distributive Law for Division.

4. If the indicated operation within the parentheses in the quotient

$$(8 + 6) \div 2$$

be first performed, in accordance with the meaning of parentheses, we have $(8 + 6) \div 2 = 14 \div 2 = 7$.

But if each term within the parentheses be first divided by 2 and the resulting quotients be then added, we have

$$8 \div 2 + 6 \div 2 = 4 + 3 = 7, \text{ as above.}$$

Therefore $\quad (8 + 6) \div 2 = 8 \div 2 + 6 \div 2.$

Likewise $\quad (-12 + 15) \div (-3) = 3 \div (-3) = -1,$

and $\quad -12 \div (-3) + 15 \div (-3) = 4 - 5 = -1$, as above.

Therefore, $\quad (-12 + 15) \div (-3) = -12 \div (-3) + 15 \div (-3).$

The above examples illustrate the following principle:

Distributive Law. — *The quotient of a multinomial by a monomial is obtained by dividing each term of the multinomial by the monomial and adding algebraically the resulting quotients;* that is,

$$(a + b - c) \div d = a \div d + b \div d - c \div d.$$

For, since $\div d \times d = \div 1$, we can replace, in

$$(a + b - c) \div d,$$

a by $a \div d \times d$, b by $b \div d \times d$, c by $c \div d \times d$,

We then have

$$(a + b - c) \div d = (a \div d \times d + b \div d \times d - c \div d \times d) \div d$$
$$= (a \div d + b \div d - c \div d) \times d \div d, \quad \text{by § 3, Art. 15,}$$
$$= a \div d + b \div d - c \div d, \text{ since } \times d \div d = \times 1.$$

5. It follows, conversely, from the Distributive Law that

$$a \div d + b \div d - c \div d = (a + b - c) \div d.$$

6. By definition of 0, we have

$$0 \div N = (a - a) \div N$$
$$= a \div N - a \div N, \text{ by the Distributive Law,}$$
$$= 0.$$

Therefore $\boldsymbol{0 \div N = 0}$, when $\boldsymbol{N \neq 0}$.

It is important to observe that this relation is proved only when $N \neq 0$. The consideration of $0 \div 0$ is deferred.

7. It follows, conversely, from Art. 6:

If a quotient be 0, *the dividend is* 0.

That is, if $$M \div N = 0,$$
then $$M = 0.$$

Division of a Multinomial by a Monomial.

8. The division of a multinomial by a monomial is a direct application of the Distributive Law.

Ex. 1. Divide $6x^2 - 12x$ by $3x$.

We have $(6x^2 - 12x) \div 3x = 6x^2 \div 3x - 12x \div 3x$ (1)

$$= 2x - 4. \qquad (2)$$

Ex. 2. Divide $-105a^3b^2 - 75a^2b^3 + 27a^2b^4$ by $-15a^2b$.

We have $(-105a^3b^2 - 75a^2b^3 + 27a^2b^4) \div (-15a^2b)$

$$= (-105a^3b^2) \div (-15a^2b) - 75a^2b^3 \div (-15a^2b)$$
$$+ 27a^2b^4 \div (-15a^2b) \qquad (1)$$
$$= 7ab + 5b^2 - \tfrac{9}{5}b^3. \qquad (2)$$

Ex. 3. Divide $x^2(1-m) - y^2(1-m)$ by $1-m$.

We have

$$[x^2(1-m) - y^2(1-m)] \div (1-m)$$
$$= x^2(1-m) \div (1-m) - y^2(1-m) \div (1-m) \qquad (1)$$
$$= x^2 - y^2. \qquad (2)$$

The steps indicated by lines (1) should be performed mentally, and lines (2) be at once written.

EXERCISES XVIII.

Divide

1. $5 + 10a$ by 5.
2. $4a + 8b$ by -4.
3. $ax + bx$ by x.
4. $3a^2 - 6ab$ by $-3a$.
5. $21a^2b - 14ab^2$ by $-7ab$.
6. $8am^2 - 2a^2m + 4a^3m^2$ by $2am$.
7. $12x^4y^2 - 2x^3y^3 + 4xy^5$ by $-2xy^2$.
8. $12a^4b^3x^4y^2 - 15a^2b^6xy^4 + 20ab^2xy^9$ by $4ab^2xy^2$.
9. $25(a+b)^3 - 2a(a+b)$ by $5(a+b)$.
10. $2(x-y)^3 - 2a(x-y)^4 - 6(x-y)^6$ by $2(x-y)^2$.

Simplify

11. $2a^2 - (a^3 - 3a) \div a$.
12. $2x(1-2x) - (12a^2x^3 + 6a^2x^2) \div (-3a^2x)$.
13. $(6x - 4x^2) \div 2x - (-2x^2y + 3xy) \div xy$.
14. $(ab - a^2b + 3a^3b) \div ab - (4a^3 - 4a^2) \div 2a$.

Divide $9a^2x^6 - 6a^3x^4 + 12a^5x^3$ by

15. $3a^2$.
16. $-3x^3$.
17. ax^2.
18. $-\frac{3}{2}a^2x^3$.

Divide $35a^3b^2c^4 - 21a^4b^3c^3 + 14a^5b^4c^3$ by

19. $7a^3$.
20. $-3a^3b^2$.
21. $-5a^2bc^3$.
22. $\frac{3}{4}a^2b^2c^2$.

Divide

23. $4a^{3n} - 6a^{2n}$ by $2a^n$.
24. $5x^{n+1}y^2 - 6x^{n+2}y$ by $-5x^ny$.

Divide $15x^{2n+1}y^5 - 12x^{2n+3}y^3 - 18x^{2n+5}y^4$ by

25. $3x^n$.
26. $-5x^{n+1}y^2$.
27. $-3x^{2n+1}y$.
28. $\frac{1}{2}x^{2n-5}y^3$.

29. Divide $2n^r(x-y)^5 + 3n^{2r}(x-y)^3 - 4n^{3r}(x-y)^6$
by $-6n^r(x-y)^2$.

30. Prove that the sum, or the difference, of two even numbers is exactly divisible by 2, and is therefore an even number.

31. Prove that the sum, or the difference, of two odd numbers is even.

32. Prove that the sum, or the difference, of an even and an odd number is odd.

33. If m be an even number, what is $m+1$? $m+4$?

34. If m be an odd number, what is $m+3$? $m+6$?

Division of a Multinomial by a Multinomial.

9. The division of one multinomial by another is performed in a way similar to that of dividing one number by another in Arithmetic. The division depends upon the following principle:

The quotient of dividing one number (**dividend**) *by another* (**divisor**) *is equal to any number whatever* (**partial quotient**), *plus the quotient of dividing the dividend minus the partial quotient times the divisor, by the divisor.*

If D be the given *dividend*, d the given *divisor*, and q *any assumed number*, the principle enunciated above, stated symbolically, is:

$$D \div d = q + (D - qd) \div d.$$

Ex. 1. $105 \div 15 = 7.$ (1)

$$105 \div 15 = 4 + (105 - 4 \times 15) \div 15$$
$$= 4 + 45 \div 15 = 4 + 3 = 7.$$
$$105 \div 15 = 9 + (105 - 9 \times 15) \div 15$$
$$= 9 + (-30) \div 15 = 9 - 2 = 7.$$
$$105 \div 15 = N + (105 - N \times 15) \div 15.$$

Ex. 2. $105 \div 18 = 5 + (105 - 5 \times 18) \div 18$ (1)

$$= 5 + 15 \div 18 = 5\tfrac{5}{6}.$$
$$105 \div 18 = 3 + (105 - 3 \times 18) \div 18$$
$$= 3 + 51 \div 18 = 3 + 2\tfrac{5}{6} = 5\tfrac{5}{6}.$$
$$105 \div 18 = 8 + (105 - 8 \times 18) \div 18$$
$$= 8 + (-39) \div 18 = 8 - 2\tfrac{1}{6} = 5\tfrac{5}{6}.$$
$$105 \div 18 = N + (105 - N \times 18) \div 18.$$

Although the partial quotient may be *any number whatever*, yet in practice we should take the greatest number whose product by the divisor is equal to or less than the dividend; as 7 in Ex. 1, and 5 in Ex. 2. The work in these examples, except lines (1), was given simply to illustrate the principle enunciated.

A quotient consisting of more than one figure is obtained by successive applications of the same principle.

Ex. 3. $8675 \div 347 = 20 + (8675 - 20 \times 347) \div 347$
$= 20 + 1735 \div 347 = 20 + 5 = 25.$

Ex. 4. $8675 \div 25 = 300 + (8675 - 300 \times 25) \div 25$
$= 300 + 1175 \div 25$
$= 300 + 40 + (1175 - 40 \times 25) \div 25$
$= 300 + 40 + 175 \div 25 = 300 + 40 + 7 = 347.$

The work may be arranged differently:

```
8675 | 25
7500 | 300 + 40 + 7, = 347.
----
1175
1000
----
 175
 175
----
```

In practice the work last given is abbreviated by omitting the ciphers:

```
8675 | 25
75     347
---
117
100
---
 175
 175
 ---
```

The proof of the principle is as follows:

We have,

$[q + (D - qd) \div d] = [q + (D - qd) \div d] \times d \div d$, since $\times d \div d = \times 1$,
$= [qd + (D - qd) \div d \times d] \div d$, by the Distr. Law,
$= [qd + (D - qd)] \div d$, since $\div d \times d = +1$,
$= D \div d$, since $qd - qd = 0$.

10. The principle of Art. 9 evidently holds when the dividend, D, and the divisor, d, are algebraic expressions (multinomials).

Ex. 1. Divide x^2+3x+2 by $x+1$.

We have

$$(x^2+3x+2)\div(x+1)=x+[(x^2+3x+2)-x(x+1)]\div(x+1) \quad (1)$$
$$=x+(x^2+3x+2-x^2-x)\div(x+1) \quad (2)$$
$$=x+(2x+2)\div(x+1) \quad (3)$$
$$=x+2+[(2x+2)-2(x+1)]\div(x+1) \quad (4)$$
$$=x+2+0\div(x+1)$$
$$=x+2,\text{ since } 0\div(x+1)=0.$$

It is advisable to take the quotient of the term containing the highest power of x in the dividend by the term containing the highest power of x in the divisor as the partial quotient at each step.

In practice the work may be arranged more conveniently thus:

x^2+3x+2 | $x+1$
$x+2$, quotient.

$x^2+\ \ x$ ... $x(x+1)$ to be subtracted from x^2+3x+2; see (1) and (2) above.

$2x+2$ ... Remainder to be divided by $x+1$; see (3) above.

$2x+2$ ... $2(x+1)$ to be subtracted from $2x+2$; see (4) above.

0

Ex. 2. Divide $x^3+3ax^2+5a^2x+6a^3$ by $x^2+ax+3a^2$.

We have

$$(x^3+3ax^2+5a^2x+6a^3)\div(x^2+ax+3a^2)$$
$$=x+[(x^3+3ax^2+5a^2x+6a^3)-x(x^2+ax+3a^2)]\div(x^2+ax+3a^2) \quad (1)$$
$$=x+(x^3+3ax^2+5a^2x+6a^3-x^3-ax^2-3a^2x)\div(x^2+ax+3a^2) \quad (2)$$
$$=x+(2ax^2+2a^2x+6a^3)\div(x^2+ax+3a^2) \quad (3)$$

$$= x + 2a + [(2ax^2 + 2a^2x + 6a^3) - 2a(x^2 + ax + 3a^2)] \div (x^2 + ax + 3a^2) \quad (4)$$

$$= x + 2a + (2ax^2 + 2a^2x + 6a^3 - 2ax^2 - 2a^2x - 6a^3) \div (x^2 + ax + 3a^2) \quad (5)$$

$$= x + 2a + 0 \div (x^2 + ax + 3a^2) \quad (6)$$

$$= x + 2a, \text{ since } 0 \div (x^2 + ax + 3a^2) = 0.$$

Arranged differently, the work is:

$$\begin{array}{l|l} x^3 + 3ax^2 + 5a^2x + 6a^3 & x^2 + ax + 3a^2 \\ \hline & x + 2a, \text{ quotient.} \end{array}$$

$x^3 + ax^2 + 3a^2x$ ... $x(x^2 + ax + 3a^2)$ to be subtracted from the dividend, see (1) and (2) above.

$2ax^2 + 2a^2x + 6a^3$ ... Remainder to be divided by $x^2 + ax + 3a^2$, see (3) above.

$2ax^2 + 2a^2x + 6a^3$ ... $2a(x^2 + ax + 3a^2)$ to be subtracted from $2ax^2 + 2a^2x + 6a^3$, see (4) and (5) above.

11. The method of applying the principle of Art. 9 to the division of multinomials, as illustrated by Exx. 1 and 2, Art. 10, may be stated as follows:

Arrange the dividend and divisor to ascending or descending powers of some common letter, the letter of arrangement.

Divide the first term of the dividend by the first term of the divisor, and write the result as the first term of the quotient.

Multiply the divisor by this first term of the quotient, and subtract the resulting product from the dividend.

Divide the first term of the remainder by the first term of the divisor, and write the result as the second term of the quotient.

Multiply the divisor by this second term of the quotient, and subtract the resulting product from the remainder previously obtained.

If there be a second remainder, proceed with it and all subsequent remainders in like manner until a remainder zero is obtained, or until the highest power of the letter of arrangement in the remainder is less than the highest power of that letter in the divisor.

In the first case the division is exact; in the second case the quotient at this stage of the work is called the quotient of the division, and the remainder the remainder of the division.

Ex. 1. Divide

$$a^3b - 15\,b^4 + 19\,ab^3 + a^4 - 8\,a^2b^2 \text{ by } a^2 - 5\,b^2 + 3\,ab.$$

Arranging dividend and divisor to descending powers of a, we have

$$\begin{array}{l|l}
a^4 + a^3b - 8\,a^2b^2 + 19\,ab^3 - 15\,b^4 & a^2 + 3\,ab - 5\,b^2 \\ \cline{2-2}
a^4 + 3\,a^3b - 5\,a^2b^2 & a^2 - 2\,ab + 3\,b^2 \\ \cline{1-1}
\quad - 2\,a^3b - 3\,a^2b^2 + 19\,ab^3 & \\
\quad - 2\,a^3b - 6\,a^2b^2 + 10\,ab^3 & \\ \cline{1-1}
\qquad\quad 3\,a^2b^2 + 9\,ab^3 - 15\,b^4 & \\
\qquad\quad 3\,a^2b^2 + 9\,ab^3 - 15\,b^4 & \\ \cline{1-1}
\end{array}$$

Ex. 2. Divide $8\,x^3 - y^3$ by $2\,xy + 4\,x^2 + y^2$.

Arranging the divisor to descending powers of x, we have

$$\begin{array}{l|l}
8\,x^3 - y^3 & 4\,x^2 + 2\,xy + y^2 \\ \cline{2-2}
8\,x^3 + 4\,x^2y + 2\,xy^2 & 2\,x - y \\ \cline{1-1}
\quad - 4\,x^2y - 2\,xy^2 - y^3 & \\
\quad - 4\,x^2y - 2\,xy^2 - y^3 & \\ \cline{1-1}
\end{array}$$

Observe that the remainder after the first partial division is arranged to descending powers of x.

Ex. 3. Divide $12\,a^{n+1} + 8\,a^n - 45\,a^{n-1} + 25\,a^{n-2}$ by $6\,a - 5$.

We have

$$\begin{array}{l|l}
12\,a^{n+1} + 8\,a^n - 45\,a^{n-1} + 25\,a^{n-2} & 6\,a - 5 \\ \cline{2-2}
12\,a^{n+1} - 10\,a^n & 2\,a^n + 3\,a^{n-1} - 5\,a^{n-2} \\ \cline{1-1}
\qquad 18\,a^n - 45\,a^{n-1} & \\
\qquad 18\,a^n - 15\,a^{n-1} & \\ \cline{1-1}
\qquad\quad - 30\,a^{n-1} + 25\,a^{n-2} & \\
\qquad\quad - 30\,a^{n-1} + 25\,a^{n-2} & \\ \cline{1-1}
\end{array}$$

Ex. 4. Divide $x^3 + (a + b + c)x^2 + (ab + ac + bc)x + abc$ by $x^2 + (a + b)x + ab$.

We have

$$\begin{array}{r|l}
x^3 + (a + b + c)x^2 + (ab + ac + bc)x + abc & x^2 + (a + b)x + ab \\
\underline{x^3 + (a + b \qquad)x^2 + \qquad\qquad ab\ x \qquad\quad} & \overline{x + c} \\
cx^2 + (ac + bc)x + abc & \\
\underline{cx^2 + (ac + bc)x + abc} &
\end{array}$$

EXERCISES XIX.

Find the values of the following indicated divisions:

1. $(x^2 + 2x + 1) \div (x + 1)$.
2. $(x^2 + 11x + 30) \div (x + 5)$.
3. $(x^2 - x - 90) \div (x + 9)$.
4. $(x^2 - 5x + 6) \div (x - 3)$.
5. $(4x^2 - 12x + 9) \div (2x - 3)$.
6. $(2m^2 - 3m + 1) \div (m - 1)$.
7. $(2a^2 + a - 6) \div (2a - 3)$.
8. $(3x^2 - 13x - 10) \div (3x + 2)$.
9. $(6x^4 - 10 - 11x^2) \div (2x^2 - 5)$.
10. $(2x^2 + 6a^2 + 7ax) \div (2x + 3a)$.
11. $(a^2 - 2ab + b^2) \div (a - b)$.
12. $(35x^2 + xy - 88y^2) \div (7x - 11y)$.
13. $(x^2 + 5\frac{2}{3}xy + 3\frac{1}{3}y^2) \div (x + 5y)$.
14. $(\frac{1}{9}a^2 + \frac{2}{5}ab + \frac{8}{25}b^2) \div (\frac{1}{3}a + \frac{4}{5}b)$.
15. $(a^2 - 18axy - 243x^2y^2) \div (a + 9xy)$.
16. $(8x^2y^2 - 65xyz^2 - 63z^4) \div (xy - 9z^2)$.
17. $(6n^3 - 7n^2x + 2nx^2) \div (-x + 2n)$.
18. $(x^4y + 6x^5 - 2x^3y^2) \div (3x^2 + 2xy)$.
19. $(-19a^2x^2 + 3x^4 + \frac{7}{2}ax^3) \div (\frac{1}{2}x - a)$.
20. $(4x^3 - 3x^2 - 24x - 9) \div (x - 3)$.
21. $(3x^3 - 13x^2 + 23x - 21) \div (3x - 7)$.
22. $(3x^4 - 3x^3 - 2x^2 - x - 1) \div (3x^2 + 1)$.
23. $(a^3 - 3a^2b + 3ab^2 - b^3) \div (a - b)$.
24. $(a^6 - 6a^4 + 9a^2 - 4) \div (a^2 - 1)$.
25. $(21a^6b + 20b^4 - 22a^2b^3 - 29a^4b^2) \div (3a^2b - 5b^2)$.
26. $(4x^4y^5 - \frac{1}{2}x^2y^6 + 12x^8y^3 - 11x^6y^4) \div (4x^3y^2 - xy^3)$.

27. $(x^3 + 8x^2 + 9x - 18) \div (x^2 + 5x - 6)$.

28. $(x^4 + x^3 - 4x^2 + 5x - 3) \div (x^2 + 2x - 3)$.

29. $(6x^4 - x^3 - 11x^2 - 10x - 2) \div (2x^2 - 3x - 1)$.

30. $(x^3 - 1) \div (x^2 + x + 1)$. 31. $(a^3 + 8) \div (a^2 - 2a + 4)$.

32. $(125x^6 - 64y^3) \div (5x^2 - 4y)$.

33. $(a^5x^5 + y^5) \div (ax + y)$. 34. $(x^4 + x^2 + 1) \div (x^2 - x + 1)$.

35. $(a^4x^5 + 64x) \div (4ax + a^2x^2 + 8)$.

36. $(24x^3 + 25x - x^5) \div (5 + x + x^3 + 5x^2)$.

37. $(4a^4 - 25c^4 - 30b^2c^2 - 9b^4) \div (2a^2 + 5c^2 + 3b^2)$.

38. $(27x^4 - 6c^2x^2 + \frac{1}{3}c^4) \div (c^2 - 6cx + 9x^2)$.

39. $(8a^3n^3 + 32a^6 + \frac{1}{2}n^6) \div (4an + n^2 + 4a^2)$.

40. $(16a^4b^2 + 9a^2b^4 - 12a^3b^3 - 8a^5b + 3a^6) \div (a^4 + 3a^2b^2 - 2a^3b)$.

41. $(28a^5c - 26a^3c^3 - 13a^4c^2 + 15a^2c^4) \div (2a^2c^2 + 7a^3c - 5ac^3)$.

42. $(81z^8 - 90b^4z^4 + 81b^6z^2 - 20b^8) \div (9z^4 + 9b^2z^2 - 5b^4)$.

43. $(32a^{10} - 80a^8b^2 + 80a^6b^4 - 40a^4b^6 + 10a^2b^8 - b^{10})$
$\div (4a^4 - 4a^2b^2 + b^4)$.

44. $(x^3 + y^3 + 3xy - 1) \div (x + y - 1)$.

45. $(a^3 + b^3 + c^3 - 3abc) \div (a + b + c)$.

46. $(a^2 + 2ab + b^2 - x^2 + 4xy - 4y^2) \div (a + b - x + 2y)$.

47. $(a^2 + 2ac - b^2 - 2bd + c^2 - d^2) \div (a + c - b - d)$.

48. $(32a^5 + b^5) \div (16a^4 - 8a^3b + 4a^2b^2 - 2ab^3 + b^4)$.

49. $(81x^8 - 16y^8) \div (27x^6 + 18x^4y^2 + 12x^2y^4 + 8y^6)$.

50. $(\frac{15}{2}a^2 - \frac{113}{6}ab + 9ac + 2b^2 - bc) \div (\frac{5}{4}a - 3b + \frac{3}{2}c)$.

51. $(28x^2 - 43\frac{3}{4}y^2 + 140yz - 112z^2) \div (7x + 8\frac{3}{4}y - 14z)$.

52. $(\frac{3}{8}a^2b + 6acd - \frac{8}{3}bc^2 + 16c^2d - \frac{1}{2}abd + \frac{4}{3}bcd - 8cd^2)$
$\div (\frac{3}{2}a + 4c - 2d)$.

53. $(\frac{2}{5}a^4x - 1\frac{61}{75}a^3x^2 + 1\frac{3}{5}a^2x^3 + \frac{3}{10}ax^4 - x^5) \div (\frac{2}{3}a^3 - \frac{4}{5}a^2x + \frac{1}{2}x^3)$.

54. $(2x^{10} - .075x^5 + 9.65x^7 - 1.05x^6 - 19.25x^8 + 8.5x^9)$
$\div (2.5x^6 - 3x^5 + .5x^7 - .15x^4)$.

Find the values of the following indicated divisions:

55. $[(b+c)x^2 - bcx + x^3 - bc(b+c)] \div (x^2 - bc)$.

56. $[x^3 + (a+b+c)x^2 + (ab+ac+bc)x + abc] \div (x+b)$.

57. $[x^3 + (a+b-c)x^2 + (ab-ac-bc)x - abc] \div (x-c)$.

58. $[abc - b^2(a+c) + a^2(b+c) + c^2(a+b)] \div (ab+ac+bc)$.

59. $[a(a-1)x^3 + (a^3 + 2a - 2)x^2 + (3a^2 - a^3)x - a^4]$
$\div (ax^2 + 2x - a^2)$.

60. $[x^5 - (1+m)x^4 + (1+m+n)x^3 - (m+n+p)x^2 + (p+n)x - p]$
$\div [x^2 - (x-1)]$.

61. $[(10a^2 + 29 - 34a)x + (5 - 2a^2 + 3a^3 - 8a)x^3 + 8a^2 + 21 - 26a$
$+ (17 - 22a + 4a^3)x^2] \div [(a-2)x + (a^2 - 1 + a)x^2 + 2a - 3]$.

Find the values of the following indicated divisions:

62. $(6x^{3n} - 25x^{2n} + 27x^n - 5) \div (2x^n - 5)$,

63. $(6x^{5n} - 11x^{4n} + 23x^{3n} + 13x^{2n} - 3x^n + 2) \div (3x^n + 2)$.

64. $(6x^{2n+1} - 29x^{2n} + 43x^{2n-1} - 20x^{2n-2}) \div (2x^n - 5x^{n-1})$.

65. $(a^{4n} - a^{3n}b^m + a^nb^{3m} - b^{4m}) \div (a^n - b^m)$.

66. $(64a^{4n} - 18a^{3n}b^m + 24a^{2n}b^{2m} - 18a^nb^{3m} - 52b^{4m}) \div (2a^n - 2b^m)$.

67. $(\frac{1}{8} - 6y^{2m+4n} + 27y^{4m+8n}) \div (\frac{1}{8} + 2y^{m+2n} + 3y^{2m+4n})$.

68. $6a^{4n+1}b^{m+6} - \frac{5}{2}a^{4n}b^{m+5} + \frac{1}{6}a^{4n-1}b^{m+4} + \frac{9}{4}a^{4n-2}b - \frac{3}{16}a^{4n-3})$
$\div (3a^{2n+1}b^{m+1} - \frac{1}{4}a^{2n}b^m)$.

69. $(15a^{3n-2m-4}b^{2p+7} + 14a^{6n-m-4}b^{4p+4} - \frac{24}{5}a^{9n-4}b^{6p+1})$
$\div (5a^{2n-4}b^{3-p} + 6a^{5n+m-4}b^p)$.

70. $(1 + a^{6x} - 2a^{3x}) \div (3a^{2x} + 2a^{3x} + 2a^x + a^{4x} + 1)$.

71. $[6x^3 + 16(a-x)^{3r} - 17(a-x)^rx^2] \div -[4(a-x)^r - 3x]$.

72. $[2a^2(b+c)^{2n} - \frac{1}{2}] \div [a(b+c)^n + \frac{1}{2}]$.

73. $[8(x-y)^{3n} - x^3] \div [4(x-y)^{2n} + 2x(x-y)^n + x^2]$.

12. In the equation of Art. 9,

$$D \div d = q + (D - qd) \div d,$$

$D - qd$ is the remainder at any stage of the work, and q is the corresponding partial quotient. If, for brevity, we let R stand for the remainder at any stage, we have

$$\boldsymbol{D \div d = q + R \div d.} \qquad (1)$$

That is, *the result of dividing one number by another is equal to the partial quotient at any stage, plus the remainder at this stage divided by the given divisor.*

E.g., $29 \div 6 = 4 + 5 \div 6 = 4 + \frac{5}{6}$;

$$(x^2 - x + 2) \div (x + 1) = (x - 2) + 4 \div (x + 1).$$

13. If both members of the equation

$$D \div d = q + R \div d$$

be multiplied by d, we have, by Ch. II., § 3, Art. 17,

$$D \div d \times d = (q + R \div d)d$$
$$= qd + R \div d \times d$$
$$= qd + R, \text{ since } \div d \times d = \div 1.$$

Therefore, $\boldsymbol{D = qd + R}$.

That is, *the dividend is equal to the product of the quotient at any stage by the divisor; plus the remainder at this stage.*

E.g., $29 = 4 \times 6 + 5$, from the first example of Art. 12;

and $x^2 - x + 2 = (x - 2)(x + 1) + 4$,

from the second example of Art. 12.

EXERCISES XX.

Find the remainder of each of the following indicated divisions, and verify the work by applying the principle of Art. 13:

1. $(x^2 - 7x + 11) \div (x - 2)$.
2. $(3x^2 + 5x - 9) \div (x - 4)$.
3. $(x^3 - 17x^2 + 15x - 13) \div (2x - 5)$.
4. $(x^4 - 19x^2 + 3x - 2) \div (5x - 7)$.
5. $(5x^5 - 7x^3 + 2x - 1) \div (x^2 - 7x + 3)$.
6. $6n^5x^3 + 12n^3x^9 - 14n^4x^6 + n^6 - 1) \div (2x^3 - n)$.
7. $(12b^5 + 8b^2c^3 - 2b^4c - 4bc^4 - 38b^3c^2) \div -(2c^2 + 6bc - 4b^2)$.
8. $(32n^4x^8 - 22n^2x^4y^3 - 3y^4 + 4nx^2y^3) \div (8n^2x^4 + 6nx^2y - y^2)$.
9. $(24a^6 + 22a^3b^2 - 20a^3x - 10b^4 - 19b^2x - 4)$
   $\div (6a^3 - 2b^2 - 5x)$.
10. $(4c^{2n}x^{3n} - 13c^{3n}x^{2n} + 14c^{4n}x^n - 2c^{5n}) \div (c^nx^{2n} - 2c^{2n}x^n + c^{3n})$.
11. $(a^{n+4}x^{n-3} + a^nx^{n+2} + a^{n+2}x^n - a^{n-2}x^{n+4})$
    $\div (a^3x^{n-2} + a^4x^{n-3} + a^2x^{n-1} + ax^n)$.

14. We have assumed in the preceding work that the dividend is of higher degree than the divisor in some one letter, and that consequently the process of division comes to an end whenever a zero remainder, or a remainder of lower degree than the divisor in the letter of arrangement is obtained. In order that algebraic division (like algebraic subtraction) may be perfectly general, we should also be able to divide an expression of lower degree by an expression of higher degree.

In illustration of this statement we add a few examples.

Ex. 1. Divide 1 by $1 + x$.

We have

$$
\begin{array}{l|l}
1 & 1 + x \\ \cline{2-2}
\underline{1 + x} & 1 - x + x^2 - x^3 \\
\quad -x & \\
\quad \underline{-x - x^2} & \\
\qquad\quad x^2 & \\
\qquad\quad \underline{x^2 + x^3} & \\
\qquad\qquad -x^3 & \\
\qquad\qquad \underline{-x^3 - x^4} & \\
\qquad\qquad\qquad x^4 &
\end{array}
$$

If we stop at this stage of the work, we have, by Art. 12,

$$1 \div (1 + x) = 1 - x + x^2 - x^3 + x^4 \div (1 + x).$$

Evidently the division can be continued indefinitely, the terms in the quotient alternating in sign.

Ex. 2. Divide $2 + x$ by $3 - x + 2x^2$.

We have

$$
\begin{array}{l|l}
2 + x & 3 - x + 2x^2 \\ \cline{2-2}
\underline{2 - \frac{2}{3}x + \frac{4}{3}x^2} & \frac{2}{3} + \frac{5}{9}x - \frac{7}{27}x^2 \\
\quad \frac{5}{3}x - \frac{4}{3}x^2 & \\
\quad \underline{\frac{5}{3}x - \frac{5}{9}x^2 + \frac{10}{9}x^3} & \\
\qquad -\frac{7}{9}x^2 - \frac{10}{9}x^3 & \\
\qquad \underline{-\frac{7}{9}x^2 + \frac{7}{27}x^3 - \frac{14}{27}x^4} & \\
\qquad\qquad -\frac{37}{27}x^3 + \frac{14}{27}x^4 &
\end{array}
$$

If we stop at this stage of the work, we have, by Art. 12,

$$(2 + x) \div (3 - x + 2x^2) = \frac{2}{3} + \frac{5}{9}x - \frac{7}{27}x^2 + \left(-\frac{37}{27}x^3 + \frac{14}{27}x^4\right) \div (3 - x + 2x^2).$$

EXERCISES XXI.

Find the quotient of each of the following indicated divisions, to include five terms:

1. $1 \div (1 - a)$.
2. $x \div (1 + x)$.
3. $(1 + x) \div (1 - x)$.
4. $(1 + 2x) \div (1 - 3x)$.
5. $(1 - x) \div (1 + 2x + x^2)$.
6. $(1 + 2x) \div (1 - x - x^2)$.
7. $1 \div (1 + x + x^2)$.

Infinity.

15. The following considerations lead to an important mathematical concept.

Observe that the quotients

$$
\begin{aligned}
3 \div (1 - .9) &= 3 \div .1 = 30, \\
3 \div (1 - .99) &= 3 \div .01 = 300, \\
3 \div (1 - .999) &= 3 \div .001 = 3000, \\
3 \div (1 - .9999) &= 3 \div .0001 = 30000, \text{ etc.},
\end{aligned}
$$

increase as the divisors decrease, the dividend remaining the same.

If the divisor be still further decreased, the dividend remaining the same, the quotient will be still further increased.

Thus, $3 \div (1 - .999999) = 3 \div .000001 = 3000000,$

$$3 \div (1 - .999999999) = 3 \div .000000001 = 3000000000, \text{ etc.}$$

It is evident that, by taking the divisor sufficiently small (and positive), we can make the quotient as great as we please. If the divisor become less than any assigned number, however small, the quotient will become greater than any assigned number, however great. From these considerations we derive the following principle:

If the dividend be positive, and remain the same, as the divisor decreases below any assigned positive number, however small, i.e., *becomes more and more nearly equal to* 0, *the quotient increases beyond any assigned positive number, however great.*

The symbol, $+\infty$, read a **Positive Infinite Number, or Positive Infinity,** is used as an abbreviation for the words, *a number greater than any assigned positive number, however great.*

The principle enunciated above can be expressed symbolically thus:

$$+N \div 0 = +\infty. \qquad (1)$$

16. It is important to observe that the symbol, $+\infty$, does not stand for one definite number. It stands for *any* number which is greater than any assigned positive number, however great, *but which can be still further increased.* Therefore one infinite number can be greater or less than another infinite number.

Likewise, equation (1), Art. 15, is to be understood only as expressing the fact that, as the divisor becomes more and more nearly equal to 0, the quotient increases beyond any assigned positive number, however great.

17. We can also arrive at the conception of a positive infinite number by taking the dividend and the divisor both negative. Thus,

$$-3 \div (1 - 1.1) = -3 \div (-.1) = 30,$$
$$-3 \div (1 - 1.0001) = -3 \div (-.0001) = 30000, \text{ etc.}$$

18. In a similar manner, we arrive at the conception of a **Negative Infinite Number**, or **Negative Infinity**.

Thus,
$$-3 \div (1 - .9) = -3 \div .1 = -30,$$
$$-3 \div (1 - .9999) = -3 \div .0001 = -30000, \text{ etc.}$$

We therefore have $-\boldsymbol{N} \div \mathbf{0} = -\boldsymbol{\infty}$.

19. The numbers which we have hitherto used in this book are, for the sake of distinction, called *finite* numbers. In subsequent work we shall assume that the numbers involved are finite, unless the contrary is expressly stated.

20. Observe that the quotients

$$3 \div 10 = .3,\quad 3 \div 100 = .03,\quad 3 \div 1000000 = .000003, \text{ etc. };$$

decrease as the divisors increase. It is evident that if the divisor become greater than any assigned number, however great, the quotient will become less than any assigned number, however small.

We therefore have the following principle:

If the dividend remain the same, as the absolute value of the divisor increases beyond any assigned number, however great, the quotient decreases in absolute value below any assigned number, however small, i.e. *becomes more and more nearly equal to* 0.

This principle can be stated symbolically thus:

$$\boldsymbol{N} \div (\pm\boldsymbol{\infty}) = \mathbf{0}.$$

21. If $N \div M = 0$, and $M \neq \infty$, then by Art. 7, $N = 0$.

22. The consideration of other relations which involve 0 and ∞ is deferred.

It is important to notice that the relation

$$0 \times a = 0$$

of § 3, Art. 18, was proved only for the case in which a is *finite*.

CHAPTER IV.

INTEGRAL ALGEBRAIC EQUATIONS.

An equation has been defined (Ch. I., § 1, Art. 12) as a statement that two numbers or expressions are equal.

We must now distinguish between two kinds of equations.

§ 1. IDENTICAL EQUATIONS.

1. Examples of the one kind are:

$$(a + b)(a - b) = a^2 - b^2. \qquad (1)$$

$$(a^2 - b^2) \div (a - b) = (a + b)^2 \div (a + b). \qquad (2)$$

The first member of (1) is reduced to the second member by performing the indicated multiplication. Both members of (2) are reduced to the common form, $a + b$, by performing the indicated divisions. Such equations are called *identical equations.*

2. An **Identical Equation**, or simply an **Identity**, is an equation one of whose members can be reduced to the other, or both of whose members can be reduced to a common form, by performing the indicated operations.

3. Notice that identical equations are true for all values that may be substituted for the literal numbers involved.

E.g., If $a = 5$ and $b = 3$, equation (1) becomes

$$8 \times 2 = 25 - 9, \text{ or } 16 = 16;$$

and equation (2) becomes

$$16 \div 2 = 64 \div 8, \text{ or } 8 = 8.$$

If $a = 7$ and $b = 4$, equation (1) becomes

$$11 \times 3 = 49 - 16, \text{ or } 33 = 33;$$

and equation (2) becomes

$$33 \div 3 = 121 \div 11, \text{ or } 11 = 11.$$

More generally, if $a = 3u$ and $b = u$, equation (1) becomes

$$4u \times 2u = 9u^2 - u^2, \text{ or } 8u^2 = 8u^2;$$

and equation (2) becomes

$$8u^2 + 2u = 16u^2 + 4u, \text{ or } 4u = 4u.$$

We need not further discuss identical equations, since we have constantly dealt with them in the preceding chapters.

§ 2. CONDITIONAL EQUATIONS.

1. Examples of the second kind are:

$x + 1 = 3.$ (1) $x^2 - 1 = 8.$ (2) $x + y = 5.$ (3)

The first member, $x + 1$, of (1) reduces to the second member, 3, when $x = 2$, as can be seen by inspection. It seems evident, and we shall later prove, that $x + 1$ reduces to 3 *only* when $x = 2$.

The first member, $x^2 - 1$, of (2) reduces to the second member, 8, when $x = +3$ and when $x = -3$, as can be seen by inspection. It seems evident, and we shall later prove, that $x^2 - 1$ reduces to 8 *only* when $x = +3$ or -3.

The first member, $x + y$, of (3) reduces to the second member, 5, when $x = 1$ and $y = 4$, when $x = -3$ and $y = 8$; but *not* when $x = 5$ and $y = 6$, when $x = -4$ and $y = 8$. Therefore, equation (3) is true for many pairs of values of x and y, but not for all pairs of values chosen at random.

2. Such equations *impose conditions* upon the values of the literal numbers involved. Equation (1) imposes the condition that if 1 be added to the value of x, the sum will be 3; equation (2) imposes the condition that if 1 be subtracted from the square of the value of x, the remainder will be 8; equation (3) imposes the condition that x and y must have such values that their sum shall be 5. Such equations are called *conditional equations.*

A Conditional Equation is an equation one of whose members can be reduced to the other only for certain definite values of one or more letters contained in it.

Whenever the word *equation* is used in subsequent work, we shall understand by it a *conditional equation*, unless the contrary is expressly stated.

3. The **Unknown Numbers** of an equation are the numbers whose values are fixed or determined by the equation.

The **Known Numbers** of an equation are the numbers whose values are given or known.

Thus, in the equations

$$x + 1 = 3 \text{ and } x^2 - 1 = 8$$

the unknown number is x, and the known numbers are 1 and 3, and 1 and 8, respectively.

In the equation $x + y = 3$

the unknown numbers are x and y; the known number is 3.

The unknown numbers are usually represented by the final letters of the alphabet, x, y, z, etc., as in the above examples.

4. An **Integral Algebraic Equation** is an equation whose members are integral algebraic expressions in the unknown number or numbers.

The known numbers may enter in any way whatever.

E.g., $3x^2 - 4 = 2x$, and $\frac{2}{3}x + 5y = \frac{4}{5}$, are integral equations.

5. The **Degree** of an integral equation is the degree of its term of highest degree in the unknown number or numbers.

6. A **Linear** or **Simple Equation** is an equation of the *first* degree.

E.g., $x + 1 = 6$ is a linear equation in one unknown number, x; $2x + 3y = 5$ is a linear equation in two unknown numbers, x and y.

7. A **Solution** of an equation is a value of the unknown number, or a set of values of the unknown numbers, which, if substituted in the equation, converts it into an identity.

E.g., 2 is a solution of the equation

$$x + 1 = 3,$$

since, when substituted for x in the equation, it converts the equation into the identity

$$2 + 1 = 3.$$

Likewise, 3 and -3 are solutions of the equation

$$x^2 - 1 = 8,$$

since, when substituted for x in the equation, they both convert the equation into the identity

$$9 - 1 = 8.$$

The set of values 1 and 2, of x and y, respectively, is a solution of the equation

$$x + y = 3,$$

since, when this set of values of x and y is substituted in the equation, it converts the equation into the identity

$$1 + 2 = 3.$$

8. To **Solve** an equation is to find its solution. The process of solving an equation is also frequently called *the solution* of the equation.

An equation is said *to be satisfied by its solution*, or *the solution is said to satisfy the equation*, since it converts the equation into an identity.

9. When the equation contains only one unknown number, a solution is frequently called a **Root** of the equation.

E.g., 2 is a root of the equation $x + 1 = 3$.

Likewise, 3 and -3 are roots of the equation $x^2 - 1 = 8$.

10. We shall now give some principles upon which the solution of integral equations depends. But it is to be kept in mind that the final test of the correctness of a solution, no matter how obtained, *is that it shall satisfy the given equation.*

§ 3. EQUIVALENT EQUATIONS.

1. *Two equations are* **equivalent** *when every solution of the first is a solution of the second, and every solution of the second is a solution of the first.* Thus, the equations

$$5x + 2 = 3x + 6;\ 2x + 2 = 6;\ x + 1 = 3$$

are equivalent equations, since they are all three satisfied by the same root 2, and, as we shall later prove, *only* by 2.

2. The methods of solving integral equations depend upon principles which enable us to change a given equation into an equivalent equation whose solution is more easily obtained than that of the given one. This process is called *transforming the equation*, or *the transformation of the equation*.

The equations $3x - 7 = 17 - 5x$ (1)

and $2x = 6$ (2)

are equivalent equations, since they are satisfied by the same root 3. But the root 3 can evidently be more easily obtained from (2) than from (1).

Fundamental Principles for Solving Integral Equations.

3. In the principles of equivalent equations which we shall now prove, the solutions are limited to finite values.

4. Addition and Subtraction. — If 3 be added to both members of the equation $3x + 1 = 7$, we obtain the equivalent equation

$$3x + 4 = 10.$$

For evidently both equations are satisfied by the same root 2.

If $2 - x$ be added to both members of the same equation, we obtain the equivalent equation

$$3x + 1 + 2 - x = 7 + 2 - x, \text{ or } 2x + 3 = 9 - x.$$

For the latter equation is also satisfied by the same root 2.

The preceding examples illustrate the following principle:

If the same number or expression be added to, or subtracted from, both members of an equation, the equation thus derived will be equivalent to the given one.

Let $P = Q$

be the given equation, and N be any number or expression. Then the equation

$$P \pm N = Q \pm N,$$

wherein the upper signs, +, go together and the lower signs, −, go together, is equivalent to the given one.

For any solution of the given equation makes P equal to Q. Therefore, by Ch. II., § 1, Art. 19 and § 2, Art. 21, that solution makes $P \pm N$

equal to $Q \pm N$, and hence is a solution of the derived equation. Consequently no solution is lost by the transformation.

But the given equation is obtained from the derived equation by subtracting the number or expression which was added, or by adding the number or expression which was subtracted, in forming the derived equation. Therefore any solution of the derived equation is a solution of the given equation, and no solution is gained by the transformation. Consequently, the two equations are equivalent.

Applications.

5. *If any term be transferred from one member of an equation to the other, its sign being reversed from + to −, or from − to +, the derived equation will be equivalent to the given one.*

E.g., $2x - 4 = x + 1$ and $2x - x = 1 + 4$

are equivalent equations. This step is equivalent to adding 4 to, and subtracting x from, both members of the given equation.

From the derived equation we obtain $x = 5$.

It is often convenient to have all the terms of an equation in one and the same member, usually the first, the second member then being 0.

Thus, the equation $$x - 2 = -3x + 6$$

becomes $$x + 3x - 2 - 6 = 0,$$

or $$4x - 8 = 0,$$

when the terms in the second member, $-3x$ and $+6$, are transferred to the first member and their signs are reversed.

6. *If equal terms be dropped from both members of an equation, the derived equation will be equivalent to the given one.*

E.g., $2x - 3 + 8 = x - 3$ and $2x + 8 = x$

are equivalent equations.

Likewise $5x - 4x = 10 - 4x$ and $5x = 10$ are equivalent equations.

This step is called *cancellation of equal terms.*

7. *If the signs of the terms of an equation be reversed, the derived equation will be equivalent to the given one.*

E.g., $5x - 3 = 9 - x$ becomes $-5x + 3 = -9 + x$,

when each term of the first member is transferred to the second member, and each term of the second member is transferred to the first member (the signs of the terms being reversed), and when, as is evidently legitimate, the two members are interchanged. This step is also equivalent to multiplying both members of the equation by -1, which the next principle will prove to be legitimate.

3. Multiplication. — If both members of the equation

$$\tfrac{1}{2}x - 3 = 6$$

be multiplied by 2, we obtain the equivalent equation

$$x - 6 = 12.$$

For evidently both equations are satisfied by the root 18. If both members of the equation

$$x^2 - 5x + 6 = 0$$

be multiplied by 7, we obtain the equivalent equation

$$7(x^2 - 5x + 6) = 7.0 = 0.$$

For the given equation and the derived equation are satisfied by the roots 2 and 3.

The preceding examples illustrate the following principle:

If both members of an equation be multiplied by one and the same number, not 0, or by an expression which does not contain the unknown number or numbers, the resulting equation will be equivalent to the given one.

It is more convenient to prove this principle when all the terms of the equation are in the same member, say the first. The latter equation is, as we have seen, equivalent to the given one. Then any solution must make the two members identical in value, *i.e.* must reduce the first member to 0.

Let $$P = 0$$

be the given equation, and N be any number, not 0, or any expression which does not contain the unknown number or numbers. Then the equation

$$N \cdot P = N \cdot 0 = 0$$

is equivalent to the given one.

For any solution of the given equation must reduce P to 0, and, therefore, by Ch. III., § 3, Art. 18, must also reduce $N \cdot P$ to 0. Hence it is also a solution of the derived equation. That is, no solution is lost by the transformation.

Any solution of the derived equation must reduce $N \cdot P$ to 0. But N is not 0, and, since it does not contain the unknown number or numbers, it cannot reduce to 0 for any value of the unknown number or numbers. Consequently, by Ch. III., § 3, Art. 20, any solution of the derived equation must reduce P to 0, and hence is a solution of the given equation. That is, no solution is gained by the transformation. Consequently, the two equations are equivalent.

Notice that if the multiplier were 0, any value of the unknown number would be a solution of the derived equation, but not of the given equation.

E.g., $2x - 6 = 0$ has the root 3, while

$$(2x - 6) \times 0 = 0$$

is evidently satisfied by 1, 2, 3, 4, etc., without end.

If the multiplier contain the unknown number or numbers, values of the unknown number or numbers will reduce the multiplier to 0, and therefore the first member of the derived equation to 0, without reducing the first member of the given equation to 0.

E.g., $2x - 6 = 0$ has the root 3, while

$$(2x - 6)(x - 2) = 0$$

is satisfied not only by 3, since $(6 - 6) \times 1 = 0 \times 1 = 0$, but also by 2, since $(4 - 6)(2 - 2) = (-2) \times 0 = 0$.

But 2 is not a solution of the given equation. That is, in multiplying both members of the given equation by $x - 2$, we have gained a root 2.

The derived equation is, therefore, not equivalent to the given one, since it has the additional root 2.

9. Division. — If both members of the equation

$$3x - 6 = 15$$

be divided by 3, we obtain the equivalent equation

$$x - 2 = 5.$$

For evidently both equations are satisfied by the root 7.

If both members of the equation

$$x^2 - 5x + 6 = 0$$

be divided by 2, we obtain the equivalent equation

$$\tfrac{1}{2}(x^2 - 5x + 6) = \tfrac{1}{2} \times 0 = 0.$$

For the given equation and the derived equation are evidently satisfied by the roots 2 and 3.

The preceding examples illustrate the following principle:

If both members of an equation be divided by one and the same number, not 0, or by an expression which does not contain the unknown number or numbers, the derived equation is equivalent to the given one.

Let us assume that all the terms of the given equation have been transferred to the first member. Then any solution must reduce this first member to 0.

Let $$P = 0$$

be the given equation, and N be any number, not 0, or an expression which does not contain the unknown number or numbers. Then the equation

$$P \div N = 0 \div N = 0$$

is equivalent to the given one.

For any solution of the given equation must reduce P to 0, and therefore, by Ch. III., § 4, Art. 6, must also reduce $P \div N$ to 0. Hence it is also a solution of the derived equation. That is, no solution is lost by the transformation.

Any solution of the derived equation must reduce $P \div N$ to 0. But N is finite and not 0, and, since it does not contain the unknown number, it cannot reduce to 0, or become infinite, for any value of the unknown number or numbers. Consequently, by Ch. III., § 4, Art. 7, any solution of the derived equation must reduce P to 0, and hence is a solution of the given equation. That is, no solution is gained by the transformation. Consequently the two equations are equivalent.

Notice that if the divisor be an expression which contains the unknown number or numbers, one or more solutions are lost.

E.g., the equation $x^2 - 1 = 2(x + 1)$

is satisfied by the two roots -1 and 3.

But if both members be divided by $x+1$, we obtain

$$x-1=2.$$

This equation is satisfied by 3, but not by -1.

The derived equation is, therefore, not equivalent to the given one, since it does not have the root -1.

Applications.

10. *An equation having fractional coefficients can be transformed into an equivalent one with integral coefficients, by multiplying both members by the L.C.M. of the denominators of the fractions.* This step is called *clearing the equation of fractions.*

E.g., if both members of the equation

$$\tfrac{1}{2}x-\tfrac{3}{4}=\tfrac{1}{3}x+\tfrac{1}{6}$$

be multiplied by 12, we obtain the equivalent equation

$$6x-9=4x+2.$$

11. *If all the terms of an equation have a common factor which does not contain the unknown number or numbers, the equation derived by dividing both members by this factor is equivalent to the given one.*

E.g., if both members of the equation

$$3x+6=9x-30$$

be divided by 3, we obtain the equivalent equation

$$x+2=3x-10.$$

12. *Any equation can be transformed into an equivalent one in which the coefficient of any term is 1, by dividing both members of the equation by the coefficient of that term.*

E.g., if both members of the equation

$$5x^2-11x+6=0$$

be divided by 5, we obtain the equivalent equation

$$x^2-\tfrac{11}{5}x+\tfrac{6}{5}=0;$$

if both members be divided by -11, we obtain the equivalent equation

$$-\tfrac{5}{11}x^2+x-\tfrac{6}{11}=0.$$

13. The preceding principles apply to integral equations of any degree. In this chapter we shall confine our attention to linear equations, in one unknown number.

§ 4. LINEAR EQUATIONS, IN ONE UNKNOWN NUMBER.

1. Ex. 1. Solve the equation $17x + 6 = 10x + 27$.

Transferring $10x$ to the first member and 6 to the second member, we have
$$17x - 10x = 27 - 6.$$

Uniting like terms, we obtain $7x = 21$.

Dividing both members by 7, we have $x = 3$.

Check. — Substituting 3 for x in the given equation, we obtain the identity
$$51 + 6 = 30 + 27.$$

Ex. 2. Solve the equation $14 - 8x = 19 - 3x$.

Transferring terms, $-8x + 3x = 19 - 14$.

Uniting like terms, $-5x = 5$.

Dividing by -5, $x = -1$.

Check. — Substituting -1 for x in the given equation, we obtain the identity
$$14 + 8 = 19 + 3.$$

Ex. 3. Solve the equation
$$15x - 14(10 - 7x) = 5x + 7(14x - 25).$$

Removing parentheses, $15x - 140 + 98x = 5x + 98x - 175$.

Canceling equal terms, $15x - 140 = 5x - 175$.

Transferring terms, $15x - 5x = -175 + 140$.

Uniting like terms, $10x = -35$.

Dividing by 10, $x = -\frac{7}{2}$.

Check. — Substituting $-3\frac{1}{2}$ for x in the given equation, we obtain the identity
$$-\tfrac{105}{2} - 14(10 + \tfrac{49}{2}) = -\tfrac{35}{2} + 7(-49 - 25), \text{ or } -535\tfrac{1}{2} = -535\tfrac{1}{2}.$$

In thus solving equations, the student should accustom himself to perform the simpler steps mentally. In the preceding examples, the steps of transferring terms and uniting like terms should be performed simultaneously.

Ex. 4. Solve the equation $2(x+1)=3[2-4(x-1)]+6x$.

Removing parentheses, $2x+2=6-12x+12+6x$.

Transferring and uniting terms, $8x=16$.

Dividing by 8, $x=2$.

Check. — Substituting 2 for x in the given equation, we obtain the identity

$$2(2+1)=3[2-4(2-1)]+12, \text{ or } 6=6.$$

Ex. 5. Solve the equation $\frac{1}{2}(x+5)-\frac{1}{3}x=\frac{1}{4}(3x-1)+1$.

Multiplying both members by 12, the lowest common multiple of the fractional coefficients, we obtain

$$6(x+5)-4x=3(3x-1)+12.$$

Removing parentheses, $6x+30-4x=9x-3+12$.

Transferring and uniting terms, $-7x=-21$.

Dividing by -7, $x=3$.

Check. — Substituting 3 for x in the given equation, we obtain the identity

$$\tfrac{1}{2}(3+5)-\tfrac{1}{3}\times 3=\tfrac{1}{4}(9-1)+1, \text{ or } 3=3.$$

As the student advances in his work he will be able, in examples like the above, to clear of fractions and remove parentheses simultaneously. He will also learn many devices for shortening his work.

Ex. 6. Solve the equation $3\frac{1}{2}(x+1)+4\frac{1}{2}(x+1)=16$.

Uniting terms in the first member, without clearing of fractions or removing parentheses, we have

$$8(x+1)=16.$$

Dividing by 8, $x+1=2$;

whence $x=1$.

Check. — Substituting 1 for x in the given equation, we obtain the identity

$$3\tfrac{1}{2}\times 2+4\tfrac{1}{2}\times 2=16, \text{ or } 16=16.$$

The beginner should check his work until he acquires some degree of confidence in his ability to solve equations readily and accurately.

2. The following general directions will be found useful in preparing an equation for solution:

(i.) *If there be any fractional coefficients in the equation, remove them by multiplying both sides of the equation by the L.C.M. of the denominators of the fractions.*

(ii.) *If there be any parentheses, remove them, especially such as contain the unknown number.*

(iii.) *Transfer all terms containing unknown numbers to one member of the equation, usually to the first member, and all the terms containing known numbers to the other member.*

(iv.) *Unite like terms. An equation thus prepared for solution is called the* **Normal Form** *of that equation.*

The preceding suggestions apply also to an integral equation of any degree. If the equation be linear in one unknown number, the solution is completed by dividing both members by the coefficient of the unknown number.

EXERCISES.

Solve each of the following equations:

1. $x + 2 = 3.$ **2.** $x - 4 = 7.$ **3.** $7 - x = 4.$

4. $15 - x = -27.$ **5.** $9 = 4 + x.$ **6.** $17 = 9 - x.$

7. $5x + 7 = 11 + 4x.$ **8.** $5x - 7 = 4x + 3.$

9. $\frac{1}{2}x + 8 = -\frac{1}{2}x - 1.$ **10.** $7x + 8 = 4x + 15 + 2x.$

11. $12x + 12 = 13x + 15.$ **12.** $8x + 19 = -5x - 4x + 11.$

13. $5x + 7 - 3x = 8x - 5x + 9.$ **14.** $-7x - 2 + 3x = -x - 4x + 3.$

15. $15x + 4 + 7x = 14x + 5 + 7x.$ **16.** $3x - 5 - 9x = 2x - 7 - 9x.$

17. $-3x - 7 = -4x - 7.$ **18.** $15x - 8 = 20x - 8 - 4x.$

19. $\frac{1}{4}x = 6.$ **20.** $\frac{1}{5}x = -4.$ **21.** $3 = -\frac{1}{2}x.$

22. $-7 = \frac{1}{3}x.$ **23.** $-2 = -\frac{1}{5}x.$ **24.** $\frac{1}{8}x = 0.$

25. $\frac{1}{8}x + 8 = 5.$ **26.** $\frac{1}{7}x - 5 = -8.$ **27.** $-\frac{1}{11}x - 12 = 4.$

28. $\frac{1}{3}(x+5)=4.$

29. $\frac{1}{2}(x-8)=7.$

30. $-\frac{1}{5}(x-6)=7.$

31. $\frac{5}{3}x=5+2x.$

32. $-\frac{15}{8}x=-3-2x.$

33. $5-2x=-\frac{15}{7}x.$

34. $\frac{2x}{3}-5=\frac{1}{3}x+2.$

35. $\frac{1}{2}x+2=-\frac{2}{3}x-5+x.$

36. $-\frac{3}{5}x+4+\frac{5}{8}x=4x-7-3x.$

37. $5x=15.$

38. $24=3x.$

39. $4x=-16.$

40. $11=-22x.$

41. $7x=0.$

42. $28x=7.$

43. $8=-56x.$

44. $-9x=36.$

45. $6x=4.$

46. $16x-11=7x+70.$

47. $3x+10=5x-70.$

48. $36-9z=116+11z.$

49. $61-5y=7y+85.$

50. $18x+4=-4+34x.$

51. $7x+4=-4-5x.$

52. $8x-18=x+12-3x.$

53. $5x+11=16-3x-4x.$

54. $37+13x-15+2x=19+36x-25.$

55. $17x-23+14x+13=29x-18-5x+127.$

56. $204-3x+47+28x=25x-19+13x-16.$

57. $x-7=\frac{1}{5}x+\frac{1}{3}x.$

58. $\frac{1}{2}x+\frac{1}{3}x=\frac{1}{4}x-7.$

59. $36-\frac{4}{9}x=8.$

60. $-x+\frac{1}{2}x+\frac{1}{3}x=11.$

61. $3x+16=\frac{5}{3}x.$

62. $\frac{1}{2}x+\frac{3}{4}x-\frac{5}{6}x=15.$

63. $\frac{3}{4}x-5=\frac{5}{6}x+2.$

64. $\frac{7}{8}x-3=\frac{9}{10}x-8.$

65. $\frac{1}{2}x+\frac{1}{3}x+\frac{1}{4}x+\frac{1}{5}x=x-17.$

66. $2x-(5x+5)=7.$

67. $7x-(3x-11)=4.$

68. $3x-7-(5x+17)=0.$

69. $3(x+1)=-5(x-1).$

70. $7(x-18)=3(x-14).$

71. $7x-3(2x-3)=2(-x-18).$

72. $5(x+1)+6(x+2)=9(x+3).$

73. $4(x-3)-7(x-4)=6-x.$

74. $\frac{1}{4}(x+3)=\frac{1}{16}(3x+16).$

75. $\frac{1}{3}(5x-2)-6=\frac{1}{5}(4x-3).$

76. $\frac{1}{2}(7-x)-3=\frac{1}{5}(3-4x)-4.$

77. $\frac{1}{4}(x-5)+6x=\frac{1}{5}(284-x).$

78. $\frac{1}{2}(19+x)=-x+\frac{1}{3}(11+x).$

79. $\frac{1}{3}(x-2)-\frac{1}{2}(12-x)=\frac{1}{4}(5x-36)-1.$

80. $\frac{1}{4}(x+8)-\frac{1}{3}(5x+2)=\frac{1}{2}(14-x)-2.$

81. $\frac{1}{10}(1-x)-\frac{1}{12}(5x-1)=2-\frac{3}{4}(x-1).$

82. $\frac{1}{12}(7+3x)-\frac{2}{27}(x-5)=1-\frac{1}{18}x.$

83. $\frac{1}{4}(x-15)-\frac{1}{21}(7-2x)=\frac{3}{14}x+\frac{1}{2}.$

84. $4x - 2(2 - x) = 6.$ 85. $6x - [7x - (8x - 18)] = 16.$

86. $4[4(x - 4) - 3] = 4.$ 87. $5[4 - (3x - 1)] = 6(x - 11) + 49.$

88. $\frac{1}{4}(x - 2) + \frac{1}{8} - [x - \frac{1}{8}(2x - 1)] = 0.$

89. $\frac{1}{9}[3x - 6 - 5(\frac{7}{2}x - 5)] + \frac{1}{4}[52(x - 5) + 1] = 0.$

90. $3\frac{1}{8}[28 - (\frac{1}{8}x + 24)] = 3\frac{1}{2}(2\frac{1}{8} + \frac{1}{4}x).$

91. $\frac{1}{8}[x - \frac{1}{9}(2x - 36)] - \frac{1}{6}(x - 18) = x + 9 - \frac{1}{4}[5x - \frac{2}{13}(x - 10)].$

92. $2(x + 1) - 3(x + 1) + 9(x + 1) + 18 = 7(x + 1).$

93. $5(3x - 5) - 17 - 8(3x - 5) - 2(3x - 5) = 3.$

94. $-17(7x - 83) + 28(7x - 83) - 34 = 12(7x - 83).$

95. $(2x + 7)(x - 3) = (x - 3)(2x + 8).$

96. $(x + 1)(x + 2) = (x - 3)(x - 4).$

97. $(16x + 5)(9x + 31) = (4x + 14)(36x + 10).$

98. $(5x - 2)(3x - 4) = (3x + 5)(5x - 6).$

99. $x(x + 2) + x(x + 1) = (2x - 1)(x + 3).$

100. $x^2 - x[1 - x - 2(3 - x)] = x + 1.$

101. $(x + 1)(x + 1) = [111 - (1 - x)]x - 80.$

102. $3 - x = 2(x - 1)(x + 2) + (x - 3)(5 - 2x).$

103. $4[8x - 5(7 - 4x) + 9(6 - 3x) + 12x]$
$= 7[20x - 2(7x - 10) - 2].$

Find the remainder of each of the following divisions, and hence the value of m which will make the dividend exactly divisible by the divisor:

104. $(9x^2 - 3x + m) \div (x - 1).$

105. $[4x^3 - 2x^2 + x - \frac{3}{8}(m + 1)] \div (2x + 3).$

106. $[7x^2 - (m - 1)x + 3] \div (x + 2).$

CHAPTER V.

PROBLEMS.

1. A **Problem** is a question proposed for solution.

Pr. 1. The greater of two numbers is three times the less, and their sum is 84. What are the numbers?

This problem involves the *given* number 84 and two *required* numbers. The statements of the problem impose two conditions upon the values of the required numbers:

(i.) *The greater number is three times the less.*

(ii.) *The sum of the two numbers is* 84.

To solve the problem, it is necessary first to translate these relations or conditions from the verbal language of the problem into the symbolic language of Algebra, *i.e.* to express them by means of algebraic signs and symbols.

We let one of the required numbers be represented by a letter, say x. We then use one of the conditions of the problem to express the second required number in terms of x and the given number 84.

Let x stand for the less required number.

Then, by the first condition, the greater number is,

in *verbal* language: *three times the less;*

in *algebraic* language: $3x$.

Consequently, the required numbers are represented by x (the less) and $3x$ (the greater).

The second condition is,

in *verbal* language: *the less number plus the greater is equal to* 84;

in *algebraic* language: $x + 3x = 84$.

This equation is called the *equation of the problem.*

From this equation we obtain $x = 21$, the less number.

Therefore $3x = 63$, the greater number.

Notice that this problem could have been solved by letting x stand for the greater number, and consequently $\frac{1}{3}x$ for the less. The resulting equation would then have been

$$\frac{1}{3}x + x = 84.$$

Whence $x = 63$, the greater number; and therefore, $\frac{1}{3}x = 21$, the less.

It is strongly recommended that at first the student write, in as concise verbal language as possible, each condition of the problem, especially when the wording is in any way involved. Then underneath this write the corresponding algebraic statement, as in the above solution.

This method leads to an equation in which the *unknown* number is one of the *required* numbers of the problem.

Pr. 2. Find two consecutive integers whose sum is 163.

In this problem the conditions are not both *explicitly* stated. The first condition is contained in the words, *two consecutive integers.*

Let x stand for the less number.

Then, by the first condition, the greater number is,

in *verbal* language: *the less number plus* 1;

in *algebraic* language: $x + 1$.

The required numbers are thus represented by x (the less) and $x + 1$ (the greater).

The second condition is,

in *verbal* language: *the less number plus the greater is equal to* 163;

in *algebraic* language: $x + (x + 1) = 163$, the equation of the problem.

From this equation we obtain $x = 81$, the less number; and therefore $x + 1 = 82$, the greater.

Let the student solve this problem, taking x as the greater number.

Pr. 3. A is 40 years old and B is 10 years old. After how many years will A be three times as old as B?

Let x stand for the required number of years, after which A will be three times as old as B.

The condition of the problem involves other unknown numbers than the required number. These we first express in terms of the required and given numbers.

In x years the *number of years in A's age* will be $40 + x$; the *number of years in B's age* will be $10 + x$.

The condition of the problem is,

in *verbal* language: *the number of years in A's age x years hence is equal to three times the number of years in B's age x years hence;*

in *algebraic* language: $40 + x = 3(10 + x)$, the equation of the problem.

From this equation we obtain $x = 5$, the required number. In 5 years A will be 45 years old, and B will be 15 years old; and $45 = 3 \times 15$.

Notice that the numbers used in the solution are *abstract* numbers. Thus, 40 is the *number of years* in A's age, not A's age.

Let the student solve the problem, taking x as the number of years in the father's age.

Pr. 4. A father divided $ 5000 between his two sons. Four times the amount received by the older son exceeds five times the amount received by the younger by $ 2000. How much did each son receive?

Let x stand for the number of dollars received by the older son.

Then, by the first condition, the number of dollars received by the younger son is,

in *verbal* language: 5000 *minus the number received by the older son;*

in *algebraic* language: $5000 - x$.

The unknown numbers involved in the second condition, expressed in terms of the known and the required numbers, are: *four times the number of dollars received by the older son*, or $4x$; and *five times the number of dollars received by the younger son*, or $5(5000 - x)$.

The second condition is,

in *verbal* language: *four times the number of dollars received by the older son minus five times the number of dollars received by the younger is* 2000;

in *algebraic* language: $4x - 5(5000 - x) = 2000$, the equation of the problem.

From this equation we obtain $x = 3000$; whence

$$5000 - x = 2000.$$

Therefore the older son received \$3000, and the younger \$2000.

Pr. 5. At an election at which 943 votes were cast, A and B were candidates. A received a majority of 65 votes. How many votes were cast for each candidate?

Let x stand for number of votes cast for A.

Then, by the first condition, the number of votes cast for B is,

in *verbal* language: 943 *minus the number cast for A*;

in *algebraic* language: $943 - x$.

The second condition is,

in *verbal* language: *the number of votes cast for A exceeds the number cast for B by* 65;

in *algebraic* language: $x - (943 - x) = 65$, the equation of the problem.

From this equation we obtain $x = 504$, whence $943 - x = 439$. A therefore received 504 votes, and B 439 votes.

This problem can also be solved by using the second condition to express the second required number in terms of the first.

Let x, as before, stand for the number of votes cast for A. Then, by the second condition, the number of votes cast for *B* is,

in *verbal* language: *the number of votes cast for A minus* 65;

in *algebraic* language: $x - 65$.

The first condition is,

in *verbal* language: *the number of votes cast for A plus the number of votes cast for B is 943*;

in *algebraic* language: $x + x - 65 = 943$, the equation of the problem.

Whence, $x = 504$, as before.

Pr. 6. Fifteen coins, dollars and quarter-dollars, amount to \$ 7.50. How many coins of each kind are there?

We take one dollar as the unit, and express parts of dollars as fractional parts of this unit.

Let x stand for the number of dollars.

Then, by the first condition, the number of quarter-dollars is,

in *verbal* language: 15 *minus the number of dollars;*

in *algebraic* language: $15 - x$.

The second condition is,

in *verbal* language: *the number of dollars plus one-fourth of the number of quarter-dollars is* $7\frac{1}{2}$;

in *algebraic* language: $x + \frac{1}{4}(15 - x) = 7\frac{1}{2}$, the equation of the problem.

From this equation we obtain $x = 5$; whence $15 - x = 10$. Therefore there are 5 dollars and 10 quarter-dollars. Evidently the total value of the coins is $5 + \frac{10}{4}$ dollars, or \$ $7\frac{1}{2}$.

Notice that both conditions refer to abstract numbers; the first condition to the number of coins, the second to the number of dollars.

Pr. 7. A drove of sheep and goats, 200 animals in all, is to be sold. A offers to pay \$ 1.25 for each sheep and \$ 1.60 for each goat; B offers to pay \$ 1.50 for each animal. The owner of the drove accepts B's offer because he finds that it will net him \$ 22 more than A's offer. Find the number of sheep and goats in the drove.

Let x stand for the number of sheep.

Then, by the first condition, the number of goats is,

in *verbal* language: 200 *minus the number of sheep;*

in *algebraic* language: $200 - x$.

The second condition involves other *unknown* numbers than the required numbers. We must express the number of dollars in A's offer and the number of dollars in B's offer in terms of the required and given numbers.

The number of dollars in A's offer is,

in *verbal* language: *the number of sheep multiplied by the number of dollars offered for each sheep, plus the number of goats multiplied by the number of dollars offered for each goat;*

in *algebraic* language: $1.25\,x + 1.6\,(200 - x)$.

The number of dollars in B's offer is,

in *verbal* language: *the number of animals multiplied by the number of dollars offered for each animal;*

in *algebraic* language: 200×1.5.

The second condition is,

in *verbal* language: *the number of dollars in B's offer minus the number of dollars in A's offer is 22;*

in *algebraic* language: $200 \times 1.5 - [1.25x + 1.6(200 - x)] = 22$, the equation of the problem.

From this equation we obtain $x = 120$, the number of sheep; whence $200 - x = 80$, the number of goats.

$$\text{A's offer} = 1.25 \times 120 + 1.6 \times 80, = 278 \text{ dollars};$$

and $$\text{B's offer} = 200 \times 1.5, = 300 \text{ dollars}.$$

Notice again that both conditions refer to abstract numbers; the first to the number of animals, the second to the number of dollars.

Pr. 8. A box contains a certain number of pencils, of which one-third are red, one-sixth are blue, and 15 are black. How many of the pencils are red, and how many are blue?

Let x stand for the number of red pencils.

Then, by the first condition, the total number of pencils is,

in *verbal* language: *three times the number of red pencils;*

in *algebraic* language: $3\,x$.

By the second condition, the number of blue pencils is,

in *verbal* language: $\frac{1}{6}$ *the total number of pencils;*

in *algebraic* language: $\frac{1}{6} \times 3\,x = \frac{1}{2}\,x$.

The third condition is not stated explicitly. It is, however,
in *verbal* language: *the number of red pencils, plus the number of blue pencils, plus the number of black pencils, is equal to the total number of pencils;*
in *algebraic* language: $x + \frac{1}{2}x + 15 = 3x$, the equation of the problem.

From this equation, we obtain $x = 10$, the number of red pencils; whence $\frac{1}{2}x = 5$, the number of blue pencils.

This problem could have been solved more readily by assuming for the *unknown* number of the equation another number than one of the required numbers.

Let x stand for the total number of pencils.

Then, by the first condition, the number of red pencils is $\frac{1}{3}x$, and by the second condition, the number of blue pencils is $\frac{1}{6}x$.

Finally, by the third condition,

$$\frac{1}{3}x + \frac{1}{6}x + 15 = x.$$

Whence $x = 30$. Therefore

$\frac{1}{3}x = 10$, the number of red pencils,

and $\frac{1}{6}x = 5$, the number of blue pencils.

Pr. 9. A number is composed of two digits whose sum is 8. If the digits be interchanged, the resulting number will be greater by 18 than the original number. What is the number?

In accordance with the suggestion in Pr. 8, we assume one of the digits of the required number, not the required number, as the unknown number of the equation of the problem.

Let x stand for the digit in the units' place.

Then, by the first condition, the digit in the tens' place is,

in *verbal* language: 8 *minus the digit in the units' place;*
in *algebraic* language: $8 - x$.

Notice that the value of a digit in the tens' place is the digit multiplied by 10.

Therefore the original number is $10(8 - x) + x$; the second number (when the digits are interchanged) is $10x + (8 - x)$.

The second condition of the problem then is,

in *verbal* language: *the second number is equal to the original number plus* 18;

in *algebraic* language: $10x+(8-x)=10(8-x)+x+18$, the equation of the problem.

From this equation, we obtain

$x=5$, the digit in the units' place;

whence $8-x=3$, the digit in the tens' place.

The original number is $10(8-x)+x$, $=35$; the second number is $10x+8-x$, $=53$, and $53-35=18$.

Pr. **10.** A carriage, starting from a point A, travels 35 miles daily; a second carriage, starting from a point B, 84 miles behind A, travels in the same direction 49 miles daily. After how many days will the second carriage overtake the first? At what distance from B will the meeting take place?

Let x stand for the number of days after which they meet. Then the number of miles traveled by the first carriage is $35x$, and the number of miles traveled by the second carriage is $49x$.

The condition of the problem is,

in *verbal* language: *the number of miles traveled by the first carriage is equal to the number of miles traveled by the second carriage minus* 84;

in *algebraic* language: $35x=49x-84$, the equation of the problem.

From this equation, we obtain $x=6$.

The distance traveled by the first carriage is 210 miles, and the distance traveled by the second carriage is 294 miles. They therefore meet 294 miles from B.

Pr. **11.** One man asked another what time it was, and received the answer: "It is between 5 and 6 o'clock, and the hour-hand is directly over the minute-hand." What time was it?

At 5 o'clock, the minute-hand points to 12 and the hour-hand to 5. The hour-hand is therefore 25 minute-divisions in advance of the minute-hand.

Let x stand for the number of minute-divisions passed over by the minute-hand from 5 o'clock until it is directly over the hour-hand between 5 and 6 o'clock.

By the first condition, which is implied in the problem, the number of minute-divisions passed over by the hour-hand is,

in *verbal* language: *the number of minute-divisions passed over by the minute-hand minus* 25;

in *algebraic* language: $x - 25$.

The second condition, which is also implied in the problem is,

in *verbal* language: *the number of minute-divisions passed over by the minute-hand is* 12 *times the number of minute-divisions passed over by the hour-hand;*

in *algebraic* language: $x = 12(x - 25)$, the equation of the problem.

From this equation we obtain $x = 27\frac{3}{11}$. Consequently, the two hands coincide at $27\frac{3}{11}$ minutes past 5 o'clock.

Observe that in this problem both conditions are implied in the problem, and are not explicitly stated.

2. The translation of the conditions of a problem into algebraic language is usually attended with some difficulty to the beginner. One reason for this is that the conditions in the problem are of necessity often stated in a more or less complicated way, or are not explicitely stated (see Pr. 2 and Pr. 11). The beginner will therefore find some suggestions useful.

(i.) *Read the problem thoughtfully to get an intelligent understanding of it as a whole.*

(ii.) *Observe what are the numbers whose values are required. It will, in general, be possible to continue the solution of the problem by representing one of these numbers by a letter, and operating upon or by that letter as if it were a known number.* See Prs. 1, 2, 3, 4, 5, 6, 7, 8 (first solution), 10, and 11.

This number is called the *unknown* number of the equation of the problem.

(iii.) *Every problem which can be solved must state, implicitly or explicitly, as many conditions as there are required numbers in the problem.*

(iv.) *Use all but one of the conditions to express the other required numbers in terms of the one selected as the unknown number, as in the problems solved.*

(v.) *The numbers involved in the statements will, in general, be not only the required numbers, but also other unknown numbers which must be expressed in terms of the required numbers.* See, in particular, Prs. 3, 4, 7, 9.

(vi.) *Express concisely in verbal language each given condition. It is frequently necessary to modify the statement in order to adapt it to translation into algebraic language.* See Prs. 2 and 11.

(vii.) *Translate each verbal statement of a condition into algebraic language. All but one of the conditions will give expressions for required numbers according to* (iv.). *The last condition will give the equation of the problem.*

(viii.) *There are problems in which other numbers than the required numbers can be used to better advantage in applying the conditions of the problems, and from which the required numbers can be readily found. In such problems it is better to take one of these numbers as the unknown number of the equation of the problem.* See Pr. 8 and Pr. 9.

3. In verbal language the relations existing between the different quantities can be expressed in a great variety of ways. Sometimes these relations are based upon the principles of other sciences (Geometry, Physics, Astronomy, etc.); or these relations are indicated by some technical expression (interest, discount, velocity, etc.). All this may increase the difficulty of forming an equation for the given problem.

4. In applying the suggestions of Art. 2, it is important to remember that the letter x always represents an abstract number. The beginner must never put x for distance, time, weight, etc., but for the *number* of miles, of hours, of pounds, etc.

Keep in mind also that in any one equation the magnitudes of all concrete quantities of the same kind must be referred to the same unit; if x refer to a certain number of yards, then all other distances must likewise represent numbers of yards, not of miles or of feet. So, if x stand for a number of dollars, all other numbers must denote numbers of dollars, not of dimes or of quarter-dollars.

The student should study the problems which have been worked in connection with the suggestions just made, and not be satisfied until he understands clearly the reason for each step taken.

EXERCISES.

1. If twice a number be added to 18, the sum will be 82. What is the number?

2. If 4 be subtracted from five times a number, the remainder will be 11. What is the number?

3. If one-fourth of a number be diminished by 5, the remainder will be 2. What is the number?

4. If three times a number be added to 12, the sum will be equal to five times the number. What is the number?

5. The sum of two consecutive even numbers is 34. What are the numbers?

6. Find the number whose double exceeds its half by 6.

7. Find three consecutive odd numbers whose sum is 57.

8. Divide 75 into two parts so that the greater shall exceed the less by 3.

9. If 48 be added to a number, the sum will be equal to nine times the number. What is the number?

10. If \$120 be divided between A and B so that A shall receive \$ 20 more than B, how many dollars will each receive?

11. A sum of \$2500 is divided between A and B. B receives \$ 4 as often as A receives \$ 1. How much does each receive?

12. Two buildings are worth $ 33,000. One-fourth and one-third of the value of the first is equal to seven-tenths of the value of the second. How much is each building worth?

13. Divide 80 into two parts, so that the second shall exceed three times the first by 4.

14. Divide 75 into two parts, such that three times the first part shall be 15 greater than seven times the second.

15. Divide 190 into three parts so that the second shall be three times the first, and the third five times the second.

16. A father's age exceeds his son's by 18 years, and the sum of their ages is four times the son's age. What are their ages?

17. A man bought a horse, a carriage, and harness for $ 320. The horse cost five times as much as the harness, and the carriage cost twice as much as the horse. How much did each cost?

18. The deposits in a bank during three days amounted to $ 16,900. If the deposits each day after the first were one-third of the deposits of the preceding day, how many dollars were deposited each day?

19. A merchant, after selling one-third, one-fourth, and one-sixth of a piece of silk, has 15 yards left. How many yards were there in the piece?

20. If two trains start together and run in the same direction, one at the rate of 20 miles an hour, and the other at the rate of 30 miles an hour, after how many hours will they be 250 miles apart?

21. A teacher proposes 16 problems to a pupil. The latter is to receive 5 marks in his favor for each problem solved, and 3 marks against him for each problem not solved. If the number of marks in his favor exceed those against him by 32, how many problems will he have solved?

22. A father divides $ 11,000 among his four sons. The second receives twice as much as the first; the third receives as much as the first two together, and the fourth receives as

much as the second and third together. How many dollars does each son receive?

23. A merchant paid 30 cents a yard for a piece of cloth. He sold one-half for 35 cents a yard, one-third for 29 cents a yard, and the remainder for 32 cents a yard, gaining $18.15 by the transaction. How many yards did he buy?

24. Two men start from points 100 miles apart and travel toward each other, one at the rate of 15 miles an hour, and the other at the rate of 10 miles an hour. After how many hours will they meet, and how far will their point of meeting be from the starting point of the first?

25. A father is 32 years old, and his son is 8 years old. After how many years will the father's age be twice the son's?

26. Divide 130 into five parts so that each part shall be 12 greater than the next less part.

27. A, traveling at the rate of 20 miles a day, has four days' start of B, who travels at the rate of 25 miles a day in the same direction. After how many days will B overtake A?

28. A sum of money is divided equally among four persons. If $60 more be divided equally among six persons, the shares will be the same as before. How many dollars are divided?

29. Atmospheric air is a mixture of four parts of nitrogen with one of oxygen. How many cubic feet of oxygen are there in a room 10 yards long, 5 yards wide, and 12 feet high?

30. The perimeter of a triangle measures 32 inches. The longest side is 8 inches longer than the shortest side, and the shortest side is 3 inches shorter than the third side. What are the lengths of the three sides?

31. A merchant paid $6.15 in an equal number of dimes and five-cent pieces. How many coins of each kind did he pay?

32. A man has $4.75 in dimes and quarters, and he has 5 more quarters than dimes. How many coins of each kind has he?

33. A leaves a certain town P, traveling at the rate of 21 miles in 5 hours; B leaves the same town 3 hours later and travels in the same direction at the rate of 21 miles in 4 hours. After how many hours will B overtake A, and at what distance from P?

34. The circumference of the front and hind wheels of a wagon are 2 and 3 yards, respectively. What distance has the wagon moved when the front wheel has made 10 revolutions more than the hind wheel?

35. The sum of two numbers is 47, and their difference increased by 7 is equal to the less. What are the numbers?

36. The sum of the ages of a father and his two sons is 100 years. The older son is 10 years older than the younger, and the father's age is equal to the sum of the ages of the two sons. What is the age of each?

37. The sum of three consecutive even numbers exceeds the least by 42. What are the numbers?

38. The sum of the two digits of a number is 4. If the digits be interchanged, the resulting number will be equal to the original one. What is the number?

39. A father is three times as old as his son, and 10 years ago he was five times as old as his son. What is the present age of each?

40. One barrel contained 48 gallons, and another 88 quarts of wine. From the first twice as much wine was drawn as from the second; the first then contained three times as much wine as the second. How much wine was drawn from each?

41. A child was born in November. On the 10th of December the number of days in its age was equal to the number of days from the 1st of November to the day of its birth, inclusive. What was the date of its birth?

42. A father has saved \$600 yearly for the last 50 years. Each of his four sons has saved the same amount yearly; the oldest for the last 27 years, the second for the last 24 years, the third for the last 19 years, and the youngest for the last

17 years. How many years ago had the father saved an amount equal to the joint savings of his sons?

43. A regiment moves from A to B, marching 20 miles a day. Two days later a second regiment leaves B for A, and marches 30 miles a day. At what distance from A do the regiments meet, A being 350 miles from B?

44. The sum of two digits of a number is 12. If the digits be interchanged, the resulting number exceeds the original one by three-fourths of the original number. What is the number?

45. Three boys, A, B, and C, have a number of marbles. A and B have 44, B and C have 43, and A and C have 39. How many marbles have they all, and how many marbles has each?

46. The tail of a fish is 4 inches long. Its head is as long as its tail and one-seventh of its body, and its body is as long as its head and one-half of its tail. How long is the fish, and how long are its head and its body?

47. A father divided his property equally among his sons. To the oldest son he gave $ 1000 and one-seventh of what remained; to the second son he gave $ 2000 and one-seventh of what was then left; to the third son he gave $ 3000 and one-seventh of the remainder; and so on. What was the amount of his property, and how many sons had he?

48. A man, wishing to give alms to several beggars, finds that in order to give 15 cents to each one, he must have 10 cents more than he has; but that if he were to give 12 cents to each one, he would have 14 cents left. How many beggars are there?

49. A train runs from A to B at the rate of 30 miles an hour; and returning runs from B to A at the rate of 28 miles an hour. The time required to go from A to B and return is 15 hours, including 30 minutes stop at B. How far is A from B?

50. A cistern has 3 taps. By the first it can be emptied in 80 minutes, by the second in 200 minutes, and by the third in 5 hours. After how many hours will the cistern be emptied, if all the taps be opened?

51. A cistern has 3 taps. By the first it can be filled in 6 hours, by the second in 8 hours, and by the third it can be emptied in 12 hours. In what time will it be filled if all the taps be opened?

52. A number of men wished to divide equally a sum of money. The amount to be divided is too small by $21.60 to give each man $1; but if each were to receive 70 cents, there would be left $10.80 undivided. How many dollars were to be divided?

53. An inlet pipe can fill a cistern in 3 hours, and an outlet pipe can empty it in 9 hours. After how many hours will the cistern be filled if both pipes be open half the time, and the outlet pipe be closed during the second half of the time?

54. In my right pocket I have as many dollars as I have cents in my left pocket. If I transfer $6.93 from my right pocket to my left, I shall have as many dollars in my left pocket as I shall have cents in my right. How much money have I in my left pocket?

55. A servant is to receive $170 and a dress for one year's services. At the end of 7 months she leaves her place and receives $95 and the dress. What is the value of the dress?

56. A farmer found that his supply of feed for his cows would last only 14 weeks. He therefore sold 60 cows, and his supply then lasted 20 weeks. How many cows had he?

57. At 6 o'clock the hands of a clock are in a straight line. At what time between 7 and 8 o'clock will they be again in a straight line? At what time between 9 and 10 o'clock?

58. A cistern has 3 pipes which can empty it in 6, 8, and 10 hours respectively. After all three pipes have been open for 2 hours they have discharged 94 gallons. What is the capacity of the cistern?

59. At what time between 1 and 2 o'clock is the minute-hand of a clock directly over the hour-hand? At what time between 6 and 7 o'clock?

60. At what time between 10 and 11 o'clock are the minute-hand and the hour-hand of a clock at right angles to each other? Find two solutions. At what time between 12 and 1 o'clock?

61. At what time between 3 and 4 o'clock will the minute-hand of a clock be 5 minute-divisions in advance of the hour hand? At what time 17 minute-divisions?

62. At what time between 9 and 10 o'clock will the hour-hand of a clock be 25 minute-divisions in advance of the minute-hand? At what time 12 minute-divisions?

A watch has the second-hand attached at the same point as the hour- and the minute-hand:

63. At what time between 1 and 2 o'clock is the second-hand over the minute-hand? At what time between 8 and 9 o'clock?

64. At what time between 4 and 5 o'clock is the second-hand of a watch over the hour-hand? At what time between 1 and 2 o'clock?

65. At what time between 8 and 9 o'clock are the second- and the minute-hand of a watch in a straight line? At what time between 4 and 5 o'clock?

66. At what time between 11 and 12 o'clock does the second-hand of a watch bisect the angle between the hour- and the minute-hand? At what time between 4 and 5 o'clock?

67. A woman sells $\frac{1}{2}$ an apple more than one-half of her apples. She next sells $\frac{1}{2}$ an apple more than one-half of the apples not yet sold, and then has 6 apples left. How many apples had she at first?

68. A steamer and a sailing vessel are both to sail from M to N. The steamer sails 40 miles every 3 hours, and the sailing vessel 24 miles in the same time. The sailing vessel has traveled $13\frac{1}{2}$ miles when the steamer sails, and arrives at N 5 hours later than the steamer. How long is the steamer in sailing from M to N, and how far is M from N?

69. A wall can be built by 20 workmen in 11 days, or by 30 other workmen in 7 days. If 22 of the first class work together with 21 of the second class, after how many days will the work be completed?

70. A merchant finds at the end of his first year in business that if he had gained \$1500 more he would have doubled his capital. At the end of the second and of the third year he finds the same condition. His capital is then eleven-fourths of his original investment. How much did he originally invest, and what were his profits each year?

71. A deposited a sum of money in a bank and received 4% simple interest. After two years he drew one-fourth of the interest then due. Two years later he again drew one-fourth of the interest then due. Two years later his principal and interest exceeded his original deposit by \$2775. What was his original deposit?

72. In a certain family each son has as many brothers as sisters, but each daughter has twice as many brothers as sisters. How many children are in the family?

73. A ship sailed from a port A toward a port B. When it was within 4 miles of B it was driven backward by a storm a distance equal to one-nineteenth of its distance from A. After the storm, the ship sailed toward B a distance equal to one-twentieth of its distance at that time from A, when it was again driven backward by a storm a distance equal to one-twentieth of its distance from A. After the second storm its distance from B was equal to one-ninth of its distance from A. How far is B from A, and how many miles did the ship sail altogether in its passage from A to B?

74. A merchant's investment yields him yearly $33\frac{1}{3}\%$ profit. At the end of each year, after deducting \$1000 for personal expenses, he adds the balance of his profits to his invested capital. At the end of three years his capital is twice his original investment. How much did he invest?

75. In a number of four digits, the first digit on the right is 2. If this digit be moved to the last place on the left, the resulting number will be less than the original one by 2106. What is the number?

76. I have in mind a number of six digits, the last one on the left being 1. If I bring this digit to the first place on the right, I shall obtain a number which is three times the number I have in mind? What is the number?

77. In a number of six digits, the first digit on the right is 2. If this digit be moved to the last place on the left, the resulting number will be one-third of the original one. What is the number?

78. A dog caught sight of a hare at a distance of 50 dog's leaps. The dog makes 3 leaps while the hare makes 4 leaps, but the length of two dog's leaps is equal to the length of 3 hare's leaps. How many leaps will the hare make before the dog overtakes him?

CHAPTER VI.

TYPE-FORMS.

We shall in this chapter consider a number of products and quotients which are of frequent occurrence. They are of such importance that the student should memorize them, and acquire facility in their applications. They are called **Type-Forms.**

§ 1. TYPE-FORMS IN MULTIPLICATION.

The Square of an Algebraic Expression.

1. By actual multiplication, we have

$$(a+b)^2 = (a+b)(a+b) = a^2 + 2ab + b^2.$$

That is, *the square of the sum of two numbers is equal to the square of the first number, plus twice the product of the two numbers, plus the square of the second number.*

E.g.,
$$(2x+5y)^2 = (2x)^2 + 2(2x)(5y) + (5y)^2$$
$$= 4x^2 + 20xy + 25y^2.$$

2. By actual multiplication, we have

$$(a-b)^2 = (a-b)(a-b) = a^2 - 2ab + b^2.$$

That is, *the square of the difference of two numbers is equal to the square of the first number, minus twice the product of the two numbers, plus the square of the second number.*

E.g.,
$$(3x-7y)^2 = (3x)^2 - 2(3x)(7y) + (7y)^2$$
$$= 9x^2 - 42xy + 49y^2.$$

Observe that this type-form is equivalent to that of Art. 1, since $a - b = a + (-b)$.

E.g.,
$$(3x-7y)^2 = (3x)^2 + 2(3x)(-7y) + (-7y)^2$$
$$= 9x^2 - 42xy + 49y^2, \text{ as above.}$$

It is also important to notice that the signs of all the terms of an expression which is to be squared may be changed, without changing the result.

For, $(-a-b)^2=[-(a+b)]^2=(a+b)^2,$

and $(a-b)^2=[-(b-a)]^2=(b-a)^2.$

3. If $a+b+c$ be regarded as the sum of $a+b$ and c, we have, by Art. 1,

$$(a+b+c)^2=[(a+b)+c]^2=(a+b)^2+2(a+b)c+c^2$$
$$=a^2+2ab+b^2+2ac+2bc+c^2.$$

Therefore $\boldsymbol{(a+b+c)^2=a^2+b^2+c^2+2ab+2ac+2bc.}$

In like manner

$$\boldsymbol{(a+b-c)^2=a^2+b^2+c^2+2ab-2ac-2bc.}$$
$$\boldsymbol{(a-b-c)^2=a^2+b^2+c^2-2ab-2ac+2bc.}$$

That is, *the square of an algebraic sum of three numbers is equal to the sum of the squares of the numbers, plus the algebraic sum of twice the product of each number by each number which follows it.*

$$(3x+5y-7z)^2=(3x)^2+(5y)^2+(-7z)^2+2(3x)(5y)$$
$$+2(3x)(-7z)+2(5y)(-7z)$$
$$=9x^2+25y^2+49z^2+30xy-42xz-70yz.$$

Observe that

$$(a+b-c)^2=(-a-b+c)^2;$$
$$(a-b-c)^2=(-a+b+c)^2.$$

4. By repeated application of the preceding articles, we can obtain the square of a multinomial of any number of terms. We have

$$(a+b+c+d)^2=[(a+b+c)^2+d]^2$$
$$=a^2+b^2+c^2+2ab+2ac+2bc+2(a+b+c)d+d^2$$
$$=a^2+b^2+c^2+d^2+2ab+2ac+2ad+2bc+2bd+2cd.$$

We therefore conclude that

The square of any multinomial is equal to the sum of the squares of the terms, plus the algebraic sum of twice the product of each term by each term which follows it.

E.g.,

$$(2x-3y+z-5u)^2=(2x)^2+(-3y)^2+z^2+(-5u)^2+2(2x)(-3y)$$
$$+2(2x)z+2(2x)(-5u)+2(-3y)z$$
$$+2(-3y)(-5u)+2z(-5u).$$
$$=4x^2+9y^2+z^2+25u^2-12xy+4xz-20xu-6yz+30yu-10zu.$$

EXERCISES I.

Write, without performing the actual multiplications, the values of

1. $(a+1)^2$.
2. $(2-b)^2$.
3. $(x+5)^2$.
4. $(a-3)^2$.
5. $(x+2a)^2$.
6. $(y-3z)^2$.
7. $(2a+3b)^2$.
8. $(5x-7y)^2$.
9. $(3ax+2by)^2$.
10. $(a^2+\frac{1}{3}ab)^2$.
11. $(\frac{1}{2}x^2+\frac{2}{3}y)^2$.
12. $(2xy-4)^2$.
13. $(2m^2-3n^2)^2$.
14. $(ab-2c^3)^2$.
15. $(4xy^2+7x^2y)^2$.
16. $(5x^2y^2-3z^3)^2$.
17. $(8a^2b^3-5c^2d^3)^2$.
18. $(\frac{1}{2}a^2b^4+\frac{3}{5}a^4b^2)^2$.
19. $(2a^n-9)^2$.
20. $(5x^{n+1}+7x)^2$.
21. $(6a^nb^m-5a^mb^n)^2$.
22. $(2a^{m+n}-3a^{m-n})^2$.
23. $(5x^py^r-6x^{p-1}y^{r+1})^2$.
24. $(7a^{n-3}b^{m+5}-3a^{n+5}b^{m-3})^2$.
25. $[a+b(x-1)]^2$.
26. $[2(x+1)-3(y+z)]^2$.
27. $[5(a+b)^n+2(a+b)^{n+1}]^2$.
28. $[7(a-b)^{2n}-4(a-b)^{m-2n}]^2$.
29. $(2x-3y+7)^2$.
30. $(m^4+n^2-1)^2$.
31. $(a^3+a^2+1)^2$.
32. $(2x^2-x+5)^2$.
33. $(2ab+3a^2+4b^2)^2$.
34. $(x^2-3xy+y^2)^2$.
35. $(xy-xz+yz-3)^2$.
36. $(x^3+2x^2y-3xy^2-5y^3)^2$.

Simplify the following expressions:

37. $a^2+b^2-(a-b)^2$.
38. $x^2+y^2-4x+6y+3$, when $x=a+1$, $y=a-2$.
39. $(a+b+c)^2+(a-b-c)^2+(a-b+c)^2+(a+b-c)^2$.
40. $(a+b-c)(a+b)+(a-b+c)(a+c)+(b+c-a)(b+c)$.

Verify the following identities by performing the indicated operations:

41. $(a^2+b^2)(x^2+y^2)-(ax+by)^2=(ay-bx)^2$.

42. $(a+b+c)^2+(a-b)^2+(a-c)^2+(b-c)^2=3(a^2+b^2+c^2)$.

43. $(a+b+c+d)^2+(a-b)^2+(a-c)^2+(a-d)^2+(b-c)^2$
$+(b-d)^2+(c-d)^2=4(a^2+b^2+c^2+d^2)$.

44. $a^2+b^2+4c^2+2ab+8bc=4(a+c)^2$, when $b=a$.

45. $s^2+(s-a)^2+(s-b)^2+(s-c)^2=a^2+b^2+c^2$, when
$2s=a+b+c$.

46. $(a^2+b^2+c^2)^2+(a+b+c)(a+b-c)(a+c-b)(b+c-a)$
$=4(a^2b^2+a^2c^2+b^2c^2)$.

47. $(a+b+c+d)^2+(a-b-c+d)^2+(a-b+c-d)^2$
$+(a+b-c-d)^2=4(a^2+b^2+c^2+d^2)$.

48. $(ap+bq+cr+ds)^2+(aq-bp+cs-dr)^2+(ar-cp+dq-bs)^2$
$+(br-cq+as-dp)^2=(a^2+b^2+c^2+d^2)(p^2+q^2+r^2+s^2)$.

49. $2(s-a)(s-b)(s-c)+a(s-b)(s-c)+b(s-a)(s-c)$
$+c(s-a)(s-b)=abc$, when $2s=a+b+c$.

Product of the Sum and Difference of Two Numbers.

5. By actual multiplication, we have

$$(a+b)(a-b)=a^2-b^2.$$

That is, *the product of the sum of two numbers and the difference of the same numbers, taken in the same order, is equal to the square of the first, minus the square of the second.*

Ex. 1. $(2x+3y)(2x-3y)=(2x)^2-(3y)^2=4x^2-9y^2$.

The product of two multinomials can frequently be brought under this type-form by properly grouping terms.

Ex. 2.
$$\begin{aligned}(x^2+x+1)(x^2-x+1)&=[(x^2+1)+x][(x^2+1)-x]\\&=(x^2+1)^2-x^2\\&=x^4+2x^2+1-x^2\\&=x^4+x^2+1.\end{aligned}$$

Ex. 3. $(x - y + z)(x + y - z) = [x - (y - z)][x + (y - z)]$
$$= x^2 - (y - z)^2$$
$$= x^2 - (y^2 - 2yz + z^2)$$
$$= x^2 - y^2 - z^2 + 2yz.$$

Observe that in thus grouping the terms a binomial or a multinomial takes the place of a single term of the type-form; as $x^2 + 1$ in Ex. 2, and $y - z$ in Ex. 3. The parentheses were unnecessary in Ex. 2, except to emphasize the mode of grouping the terms. The insertion of the parentheses, preceded by the sign $-$, in the first factor of Ex. 3, was necessary in order to bring the product to the required type-form.

EXERCISES II.

Find, without performing the actual multiplications, the values of

1. $(x + 2)(x - 2)$.
2. $(2a - 3)(2a + 3)$.
3. $(a - \frac{2}{3})(a + \frac{2}{3})$.
4. $(5x + 4y)(5x - 4y)$.
5. $(3a^2 + \frac{1}{2}ab)(3a^2 - \frac{1}{2}ab)$.
6. $(2x^3y^2 - 5x^2y^3)(2x^3y^2 + 5x^2y^3)$.
7. $(-3x^2 + 7)(3x^2 + 7)$.
8. $(5mn^2 + 2m^2n)(-5mn^2 + 2m^2n)$.
9. $(2ax^2 - 3a^2x)(2ax^2 + 3a^2x)$.
10. $(3a^n + 7b^m)(3a^n - 7b^m)$.
11. $(-5x^{n+1} + 9x^{n-1})(5x^{n+1} + 9x^{n-1})$.
12. $7x^2y^{n-1} + 5y^3x^m)(7x^2y^{n-1} - 5y^3x^m)$.
13. $[a^2 + 6(a + b)][a^2 - 6(a + b)]$.
14. $[3(m + n) - 5][-3(m + n) - 5]$.
15. $(x + y + 5)(x + y - 5)$.
16. $(4a - 3b - 7)(4a - 3b + 7)$.
17. $(x^2 + y^2 + z^2)(-x^2 + y^2 + z^2)$.
18. $(a^2 - ab + b^2)(a^2 + ab + b^2)$.
19. $(5a - 7b - 6c)(5a + 7b - 6c)$.
20. $(-2x + 3y - 5z)(2x + 3y - 5z)$.
21. $(x^2 + 2x - 1)(x^2 - 2x - 1)$.
22. $(x^4 - x^2 + 1)(x^4 + x^2 - 1)$.
23. $(-a^2 - b^2 + 3)(a^2 - b^2 + 3)$.
24. $(a^2 - b^2 - c^2)(a^2 + b^2 + c^2)$.
25. $(4x - 5y + 7)(4x + 5y - 7)$.
26. $(ax - by + cz)(ax + by - cz)$.

27. $(1+2a+3b+4c)(1+2a-3b-4c)$.

28. $(a+b+c-d)(a+b-c+d)$.

29. $(a^3-3a^2x+3ax^2-x^3)(a^3+3a^2x+3ax^2+x^3)$.

30. $(x^2-y^2-z^2-w^2)(x^2+y^2+z^2-w^2)$.

31. $(-x^3+x^2-2x-1)(x^3+x^2+2x-1)$.

32. $(x^2+y^2-2x-3y+1)(x^2+y^2+2x+3y+1)$.

33. $(a^4-2a^3+3a^2-2a+1)(a^4+2a^3+3a^2+2a+1)$.

Simplify the following expressions:

34. $(1+x)^2-(1-x)(1+x)$. 35. $(2x+3y)^2(2x-3y)^2$.

36. $(1-ab)^2(1+ab)^2$. 37. $(x-3)(x-1)(x+1)(x+3)$.

38. $(a^2-b)(a^2-b)(a^2+b)(a^2+b)$.

39. $(a-x)(a+x)(a^2+x^2)(a^4+x^4)$.

40. $(x^2-1)(x^8+1)(x^4+1)(x^2+1)$.

41. $(a^2+2ab)(a^2-2ab)(a^8+16a^4b^4)(a^4+4a^2b^2)$.

42. $(a^2x^3-2xy^3)(a^2x^3+2xy^3)(a^4x^6+4x^2y^6)(a^8x^{12}+16x^4y^{12})$.

43. $(x^2-x+1)(x^2+x+1)(x^4-x^2+1)$.

44. $(a+b-c)(a+c-b)(b+c-a)(a+b+c)$.

45. $(a-b)(a+b-c)+(b-c)(b+c-a)+(c-a)(c+a-b)$.

Multiply

46. $(a+b)x^2+(a-b)x+2$ by $(a-b)x^2+(a+b)x+1$.

47. $a^3+(n+1)a^2+(n^2-n+1)a+3$ by
$a^2-(n-1)a-(n^2+n+1)$.

What is the value of a^2-ab+3,

48. When $a=x+1$, $b=x-1$?

49. When $a=x+y$, $b=x-y$?

50. When $a=m+n-1$, $b=m-n+1$?

The Product $(x+a)(x+b)$.

6. By actual multiplication, we have

$$(x+a)(x+b)=x^2+(a+b)x+ab;$$
$$(x+a)(x-b)=x^2+(a-b)x-ab;$$
$$(x-a)(x-b)=x^2-(a+b)x+ab.$$

That is, *the product of two binomials, having the same first term, is equal to the square of this term, plus the product of the algebraic sum of the second terms by the common first term, plus the product of the second terms.*

Ex. 1. $(x+3)(x+5) = x^2 + 8x + 15.$

Ex. 2. $(x^2-7)(x^2+4) = x^4 - 3x^2 - 28.$

Ex. 3. $(ax-b)(ax-c) = a^2x^2 - (b+c)ax + bc.$

Ex. 4. $(a+b+5)(a+b-3) = [(a+b)+5][(a+b)-3]$
$= (a+b)^2 + 2(a+b) - 15.$

Observe that in applying the type-form to the last example the binomial $a+b$ is regarded as the first term.

EXERCISES III.

Find, without performing the actual multiplications, the values of the following indicated products:

1. $(x+7)(x+4)$.
2. $(m+3)(m+9)$.
3. $(x-5)(x-4)$.
4. $(y-1)(y-11)$.
5. $(a-6)(a+8)$.
6. $(x-3)(x+2)$.
7. $(2+x)(1+x)$.
8. $(5+b)(7+b)$.
9. $(3-x)(4-x)$.
10. $(7-x)(5-x)$.
11. $(6+y)(-3+y)$.
12. $(7-z)(-5-z)$.
13. $(2x+1)(2x+3)$.
14. $(ay+1)(ay+5)$.
15. $(3xy-8)(3xy-7)$.
16. $(5a-3b)(5a-7b)$.
17. $(2xy+3z)(2xy-4z)$.
18. $(x^2-7)(x^2-5)$.
19. $(2x^3-11)(2x^3+4)$.
20. $(5x^2-3yz)(5x^2-2yz)$.
21. $(a^n+3)(a^n-5)$.
22. $(x^{m+1}-4)(x^{m+1}-7)$.
23. $(2a^{2m-1}-b^{m+1})(2a^{2m-1}+3b^{m+1})$.
24. $(x+y-3)(x+y-5)$.
25. $(a+3b+7)(a+3b-11)$.
26. $(m-n-3p)(m-n+5p)$.
27. $(ax+by-c)(ax+by-2c)$.

The Cube of an Algebraic Expression.

7. By actual multiplication, we have

$$(a+b)^3 = a^3 + 3a^2b + 3ab^2 + b^3.$$

That is, *the cube of the sum of two numbers is equal to the cube of the first number, plus three times the product of the square of the first number by the second, plus three times the product of the first number by the square of the second, plus the cube of the second.*

E.g., $(3x+2y)^3 = (3x)^3 + 3(3x)^2(2y) + 3(3x)(2y)^2 + (2y)^3$
$= 27x^3 + 54x^2y + 36xy^2 + 8y^3.$

8. By actual multiplication, we have

$$(a-b)^3 = a^3 - 3a^2b + 3ab^2 - b^3.$$

That is, *the cube of the difference of two numbers is equal to the cube of the first number, minus three times the product of the square of the first number by the second, plus three times the product of the first number by the square of the second, minus the cube of the second.*

E.g., $(5x-3y)^3 = (5x)^3 - 3(5x)^2(3y) + 3(5x)(3y)^2 - (3y)^3$
$= 125x^3 - 225x^2y + 135xy^2 - 27y^3.$

Observe that this type-form is equivalent to that of Art. 7, since $(a-b)^3 = [a+(-b)]^3$.

It is important to notice also that

$$(a-b)^3 = -(b-a)^3.$$

9. *The cube of a trinomial can be obtained by regarding the sum of any two of its terms as one number.*

E.g., in $(x+2y-3z)^3$,

if $x+2y$ be regarded as one number, and $3z$ as the other, we have

$$(x+2y-3z)^3 = [(x+2y)-(3z)]^3 = (x+2y)^3 - 3(x+2y)^2(3z) + 3(x+2y)(3z)^2 - (3z)^3$$
$$= x^3 + 6x^2y + 12xy^2 + 8y^3 - 9x^2z - 36xyz - 36y^2z + 27xz^2 + 54yz^2 - 27z^3.$$

EXERCISES IV.

Write, without performing the actual multiplications, the values of

1. $(a+1)^3$.
2. $(a-2)^3$.
3. $(2x+3)^3$.
4. $(3-4y)^3$.
5. $(a+2b)^3$.
6. $(3a-b)^3$.
7. $(ax+by)^3$.
8. $(2x-3yz)^3$.
9. $(x^2+3x)^3$.
10. $(a^2n^2-6an)^3$.
11. $(2xm^3+5x^2m^2)^3$.
12. $(\frac{5}{3}a^2x-\frac{1}{5}ax^2)^3$.
13. $(4xy^2u-\frac{1}{2}x^2yu^2)^3$.
14. $(7x^n+2x^{n-1})^3$.
15. $(a^nb-5ab^{n+1})^3$.
16. $(5x^{m+1}y^{n-1}+1)^3$.
17. $(3a^{r+1}b^{r-1}c^p-7a^{r-1}b^{r+1}c^{1+p})^3$.
18. $(a^2+a+1)^3$.
19. $(3-x-x^2)^3$.
20. $(\frac{5}{3}x^2-x+\frac{1}{2})^3$.
21. $(2-ab+a^2b^2)^3$.
22. $(2a^3x-a^2x^2+3ax^2)^2$.
23. $(a^{n-1}-3b^{m+1}+2a^{n-1}b^{m-1})^3$.
24. What is the value of x^3-3x^2+3x-1, when $x=m+1$?
25. What is the value of x^3-2x^2+3x-1, when $x=y-2$?
26. What is the value of a^3-3a^2b+2ab, when $a=x+y$, $b=x-y$?

Verify the following identities:

27. $(a+b+c)^3-3(a+b)(b+c)(c+a)=a^3+b^3+c^3$.
28. $(a-b)^3+3(a-b)^2(a+b)+(a+b)^3+3(a-b)(a+b)^2=8a^3$.
29. $(a+b+c)^3-(b+c-a)^3-(c+a-b)^3-(a+b-c)^3=24abc$.
30. Multiply $x^3+(a+b)x^2+(a-b)x+(a-b)^2$
 by $x^2-(a-b)x+(a+b)$.

Higher Powers of a Binomial.

10. By actual multiplication, we have

$$(a+b)^4=a^4+4a^3b+6a^2b^2+4ab^3+b^4;$$
$$(a-b)^4=a^4-4a^3b+6a^2b^2-4ab^3+b^4;$$
$$(a+b)^5=a^5+5a^4b+10a^3b^2+10a^2b^3+5ab^4+b^5;$$
$$(a-b)^5=a^5-5a^4b+10a^3b^2-10a^2b^3+5ab^4-b^5;$$
$$(a+b)^6=a^6+6a^5b+15a^4b^2+20a^3b^3+15a^2b^4+6ab^5+b^6;$$
$$(a-b)^6=a^6-6a^5b+15a^4b^2-20a^3b^3+15a^2b^4-6ab^5+b^6.$$

The result of performing the indicated operation in a power of a binomial is called the **Expansion** of that power of the binomial. Thus, the second members of the above identities are the expansions of the corresponding powers of $a + b$.

In the preceding powers the following laws are evident:

(i.) *The number of terms in the expansion exceeds the binomial exponent by* 1.

(ii.) *The exponent of* a *in the first term is equal to the binomial exponent, and decreases by* 1 *from term to term.*

(iii.) *The exponent of* b *in the second term is* 1 *and increases by* 1 *from term to term, and in the last term is equal to the binomial exponent.*

(iv.) *The coefficient of the first term is* 1, *and that of the second term is equal to the binomial exponent.*

(v.) *The coefficient of any term after the second is obtained by multiplying the coefficient of the preceding term by the exponent of* a *in that term, and dividing the product by a number greater by* 1 *than the exponent of* b *in that term.*

E.g., the coefficient of the *fifth* term in the expansion of

$$(a + b)^5 \text{ is } 10 \times 2 \div 4 = 5.$$

(vi.) *The signs of the terms are all positive when the terms of the binomial are both positive; the signs of the terms alternate,* + *and* −, *when one of the terms of the binomial is negative.*

Observe, as a check:

(vii.) *The sum of the exponents of* a *and* b *in any term is equal to the binomial exponent.*

(viii.) *The coefficients of two terms equally distant from the beginning and the end of the expansion are equal.*

In a subsequent chapter the above laws will be proved to hold for any positive integral power of the binomial.

Ex. 1. $(2x - 3y)^4 = (2x)^4 - 4(2x)^3(3y) + 6(2x)^2(3y)^2$

$$- 4(2x)(3y)^3 + (3y)^4$$

$$= 16x^4 - 96x^3y + 216x^2y^2 - 216xy^3 + 81y^4.$$

Ex. 2. $(3a^2+ab)^6=(3a^2)^6+6(3a^2)^5(ab)+15(3a^2)^4(ab)^2$
$+20(3a^2)^3(ab)^3+15(3a^2)^2(ab)^4$
$+6(3a^2)(ab)^5+(ab)^6$
$=729a^{12}+1458a^{11}b+1215a^{10}b^2+540a^9b^3$
$+135a^8b^4+18a^7b^5+a^6b^6.$

EXERCISES V.

Find the values of the following powers:

1. $(1+x)^4$.
2. $(3-4a)^4$.
3. $(2x+3y)^5$.
4. $(3m-2n)^5$.
5. $(2a+1)^6$
6. $(3x-2y)^6$.
7. $(a^2+ab)^4$.
8. $(x^3-x)^5$.
9. $(2m^3n-mn^3)^6$.
10. $(a^{n+1}-b)^4$.
11. $(a^{n-1}b+ab^n)^5$.
12. $(a^{m+1}x^{n-1}-a^{n-1}x^{m+1})^6$.
13. Verify the identity

$$[(a-b)^2+(b-c)^2+(c-a)^2]^2=2[(a-b)^4+(b-c)^4+(c-a)^4].$$

§ 2. TYPE-FORMS IN DIVISION.

Quotient of the Sum or the Difference of Like Powers of Two Numbers by the Sum or the Difference of the Numbers.

1. By actual division, we have

$(a^2-b^2)\div(a+b)=a-b$ and $(a^2-b^2)\div(a-b)=a+b$.

That is, *the difference of the squares of two numbers is exactly divisible by the sum of the numbers, and also by the difference of the numbers, taken in the same order; the quotient in the first case is the difference of the two numbers, taken in the same order, and in the second case is the sum of the two numbers.*

Ex. 1. $(9-25x^2)\div(3+5x)=3-5x$.

Ex. 2. $(16x^4-81y^6)\div(4x^2-9y^3)=4x^2+9y^3$.

Ex. 3. $(x^2+y^2-z^2-2xy)\div(x-y-z)$
$=(x^2-2xy+y^2-z^2)\div(x-y-z)$
$=[(x-y)^2-z^2]\div[(x-y)-z]$
$=x-y+z.$

2. By actual division, we have

$$\text{(i.)}\ \begin{cases} (a^3+b^3)\div(a+b)=a^2-ab+b^2; \\ (a^5+b^5)\div(a+b)=a^4-a^3b+a^2b^2-ab^3+b^4. \end{cases}$$

$$\text{(ii.)}\ \begin{cases} (a^3-b^3)\div(a-b)=a^2+ab+b^2; \\ (a^5-b^5)\div(a-b)=a^4+a^3b+a^2b^2+ab^3+b^4. \end{cases}$$

$$\text{(iii.)}\ \begin{cases} (a^4-b^4)\div(a+b)=a^3-a^2b+ab^2-b^3; \\ (a^4-b^4)\div(a-b)=a^3+a^2b+ab^2+b^3. \end{cases}$$

The above identities, and the identities in Art. 1, illustrate the following principles, which will be proved in Art. 6:

(i.) *The sum of the like odd powers of two numbers is exactly divisible by the sum of the numbers.*

(ii.) *The difference of the like odd powers of two numbers is exactly divisible by the difference of the numbers, taken in the same order.*

(iii.) *The difference of the like even powers of two numbers is exactly divisible by the sum, and also by the difference of the numbers, taken in the same order.*

(iv.) *When the divisor is a sum, the signs of the terms of the quotient alternate,* + *and* −.

(v.) *When the divisor is a difference, the signs of the terms of the quotient are all* +.

(vi.) *In the first term of the quotient the exponent of* a *is less by* 1 *than its exponent in the dividend, and decreases by* 1 *from term to term.*

(vii.) *The exponent of* b *is* 1 *in the second term of the quotient, and increases by* 1 *from term to term.*

Observe that the quotient is homogeneous in a and b, of degree less by 1 than the degree of the dividend.

It is important to notice that, as we shall prove in Art. 6, *the sum of the like even powers of two numbers is not exactly divisible by either the sum or the difference of the numbers.*

E.g., a^4+b^4 is not divisible either by $a+b$ or by $a-b$.

Ex. 1. $(8x^3+\frac{1}{125})\div(2x+\frac{1}{5})=(2x)^2-(2x)(\frac{1}{5})+(\frac{1}{5})^2$

$$=4x^2-\tfrac{2}{5}x+\tfrac{1}{25}.$$

Ex. 2. $(625\,a^4b^4 - c^8) \div (5\,ab - c^2)$

$$= (5\,ab)^3 + (5\,ab)^2c^2 + 5\,ab\,(c^2)^2 + (c^2)^3$$
$$= 125\,a^3b^3 + 25\,a^2b^2c^2 + 5\,abc^4 + c^6.$$

Ex. 3. $(32\,a^{10} - x^{15}) \div (2\,a^2 - x^3)$

$$= (2\,a^2)^4 + (2\,a^2)^3(x^3) + (2\,a^2)^2(x^3)^2 + 2\,a^2\,(x^3)^3 + (x^3)^4$$
$$= 16\,a^8 + 8\,a^6x^3 + 4\,a^4x^6 + 2\,a^2x^9 + x^{12}.$$

Notice that in the type-forms each term, *beginning with the second*, is equal to the preceding term multiplied by $b \div a$ when the divisor is $a - b$, and by $-b \div a$ when the divisor is $a + b$. This gives a convenient working rule for more complicated examples.

By this rule, the terms of the quotient in Ex. 3 are formed as follows:

$$16\,a^8 = 32\,a^{10} \div 2\,a^2;\ 8\,a^6x^3 = 16\,a^8 \times x^3 \div 2\,a^2;$$
$$4\,a^4x^6 = 8\,a^6x^3 \times x^3 \div 2\,a^2;\ 2\,a^2x^9 = 4\,a^4x^6 \times x^3 \div 2\,a^2;$$
$$x^{12} = 2\,a^2x^9 \times x^3 \div 2\,a^2.$$

EXERCISES VI.

Find the values of the following quotients, without performing the actual divisions:

1. $(x^2 - 1) \div (x - 1)$.
2. $(25 - x^2) \div (5 + x)$.
3. $(4\,a^2 - 9) \div (2\,a - 3)$.
4. $(\frac{1}{9} - x^2y^2) \div (\frac{1}{3} + xy)$.
5. $(16\,x^2 - 9\,y^2) \div (4\,x - 3\,y)$.
6. $(64\,a^2b^2 - 121\,c^2) \div (8\,ab + 11\,c)$.
7. $(16\,x^2y^2 - 81\,a^2b^2c^2) \div (4\,xy - 9\,abc)$.
8. $(x^4 - 1) \div (x^2 + 1)$.
9. $(4\,a^4 - b^2) \div (2\,a^2 - b)$.
10. $(81\,a^8b^2 - 25\,x^{10}) \div (9\,a^4b + 5\,x^5)$.
11. $(81\,x^4 - n^8y^{12}) \div (9\,x^2 - n^4y^6)$.
12. $(x^{2n} - 1) \div (x^n - 1)$.
13. $(a^{4n} - 16\,b^{16}) \div (a^{2n} + 4\,b^8)$.
14. $(x^{2n+2} - 4) \div (x^{n+1} + 2)$.
15. $(a^{8n} - b^{4n+4}) \div (a^{4n} - b^{2n+2})$.
16. $(\frac{1}{25}\,x^{2m+2}y^{6m-2} - 9\,z^{8m+8n}) \div (\frac{1}{5}\,x^{m+1}y^{3m-1} + 3\,z^{4m+4n})$.
17. $[(a+b)^2 - 1] \div (a+b+1)$.
18. $[4 - (a+b)^2] \div (2 - a - b)$.
19. $(a^2 - 2\,ab + b^2 - 1) \div (a - b + 1)$.

20. $(a^2 - n^2 - p^2 + 2np) \div (a - n + p)$.

21. $(p^2 - r^2 - 4 - 4r) \div (p - r - 2)$.

22. $[(a^2 + 2ab + b^2)x^6 - y^4] \div [(a + b)x^3 + y^2]$.

23. $(x^4 + 2x^2y^2 + y^4 - z^2 - 2zu - u^2) \div (x^2 + y^2 + u + z)$.

24. $(a^2 - b^2 + 2bz - 2ax + x^2 - z^2) \div (a - x - b + z)$.

25. $(a^4 + 4a^2c - 4b^2 + 4bd^2 + 4c^2 - d^4) \div (a^2 + d^2 + 2c - 2b)$.

26. $(x^3 + 1) \div (x + 1)$. **27.** $(1 - a^3) \div (1 - a)$.

28. $(x^3 + 125) \div (x + 5)$. **29.** $(8a^3 - 27) \div (3 - 2a)$.

30. $(x^3y^3 + 1) \div (xy + 1)$.

31. $(\frac{1}{8}m^3n^3 - 125x^3y^3z^3) \div (\frac{1}{2}mn - 5xyz)$.

32. $(a^6 - 1) \div (a^2 - 1)$. **33.** $(a^{12} + 27) \div (a^4 + 3)$.

34. $(a^9x^6 + 64b^6y^9) \div (a^3x^2 + 4b^2y^3)$.

35. $(8m^{15}n^3 - p^{12}) \div (2m^5n - p^4)$. **36.** $(a^{3n} - 1) \div (a^n - 1)$.

37. $(343x^{3m-3} - y^{6n}) \div (7x^{m-1} - y^{2n})$.

38. $[(x + y)^3 - z^6] \div (x + y - z^2)$.

39. $[8a^6b^6 - (ab + 1)^3] \div (2a^2b^2 - ab - 1)$.

40. $(x^3 + 8y^3 + z^3 + 6x^2y + 12xy^2) \div (x + 2y + z)$.

41. $(x^4 - 1) \div (x - 1)$. **42.** $(1 - 16a^4) \div (1 + 2a)$.

43. $(\frac{1}{81}x^4 - 16y^4) \div (\frac{1}{3}x - 2y)$. **44.** $(81x^8 - 16y^8) \div (3x^2 + 2y^2)$.

45. $(x^8y^{12} - 256z^{16}) \div (x^2y^3 - 4z^4)$. **46.** $(x^{4n} - y^{4n}) \div (x^n + y^n)$.

47. $(x^{8n+4} - 16y^{12n}) \div (x^{2n+1} - 2y^{3n})$.

48. $(81a^{4n}b^{12n-4} - c^{8m}) \div (3a^nb^{3n-1} - c^{2m})$.

49. $(a^5 + 1) \div (a + 1)$. **50.** $(32x^5 + y^5) \div (2x + y)$.

51. $(1 - x^5) \div (x - 1)$. **52.** $(243a^5b^5 - c^5) \div (3ab - c)$.

53. $(x^{10}y^5 + 32z^{15}) \div (x^2y + 2z^3)$. **54.** $(a^{10}b^{15}c^{20} - d^{25}) \div (a^2b^3c^4 - d^5)$.

55. $(x^{5n} - 1) \div (x^n - 1)$. **56.** $(1 + a^{5m}) \div (1 + a^m)$.

57. $(a^6 - b^6) \div (a - b)$. **58.** $(a^7 + b^7) \div (a + b)$.

59. $(a^{11} + 1) \div (a + 1)$. **60.** $(a^{12} - 1) \div (a - 1)$.

61. $(a^{10} - \frac{1}{32}b^5) \div (a^2 - \frac{1}{2}b)$. **62.** $(a^7 - x^{14}) \div (a - x^2)$.

63. $(x^{30} - 1) \div (x^5 + 1)$. **64.** $(a^{14}x^{7n} + b^{14m}) \div (a^2x^n + b^{2m})$.

Of what divisions are the following expressions the quotients:

65. x^2+x+1.

66. a^2-ab+b^2.

67. x^3-x^2+x-1.

68. $a^4+a^3+a^2+a+1$.

69. $x^3y^3+mx^2y^2+m^2xy+m^3$.

70. $x^{m-1}+x^{m-2}+x^{m-3}+\cdots+x+1$.

3. If $3x^3-4x^2-6x+7$ be divided by $x-2$, we have

$$\begin{array}{l|l} 3x^3-4x^2-6x+7 & x-2 \\ \underline{3x^3-6x^2} & \overline{3x^2+2x-2} \\ \quad 2x^2 & \\ \quad \underline{2x^2-4x} & \\ \qquad -2x & \\ \qquad \underline{-2x+4} & \\ \qquad\qquad 3 & \end{array}$$

The division is not exact, and the remainder is 3. If now 2 be substituted for x in the given expression, we obtain

$$3\times 2^3-4\times 2^2-6\times 2+7=3, \text{ the above remainder.}$$

This example illustrates the following principle:

If an expression, arranged to ascending or descending powers of a letter of arrangement, say x, be not exactly divisible by $x-a$, the remainder of the division is equal to the result of substituting a for x in the given expression.

The method of proving the principle enunciated will be first illustrated by the above example.

By Ch. III., § 4, Art. 13, we have

$$3x^3-4x^2-6x+7=(3x^2+2x-2)(x-2)+3.$$

If 2 be substituted for x in this equation, we obtain

$$\begin{aligned} 3\times 2^3-4\times 2^2-6\times 2+7 &= (3\times 2^2+2\times 2-2)(2-2)+3 \\ &= (3\times 2^2+2\times 2-2)\times 0+3 \\ &= 3, \text{ since } N\times 0=0, \text{ and } 0+3=3. \end{aligned}$$

We thus see that

$$3\times 2^3-4\times 2^2-6\times 2+7 \text{ is equal to } 3,$$

without performing the indicated operations.

In general, let the given expression be of the form

$$Ax^n+Bx^{n-1}+\cdots+Ux+V,$$

in which n is a positive integer.

Let Q stand for the quotient of the division by $x-a$, and R for the remainder.

Then, by Ch. III., § 4, Art. 13, we have

$$Ax^n + Bx^{n-1} + \cdots + Mx + V = Q(x-a) + R.$$

If now a be substituted for x in the last equation, we obtain

$$Aa^n + Ba^{n-1} + \cdots + Ma + V = Q(a-a) + R$$
$$= Q \cdot 0 + R = R.$$

That is, the remainder, R, of dividing the given expression by $x - a$ is equal to the result of substituting a for x in the expression.

4. From the principle of the preceding article we derive the following:

If an expression, arranged to ascending or descending powers of a letter of arrangement, say x, be exactly divisible by $x - a$, the result of substituting a for x in the given expression is 0; and conversely.

For if the division be exact, the remainder is 0, and therefore the result of the substitution is 0.

E.g., $3x^3 - 4x^2 - 6x + 4$ is exactly divisible by $x - 2$.

Substituting 2 for x, we obtain

$$3 \times 2^3 - 4 \times 2^2 - 6 \times 2 + 4 = 0.$$

5. If an expression be arranged to descending powers of a letter of arrangement, the following is a convenient method of substituting a particular value for the letter of arrangement.

Ex. 1. Substitute 2 for x in $3x^3 - 4x^2 - 6x + 7$.

We have $3x^3 = 3x \cdot x^2 = 6x^2$, when $x = 2$;

therefore $3x^3 - 4x^2 = 6x^2 - 4x^2 = 2x^2$, when $x = 2$;

then $2x^2 = 2x \cdot x = 4x$, when $x = 2$;

and $4x - 6x = -2x = -4$, when $x = 2$;

finally $-4 + 7 = 3$, the result of the substitution.

Ex. 2. Substitute -3 for x in $12x^6 + 5x^5 - 3x^4 - 11x^2 + 6x + 3$.

When $x = -3$, we have

$$12x^6 = -36x^5;$$
$$-36x^5 + 5x^5 = -31x^5 = 93x^4;$$
$$93x^4 - 3x^4 = 90x^4 = -270x^3 = 810x^2;$$
$$810x^2 - 11x^2 = 799x^2 = -2397x;$$
$$-2397x + 6x = -2391x = 7173;$$

$7173 + 3 = 7176$, the result of the substitution.

Ex. 3. Is $x^3 + x^2 - x + 7$ exactly divisible by $x + 2$?

Since $x + 2 = x - (-2)$, we substitute -2 for x in the given expression.

We then have

$$x^3 = -2x^2;$$
$$-2x^2 + x^2 = -x^2 = 2x;$$
$$2x - x = x = -2;$$
$$-2 + 7 = 5.$$

Therefore $x^3 + x^2 - x + 7$ is not exactly divisible by $x + 2$, and the remainder of the division is 5.

Ex. 4. Is $x^5 - 2x^4 - 6x^3 + 8x^2 + 5x - 6$ exactly divisible by

$$(x + 1)(x - 1)(x - 3)?$$

If the given expression be divisible by $(x + 1)(x - 1)(x - 3)$, it must be divisible by each factor of that product.

Since $x + 1 = x - (-1)$, we substitute -1 for x in the given expression, and have

$$x^5 = -x^4;$$
$$-x^4 - 2x^4 = -3x^4 = 3x^3;$$
$$3x^3 - 6x^3 = -3x^3 = 3x^2;$$
$$3x^2 + 8x^2 = 11x^2 = -11x;$$
$$-11x + 5x = -6x = 6;$$
$$6 - 6 = 0, \text{ the result of the substitution.}$$

In like manner it can be shown that the given expression is exactly divisible by $x - 1$ and $x - 3$.

EXERCISES VII.

Prove that the following dividends are exactly divisible by the corresponding divisors, without performing the divisions:

1. $(x^2 - 3x + 2) \div (x - 2)$. **2.** $(x^2 - 3x + 2) \div (x - 1)$.

3. $(x^3 - 18x - 35) \div (x - 5)$. **4.** $(x^3 + 2x^2 - x - 2) \div (x + 1)$.

5. $(x^3 + 21x + 342) \div (x + 6)$. **6.** $(2x^3 + 3x^2 + 3x + 1) \div (x + \frac{1}{2})$.

7. $(x^6 - 6x^4 - 19x^2 + 84) \div (x^2 - 3)(x^2 + 4)$.

8. $(x^6 + 4x^4 + x^2 - 6) \div (x^2 - 1)(x^2 + 2)$.

9. $(75x^4 + 140x^3 - 223x^2 + 92x - 12) \div (x - \frac{2}{5})(x - \frac{1}{3})$.

10. $(x^4 + 2x^3 - 7x^2 - 8x + 12) \div (x - 2)(x + 2)(x + 3)$.

11. $(2x^4 + 4ax^3 - 5a^2x^2 - 3a^3x + 2a^4) \div (x - a)$.

12. $(x^4 + 3a^2x^2 + 5a^3x + a^4) \div (x + a)$.

Find the remainders of the following indicated divisions, without performing the divisions:

13. $(2x^3 - 7x^2 + 6x - 15) \div (x + 4)$.

14. $(5x^4 - 11x^2 + 2x - 7) \div (x - 2)$.

15. $(17x^3 - 2x^2 + 4x - 3) \div (x - \frac{3}{5})$.

16. Prove that $(x + 1)^m + (x - 1)^m$ is exactly divisible by x when m is odd.

17. Prove that $(a + b + c)^3 - a^3 - b^3 - c^3$ is exactly divisible by $(a + b)(a + c)(b + c)$.

18. Prove that $x^q y^r + y^q z^r + z^q x^r - x^r y^q - y^r z^q - z^r x^q$ is exactly divisible by $(x - y)(x - z)(y - z)$.

6. We are now prepared to prove the following principles enunciated in Art. 2:

(i.) *$a^n + b^n$ is exactly divisible by $a + b$, but not by $a - b$, when n is odd.*

(ii.) *$a^n - b^n$ is exactly divisible by $a - b$, but not by $a + b$, when n is odd.*

(iii.) *$a^n - b^n$ is exactly divisible by $a + b$, and by $a - b$, when n is even.*

(iv.) *$a^n + b^n$ is not exactly divisible by either $a + b$ or $a - b$, when n is even.*

For if $-b$ be substituted for a in $a^n + b^n$, we obtain

$$(-b)^n + b^n = 0, \text{ only when } n \text{ is odd.}$$

Therefore, $a^n + b^n$ is exactly divisible by $a + b$, only when n is odd.

If b be substituted for a in $a^n + b^n$, we obtain

$$b^n + b^n, \neq 0.$$

Therefore $a^n + b^n$ is not divisible by $a - b$.

In like manner the other principles can be proved.

It is evident that the forms of the quotients considered in this article, obtained by actual division, conform to principles (iv.)-(vii.), Art. 2.

CHAPTER VII.

PARENTHESES.

1. The use of parentheses to inclose an expression which is to be treated as a whole in subsequent operations has been briefly discussed in Ch. II., § 2, Arts. 12–15. It is frequently necessary to employ more than two sets of parentheses in the same chain of operations, and in order to distinguish them the following forms are used:

Parentheses, ( ); **Brackets,** [ ]; **Braces,** { }.

A **Vinculum** is a line drawn over an expression, and is equivalent to parentheses inclosing it.

E.g., $(a+b)(c-d) = \overline{a+b} \cdot \overline{c-d}.$

If more forms of parentheses than the above are needed in any operations, one or more of them is made larger and heavier.

Removal of Parentheses.

2. The principles given in Ch. II., § 2, Arts. 12–14, are to be applied successively when several sets of parentheses are to be removed from a given expression.

In thus removing parentheses we may begin either with the inmost or with the outmost.

3. The following examples will illustrate the method of removing parentheses, beginning with the inmost.

Ex. 1.

$$\begin{aligned} &4a - \{3a + [2a - (a-1)]\} \\ &= 4a - \{3a + [2a - a + 1]\} \\ &= 4a - \{3a + [a+1]\} \\ &= 4a - \{3a + a + 1\} \\ &= 4a - 3a - a - 1 = -1. \end{aligned}$$

Ex. 2. $x^2 - \{2x^2 - [3x^2 + (4x^2 - \overline{5x^2 - 1})]\}$

$= x^2 - \{2x^2 - [3x^2 + (4x^2 - 5x^2 + 1)]\}$

$= x^2 - \{2x^2 - [3x^2 + (-x^2 + 1)]\}$

$= x^2 - \{2x^2 - [3x^2 - x^2 + 1]\}$

$= x^2 - \{2x^2 - [2x^2 + 1]\}$

$= x^2 - \{2x^2 - 2x^2 - 1\}$

$= x^2 - \{-1\} = x^2 + 1.$

Ex. 3. $[b^2 - \{(a^2 + b)a - (a^2 - b)b - a^2(a - b)\}]^3$

$= [b^2 - \{a^3 + ab - a^2b + b^2 - a^3 + a^2b\}]^3$

$= [b^2 - \{ab + b^2\}]^3$

$= [b^2 - ab - b^2]^3 = (-ab)^3 = -a^3b^3.$

Ex. 4.

$60x + 5\left(1 - x^2 + \{x^2 + 3 - 3[4x^2 + 1 + 2(x - x \cdot \overline{2x - 1})]\}\right)$

$= 60x + 5\left(1 - x^2 + \{x^2 + 3 - 3[4x^2 + 1 + 2(x - 2x^2 + x)]\}\right)$

$= 60x + 5\left(1 - x^2 + \{x^2 + 3 - 3[4x^2 + 1 + 4x - 4x^2]\}\right)$

$= 60x + 5\left(1 - x^2 + \{x^2 + 3 - 3 - 12x\}\right)$

$= 60x + 5\left(1 - x^2 + x^2 - 12x\right)$

$= 60x + 5 - 60x = 5.$

4. The method of removing parentheses, beginning with the outmost, may be illustrated by reworking Ex. 2, Art. 3.

Notice that the part which is free from parentheses is simplified at the same time that the next parentheses are removed.

$x^2 - \{2x^2 - [3x^2 + (4x^2 - \overline{5x^2 - 1})]\}$

$= x^2 - 2x^2 + [3x^2 + (4x^2 - \overline{5x^2 - 1})]$, removing braces,

$= -x^2 + 3x^2 + (4x^2 - \overline{5x^2 - 1})$, removing brackets,

$= 2x^2 + 4x^2 - \overline{5x^2 - 1}$, removing parentheses,

$= 6x^2 - 5x^2 + 1$, removing vinculum,

$= x^2 + 1.$

EXERCISES I.

Simplify the following expressions by removing parentheses:

1. $a + 2b - [6a - \{3b - (6a - 6b)\}]$.
2. $7a - \{3a - [4a - (5a - 2a)]\}$.
3. $2x - \{3y - [4x - (5y - 6x)]\}$.
4. $\{[(\overline{x + 4} \cdot x - 3)x + 7]x + 8\}x$.
5. $\{50 - [35 - (10 - x)x]x\}x$.
6. $4x - \{[x - 3(2 - x)]x - 4\}2$.
7. $6a - [7a - \{8a - (9a - \overline{10a - b})\}]$.
8. $a - \{2b + [3c - 3a - (a + b) + 2a - (b + 3c)]\}$.
9. $a - \{5b - [a - (3c - 3b) + 2c - (a - 2b - c)]\}$.
10. $12 - 13[10(7 \cdot \overline{4x - 3} - 6) - 9]$.
11. $12a - 13\{10[7(4a - 3) - 6] - 9a\}$.
12. $x - \{x + y - [x + y + z - (x + y + z + v)]\}$.
13. $10 - 2\{x - 5[3 - 2x - 6(4x - 7)] - 3(5 - 2x)\}$.
14. $7a^m - \{2a^m - [a^n - 3a^m + (5a^m - 2a^n) - 4a^m] - 2a^n\}$.
15. $5a - \{4a - [2a - (3a + 2b + 3c) + 4b] + 2b\} + 4c$.
16. $3m - \Big[4m - \{2n - [2m + 3n - (3m + n)] + 3m\}$
    $- (2n - 3m)\Big]$.
17. $\{[(x + y^2)x - (2y - 1)]x - (x^2 - 2y)x - x^2y^2\}^2$.
18. $[(x - y)^2 + 6xy] - \Big[(x^2 + 2xy) - \{x^2 - [2xy - (4xy - y^2)]\}$
    $- (- x^2 - 2xy)\Big]$.
19. $\frac{2}{5}a^2x + \frac{3}{4}ax^2 - \{\frac{1}{4}ax^2 - [-\frac{2}{5}a^2x - (\frac{1}{2}ax^2 - a)]\}$.

Find the values of the expressions in Exx. 1–9 and 14–19,

20. When $a = -3$, $b = 4$, $c = -5$, $m = 2$, $n = 1$, $x = 8$, $y = -9$.
21. When $a = 7$, $b = -5$, $c = 3$, $m = 4$, $n = 3$, $x = -11$, $y = 0$.

Solve the following equations:

22. $(x + 3)^2 = x^2 + 15$.
23. $(2x + 1)^2 - 8 = (2x - 1)^2$.
24. $(5 + x)^2 + 9 = 3x(9 + \frac{1}{3}x)$.
25. $(x + 1)^2 = [6 - (1 - x)]x - 2$.

26. $(4x - 11)^2 = (4x + 3)(4x - 3)$.

27. $(3 - \frac{1}{2}x)^2 + (\frac{2}{3}x - 5)^2 = (\frac{5}{6}x)^2$.

28. $(6x - 1)^2 + (8x - 3)^2 = (10x - 7)^2$.

29. $2x^2 + 17x = (8 + 2x)^2 - 67 - x(3 + 2x)$.

30. $(x + 1)(x - 1) = x(x - 2)$.

31. $(2x^2 + 1 - x)(2x^2 - 1 + x) = 1 + x^2(2x + 1)(2x - 1)$.

32. $(x - 1)(x^2 + x + 1) - 6(x^2 - 1) = -(2 - x)^3$.

33. $(x + 4)^3 - 6x^2 = x(x^2 + 6x + 20) + 68$.

34. $56x^3 - 4x(4x - 7)^2 = 164x^2 - (2x - 5)^3$.

35. $6 - [3 + \{2 - (5 - x)\}] = 8$. 36. $\frac{1}{5}\left(\frac{1}{5}\{\frac{1}{5}[\frac{1}{5}(\frac{1}{5}x - 1)]\}\right) = 1$.

37. $6 - \left\{5 - \left(4 - \{3 - [2 - (1 - x)]\}\right)\right\}$.

38. $2[8 - 2\{6 - 2(5 - \overline{2x - 1})\}] = 8$.

39. $\frac{1}{2}\left(\frac{3}{4} + \frac{1}{2}\{\frac{5}{8} + \frac{1}{2}[\frac{3}{4} + \frac{1}{2}(5 + x)]\}\right) = 1$.

40. $\frac{2}{3}[\frac{3}{2}\{\frac{3}{2}(6 - x) - x\} - x] - x = 4$.

41. $\frac{1}{3}\left(\frac{1}{3}\{\frac{1}{3}[\frac{1}{3}(\frac{1}{3}x + 2) + 2] + 2\} + 2\right) = 1$.

42. $4\{4[4(4x - 3) - 3] - 3\} - 3 = 1$.

43. $13\left(10\{[7(4x - 3) - 6] - 9\} - 12\right) = 1$.

44. $\frac{1}{9}\left(\frac{1}{7}\{\frac{1}{5}[\frac{1}{3}(x + 2) + 4] + 6\} + 8\right) = 1$.

45. $-4 - 4\{4 - 4[4 - 4(4 - x)]\} = 44$.

46. $-4x - \left(5x - [6x - \{7x - (8x - 9)\}]\right) = -10$.

47. $\frac{1}{3}[\frac{1}{3}\{\frac{1}{3}(\frac{1}{3}x - \frac{1}{3}) - \frac{1}{3}\} - \frac{1}{3}] - \frac{1}{3} = 0$.

48. $4x + \frac{1}{2}(x - 2) - 2\left(2x - [\frac{1}{4}x - \frac{1}{18}\{16 - \frac{1}{2}(x + 4)\}]\right) = \frac{2}{3}(x + 2)$.

49. $10 - \{x - [3 - 2x - (4x - 7)]\} = x - \{10 + 3x - [5 - (5 - 3x)]\}$.

What is the value of the product

$$(a + b - c)(a - b + c)(a + b + c),$$

50. When $a = 4,\ b = 3,\ c = -2$?

51. When $a = -5,\ b = -\frac{2}{5},\ c = 8$?

52. When $a = m + 1,\ b = m + 2,\ c = 2m + 1$?

53. When $a = b$? 54. When $a = -b$, and $c = -b$?

Insertion of Parentheses.

5. The principles for inserting parentheses in a given expression were proved in Ch. II., § 2, Art. 15.

Ex. **1.** Express $4(x-y)+y-x$ as a product, of which one factor is $x-y$.

We have $$4(x-y)+y-x=4(x-y)-(x-y)$$ $$=3(x-y).$$

Ex. **2.** Express $a^2+b^2+2ab+c^2-2cd$ as a difference in which the minuend is $a^2+b^2+c^2$.

We have

$$a^2+b^2+2ab+c^2-2cd=a^2+b^2+c^2-(2cd-2ab).$$

The sign + or − before a pair of parentheses can evidently be reversed from + to −, or from − to +, if the signs of the terms within the parentheses be reversed.

Ex. **3.** $$7(x-1)-3(1-x)=7(x-1)+3(x-1)$$ $$=10(x-1).$$

EXERCISES II.

Write each of the following expressions as a product, of which the expression within the parentheses is one of the factors:

1. $3(a-b)-a+b$.

2. $5(x^2-y)-x^2+y$.

3. $3m-5n-4(5n-3m)$.

4. $1-a^n+3(a^n-1)$.

5. $a^n-b^n-2(b^n-a^n)$.

6. $x^2-2-7(2-x^2)$.

7. $5(x^2-x+1)-x^2+x-1$.

8. $x-y-z-6(y+z-x)$.

9. $6(5m-3n-2)+2-5m+3n$.

10. Write each one of the expressions in Exx. 1–9 as a product, of which one of the factors is the expression without the parentheses.

Write each of the following expressions as a single product, of which the expression within the first parentheses is a factor:

11. $(2x-1)-3(1-2x)$.

12. $2(2m-3n)+(3n-2m)$.

13. $5(x^2-y^2)+2(y^2-x^2)$.

14. $7(xy-z)-(z-xy)$.

15. $3(x^2 + x - 1) - 2(1 - x^2 - x)$.

16. $6(a^2 - ab + b^2) + 5(ab - b^2 - a^2)$.

17. Write each of the expressions in Exx. 11–16 as a single product, of which the expression within the second parentheses is a factor.

Simplify the following expressions without removing the parentheses:

18. $(a - b)c + (b - a)c$.

19. $5(x - y)z + 5(y - x)z$.

20. $(1 - x)(1 + x^2) + (x - 1)(1 + x^2)$.

21. $(m - n)(m + n) - 3(n + m)(n - m)$.

22. $9(xy + 3)(z - 5) + 7(xy + 3)(5 - z)$.

CHAPTER VIII.

FACTORS AND MULTIPLES OF INTEGRAL ALGEBRAIC EXPRESSIONS.

§ 1. INTEGRAL ALGEBRAIC FACTORS.

1. Factors have already been defined in multiplication (Ch. II., § 3, Art. 15). The factors were there given, and their product was required.

The converse process, given a product to find its factors, is equally important, and to that we now turn.

2. A product of two or more factors is, by the definition of division, exactly divisible by any one of the factors.

An **Integral Algebraic Factor** of an integral expression is an integral expression by which the given one is exactly divisible.

E.g., integral factors of $6a^2x$ are 6, a^2x, $3x$, $2a^2$, etc.;
integral factors of $a^2 - b^2$ are $a + b$ and $a - b$;
integral factors of $ac + bc$ are c and $a + b$.

The word integral, here as in Ch. III., § 1, Art. 1, refers only to the *literal* parts of the expression.

E.g., $\frac{1}{2}a$ and $\frac{3}{4}x$ are integral algebraic factors of $6a^2x$.

We shall, however, not introduce numerical fractions as factors of integral algebraic expressions whose numerical coefficients are integers, unless there is some special reason for so doing.

3. Since any expression is divisible without a remainder by itself and by unity, it has itself and unity as factors.

Thus, 1 and a are factors of a; 1 and $x + y$ are factors of $x + y$.

4. A **Prime Factor** is one which is exactly divisible only by itself and unity.

E.g., the prime factors of $6a^2x$ are 2, 3, a, a, x.

A **Composite Factor** is one which is not prime, *i.e.*, which is itself the product of two or more prime factors.

E.g., composite factors of $6a^2x$ are 6, ax, $2a$, $3ax$, etc.

5. Any monomial can easily be resolved into its prime factors.

E.g., the prime factors of $4a^3b^2$ are 2, 2, a, a, a, b, b;
the prime factors of $3x(a+b)$ are 3, x, $a+b$.

The Fundamental Formula for Factoring.

6. A multinomial whose terms contain a common factor can be factored by applying the converse of the Distributive Law for Multiplication. From Ch. III., § 3, Art. 15, we have

$$ab + ac - ad = a(b + c - d).$$

That is, if the terms of a multinomial contain a common factor, the multinomial can be written as the product of the common factor and the algebraic sum of the remaining factors of the terms.

As we shall have occasion to refer frequently to the above relation, let us call it the *Fundamental Formula for Factoring.*

Ex. 1. Factor $2x^2y - 2xy^2$.

The factor $2xy$ is common to both terms; the remaining factor of the first term is x, that of the second term is $-y$, and their algebraic sum is $x - y$.

Consequently $2x^2y - 2xy^2 = 2xy(x - y)$.

Ex. 2. $ab^2 + abc + b^2c = b(ab + ac + bc)$.

7. In the fundamental formula the letters a, b, c, d may stand for binomial or multinomial expressions.

Ex. 1. Factor $a(x - 2y) + b(x - 2y)$.

The factor $x - 2y$ is common to both terms; the remaining factor of the first term is a, that of the second term is b, and their algebraic sum is $a + b$.

Consequently $a(x - 2y) + b(x - 2y) = (x - 2y)(a + b)$.

Ex. 2. Factor $1 - a + x(1 - a)$.

The factor $1 - a$ is common to both parts; the remaining factor of the first part is 1, that of the second part is x.

Hence $1 - a + x(1 - a) = (1 - a)(1 + x)$.

Ex. 3. Factor $(x - y)(a^2 + b^2) - (x + y)(a^2 + b^2)$.

The factor $a^2 + b^2$ is common to both parts; the remaining factor of the first part is $x - y$, that of the second part is $-(x + y)$.

Consequently

$$(x-y)(a^2+b^2)-(x+y)(a^2+b^2)=(a^2+b^2)[(x-y)-(x+y)]$$
$$= -2y(a^2 + b^2).$$

The beginner should at first, in examples like the last, write parentheses within parentheses, simplifying afterwards. As he advances in factoring he can shorten his work by performing some of the reductions mentally.

The following method of procedure is recommended:

Determine by inspection the common factor of the parts of the given expression; write this factor on the left of a parenthesis, (; *after this parenthesis write each term as it is obtained by dividing the corresponding term of the given expression by the common factor, followed by a closing parenthesis,*).

This division can usually be done mentally.

Ex. 4. $x^2(x-a)+x(x-a)^2+x-a=(x-a)(x^2+x[x-a]+1)$
$$=(x-a)(2x^2-ax+1).$$

Ex. 5. $x(a^2 + b^2) - x(a^2 - b^2) = x[(a^2 + b^2) - (a^2 - b^2)]$
$$= x[2b^2] = 2b^2x.$$

It frequently happens that the parts of a given expression have a common factor except for sign.

Ex. 6. Factor $x^2(1-m)-y^2(m-1)$.

Since $1-m$ and $m-1$ differ only in sign, *i.e.*,

$$m-1=-(1-m),$$

we may take either as the common factor.

Taking $1-m$ as the common factor, we have

$$x^2(1-m)-y^2(m-1)=(1-m)(x^2+y^2).$$

Taking $m-1$ as the common factor, we have

$$\begin{aligned} x^2(1-m)-y^2(m-1)&=(m-1)(-x^2-y^2)\\ &=-(m-1)(x^2+y^2). \end{aligned}$$

Ex. 7. $$\begin{aligned} a(m-n)+b(n-m)&=(m-n)(a-b)\\ &=(n-m)(-a+b). \end{aligned}$$

The fundamental formula must often be applied more than once.

Ex. 8. Factor $by(x-a)-bx(y-a)$.

Taking out the factor b, we obtain

$$b[y(x-a)-x(y-a)];$$

reducing the expression within the brackets, we obtain

$$b[ax-ay].$$

Taking out the factor a, we have

$$ab(x-y).$$

Ex. 9. $$\begin{aligned} &x(y-z)(a+b)+y(z-x)(a+b)\\ &=(a+b)[x(y-z)+y(z-x)]\\ &=(a+b)(-xz+yz)=z(a+b)(y-x). \end{aligned}$$

EXERCISES I.

Factor the following expressions:

1. $5x+5$.
2. $3x-3$.
3. $4m+4n$.
4. $ab+bc$.
5. $ax-a$.
6. $4a^3-6$.
7. $-x^3-x^2$.
8. a^2b-ab^2.
9. $2an-4n^2$.
10. $3x^4-2x^3$.
11. $12a^3b^3-3a^2b^2$.
12. $10a^4x^2-15a^2x^4$.
13. $30a^2b^3d+12a^5b^2c^4d^3$.
14. $15+20a-30a^2$.
15. $12a^2-60ab+24ad$.
16. $3ab+6ac-12ad$.

17. $12\,abd - 6\,bcd + 3\,bd^2$. **18.** $70\,xy - 98\,y^2 - 140\,yz$.

19. $\frac{15}{16}\,ax + \frac{15}{16}\,bx^2 + \frac{1}{4}\,x$. **20.** $6\,ax^4 - 15\,a^3bx^5 + 18\,a^2b^2x^6$.

21. $8\,a^2n^5x^5 - 10\,an^4x^7 + 4\,a^2n^3x^8$. **22.** $7\,(a + b) - 14$.

23. $16\,(m^2 + n^2) - 8$. **24.** $a^2(a + x) + x^2(a + x)$.

25. $a\,(n + 1) - b\,(n + 1)$. **26.** $5\,a^3(a - 2\,x) + 2\,(a - 2\,x)$.

27. $3\,a\,(a - 1) - 3\,(a - 1)$. **28.** $2\,(n + 1)^2 - 4\,(n + 1)$.

29. $a\,(x - 1) - x + 1$. **30.** $5\,ax^2(x - 2\,a)^2 + 15\,a^3x\,(x - 2\,a)$.

31. $4\,x\,(x - m) + m - x$. **32.** $m\,(q - p) - (p - q)$.

33. $2\,x\,(x^2 - x + 1) - 3\,(x^2 - x + 1)$.

34. $9\,a\,(a - 1)^3 + 36\,b\,(a - 1)^2 - 63\,(a - 1)^2$.

35. $-2\,a^x + 4\,a^{2x}$. **36.** $a^{n+1} - a$.

37. $5^{n+1} - 5$. **38.** $5^{n+3} - 125$.

39. $6\,m^{n+1} - 3\,m^n$. **40.** $8\,a^{m+n}b^{r+1} - 12\,a^{m+1}b^{r+p}$.

41. $ax^n - bx^{n+1} + cx^{n+2}$. **42.** $a^{n-1} + 2\,a^{n+1} - 3\,a^{n+3}$.

43. $3\,a^m - 4\,a^{m-2} - 5\,a^{m+1}$. **44.** $2^{n+4} - 8 \times 2^{n-1} + 16$.

8. When all the terms of a given expression do not contain a common factor, it is sometimes possible to group the terms in such a way that all the groups shall contain a common factor.

Ex. 1. Factor $2\,a + 2\,b + ax + bx$.

Factoring the first two terms by themselves, and the last two terms by themselves, we obtain

$$2\,(a + b) + x\,(a + b) = (a + b)\,(2 + x).$$

Ex. 2. $x^2 - xy - xz + yz = x\,(x - y) - z\,(x - y) = (x - y)\,(x - z)$.

Ex. 3. $(2\,a - b)^2 + 4\,ax - 2\,bx = (2\,a - b)^2 + 2\,x\,(2\,a - b)$
$= (2\,a - b)\,[2\,a - b + 2\,x]$.

EXERCISES II.

Factor the following expressions:

1. $ac + ad + bc + bd$. **2.** $2\,ax - 3\,by - 2\,ay + 3\,bx$.

3. $6\,ad + 10\,bd - 18\,af - 30\,bf$. **4.** $20\,ad - 35\,bd - 8\,ax + 14\,bx$.

5. $24\,ag - 32\,bg - 6\,ah + 8\,bh$. **6.** $5\,ax - cx - 5\,ay + cy$.

7. $33\,abdf - 21\,d^2f + 22\,abg - 14\,dg.$

8. $a^3 + a^2b + 2\,ab^2 + 2\,b^3.$

9. $a^3 - a^2c + ac^2 - c^3.$

10. $x^3 - x^2 + x - 1.$

11. $18\,n^2x - 12\,x - 9\,n^2 + 6.$

12. $3\,x^4 - x^3 + 6\,x - 2.$

13. $3\,c^4 - 3\,c^3n + cn^3 - n^3.$

14. $3\,r^3 + nx^2 - 6\,n^2x - 2\,x^3.$

15. $6\,n^3 - 9\,a^2n - 2\,an^2 + 3\,a^3.$

16. $6\,a^3 - 6\,a^2y + 2\,ay^2 - 2\,y^3.$

17. $\frac{9}{20}\,ad - \frac{12}{25}\,bd - \frac{21}{44}\,af + \frac{28}{55}\,bf.$

18. $\frac{1}{6}\,ty + \frac{2}{9}\,xy - \frac{1}{8}\,tz - \frac{1}{6}\,xz.$

19. $12\,a^3b^4 - 4\,a^2b^4 - 4\,a^2b^3 + 12\,a^3b^3.$

20. $\frac{4}{5}\,a^2b^2x^3 + \frac{3}{5}\,ab^3x^3 - \frac{4}{9}\,ab^2x^3 - \frac{1}{3}\,b^3x^3.$

21. $a^4 - a^3n^2 + a^2n - an^3 + n^5 - an^3.$

22. $ax^3 - bx^3 + ax - cx^2 - bx - cx.$

23. $ad - ah + bd - bh - ch + cd.$

24. $2\,ag + 3\,bg - 5\,gf + 8\,ah + 12\,bh - 20\,fh.$

25. $x^4 - ax^3 + 3\,a^2x^2 - 2\,a^2bx^2 + 2\,a^3bx - 6\,a^4b.$

26. $8\,x^{r+2} - 4\,x^{r+1}y - 2\,x^2y + 20\,x^{r+1}z + xy^2 - 5\,xyz.$

27. $ax + by + cz + bx + cy + az + cx + ay + bz.$

Use of Type-Forms in Factoring.

9. If an expression be in the form of one of the type-forms considered in Ch. VI., or if it can be reduced to such a form, its factors can be written by inspection.

We shall now consider the most important type-forms in connection with the expressions which they enable us to factor.

Trinomial Type-Forms.

10. From Ch. VI., § 1, Arts. 1 and 2, we have

$$a^2 + 2\,ab + b^2 = (a + b)^2,$$

$$a^2 - 2\,ab + b^2 = (a - b)^2.$$

From these identities we see that a trinomial which is the square of a binomial must satisfy the following three conditions:

(i.) *One term of the trinomial is the square of the first term of the binomial.*

(ii.) *A second term of the trinomial is the square of the second term of the binomial.*

(iii.) *The remaining term of the trinomial is twice the product of the two terms of the binomial.*

Ex. 1. Factor $x^2 + 6x + 9$.

x^2 is the square of x, 9 is the square of 3, and $6x = 2 \cdot x \cdot 3$.

Therefore $$x^2 + 6x + 9 = (x + 3)^2.$$

Notice that the following steps would have led to an equally correct result:

x^2 is the square of $-x$, 9 is the square of -3, and

$$6x = 2(-x)(-3).$$

Therefore $$x^2 + 6x + 9 = (-x - 3)^2.$$

Although the last result is equally correct, we naturally take the first form, which is simpler, unless there is some reason to the contrary.

Ex. 2. Factor $-4xy + 4x^2 + y^2$.

$4x^2$ is the square of $2x$, or of $-2x$; y^2 is the square of y, or of $-y$.

Since the middle term in the given expression is negative, we must take one of the terms of the binomial negative, the other positive.

Therefore $-4xy + 4x^2 + y^2 = (2x - y)^2 = (-2x + y)^2$.

Ex. 3. $4a^2x^2 - 12ab^2xy + 9b^4y^2 = (2ax - 3b^2y)^2$.

Ex. 4. $(n^2 - 2nx)^2 + 2(n^2x^2 - 2nx^3) + x^4$.

$$\begin{aligned} &= (n^2 - 2nx)^2 + 2x^2(n^2 - 2nx) + x^4 \\ &= (n^2 - 2nx + x^2)^2 \\ &= [(n - x)^2]^2 = (n - x)^4. \end{aligned}$$

Ex. 5. $60xy - 36x^2 - 25y^2 = -(36x^2 - 60xy + 25y^2)$

$$= -(6x - 5y)^2.$$

EXERCISES III.

Factor the following expressions:

1. $x^2 - 2x + 1$.
2. $a^2 + 6a + 9$.
3. $y^2 + 12y + 36$.
4. $a^2 - 10a + 25$.
5. $x^2 - 14x + 49$.
6. $4x^2 - 12x + 9$.
7. $9a^2 + 30a + 25$.
8. $20x - 4x^2 - 25$.
9. $36x - 4x^2 - 81$.
10. $-90x - 25x^2 - 81$.
11. $4x^2 - 12xy + 9y^2$.
12. $16a^2 + 40ab + 25b^2$.
13. $49x^2 - 28xy + 4y^2$.
14. $a^2b^2 + 2abcd + c^2d^2$.
15. $9x^2y^2 - 30xyz^2 + 25z^4$.
16. $24xy - 9x^2 - 16y^2$.
17. $a^4 - 2a^2x + x^2$.
18. $x^4 - 2x^2y^2 + y^4$.
19. $a^2x^2 - 4ac^3x + 4c^6$.
20. $2a^2x^2 - a^4 - x^4$.
21. $x^8 - 2x^4y^4 + y^8$.
22. $(a + x)^2 + 2(a + x) + 1$.
23. $(x - 4)^2 - 4(x - 4) + 4$.
24. $(2x - 9)^2 - 6(9 - 2x) + 9$.
25. $4x^{2n} - 12x^n + 9$.
26. $a^{2n} + 2a^np^n + p^{2n}$.
27. $25x^{4p} - 8x^{2p} + \frac{16}{25}$.
28. $36a^{n+2} - 48a^n + 16a^{n-2}$.
29. $4ax + 2a^2 + 2x^2$.
30. $2a^4n + a^5 + a^3n^2$.
31. $6a^2x^2 - 3a^2x^3 - 3a^2x$.
32. $8a^3x + 18ax^3 - 24a^2x^2$.
33. $16a^2b^6 + 9c^8 + 24ab^3c^4$.
34. $27a^4 + 3n^4x^6 - 18a^2n^2x^3$.
35. $a^5x^3 - 18a^3x^5 + 81ax^7$.
36. $a^{2n-2} - 2a^{n-1}x^{n+1} + x^{2n+2}$.
37. $a^{n+6} + a^nb^{12} - 2a^{n+3}b^6$.
38. $x^2(x + 2)^2 + 2(x + 2)^3 + 2x(x + 2)^2$.
39. $(a^2 + 2ab + b^2)c + (a + b)d^2$.
40. $xy - xz - (y^2 - 2yz + z^2)$.
41. $a^2 + 2an + n^2 - ap - pn$.
42. $2a + ad - d^2 - 4d - 4$.
43. $a^2 + 2ab - 4ac - 4bc + 4c^2$.
44. $n^2x - xy - n^4y + 2n^2y^2 - y^3$.
45. $8a^{n+5}x^{n+8} + 4a^{n+10}x^{n-2} + 4a^nx^{n+18}$.
46. $2a^3 - a^2n + (n - 2)(an - a)^2$.
47. $(a - c)^3 + 2a^2c - 4ac^2 + 2c^3$.

11. From Ch. VI., § 1, Art. 6, we have

$$x^2 + (a + b)x + ab = (x + a)(x + b).$$

In this identity a and b, either or both, may be positive or negative.

When a trinomial, arranged to descending powers of some letter, say x, can be factored into two binomials, in both of which the first term is the letter of arrangement, it must satisfy the following three conditions:

(i.) *One term of the trinomial is the square of the letter of arrangement,* i.e., *of the common first term of the binomial factors.*

(ii.) *The coefficient of the first power of the letter of arrangement in the trinomial is the algebraic sum of two numbers whose product is the remaining term of the trinomial.*

(iii.) *These two numbers are the second terms of the binomial factors.*

Ex. 1. Factor $x^2 + 8x + 15$.

The common first term of the binomial factors is evidently x. The second terms are two numbers whose product is 15, and whose sum is 8.

By inspection we see that

$$3 + 5 = 8 \text{ and } 3 \times 5 = 15;$$

that is, the second terms of the binomial factors are 3 and 5.

Consequently, $x^2 + 8x + 15 = (x + 3)(x + 5)$.

Ex. 2. Factor $x^2 - 7x + 12$.

The common first term of the binomial factors is x. The second terms are two numbers whose product is 12, and whose sum is -7. Since their product is *positive*, they must be *both positive* or *both negative;* and since their sum is negative, they must be *both negative.*

The possible pairs of negative factors of 12 are -1 and -12, -2 and -6, -3 and -4.

Since $$-3 + (-4) = -7,$$

the required second terms of the binomial factors are -3 and -4.

Consequently $x^2 - 7x + 12 = (x - 3)(x - 4)$.

Ex. 3. Factor $a^2x^2 + 5ax - 24$.

The common first term of the binomial factors is ax. The second terms are two numbers whose product is -24, and whose sum is 5. Since their product is negative, one must be positive and the other negative; and since their sum is positive, the positive number must have the greater absolute value. The possible pairs of factors of -24 are -1 and 24, -2 and 12, -3 and 8, -4 and 6.

Since $$-3+8=5,$$

the required second terms of the binomial factors are -3 and 8.

Consequently $a^2x^2+5ax-24=(ax-3)(ax+8)$.

Ex. 4. Factor $x^2-3xy-28y^2$.

The common first term of the binomial factors is x. The second terms are two numbers whose product is $-28y^2$, and whose sum is $-3y$. It is evident that both of these terms contain y as a factor. Therefore we have only to find their numerical coefficients.

Since their product is negative, one must be negative and the other positive; and since their sum is negative, the negative number must have the greater absolute value. The possible pairs of factors of -28 are 1 and -28, 2 and -14, 4 and -7.

Since $$4+(-7)=-3,$$

the required second terms of the binomial factors are $4y$ and $-7y$.

Consequently $x^2-3xy-28y^2=(x+4y)(x-7y)$.

EXERCISES IV.

Factor the following expressions:

1. x^2+3x+2.
2. x^2+x-2.
3. x^2-x-6.
4. x^2-5x+6.
5. x^2-6x+5.
6. $x^2-4x-60$.
7. $x^2+7x-30$.
8. $x^2+7x+12$.
9. $x^2+12x+32$.
10. $x^2+10x-11$.
11. $x^2-3x-40$.
12. $x^2-12x+35$.
13. x^3-17x^2+72x.
14. $x^2-6x-27$.
15. $x^2+13x-30$.
16. $x^2+18x+72$.
17. $120x^2+25x^3-5x^4$.
18. $6x-x^2-x^3$.
19. $35+2x-x^2$.
20. $24-11x+x^2$.
21. x^4+8x^2+15.
22. x^4+2x^2-24.
23. x^4-5x^2-24.

24. $x^4 - 24x^2 + 63$. **25.** $3x^6 + 39x^3 + 66$. **26.** $x^6 - x^3 - 56$.

27. $x^6 - 9x^3 - 22$. **28.** $x^6 - 28x^3 + 75$. **29.** $x^{2n} + 15x^n + 36$.

30. $x^{2m} + 10x^m - 24$. **31.** $x^{2n} + 6x^n - 112$. **32.** $x^{2n} - 16x^n + 55$.

33. $x^2 + (a + b)x + ab$. **34.** $x^2 - (m + n)x + mn$.

35. $x^2 + (p - q)x - pq$. **36.** $x^2 + (3r - 2s)x - 6rs$.

37. $ax^2 + 7a^2x + 6a^3$. **38.** $x^2 + 2xy - 15y^2$.

39. $x^2 - 4ax - 12a^2$. **40.** $x^2 - 7ax + 12a^2$.

41. $2ax^3y^2 - 26ax^2y^3 + 84axy^4$. **42.** $x^2 - 11xm + 30m^2$.

43. $x^2z^2 + 12xz - 13$. **44.** $a^2b^2 - 7ab + 10$.

45. $m^2n^2 - 20mn + 99$. **46.** $(a + b)^2 + 7(a + b) + 6$.

47. $(a - b)^2 + 7(a - b) + 12$. **48.** $(m + n)^2 + 2(m + n) - 15$.

49. $m(m - n)^2 - 4m(m - n) - 12m$.

12. By actual multiplication, we obtain

$$(ax + b)(cx + d) = acx^2 + (ad + bc)x + bd.$$

A trinomial which can be factored by this type-form must satisfy the following three conditions:

(i.) *One term of the trinomial is the product of the first terms of its binomial factors.*

(ii.) *A second term of the trinomial is the product of the second terms of its binomial factors.*

(iii.) *The remaining term of the trinomial is the sum of the products of the first term of each binomial factor by the second term of the other.*

In this type-form that part of the multiplication which gives the middle term of the type-form may be represented concisely by the following arrangement:

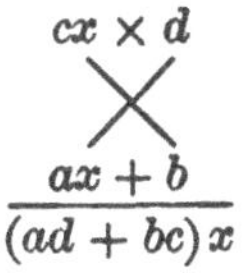

The products of the terms connected by the crossed lines are called *cross-products*, and their sum is the middle term of the given trinomial.

Ex. 1. Factor $6x^2 + 19x + 10$.

The first terms of the required binomial factors are factors of $6x^2$, the second terms are factors of 10; and the sum of the cross-products is $19x$.

The factors of $6x^2$ are x and $6x$, $2x$ and $3x$; and the factors of 10 are 1 and 10, 2 and 5.

The following arrangements represent possible pairs of factors:

$x+1$	$x+10$	$x+2$	$x+5$
$\times$	$\times$	$\times$	$\times$
$6x+10$	$6x+1$	$6x+5$	$6x+2$
$16x$	$61x$	$17x$	$32x$
$2x+1$	$2x+10$	$2x+2$	$2x+5$
$\times$	$\times$	$\times$	$\times$
$3x+10$	$3x+1$	$3x+5$	$3x+2$
$23x$	$32x$	$16x$	$19x$

Since the sum of the cross-products in the last arrangement is equal to the middle term of the given trinomial, we have

$$6x^2 + 19x + 10 = (2x+5)(3x+2).$$

Ex. 2. Factor $5x^2 - 6xy - 8y^2$.

The factors of $5x^2$ are x and $5x$, and the factors of $-8y^2$ are y and $-8y$, $-y$ and $8y$, $2y$ and $-4y$, $-2y$ and $4y$.

The following arrangements represent possible pairs of factors:

$x+y$	$x-8y$	$x-y$	$x+8y$
$\times$	$\times$	$\times$	$\times$
$5x-8y$	$5x+y$	$5x+8y$	$5x-y$
$-3xy$	$-39xy$	$3xy$	$39xy$
$x+2y$	$x-4y$	$x-2y$	$x+4y$
$\times$	$\times$	$\times$	$\times$
$5x-4y$	$5x+2y$	$5x+4y$	$5x-2y$
$6xy$	$-18xy$	$-6xy$	$18xy$

Since the sum of the cross-products in the next to the last arrangement is equal to the middle term of the given trinomial, we have

$$5x^2 - 6xy - 8y^2 = (x - 2y)(5x + 4y).$$

Observe that the reason given in Art. 11, Ex. 4, for rejecting at sight some factors of the last term of the trinomial does not hold in the above example.

For, although the middle term, $-6xy$, is negative, the negative factor of $-8y^2$ is less in absolute value than the positive factor.

The student should accustom himself to determine the factors of a given trinomial without the continued use of such arrangements to represent possible pairs of factors. They have been given here because the beginner does not always find it easy to choose the proper factors and needs some systematic guidance. He will soon learn to reject some pairs of factors at sight, and to test other pairs of factors by performing mentally the cross-multiplications and additions.

Ex. 3. Factor $6a^2 - 17ab + 10b^2$.

The factors of $6a^2$ are a and $6a$, $2a$ and $3a$; and the factors of $10b^2$ are b and $10b$, $-b$ and $-10b$, $2b$ and $5b$, $-2b$ and $-5b$.

Since the second terms of the required binomial factors are either both positive or both negative, and since the sum of the cross-products, $-17ab$, is negative, the positive factors of $10b^2$ can be rejected at sight. By trial we then find that the arrangement

$$\begin{array}{c} a - 2b \\ \times \\ 6a - 5b \\ \hline -17ab \end{array}$$

gives the middle term of the trinomial.

Consequently, $6a^2 - 17ab + 10b^2 = (a - 2b)(6a - 5b)$.

Ex. 4. Factor $10a^4 + a^2b - 21b^2$.

Notice that this example is arranged to descending powers of a^2.

The factors of $10a^4$ are a^2 and $10a^2$, $2a^2$ and $5a^2$; and the factors of $-21b^2$ are b and $-21b$, $-b$ and $21b$, $3b$ and $-7b$, $-3b$ and $7b$.

By trial we find that the arrangement

$$\begin{array}{c} 2a^2 + 3b \\ \times \\ \underline{5a^2 - 7b} \\ a^2b \end{array}$$

gives the required middle term of the trinomial.

Consequently, $10a^4 + a^2b - 21b^2 = (2a^2 + 3b)(5a^2 - 7b)$.

If the first term of the given trinomial be negative, either of its factors, *i.e.*, the first term of either of the required binomial factors, may be taken positively, and the other negatively.

Thus, since

$$(2x - 5)(-3x + 4) = -6x^2 + 23x - 20,$$

and

$$(-2x + 5)(3x - 4) = -6x^2 + 23x - 20,$$

the factors of $-6x^2 + 23x - 20$ are

$$2x - 5 \text{ and } -3x + 4, \text{ or } -2x + 5 \text{ and } 3x - 4.$$

Ex. 5. Factor $-15x^2 + 22x - 8$.

The factors of $-15x^2$ are x and $-15x$, $3x$ and $-5x$; the factors $-x$ and $15x$, $-3x$ and $5x$, need not be considered, since, as we have seen, either factor may be taken positively and the other negatively.

The factors of -8 are 1 and -8, -1 and 8, 2 and -4, -2 and 4.

By trial we find that the arrangement

$$\begin{array}{c} 3x - 2 \\ \times \\ \underline{-5x + 4} \\ 22x \end{array}$$

gives the middle term of the trinomial.

Consequently, $-15x^2+22x-8=(3x-2)(-5x+4)$.

If we had taken the factors $-3x$ and $5x$ of $-15x^2$, we should have obtained

$$-15x^2+22x-8=(-3x+2)(5x-4).$$

13. The following directions may be observed in factoring trinomials which come under this type-form:

(i.) *When all the terms of the trinomial are positive, only positive factors of the last term are to be tried.*

(ii.) *When the middle term of the trinomial is negative and the last term is positive, the factors of the last term must be both negative.*

(iii.) *When the middle term and the last term of the trinomial are both negative, one of the factors of the last term must be positive and the other negative.*

(iv.) *Select that pair of factors of the last term which, by cross-multiplication and addition, gives the middle term of the trinomial.*

EXERCISES V.

Factor the following expressions:

1. $2x^2+5x+2$. **2.** $10+16x+6x^2$. **3.** $4x^2+8x+3$.

4. $9x^2+36x+20$. **5.** $6+13x-63x^2$. **6.** $3x^2+13x+12$.

7. $6x^2+19x-36$. **8.** $40+2x-2x^2$. **9.** $25x^3+25x^2-6x$.

10. $36x^4-18x^2-10$. **11.** $12x-6x^2-90x^3$.

12. $10x^2+7x-33$. **13.** $8x^4-19x^2-15$.

14. $40+6x-27x^2$. **15.** $49x^2-35x+6$.

16. $64x^2-92x+30$. **17.** $6-19x+15x^2$.

18. $6x^2-41x-56$. **19.** $14x^2-39x+10$.

20. $30x^2-89x+35$. **21.** $18x^2-3xy-45y^2$.

22. $3a^2-5ab-2b^2$. **23.** $18x^2y^2-71xyz-45z^2$.

24. $18x^4+3x^2y-10y^2$. **25.** $abx^2-(a^2-b^2)x-ab$.

26. $5a^4x^2-4a^2xz-96z^2$. **27.** $-10a^4+7a^2b^2+12b^4$.

28. $4x^2-xy-3y^2$. **29.** $10a^2+11ab-6b^2$.

30. $9x^{2n}-4x^n-5$. **31.** $2x^{2r+2}-3x^{r+1}-2$.

32. $3x^{2m-4} + 4x^{m-2} - 4$. **33.** $6x^{2m} + x^m y^n - 15y^{2n}$.

34. $10(a+b)^2 + 7c(a+b) - 6c^2$. **35.** $7(x-y)^2 - 37z(x-y) + 10z^2$.

36. $6(x^2 + y^2)^2 - 9(x^2 + y^2)z^2 - 15z^4$.

37. $2(a^2 - c^2)^2 - 4b(a^2 - c^2) - 6b^2$.

Binomial Type-Forms.

14. From Ch. VI., § 1, Art. 5, we have

$$a^2 - b^2 = (a + b)(a - b).$$

That is, *the difference of the squares of two numbers can be written as the product of the sum and the difference of the numbers.*

Ex. 1.
$$a^2x^2 - \tfrac{1}{4}b^2 = (ax)^2 - (\tfrac{1}{2}b)^2$$
$$= (ax + \tfrac{1}{2}b)(ax - \tfrac{1}{2}b).$$

Ex. 2.
$$32m^4n - 2n^3 = 2n(16m^4 - n^2)$$
$$= 2n[(4m^2)^2 - n^2]$$
$$= 2n(4m^2 + n)(4m^2 - n).$$

As in the preceding type-forms, the letters a and b may stand for multinomials.

Ex. 3.
$$(a+b)^2 - (a-b)^2 = [(a+b) + (a-b)][(a+b) - (a-b)]$$
$$= (2a)(2b) = 4ab.$$

Ex. 4.
$$x^2 - 4xy + 4y^2 - 9z^2 = (x - 2y)^2 - (3z)^2$$
$$= (x - 2y + 3z)(x - 2y - 3z).$$

In factoring a given expression the type-form must frequently be applied more than once.

Ex. 5.
$$4a^2c^2 - (a^2 - b^2 + c^2)^2$$
$$= (2ac + a^2 - b^2 + c^2)(2ac - a^2 + b^2 - c^2)$$
$$= [(a + c)^2 - b^2][b^2 - (a - c)^2]$$
$$= (a + c + b)(a + c - b)(b + a - c)(b - a + c).$$

15. The difference of the like, or unlike, even powers of two numbers can always be written as the difference of the squares of two numbers, and should therefore first be factored by applying this type-form.

Ex. 6.
$$a^4 - b^4 = (a^2)^2 - (b^2)^2$$
$$= (a^2 + b^2)(a^2 - b^2)$$
$$= (a^2 + b^2)(a + b)(a - b).$$

EXERCISES VI.

Factor the following expressions:

1. $x^2 - 1.$ 2. $4 - a^2.$ 3. $a^2 - x^2y^2.$
4. $25\,x^2 - 9.$ 5. $36\,a^2 - 49\,b^2.$ 6. $4\,x^2 - y^4.$
7. $86^2 - 14^2.$ 8. $57^2 - 43^2.$ 9. $37^2 - 27^2.$
10. $81\,a^4 - 16.$ 11. $25\,a^2x^2 - 16\,b^2y^2.$ 12. $\frac{4}{9}\,a^2b^2 - \frac{25}{49}\,c^2d^2.$
13. $16\,a^6 - 25\,b^4c^6.$ 14. $a^2b^4c^6 - \frac{1}{4}.$ 15. $\frac{1}{9}\,a^2n^4 - \frac{1}{100}\,x^6.$
16. $a^{2n} - 1.$ 17. $a^{2n} - b^{2m}.$ 18. $x^{2r+2} - 4.$
19. $x^{2m+2} - 9\,y^{6m}.$ 20. $9\,a^{2n}b^2 - 4\,c^{2m}.$ 21. $36\,a^{2m+2}b^{2n-2} - 1.$
22. $(m - n)^2 - 1.$ 23. $(a + b)^2 - c^2.$ 24. $c^2 - (a - b)^2.$
25. $4 - (2 - x)^2.$ 26. $9 - (3 - x)^2.$ 27. $(4\,x - 3)^2 - 16\,x^2.$
28. $(a - b)^2 - (c - d)^2.$ 29. $(5\,x - 2)^2 - (4\,x - 3)^2.$
30. $(7\,a + 6)^2 - (5\,a - 8)^2.$ 31. $(2\,a + 3\,b)^2 - (a + 2\,b)^2.$
32. $(3\,xy - 5)^2 - (2\,xy - 6)^2.$ 33. $(x^2 + x + 1)^2 - (x^2 - x + 1)^2.$
34. $a^4 - 1.$ 35. $7 - 112\,x^4.$
36. $16\,x^4 - y^4.$ 37. $625\,a^4x^4 - b^4y^4.$ 38. $a^8 - b^8.$
39. $1 - 256\,x^8y^8.$ 40. $x^{16} - y^{16}.$ 41. $a^{16} - 1.$
42. $5\,a^2 - 180\,b^4.$ 43. $75\,a^2b^4 - 108\,c^2d^4$
44. $300\,abc^2 - 432\,abd^2.$ 45. $243\,b^5c^6 - 75\,b^7.$
46. $18\,a^2x^3 - 98\,b^2xy^4.$ 47. $\frac{2}{4}\,ab^2 - \frac{2}{9}\,ac^4.$
48. $\frac{5}{4}\,xy^4 - \frac{5}{25}\,xz^6.$ 49. $\frac{7}{16}\,a^2b^3c^4 - \frac{7}{36}\,a^4b^5c^8.$
50. $\frac{2}{5}\,a^6n^2z^3 - \frac{8}{5}\,a^2x^6z^7.$ 51. $a^{2n} - b^{2m}.$
52. $a^{4x} - b^{4x}.$ 53. $144\,x^n - x^{n+2}.$
54. $4\,a^{3n+3} - a^{n+1}.$ 55. $a^{2n+3}b^{2n} - a^3b^{2n+2}$
56. $9\,x^{n+5}y^{2n+7} - 16\,x^{n+3}y^{2n+1}$ 57. $\frac{1}{4}\,a^{5n-3}b^{5m} - \frac{1}{9}\,a^{3n-1}b^m$
58. $m^{2n-4}n^{6m+2} - 1.$ 59. $xy^{5p+7r} - 9\,x^{2n+1}y^{p+r}.$
60. $9\,x^{n+5}y^{2n+7} - 16\,x^{n+3}y^{2n+1}.$ 61. $\frac{1}{4}\,a^{5n-3}b^{5m} - \frac{1}{9}\,a^{3n-1}b^m.$
62. $xy^{5p+7r} - 9\,x^{2n+1}y^{p+r}.$ 63. $a^2 - b^2 + (a + b)\,c.$

64. $x^2+3x^3-x^4-3x.$ **65.** $a^2-x^2+a-x.$

66. $x^2-xz-yz-y^2.$ **67.** $a^3-a^2n+an^2-n^3.$

68. $a^2+2ab+b^2-c^2.$ **69.** $x^2-2xy+y^2-z^2.$

70. $z^2-x^2+2xy-y^2.$ **71.** $a^2+2bc-b^2-c^2.$

72. $a^2-n^2+2np-p^2.$ **73.** $p^2-z^2-4z-4.$

74. $a^4-2ab^3-b^4+2a^3b.$ **75.** $a^2+b^2-c^2-d^2+2(ab+cd).$

76. $a^2+b^2-c^2-d^2-2(ab-cd).$ **77.** $2(ab+cd)-(a^2+b^2-c^2-d^2).$

78. $2(ab-cd)-(a^2+b^2-c^2-d^2).$ **79.** $a^2-b^2+2bz-2ax+x^2-z^2.$

80. $a^4+4a^2c-4b^2+4bd^2+4c^2-d^4.$ **81.** $4a^2b^2-(a^2+b^2-c^2)^2.$

82. $4(ad+bc)^2-(a^2-b^2-c^2+d^2)^2.$ **83.** $a^{2r}-a^{4r}-2a^{7r}-a^{10r}.$

84. $x^3y-xy^3+x^2y+xy^2.$ **85.** $(a+n)(a^2-x^2)-(a-x)(a^2-n^2).$

86. $(n-x)(5n^2-4x^2)-(3x^2-4n^2)(x-n).$

16. From Ch. VI., § 2, Art. 2 (i.) and (ii.), we derive

$$\boldsymbol{a^3+b^3=(a+b)(a^2-ab+b^2)}$$
$$\boldsymbol{a^3-b^3=(a-b)(a^2+ab+b^2).}$$

Ex. **1.**
$$\begin{aligned} x^3+8y^3 &= x^3+(2y)^3 \\ &=(x+2y)[x^2-x(2y)+(2y)^2] \\ &=(x+2y)(x^2-2xy+4y^2). \end{aligned}$$

Ex. **2.**
$$\begin{aligned} 27-125x^6y^6 &= 3^3-(5x^2y^2)^3 \\ &=(3-5x^2y^2)[3^2+3(5x^2y^2)+(5x^2y^2)^2] \\ &=(3-5x^2y^2)(9+15x^2y^2+25x^4y^4]. \end{aligned}$$

Ex. **3.**
$$\begin{aligned} (1-x)^3-8x^3 &=(1-x)^3-(2x)^3 \\ &=(1-x-2x)[(1-x)^2+(1-x)(2x)+(2x)^2] \\ &=(1-3x)(1+3x^2). \end{aligned}$$

Ex. **4.**
$$\begin{aligned} 512x^9+y^9 &=(8x^3)^3+(y^3)^3 \\ &=(8x^3+y^3)[(8x^3)^2-(8x^3)(y^3)+(y^3)^2] \\ &=[(2x)^3+y^3](64x^6-8x^3y^3+y^6) \\ &=(2x+y)(4x^2-2xy+y^2)(64x^6-8x^3y^3+y^6). \end{aligned}$$

Ex. **5.**
$$\begin{aligned} a^6-729b^6 &=(a^3)^2-(27b^3)^2 \\ &=(a^3+27b^3)(a^3-27b^3) \\ &=(a+3b)(a^2-3ab+9b^2)(a-3b)(a^2+3ab+9b^2). \end{aligned}$$

17. From Ch. VI., § 2, Art. 2 (i.) and (ii.), we infer:

(i.) *The sum of the like odd powers of two numbers contains the sum of the numbers as a factor.*

(ii.) *The difference of the like odd powers of two numbers contains the difference of the numbers as a factor.*

Ex. 1. $x^5 + y^5 = (x + y)(x^4 - x^3y + x^2y^2 - xy^3 + y^4)$.

Ex. 2. $x^7 - y^7 = (x - y)(x^6 + x^5y + x^4y^2 + x^3y^3 + x^2y^4 + xy^5 + y^6)$.

18. The sum of the like even powers of two numbers, whose exponents are divisible by an odd number, except 1, can be factored by applying the type-forms of Arts. 16 and 17.

Ex.
$$\begin{aligned} x^{12} + y^{12} &= (x^4)^3 + (y^4)^3 \\ &= (x^4 + y^4)[(x^4)^2 - (x^4)(y^4) + (y^4)^2] \\ &= (x^4 + y^4)(x^8 - x^4y^4 + y^8). \end{aligned}$$

EXERCISES VII.

Factor the following expressions:

1. $x^3 + 1$. **2.** $x^3 - 8$. **3.** $a^3 + 27$.

4. $64x^3 - 1$. **5.** $a^3b^3 + 1$. **6.** $8x^3y^3 - 27$.

7. $27a^3 + \frac{64}{1000}b^3$. **8.** $8x^3 - y^6$. **9.** $64a^3b^3 - 1$.

10. $125x^3y^6 + 8$. **11.** $x^4 + x$. **12.** $3a^2 - 24a^5$.

13. $4x^2 - 32x^5y^3$. **14.** $27a - a^4b^6$. **15.** $3a^3b^4c^5 - 375bc^8$.

16. $2x^3y^5 + 432y^2$. **17.** $a^5 + 243$. **18.** $x^6 + y^6$.

19. $x^6 - 64$. **20.** $x^6 - 64y^6$. **21.** $x^9 + y^9$. **22.** $x^9 - 1$.

23. $a^{10} - b^{10}$. **24.** $x^{10} + y^{10}$. **25.** $x^{12} - 1$. **26.** $x^{12}y^{12} - z^{12}$.

27. $n^{12} - 1$. **28.** $x^{14}y^{14} - 1$. **29.** $x^{14} + y^{14}$. **30.** $a^{15} + b^{15}$.

31. $1 - x^{15}$. **32.** $1 - a^{18}$. **33.** $x^{18} + y^{18}$. **34.** $a^{3n} - b^{3n}$.

35. $8x^{3n}y^n - 729y^{n+3}z^6$. **36.** $81a^9x^{4n} - 648x^{n+6}y^9$.

37. $1 - (x + y)^3$. **38.** $8 - (a + b + c)^3$.

39. $27 - (3 + 2x)^3$. **40.** $(a + b)^3 - (c + d)^3$.

41. $(2a + x)^3 + (a - 2x)^3$. **42.** $x^5 - x^3 - x^2 + 1$.

43. $4 - x^2 + 4x^3 - x^5$. **44.** $x^3 - 8 - 6x^2 + 12x$.

45. $x^3 - y^3 - 2x^2y + 2xy^2$. **46.** $a^3 - 4a^2c - 4ac^2 + c^3$.

47. $n^6 + 5n^4x^2 + 5n^2x^4 + x^6$.

Multinomial Type-Forms.

19. From Ch. VI., § 1, Art. 3, we derive

$$a^2 + b^2 + c^2 + 2ab + 2ac + 2bc = (a + b + c)^2,$$

wherein a, b, and c may be positive or negative.

That is, *if three terms of a multinomial of six terms be the squares of three numbers, respectively, and if the three remaining terms be equal to twice the products of these numbers, taken in pairs, the multinomial is the square of the sum of the three numbers.*

Ex. Factor $4x^2 - 12xy + 9y^2 + 16xz - 24yz + 16z^2$.

Since some of the terms of the given multinomial are negative, one or more of the terms of the factors must be negative.

We have

$$4x^2 = (2x)^2, \text{ or } (-2x)^2;$$
$$9y^2 = (3y)^2, \text{ or } (-3y)^2;$$
$$16z^2 = (4z)^2, \text{ or } (-4z)^2.$$

The term $+16xz$ shows that the terms in x and z have the same sign; the terms $-12xy$ and $-24yz$ show that when the terms in x and z have the sign $+$, the term in y has the sign $-$, and when the terms in x and z have the sign $-$, the term in y has the sign $+$.

If the terms in x and z be taken with the sign $+$, we have

$$-12xy = 2(2x)(-3y),$$
$$16xz = 2(2x)(4z),$$
$$-24yz = 2(-3y)(4z).$$

Consequently,

$$4x^2 - 12xy + 9y^2 + 16xz - 24yz + 16z^2 = (2x - 3y + 4z)^2.$$

If the terms in x and z be taken with the sign $-$, we have

$$-12xy = 2(-2x)(3y),$$
$$16xz = 2(-2x)(-4z),$$
$$-24yz = 2(3y)(-4z).$$

Consequently,

$$4x^2 - 12xy + 9y^2 + 16xz - 24yz + 16z^2 = (-2x + 3y - 4z)^2.$$

Notice that the trinomials

$$2x - 3y + 4z \text{ and } -2x + 3y - 4z, = -(2x - 3y + 4z),$$

differ only in sign, and that therefore their squares are equal.

EXERCISES VIII.

Factor the following expressions:

1. $a^2 + b^2 + 2a + 2b + 2ab + 1.$
2. $4a^2 + b^2 + c^2 + 4ab + 4ac + 2bc.$
3. $1 + 4a + 4a^2 - 6b + 9b^2 - 12ab.$
4. $a^4 + 4b^2 + 9c^2 - 4ab + 6ac - 12bc.$
5. $4x^2 + 4y^2 + z^2 - 8xy - 4yz + 4xz.$
6. $36x^2 + 60xy + 25y^2 + 48x + 40y + 16.$
7. $16x^2 + \frac{1}{4}y^2 + 4z^2 - 4xy - 16xz + 2yz.$
8. $25x^2 + \frac{1}{4}y^2 + \frac{1}{4}z^2 + 5xy - 5xz - \frac{1}{2}yz.$
9. $a^2 + b^2 + c^2 + 2ab + 2ac + 2bc - 4.$
10. $x^2 + \frac{1}{4}y^2 + z^2 + xy + 2xz + yz - a^2 - 2ab - b^2.$

20. From Ch. VI., § 1, Arts. 7 and 8, we derive

$$a^3 + 3a^2b + 3ab^2 + b^3 = (a + b)^3,$$

wherein a and b may be positive or negative.

From this identity we see that a multinomial of four terms which is the cube of a binomial must satisfy the following three conditions:

(i.) *One term of the multinomial is the cube of the first term of the binomial.*

(ii.) *A second term of the multinomial is the cube of the second term of the binomial.*

(iii.) *The two remaining terms of the multinomial are three times the product of the square of the first term of the binomial by the second, and three times the product of the first term of the binomial by the square of the second, respectively.*

Ex. Factor $8 - 36x + 54x^2 - 27x^3$.

8 is the cube of 2, $-27x^3$ is the cube of $-3x$, and

$$-36x = 3 \times 2^2 \times (-3x), \quad 54x^2 = 3 \times 2 \times (-3x)^2.$$

Therefore, $8 - 36x + 54x^2 - 27x^3 = (2 - 3x)^3$.

EXERCISES IX.

Factor the following expressions:

1. $27a^3 - 27a^2b + 9ab^2 - b^3$.
2. $8a^3 - 12a^2b + 6ab^2 - b^3$.
3. $8x^3 - 36x^2y + 54xy^2 - 27y^3$.
4. $125a^3 - 525a^2b - 735ab^2 - 343b^3$.
5. $x^6 - 12x^4 + 48x^2 - 64$.
6. $a^6 - 15a^4b^2 + 75a^2b^4 - 125b^6$.
7. $x^9 - 18x^6y^2 + 108x^3y^4 - 216y^6$.
8. $x^{3n} - 3x^{2n} + 3x^n - 1$.
9. $1 - 6x + 12x^2 - 8x^3 - a^3 + 3a^2b - 3ab^2 + b^3$.

21. The method of Art. 12 can be extended to factor multinomials of the second degree whose factors contain three or more terms.

Ex. 1. Factor $ax^2 - ay^2 + x + y - ax + ay - 1$.

The terms $ax^2 - ay^2$ are evidently the product of the terms which contain x and y in the two factors. These may be either

$$ax + ay \text{ and } x - y, \text{ or } ax - ay \text{ and } x + y.$$

Since the last term of the multinomial does not contain either x or y, it must be the product of terms in the factors which do not contain x or y. These can be only $+1$ and -1. Therefore, the following arrangements represent possible pairs of factors:

$(ax+ay)+1$	$(ax+ay)-1$	$(ax-ay)+1$	$(ax-ay)-1$
$\times$	$\times$	$\times$	$\times$
$(x-y)-1$	$(x-y)+1$	$(x+y)-1$	$(x+y)+1$
$x-y-ax-ay$	$-x+y+ax+ay$	$x+y-ax+ay$	$-x-y+ax-ay$

Only the sums of the cross-products are given above. Since the third arrangement gives the remaining terms, $x+y-ax+ay$, of the multinomial, we have

$$ax^2-ay^2+x+y-ax+ay-1=(x+y-1)(ax-ay+1).$$

Ex. 2. Factor $x^2+3xy+2y^2-x-3y-2$.

The factors of the part of the second degree are $x+y$ and $x+2y$; the factors of the last term are 1 and -2, or -1 and 2.

Therefore the following arrangements represent possible pairs of factors:

$$\begin{array}{c} (x+y)+1 \\ \underline{(x+2y)-2} \\ -x \end{array} \qquad \begin{array}{c} (x+y)-2 \\ \underline{(x+2y)+1} \\ -x-3y \end{array} \qquad \begin{array}{c} (x+y)-1 \\ \underline{(x+2y)+2} \\ x \end{array} \qquad \begin{array}{c} (x+y)+2 \\ \underline{(x+2y)-1} \\ x+3y \end{array}$$

Since the second arrangement gives the remaining terms, $-x-3y$, of the multinomial, we have

$$x^2+3xy+2y^2-x-3y-2=(x+2y+1)(x+y-2).$$

Ex. 3. Factor $2x^2-12y^2+4z^2-5xy-8yz-9xz$.

The given expression is of the form of the product of two trinomials, both of which contain terms in x, y, and z. The part $2x^2-5xy-12y^2$ is evidently the product of the parts of the factors which contain terms in x and y, and the term $4z^2$ is evidently the product of the terms in z in the factors.

The factors of $2x^2-5xy-12y^2$ are found to be $2x+3y$ and $x-4y$; the factors of $4z^2$ are z and $4z$, $-z$ and $-4z$, $2z$ and $2z$, $-2z$ and $-2z$. By trial we find that the following arrangement

$$\begin{array}{c} (2x+3y)-\ \ z \\ \underline{(x-4y)-4z} \\ -9xz-8yz \end{array}$$

gives the remaining terms of the multinomial.

Consequently,

$$2x^2-12y^2+4z^2-5xy-8yz-9xz$$
$$=(2x+3y-z)(x-4y-4z).$$

EXERCISES X.

Factor the following expressions:

1. $y^2-4x^2+4x-1.$
2. $x^2-y^2-2y-1.$
3. $a^2+b^2+2ab+8a+8b-9.$
4. $x^2+xy-2x-6y^2-16y-8.$
5. $x^2+2xy+3x+y^2+3y+2.$
6. $3a^2-7ab-6b^2-bc+4ac+c^2.$
7. $2x^2-8xy+23x-12y+30.$
8. $2x^2-3y^2-z^2+xy+xz+4yz.$
9. $x^2+3xy+5x+2y^2+8y+6.$
10. $2a^2-9ac-5ab+4c^2-8bc-12b^2.$
11. $x^2+3xy+3x+2y^2+5y+2.$
12. $x^2-3xy+3x+2y^2-5y+2.$
13. $x^2+4y^2-4xy-10x+20y-56.$
14. $4x^2+9y^2+19-12xy+40x-60y.$
15. $6x^6-x^3y^3-6x^3z^3-2y^6+11y^3z^3-12z^6.$
16. $56x^2-6y^2-12z^2-5xy+34xz-22yz.$
17. $3x^2-4ax+2bx-2ab+a^2.$
18. $28a^4-21a^3b+23a^2b^2-12ab^3+4b^4.$

22. By actual multiplication, we can verify the following identities:

$$a^3+b^3+c^3-3abc=(a+b+c)(a^2+b^2+c^2-ab-ac-bc) \quad \text{(i.)}$$

$$(a+b+c)^3-(a^3+b^3+c^3)=3(a+b)(a+c)(b+c) \quad \text{(ii.)}$$

$$2(a^2b^2+a^2c^2+b^2c^2)-(a^4+b^4+c^4)$$
$$=(a+b+c)(a+b-c)(a-b+c)(b+c-a) \quad \text{(iii.)}$$

Ex. 1. Factor $8x^3-y^3-125z^3-30xyz.$

Since $8x^3=(2x)^3,\ -y^3=(-y)^3,\ -125z^3=(-5z)^3,$

and $-30xyz=-3(2x)(-y)(-5z),$

we have

$$8x^3-y^3-125z^3-30xyz=(2x-y-5z)(4x^2+y^2+25z^2+2xy+10xz-5yz).$$

Ex. 2. Factor $(2x+3y-5z)^3-(27y^3+8x^3-125z^3)$.

Since $8x^3=(2x)^3,\ 27y^3=(3y)^3,\ -125z^3=(-5z)^3,$

we have

$$(2x+3y-5z)^3-(27y^3+8x^3-125z^3)=3(2x+3y)(2x-5z)(3y-5z).$$

Ex. 3. Factor $8a^2+18b^2+72a^2b^2-(1+16a^4+81b^4)$.

Since $1=1^4,\ 16a^4=(2a)^4,\ 81b^4=(3b)^4,$

and $8a^2=2(1)^2(2a)^2,\ 18b^2=2(1)^2(3b)^2,\ 72a^2b^2=2(2a)^2(3b)^2,$

we have $8a^2+18b^2+72a^2b^2-(1+16a^4+81b^4)$

$$=(1+2a+3b)(1+2a-3b)(1-2a+3b)(2a+3b-1).$$

EXERCISES XI.

Factor the following expressions:

1. $a^3+b^3-c^3+3abc.$
2. $(5x-2y+7z)^3+8y^3-343z^3-125x^3.$
3. $8x^2y^2+2x^2z^2+8y^2z^2-x^4-16y^4-z^4.$
4. $72x^2y^2+18x^2z^2+8y^2z^2-81x^4-16y^4-z^4.$
5. $1-8m^3-512n^3+(2m-1+8n)^3.$
6. $a^3-b^3-c^3-3abc.$
7. $x^3+y^3-1+3xy.$
8. $(ab-3+5cd)^3-(a^3b^3-27+125c^3d^3).$
9. $8x^2+72x^2y^2+18y^2-16x^4-81y^4-1.$
10. $72x^2+128y^2+288x^2y^2-81x^4-256y^4-16.$
11. $x^3y^6-x^6y^3+x^3y^3z^3+(x^2y-xy^2-xyz)^3.$
12. $x^3-8y^3-27z^3-18xyz.$
13. $\frac{1}{8}x^3+\frac{1}{27}y^3+\frac{1}{64}z^3-\frac{1}{8}xyz.$
14. $8x^2+32y^2+128x^2y^2-16x^4-256y^4-1.$
15. $(5m-3n+2)^3-8+27n^3-125m^3.$
16. $343a^3b^3-512d^3e^3-1000f^3g^3+(10fg+8ed-7ab)^3.$
17. $2x^2y^2+\frac{9}{2}x^2z^2+72y^2z^2-\frac{1}{16}x^4-16y^4-81z^4.$
18. $\frac{1}{18}x^2+\frac{1}{72}y^2+\frac{1}{32}x^2y^2-\frac{1}{16}x^4-\frac{1}{256}y^4-\frac{1}{81}.$
19. $27x^3+\frac{1}{8}y^3-8z^3+9xyz.$
20. $\frac{1}{8}x^3-125y^3-\frac{8}{27}z^3-5xyz.$
21. $72x^2+2y^2+\frac{9}{2}x^2y^2-16-81x^4-\frac{1}{16}y^4.$

Special Devices for Factoring.

23. The reduction of factorable expressions to known type-forms will now be taken up. Only general suggestions, not a definite mode of procedure for all expressions, can be given. Much will in the end depend upon the ingenuity of the student.

24. A factorable expression can frequently be brought to some known type-form by adding to, or subtracting from it one or more terms.

Ex. 1. Factor $x^4 + x^2y^2 + y^4$.

This expression would be the square of $x^2 + y^2$, if the coefficient of x^2y^2 were 2. We therefore add x^2y^2, and, in order that the value of the expression may remain the same, we subtract x^2y^2. We then have

$$\begin{aligned} x^4 + 2x^2y^2 + y^4 - x^2y^2 &= (x^2 + y^2)^2 - x^2y^2 \\ &= (x^2 + y^2 + xy)(x^2 + y^2 - xy). \end{aligned}$$

Ex. 2. Factor $x^3 + 3x^2 - 2$.

The terms $x^3 + 3x^2$ suggest the cube of $x + 1$.

To complete the cube of $x + 1$ we must add, and therefore also subtract, $3x + 1$.

Doing so, we obtain

$$\begin{aligned} x^3 + 3x^2 + 3x + 1 - 3x - 3 &= (x + 1)^3 - 3(x + 1) \\ &= (x + 1)[(x + 1)^2 - 3] \\ &= (x + 1)(x^2 + 2x - 2). \end{aligned}$$

Ex. 3. Factor $x^3 - 3x + 2$.

Subtracting 1 from, and adding 1 to the given expression, we obtain

$$\begin{aligned} x^3 - 3x + 2 &= x^3 - 1 - 3x + 3 \\ &= (x^3 - 1) - 3(x - 1) \\ &= (x - 1)[(x^2 + x + 1) - 3] \\ &= (x - 1)(x^2 + x - 2) \\ &= (x - 1)(x - 1)(x + 2). \end{aligned}$$

The method of adding or subtracting terms could have been applied to factor the type-forms, without assuming that they are the products of known factors.

Ex. 4.
$$\begin{aligned} a^2 - b^2 &= a^2 - ab + ab - b^2 \\ &= a(a-b) + b(a-b) \\ &= (a+b)(a-b). \end{aligned}$$

25. Another device consists in separating a term into two or more terms, and grouping these component terms with others of the given expression.

Ex. 1. Factor $x^3 + x + 2$.

Separating $+2$ into $+1+1$, and rearranging terms, we obtain

$$\begin{aligned} x^3 + x + 2 &= (x^3+1) + (x+1) \\ &= (x+1)(x^2 - x + 1) + (x+1) \\ &= (x+1)[(x^2 - x + 1) + 1] \\ &= (x+1)(x^2 - x + 2). \end{aligned}$$

Ex. 2. Factor $x^3 - 3x^2 + 4$.

Separating $-3x^2$ into $-2x^2$ and $-x^2$, we obtain

$$\begin{aligned} x^3 - 3x^2 + 4 &= x^3 - 2x^2 - x^2 + 4 \\ &= x^2(x-2) - (x^2 - 4) \\ &= (x-2)[x^2 - (x+2)] \\ &= (x-2)(x^2 - x - 2) \\ &= (x-2)(x-2)(x+1) \\ &= (x-2)^2(x+1). \end{aligned}$$

The trinomials considered in Arts. 10–12 could have been factored by the methods of this and the preceding article.

Ex. 3.
$$\begin{aligned} x^2 + 5x + 6 &= x^2 + 2x + 3x + 6 \\ &= x(x+2) + 3(x+2) \\ &= (x+2)(x+3). \end{aligned}$$

Ex. 4.
$$\begin{aligned} 10x^2 - 7xy - 12y^2 &= 10x^2 - 15xy + 8xy - 12y^2 \\ &= 5x(2x-3y) + 4y(2x-3y) \\ &= (2x-3y)(5x+4y). \end{aligned}$$

26. Symmetry. — An expression is *symmetrical* with respect to two letters if it remain the same when these letters are interchanged.

E.g., ab, $a^3 + b^3$, $a^2 + 2ab + b^2$ are symmetrical with respect to a and b, since, when a and b are interchanged, they become

$$ba,\quad b^3 + a^3,\quad b^2 + 2ba + a^2, \text{ as above.}$$

27. An expression is symmetrical with respect to three or more letters, if it be symmetrical with respect to any two of them.

E.g., $a(b + c) + b(c + a) + c(a + b)$ is symmetrical with respect to the three letters a, b, and c. For, if any two letters, say a and c, be interchanged, we obtain

$$c(b + a) + b(a + c) + a(c + b), \text{ as above.}$$

28. Cyclo-symmetry. — An expression is cyclo-symmetrical with respect to three or more letters it it remain the same when the first letter is changed into the second, the second into the third, and so on, and the last into the first.

Such an interchange of letters is called a *cyclic interchange.*

Thus, abc becomes bca by a first cyclic interchange;
becomes cab by a second cyclic interchange;
becomes abc by a third cyclic interchange.

Therefore abc is a cyclo-symmetrical expression with respect to a, b, and c.

The expression $(a - d)(b^2 - c^2) + (b - d)(c^2 - a^2) + (c - d)(a^2 - b^2)$ is cyclo-symmetrical with respect to a, b, and c. For, after making a cyclic interchange of these letters, we have

$$(b - d)(c^2 - a^2) + (c - d)(a^2 - b^2) + (a - d)(b^2 - c^2), \text{ as above.}$$

Observe that this expression is *not* cyclo-symmetrical with respect to all four letters, a, b, c, and d.

29. A symmetrical or cyclo-symmetrical expression can frequently be factored by arranging its terms to powers of one of the letters with respect to which it is symmetrical.

Ex. Factor $\quad b^3(c - a) + c^3(a - b) + a^3(b - c)$.

Arranging the given expression to descending powers of a, we have

$$\begin{aligned} &a^3(b - c) - a(b^3 - c^3) + bc(b^2 - c^2) \\ &= (b - c)[a^3 - a(b^2 + bc + c^2) + bc(b + c)] \\ &= (b - c)[a^3 - ab^2 - abc - ac^2 + b^2c + bc^2] \\ &= (b - c)[a(a^2 - c^2) - b^2(a - c) - bc(a - c)] \\ &= (b - c)(a - c)[a(a + c) - b^2 - bc] \\ &= (b - c)(a - c)[a^2 + ac - b^2 - bc] \\ &= (b - c)(a - c)[(a^2 - b^2) + c(a - b)] \\ &= (b - c)(a - c)(a - b)(a + b + c). \end{aligned}$$

EXERCISES XII.

Factor the following expressions:

1. $1 + 4x^4$.
2. $m^4 + 4n^4$.
3. $1 + 64x^4$.
4. $1 + 4a^{4r}$.
5. $x^{4n} + 4y^{4n}$.
6. $a^4b^8 + 64c^8d^4$.
7. $1 + 3a^2 + 4a^4$.
8. $1 - 7a^2 + a^4$.
9. $1 + 2x^2y^2 + 9x^4y^4$.
10. $x^4 - x^2y^2 + 16y^4$
11. $x^4 + y^4 - 11x^2y^2$.
12. $x^4 + y^4 - 6x^2y^2$
13. $x^4 + 2x^2y^2 + 9y^4$
14. $16x^4 - x^2y^2 + y^4$.
15. $x^4y^4 + 13x^2y^2 + 49y^4$
16. $25a^4b^4 + 5a^2b^2c^2d^2 + 9c^4d^4$.
17. $x^4 + 4y^4 - 12x^2y^2$.
18. $36x^8y^8 + 3x^4y^4 + y^8$.
19. $x^4 + y^8 + x^2y^4$
20. $x^8 + y^8 - 142x^4y^4$.
21. $x^8 - 3x^4y^2 + 9y^4$
22. $4m^4n^8 + 8m^2n^4p^4q^2 + 121p^8q^4$.
23. $25(x - y)^4 + 11(x^2 - y^2)^2 + 36(x + y)^4$.
24. $1 + 2(a^2 + b^2)^2 + 9(a^2 + b^2)^4$.
25. $a^{4n} + a^{2n} + 1$.
26. $49x^{4n}y^{4n} + 10x^{2n}y^{2n} + 1$.
27. $x^3 - 6x^2 + 16$.
28. $x^3 - 15x^2 + 250$.
29. $x^3 + 6x^2 + 10x + 4$.
30. $x^3 - 9x^2 + 32x - 42$.
31. $x^3 - 15x^2 + 72x - 110$.
32. $8x^3 - 36x^2 + 48x - 18$.
33. $27x^3 - 27x^2 - 6x + 4$.
34. $ac(a - c) - ab(a - b) - bc(b - c)$.
35. $a^2(b - c) + b^2(c - a) + c^2(a - b)$.
36. $a^2(b + c) + b^2(c + a) + c^2(a + b) + 2abc$.
37. $(a - d)(b^2 - c^2) + (b - d)(c^2 - a^2) + (c - d)(a^2 - b^2)$.
38. $a^3(b^2 - c^2) + b^3(c^2 - a^2) + c^3(a^2 - b^2)$.
39. $a^4(b^2 - c^2) + b^4(c^2 - a^2) + c^4(a^2 - b^2)$.

Method of Substitution.

30. In Ch. VI., § 2, Art. 4, it was proved that if an expression be divisible by $x - a$ (*i.e.*, if $x - a$ be a *factor* of the expression), the result of substituting a for x in the given expression is 0.

The following example will illustrate the method of applying this principle in factoring certain symmetrical expressions.

Ex. 1. Factor $(x-y)^3+(y-z)^3+(z-x)^3$.

The given expression has the factor $x-y$, if it be divisible by $x-y$; that is, if the result of substituting y for x be 0. Making this substitution, we have

$$0^3+(y-z)^3+(z-y)^3=(y-z)^3-(y-z)^3$$
$$=0.$$

Therefore, $x-y$ is a factor of the given expression.

In like manner it can be shown that $y-z$ and $z-x$ are factors. It is evident that the given expression, which is of the *third* degree, cannot have a fourth *literal* factor, since the product of four literal factors is an expression of the *fourth* degree. But it may possibly be equal to the product of the three factors $x-y$, $y-z$, $z-x$, and a numerical coefficient.

Let us assume, therefore,

$$(x-y)^3+(y-z)^3+(z-x)^3=C(x-y)(y-z)(z-x), \tag{1}$$

wherein C is some numerical coefficient yet to be determined.

Now (1) is an identity, and hence its first member must be equal to its second member for all values of the letters x, y, and z (Ch. IV., § 1, Art. 3).

If we substitute $x=0$, $y=1$, $z=2$, in both members of (1), we have

$$(-1)^3+(1-2)^3+2^3=C(-1)(1-2)\times 2,$$

or

$$6=2C \tag{2}$$

Solving (2) for C, we have $\quad C=3.$

Consequently,

$$(x-y)^3+(y-z)^3+(z-x)^3=3(x-y)(y-z)(z-x),$$

and the given expression is completely factored.

This method evidently depends upon the possibility of forming in advance of factoring a given expression some idea of the nature of its factors.

Ex. 2. Factor $(a+b-c)^2(a-b+c)+(a+b+c)(a+b-c)(b+c-a)$.

The given expression has the factor b if it be exactly divisible by $b-0$; that is, if the result of substituting 0 for b be 0. Making this substitution, we have

$$(a-c)^2(a+c)+(a+c)(a-c)(c-a)=(a-c)^2(a+c)-(a+c)(a-c)^2=0.$$

Therefore, b is a factor of the given expression.

In like manner, it can be shown that c is a factor, and that a is not a factor.

It is evident that the given expression reduces to 0, when $a+b-c=0$. Therefore, $a+b-c$ is a factor.

We now assume

$$(a+b-c)^2(a-b+c)+(a+b+c)(a+b-c)(b+c-a)=Cbc(a+b-c).$$

In this identity let $a = 0$, $b = 1$, $c = 2$. By this substitution we obtain $C = 4$.

Therefore,

$$(a+b-c)^2(a-b+c)+(a+b+c)(a+b-c)(b+c-a)=4\,bc\,(a+b-c).$$

In factoring expressions like the above, it is advisable to test such expressions as $a + b$, $a - b$, $a + b + c$, $a + b - c$, a, b, etc., according as any one of them is suggested by the form of the expression to be factored.

EXERCISES XIII.

1–6. Factor the expressions given in Exercises XII., Exx. 34–39.

Factor the following expressions:

7. $a(b+c)^2 + b(c+a)^2 + c(a+b)^2 - 4\,abc$.
8. $(b+c-a)^2(a-b+c)+(a+b+c)(b+c-a)(a+b-c)$.
9. $(a+b)^2+(a+c)^2-(c+d)^2-(b+d)^2$.
10. $(x+y+z)(xy+yz+zx)-xyz$.
11. $(a^2-b^2)^3+(b^2-c^2)^3+(c^2-a^2)^3$.
12. $(x+y+z)^3-(x+y-z)^3-(x+z-y)^3-(z+y-x)^3$.
13. $a(b-c)^3+b(c-a)^3+c(a-b)^3$.
14. $x^2y+y^2z+z^2x-xy^2-yz^2-x^2z$.
15. $(a+b+c)^3-(a^3+b^3+c^3)$.
16. $(a+b+c)^4+a^4+b^4+c^4-(b+c)^4-(c+a)^4-(a+b)^4$.
17. $2(a^2b^2+a^2c^2+b^2c^2)-(a^4+b^4+c^4)$.

EXERCISES XIV.

Factor the following expressions by the methods given in this chapter:

1. $a^4 + 2\,a^3b - 2\,ab^3 - b^4$.
2. $ax^2 + (a+b+c)x + b + c$.
3. $10\,c^{4n+1} - 5\,c^{7n+1} - 5\,c^{n+1}$.
4. $x^2y^2 + 17\,xy + 16$.
5. $6\,a^4 - 6\,a^3c + 2\,a^3x^2 - 2\,a^2cx^2 + 6\,a^3x + 2\,a^2x^3$.
6. $a^6 - a^6z^4 + 3\,a^4z^2 - a^4z^6 + 3\,a^2z^4 + z^6$.
7. $2(x^6+y^6) - xy(x^2+y^2)(2\,xy - 3\,x^2 + 3\,y^2)$.
8. $x^6 + 64$.
9. $a^{10} - a$.
10. $36\,a^x - a^{x+2}$.
11. $x^4 + 2\,x^2 + 9$.
12. $24\,x^2 - (3\,b - 8\,a)x - ab$.
13. $b^2 - c^2 + a(a - 2\,b)$.
14. $x^{2m-2} + 2\,x^{m+n} + x^{2n+2}$.
15. $x^4 - 2\,x^3 - 1 + 2\,x$.

16. $3x^2 + 4y^2 + z^2 - 8xy - 4yz + 4xz$.

17. $x^{3n+3} - 3x^{2n+2}y^2 + 3x^{n+1}y^4 - y^6$.

18. $ax^5 + bx^4 + cx^3 - ax^2 - bx - c$.

19. $(a+b)x^2 + (a-2b)x - 3b$.

20. $a^2 - b^2 - c^2 - 2a + 2bc + 1$.

21. $49x^4y^6 + 42x^7y^9 + 9x^{10}y^{12}$.

22. $ab(x^2 + y^2) + xy(a^2 + b^2)$.

23. $27x^3 + 27x^2y + 9xy^2 + y^3 - x^6 - 3x^4 - 3x^2 - 1$.

24. $(ax + by)^2 - (a - b)(x + z)(ax + by) + (a - b)^2xz$.

25. $28(x + 3)^2 - 23(x^2 - 9) - 15(x - 3)^2$.

26. $x^2 - 2x + 1 - y^2$.

27. $15x^2 + x - 40$.

28. $x^3 - x^2z + xz^2 - z^3$.

29. $a^3 - 1 + c - ac$.

30. $a^6b^6 + 1$.

31. $x^7 + y^7$.

32. $a^2 - a - 1 - a^2c + ac + c$.

33. $2a^2 + a - 4ax - x + 2x^2$.

34. $9ac + 2a^2 - 5ab + 4c^2 + 8bc - 12b^2$.

35. $(ax - by)^2 - (a + b)(x + z)(ax - by) + (a + b)^2xz$.

36. $a^7 - 1$.

37. $2^{3x+3} - 64$.

38. $20x^2 - 123x + 180$.

39. $x^3 - 5x^2 - x + 5$.

40. $x^3(x + 1) - b^3(b + 1)$.

41. $25a^4b^4 + 70a^2b^2c^2 + 49c^4$.

42. $x^4y + zx^3 - xy - z$.

43. $x^2 - 9z^2 - 4y(y + 3z)$.

44. $x^8 - 2x^4y^4 + y^8 - 4x^2y^2(x^2 - y^2)^2$.

45. $a^3 + a^2c + abc + b^2c - b^3$.

46. $x^6 - 2x^5y + 2x^3y^3 - 2xy^5 + y^6$.

47. $3(a - 1)^3 - (1 - a)$.

48. $x^6 - y^6 + 1 - 2x^3$.

49. $x^2 - ax - bx + ab$.

50. $x^2y^2 + 25 - 9z^2 - 10xy$.

51. $x^2 + 9 - 2x(3 + 2xy^2)$.

52. $abx^2 - (a + b)(ab + 1)x + (ab + 1)^2$.

53. $ab + 2a^2 - 3b^2 - 4bc - ac - c^2$.

54. $3x^6 + 8x^4 - 8x^2 - 3$.

55. $7a^3x^2 + 49a^2x + 84a$.

56. $\frac{2}{3}a^3n^8x + \frac{2}{3}a^3n^2x^5 - \frac{4}{3}a^3n^5x^3$.

57. $(x^2 + xy + y^2)^2 - (x^2 - xy + y^2)^2$.

58. $a^3 + 8b^3 - 27c^3 + 18abc$.

59. $25a^4b^{2n+2} + 9a^{2n-2}b^4 - 30a^{n+1}b^{n+3}$.

60. $49x^8 + 121y^4 - 155x^4y^2$.

61. $cd - bd + a(b - c)$.

62. $8x^3 - 60x^2 + 140x - 100$.

63. $(x^2 + 1)^3 - (y^2 + 1)^3$.

64. $x^6 + y^3 + 1 - 3x^2y$.

65. $abx^2 + x + ab + 1$.

66. $36a^4 - 21a^2 + 1$.

67. $10x^4 - 47x^2 + 42$.

68. $(x^2+xy-y^2)^2-(x^2-xy-y^2)^2$.

69. $x^2+c(a+b)x+ab(a+c)(c-b)$.

70. $4+a^2-(b^2+c^2)-2(bc-2a)$.

71. $x^4-2bx^3-(a^2-b^2)x^2+2a^2bx-a^2b^2$.

72. $5a^2-180b^2$.

73. $\frac{7}{16}abc^2-\frac{7}{36}abd^2$.

74. $10x^2+3x-18$.

75. $x^{2n}-y^{2n}+4y^n-4x^n$.

76. $ab(x^2-y^2)+xy(a^2-b^2)$.

77. $125x^3-150x^2+45x-2$.

78. $36a^4b^2-60a^3b^3+25a^2b^4$.

79. $a^2(a^2-1)-b^2(b^2-1)$.

80. $(m-n)^2-12(m-n)+27$.

81. $a^4x^5(a^3-x)-a^5x^2(x^3-a)$.

82. $33a^2-36ab+66ad-220ag+240bg-440dg$.

83. $3(a-1)(a^5+7)^2-12(4a^5+28)(a-1)+192(a-1)$.

84. x^6+32xy^5.

85. a^5-1.

86. $2a^4-16ab^3$.

87. x^5y^5-32.

88. $(x+y)^2-18(x+y)+77$.

89. $(a^2-b^2)x^2-(a^2+b^2)x+ab$.

90. $300abc^2-432abd^2$.

91. $\frac{1}{2}ab^2-\frac{2}{9}ac^2$.

92. $(x^2+x-1)^3-(y^2+x-1)^3$.

93. $80a^6c^{15}n^{13}+5a^{12}c^6n+320c^{24}n^{25}$.

94. $75a^2b^2-108c^2d^2$.

95. $\frac{16}{25}abx^2y^2-\frac{16}{49}abz^2$.

96. $18a^2x^2-98b^2y^2$.

97. $\frac{5}{4}xy^2-\frac{5}{25}xz^2$.

98. $\frac{8}{9}ab^2-\frac{8}{25}ac^2$.

99. x^5-2x^3+x.

100. $18(x+y)^2+23(x^2-y^2)-6(x-y)^2$.

101. Express $(a^2-b^2)(c^2-d^2)$ as the difference of two squares.

§ 2. HIGHEST COMMON FACTORS.

1. If two or more integral algebraic expressions have no common factor except 1, they are said to be *prime to one another*, or are called *relatively prime expressions*.

E.g., ab and cd; $5x^2y$ and $8z^3$; a^2+b^2 and a^2-b^2.

2. The **Highest Common Factor** of two or more integral algebraic expressions is the expression of highest degree which exactly divides each of them.

E.g., the highest common factor of ax^2, bx^3, and cx^4 is evidently x^2.

The words Highest Common Factor are frequently abbreviated to H. C. F.

H. C. F. by Factoring.

3. Monomial Expressions. — The H. C. F. of monomials can be found by inspection.

Ex. 1. Find the H. C. F. of x^2y^5z, $x^4y^3z^2$, and $x^3y^4z^4$.

The expression of highest degree which exactly divides each of the given expressions evidently cannot contain a higher power of x than x^2, a higher power of y than y^3, and a higher power of z than z. Therefore the required H. C. F. is x^2y^3z.

Observe that the power of each letter in the H. C. F. is the *lowest* power to which it occurs in any of the given expressions.

If the monomials contain numerical factors, the Greatest Common Measure (G. C. M.) of these factors should be found as in Arithmetic.

Ex. 2. Find the H. C. F. of $18\,a^4b^5c^3d$, $42\,a^3bc^4$, and $30\,a^2b^2c^2$.

The G. C. M. of the numerical coefficients is evidently 6. The lowest power of a in any of the given expressions is a^2; the lowest power of b is b; the lowest power of c is c^2; and d is not a common factor.

Therefore the required H. C. F. is $6\,a^2bc^2$.

In general, *the H. C. F. of two or more monomials is obtained by multiplying the G. C. M. of their numerical coefficients by the product of their common literal factors, each to the lowest power to which it occurs in any of the given monomials.*

EXERCISES XV.

Find the H. C. F. of the following expressions:

1. ax^3, a^2x.
2. $15\,a$, $20\,a^2$, $10\,b$, 5.
3. a^2bx^2, ab^2x^2, a^2b^2x.
4. $2\,x^4$, $3\,x^5$, x^3, $5\,x^6$.
5. $56\,xy^3$, $70\,x^3y$, $98\,x^3y^3$.
6. $20\,a^2x^4b$, $40\,ax^3$, $10\,a^2x^3$.
7. $20\,a^4b^2$, $12\,a^3$, $10\,a^2b$.
8. $55\,x^2b^4$, $20\,x^2b^3$, $15\,a^2b^3x$, $5\,a^4b^4$.
9. $100\,x^4y$, $120\,y^3$, $60\,xz^3$, $80\,x^2y^2z^2$.
10. $105\,a^3b^3c^4x^2y^5$, $35\,a^2b^4c^5xy^4$, $28\,a^4b^2c^3x^4y^2$, $49\,a^5b^8c^4x^3y^6$.
11. $15\,a^2x^n$, $20\,a^3x^{n-1}$, $10\,ax^{n+1}$, $5\,a^2x^{n+2}$, $25\,a^3x^n$.
12. $(x-y)^2(x+z)^3$, $(x-y)^3(x+z)^2$.
13. $27\,(x-y)^3(a+x)^2(b+x)$, $18\,(x-y)^2(x+b)(x+a)$.

4. Multinomial Expressions. — The method of finding the H. C. F. of multinomials by factoring is similar to that of finding the H. C. F. of monomials.

Ex. 1. The expressions

$$x^2 - 1 = (x-1)(x+1),$$

and

$$x^2 + x - 2 = (x-1)(x+2),$$

have only the common factor $x - 1$. This is therefore their H. C. F.

Ex. 2. The expressions

$$x^4 - y^4 = (x^2 + y^2)(x+y)(x-y),$$

and

$$x^3 - 3xy^2 - 2y^3 = (x+y)^2(x-2y),$$

have only the common factor $x + y$. This is therefore their H. C. F.

Ex. 3. The expressions

$$4x^4y^2 - 4x^2y^4 + 8x^3y^3 - 8xy^5 = 4xy^2(x^3 - xy^2 + 2x^2y - 2y^3)$$
$$= 4xy^2(x+2y)(x+y)(x-y),$$
$$2x^4y - 2x^2y^3 + 2x^3y - 2xy^3 = 2xy(x^3 - xy^2 + x^2 - y^2)$$
$$= 2xy(x+1)(x+y)(x-y),$$

have the common factors $2xy$, $x+y$, $x-y$.

Therefore their H. C. F. is

$$2xy(x+y)(x-y) = 2xy(x^2 - y^2).$$

Ex. 4. The expressions

$$x^5y^3 - 2x^4y^3 + x^3y^3 = x^3y^3(x^2 - 2x + 1)$$
$$= x^3y^3(x-1)^2,$$

and

$$x^5y^2 - 3x^4y^2 + 3x^3y^2 - x^2y^2 = x^2y^2(x^3 - 3x^2 + 3x - 1)$$
$$= x^2y^2(x-1)^3,$$

have the common factors x^2y^2 and $(x-1)^2$.

Therefore, their H. C. F. is $x^2y^2(x-1)^2$.

In general, *the H. C. F. of two or more multinomial expressions is the product of their common factors, each to the lowest power to which it occurs in any of them.*

EXERCISES XVI.

Find the H. C. F. of the following expressions:

1. $x^2 + xy$, $x^2 + 2xy + y^2$.
2. $a^2 - ab$, $a^2 - 2ab + b^2$.
3. $x^2 - 1$, $x^2 + 2x + 1$.
4. $x^3 - 1$, $x^2 + x + 1$.
5. $a^3 + b^3$, $a^2 - b^2$.
6. $16a^4 - 81b^4$, $4a - 6b$.
7. $8 - a^3$, $a^2 - 4$.
8. $x^2 + 2x - 8$, $x^2 + 4x - 12$.
9. $x^2+7x-30$, x^2-5x+6.
10. $x^3 + 27$, $x^2 + 6x + 9$.
11. $x^4 - y^4$, $(x^2 - y^2)^2$.
12. $x^2 - 4x + 3$, $x^2 - 7x + 12$.
13. $x^2+2xy+y^2-a^2$, $2x+2y+2a$.
14. $x^2 + xy$, $x^5 + y^5$.
15. $x^5 - y^5$, $ax - ay$.
16. $x^5 - y^5$, $x^2 - y^2$.
17. $3x^2+16x-35$, $5x^2+33x-14$.
18. x^3+1, x^3+mx^2+mx+1.
19. $x^4 + 9x^2 + 20$, $x^4 + 7x^2 + 10$.
20. $2x^2 + 5ax + 3a^2$, $3x^2 - ax - 4a^2$.
21. $3x^3 - 8x^2 + 4x$, $x^3 - 6x^2 + 12x - 8$.
22. $2x^4 + 5a^2x^2 + 3a^4$, $6x^4 - 3a^2x^2 - 9a^4$.
23. $3x^2 - ax - 4a^2$, $6x^2 - 17ax + 12a^2$.
24. $a^3 + 2a^2 + 2a + 1$, $a^3 + 1$.
25. $ax^4 - 3a^2x^3 + 3x^2a^3 - xa^4$, $5x^3 - 10ax^2 + 5a^2x$.
26. $x^3 - 1$, $x^2 - 1$, $(x - 1)^2$.
27. $a^2 - b^2$, $3(a - b)$, $5(a - b)^2$.
28. $x^2 + 5x + 4$, $x^2 + 2x - 8$, $x^2 + 7x + 12$.
29. $x^2 - 2a^2 - ax$, $x^2 - 4a^2$, $x^2 - 6a^2 + ax$.
30. $2x^2 + 9xy + 7y^2$, $2x^2 - 2y^2$, $3x^2 - 2xy - 5y^2$.
31. $x^2 - 2x - 3$, $x^2 - 7x + 12$, $x^2 - x - 6$.
32. $x^6 - y^6$, $x^4 + xy^3$, $x^6 + 2x^3y^3 + y^6$.
33. $a^5 + b^5$, $a^4 - b^4$, $a^3 + b^3$.
34. $ab(x + a)$, $a[x^2 + (a - b)x - ab]$, $b(x^2 + ax)$.

H. C. F. by Division.

5. **A Multiple** of an integral algebraic expression is an expression which is exactly divisible by the given one.

E.g., multiples of $a + b$ are $2(a + b)$, $(x - y)(a + b)$, etc.

6. If the given expressions cannot be readily factored, their H. C. F. can be obtained by a method analogous to that used in Arithmetic to find the G. C. M. of numbers.

This method depends upon the following principles:

(i.) *A factor of an integral algebraic expression is also a factor of any multiple of the expression.*

E.g., $x-y$ is a factor of x^2-y^2, and also of the multiples $2(x^2-y^2)$, $x^3(x^2-y^2)$, etc.

(ii.) *A common factor of two integral algebraic expressions is also a factor of the sum or the difference of any multiples of the expressions (including simply the sum or the difference of the expressions).*

E.g., $x-y$ is a common factor of x^2-y^2 and $x^2-2xy+y^2$; and $x-y$ is also a factor of

$$(x^2-y^2)+(x^2-2xy+y^2)=2x^2-2xy=2x(x-y),$$

of $$(x^2-y^2)-(x^2-2xy+y^2)=-2y^2+2xy=2y(x-y),$$

of $$3(x^2-y^2)+2(x^2-2xy+y^2)=5x^2-4xy-y^2=(5x+y)(x-y).$$

The proofs of the principles enunciated are as follows:

(i.) Let E stand for any integral algebraic expression, and F for any factor of E; then we are to prove that F is also a factor of ME, wherein M stands for any number or integral algebraic expression.

Since F is a factor of E, E is the product of F and some other algebraic expression, say Q; or,

$$E=QF.$$

Multiplying both members of this equation by M, we have, by Ch. II., § 3, Art. 17,

$$ME=MQF.$$

The latter equation shows that ME, $=MQ\cdot F$, contains F as a factor, the other factor being MQ.

(ii.) Let E_1 and E_2 stand for the two expressions, and F for their common factor. Then we are to prove that F is a factor of

$$ME_1+NE_2,$$

wherein M and N stand for two numbers or integral algebraic expressions.

Since F is a factor of E_1, E_1 is the product of F and some other expression, say Q_1; or

$$E_1=Q_1F. \quad (1)$$

For a similar reason $$E_2=Q_2F. \quad (2)$$

Multiplying both members of (1) by M, and both members of (2) by N, we have, by Ch. II., § 3, Art. 17,

$$ME_1 = MQ_1F, \text{ and } NE_2 = NQ_2F.$$

Adding corresponding members of the last equations, we have, by Ch. II., § 1, Art. 19,

$$\begin{aligned} ME_1 + NE_2 &= MQ_1F + NQ_2F \\ &= (MQ_1 + NQ_2)F. \end{aligned}$$

The last equation shows that $ME_1 + NE_2$ contains F as a factor, the remaining factor being $MQ_1 + NQ_2$.

7. The expressions whose H. C. F. is required should be arranged to descending (or ascending) powers of some common letter of arrangement.

If one of two expressions be divisible without a remainder by the other, which must be of the same or lower degree in the letter of arrangement, then the latter (the divisor) is the required H. C. F. For it is a factor of the other expression.

But if the one expression be not divisible without a remainder by the other, their H. C. F. is found by repeated applications of the following principle:

If an integral algebraic expression be divided by another (of the same or lower degree in a common letter of arrangement) and if there be a remainder, then the H. C. F. of this remainder and the divisor is the H. C. F. of the given expressions.

E.g., the H. C. F. of

$$x^4 - 10x^3 + 35x^2 - 50x + 24, = (x-1)(x-2)(x-3)(x-4), \quad (1)$$

and $$x^3 - 7x^2 + 11x - 5, = (x-1)(x-1)(x-5) \quad (2)$$

is evidently $x - 1$. The remainder obtained by dividing (1) by (2) is found to be

$$3x^2 - 12x + 9, = 3(x-1)(x-3). \quad (3)$$

The H. C. F. of this remainder and the divisor (2) is evidently also $x - 1$, the H. C. F. of (1) and (2).

It is important to notice that the H. C. F. of the remainder and the dividend (1) is $(x-1)(x-3)$, and is *not* the H. C. F. of (1) and (2).

The proof of the principle is as follows:

Let E_1 and E_2 stand for two integral algebraic expressions, which have a common factor, and let E_1 be of the same or higher degree than E_2 in some letter of arrangement.

Let Q be the quotient and R the remainder of dividing E_1 by E_2. Then since, by Ch. III., § 4, Art. 13, *the dividend is equal to the product of the quotient and the divisor, plus the remainder*, we have

$$E_1 = QE_2 + R. \qquad (1)$$

From this equation we infer, by Art. 6 (ii.), that any common factor of R and E_2, *and therefore their H. C. F.*, is also a factor of $QE_2 + R$; that is, of E_1.

Transferring QE_2 to the first member of (1), we obtain

$$E_1 - QE_2 = R. \qquad (2)$$

From the last equation we infer that any common factor of E_1 and E_2, *and therefore their H. C. F.*, is also a factor of $E_1 - QE_2$; that is, of R.

If now the H. C. F. of E_1 and E_2 be not the H. C. F. of R and E_2, then E_2 must have in common with R some factor of higher degree than is contained in E_1. But this contradicts the first part of the proof, that the H. C. F. of R and E_2 is also a factor of E_1.

Hence the truth of the principle enunciated.

8. The following example will illustrate the method of applying the principle of Art. 7 to find the H. C. F. of two given expressions.

Ex. Find the H. C. F. of

$$x^3 - 4x^2 + 4x - 1 \qquad (1)$$

and

$$x^2 - 3x + 2. \qquad (2)$$

Dividing (1) by (2), we have

$$\begin{array}{l|l} x^3 - 4x^2 + 4x - 1 & x^2 - 3x + 2 \\ \underline{x^3 - 3x^2 + 2x} & x - 1 \\ \quad -x^2 + 2x & \\ \quad \underline{-x^2 + 3x - 2} & \\ \qquad -x + 1 & \end{array}$$

By Art. 7, the H. C. F. of (1) and (2) is the H. C. F. of

$$x^2 - 3x + 2, \qquad (2)$$

and the remainder

$$-x + 1. \qquad (3)$$

Dividing (2) by (3), we have

$$\begin{array}{r|l} x^2 - 3x + 2 & -x + 1 \\ \underline{x^2 - x} & -x + 2 \\ -2x + 2 & \\ \underline{-2x + 2} & \end{array}$$

Since the remainder of this division is 0, the divisor $-x+1$ (*i.e.*, the remainder of the first division) is the H. C. F. of itself and (2), and therefore of (1) and (2).

9. The following principle will simplify the work of finding the H. C. F. of two expressions:

Either of the expressions can be multiplied or divided by any number which is not already a factor of the other expression.

For a factor introduced by multiplication into one expression will not be common to both of them, and therefore will not be introduced into their H. C. F.

In like manner, a factor removed by division from one expression was not common to both of them, and therefore would not have been a factor of their H. C. F.

Ex. 1. Find the H. C. F. of

$$x^4 - 10x^3 + 35x^2 - 50x + 24 \qquad (1)$$

and

$$x^3 - 7x^2 + 11x - 5. \qquad (2)$$

If (1) be divided by (2), we have

$$\begin{array}{r|l} x^4 - 10x^3 + 35x^2 - 50x + 24 & x^3 - 7x^2 + 11x - 5 \\ \underline{x^4 - 7x^3 + 11x^2 - 5x} & x - 3 \\ -3x^3 + 24x^2 - 45x & \\ \underline{-3x^3 + 21x^2 - 33x + 15} & \\ 3x^2 - 12x + 9 & \end{array}$$

But, by Art. 7, the H. C. F. of (1) and (2) is the H. C. F. of

$$x^3 - 7x^2 + 11x - 5, \qquad (2)$$

and the remainder

$$3x^2 - 12x + 9. \qquad (3)$$

The work of finding the H. C. F. of (2) and (3) by division will introduce fractional coefficients. To avoid these, we divide (3) by 3, *since* 3 *is not a factor of* (2), and obtain

$$x^2 - 4x + 3. \qquad (4)$$

The second stage of the work is then as follows:

$$\begin{array}{l|l} x^3-7x^2+11x-5 & x^2-4x+3 \\ \cline{2-2} \underline{x^3-4x^2+\ \ 3x} & x-3 \\ -3x^2+\ \ 8x & \\ \underline{-3x^2+12x-9} & \\ -\ \ 4x+4 & \end{array}$$

But, by Art. 7, the H. C. F. of (2) and (3), and therefore the H. C. F. of (1) and (2), is also the H. C. F. of the first remainder (divided by 3)

$$x^2-4x+3, \qquad (4)$$

and the second remainder $-4x+4$. (5)

To avoid fractional coefficients, $-4x+4$ is divided by -4, since -4 *is not a factor of* (4). The third stage of the work is as follows:

$$\begin{array}{l|l} x^2-4x+3 & x-1 \\ \cline{2-2} \underline{x^2-\ \ x} & x-3 \\ -3x+3 & \\ \underline{-3x+3} & \end{array}$$

Since the remainder of this division is 0, the last divisor, $x-1$, is the H. C. F. of (4) and (5), and therefore of (2) and (3), and hence of (1) and (2).

The work of this example may be arranged more compactly thus:

1st divisor,

$$\begin{array}{l|l|l} x^3-7x^2+11x-5 & x^4-10x^3+35x^2-50x+24 & x-3 \\ & \underline{x^4-\ \ 7x^3+11x^2-\ \ 5x} & \\ & -\ \ 3x^3+24x^2-45x & \\ & \underline{-\ \ 3x^3+21x^2-33x+15} & \\ & \div 3 \qquad \lfloor 3x^2-12x+\ \ 9 & \end{array}$$

2d divisor,

$$\begin{array}{l|l|l} x^2-\ \ 4x+\ \ 3 & x^3-7x^2+11x-5 & x-3 \\ & \underline{x^3-4x^2+\ \ 3x} & \\ & -3x^2+\ \ 8x & \\ & \underline{-3x^2+12x-9} & \\ & \div(-4)\ \lfloor -\ \ 4x+4 & \end{array}$$

3d divisor and H.C.F.,

$$\begin{array}{l|l|l} x-1 & x^2-4x+3 & x-3 \\ & \underline{x^2-x} & \\ & -3x & \\ & \underline{-3x+3} & \end{array}$$

Observe that the divisor is, for convenience, placed on the left of the dividend and the quotient on the right.

Ex. 2. Find the H.C.F. of

$$x^4 - 6x^3 + 6x^2 - 3x + 2, \qquad (1)$$

$$2x^3 - 5x^2 + 7x - 4. \qquad (2)$$

The first of these expressions cannot be divided by the second without introducing fractional coefficients. To avoid these we multiply the first by 2, *since 2 is not a factor of the other expression.*

The work then proceeds as follows:

$$\begin{array}{l}
2x^3-5x^2+7x-4 \,|\, 2x^4-12x^3+12x^2-6x+4 \,|\, x+7 \\
\qquad\qquad\qquad\quad \underline{2x^4-5x^3+7x^2-4x} \\
\qquad\qquad\quad \times(-2) \,|\, -7x^3+5x^2-2x+4 \\
\qquad\qquad\qquad\qquad 14x^3-10x^2+4x-8 \\
\qquad\qquad\qquad\qquad \underline{14x^3-35x^2+49x-28} \\
\qquad\qquad\qquad\quad +5 \,|\, 25x^2-45x+20 \\
\qquad\qquad\qquad 5x^2-9x+4 \,|\, 10x^3-25x^2+35x-20 \,|\, 2x-7 \\
\qquad\qquad\qquad\qquad\qquad\quad \underline{10x^3-18x^2+8x} \\
\qquad\qquad\qquad\qquad\qquad \times 5 \,|\, -7x^2+27x-20 \\
\qquad\qquad\qquad\qquad\qquad\qquad -35x^2+135x-100 \\
\qquad\qquad\qquad\qquad\qquad\qquad \underline{-35x^2+63x-28} \\
\qquad\qquad\qquad\qquad\qquad\qquad +72 \,|\, 72x-72 \\
\qquad\qquad\qquad\qquad\qquad\qquad\qquad x-1 \,|\, 5x^2-9x+4 \,|\, 5x-4 \\
\qquad\qquad\qquad\qquad\qquad\qquad\qquad\qquad \underline{5x^2-5x} \\
\qquad\qquad\qquad\qquad\qquad\qquad\qquad\qquad\quad -4x \\
\qquad\qquad\qquad\qquad\qquad\qquad\qquad\qquad\quad \underline{-4x+4}
\end{array}$$

Observe that in beginning the second stage of the work the dividend is the first divisor multiplied by 5.

The required H.C.F. is $x - 1$.

10. If either or both of the given expressions contain monomial factors, these should first be removed, and their H.C.F. set aside as a factor of the H.C.F. of the given expressions. The remaining factor of the required H.C.F. will then be obtained by the method of division. Notice in particular that at any stage of the work either the divisor or the dividend may be multiplied or divided by any positive or negative number with the limitations stated in Art. 7.

Ex. 4. Find the H.C.F. of

$$2x^4y - 2(a+b)x^3y + 2(ab+1)x^2y - 2axy$$
$$= 2xy[x^3 - (a+b)x^2 + (ab+1)x - a]$$

and

$$x^4y^2 - (a+1)x^3y^2 + (a+b)x^2y^2 - abxy^2$$
$$= xy^2[x^3 - (a+1)x^2 + (a+b)x - ab].$$

We set aside xy, the H.C.F. of $2xy$ and xy^2, as a factor of the required H.C.F., and find the H.C.F. of the remaining factors by division.

$$\begin{array}{r|l|l}
x^3-(a+1)x^2+(a+b)x-ab & x^3-(a+b)x^2+(ab+1)x-a & 1 \\
 & \underline{x^3-(a+1)x^2+(a+b)x-ab} & \\
\div(1-b) & \underline{(1-b)x^2+(1-a)(1-b)x-a(1-b)} & \\
 & x^2+(1-a)x-a &
\end{array}$$

$$\begin{array}{r|l|l}
x^2+(1-a)x-a & x^3-(a+1)x^2+(a+b)x-ab & x-2 \\
 & \underline{x^3+(1-a)x^2-ax} & \\
 & -2x^2+(2a+b)x-ab & \\
 & \underline{-2x^2-2(1-a)x+2a} & \\
\div(b+2) & \underline{(b+2)x-a(b+2)} &
\end{array}$$

$$\begin{array}{r|l|l}
x-a & x^2+(1-a)x-a & x+1 \\
 & \underline{x^2-ax} & \\
 & x-a & \\
 & \underline{\underline{x-a}} &
\end{array}$$

The required H.C.F. is $xy(x-a)$.

If the divisor and dividend at any stage can be factored readily, it is better to find their H.C.F. by factoring than by continuing the method of division.

Thus, in the last example, after the first partial division in the second stage of the work, we have to find the H.C.F. of the two expressions

$$x^2 + (1-a)x - a = (x-a)(x+1)$$

and

$$-2x^2 + (2a+b)x - ab = (b-2x)(x-a).$$

The factor, $x - a$, is thus obtained with less work.

11. The examples worked in the preceding articles illustrate the following method of finding the H.C.F. of two expressions:

(i.) *Remove from the given expressions any monomial factors, and set aside their H. C. F. as a factor of the required H. C. F.*

(ii.) *Divide the expression of higher degree in a common letter of arrangement by the one of lower degree; if the expressions be of the same degree, either may be taken as the first divisor.*

(iii.) *Divide the first divisor by the first remainder, the first remainder (second divisor) by the second remainder, and so on, until a remainder* 0 *is obtained. The last divisor will be the required H. C. F.*

If a remainder which does not contain the letter of arrangement, and which is not 0, is obtained, the given expressions do not have a H. C. F. in this letter of arrangement.

(iv.) *At any stage of the work the dividend may be multiplied by any number which is not a factor of the corresponding divisor; or the divisor may be divided by any number which is not a factor of the corresponding dividend.*

12. To find the H. C. F. of three or more integral algebraic expressions find the H. C. F. of any two of them, next the H. C. F. of that H. C. F. and the third expression, and so on.

For any common factor of three or more expressions, and hence their H. C. F., must be a factor of the H. C. F. of any two of them.

EXERCISES XVII.

Find the H. C. F. of the following expressions:

1. $x^2 - 5x + 6,\ x^2 + 3x - 10.$
2. $18a^2 - 9a - 14,\ 30a^2 - 59a + 28.$
3. $18x^4 - 33x^2 + 14,\ 3x^4 + x^2 - 2.$
4. $x^3 + 4x - 5,\ x^3 - 2x^2 + 6x - 5.$
5. $2x^3 + 3x^2 - x - 12,\ 6x^3 - 17x^2 + 2x + 15.$
6. $x^3 - 3x^2 + 4,\ x^3 - 2x^2 - 4x + 8.$
7. $x^3 - 3x + 2,\ x^4 - 6x^2 + 8x - 3.$

8. $2x^2 + 3x - 2,\ 4x^3 + 16x^2 - 19x + 5.$

9. $x^3 - 3x^2 + 4,\ 3x^3 - 18x^2 + 36x - 24.$

10. $x^3 - (a + b - c)x^2 + (ab - ac - bc)x + abc,$
$x^3 - (a - b + c)x^2 + (ac - ab - bc)x + abc.$

11. $x^3 + x^2 - 5x + 3,\ 2x^3 + 7x^2 - 9.$

12. $3x^3 - 8x^2 - 36x + 5,\ 9x^3 - 50x^2 + 27x - 10.$

13. $3a^3 - 23a^2 + 15a - 7,\ 2a^3 - 11a^2 - 25a + 28.$

14. $4x^3y^3 - 3x^2y^2 - 4xy + 3,\ 5x^3y^3 + 8x^2y^2 + xy - 14.$

15. $x^3 - 3xy^2 - 2y^3,\ 2x^3 - 5x^2y - xy^2 + 6y^3.$

16. $7x^3 + 34x^2y - 102xy^2 + 21y^3,\ 4x^3 + 33x^2y + 29xy^2 - 42y^3.$

17. $a^3 - a^2 - 5a + 2,\ 3a^3 - a^2 - 8a + 12.$

18. $6x^3 + 4x^2 - 51x + 21,\ 21x^3 - 94x^2 + 144x - 91.$

19. $14x^3 - 41x^2y + 17xy^2 - 5y^3,\ 10x^3 - 31x^2y + 23xy^2 - 20y^3.$

20. $6x^3 - x^2 + 16,\ 10x^3 - 19x^2 + 26x - 8.$

21. $x^3 + 2x^2 + 2x + 1,\ x^3 - 4x^2 - 4x - 5.$

22. $2a^3x^3 + a^2x^2 - ax + 10,\ 3a^3x^3 + 11a^2x^2 + 3ax - 14.$

23. $a^3x^3 - 5a^2x^2 - 5ax - 6,\ 2a^3x^3 - 3a^2x^2 - 3ax - 5.$

24. $30x^3 - 25ax^2 + 8a^2x - a^3,\ 18x^3 - 24ax^2 + 15a^2x - 3a^3.$

25. $5x^3 - 31x^2 + 7x - 6,\ 15x^3 - 13x^2 + 5x - 2.$

26. $36a^6 + 9a^3 - 27a^4 - 18a^5,\ 27a^5b^2 - 9a^3b^2 - 18a^4b^2.$

27. $3x^5 - 10x^3 + 15x + 8,\ x^5 - 2x^4 - 6x^3 + 4x^2 + 13x + 6.$

28. $2a^3x^4 - 2a^2bx^3y + 2ab^2x^2y^2 - 2b^3xy^3,$
$4a^2b^2x^3y^2 - 2ab^3x^2y^3 - 2b^4xy^4.$

29. $2x^3 - 9x^2 - 25x + 42,\ 4x^3 - 29x^2 + 37x - 42.$

30. $9x^3 - 9x^2 - 4x + 4,\ 15x^3 - 19x^2 + 4.$

31. $3x^3 - 2x^2 - 36x - 35,\ 2x^3 - 17x^2 + 23x + 55.$

32. $x^3 + 4x^2 - 4x + 5,\ x^4 + 4x^3 - 5x^2 + 3x + 15.$

33. $2x^3 - 3x^2 - 8x - 3,\ 2x^4 - 9x^3 + 13x^2 - 23x - 16.$

34. $2a^3x^3 - 7a^2x^2 + 11ax - 15,\ 2a^4x^4 - 7a^3x^3 + 8a^2x^2 - 12ax - 9.$

35. $x^5 + x^3 - 8x^2 - 8,\ x^4 - 2x^3 + x^2 - 2x.$

36. $75x^4 + 140x^3 - 223x^2 + 92x - 12,$
$45x^4 - 93x^3 + 65x^2 - 19x + 2.$

37. $x^2+4x+3,\ x^2-4x-5,\ 2x^2-5x-7.$

38. $2x^2+5x-12,\ 2x^2-13x+15,\ 10x^2-23x+12.$

39. $x^2-4x+3,\ 2x^3+x^2-7x+4,\ x^3-2x^2+1.$

40. $2x^3+5x^2-4x-10,\ 2x^3+5x^2+2x+5,\ 2x^3+7x^2+7x+5.$

41. $x^3+x^2-5x+3,\ x^3-7x+6,\ x^3-x^2-10x+10.$

42. $x^3-2x^2-34x+5,\ x^3+5x^2+x+5,\ 3x^3+16x^2+6x+5.$

43. $3x^4-14x^3-9x+2,\ 2x^4-9x^3+x^2-19x+4,$
$5x^3-23x^2-5x+2.$

44. $2x^4+6x^3+4x^2,\ 3x^3+9x^2+9x+6,\ 3x^3+8x^2+5x+2.$

45. $10a^5+10a^3b^2+20a^4b,\ 2a^3+2b^3,\ 4b^4+12a^2b^2+4a^3b+12ab^3.$

46. $x^3-3x^2-4x+12,\ x^3-7x^2+16x-12,\ 2x^3-9x^2+7x+6.$

47. $a^3x^3-a^2x^2-11ax-10,\ a^3x^3-5a^2x^2+ax+10,$
$2a^3x^3-5a^2x^2-13ax-5.$

48. $2x^4-x^3+3x^2+x+4,\ 2x^4-3x^3-2x^2+9x-12,$
$4x^4-16x^3+25x^2-23x+4.$

13. The words *Highest Common Factor* in Algebra refer to the degree of the common factor. Thus, the H. C. F. of

$$x^3-2x^2-x+2=(x^2-1)(x-2),$$

and

$$x^3-4x^2-x+4=(x^2-1)(x-4)$$

is evidently x^2-1.

That factor is of higher degree in x than any other common factor.

E.g., than $x-1$ or $x+1$.

The words *Greatest Common Measure* refer to the greatest numerical common measure when particular numerical values are substituted for the letters.

Thus, if we substitute 6 for x in the expressions given above, we have

$$x^3-2x^2-x+2=(x^2-1)(x-2)=35\times 4=140,$$

and $$x^3-4x^2-x+4=(x^2-1)(x-4)=35\times 2=70.$$

The arithmetical G. C. M. of 70 and 140 is evidently 70.

Now notice that when $x = 6$, the G. C. M. of the expressions is not the same in numerical value as the H. C. F.; for, when $x = 6$,

$$x^2 - 1 = 35, \text{ not } 70.$$

The reason for this is that while $x - 2$ and $x - 4$ do not have an algebraic common factor, their numerical values for particular values of x may have a common numerical factor.

Thus, when $x = 6$, $x - 2$ and $x - 4$ have the values 4 and 2 respectively, and therefore have the common factor 2.

The words *Greatest Common Measure* should not therefore be used in the same sense as the words *Highest Common Factor.*

14. The following principles will be of use in subsequent work:

(i.) *In the process for finding the arithmetical G. C. M. of two integers, M and N, the remainder at any stage of the work can be expressed in the form*

$$\pm(mM - nN),$$

wherein m and n are positive integers, and the upper sign goes with the first, third, etc., remainders, and the lower sign with the second, fourth, etc.

Let M be greater than N. Then, in the process for finding the G. C. M., let Q_1 be the quotient and R_1 the remainder of the first division, Q_2 and R_2 the quotient and the remainder, respectively, of the second division, and so on. It is to be kept in mind that the Q's and the R's are positive integers.

Then, by Ch. III., § 4, Art. 13, we have

$$M = Q_1N + R_1, \qquad (1)$$

$$N = Q_2R_1 + R_2, \qquad (2)$$

$$R_1 = Q_3R_2 + R_3, \qquad (3)$$

etc.

We now have,

from (1): $R_1 = M - Q_1N;$ (4)

from (2): $R_2 = -Q_2R_1 + N$

$= -Q_2(M - Q_1N) + N,$

substituting the value of R_1 from (4)

$= -[Q_2M - (Q_1Q_2 + 1)N];$ (5)

from (3): $R_3 = -Q_3R_2 + R_1$

$= Q_3[Q_2M - (Q_1Q_2 + 1)N] + M - Q_1N$

$= (Q_2Q_3 + 1)M - (Q_1Q_2Q_3 + Q_1 + Q_3)N.$ (6)

In like manner, the value of each succeeding remainder, in terms of M and N, can be derived.

In (4), $m = 1$, $n = Q_1$; in (5), $m = Q_2$, $n = Q_1Q_2 + 1$, and so on. Since the Q's are positive integers, these values of m and n must be positive integers. Hence the truth of the principle enunciated.

(ii.) *If M and N be two positive integers, prime to each other, then two positive integers, m and n, can be found, such that*

$$mM - nN = \pm 1.$$

Since M and N are prime to each other, 1 is their G. C. M. Therefore, the next to the last remainder will be 1 (the last being 0). Consequently, by (i.) two positive integers, m and n, can be found, such that

$$\pm(mM - nN) = 1;$$

whence, $$mM - nN = \pm 1.$$

(iii.) *If M and N be two positive integers, prime to each other, then any common factor of M and NR must be a factor of R.*

For by (ii.), $$mM - nN = \pm 1.$$

Therefore, $mMR - nNR = \pm R$, or $mR \cdot M - n \cdot NR = \pm R$.

Since, by Art. 6 (ii.), any common factor of M and NR is a factor of $mR \cdot M - n \cdot NR$, the last equation shows that this factor is a factor of R.

The following principles follow directly from (iii.):

(iv.) *If M be a factor of NR and be prime to N, it is a factor of R.*

(v.) *If M be prime to R, S, etc., it is prime to $RS \cdots$.*

(vi.) *If each of the integers M, N, P be prime to each of the integers R, S, T, then MNP is prime to RST.*

(vii.) *If M be prime to N, then M^p is prime to N^p, wherein p is a positive integer.*

§ 3. LOWEST COMMON MULTIPLES.

1. The **Lowest Common Multiple** of two or more integral algebraic expressions is the integral expression of lowest degree which is exactly divisible by each of them.

E.g., the lowest common multiple of ax^2, bx^3, and cx^4 is evidently $abcx^4$.

The words *Lowest Common Multiple* are frequently abbreviated to L. C. M.

L. C. M. by Factoring.

2. The following examples will illustrate the method of finding the L. C. M. of two or more expressions which can be readily factored.

Ex. 1. Find the L. C. M. of a^3b, a^2bc^2, and ab^2c^4.

The expression of lowest degree which is exactly divisible by each of the given expressions evidently cannot contain a lower power of a than a^3, a lower power of b than b^2, and a lower power of c than c^4. Therefore, the required L. C. M. is $a^3b^2c^4$.

Observe that the power of each letter in the L. C. M. is the *highest* power to which it occurs in any of the given expressions.

If the expressions contain numerical factors, the L. C. M. of these factors should be found as in Arithmetic.

Ex. 2. Find the L. C. M. of

$$3ab^2,\ 6b(x+y)^2, \text{ and } 4a^2b(x-y)(x+y).$$

The L. C. M. of the numerical coefficients is 12.

The highest power of a in any of the expressions is a^2; of b is b^2; of $x+y$ is $(x+y)^2$; and of $x-y$ is $x-y$.

Consequently the required L. C. M. is $12a^2b^2(x+y)^2(x-y)$.

In general, *the L. C. M. of two or more expressions is obtained by multiplying the L. C. M. of their numerical coefficients by the product of all the different prime factors of the expressions, each to the highest power to which it occurs in any of them.*

EXERCISES XVIII.

Find the L. C. M. of the following expressions:

1. $2a,\ 3b$.
2. $4a^3b,\ 2ab^3,\ 3ax$.
3. $4ab,\ 2a^2b^2,\ 12a^3b$.
4. $14a^3,\ 21a^2,\ 5b,\ 7a$.
5. $7a^2b,\ 3a^3bx,\ 2ab,\ 2a^2x^3$.
6. $7a^3m^2,\ 21x^2m^3,\ 343xm$.
7. $12a^3b^2x,\ 18x^3a^2b,\ 36ab^3x$.
8. $20m^4l^2,\ 12m^3,\ 10m^2l$.
9. $8a^2x^3,\ 30a^3x^3,\ 4a^2x^4,\ 10ax$.
10. $90xy^3,\ 50x^3m^4,\ 6x^2y^2m^2$.

11. $4a^4x^n,\ 6a^2x^{n+1},\ 3a^5x^{n-1}$.

12. $10a^4b^{r+1}x^3,\ 15a^6b^{r-1}x^2$.

13. $20a^2x^n,\ 15a^3x^{n-1},\ 10ax^{2n+1}$.

14. $5x+11,\ 10x-33$.

15. $a+1,\ a-1,\ a$.

16. $a+b,\ a^2+2ab+b^2$.

17. $x+1,\ x^2-2x-3$.

18. $a+x,\ a^2-x^2$.

19. $x-5,\ x^2-3x-10$.

20. $3x-3,\ a^2-2ax+x^2$.

21. $8a^2+16a,\ a^3+4a^2+4a,\ a^3$.

22. $a^2-b^2,\ 4a+4b,\ a^3-b^3-3a^2b+3ab^2$.

23. $x+1,\ x^2-1,\ x^3-1$.

24. $a^3-x^3,\ a^2-x^2,\ x-a$.

25. $x^2-y^2,\ (x-y)^2,\ x^3-y^3$.

26. $x-a,\ a^2-x^2,\ x^4-a^4$.

27. $1-2x,\ 4x^2-1,\ 1+4x^2$.

28. $1-x,\ x^2-1,\ x-2,\ x^2-4$.

29. $x^3-x,\ x^3-1,\ x^3+1$.

30. $x^2-4a^2,\ (x+2a)^3,\ (x-2a)^3$.

31. $4(1-x)^2,\ 8(1-x),\ 8(1+x),\ 4(1-x^2)$.

32. $9a^4b^2-4c^2d^4,\ 9a^4b^2-12a^2bcd^2+4c^2d^4$.

33. $a^2-9a+8,\ a^2-10a+16$.

34. $6x^2-x-1,\ 2x^2+3x-2$.

35. $3x^2-5x+2,\ 4x^3-4x^2-x+1$.

36. $x^2-4a^2,\ x^3+2ax^2+4a^2x+8a^3,\ x^3-2ax^2+4a^2x-8a^3$.

37. $6(a^3-b^3)(a-b)^3,\ 9(a^4-b^4)(a-b)^2,\ 12(a^2-b^2)$.

Lowest Common Multiple by Means of H. C. F.

3. If the given expressions cannot be readily factored, their L. C. M. can be obtained by first finding their H. C. F. The method depends upon a principle which will be first illustrated and then proved.

Ex. 1. Find the L. C. M. of

$$x^3-2x^2-2x^2y+4xy+x-2y \text{ and } x^3-2x^2y+xy^2-2y^3.$$

The H. C. F. of these expressions is found to be $x-2y$.

Consequently the other factors of the given expressions can be found by dividing each of them by their H. C. F. We thus obtain

$$x^3-2x^2-2x^2y+4xy+x-2y=(x-2y)(x^2-2x+1),$$

$$x^3-2x^2y+xy^2-2y^3=(x-2y)(x^2+y^2).$$

From the definition of the H. C. F. we know that these second factors, $x^2 - 2x + 1$ and $x^2 + y^2$, have no common factor, and therefore that the L. C. M. of the given expressions must contain both of them as factors.

Consequently the required L. C. M. is

$$(x - 2y)(x^2 + y^2)(x - 1)^2.$$

4. The example of Art. 3 illustrates the following principle:

The L. C. M. of two integral algebraic expressions is the product of their H. C. F. by the remaining factors of the expressions.

Let E_1 and E_2 stand for the two expressions, and let F stand for their H. C. F.

Then $$E_1 = FQ_1 \text{ and } E_2 = FQ_2,$$
wherein Q_1 and Q_2 are the remaining factors of E_1 and E_2, respectively.

Since Q_1 and Q_2 have no common factor, the expression of lowest degree which is exactly divisible by FQ_1 and FQ_2 must be FQ_1Q_2; hence the principle enunciated.

5. To find the L. C. M. of three or more integral algebraic expressions, find the L. C. M. of any two of them; next, the L. C. M. of a third and the L. C. M. already found, and so on.

Let E_1, E_2, E_3 stand for three integral algebraic expressions, and let M_1 stand for the L. C. M. of E_1 and E_2, and M_2 for the L. C. M. of M_1 and E_3. Then M_2 is the expression of lowest degree which is exactly divisible by M_1 and E_3; but M_1 is the expression of lowest degree which is exactly divisible by E_1 and E_2. Hence M_2 is the expression of lowest degree which is exactly divisible by E_1, E_2, E_3.

Ex. Find the L. C. M. of

$$(x-1)(x-a)(x-b), \quad (1)$$
$$(x-1)(x-a)(x+b), \quad (2)$$
$$(x-1)(x-a)(x+c). \quad (3)$$

The L. C. M. of (1) and (2) is

$$(x-1)(x-a)(x-b)(x+b), \quad (4)$$

and the L. C. M. of (3) and (4) is

$$(x-1)(x-a)(x-b)(x+b)(x+c).$$

Consequently

$$(x-1)(x-a)(x-b)(x+b)(x+c)$$

is the required L. C. M.

EXERCISES XIX.

Find the L. C. M. of the following expressions:

1. $x^3 - x$, $x^3 - 1$.
2. $x^2 - 1$, $x^3 + 1$.
3. $x^3 - 3x + 2$, $x^3 + 2x^2 - x - 2$.
4. $x^2 + x - 2$, $x^3 + 2x^2 + 2x + 1$.
5. $2x^3 - 17x^2 + 19x - 4$, $3x^3 - 20x^2 - 10x + 27$.
6. $6x^3 - x^2 + 11x + 4$, $3x^3 + 13x^2 + x - 3$.
7. $x^3 - 5x^2 + 9x - 9$, $x^4 - 4x^2 + 12x - 9$.
8. $x^3 - x^2 - 9x + 9$, $x^4 - 4x^2 + 12x - 9$.
9. $14x^3 - 17x^2 + 11x - 3$, $6x^4 - 3x^3 + 4x^2 - 1$.
10. $2x^4 - 3x^3 + 4x^2 - 5x - 4$, $2x^4 - x^3 + x - 12$.
11. $4x^3 - 8x^2 + 5x - 3$, $2x^4 - 3x^3 + 6x^2 - 3x + 2$.
12. $4x^4 - 8x^3 - 3x^2 + 7x - 2$, $3x^3 - 11x^2 + 2x + 16$.
13. $2x - 1$, $4x^2 - 1$, $4x^2 + 1$.
14. $x^3 - 6x^2 + 11x - 6$, $x^3 - 9x^2 + 26x - 24$, $x^3 - 8x^2 + 19x - 12$.
15. $x^3 - 5x^2 + 9x - 9$, $x^3 - x^2 - 9x + 9$, $x^4 - 4x^2 + 12x - 9$.

6. The following example illustrates an important relation between the H. C. F. and the L. C. M. of two integral algebraic expressions.

Ex. The H. C. F. of

$$x^3 - 1 = (x - 1)(x^2 + x + 1)$$

and

$$x^2 - 1 = (x - 1)(x + 1)$$

is

$$(x - 1).$$

The L. C. M. of the same expressions is

$$(x - 1)(x + 1)(x^2 + x + 1).$$

The product of the two given expressions is

$$(x-1)(x-1)(x+1)(x^2+x+1) = (\text{H. C. F.}) \times (\text{L. C. M.}).$$

This example illustrates the principle:

The product of two integral algebraic expressions is equal to the product of their H. C. F. and their L. C. M.

Let E_1 and E_2 stand for two integral algebraic expressions, and let F stand for their H. C. F. and M for their L. C. M.

Then $$E_1 = Q_1F \text{ and } E_2 = Q_2F,$$

wherein, as before, Q_1 and Q_2 stand for the remaining factors of E_1 and E_2, respectively.

The L. C. M. of the two expressions is Q_1Q_2F.

But the product of the two expressions is

$$Q_1FQ_2F=(Q_1Q_2F)F=M\cdot F.$$

It follows directly from this principle that the L. C. M. of two integral algebraic expressions can be found by dividing their product by their H. C. F.

§ 4. SOLUTION OF EQUATIONS BY FACTORING.

1. The roots of the equation

$$(x-1)(x-2)=0 \qquad (1)$$

are evidently 1 and 2. For 1 reduces the first member to $0\times(-1)$, $=0$; and 2 reduces the first member to 1×0, $=0$. Therefore equation (1) is equivalent to the equations

$$x-1=0 \text{ and } x-2=0, \text{ jointly.}$$

This example illustrates the following principle:

If all the terms of an integral equation be transferred to the first member, and if this first member be factored, the given equation is equivalent to the set of equations obtained by equating to 0 each factor of its first member.

Let

$$P\times Q\times R=0 \qquad (1)$$

be the given equation. Then we are to prove that the equation is equivalent to the set of equations

$$P=0,\quad Q=0,\quad R=0. \qquad (2)$$

For any solution of (1) must reduce $P\times Q\times R$ to 0, and, therefore, by Ch. III., § 3, Art. 20, either P, or Q, or R to 0. That is, every solution of (1) is a solution of one of equations (2).

Any solution of $P=0$ must reduce P to 0, and, therefore, by Ch. III., § 3, Art. 18, $P\times Q\times R$ to 0.

That is, every solution of $P=0$ is a solution of $P\times Q\times R=0$.

In like manner, it can be shown that the solutions of $Q=0$ and $R=0$ are solutions of (1).

The proof can be easily extended to an equation whose second member is 0, and whose first member is the product of any number of factors.

Ex. 1. Solve the equation $x(x-2)(x+5)=0$.

The given equation is equivalent to the equations

$$x=0,\quad x-2=0,\quad \text{and}\quad x+5=0.$$

The roots are therefore 0, 2, and -5.

Ex. 2. Solve the equation $x^2 - 9x + 20 = 0$.

The factors of the first member are $x - 4$ and $x - 5$.

Therefore the equations

$$x - 4 = 0 \text{ and } x - 5 = 0$$

are jointly equivalent to the given equation.

The required roots are 4 and 5.

EXERCISES XX.

Solve the following equations:

1. $x(x-1)=0$.
2. $5y(y+11)=0$.
3. $(x+2)(2x-3)=0$.
4. $(5x+4)(9-3x)=0$.
5. $x(x-5)(3-2x)=0$.
6. $5x(6x-7)(2-4x)=0$.
7. $x^2-5x+6=0$.
8. $x^2+7x+10=0$.
9. $10x^2+7x-12=0$.
10. $10x^2+x-21=0$.
11. $x^3+6x^2-16x=0$.
12. $x^3-3x^2-10x=0$.
13. $(x^2-4)(x^2-9)=0$.
14. $(9x^2-25)(12-5x-2x^2)=0$

2. The expression $(x-1)(x-2)$

reduces to 0 for $x=1$ and $x=2$, and the expression

$$(x-5)(x+4)$$

reduces to 0 for $x=5$ and $x=-4$.

Observe that the two expressions do not have a common factor, and do not reduce to 0 for the same values of x. This example illustrates the following principle:

If two expressions in one and the same unknown number do not have a common factor, they cannot reduce to 0 for the same value of the unknown number.

Let E_1 and E_2 be two integral expressions in x which do not have a common factor. Then we are to prove that E_1 and E_2 cannot reduce to 0 for the same value of x.

For if E_1 and E_2 do reduce to 0 for the same value of x, say a, they must both be divisible by $x-a$ without a remainder (Ch. VI., § 2, Art. 4). That is, $x-a$ must be a factor of both expressions. But this contradicts the hypothesis that E_1 and E_2 do not have a common factor. Hence the truth of the principle enunciated.

CHAPTER IX.

FRACTIONS.

1. The quotient of a division can be expressed as an integer or an integral expression only when the dividend is a multiple of the divisor.

E.g., $a^2b \div ab = a$; $(ax^2 + 2\,bx) \div x = ax + 2\,b$.

If the dividend be not a multiple of the divisor, the quotient is called a **Fraction**.

E.g., $a \div b$; $(ax^2 + 2\,bx) \div x^3$.

2. The notation for a fraction in Algebra is the same as in ordinary Arithmetic.

Thus, $a \div b$ is written $\frac{a}{b}$, and $(ax^2 + 2\,bx) \div x^3$ is written $\frac{ax^2 + 2\,bx}{x^3}$.

The **Solidus**, /, is frequently used instead of the horizontal line to denote a fraction; as a/b for $\frac{a}{b}$, and $(ax^2 + bx)/x^3$ for $\frac{ax^2 + bx}{x^3}$.

3. As in Arithmetic, the dividend is called the **Numerator** of the fraction, the divisor the **Denominator**, and the two are called the **Terms** of the fraction.

4. An integer or an integral expression can be written in a *fractional form* with a denominator 1.

E.g., $7 = \frac{7}{1}$, $a + b = \frac{a + b}{1}$.

It is important to notice that an algebraic fraction may be *arithmetically* integral for certain values of its terms.

E.g., when $a = 4$ and $b = 2$, the fraction a/b becomes $4/2 = 2$.

5. By the definition of a fraction, a/b is a number which, multiplied by b, becomes a; that is,

$$(a/b)\times b=a,\quad\text{or}\quad \frac{a}{b}\times b=a. \tag{1}$$

6. The definition of an algebraic fraction given in Art. 1 is an extension of the meaning of division to the case in which the dividend is not exactly divisible by the divisor. That this definition is consistent with the meaning of an arithmetical fraction as hitherto used in this book is evident from the following considerations.

If the unit be divided into 2, 3, 4, ⋯, b equal parts, the parts were designated by $\frac{1}{2}, \frac{1}{3}, \frac{1}{4}, \cdots, \frac{1}{b}$, respectively. These fractions were called *fractional units.*

The fraction $\frac{a}{b}$, in which a and b are *positive* integers, was defined as the sum of a fractional units, each equal to $\frac{1}{b}$; or

$$\frac{a}{b}=\frac{1}{b}+\frac{1}{b}+\frac{1}{b}+\cdots a \text{ terms}. \tag{1}$$

And
$$\frac{a}{b}\times b=\frac{a}{b}+\frac{a}{b}+\frac{a}{b}+\cdots b \text{ terms}. \tag{2}$$

Now if each fraction $\frac{a}{b}$, in the second member of identity (2), be replaced by its value given in (1), written in a vertical column, we have

$$\begin{aligned}\frac{a}{b}\times b=&\frac{1}{b}+\frac{1}{b}+\frac{1}{b}+\cdots b \text{ terms}\\ &+\frac{1}{b}+\frac{1}{b}+\frac{1}{b}+\cdots b \text{ terms}\\ &+\frac{1}{b}+\frac{1}{b}+\frac{1}{b}+\cdots b \text{ terms}\\ &\cdot\quad\cdot\quad\cdot\quad\cdot\quad\cdot\quad\cdot\quad\cdot\quad\cdot\quad\cdot\quad\cdot\end{aligned}$$

a rows.

The sum of the fractional units in each row is 1; and, since there are a rows, the sum of all the fractional units is $a\times 1=a$.

Consequently $$\frac{a}{b} \times b = a.$$

This relation agrees with that obtained in Art. 5 from the definition of an algebraic fraction.

7. The Sign of a Fraction. — The sign of a fraction is written before the line separating its numerator from its denominator; as $+\frac{a}{b}$, $-\frac{a}{b}$.

Since a fraction is a quotient, the sign of a fraction is determined by the rule of signs in division.

E.g., $$\frac{+a}{+b} = +\frac{a}{b},\quad \frac{-a}{-b} = +\frac{a}{b},\quad \frac{+a}{-b} = -\frac{a}{b},\quad \frac{-a}{+b} = -\frac{a}{b}.$$

8. From the rule of signs we derive:

(i.) *If the signs of the numerator and the denominator of a fraction be reversed, the sign of the fraction is unchanged.*

E.g., $$\frac{-7}{3} = \frac{7}{-3};\quad \frac{x}{x-1} = \frac{-x}{1-x}.$$

This step is equivalent to multiplying or dividing both terms of the fraction by -1, which in Art. 12 will be proved to be legitimate.

(ii.) *If the sign of either the numerator or the denominator of a fraction be reversed, the sign of the fraction is reversed; and conversely.*

E.g., $$\frac{7}{3} = -\frac{-7}{3};\quad \frac{-x}{x-1} = -\frac{x}{x-1};\quad -\frac{x-a}{b-x} = \frac{x-a}{x-b}.$$

(iii.) *If, in a fraction whose numerator or denominator is a product of two or more factors, the signs of an even number of factors be reversed, the sign of the fraction is unchanged; but, if the signs of an odd number of factors be reversed, the sign of the fraction is reversed.*

E.g., $$\frac{(x-a)}{(a-b)(b-c)(c-a)} = -\frac{(x-a)}{(a-b)(b-c)(a-c)}$$
$$= \frac{(x-a)}{(b-a)(b-c)(a-c)}$$
$$= \frac{a-x}{(a-b)(b-c)(a-c)}.$$

9. Observe that the sign of a fraction affects each term of the numerator (or each term of the denominator); or, the dividing line between the numerator and the denominator has the same effect as parentheses.

E.g.,
$$-\frac{a-b+c}{d} = -(a-b+c) \div d$$
$$= (-a+b-c) \div d$$
$$= \frac{-a+b-c}{d};$$

and
$$-\frac{x+y}{a-b} = \frac{x+y}{-a+b} = \frac{x+y}{b-a}.$$

EXERCISES I.

Change each of the following fractions into an equivalent fraction in which the sign of the first term in the numerator is reversed:

1. $\frac{-x+1}{x-3}$.
2. $\frac{2x-1}{4x+5}$.
3. $\frac{-5x+6}{x^2-3x+1}$.
4. $\frac{-x-1}{x^2+1}$.
5. $\frac{a-b}{c+d}$.
6. $\frac{a-b-c}{x-y}$.

Change each of the following fractions into an equivalent fraction with sign reversed, leaving the denominator unchanged:

7. $-\frac{a-b}{x-y}$.
8. $\frac{1-x}{y-1}$.
9. $-\frac{a-b+c}{x+y-z}$.
10. $\frac{x^3-x-1}{x^2+x-1}$.
11. $-\frac{2a+b-3c}{6x-3y+z}$.
12. $\frac{(x-a)(b-c)}{(x-b)(a-c)}$.

13.–18. Change each of the fractions in Exx. 7–12 into an equivalent fraction with sign reversed, leaving the numerator unchanged.

Change each of the following pairs of fractions into two equivalent fractions whose denominators are equal:

19. $\frac{1}{x-1}, \frac{1}{1-x}$.
20. $\frac{x}{x^2-y^2}, \frac{y}{y^2-x^2}$.

21. $-\frac{a}{a^3-1},\ \frac{1}{1-a^3}.$ **22.** $\frac{1}{1+x-x^4},\ -\frac{x}{x^4-x-1}.$

23. $-\frac{a}{b+c},\ \frac{1}{-b-c}.$ **24.** $\frac{1}{a+b-c},\ \frac{1}{c-a-b}.$

Change each of the following pairs of fractions into two equivalent fractions whose denominators have a common factor:

25. $\frac{1}{(a-b)(a-c)},\ \frac{1}{(b-a)(b-c)}.$

26. $\frac{x-a}{(x-b)(x-c)},\ \frac{b-x}{(a-x)(c-x)}.$

27. $\frac{a}{(a-b)(a-c)(x-c)},\ \frac{b}{(b-c)(c-a)(x-b)}.$

Change each of the following sets of fractions into three equivalent fractions whose denominators, in pairs, have a common factor:

28. $\frac{1}{(a-b)(a-c)},\ \frac{1}{(b-a)(b-c)},\ \frac{1}{(c-a)(c-b)}.$

29. $\frac{x-a}{(x-y)(y-z)},\ \frac{x-b}{(z-y)(x-z)},\ \frac{x-c}{(x-y)(z-x)}.$

10. A **Proper Fraction** is one whose numerator is of lower degree than its denominator in a common letter of arrangement.

E.g., $\frac{1}{x+1},\ \frac{x-2}{x^2+2x-1}.$

An **Improper Fraction** is one whose numerator is of the same or of a higher degree than its denominator in a common letter of arrangement.

E.g., $\frac{x}{x+1},\ \frac{x^3+3x^2+x-1}{x^2+2x-1}.$

A **Fractional Expression** is an expression which has one or more fractional terms.

E.g., $a+\frac{b}{c},\ ax+by-\frac{c}{x+y},\ \frac{x-y}{a+b}.$

11. If the numerator and denominator of a fraction be multinomials, arranged to ascending or descending powers of a common letter of arrangement, and if the numerator be of the same or of a higher degree than the denominator in this letter, the fraction can be reduced to the sum of an integral expression and a fraction whose numerator is of a lower degree than its denominator.

Ex. 1. $$\frac{x}{x+1}=1+\frac{-1}{x+1}, \text{ by Ch. III., § 4, Art. 12,}$$
$$=1-\frac{1}{x+1}, \text{ by Art. 8, (ii.).}$$

Ex. 2.
$$\frac{x^3+x^2-4x+3}{x^2+2x-1}=x-1+\frac{-x+2}{x^2+2x-1}=x-1-\frac{x-2}{x^2+2x-1}.$$

EXERCISES II.

Reduce each of the following fractions to equivalent fractional expressions, containing only proper fractions:

1. $\frac{x^3+x^2-1}{x^2}$.
2. $\frac{x^2-x-1}{x^2}$.
3. $\frac{10a^2-3a+4}{5a^2}$.
4. $\frac{6a^3-9a^2b+5b}{3a}$.
5. $\frac{x^2+x-xy}{x-y}$.
6. $\frac{a^2-b^2-a}{a-b}$.
7. $\frac{9x^2-9x+3}{x-1}$.
8. $\frac{2x^2+x-5}{x+1}$.
9. $\frac{21x^2+20x-1}{3x+2}$.
10. $\frac{m^3-n^3-1}{m-n}$.
11. $\frac{x^3-3x^2+2x-3}{x-1}$.
12. $\frac{m^3-mn^2-m^2n+n^3+1}{m-n}$.
13. $\frac{3x^2+4x+1}{x^2+1}$.
14. $\frac{5x^2-3x-14}{x^2-2}$.
15. $\frac{12x^3+3x^2-9}{4x^2+x}$.
16. $\frac{4x^3+21x+9}{x^2+7}$.
17. $\frac{x^3+x^2-2}{x^2-1}$.
18. $\frac{6x^4-2x^3-5x^2-2x-5}{3x^2-x+2}$.

Reduction of Fractions.

12. The reduction of fractions is based upon the following principles:

(i.) *If both numerator and denominator of a fraction be multiplied by one and the same number or expression, the value of the fraction is not changed;* or, stated symbolically,

$$\frac{a}{b}=\frac{am}{bm}.$$

E.g., $\frac{2}{3}=\frac{2\times 5}{3\times 5}=\frac{10}{15}$; $\frac{a-x}{a+x}=\frac{(a-x)\times(a+x)}{(a+x)\times(a+x)}=\frac{a^2-x^2}{(a+x)^2}$.

(ii.) *If both numerator and denominator of a fraction be divided by one and the same number or expression, the value of the fraction is not changed;* or, stated symbolically,

$$\frac{a}{b}=\frac{a\div m}{b\div m}.$$

E.g., $\frac{6}{8}=\frac{6\div 2}{8\div 2}=\frac{3}{4}$; $\frac{a+ab}{a+ac}=\frac{(a+ab)\div a}{(a+ac)\div a}=\frac{1+b}{1+c}$.

Let the fraction $\frac{a}{b}$ be denoted by q, or

$$q=\frac{a}{b}. \qquad (1)$$

Multiplying both members of (1) by b, we have, by Ch. II., § 3, Art. 17,

$$qb=\frac{a}{b}\times b;$$

or

$$qb=a, \text{ by Art. 5.} \qquad (2)$$

Multiplying both members of (2) by m, we have

$$qbm=am. \qquad (3)$$

Dividing both members of (3) by bm, we have, by Ch. II., § 4, Art. 13,

$$q=am\div bm$$

$$=\frac{am}{bm}, \text{ by definition of a fraction.} \qquad (4)$$

From (1) and (4), we have, by Axiom (iv.),

$$\frac{a}{b}=\frac{am}{bm}.$$

The principle enunciated in (ii.) can be proved in a similar way.

Reduction of Fractions to Lowest Terms.

13. A fraction is said to be *in its lowest terms* when its numerator and denominator have no common integral factor.

E.g., $$\frac{2}{3}, \ \frac{x-1}{x^2+1}.$$

If the numerator and denominator of a given fraction have common factors, the fraction can be reduced to an equivalent fraction in its lowest terms by dividing its numerator and denominator by all their common factors [Art. 12 (ii.)].

This step is called *canceling common factors.*

Ex. 1. Reduce $\frac{6a^3b^2c^4}{8a^2b^5c^6}$ to its lowest terms.

The factor $2a^2b^2c^4$ is common to both numerator and denominator. We, therefore, have

$$\frac{6a^3b^2c^4}{8a^2b^5c^6} = \frac{3a}{4b^3c^2}, \text{ canceling the common factor.}$$

Ex. 2.
$$\frac{a^2-x^2}{(a+x)^2} = \frac{(a+x)(a-x)}{(a+x)(a+x)}$$
$$= \frac{a-x}{a+x}, \text{ canceling common factor.}$$

Ex. 3.
$$\frac{a^4-1}{a^6-a^5-a^2+a} = \frac{a^4-1}{a(a^5-a^4-a+1)}$$
$$= \frac{a^4-1}{a[a^4(a-1)-(a-1)]}$$
$$= \frac{a^4-1}{a(a^4-1)(a-1)} = \frac{1}{a(a-1)}.$$

Ex. 4.
$$\frac{8a^3b^3+8a^2b^4}{12a^5b^4+24a^4b^5+12a^3b^6} = \frac{8a^2b^3(a+b)}{12a^3b^4(a^2+2ab+b^2)}$$
$$= \frac{2}{3ab(a+b)}.$$

EXERCISES III.

Reduce each of the following fractions to its lowest terms:

1. $\frac{ab}{ac}$. 2. $\frac{a^2x}{ax^2}$. 3. $\frac{a^2x^3}{5a^3x^2}$.

4. $\frac{4x^4m^2n^3}{8x^3m^2n^6}$. 5. $\frac{2a^2b^3c^4}{5a^3b^2c^5}$. 6. $\frac{150a^3x^4z^7}{48a^4x^7}$.

7. $\frac{4(a+b)}{2(a+b)}$. 8. $\frac{5(x+y)^3}{15(x+y)^2}$. 9. $\frac{44(a+c)^r}{66(a+c)^{r-2}}$.

10. $\frac{a^{n+1}b}{a^{n-1}b^m}$. 11. $\frac{54a^nb^{n-2}y^{n+1}}{72ab^{n-1}y^n}$.

12. $\frac{360(a-b)^{n+1}(x+y)^{n-3}}{75(a-b)^n(x+y)^{n-5}}$. 13. $\frac{m-n}{2m-2n}$.

14. $\frac{a^2+ab}{a^2-ab}$. 15. $\frac{6a-9b}{8a-12b}$. 16. $\frac{2ac-2a}{3ac-3a}$.

17. $\frac{3ab+6ac}{5bd+10dc}$. 18. $\frac{12a^2-60ab}{36a^2+12ab}$. 19. $\frac{a^n+a^{n+2}}{a^{n+1}+a^{n+3}}$.

20. $\frac{x^{n+2}-x^ny^4}{x^{n+5}-x^{n+3}y^4}$. 21. $\frac{ax-ab}{ax+3x-3b-ab}$.

22. $\frac{an+2a-cn-2c}{2an-2a-2cn+2c}$. 23. $\frac{x^3-ax^2+b^2x-ab^2}{x^3-ax^2-b^2x+ab^2}$.

24. $\frac{x^2+2x-3}{x^2+5x+6}$. 25. $\frac{x^2+x-6}{(x-2)^3}$. 26. $\frac{x^2-x-12}{(x+3)^2}$.

27. $\frac{x^2-5x-14}{x^2+10x+16}$. 28. $\frac{3x^2+x-2}{3x^2+4x-4}$. 29. $\frac{5x^2+4x-1}{5x^2+19x-4}$.

30. $\frac{3x^2+16x-35}{5x^2+33x-14}$. 31. $\frac{x^4+x^2-2}{x^4+5x^2+6}$. 32. $\frac{x^{2n}+2x^n+1}{x^{2n}+3x^n+2}$.

33. $\frac{x^2-(a-b)x-ab}{x^2-(a+c)x+ac}$. 34. $\frac{x-1}{x^2-1}$. 35. $\frac{3-2a}{4a^2-9}$.

36. $\frac{2-x}{x^2-4}$. 37. $\frac{5a^2+5ax}{a^2-x^2}$. 38. $\frac{ax+bx}{na^2-nb^2}$.

39. $\frac{3x^2-12a^2}{3x+6a}$. 40. $\frac{a^2x^2-a^2}{2a^2x-2a^2x^2}$. 41. $\frac{4a^2-12ab+9b^2}{4a^2-9b^2}$.

42. $\dfrac{x^2 - 12xy + 36y^2}{x^2 - 36y^2}$.

43. $\dfrac{2ay - b + a - 2by}{a^2 - b^2}$.

44. $\dfrac{a - b}{a^3 - b^3}$.

45. $\dfrac{a^2 - ab + b^2}{a^3 + b^3}$.

46. $\dfrac{2a - 3b}{8a^3 - 27b^3}$.

47. $\dfrac{18abx + 30bdx - 24aby - 40bdy}{18a^2b^2 - 50b^2d^2}$.

48. $\dfrac{54a^2bx + 63a^2by - 126acdx - 147acdy}{27a^3b^2 - 147ac^2d^2}$.

49. $\dfrac{bx + 2}{2b + (b^2 - 4)x - 2bx^2}$.

50. $\dfrac{1 - a^2}{(1 + ax)^2 - (a + x)^2}$.

51. $\dfrac{x^5 - x^4y - xy^4 + y^5}{x^4 - x^3y - x^2y^2 + xy^3}$.

52. $\dfrac{ab(x^2 + y^2) + xy(a^2 + b^2)}{ab(x^2 - y^2) + xy(a^2 - b^2)}$.

53. $\dfrac{a^6 + a^4 - a^2 - 1}{a^8 - a^6 + a^2 - 1}$.

54. $\dfrac{n^4 - 16}{n^4 - 4n^3 + 8n^2 - 16n + 16}$.

55. $\dfrac{x(b - c)}{(a - b)(c - b)(a - c)}$.

56. $\dfrac{(a + b + c)^2 - (a - b - c)^2}{3a(b^2 + 2bc + c^2)}$.

57. $\dfrac{a^2 - (b - c)^2}{(a + c)^2 - b^2}$.

58. $\dfrac{bc(b - c) + ca(c - a) + ab(a - b)}{(b - c)(c - a)(a - b)}$.

59. $\dfrac{(y - z)^3 + (z - x)^3 + (x - y)^3}{(x - y)(y - z)(z - x)}$.

60. $\dfrac{a^4(b^2 - c^2) + b^4(c^2 - a^2) + c^4(a^2 - b^2)}{a^2(b - c) + b^2(c - a) + c^2(a - b)}$.

14. If the numerator and denominator of a fraction cannot be readily factored, we find their H. C. F. by the method of division. When the terms of the fraction are divided by this H. C. F., the fraction is in its lowest terms.

Ex. The H. C. F. of the numerator and denominator of the fraction

$$\frac{3x^2 - 14x + 16}{6x^3 - x^2 - 61x + 56}$$

is $3x - 8$. Dividing both terms of the fraction by $3x - 8$, we have

$$\frac{x - 2}{2x^2 + 5x - 7}.$$

EXERCISES IV.

Reduce the following fractions to their lowest terms:

1. $\dfrac{3x^3-8x^2+8x-5}{2x^3+5x^2-5x+7}$.

2. $\dfrac{6x^3+11x^2-6x-5}{3x^3+10x^2+3x-10}$.

3. $\dfrac{x^3-x^2+2}{x^3-3x^2+4x-2}$.

4. $\dfrac{2x^3-13x^2+19x-20}{2x^3+9x^2-14x+24}$.

5. $\dfrac{x^3-5x^2+13x-14}{x^3-x^2+x+14}$.

6. $\dfrac{x^3-3x^2+4}{x^3-2x^2-4x+8}$.

7. $\dfrac{x^3-3x^2+4}{3x^3-18x^2+36x-24}$.

8. $\dfrac{8x^3+2x^2-5x+1}{8x^3+10x^2-11x+2}$.

9. $\dfrac{4x^3-13x^2+11x-2}{8x^3-22x^2+13x-2}$.

10. $\dfrac{2x^5-21x^4+68x^3-84x^2+32x}{2x^5-14x^4+31x^3-24x^2+12x-16}$.

11. $\dfrac{a^3x^3-ax-6}{a^3x^3+4a^2x^2+7ax+6}$.

12. $\dfrac{2x^3y^3-17x^2y^2+27xy-9}{2x^3y^3-11x^2y^2-15xy+9}$.

13. $\dfrac{3a^3x^3-7a^2bx^2+3ab^2x-2b^3}{a^3x^3-4a^2bx^2+7ab^2x-6b^3}$.

Reduction of Two or More Fractions to a Lowest Common Denominator.

15. Two or more fractions are said to have a common denominator when their denominators are the same.

E.g., $\dfrac{a}{b}$ and $\dfrac{c}{b}$; $\dfrac{x}{a^2-x^2}$ and $\dfrac{x-y}{(a+x)(a-x)}$.

The **Lowest Common Denominator** (L. C. D.) of two or more fractions is the L. C. M. of their denominators.

E.g., the L. C. D. of $\dfrac{x}{x^2-1}$ and $\dfrac{2x}{(x+1)^2}$

is $(x+1)^2(x-1)$, the L. C. M. of x^2-1 and $(x+1)^2$.

To reduce two or more fractions to equivalent fractions with a lowest common denominator, multiply both numerator and denominator of each fraction by the quotient obtained by dividing their L. C. D. by the denominator of the fraction.

Ex. 1. Reduce $\frac{a}{b^2c}$ and $\frac{d}{bc^2}$ to equivalent fractions having a least common denominator.

Their required L. C. D. is b^2c^2. Multiplying both terms of $\frac{a}{b^2c}$ by $b^2c^2 \div b^2c, = c$, and both terms of $\frac{d}{bc^2}$ by $b^2c^2 \div bc^2, = b$, we have

$$\frac{ac}{b^2c^2} \text{ and } \frac{bd}{b^2c^2}.$$

Ex. 2. Reduce $\frac{x}{a^2 - x^2}$ and $\frac{y}{(a+x)^2}$ to equivalent fractions having a least common denominator.

The required L. C. D. is $(a-x)(a+x)^2$. Multiplying both terms of $\frac{x}{a^2-x^2}$ by $(a-x)(a+x)^2 \div (a^2 - x^2), = a + x$, and both terms of

$$\frac{y}{(a+x)^2} \text{ by } (a-x)(a+x)^2 \div (a+x)^2, = a - x,$$

we have $$\frac{x(a+x)}{(a-x)(a+x)^2} \text{ and } \frac{y(a-x)}{(a-x)(a+x)^2}.$$

Ex. 3. Reduce $x, = \frac{x}{1}$, and $\frac{y}{x-y}$ to equivalent fractions having a least common denominator.

The required L. C. D. is $x - y$. Multiplying both terms of $\frac{x}{1}$ by $x - y$, and both terms of $\frac{y}{x-y}$ by 1, we have

$$\frac{x^2 - xy}{x-y} \text{ and } \frac{y}{x-y}.$$

Ex. 4. Reduce $\frac{1}{x-a}$ and $\frac{1}{a^2 - x^2}$ to equivalent fractions having a least common denominator.

Observe that the denominator of the first fraction is, except for sign, a factor of the denominator of the second fraction.

In such examples it is advisable first to change the fractions into equivalent fractions whose denominators are arranged to ascending or descending powers of some letter, with the term of highest degree in this letter in each denominator positive.

If in this example the denominators be arranged to descending powers of a, the first fraction becomes

$$\frac{-1}{a-x}, \text{ by Art. 8 (i.).}$$

The required L. C. D. is now $a^2-x^2, =(a-x)(a+x)$. Multiplying both terms of $\frac{-1}{a-x}$ by $(a-x)(a+x)\div(a-x), =(a+x)$, and both terms of $\frac{1}{a^2-x^2}$ by 1, we have

$$\frac{-(a+x)}{a^2-x^2}, =-\frac{a+x}{a^2-x^2}, \text{ and } \frac{1}{a^2-x^2}.$$

EXERCISES V.

Reduce the following fractions to equivalent fractions having a least common denominator:

1. $1-a, \frac{a^2}{a+1}.$

2. $m, \frac{1+4m}{m-4}.$

3. $\frac{15}{14xy^2}, \frac{2x}{3y^3}.$

4. $\frac{3}{5a^2b}, \frac{7}{15abx}, \frac{1}{10b^2x}.$

5. $\frac{x}{x-1}, \frac{1}{x+1}, \frac{1}{1-x^2}.$

6. $\frac{m}{y(x-y)}, \frac{y}{m(y-x)}, \frac{1+m}{my}.$

7. $\frac{ax-b}{ax+ab}, \frac{a-bx}{bx+b^2}, \frac{1}{a^2b^2}.$

8. $\frac{a}{1-a}, \frac{1}{a^2-a}, \frac{3a+1}{a^2-1}.$

9. $\frac{3}{2x-2}, \frac{5}{x-2x+1}, \frac{x}{1-x^2}.$

10. $\frac{1}{n-m}, \frac{3nm}{n^3-m^3}, \frac{m-n}{m^2+mn+n^2}.$

11. $\dfrac{1}{2x^2-4x+2}$, $\dfrac{1}{2x^2+4x+2}$, $\dfrac{1}{1-x^2}$.

12. $\dfrac{1}{(a-c)(a-b)}$, $\dfrac{1}{(b-a)(b-c)}$, $\dfrac{1}{(c-a)(c-b)}$.

13. $\dfrac{1}{x^2-3x+2}$, $\dfrac{1}{x^2-5x+6}$. 14. $\dfrac{1}{x^2-9x+20}$, $\dfrac{1}{9x-20-x^2}$.

15. $\dfrac{1}{x^2-8x+7}$, $\dfrac{1}{x^2+7x-8}$.

Addition and Subtraction of Fractions.

16. *The sum, or the difference, of two fractions having a common denominator is a fraction whose numerator is the sum, or the difference, of the numerators of the given fractions, and whose denominator is their common denominator;* or, stated symbolically,

$$\frac{a}{c}+\frac{b}{c}=\frac{a+b}{c}, \text{ and } \frac{a}{c}-\frac{b}{c}=\frac{a-b}{c}.$$

E.g.,
$$\frac{x}{x^2+1}+\frac{1-x}{x^2+1}=\frac{x+(1-x)}{x^2+1}=\frac{1}{x^2+1};$$

$$\frac{2x}{x-1}-\frac{1+x}{x-1}=\frac{2x-(1+x)}{x-1}=\frac{x-1}{x-1}=1.$$

We have $\frac{a}{c}+\frac{b}{c}=a\div c+b\div c$, by definition of a fraction,

$=(a+b)\div c$, by the Distributive Law,

$=\frac{a+b}{c}$, by definition of a fraction.

In like manner, the principle can be proved for the difference of two fractions.

17. If the fractions to be added or subtracted do not have a common denominator, they should first be reduced to equivalent fractions having a least common denominator.

Ex. 1. Simplify $\frac{a}{b^2c}+\frac{d}{bc^2}$.

We have $$\frac{a}{b^2c}+\frac{d}{bc^2}=\frac{ac}{b^2c^2}+\frac{bd}{b^2c^2},$$
reducing to L. C. D., (1)

$$=\frac{ac+bd}{b^2c^2}, \text{ by Art. 16.} \qquad (2)$$

Ex. 2. Simplify $\frac{x}{a-x}+\frac{y}{a+y}$.

We have $$\frac{x}{a-x}+\frac{y}{a+y}=\frac{x(a+y)}{(a-x)(a+y)}+\frac{y(a-x)}{(a-x)(a+y)},$$
reducing to L. C. D., (1)

$$=\frac{x(a+y)+y(a-x)}{(a-x)(a+y)} \qquad (2)$$

$$=\frac{a(x+y)}{(a-x)(a+y)}.$$

The student should accustom himself to write at once the results indicated by (2) in the above examples, omitting (1).

Ex. 3. Simplify $\frac{1}{1-x}-\frac{2}{x+1}+\frac{3x}{x^2-1}$.

We have $$\frac{1}{1-x}-\frac{2}{x+1}+\frac{3x}{x^2-1}=\frac{-1}{x-1}-\frac{2}{x+1}+\frac{3x}{x^2-1}$$

$$=\frac{-(x+1)-2(x-1)+3x}{x^2-1}$$

$$=\frac{1}{x^2-1}.$$

Observe that in this example the denominators of the given fractions were first arranged to descending powers of x, in accordance with the suggestion given in Art. 15, Ex. 4.

Ex. 4. Simplify $\frac{3}{x+4}+\frac{4}{x-5}-\frac{3}{x-4}-\frac{4}{x+5}$.

The character of the denominators in this example suggests that it is better first to unite the first and third fractions, and the second and fourth fractions separately, and then to unite these results.

We have $\dfrac{3}{x+4}-\dfrac{3}{x-4}+\dfrac{4}{x-5}-\dfrac{4}{x+5}$

$$=\frac{3(x-4)-3(x+4)}{x^2-16}+\frac{4(x+5)-4(x-5)}{x^2-25}$$

$$=\frac{-24}{x^2-16}+\frac{40}{x^2-25}$$

$$=\frac{-24(x^2-25)+40(x^2-16)}{(x^2-16)(x^2-25)}$$

$$=\frac{16x^2-40}{(x^2-16)(x^2-25)}=\frac{8(2x^2-5)}{(x^2-16)(x^2-25)}.$$

Ex. 5. Simplify $\dfrac{1}{(a-b)(a-c)}+\dfrac{1}{(b-a)(b-c)}+\dfrac{1}{(c-a)(c-b)}$.

We have $\dfrac{1}{(a-b)(a-c)}+\dfrac{1}{(b-a)(b-c)}+\dfrac{1}{(c-a)(c-b)}$

$$=\frac{1}{(a-b)(a-c)}-\frac{1}{(a-b)(b-c)}+\frac{1}{(a-c)(b-c)}$$

$$=\frac{(b-c)-(a-c)+(a-b)}{(a-b)(a-c)(b-c)}, \text{ reducing to L. C. D.,}$$

$$=\frac{0}{(a-b)(a-c)(b-c)}=0, \text{ since } 0\div N=0.$$

Observe that in this example the fractions are first changed into equivalent fractions, whose denominators, taken in pairs, have one common factor.

Ex. 6. Simplify $1-x+x^2-\dfrac{x^3}{1+x}$.

We have $1-x+x^2-\dfrac{x^3}{1+x}=\dfrac{(1-x+x^2)(1+x)-x^3}{1+x}$

$$=\frac{1+x^3-x^3}{1+x}=\frac{1}{1+x}.$$

EXERCISES VI.

Simplify the following expressions:

1. $\dfrac{5b}{a}-\left(\dfrac{b-a}{a}+\dfrac{3b+2a}{a}\right)$.

2. $\dfrac{6ab-2ad}{n}-\left(\dfrac{3ab-5ad}{n}-\dfrac{7ad-2ab}{n}\right)$.

3. $\dfrac{3a-5b}{3a}+\dfrac{-a+2b}{3a}-\dfrac{4b-a}{3a}$.

4. $1+\dfrac{1}{x-1}$.

5. $2m-\dfrac{3m-5n}{4}$.

6. $a-\dfrac{a^2}{a+b}$.

7. $3a+\dfrac{1-8a}{3}$.

8. $x-\dfrac{3x-4}{3-x}$.

9. $a^2+ax+x^2+\dfrac{x^3}{a-x}$.

10. $1+\dfrac{(a-b)^2}{4ab}$.

11. $1-\left(a-\dfrac{a^2}{1+a}\right)$.

12. $a+b-\dfrac{2ab}{a+b}$.

13. $4a^2-\left[2a-\left(1+\dfrac{5-8a^3}{2a+1}\right)\right]$.

14. $1+a+a^2+\dfrac{a^3}{1-a}$.

15. $\dfrac{b}{2a}+\dfrac{3b}{4a}-\dfrac{5b}{6a}$.

16. $\dfrac{5ab-3y}{20y}+\dfrac{3ab-5x}{4x}$.

17. $\dfrac{1}{x^3}-\dfrac{x^2-1}{x^5}$.

18. $\dfrac{1}{a^n}+\dfrac{1}{a^{n-1}}$.

19. $\dfrac{1-x}{x^n}+\dfrac{1}{x^{n-1}}$.

20. $\dfrac{3}{a^n}-\dfrac{4}{a^{n-1}}+\dfrac{5}{a^{n-2}}$.

21. $\dfrac{1}{xy}+\dfrac{1}{xz}-\dfrac{1}{yz}$.

22. $\dfrac{x^4}{(1-x)^4}+\dfrac{2x^5-x^4}{(1-x)^5}+\dfrac{x^6-x^5+x^4}{(1-x)^6}$.

23. $\dfrac{1}{(x-y)^3}-\dfrac{1}{(y-x)^2}$.

24. $\dfrac{1}{(a-1)^3}-\dfrac{2}{(1-a)^3}$.

25. $\dfrac{3(2a^3+1)}{2a^3n^4}-\dfrac{2(3a^3x^4+2)}{5a^3x^5}-\dfrac{3(5x-2n^4)}{5n^4x}$.

26. $a^{n-2}-\dfrac{a^{n-1}}{2b^3}+\dfrac{1}{3a^2b}$.

27. $\dfrac{9a^n}{14b^6c^4}-\dfrac{5b^{n-4}}{21ac^2}-\dfrac{2c^{n-5}}{15ab^5}$.

28. $\dfrac{x}{x-1}+\dfrac{1}{2x-1}.$

29. $\dfrac{3}{x-3}-\dfrac{4}{x+4}.$

30. $\dfrac{1}{x-y}-\dfrac{1}{x+y}.$

31. $\dfrac{2a+3x}{2a-3x}-\dfrac{2a-3x}{2a+3x}.$

32. $\dfrac{m+n}{m-n}-\dfrac{m-n}{m+n}.$

33. $\dfrac{x^m+y^m}{x^m-y^m}-\dfrac{x^m-y^m}{x^m+y^m}.$

34. $\dfrac{1+x}{1+x+x^2}+\dfrac{1-x}{1-x+x^2}.$

35. $\dfrac{1}{ac+c^2}-\dfrac{1}{a^2+ac}.$

36. $\dfrac{a^2}{ax+x^2}-\left(\dfrac{a^2-x^2}{ax}+\dfrac{x^2}{a^2+ax}\right).$

37. $\dfrac{n}{a^n+1}-\dfrac{n}{a^n-1}.$

38. $\dfrac{a}{3-a}-\dfrac{9}{a^2-3a}.$

39. $\dfrac{2}{x}+\dfrac{x-6}{3x+6}-\dfrac{1}{x^2+2x}.$

40. $\dfrac{3}{2x^3+2x^2}-\dfrac{1}{(x+1)^2}+\dfrac{5}{4x(x+1)}.$

41. $\dfrac{a^2+3n^2+4an}{a^2+n^2+2an}-2.$

42. $\dfrac{5n}{6x}-\left(\dfrac{10n^2-17nx}{12nx-6x^2}+\dfrac{x}{x-2n}\right).$

43. $\dfrac{a-b}{a+b}-\left(\dfrac{a^2-b^2}{a^2+b^2}-\dfrac{a+b}{a-b}\right).$

44. $\dfrac{1}{(x-1)^2}+\dfrac{2}{x-1}-\dfrac{2x}{x^2+1}.$

45. $\dfrac{1}{1+x}+\dfrac{1}{1-x}-\dfrac{2x}{1-x^2}.$

46. $\dfrac{2b}{a^2-b^2}+\dfrac{1}{a+b}.$

47. $\dfrac{2a}{a^2-1}-\dfrac{1}{a+1}.$

48. $\dfrac{ac}{a^2-4y^2}+\dfrac{bd}{ac+2cy}.$

49. $\dfrac{5xy}{4a^2-9b^2}-\dfrac{xy}{9bd-6ad}.$

50. $\dfrac{m}{m-n}+\dfrac{2mn}{n^2-m^2}-\dfrac{2m}{m+n}.$

51. $\dfrac{3}{2x-1}+\dfrac{7}{2x+1}-\dfrac{4-20x}{1-4x^2}.$

52. $\dfrac{3a}{a+x}+\dfrac{a}{a-x}-\dfrac{2ax}{a^2-x^2}.$

53. $\dfrac{30a}{9a^2-1}-\dfrac{4}{3a-1}-\dfrac{5}{3a+1}.$

54. $\dfrac{a(16-a)}{a^2-4}+\dfrac{3+2a}{2-a}-\dfrac{2-3a}{a+2}.$

55. $\dfrac{2m-3}{1-4m^2}+\dfrac{3}{1-2m}+\dfrac{2}{m}.$

56. $\dfrac{a-1}{a+1}-\left(\dfrac{a+1}{1-a}+\dfrac{a^2+1}{a^2-1}\right).$

57. $\dfrac{a+x}{a-x}-\left[\dfrac{x-a}{x+a}-\left(\dfrac{a^2+x^2}{a^2-x^2}+\dfrac{4ax}{a^2+x^2}\right)\right].$

58. $\dfrac{a+x}{2a+2x+4}-\dfrac{2}{a^2+2ax+2a+2x+x^2}$.

59. $\dfrac{ab+x}{bn-dn+bd-b^2}+\dfrac{ad+x}{dn+bd-bn-d^2}$.

60. $\dfrac{5a}{9a^2-25b^2}-\dfrac{2a+3b}{6ad+10bd}-\dfrac{4a-b}{6ad-10bd}$.

61. $\dfrac{3a-5bx}{12a^2-300b^2}-\dfrac{6+x}{60b-12a}+\dfrac{-3d}{4ad+20bd}$.

62. $\dfrac{1}{a-1}-\dfrac{a^2+2a}{a^3-1}$.

63. $\dfrac{1}{x+1}+\dfrac{x^2+x}{x^3+1}$.

64. $\dfrac{1}{(a-1)^2+3a}-\dfrac{1}{1-a^3}-\dfrac{1}{a-1}$.

65. $\dfrac{a-2}{a^2-a+1}-\dfrac{1}{a+1}+\dfrac{a^2+a+3}{a^3+1}$.

66. $\dfrac{a-2n}{a^3+n^3}-\dfrac{a-n}{a^2n-an^2+n^3}-\dfrac{1}{an+n^2}$.

67. $\dfrac{1}{n-m}-\dfrac{3nm}{n^3-m^3}-\dfrac{m-n}{m^2+mn+n^2}$.

68. $\dfrac{1}{x^4+x^2+1}-\dfrac{1}{x-1-x^2}+\dfrac{1}{x+1+x^2}$.

69. $\dfrac{1}{2x^2-4x+2}+\dfrac{1}{2x^2+4x+2}-\dfrac{1}{1-x^2}$.

70. $\dfrac{1}{x-3}-\dfrac{1-x}{x^2+3x+9}+\dfrac{x^2+x-3}{x^3-27}$.

71. $\dfrac{2}{x+a}+\dfrac{3a}{(x+a)^2}-\dfrac{2x-3a}{x^2-2ax-3a^2}$.

72. $\dfrac{1}{a^2+5a+6}+\dfrac{2a}{a^2+4a+3}+\dfrac{1}{(a+1)^2+(a+1)}-\dfrac{2}{a+3}$.

73. $\dfrac{x^2+x-1}{x^3-x^2+x-1}+\dfrac{x^2-x-1}{x^3+x^2+x+1}-\dfrac{x}{1-x^2}-\dfrac{2x^3}{x^4-1}$.

74. $\dfrac{x+3}{2x-1}-\dfrac{x^2-5}{4x^2-4x+1}-\dfrac{2x^3-x(1-5x)-1}{8x^3-12x^2+6x-1}$.

75. $\dfrac{b}{b+x}-\dfrac{bx}{b^2+x^2}-\dfrac{x^2}{b^2-x^2}-\dfrac{2\,bx^3}{b^4-x^4}.$

76. $\dfrac{3}{4\,x^2-8\,x+4}-\left[\dfrac{1-x}{4-4\,x^2}-\left(\dfrac{1}{8+8\,x}-\dfrac{3}{8\,x-8}\right)\right].$

77. $\dfrac{1}{x^2+3\,x+2}-\dfrac{3}{(x+1)(x+2)(x+3)}.$

78. $\dfrac{5}{x-2}+\dfrac{7}{x-1}-\dfrac{5}{x+2}-\dfrac{7}{x+1}.$

79. $\dfrac{4}{x+7}-\dfrac{1}{x-8}-\dfrac{4}{x-7}+\dfrac{1}{x+8}.$

80. $\dfrac{a+b}{(b-c)(c-a)}+\dfrac{b+c}{(c-a)(a-b)}+\dfrac{c+a}{(a-b)(b-c)}.$

81. $\dfrac{ab}{(b-c)(c-a)}+\dfrac{bc}{(c-a)(a-b)}+\dfrac{ca}{(a-b)(b-c)}.$

82. $\dfrac{1}{a(a-b)(a-c)}+\dfrac{1}{b(b-a)(b-c)}+\dfrac{1}{c(c-a)(c-b)}.$

83. $\dfrac{a}{(a-b)(a-c)}+\dfrac{b}{(b-a)(b-c)}+\dfrac{c}{(c-a)(c-b)}.$

84. $\dfrac{a^2}{(a-b)(a-c)}+\dfrac{b^2}{(b-a)(b-c)}+\dfrac{c^2}{(c-a)(c-b)}.$

85. $\dfrac{bc}{a(a^2-b^2)(a^2-c^2)}+\dfrac{ac}{b(b^2-a^2)(b^2-c^2)}+\dfrac{ab}{c(c^2-a^2)(c^2-b^2)}.$

86. $\dfrac{a^2-bc}{(a-b)(a-c)}+\dfrac{b^2+ac}{(b+c)(b-a)}+\dfrac{c^2+ab}{(c-a)(c+b)}.$

87. $\dfrac{a^2-bc}{(a+b)(a+c)}+\dfrac{b^2-ca}{(b+c)(b+a)}+\dfrac{c^2-ab}{(c+a)(c+b)}.$

88. $\dfrac{a-b}{a+b}+\dfrac{b-c}{b+c}+\dfrac{c-a}{c+a}+\dfrac{(a-b)(b-c)(c-a)}{(a+b)(b+c)(c+a)}.$

89. $\dfrac{2}{a-b}+\dfrac{2}{b-c}+\dfrac{2}{c-a}+\dfrac{(a-b)^2+(b-c)^2+(c-a)^2}{(a-b)(b-c)(c-a)}.$

90. $\dfrac{x^2y^2}{b^2c^2}+\dfrac{(x^2-b^2)(b^2-y^2)}{b^2(c^2-b^2)}+\dfrac{(c^2-x^2)(c^2-y^2)}{c^2(c^2-b^2)}.$

91. $\dfrac{x^2-(y-z)^2}{(x+z)^2-y^2}+\dfrac{y^2-(x-z)^2}{(x+y)^2-z^2}+\dfrac{z^2-(x-y)^2}{(y+z)^2-x^2}.$

18. *The product of two fractions is a fraction whose numerator is the product of the numerators of the given fractions, and whose denominator is the product of their denominators;* or, stated symbolically,

$$\frac{a}{c} \times \frac{b}{d} = \frac{ab}{cd}.$$

E.g.,
$$\frac{a+x}{b+x} \times \frac{a-x}{b-x} = \frac{a^2-x^2}{b^2-x^2}.$$

We have $\frac{a}{c} \times \frac{b}{d} = (a \div c) \times (b \div d)$, by definition of a fraction,

$= a \times b \div c \div d$, by Commutative Law,

$= (ab) \div (cd)$, by Ch. II., § 4, Art. 7 (ii.),

$= \frac{ab}{cd}$, by definition of a fraction.

It follows at once that the product of three or more fractions is a fraction whose numerator is the product of the numerators of the given fractions, and whose denominator is the product of their denominators.

It is assumed that the fractions to be multiplied are in their lowest terms. It frequently happens, however, that the numerator of one fraction and the denominator of another have a common factor. In such cases, all common factors should be canceled before the multiplications are performed.

Ex. 1. Simplify $\frac{6(a^2-b^2)}{x^2-y^2} \times \frac{(x+y)^2}{3(a-b)^2}$.

The factor $3(a-b)$ is common to the numerator of the first fraction and the denominator of the second; and the factor $x+y$ is common to the denominator of the first fraction and the numerator of the second.

Canceling these common factors, we have

$$\frac{6(a^2-b^2)}{x^2-y^2} \times \frac{(x+y)^2}{3(a-b)^2} = \frac{2(a+b)}{x-y} \times \frac{x+y}{a-b} = \frac{2(a+b)(x+y)}{(x-y)(a-b)}.$$

Ex. 2. $\frac{2x}{x^2-3x+2} \times \frac{x-1}{x+1} \times \frac{x+1}{x(x^2+1)^2}$.

The factor x is common to the numerator of the first fraction and the denominator of the third; the factor $x-1$ is common to the numerator of the second and the denominator of the first; and the factor $x+1$ is common to the numerator of the third and the denominator of the second. Canceling these common factors, we have

$$\frac{2x}{x^2-3x+2}\times\frac{x-1}{x+1}\times\frac{x+1}{x(x^2+1)^2}=\frac{2}{x-2}\times\frac{1}{1}\times\frac{1}{(x^2+1)^2}$$
$$=\frac{2}{(x-2)(x^2+1)^2}.$$

Ex. 3. $$\frac{1}{ab}\times\frac{b-a^4b}{1+a+a^2}\times\frac{a-a^4}{(1+a^2)(1+a)}=\frac{1}{ab}\times\frac{b(1-a^2)(1+a^2)}{1+a+a^2}$$
$$\times\frac{a(1-a)(1+a+a^2)}{(1+a^2)(1+a)}=(1-a)(1-a)=(1-a)^2.$$

19. The principle proved in Ch. III., § 4, Art. 2, namely,

$$a^m \div a^n = a^{m-n},\ \text{when } m>n,$$

can now be extended to the case in which the dividend (numerator) is of a lower power than the divisor (denominator).

E.g., $$\frac{a^3}{a^5}=\frac{a^3}{a^3\times a^{5-3}}=\frac{1}{a^{5-3}}=\frac{1}{a^2}.$$

In general, $$\frac{a^m}{a^n}=\frac{1}{a^{n-m}},\ \text{when } m<n.$$

When $m<n$, we have

$$\frac{a^m}{a^n}=\frac{aaa\cdots\text{ to } m \text{ factors}}{aaa\cdots\text{ to } n \text{ factors}}$$
$$=\frac{aaa\cdots\text{ to } m \text{ factors}}{aaa\cdots\text{ to } m\text{ factors}\times aaa\cdots\text{ to } (n-m)\text{ factors}}$$
$$=\frac{1}{aaa\cdots\text{ to } (n-m)\text{ factors}}$$
$$=\frac{1}{a^{n-m}}.$$

EXERCISES VII.

Simplify the following expressions:

1. $\frac{7x}{a^3} \times \frac{5ab^3}{14x^3}$.

2. $\frac{15a^3b^2}{22x^2y^5} \times \frac{14xy^2}{25a^2b}$.

3. $\frac{33b^4x^2z^5}{75a^3y^4} \times \left(-\frac{15ay^2}{22b^3z^4}\right)$.

4. $\frac{8a^2}{15b^3x^{n-1}} \times \frac{x^{m+n}}{a^{n+1}}$.

5. $\frac{3c^3y}{4x^{n-1}} \times \frac{2a^n}{15b^m} \times \frac{8b^{m-1}x^{n+1}}{9c^5y^n}$.

6. $\frac{8a^5b^6x^7}{15c^6y^5} \times \frac{5a^4b^3c^3}{6x^3y^3} \times \frac{3c^5y^4}{4a^{13}b^4}$.

7. $\frac{15(a^2+b^2)^2}{14a^2(a-b)} \times \frac{21a^3(a-b)^5}{10b^3(a^2+b^2)^3}$.

8. $\frac{2x-5y}{x+y} \times \frac{2x+5y}{x-y}$.

9. $\frac{5x}{15a-10b} \times (3a-2b)$.

10. $\frac{x-y}{2x} \times \frac{8x}{x^2-y^2}$.

11. $\frac{8a^2}{a^2-b^2} \times \frac{a+b}{2a}$.

12. $\frac{ab^2-b^3}{a^2+ab} \times \frac{a^3-ab^2}{2b^2}$.

13. $\frac{x-3}{x+1} \times \frac{x^2+2x+1}{x^3-27}$.

14. $\frac{2x}{a^3-b^3} \times (a^2+ab+b^2)$.

15. $\frac{a(a+b)}{a^2-2ab+b^2} \times \frac{b(a-b)}{a^2+2ab+b^2}$.

16. $\frac{6ax-15bx}{40ay+15dy} \times \frac{8ax+3dx}{4a^2-25b^2}$.

17. $\frac{10ab-6ad}{24bx+15by} \times \frac{72dx+45dy}{20b-12d} \times \frac{2b}{a}$.

18. $\frac{x^4-y^4}{(x+y)^2} \times \frac{x^2-y^2}{x^2+y^2} \times \frac{x+y}{(x-y)^2}$.

19. $\frac{x^4-y^4}{a^3+b^3} \times \frac{a^2-ab+b^2}{x-y} \times \frac{a+b}{x+y}$.

20. $\frac{ax+x^2}{2b-cx} \times \frac{2bx-cx^2}{(a+x)^2} \times \frac{(a+x)}{x^2}$.

21. $\frac{24abx-15bcy}{20xz-70yz} \times \frac{4x^2-49y^2}{88abx-55bcy}$.

22. $\frac{x^2-(a+b)x+ab}{x^2-(a+c)x+ac} \times \frac{x^2-c^2}{x^2-b^2}$.

23. $\frac{a^2-(b-c)^2}{x^2-y^2} \times \frac{(x+y)^2}{(a-b)^2-c^2}$.

24. $\frac{x^3-8y^3}{x^2-y^2} \times \frac{x+y}{x^2-2xy+4y^2}$.

25. $\frac{x^2-4}{x^2-8x+15} \times \frac{x^2-9}{x^2-8x+12}$.

26. $\dfrac{x^3+y^3}{6x^2(x^4-y^4)} \times \dfrac{(x^2+y^2)}{x^2-xy+y^2}.$

27. $\dfrac{10ab-12ad}{6dx-15dy} \times \dfrac{24dx-60dy}{35b^2-42bd} \times \dfrac{7bd}{8a}.$

28. $\dfrac{4x^2-9y^2}{22a^2-10ab} \times \dfrac{33ab-15b^2}{6ax-9ay} \times \dfrac{12a^2}{10bx+15by}.$

29. $\dfrac{x^2+x-6}{x^2-x-20} \times \dfrac{x^2+x-12}{x^2+x-6} \times \dfrac{x^2-3x-10}{x^2-4}.$

30. $\dfrac{y+x}{(m+n)^3} \times \dfrac{x^2-y^2}{12} \times \dfrac{(m+n)^2}{m-n} \times \dfrac{6(m^2-n^2)}{x+y}.$

Powers of Fractions.

20. From the principle for multiplying fractions it follows that:

A power of a fraction is a fraction whose numerator is the like power of the numerator of the given fraction, and whose denominator is the like power of the denominator; or, stated symbolically,

$$\left(\frac{a}{b}\right)^n = \frac{a^n}{b^n},$$

wherein n is as yet a positive integer.

E.g., $$\left(\frac{2a^2b^3}{c^4}\right)^3 = \frac{8(a^2)^3(b^3)^3}{(c^4)^3} = \frac{8a^6b^9}{c^{12}}.$$

The converse of the principle evidently holds; that is,

$$\frac{a^n}{b^n} = \left(\frac{a}{b}\right)^n.$$

Ex. 1. $$\frac{(x^2-5x+6)^2}{(x-3)^2} = \left(\frac{x^2-5x+6}{x-3}\right)^2 = (x-2)^2.$$

EXERCISES VIII.

Simplify the following expressions:

1. $\left(\frac{2ab^2}{3x}\right)^3$.

2. $\frac{(2ab^2)^3}{3x}$.

3. $\left(-\frac{6x^2y}{7b^3c^2}\right)^3$.

4. $\left(\frac{5a^2b^3}{3x^2y^2}\right)^4$.

5. $\left(-\frac{2a^3x^4z^2}{5b^2c^3y}\right)^6$.

6. $\left(\frac{a^2}{b}\right)^n$.

7. $\left(-\frac{2a^3x}{5b^2y^3}\right)^{2n}$.

8. $\left(\frac{2a^2x^{n-1}}{3b^3y^n}\right)^{n+1}$.

9. $\frac{(x^2-1)^5}{(x-1)^5}$.

10. $\frac{(a^2-ab)^6}{(a-b)^6}$.

11. $\left(\frac{a^2-b^2}{x+y}\right)^2 \times \left(\frac{x^3+y^3}{a+b}\right)^2$.

12. $\frac{(x^3-1)^3}{(x^2+x+1)^3}$.

13. $\frac{(x^4-y^4)^3}{(x^2+y^2)^3}$.

14. $\frac{(x^2-5x+6)^3}{(x-2)^3}$.

15. $\left(\frac{a+1}{b+1}\right)^2 \times \frac{b^2-1}{a^3+1}$.

16. $\frac{(x+7)^4}{(x^2+6x-7)^4}$.

17. $\frac{(x^4-xy^3)^{15}}{(x^3-y^3)^{15}}$.

18. $\left(\frac{a^2+ab+b^2}{x^3+y^3}\right)^3 \times \left(\frac{x^2-xy+y^2}{a^3-b^3}\right)^3$.

19. $\left(\frac{8x^2-10x-3}{x^2-2xy+y^2}\right)^2 \times \left(\frac{x^2-y^2}{2x^2-5x+3}\right)^2$.

20. $\frac{d^3-z^3}{d^3+z^3} \times \frac{d+z}{d-z} \times \left(\frac{d^2-dz+z^2}{d^2+dz+z^2}\right)^2$.

Express each of the following fractions as the square of a fraction:

21. $\frac{x^2}{y^4}$.

22. $\frac{16a^4b^6}{81x^6y^8}$.

23. $\frac{625m^{12}n^{6n}}{a^{2n-2}b^4}$.

Reciprocal Fractions.

21. The **Reciprocal** of a fraction is a fraction whose numerator is the denominator, and whose denominator is the numerator of the given fraction.

E.g., the reciprocal of $\frac{a}{b}$ is $\frac{b}{a}$.

Conversely, $\frac{a}{b}$ is the reciprocal of $\frac{b}{a}$.

22. The product of a fraction and its reciprocal is 1.

For
$$\frac{a}{b}\times\frac{b}{a}=\frac{ab}{ba}=1.$$

Division of Fractions.

23. *The quotient of one fraction divided by another is equal to the product of the dividend and the reciprocal of the divisor;* or, stated symbolically,

$$\frac{a}{b}\div\frac{c}{d}=\frac{a}{b}\times\frac{d}{c}.$$

E.g.,
$$\frac{a-x}{b-x}\div\frac{b+x}{a+x}=\frac{a-x}{b-x}\times\frac{a+x}{b+x}=\frac{a^2-x^2}{b^2-x^2}.$$

We have $\frac{a}{b}\div\frac{c}{d}=(a\div b)\div(c\div d)$, by definition of a fraction,

$= a\div b\div c\times d$, since $\div(c\div d)=\div c\times d$,

$= a\div b\times d\div c$, since $\div c\times d=\times d\div c$,

$=(a\div b)\times(d\div c)$, since $\times d\div c=\times(d\div c)$,

$=\frac{a}{b}\times\frac{d}{c}$, by definition of a fraction.

Ex. 1.
$$\frac{4(a^2-ab)}{(a+b)^2}\div\frac{6a}{a^2-b^2}=\frac{4a(a-b)}{(a+b)^2}\times\frac{(a-b)(a+b)}{6a}$$
$$=\frac{2(a-b)^2}{3(a+b)}.$$

Ex. 2.
$$\frac{4y^3}{x^3-y^3}\div\frac{2y}{y-x}=\frac{4y^3}{(x-y)(x^2+xy+y^2)}\times\frac{-(x-y)}{2y}$$
$$=\frac{-2y^2}{x^2+xy+y^2}=-\frac{2y^2}{x^2+xy+y^2}.$$

Ex. 3.
$$(a^2-b^2-c^2+2bc)\div\frac{a+b-c}{a+b+c}$$
$$=(a+b-c)(a-b+c)\times\frac{a+b+c}{a+b-c}$$
$$=(a-b+c)(a+b+c).$$

24. If the numerator and denominator of the dividend be multiples of the numerator and denominator of the divisor, respectively, the following principle should invariably be used:

The quotient of one fraction divided by another is a fraction whose numerator is the quotient of the numerator of the first fraction divided by the numerator of the second, and whose denominator is the quotient of the denominator of the first fraction divided by the denominator of the second; or, stated symbolically,

$$\frac{a}{b} \div \frac{c}{d} = \frac{a \div c}{b \div d}.$$

E.g.,

$$\frac{8}{27} \div \frac{2}{3} = \frac{8 \div 2}{27 \div 3} = \frac{4}{9};$$

$$\frac{a^2 - x^2}{b^2 - x^2} \div \frac{a - x}{b - x} = \frac{(a^2 - x^2) \div (a - x)}{(b^2 - x^2) \div (b - x)} = \frac{a + x}{b + x}.$$

We have

$$\begin{aligned}
\frac{a}{b} \div \frac{c}{d} &= (a \div b) \div (c \div d), \text{ by definition of a fraction,}\\
&= a \div b \div c \times d, \text{ since } \div (c \div d) = \div c \times d,\\
&= a \div c \div b \times d, \text{ since } \div b \div c = \div c \div b,\\
&= (a \div c) \div (b \div d), \text{ since } \div b \times d = \div (b \div d),\\
&= \frac{a \div c}{b \div d}, \text{ by definition of a fraction.}
\end{aligned}$$

Ex. 1. $\dfrac{x^2}{2x - 2} \div \dfrac{x}{x - 1} = \dfrac{x^2 \div x}{(2x - 2) \div (x - 1)} = \dfrac{x}{2}.$

Ex. 2. $\dfrac{a^2 - ab}{(a + b)^2} \div \dfrac{b - a}{b + a} = \dfrac{a(a - b) \div [-(a - b)]}{(a + b)^2 \div (a + b)} = \dfrac{-a}{a + b} = -\dfrac{a}{a + b}.$

25. Observe that a fraction is divided by an integral expression, which is a factor of its numerator, by dividing its numerator by the expression.

Ex. 1. $\dfrac{a^2 - b^2}{xy} \div (a - b) = \dfrac{(a^2 - b^2) \div (a - b)}{xy} = \dfrac{a + b}{xy}.$

Also that a fraction is divided by an integral expression, which is not a factor of its numerator, by multiplying its denominator by the expression.

Ex. 2. $\dfrac{a^2 + b^2}{xy} \div (a + b) = \dfrac{a^2 + b^2}{xy(a + b)}.$

EXERCISES IX.

Simplify the following expressions:

1. $\dfrac{27\,a^3b^4}{16\,x^5y^2} \div \dfrac{9\,a^5b^2}{4\,x^3y^6}.$

2. $\dfrac{a^5b^6}{x^7y^8} \div \dfrac{a^3b^4}{x^5y^6}.$

3. $\dfrac{12\,x^5y^6}{35\,a^7b^3} \div \dfrac{18\,x^6y^5}{7\,a^4b^6}.$

4. $\dfrac{a^{n-1}x^n}{b^3y^{n-1}} \div \dfrac{a^{n+1}x^3}{b^2y^n}.$

5. $\dfrac{a^{n+1}b^{n-1}}{c^{m+1}d^{m-1}} \div \dfrac{a^nb^n}{c^md^m}.$

6. $\dfrac{6\,(a^2-b^2)^2}{7\,(x^3-1)} \div \dfrac{3\,(a+b)}{(1-x)}.$

7. $\dfrac{(a+b)^{n+1}}{(x-y)^{n-1}} \div \dfrac{(a+b)^{n-1}}{(x-y)^{n+1}}.$

8. $\dfrac{x^2+7\,x+12}{x^2+2\,x-15} \div \dfrac{x+4}{x+5}.$

9. $\dfrac{2\,a^3-2\,ab^2}{a+2\,b} \div \dfrac{a^2-b^2}{2\,a+4\,b}.$

10. $\dfrac{12\,(a^6-b^6)}{x^4-5\,x^3+6\,x^2} \div \dfrac{3\,(a^4+a^2b^2+b^4)}{x^3-3\,x^2}.$

11. $\dfrac{x^2-6\,x+8}{x^2+2\,x+1} \div \dfrac{x-4}{x+1}.$

12. $\dfrac{x^2+y^2-2\,xy-z^2}{a^2-9+4\,b^2+4\,ab} \div \dfrac{x-y+z}{a+2\,b-3}.$

13. $\dfrac{9\,a^2-16\,b^2}{15\,ax+21\,ay} \div \dfrac{6\,ab+8\,b^2}{5\,x+7\,y}.$

14. $\dfrac{12\,ax+9\,ay}{14\,bx-35\,by} \div \dfrac{36\,dx+27\,dy}{28\,gx-70\,gy}.$

15. $\dfrac{24\,dx-60\,dy}{35\,b^2-42\,bd} \div \dfrac{6\,dx-15\,dy}{10\,ab-12\,ad}.$

16. $\dfrac{45\,dx-9\,dy}{20\,abx-10\,b^2x} \div \dfrac{30\,x-6\,y}{20\,a^2x-5\,b^2x}.$

17. $\dfrac{a^2-(b-c)^2}{(a^2-b^2)^2} \div \dfrac{a-b+c}{a^4-b^4}.$

18. $\dfrac{x^3-1}{x^2-a^2} \div \dfrac{x^2+x+1}{x-a}.$

19. $\dfrac{1+n-n^3-n^4}{1-a^2} \div \dfrac{n^2-1}{a^2-1}.$

20. $\dfrac{1-2\,x}{1-x^3} \div \dfrac{1-2\,x+x^2-2\,x^3}{1+2\,x+2\,x^2+x^3}.$

21. $\frac{a^2+ab}{a^2+b^2} \div \frac{a^3b+ab^3+2a^2b^2}{a^4-b^4}$.

22. $\frac{1-x}{x^3+x^4-x^5} \div \frac{1-x^3}{x^5-x^3-2x^2-x}$.

23. $\frac{x^4-x-3x^3+3x^2}{x^3y-y^4} \div \frac{x^4+x^2-2x^3}{x^2y^2+xy^3+y^4}$.

24. $\frac{(a+2b)a^3-(2a+b)b^3}{a^4b^4} \div \frac{(a+b)^2}{a^4b^2+a^2b^4}$.

25. $\frac{[(a+x)^2-4ax][(a-x)^2+4ax]}{a^6-x^6}$

$$\div \frac{a^2x-ax^2}{[(a+x)^2-ax][(a-x)^2+ax]}.$$

26. $\frac{6x^{2n}-24}{x^{2n+3}+6x^{n+3}+9x^3} \div \frac{3x^n+6}{x^{n+2}+3x^2}$.

27. $\frac{1-x^2}{1+y} \div \frac{y^3-1}{x^2} \times \frac{1-y^2}{x+x^2}$.

28. $\frac{x^2+2x-3}{x^2-2x-3} \div \frac{x^2+4x+3}{x^2-4x+3} \times \frac{x^3+1}{x^3-1}$.

29. $\frac{x^4+x^2y^2+y^4}{x^2+y^2} \times \frac{x^2+y(2x+y)}{x^3-y^3} \div \frac{x^3+y^3}{x^2-y(2x-y)}$.

Multiplication and Division of Fractional Expressions.

26. Multinomial expressions having fractional terms are multiplied and divided by methods similar to those used in multiplying and dividing integral expressions.

Ex. 1. $\left(\frac{a^2}{2bx}-\frac{5ac}{9dx}+\frac{25ay}{6}-\frac{ad}{6cx}\right) \div \frac{5a}{6x}$

$$=\left(\frac{a^2}{2bx}-\frac{5ac}{9dx}+\frac{25ay}{6}-\frac{ad}{6cx}\right) \times \frac{6x}{5a}$$

$$=\frac{a^2}{2bx}\times\frac{6x}{5a}-\frac{5ac}{9dx}\times\frac{6x}{5a}+\frac{25ay}{6}\times\frac{6x}{5a}-\frac{ad}{6cx}\times\frac{6x}{5a} \quad (1)$$

$$=\frac{3a}{5b}-\frac{2c}{3d}+5xy-\frac{d}{5c}.$$

The student should accustom himself to write line (1) directly from the given example.

Ex. 2. Multiply $\frac{9n^3}{10x}+\frac{x}{15n}-\frac{6n}{5}$ by $\frac{5n^2}{x}-\frac{10x}{9n^2}+\frac{15}{4}$.

Arranging the work as in multiplication of integral expressions, we have

$$\begin{array}{llllll}
\frac{9n^3}{10x} & +\frac{x}{15n} & -\frac{6n}{5} & & & \\
\frac{5n^2}{x} & -\frac{10x}{9n^2} & +\frac{15}{4} & & & \\
\hline
\frac{9n^5}{2x^2} & +\frac{n}{3} & -\frac{6n^3}{x} & & & \\
 & -n & & -\frac{2x^2}{27n^3} & +\frac{4x}{3n} & \\
 & & +\frac{27n^3}{8x} & & +\frac{x}{4n} & -\frac{9n}{2} \\
\hline
\frac{9n^5}{2x^2} & -\frac{2n}{3} & -\frac{21n^3}{8x} & -\frac{2x^2}{27n^3} & +\frac{19x}{12n} & -\frac{9n}{2}
\end{array}$$

Ex. 3. Divide

$$\frac{12a^2}{125x^2}-15b^2+\frac{15bc}{y}-\frac{15c^2}{4y^2} \text{ by } \frac{2a}{5x}+5b-\frac{5c}{2y}.$$

We have

$$\begin{array}{llll|l}
\frac{12a^2}{125x^2} & -15b^2+\frac{15bc}{y} & -\frac{15c^2}{4y^2} & & \frac{2a}{5x}+5b-\frac{5c}{2y} \\
\cline{5-5}
\frac{12a^2}{125x^2}+\frac{6ab}{5x} & -\frac{3ac}{5xy} & & & \frac{6a}{25x}-3b+\frac{3c}{2y} \\
\cline{1-2}
-\frac{6ab}{5x} & +\frac{3ac}{5xy} & & & \\
-\frac{6ab}{5x} & -15b^2+\frac{15bc}{2y} & & & \\
\cline{1-4}
 & \frac{3ac}{5xy}+\frac{15bc}{2y} & -\frac{15c^2}{4y^2} & & \\
 & \frac{3ac}{5xy}+\frac{15bc}{2y} & -\frac{15c^2}{4y^2} & & \\
\cline{2-3}
\end{array}$$

EXERCISES X.

Simplify the following expressions:

1. $\left(\frac{2\,c}{5\,d}-\frac{9\,b^2}{20\,ad^2}+\frac{3\,cb}{a}-\frac{3\,c^2b^3}{5\,ad}\right)\div\frac{3\,cb}{5\,ad}.$

2. $\left(\frac{2\,x^3}{3\,y^3}-5\frac{x^2}{y^2}+\frac{x}{9\,y}+\frac{4}{5}-\frac{7\,y}{x}\right)\times\frac{3\,y}{5\,x}.$

3. $\left(\frac{a^nb^{n-1}}{x^3y^4}-\frac{a^{n-1}b}{x^2y^5}+\frac{a^2b^3}{x^ny^{n-1}}+\frac{3}{4}\right)\div\frac{a^nb^3}{x^4y^{n-1}}.$

4. $\left(x-\frac{1}{x}\right)^2.$ 5. $\left(2\,x+\frac{1}{2\,x}\right)^3.$ 6. $\left(2\,a+\frac{3}{a}\right)^4.$

7. $\left(\frac{a}{b}+1+\frac{b}{a}\right)\left(\frac{a}{b}-1+\frac{b}{a}\right).$

8. $\left(\frac{2\,a}{3\,b}-\frac{4\,x}{5\,y}\right)\left(\frac{4\,a^2}{9\,b^2}+\frac{8\,ax}{15\,by}+\frac{16\,x^2}{25\,y^2}\right).$

9. $\left(\frac{a^3b}{12\,c^6}+\frac{4\,ab^3}{27\,c^2}-\frac{a^4}{16\,c^8}+\frac{16\,b^4}{81}+\frac{a^2b^2}{9\,c^4}\right)\left(\frac{a}{2\,c^2}-\frac{2\,b}{3}\right).$

10. $\left(\frac{2\,ab^4}{3\,nx^2}+\frac{4\,a^3b^2}{5\,n^3}-\frac{3\,a^2b^3}{2\,n^2x}-\frac{b^5}{x^3}\right)\left(\frac{2\,a^2b}{3\,n^2}-\frac{3\,b^3}{2\,x^2}-\frac{3\,ab^2}{5\,nx}\right).$

11. $\left(\frac{a^4}{b^2}-\frac{b^4}{a^2}\right)\div\left(\frac{a^3}{b}+2\,a^2+2\,ab+b^2\right).$

12. $\left(\frac{ab^3}{xy}-\frac{ax^5}{by}-\frac{b^3y}{ax}+\frac{x^5y}{ab}\right)\div\left(\frac{ab}{y}-\frac{ax^3}{by}+b-\frac{x^3}{b}\right).$

13. $\left(\frac{4\,a^6b^2}{x^2}-\frac{9\,a^6y^4}{b^2x^2}-4\,b^2x^4+\frac{9\,x^4y^4}{b^2}\right)$
$\div\left(\frac{2\,a^3b^2}{x^2}-\frac{3\,a^3y^2}{x^2}+2\,b^2x-3\,xy^2\right).$

14. $\left(\frac{a^{4n}}{b^{6n}}-\frac{4\,c^{6m}d^{8m}}{b^{10n}}+\frac{14\,c^{5m}d^{4m}}{a^{4n}b^{8n}}-\frac{49\,c^{4m}}{4\,a^{8n}b^{6n}}\right)$
$\div\left(\frac{a^{2n}}{b^{3n}}+\frac{2\,c^{3m}d^{4m}}{b^{5n}}-\frac{7\,c^{2m}}{2\,a^{4n}b^{3n}}\right).$

Complex Fractions.

27. A **Complex Fraction** is a fraction whose numerator and denominator, either or both, are fractional expressions.

E.g., $$\dfrac{\dfrac{2}{3}}{\dfrac{4}{5}},\quad \dfrac{\dfrac{a+x}{a-x}}{\dfrac{a+y}{a-y}},\quad \dfrac{1+\dfrac{1}{x}}{1-\dfrac{1}{x}}.$$

Observe that the line which separates the terms of the complex fraction is drawn heavier than the lines which separate the terms of the fractions in its numerator and denominator.

If no distinction be made between the lines of division, the indicated divisions are to be performed successively from above downward.

E.g., $\dfrac{\frac{\frac{2}{3}}{4}}{5} = 2 \div 3 \div 4 \div 5 = 2 \div (3 \times 4 \times 5)$, by Ch. II., § 4, Art. 7,

$= 2 \div 60 = \frac{1}{30}$;

while $$\dfrac{\frac{2}{3}}{\frac{4}{5}} = \tfrac{2}{3} \times \tfrac{5}{4} = \tfrac{5}{6}.$$

28. Compound fractions are simplified by applying successively the principles already established for simple fractions.

Ex. 1. $$\dfrac{\dfrac{1-x^2}{x}}{1-x} = \frac{1-x^2}{x} \div (1-x), \text{ by definition of a fraction,}$$
$$= \frac{1+x}{x}.$$

Ex. 2. $$\dfrac{\dfrac{m^2+n^2}{n} - m}{\dfrac{1}{n} - \dfrac{1}{m}} \div \frac{m^3+n^3}{m^2-n^2}$$
$$= \dfrac{\dfrac{m^2+n^2-mn}{n}}{\dfrac{m-n}{mn}} \times \frac{m^2-n^2}{m^3+n^3}$$
$$= \frac{m(m^2+n^2-mn)}{m-n} \times \frac{(m+n)(m-n)}{(m+n)(m^2-mn+n^2)}$$
$$= m.$$

In complicated examples there is often an advantage in simplifying the parts separately.

Ex. 3. Simplify

$$\frac{\left[\frac{(a+x)^2}{ax}-4\right]\left[\frac{(a-x)^2}{ax}+4\right]+(a^6-x^6)}{(a^2x-ax^2)+\{[(a+x)^2-ax][(a-x)^2+ax]\}}\times\frac{1-\frac{x}{a+x}}{1+\frac{x}{a-x}}$$

We have

$$\left[\frac{(a+x)^2}{ax}-4\right]\left[\frac{(a-x)^2}{ax}+4\right]$$
$$=\frac{a^2+2ax+x^2-4ax}{ax}\times\frac{a^2-2ax+x^2+4ax}{ax}$$
$$=\frac{(a-x)^2(a+x)^2}{a^2x^2};$$

$$a^6-x^6=(a^3-x^3)(a^3+x^3)$$
$$=(a-x)(a+x)(a^2+ax+x^2)(a^2-ax+x^2);$$
$$a^2x-ax^2=ax(a-x);$$

and $$[(a+x)^2-ax][(a-x)^2+ax]$$
$$=(a^2+ax+x^2)(a^2-ax+x^2);$$

finally $$\frac{1-\frac{x}{a+x}}{1+\frac{x}{a-x}}=\frac{\frac{a}{a+x}}{\frac{a}{a-x}}=\frac{a-x}{a+x}.$$

Consequently the given expression becomes

$$\frac{\frac{(a-x)^2(a+x)^2}{a^2x^2}\times\frac{1}{(a-x)(a+x)(a^2+ax+x^2)(a^2-ax+x^2)}}{\frac{ax(a-x)}{(a^2+ax+x^2)(a^2-ax+x^2)}}\times\frac{a-x}{a+x}$$

$$=\frac{\frac{(a-x)(a+x)}{a^2x^2(a^2+ax+x^2)(a^2-ax+x^2)}}{\frac{ax(a-x)}{(a^2+ax+x^2)(a^2-ax+x^2)}}\times\frac{a-x}{a+x}$$

$$=\frac{a+x}{a^3x^3}\times\frac{a-x}{a+x}=\frac{a-x}{a^3x^3}.$$

Continued Fractions.

29. A **Continued Fraction** is a fraction whose numerator is an integer, and whose denominator is an integer plus (or minus) another fraction whose numerator is an integer, and whose denominator is an integer plus (or minus) a third fraction, etc.

Ex. 1. $$\frac{1}{2+\dfrac{3}{4-\frac{6}{7}}}=\frac{1}{2+\dfrac{3}{\frac{22}{7}}}=\frac{1}{2+\frac{21}{22}}=\frac{1}{\frac{65}{22}}=\frac{22}{65}.$$

Observe that in this reduction the work proceeds from below upward.

Ex. 2. $$\frac{3}{x+\dfrac{1}{1+\dfrac{x+1}{3-x}}}=\frac{3}{x+\dfrac{1}{\dfrac{4}{3-x}}}$$

$$=\frac{3}{x+\dfrac{3-x}{4}}$$

$$=\frac{3}{\dfrac{3x+3}{4}}=\frac{4}{x+1}.$$

EXERCISES XI.

Simplify the following expressions:

1. $\dfrac{a+\dfrac{a^2}{c}}{b+\dfrac{bc}{a}}.$

2. $\dfrac{a-\dfrac{ax}{a+x}}{a+\dfrac{ax}{a-x}}.$

3. $\dfrac{\dfrac{x}{x-1}-\dfrac{x+1}{x}}{\dfrac{x}{x+1}-\dfrac{x-1}{x}}.$

4. $\dfrac{\dfrac{a}{a-1}+1}{1-\dfrac{a}{1-a}}.$

5. $\dfrac{x}{1-\dfrac{1}{1+x}}.$

6. $a+\dfrac{a}{a+\dfrac{1}{a}}.$

7. $\dfrac{\dfrac{1+\frac{1}{m}}{1-\frac{1}{m}}}{\dfrac{m+1}{m-1}}.$

8. $\dfrac{1-\dfrac{1-x}{1+x}}{1+\dfrac{1-x}{1+x}}.$

9. $\dfrac{\dfrac{a+b}{a}+\dfrac{a-b}{b}}{\dfrac{a-b}{a}-\dfrac{a+b}{b}}.$

10. $\dfrac{x-\dfrac{1}{1+x}}{\dfrac{1-x-x^2}{x+1}}$.

11. $\dfrac{\dfrac{a^2}{b^3}-\dfrac{b^2}{a^3}}{1-\dfrac{b}{a}}$.

12. $x-\dfrac{x}{1+x+\dfrac{2x^2}{1-x}}$.

13. $\dfrac{1}{x-1+\dfrac{1}{1+\dfrac{x}{4-x}}}$.

14. $\dfrac{\dfrac{x^2+1}{2x-1}-\dfrac{1}{2}x}{\dfrac{x+2}{1-2x}}$.

15. $\dfrac{\dfrac{a+x}{x}-\dfrac{2x}{x-a}}{\dfrac{a^2+x^2}{x-a}}$.

16. $\dfrac{\dfrac{a+x}{a}-\dfrac{x-y}{x}}{\dfrac{x^2+ay}{a^2}}$.

17. $\dfrac{1}{1+\dfrac{x}{1+x+\dfrac{2x^2}{1-x}}}$.

18. $\dfrac{1}{1-\dfrac{1}{1-\dfrac{1}{1-x}}}$.

19. $\dfrac{\dfrac{n}{n+x}-\dfrac{n}{n-x}}{\dfrac{n}{n-x}+\dfrac{n}{n+x}}$.

20. $\dfrac{\dfrac{a+x}{a-x}-\dfrac{a-x}{a+x}}{\dfrac{4ax}{a^2-x^2}}$.

21. $\dfrac{\dfrac{a+1}{a-1}+\dfrac{a-1}{a+1}}{\dfrac{a+1}{a-1}-\dfrac{a-1}{a+1}}$.

22. $\dfrac{\dfrac{n}{a}+\dfrac{a^2}{n^2}}{\dfrac{1}{a^2n}+\dfrac{1}{n^3}-\dfrac{1}{an^2}}$.

23. $\dfrac{\dfrac{1+a^2}{1-a^2}-\dfrac{1-a^2}{1+a^2}}{\left(\dfrac{1+a}{1-a}-\dfrac{1-a}{1+a}\right)\left(1+a^2\right)}$.

24. $\dfrac{x^2}{a+\dfrac{x^2}{a+\dfrac{x^2}{a}}}$.

25. $x+1-\dfrac{x}{x+2-\dfrac{x+1}{x+\dfrac{1}{x+2}}}$.

26. $\dfrac{\dfrac{x}{x-2}-\dfrac{x}{x+2}}{\dfrac{2x}{\frac{1}{2}x^4-x^3+4x-8}}$.

27. $\dfrac{\dfrac{x^4}{x+1}-\dfrac{1}{x^4+x^3}}{x^3+x+\dfrac{1}{x}+\dfrac{1}{x^3}}$.

28. $\dfrac{\dfrac{a}{n}-\dfrac{n-x}{a}+\dfrac{ax}{n^2-nx}}{\dfrac{a}{n-x}+\dfrac{n-x}{a}+2}$.

29. $\dfrac{\left(p+\dfrac{1}{q}\right)^p\left(p-\dfrac{1}{q}\right)^q}{\left(q+\dfrac{1}{p}\right)^p\left(q-\dfrac{1}{p}\right)^q}$.

30. $$\frac{\dfrac{a+2b}{ab^4}-\dfrac{2a+b}{a^4b}}{\dfrac{b^2+c^2}{b^2c^2}\left(\dfrac{1}{b^2}-\dfrac{1}{c^2}\right)-\left(\dfrac{1}{a^2}-\dfrac{1}{c^2}\right)\dfrac{a^2+c^2}{a^2c^2}}.$$

31. $$\frac{\dfrac{a^2+ab}{a^2+b^2}}{\dfrac{a^3b+ab^3+2a^2b^2}{a^4-b^4}}\times\frac{\dfrac{a^4-a-3a^3+3a^2}{a^3b-b^4}}{\dfrac{a^4+a^2-2a^3}{a^2b^2+ab^3+b^4}}.$$

32. $$\frac{\dfrac{4x^n-4}{x^{2n}-2}+\dfrac{x^{2n}-2}{x^n+1}}{\dfrac{3x^{2n-2}-12x^{n-4}}{x^{3n}+5x^{2n}-2x^n-10}\times\dfrac{x^{4n+1}+5x^{3n+1}}{x^{n+2}-4}}.$$

Factors of Fractional Expressions.

30. Fractional Expressions can be factored by the same methods as were employed in factoring integral expressions:

Ex. 1. $$\frac{a}{x}-\frac{a}{x^2}=\frac{a}{x}\left(1-\frac{1}{x}\right).$$

Ex. 2. $$k^2+k+n+\frac{n}{k}=k^2\left(1+\frac{1}{k}\right)+n\left(1+\frac{1}{k}\right)$$
$$=\left(1+\frac{1}{k}\right)\left(k^2+n\right).$$

Ex. 3. Factor $x^2+2+\dfrac{1}{x^2}.$

Since $$x^2=x^2,\ \frac{1}{x^2}=\left(\frac{1}{x}\right)^2,\ \text{and}\ 2=2\cdot x\cdot\frac{1}{x},$$

therefore, $$x^2+2+\frac{1}{x^2}=\left(x+\frac{1}{x}\right)^2.$$

Ex. 4. Factor $6x^2+\dfrac{x}{y}-\dfrac{2}{y^2}.$

The factors of the first term are x and $6x$, or $2x$ and $3x$; those of the second term are $\dfrac{1}{y}$ and $-\dfrac{2}{y}$, or $-\dfrac{1}{y}$ and $\dfrac{2}{y}$. By trial we find $6x^2+\dfrac{x}{y}-\dfrac{2}{y^2}=\left(2x-\dfrac{1}{y}\right)\left(3x+\dfrac{2}{y}\right).$

Ex. 5. $$\frac{4a^2}{9b^2}-1=\left(\frac{2a}{3b}+1\right)\left(\frac{2a}{3b}-1\right).$$

Ex. 6. $$x^2+\frac{1}{x^2}+1=x^2+2+\frac{1}{x^2}-1$$
$$=\left(x+\frac{1}{x}\right)^2-1$$
$$=\left(x+\frac{1}{x}+1\right)\left(x+\frac{1}{x}-1\right).$$

EXERCISES XII.

Factor the following expressions:

1. $a^2b^2-\frac{1}{c^2d^2}.$

2. $\frac{9a^2x^2}{b^2y^2}-\frac{m^2}{n^2}.$

3. $x^3+\frac{1}{x^3}.$

4. $27-\frac{1}{x^3}.$

5. $\frac{64x^3}{125y^3}-1.$

6. $x^2+\frac{1}{4x^2}.$

7. $x^2+1+\frac{1}{4x^2}.$

8. $4x^2+2\frac{x}{y}+\frac{1}{4y^2}.$

9. $x^3+\frac{1}{x^3}+3x+\frac{3}{x}.$

10. $x^3+x+\frac{1}{x}+\frac{1}{x^3}.$

11. $\frac{x^2}{y^2}+\frac{y^2}{x^2}+1.$

12. $x^2+x+2+\frac{1}{x}+\frac{1}{x^2}.$

13. $x^4+4x^2+6+\frac{4}{x^2}+\frac{1}{x^4}.$

14. $x^5-\frac{1}{x^5}+x-\frac{1}{x}.$

15. $a^2x^2+ax+bc+\frac{bc}{ax}.$

16. $\frac{a^3x^3}{b^3y^3}-\frac{a^2x^2}{b^2y^2}+\frac{ax}{by}-1.$

17. $\frac{a^2}{b^2}-\frac{x^2}{y^2}+\frac{c^2}{d^2}-\frac{2ac}{bd}.$

18. $\frac{a^3}{b^3}-\frac{c^3}{d^3}-\frac{2a^2c}{b^2d}+\frac{2ac^2}{bd^2}.$

19. $\frac{a^2x^2}{b^2y^2}+\frac{2anx}{by}-\frac{ax}{by}\cdot\frac{u}{v}+n^2-\frac{u}{v}\cdot n.$

Indeterminate Fractions.

31. By Art. 5, a fraction is a number which, multiplied by the denominator, gives the numerator. Therefore the fraction $\frac{0}{0}$ is a number which, multiplied by 0, gives 0. But by Ch. III., § 3, Art. 18, *any* number, multiplied by 0, gives 0.

We therefore conclude that the fraction $\frac{0}{0}$ may denote *any* number whatever. For this reason, it is called an **Indeterminate Fraction.**

But under certain conditions a fraction of the form $\frac{0}{0}$ has a definite value, which can be determined.

32. The fraction
$$\frac{x^2-9}{x-3}$$
assumes the form $\frac{0}{0}$, when $x=3$.

But for every other value of x, the fraction has a definite value, which can be determined by substituting the particular value of x either in the given fraction $\frac{x^2-9}{x-3}$, or in the equivalent expression $x+3$. For the *identity*
$$\frac{x^2-9}{x-3}=x+3 \qquad (1)$$
is known to be true for all values of x, except 3.

Thus, when $x=1$, the first member becomes $\frac{1-9}{1-3}$, $=4$, and the second member becomes $1+3$, $=4$; when $x=2$, the first member becomes $\frac{4-9}{2-3}$, $=5$, and the second member becomes $2+3$, $=5$. When $x=2.99$, the first member becomes
$$\frac{(2.99)^2-9}{2.99-3}, \quad =\frac{-.0599}{-.01}, \quad =5.99,$$
and the second member becomes $2.99+3$, $=5.99$.

That is, the identity (1) is true however little x may differ from 3. If, therefore, we assume that the identity holds when $x=3$, we can obtain the value of $\frac{0}{0}$ in this example by substituting 3 for x in $x+3$.

Therefore, $\frac{0}{0}=3+3=6$, under this condition.

33. In like manner the value of any fraction, whose numerator and denominator have a common factor of the form $x - a$, can be determined when $x = a$. Evidently the fraction then reduces to $\frac{0}{0}$.

Ex. 1. The fraction $\dfrac{x^2 - 3x + 2}{x^2 + x - 6}$

becomes $\dfrac{0}{0}$, when $x = 2$. But

$$\frac{x^2 - 3x + 2}{x^2 + x - 6} = \frac{(x - 1)(x - 2)}{(x + 3)(x - 2)} = \frac{x - 1}{x + 3}.$$

Therefore, under this condition, $\dfrac{0}{0} = \dfrac{2 - 1}{2 + 3} = \dfrac{1}{5}$.

Ex. 2. The fraction $\dfrac{a^3 - b^3}{a^2 - b^2}$

becomes $\dfrac{0}{0}$, when $a = b$. But

$$\frac{a^3 - b^3}{a \;- b^2} = \frac{(a - b)(a^2 + ab + b^2)}{(a - b)(a + b)} = \frac{a^2 + ab + b^2}{a + b}.$$

Therefore, under this condition, $\dfrac{0}{0} = \dfrac{b^2 + b^2 + b^2}{b + b} = \dfrac{3b}{2}$.

In particular, if $a = 1$, $b = 1$, then $\dfrac{0}{0} = \dfrac{3}{2}$;

if $a = 2$, $b = 2$, then $\dfrac{0}{0} = 3$; and so on.

EXERCISES XIII.

Find the values of the following indeterminate fractions:

1. $\dfrac{9n^2 - 1}{3n - 1}$, when $n = \dfrac{1}{3}$.
2. $\dfrac{a^3 - ax^2}{a - x}$, when $x = a$.
3. $\dfrac{a^2 - 4}{ac + 2c - a - 2}$, when $a = -2$.
4. $\dfrac{x^3 - 3x + 2}{x^3 - 6x + 5}$, when $x = 1$.
5. $\dfrac{x^3 - y^3}{x^4 - y^4}$, when $x = y$; when $x = 2$, $y = 2$; when $x = 3$, $y = 3$.

6. $\dfrac{a^5+n^5}{a^3+n^3}$, when $a=-n$; when $a=-1$, $n=1$;
when $a=7$, $n=-7$.

7. $\dfrac{5a^2-10ax}{a^3-2a^2x-2x^2+ax}$, when $a=2x$; when $a=6$, $x=3$.

8. $\dfrac{x^3-(2-3x)x-6}{(x+4)^2-1}$, when $x=-3$.

9. $\dfrac{a^7-64a+a^6-64}{20a^3+20a^2-80a-80}$, when $a=2$.

10. $\dfrac{x^4-5x^2-4x^2y+10y+4y^2}{2x^2-4y-10}$, when $x=1$, $y=-2$.

34. The following principles will be of use in subsequent work:

(i.) *If*
$$\frac{n_1}{d_1}=\frac{n_2}{d_2}=\frac{n_3}{d_3}=\text{etc.},$$

then each of these fractions is equal to the fraction

$$\frac{an_1+bn_2+cn_3+\cdots}{ad_1+bd_2+cd_3+\cdots}.$$

E.g.,
$$\frac{2}{3}=\frac{4}{6}=\frac{5\times 2+6\times 4}{5\times 3+6\times 6}.$$

Let the common value of the given fractions be v. Then from

$$\frac{n_1}{d_1}=v,\quad \frac{n_2}{d_2}=v,\quad \frac{n_3}{d_3}=v,\ \text{etc.},$$

we have
$$n_1=d_1v,\ n_2=d_2v,\ n_3=d_3v,\ \text{etc.}$$

Multiplying these equations by a, b, c, etc., respectively, and adding corresponding members of the resulting equations, we have

$$an_1+bn_2+cn_3+\cdots=ad_1v+bd_2v+cd_3v+\cdots$$
$$=(ad_1+bd_2+cd_3+\cdots)v.$$

Therefore

$$\frac{an_1+bn_2+cn_3+\cdots}{ad_1+bd_2+cd_3+\cdots}=v=\frac{n_1}{d_1}=\frac{n_2}{d_2}=\text{etc.}$$

(ii.) *In particular,*

$$\frac{n_1}{d_1} = \frac{n_2}{d_2} = \frac{n_3}{d_3} = \cdots = \frac{n_1 + n_2 + n_3 + \cdots}{d_1 + d_2 + d_3 + \cdots}.$$

E.g., $$\frac{2}{3} = \frac{4}{6} = \frac{2 + 4}{3 + 6}.$$

(iii.) *If the fraction* $\frac{n}{d}$ *be in its lowest terms, then* $\frac{n^p}{d^p}$, *wherein* p *is a positive integer, is in its lowest terms.*

For by Ch. VIII., § 2, Art. 14 (vii.), n^p and d^p are prime to each other when n and d are prime to each other.

(iv.) *If two fractions,* $\frac{n}{d}$ *and* $\frac{N}{D}$, *whose terms are positive integers, be equal, and if* $\frac{n}{d}$ *be in its lowest terms, then* $N = kn$, $D = kd$, *wherein* k *is a positive integer.*

E.g., $\frac{10}{15} = \frac{2}{3}$, and $10 = 5 \times 2$, $15 = 5 \times 3$.

From $$\frac{N}{D} = \frac{n}{d}, \text{ we have } N = \frac{nD}{d}. \qquad (1)$$

Since N is an integer, this equation shows that nD is exactly divisible by d; that is, that it contains d as a factor. Also, since $\frac{n}{d}$ is in its lowest terms, n and d are prime to each other. Therefore, by Ch. VIII., § 2, Art. 14 (iv.), d is a factor of D; that is, $D = kd$, wherein k is a positive integer.

Substituting kd for D in (1), we have

$$N = \frac{nkd}{d} = kn.$$

Hence the truth of the principle enunciated.

(v.) *If two fractions,* $\frac{n}{d}$ *and* $\frac{N}{D}$, *whose terms are positive integers, be equal, and each be in its lowest terms, then* $N = n$ *and* $D = d$.

This principle follows directly from (iv.).

EXERCISES XIV.

MISCELLANEOUS EXAMPLES.

Simplify the following expressions:

1. $\dfrac{1-\left(\dfrac{1-a}{1+a}\right)^2}{1+\left(\dfrac{1-a}{1+a}\right)^2}.$

2. $\dfrac{(a-b)^2-\left(\dfrac{a^2+b^2}{a+b}\right)^2}{b-a+\dfrac{a^2}{a+b}}.$

3. $\dfrac{a^2-x^2}{a+b}\times\dfrac{a^2-b^2}{ax+x^2}\left(a+\dfrac{ax}{a-x}\right).$

4. $\left(1+a-\dfrac{a^2+3}{a-1}\right)(1-a^2).$

5. $\dfrac{a+b}{ab}\left(\dfrac{1}{a}-\dfrac{1}{b}\right)-\dfrac{b+c}{bc}\left(\dfrac{1}{c}-\dfrac{1}{b}\right).$

6. $\left(\dfrac{a+b}{c+d}+\dfrac{a-b}{c-d}\right)\div\left(\dfrac{a+b}{c-d}+\dfrac{a-b}{c+d}\right).$

7. $a+b-\dfrac{1}{a+\dfrac{1}{b}}-\dfrac{1}{b+\dfrac{1}{a}}.$

8. $\dfrac{a}{1+\dfrac{a}{b}}+\dfrac{b}{1+\dfrac{b}{a}}-\dfrac{2}{\dfrac{1}{a}+\dfrac{1}{b}}.$

9. $m-\dfrac{1}{1-m+m^2+\dfrac{m^3}{1+m}}.$

10. $\left(\dfrac{x+2y}{x+y}+\dfrac{x}{y}\right)\div\left(\dfrac{x+2y}{y}-\dfrac{x}{x+y}\right).$

11. $\left(\dfrac{x}{a+x}+a\right)\left(\dfrac{a}{a-x}-x\right)-\left(\dfrac{a}{a+x}+x\right)\left(\dfrac{x}{a-x}-a\right).$

12. $\dfrac{1}{1-\dfrac{x}{x-1}}-\dfrac{1}{\dfrac{x}{x+1}-1}.$

13. $\dfrac{a^2-x^2}{\dfrac{1}{a^2}-\dfrac{2}{ax}+\dfrac{1}{x^2}}\times\dfrac{\dfrac{1}{a^2x^2}}{a+x}.$

14. $\dfrac{x-\dfrac{1}{x^2}}{x+\dfrac{1}{x}-2}\div\dfrac{\left(x+\dfrac{1}{x}\right)^2-1}{\left(1-\dfrac{1}{x}\right)\left(x-1+\dfrac{1}{x}\right)}.$

15. $$\frac{\dfrac{n+3}{n^2-4n+3}}{\dfrac{6}{n-1}-\dfrac{5n-30}{n^2-5n+6}}.$$

16. $$\left(\frac{2x+y}{x+y}+\frac{2y-x}{x-y}-\frac{x^2}{x^2-y^2}\right)\div\frac{x^2+y^2}{x^2-y^2}.$$

17. $$\frac{3x^2+3xy}{4xy+6ay}\times\left(\frac{x}{ax+ay}+\frac{3}{2x+2y}\right).$$

18. $$\left(\frac{n-1}{n+1}-\frac{n+1}{n-1}\right)\times\left(\frac{1}{2}-\frac{n}{4}-\frac{1}{4n}\right).$$

19. $$\frac{\dfrac{a^2+ax}{2x}}{a^2-x^2}\times\left(\frac{(a+x)^2}{4ax}-1\right).$$

20. $$\frac{\dfrac{a^2}{a+n}-\dfrac{a^3}{a^2+n^2+2an}}{\dfrac{a}{a+n}-\dfrac{a^2}{a^2-n^2}}.$$

21. $$\frac{\dfrac{ab+1}{b}}{a+\dfrac{1}{\dfrac{ac+1}{c}}}-\frac{1}{b(abc+a+c)}.$$

22. $$\frac{12x-12}{x^4\left[x+\dfrac{8(x+2)}{x}\right]}\times\frac{x^4-16x^2}{15x^2-15}.$$

23. $$\frac{\dfrac{c-d}{c^2+cd}-\dfrac{c}{d^2+cd}}{\dfrac{d^2}{c^3-cd^2}+\dfrac{1}{c+d}}.$$

24. $$\frac{a+\dfrac{ab}{c+d}}{c+\dfrac{cd}{a+b}}\times\frac{a-b+d}{b+c+d}\div\frac{a+b}{c+d}.$$

25. $$\frac{a+\dfrac{1}{b}}{b+\dfrac{1}{a}}\times\frac{b+\dfrac{1}{c}}{c+\dfrac{1}{b}}\times\frac{c+\dfrac{1}{a}}{a+\dfrac{1}{c}}.$$

26. $$\frac{\dfrac{1}{x}-\dfrac{x+a}{x^2+a^2}}{\dfrac{1}{a}-\dfrac{a+x}{a^2+x^2}}+\frac{\dfrac{1}{x}-\dfrac{x-a}{x^2+a^2}}{\dfrac{1}{a}-\dfrac{a-x}{a^2+x^2}}.$$

27. $$\frac{\dfrac{x}{y}+\dfrac{y}{x}+2}{x+y}\div\frac{\dfrac{x}{y}+\dfrac{y}{x}-2}{x-y}.$$

28. $$\left[\frac{1}{p^2}+\frac{1}{q^2}+\frac{2}{p+q}\left(\frac{1}{p}+\frac{1}{q}\right)\right]\div(p+q)^2.$$

29. $$\left[\left(\frac{x^2}{y^3}+\frac{1}{x}\right)\div\left(\frac{x}{y^2}-\frac{1}{y}+\frac{1}{x}\right)\right]\times\frac{-y}{x+y}.$$

30. $\left[\left(\dfrac{2x}{x^2+1}+\dfrac{2x}{x^2-1}\right)\div\left(\dfrac{x}{x^2+1}-\dfrac{x}{x^2-1}\right)\right]^2.$

31. $\left[(a^2-b^2)\div\left(\dfrac{1}{b}-\dfrac{1}{a}\right)\right]-\left[(a^2-b^2)\div\left(\dfrac{1}{b}+\dfrac{1}{a}\right)\right].$

32. $\left[\left(\dfrac{1}{a}+\dfrac{1}{b+c}\right)\div\left(\dfrac{1}{a}-\dfrac{1}{b+c}\right)\right]\times\left(1+\dfrac{b^2+c^2-a^2}{2bc}\right).$

33. $\dfrac{x^3y-y^4}{xy^3+x^2y}\div\left\{\dfrac{x^4+x^3y+x^2y^2}{(x^2-y^2)^3}\div\left(1+\dfrac{y}{x}\right)^2\right\}.$

34. $-\dfrac{a^2-1}{n^2+n}\times\left\{1-\dfrac{1}{1-\dfrac{1}{n}}\right\}\times\dfrac{1+n-n^3-n^4}{1-a^2}.$

35. $\dfrac{\dfrac{1+x}{1-x}-\dfrac{1-x}{1+x}-\dfrac{4x^2}{x^2-1}}{-2\left(\dfrac{1}{x^2+x^3}-\dfrac{1-x}{x^2}-1\right)}.$

36. $\dfrac{\dfrac{x-1}{3x+(x-1)^2}-\dfrac{1-3x+x^2}{x^3-1}-\dfrac{1}{x-1}}{\dfrac{1-2x+x^2-2x^3}{1+2x+2x^2+x^3}}.$

37. $\dfrac{1+x+x^2+\cdots+x^{n-1}+\dfrac{x^n}{1-x}}{1+2x+x^2-\dfrac{x+2}{x^2-1}\cdot x^3}.$

38. $\left[x+\dfrac{b(a+b-c)+ac}{x-b}-(c-a-b)\right]$
$\div\left[\dfrac{c-a}{(c-b)(x-c)}-\dfrac{b-a}{(c-b)(x-b)}\right].$

39. $\dfrac{\left(x^n-1-\dfrac{7-x^n}{3+x^n}\right)\times\dfrac{4}{x^{n+2}+3x^2}}{\dfrac{6x^{2n}-24}{x^{2n+3}+6x^{n+3}+9x^3}\times\dfrac{2x}{3x^n+6}}.$

40. $\dfrac{x(x+3)-a(a-1)+2}{2x}\div\left\{\dfrac{\dfrac{1}{a}+\dfrac{1}{x+1}}{\dfrac{1}{a}-\dfrac{1}{x+1}}\left[1+\dfrac{(x-a)(x+a)+1}{2x}\right]\right\}.$

In each of the following expressions make the indicated substitution, and simplify the result:

41. In $\frac{x-2x+b}{x+a-2b}$, let $x=\frac{a+b}{2}$.

42. In $\left(\frac{m-a}{m-b}\right)^3$, let $m=\frac{a+b}{2}$.

43. In $\frac{ax^2}{n-cx}+\frac{ax}{c}$, let $x=\frac{cn}{c^2-n}$.

44. In $1+\frac{b^2+c^2-a^2}{2bc}$, let $a+b+c=2s$.

45. In $\frac{m}{n}\left(1-\frac{m}{a}\right)+\frac{n}{m}\left(1-\frac{n}{a}\right)$, let $a=m+n$.

46. In $\frac{a-x}{x-b}$, let $x=\frac{(a+b)^2-(a^2+b^2)}{a+b}$.

47. In $\frac{ax+1}{x}-\left[a(x+1)-\frac{a(x^2-1)-x}{x}\right]$, let $x=\frac{a-1}{a}$.

48. In $\frac{8ax+6a-3}{x(4a^2-1)}-\frac{x(2a+1)-3}{x(2a-1)+3}$, let $x=\frac{3-6a}{3+4a}$.

Verify each of the following identities:

49. $\frac{a(a-x)}{b}-\frac{b(b+x)}{a}=x$, when $x=a-b$.

50. $\frac{a(x-a)}{b+c}+\frac{b(x-b)}{a+c}+\frac{c(x-c)}{a+b}=x$, when $x=a+b+c$.

51. $\frac{x+2a}{2b-x}+\frac{x-2a}{2b+x}=\frac{4ab}{4b^2-x^2}$, when $x=\frac{ab}{a+b}$.

52. $(1+x)(1+y)(1+z)=(1-x)(1-y)(1-z)$, when

$$x=\frac{a-b}{a+b},\ y=\frac{b-c}{b+c},\ z=\frac{c-a}{c+a}.$$

53. $b^2-x^2=\frac{4}{c^2}s(s-a)(s-b)(s-c)$, when

$$x=\frac{b^2+c^2-a^2}{2c} \text{ and } a+b+c=2s.$$

CHAPTER X.

FRACTIONAL EQUATIONS IN ONE UNKNOWN NUMBER.

1. A **Fractional Equation** is an equation whose members, either or both, are fractional expressions in the unknown number or numbers.

E.g., $\frac{3}{x+2}=\frac{2}{x+1}$, $x-2+\frac{4-2x}{x+1}=0$.

Observe that we cannot speak of the *degree* of a fractional equation. The term *degree*, as used in Ch. IV., § 2, Art. 5, applies only to *integral* equations.

2. The principles of equivalent equations established in Ch. IV., § 3, hold also for fractional equations. They can therefore be applied in this chapter.

3. If both members of a fractional equation be multiplied by the L. C. D. of the denominators of its fractional terms, an integral equation will be obtained (Ch. II., § 3, Art. 17).

This step is called *clearing the equation of fractions.*

Ex. 1. If both members of the equation

$$\frac{3}{x+2}=\frac{2}{x+1} \qquad (1)$$

be multiplied by $(x+2)(x+1)$, the L. C. D. of the denominators of its fractional terms, we obtain the integral equation

$$3(x+1)=2(x+2). \qquad (2)$$

Ex. 2. If both members of the equation

$$\frac{-2x^2}{x^2-1}+\frac{x}{1-x}=-\frac{x}{x+1}-3 \qquad (1)$$

be multiplied by x^2-1, we obtain the integral equation

$$-2x^2-x(x+1)=-x(x-1)-3(x^2-1),$$

or

$$(x+1)(x-3)=0. \tag{2}$$

Observe that, in Ex. 2, it is not necessary to multiply by $x^2-1, =(x+1)(x-1)$, in order to clear the equation of fractions. For, if the terms in the second member be transferred to the first member, we have

$$\frac{-2x^2}{x^2-1}+\frac{x}{1-x}+\frac{x}{1+x}+3=0,$$

or, uniting terms,

$$\frac{x^2-2x-3}{x^2-1}=0,$$

or

$$\frac{x-3}{x-1}=0.$$

Clearing the last equation of fractions by multiplying by $x-1$, we have the integral equation

$$x-3=0. \tag{3}$$

It is important to notice that the factor $x+1$, by which it is not necessary to multiply in order to clear the given equation of fractions, is a factor of the integral equation (2).

It can be easily seen that no unnecessary factor was used in clearing of fractions the equation in Ex. 1.

4. The roots of a fractional equation are found by solving the integral equation derived from it by clearing of fractions. The equivalence of the given fractional equation and the derived integral equation is determined by the following principle:

If both members of a fractional equation, in one unknown number, be multiplied by an integral expression which is necessary to clear the equation of fractions, the integral equation thus derived will be equivalent to the given fractional equation.

Let $$\frac{N}{D}=0 \tag{1}$$

be the given fractional equation when all its terms are transferred to the first member, added algebraically, and the resulting fraction reduced to its lowest terms.

In deriving equation (1) from the given fractional equation, terms were transferred from one member to the other, by Ch. IV., § 3, Art. 5; and then only indicated operations were performed. Therefore equation (1) is equivalent to the given fractional equation.

Clearing (1) of fractions, we have the integral equation

$$N=0. \tag{2}$$

Any root of (1) reduces $\frac{N}{D}$ to 0. But any value of x which reduces $\frac{N}{D}$ to 0 must reduce N to 0 (Ch. III., § 4, Art. 7), and hence is a root of the derived equation. That is, no solution is lost by the transformation.

Any root of the derived equation reduces N to 0. But, since $\frac{N}{D}$ is a fraction in its lowest terms, N and D have no common factor, and therefore cannot both reduce to 0 for the same value of x (Ch. VIII., § 4, Art. 2). Consequently, any value of x which reduces N to 0 must reduce $\frac{N}{D}$ to 0 (Ch. III., § 4, Art. 6). That is, no root is gained by the transformation.

Therefore the derived integral equation is equivalent to equation (1), and hence to the given fractional equation.

We will now apply this principle to complete the solutions of Exx. 1 and 2 of the preceding article.

Ex. 1. The derived integral equation

$$3(x+1)=2(x+2)$$

is equivalent to the given fractional equation, since no unnecessary factor was used in clearing of fractions. The root 1 of this derived equation is therefore the required root of the given equation.

Ex. 2. The derived integral equation

$$x-3=0$$

is equivalent to the given fractional equation, since no unnecessary factor was used in clearing of fractions. The root 3 of this derived equation is therefore the required root of the given equation.

It is important to observe that in Ex. 2 the integral equation

$$(x+1)(x-3)=0$$

is not equivalent to the given fractional equation, since the unnecessary factor $x+1$ was used in clearing of fractions. This integral equation therefore has the root -1, which is not a root of the given equation.

This root, which does not satisfy the given equation, and which *was introduced by multiplying both members of the given equation by the unnecessary factor* $x+1$, is a root of the equation obtained by equating this factor to 0.

5. If all the terms of a fractional equation be transferred to its first member, be united into a single fraction, and this fraction be reduced to its lowest terms, the integral equation obtained by then clearing of fractions is equivalent to the given fractional equation. But it is not necessary, nor advisable, to make this transformation before clearing of fractions. If any new root be introduced, it will, as we have seen, be a root of one of the factors of the L. C. D, equated to 0, and can therefore be rejected at sight.

Ex. 1. Solve the equation $\frac{7x+10}{x-2}=\frac{5x}{12}+\frac{35}{6}$.

Multiplying by $12(x-2)$,

$$84x+120=5x^2-10x+70x-140.$$

Transferring and uniting terms,

$$5x^2-24x-260=0.$$

Factoring, $$(x-10)(5x+26)=0.$$

Whence $$x=10 \text{ and } x=-\tfrac{26}{5}.$$

Since neither 10 nor $-\frac{26}{5}$ is a root of the L. C. D. equated to 0, that is, of

$$12(x-2)=0,$$

both 10 and $-\frac{26}{5}$ are roots of the given equation.

6. The following suggestions will simplify the work of solving many fractional equations:

(i.) *If any fraction be not in its lowest terms it should be reduced.*

(ii.) *The form of an equation will frequently suggest grouping and uniting some fractional terms. This reduction should always be made if two or more fractions have a common denominator.*

Ex. 1. Solve the equation $\frac{1}{x-2} - \frac{1}{x-4} = \frac{1}{x-6} - \frac{1}{x-8}$.

Uniting the fractional terms in each member separately,

$$\frac{-2}{(x-2)(x-4)} = \frac{-2}{(x-6)(x-8)}.$$

Dividing by -2,

$$\frac{1}{(x-2)(x-4)} = \frac{1}{(x-6)(x-8)}.$$

Clearing of fractions,

$$(x-6)(x-8) = (x-2)(x-4).$$

Therefore $\quad x^2 - 14x + 48 = x^2 - 6x + 8.$

Whence $\quad -8x = -40$, or $x = 5$.

Since 5 is not a root of any factor of the L. C. D. equated to 0, it is a root of the given equation.

Ex. 2. Solve the equation $\frac{x^2}{x-1} = \frac{1}{x-1} + 10. \qquad (1)$

Transferring and uniting terms, $\frac{x^2-1}{x-1} = 10.$

Reducing to lowest terms, $\quad x + 1 = 10. \qquad (2)$

Whence $\quad x = 9.$

The integral equation (2) is equivalent to the given equation, and therefore 9 is the required root.

Had the given equation been cleared of fractions by multiplying by $x-1$, the root 1 of $x-1=0$ would have been introduced.

(iii.) *An improper fraction should in some cases first be reduced to the sum of an integral expression and a proper fraction.*

Ex. 3. Solve the equation $\frac{x-2}{x-4}+\frac{x-3}{x-5}=2.$

Reducing the improper fractions, we obtain

$$1+\frac{2}{x-4}+1+\frac{2}{x-5}=2,$$

or

$$\frac{1}{x-4}+\frac{1}{x-5}=0.$$

Clearing of fractions, $x-5+x-4=0.$

Whence $x=\frac{9}{2}.$

EXERCISES I.

Solve the following equations:

1. $\frac{12}{x}=4.$

2. $\frac{7}{x}=3.$

3. $5-\frac{3}{x}=2.$

4. $\frac{3x-16}{x}=\frac{5}{3}.$

5. $\frac{5x-5}{x+1}=3.$

6. $\frac{x-1}{x+1}=\frac{2}{3}.$

7. $\frac{1}{5-\frac{1}{x}}=\frac{2}{7}.$

8. $\frac{x}{1+2x}=\frac{1-2x}{4(1-x)}.$

9. $\frac{x-1}{x-3}=\frac{x-4}{x-2}.$

10. $\frac{x-2}{2x-5}=\frac{x-5}{2x-2}.$

11. $\frac{2x-3}{3x-4}=\frac{4x-5}{6x-7}.$

12. $\frac{25}{x-\frac{7}{2}}-\frac{10}{3x-4}=0.$

13. $\frac{x}{x+1}=\frac{3x}{x+2}-2.$

14. $\frac{x-1}{x+1}+\frac{1}{x}=1.$

15. $\frac{10-7x}{x-1}=\frac{5}{x+1}-7.$

16. $\frac{2x+1}{(x+2)^2}+\frac{2x+1}{x+2}=2.$

17. $\frac{5x}{3x+1}-\frac{1}{9x+3}=\frac{7}{6}.$

18. $\frac{9x-8}{45}=\frac{x^2-1}{5x+1}-\frac{1}{9}.$

19. $\frac{5x+1}{36x-48}-\frac{x}{9x-12}=\frac{1}{8}.$

20. $\frac{1}{2}+\frac{2}{x+2}=\frac{13}{8}-\frac{5x}{4x+8}.$

21. $\frac{3}{4}-\frac{\frac{3}{4}x+\frac{3}{4}}{\frac{3}{4}+x}=\frac{\frac{3}{4}}{\frac{3}{4}+x}-\frac{3}{4}.$

22. $\frac{2x-1}{4x+2}=\frac{9}{22}+\frac{4x-2}{2x+1}.$

23. $\frac{4}{(x+1)^2}+\frac{4}{x(x+1)^2}=\frac{5}{2x(x+1)}.$

24. $\frac{3x-1}{14}+\frac{x-7}{12x-4}=\frac{12x-1}{56}.$

25. $\frac{3}{1-3x}+\frac{5}{1-5x}=-\frac{4}{2x-1}.$

26. $\frac{x-7}{x+7}-\frac{2x-15}{2x-6}=-\frac{1}{2(x+7)}.$

27. $\frac{1}{x+3}-\frac{1}{x+5}=\frac{1}{6}\cdot\frac{1}{x+3}.$

28. $\frac{2}{x-4}-\frac{5}{x-2}+\frac{3}{x-6}=0.$

29. $\frac{21}{3x-7}-\frac{5}{x+3}=\frac{2}{x-5}.$

30. $\frac{x+\frac{1}{x}}{x^2+1}-\frac{1}{x+1}=\frac{3}{x^2+2x+1}.$

31. $\frac{1}{1+x}+\frac{3}{1-x}=\frac{24}{1-x^2}.$

32. $\frac{2x+1}{2x-1}-\frac{8}{4x^2-1}=\frac{2x-1}{2x+1}.$

33. $\frac{4}{x+2}+\frac{7}{x+3}=\frac{37}{x^2+5x+6}.$

34. $\frac{1}{1-x^2}+\frac{x+2}{4-x^2}+\frac{3+x}{x^2-9}=0.$

35. $\frac{x-3}{x^2-9}-\frac{12-2x}{x^2-36}=\frac{3x-27}{x^2-81}.$

36. $\frac{2x+19}{5x^2-5}-\frac{17}{x^2-1}-\frac{3}{1-x}=0.$

37. $\dfrac{5x-8}{6x-15}-\dfrac{2x-5}{10x-4}=\dfrac{19x^2-29}{(2x-5)(15x-6)}.$

38. $\dfrac{7}{x^2-1}+\dfrac{8}{x^2-2x+1}=\dfrac{37-9x}{x^3-x^2-x+1}.$

39. $\dfrac{7}{6x+30}+\dfrac{3}{4x-20}=\dfrac{15}{2x^2-50}.$

40. $\dfrac{x^2-6}{x^3+8}+\dfrac{4}{5x^2-10x+20}-\dfrac{1}{x+2}=0.$

41. $\dfrac{1}{x^2+2x+1}+\dfrac{4}{x+2x^2+x^3}=\dfrac{5}{2x+2x^2}.$

42. $\dfrac{x-1-\dfrac{x^2-5x+4}{x-3}}{x-1}-\dfrac{x}{x^2+3x+9}=\dfrac{15}{x^3-27}.$

43. $\dfrac{3-2x}{1-2x}-\dfrac{2x-5}{2x-7}=1-\dfrac{4x^2-2}{7-16x+4x^2}.$

44. $\dfrac{1}{x-1}-\dfrac{1}{2(x+1)}-\dfrac{x+3}{2(x^2+1)}=\dfrac{6}{x^4-1}.$

45. $\dfrac{1-3x}{4-x}+\dfrac{2x-1}{x+4}-5=\dfrac{96}{x^2-16}.$

46. $\dfrac{1}{x-2}-\dfrac{1}{x-4}=\dfrac{1}{x-6}-\dfrac{1}{x-8}.$

47. $\dfrac{6}{x-5}-\dfrac{9}{x-3}=\dfrac{1}{x-7}-\dfrac{4}{x-1}.$

48. $\dfrac{x^2+9x+18}{x+6}-\dfrac{x^2-9x+18}{x-6}=22-\dfrac{x^2+3x-18}{x-3}.$

49. $\dfrac{5}{(x-1)(x+2)}-\dfrac{2}{x^2-x-2}=\dfrac{8}{x^2-1}-\dfrac{5}{x^2-4}.$

50. $\dfrac{1}{x-8}-\dfrac{1}{x-7}+\dfrac{1}{x-4}=\dfrac{1}{x-5}.$

51. $\dfrac{7}{x-9}+\dfrac{2}{x-4}=\dfrac{7}{x-7}+\dfrac{2}{x-11}.$

52. $\frac{1}{x-13}-\frac{2}{x-15}+\frac{2}{x-18}=\frac{1}{x-19}$.

53. $\frac{4}{x-8}+\frac{3}{2x-16}-\frac{29}{24}=\frac{2}{3x-24}$.

54. $\frac{x+3}{x+1}-\frac{x+4}{x+2}+\frac{x-6}{x-4}=\frac{x^2-2x-15}{x^2-9}$.

55. $\frac{x+8}{x-5}-\frac{x+6}{x-6}+\frac{x+4}{x-7}=\frac{x+5}{x-5}-\frac{x+2}{x-6}+\frac{x+3}{x-7}$.

56. $\frac{x+5}{x-7}-\frac{x+3}{x-8}+\frac{x+1}{x-9}=\frac{x+2}{x-7}-\frac{x-1}{x-8}+\frac{x}{x-9}$.

Problems.

7. We add a few problems which lead to fractional equations.

Pr. **1.** If 5 be added to a certain number, and 9 be subtracted from the same number, the quotient of the sum divided by the difference will be 3. What is the number?

Let x stand for the required number.

Then, by the condition of the problem, we have

$$\frac{x+5}{x-9}=3;$$

whence $x=16$; and $\frac{16+5}{16-9}=\frac{21}{7}=3$.

Pr. **2.** A number of men received \$120, to be divided equally. If their number had been 4 less, each one would have received three times as much. How many men were there?

Let x stand for the number of men. Then each man received $\frac{120}{x}$ dollars. If their number had been 4 less, each one would have received $\frac{120}{x-4}$ dollars.

Therefore, by the condition of the problem, we have

$$\frac{120}{x-4}=3\times\frac{120}{x}.$$

Whence $x=6$. Each man received \$20; and if their number had been 4 less, each one would have received \$60.

Pr. 3. The value of a fraction when reduced to its lowest terms is $\frac{1}{5}$. If its numerator and denominator be each diminished by 1, the resulting fraction will be equal to $\frac{1}{6}$. What is the fraction?

The numerator of the required fraction must be a multiple of 1, and the denominator the same multiple of 5.

Let x stand for this multiple. Then the required fraction is

$$\frac{x}{5x}.$$

By the condition of the problem, we have

$$\frac{x-1}{5x-1}=\frac{1}{6}.$$

Whence $x=5$. The required fraction is $\frac{5}{25}$, $=\frac{1}{5}$; and

$$\frac{5-1}{25-1}=\frac{4}{24}=\frac{1}{6}.$$

EXERCISES II.

1. What number added to the numerator and denominator of $\frac{3}{7}$ will give a fraction equal to $\frac{3}{4}$?

2. The sum of two numbers is 18, and the quotient of the less divided by the greater is equal to $\frac{1}{5}$. What are the numbers?

3. The denominator of a fraction exceeds its numerator by 2, and if 1 be added to both numerator and denominator, the resulting fraction will be equal to $\frac{2}{3}$. What is the fraction?

4. The sum of a number and seven times its reciprocal is 8. What is the number?

5. The value of a fraction, when reduced to its lowest terms, is $\frac{3}{7}$. If its numerator be increased by 7 and its denominator be decreased by 7, the resulting fraction will be equal to $\frac{2}{3}$. What is the number?

6. What number must be added to the numerator and subtracted from the denominator of the fraction $\frac{7}{18}$, to give its reciprocal?

7. If $\frac{1}{4}$ be divided by a certain number increased by $\frac{1}{4}$, and $\frac{1}{4}$ be subtracted from the quotient, the remainder will be $\frac{1}{4}$. What is the number?

8. A train runs 200 miles in a certain time. If it were to run 5 miles an hour faster, it would run 40 miles further in the same time. What is the rate of the train?

9. A number has three digits, which increase by 1 from left to right. The quotient of the number divided by the sum of the digits is 26. What is the number?

10. A number of men have $72 to divide. If $144 were divided among 3 more men, each one would receive $4 more. How many men are there?

11. It was intended to divide $\frac{1}{2}$ by a certain number, but by mistake $\frac{1}{2}$ was added to the number. The result was, nevertheless, the same. What is the number?

12. A steamer can run 20 miles an hour in still water. If it can run 72 miles with the current in the same time that it can run 48 miles against the current, what is the speed of the current?

13. A man buys two kinds of wine, 14 bottles in all, paying $9 for one kind and $12 for the other. If the price of each kind is the same, how many bottles of each does he buy?

14. The circumference of the hind wheel of a carriage exceeds the circumference of the front wheel by 4 feet, and the front wheel makes the same number of revolutions in running 400 yards that the hind wheel makes in running 500 yards. What is the circumference of each wheel?

15. In a number of two digits, the digit in the tens' place exceeds the digit in the units' place by 2. If the digits be interchanged and the resulting number be divided by the original number, the quotient will be equal to $\frac{23}{32}$. What is the number?

16. In a number of three digits, the digit in the hundreds' place is 2; if this digit be transferred to the units' place, and the resulting number be divided by the original number, the quotient will be equal to $\frac{77}{41}$. What is the number?

17. In one hour a train runs 10 miles further than a man rides on a bicycle in the same time. If it takes the train 6 hours longer to run 255 miles than it takes the man to ride 63 miles, what is the rate of the train?

18. Two engines are used in different places in a mine to pump out water. The one pumps 11 gallons every 5 minutes from a depth of 155 yards; the other pumps 31 gallons every 10 minutes from a depth of 88 yards. The engines together represent the power of 54 horses. Each engine represents the power of how many horses?

19. A cistern has three pipes. To fill it, the first pipe takes one-half of the time required by the second, and the second takes two-thirds of the time required by the third. If the three pipes be open together, the cistern will be filled in 6 hours. In what time will each pipe fill the cistern?

20. A and B ride 100 miles from P to Q. They ride together at a uniform rate until they are within 30 miles of Q, when A increases his rate by $\frac{1}{5}$ of his previous rate. When B is within 20 miles of Q, he increases his rate by $\frac{1}{2}$ of his previous rate, and arrives at Q 10 minutes earlier than A. At what rate did A and B first ride?

21. A circular road has three stations, A, B, and C, so placed that A is 15 miles from B, B is 13 miles from C in the same direction, and C is 14 miles from A in the same direction. Two messengers leaving A at the same time, and traveling in opposite directions, meet at B. The faster messenger then reaches A, 7 hours before the slower one. What is the rate of each messenger?

CHAPTER XI.

LITERAL EQUATIONS IN ONE UNKNOWN NUMBER.

1. The unknown numbers of an equation are frequently to be determined in terms of general numbers, *i.e.*, in terms of numbers represented by letters. The latter are commonly represented by the leading letters of the alphabet, a, b, c, etc.

Such general numbers, a, b, c, etc., are then to be regarded as *known*.

E.g., in the equation $x + a = b$,

a and b are the *known* numbers, and x is the *unknown* number.

From this equation we obtain $x = b - a$.

2. It is important to notice that the assumption that x, y, z, etc., are the unknown numbers of an equation, and that a, b, c, etc., are the known numbers, is arbitrary.

In the equation $x + a = b$, either a or b could be taken as the unknown number. If a be taken as the unknown number, we have

$$a = b - x;$$

if b be taken as the unknown number, we have

$$b = x + a.$$

3. A **Numerical Equation** is one in which all the known numbers are numerals.

E.g., $2x + 3 = 7$; $4x - 3y = 7$.

A **Literal Equation** is one in which some or all of the known numbers are literal.

E.g., $ax + b = 3$, $2ax + 3b = 5$; $ax + by = c$.

4. The principles of equivalent equations, of course, hold when the equations are literal.

Ex. 1. Solve the equation $\frac{x-a}{b}+\frac{x-b}{a}=-\frac{(a-b)^2}{2ab}$.

Clearing of fractions,

$$2ax-2a^2+2bx-2b^2=-a^2+2ab-b^2.$$

Transferring and uniting terms,

$$2(a+b)x=a^2+2ab+b^2.$$

Dividing by $2(a+b)$, $x=\frac{a+b}{2}$.

It is important to notice that the above equation, although algebraically fractional, is integral in the unknown number x. The equations which follow are fractional in the unknown number.

Ex. 2. Solve the equation $\frac{1-ax}{bx}=\frac{bx-1}{ax}+2$.

Clearing of fractions, $a-a^2x=b^2x-b+2abx$.

Transferring and uniting terms,

$$(a^2+2ab+b^2)x=a+b.$$

Dividing by $a^2+2ab+b^2$, $x=\frac{1}{a+b}$.

Ex. 3. Solve the equation $a+\frac{1-a^2}{x}=1$.

Transferring a to second member, $\frac{1-a^2}{x}=1-a$.

Dividing by $1-a$, $\frac{1+a}{x}=1$.

Whence $x=1+a$.

Ex. 4. Solve the equation $\frac{x+a}{x-b}+\frac{x+b}{x-a}=2$.

Reducing improper fractions, $1+\frac{a+b}{x-b}+1+\frac{a+b}{x-a}=2$.

Canceling equal terms, $\frac{a+b}{x-b}+\frac{a+b}{x-a}=0.$

Dividing by $a+b$, $\frac{1}{x-b}+\frac{1}{x-a}=0.$

Clearing of fractions, $x-a+x-b=0.$

Whence $x=\frac{a+b}{2}.$

Ex. 5. Solve the equation

$$\frac{1}{a-b}+\frac{c}{(x-b)(x-c)}=\frac{x}{(x-b)(x-c)}.$$

Uniting fractions with common denominator,

$$\frac{1}{a-b}+\frac{c-x}{(x-b)(x-c)}=0.$$

Reducing to lowest terms, $\frac{1}{a-b}-\frac{1}{x-b}=0.$

Clearing of fractions, $x-b-a+b=0.$

Whence $x=a.$

Had we cleared of fractions at once, we should have introduced the root c of the factor $x-c$ equated to 0.

5. ***A linear equation in one unknown number has one, and only one, distinct root.***

Any linear equation in one unknown number can be reduced to the form

$$ax=b,$$

in which ax is the algebraic sum of all the terms which contain x (transferred to the first member), and b is the algebraic sum of all the terms free from x (transferred to the second member).

If both members of the equation

$$ax=b$$

be divided by a, when $a\neq 0$, we obtain $x=\frac{b}{a}.$

Since this value of x satisfies the equation, we conclude that every linear equation has at least one root.

Let us assume that the equation

$$ax=b$$

has two distinct roots, and let us denote them by r_1 and r_2. Then, since they must both satisfy the given equation, we have

$$ar_1 = b \tag{1}$$

and

$$ar_2 = b. \tag{2}$$

Subtracting (2) from (1), we obtain

$$a(r_1 - r_2) = 0.$$

Since $a \neq 0$, we have, by Ch. III., § 3, Art. 20,

$$r_1 - r_2 = 0;$$

whence

$$r_1 = r_2.$$

Therefore the assumption that the equation has two distinct roots is untenable. Hence the truth of the principle enunciated.

6. Observe that, in Art. 5, we have no authority for dividing both members of the equation

$$ax = b$$

by a, when $a = 0$. But if we assume that $\frac{b}{a}$ still gives the solution of the equation when $a = 0$, the value of x will be indeterminate $(\frac{0}{0})$ or infinite (∞), according as $b = 0$ or $b \neq 0$. Evidently, when $a = 0$ and $b = 0$, any finite value of x will satisfy the equation; while, when $a = 0$ and $b \neq 0$, no finite value of x will satisfy the equation.

EXERCISES I.

Solve the following equations:

1. $a - x = c.$ **2.** $mx + a = b.$ **3.** $a - bx = c.$

4. $mx = nx + 2.$ **5.** $x - ax + 1 = bx.$ **6.** $ax + bx - x = 0.$

7. $3ax - 5ab + 6ax - 7ac = 2ax + 2ab.$

8. $4a^2 - 2abx + b^2 + 3a^2x = 5a^2 - b^2x + 2a^2x.$

9. $19a - 10ax + 15b - 8bx = 16b - 9ax + 16a - 11bx.$

10. $(2a - b)x = 4a^2 - 3a(b + x).$

11. $a(x + a) - b(x - b) = 3ax + (a - b)^2.$

12. $x(x + a) + x(x + b) - 2(x + a)(x + b) = 0.$

13. $(x + a)^2 = (x - b)^2.$ **14.** $(x + a)^2 = 5a^2 + (x - a)^2.$

15. $(a + x)(b + x) - (c - x)(d - x) = 0.$

16. $(3a - x)(a - b) + 2ax = 4b(a + x).$

17. $(2b + 2c - x)^2 + (2b - 2c + x)^2 - (2b - 2d + x)^2$
$= (2b + 2d - x)^2.$

18. $a+\frac{b}{x}=c.$

19. $\frac{a}{x}+b=\frac{b}{x}+a.$

20. $\frac{a}{x}+\frac{x-b}{x}-\frac{a-x}{x}=1.$

21. $\frac{b^2}{ax}+\frac{b}{a}-\frac{a}{b}=\frac{a}{x}.$

22. $\frac{2}{ab}-\frac{1}{bx}+\frac{1}{2ax}-\frac{1}{a^2}=0.$

23. $\frac{a^2x}{b-c}=bx+cd-ac.$

24. $\frac{x+a}{x-a}=\frac{5}{4}.$

25. $\frac{a+x}{b+x}=\frac{a+1}{b+1}.$

26. $\frac{a-bx}{ax-b}=\frac{3}{4}.$

27. $\frac{4x^2+ax-b}{6x^2+bx-a}=\frac{2}{3}.$

28. $\frac{x}{a+b}+\frac{x}{a-b}=2a.$

29. $\frac{ax}{a+b}+\frac{bx}{a-b}=x+\frac{2b}{a}.$

30. $\frac{x+a}{2}-\frac{2}{x+a}=\frac{x-a}{2}.$

31. $\frac{6x+a}{4x+b}-\frac{3x-b}{2x-a}=0.$

32. $\frac{a+x}{b+a}=\frac{a-x}{b-a}.$

33. $\frac{x-a}{x-2}=\frac{x+b}{x-3}.$

34. $\frac{a}{b}=\frac{x-b^2}{x-a^2}.$

35. $\frac{x+ab}{x-ab}=\frac{a^2+ab+b^2}{a^2-ab+b^2}.$

36. $\frac{x+a}{x-b}=\frac{(2x+a)^2}{(2x-b)^2}.$

37. $\frac{x+a}{x+b}=\frac{(2x+a+c)^2}{(2x+b+c)^2}.$

38. $\frac{a(x+1)-b(x-1)}{b(x+1)-a(x-1)}=\frac{a^3}{b^3}.$

39. $\frac{a^3-b^3}{a^3+b^3}=\frac{a(x-b^2)+b(a^2-x)}{a(x-b^2)-b(a^2-x)}.$

40. $\frac{1}{ab-ax}+\frac{1}{bc-bx}-\frac{1}{ac-ax}=0.$

41. $\frac{c}{a-b}\left(1+\frac{1}{x}\right)-\frac{b}{a-c}\left(1+\frac{1}{x}\right)=\frac{a+c}{(a-c)x}+1.$

42. $\frac{x^2+a^2}{4x^2-a^2}-\frac{x}{2x+a}=-\frac{1}{4}.$

43. $\frac{x-a}{2bc}+\frac{x-b}{2ac}+\frac{x-c}{2ab}=\frac{1}{a}+\frac{1}{b}+\frac{1}{c}.$

44. $\frac{a-b}{x-1}+\frac{b-c}{x-2}+\frac{c-a}{x-3}=0.$

45. $\frac{a+b}{x-a-b}+\frac{a-b}{x-a+b}+\frac{2a}{2a-x}=0.$

46. $\frac{x-a}{b+c}+\frac{x-b}{a+c}+\frac{x-c}{a+b}=\frac{3x}{a+b+c}.$

47. $\frac{a+x}{a^2+ax+x^2}+\frac{a-x}{a^2-ax+x^2}=\frac{3a}{x(a^4+a^2x^2+x^4)}.$

48. $\frac{1}{(x-a)(x-c)}+\frac{2}{(a-c)(a-x)}=\frac{1}{(c-a)(c-x)}.$

49. $\frac{1-2ax^2}{1+2bx^2}-\frac{1+2ax^2}{1-2bx^2}=\frac{4abx^3}{4b^2x^4-1}.$

50. $\frac{a^3+4a}{x^2+x-a^2+a}-\frac{a}{x+a}=\frac{1}{x-a+1}.$

51. $\frac{x+4a+3}{1+a}-\frac{1-x}{1+a^2}-\frac{1-x}{a^3-a^2+a-1}=\frac{1-x}{1+a+a^2+a^3}$
$-\frac{1-x}{1-a^2}+\frac{1-x}{1-a}.$

52. $\frac{x-1}{a-1}+\frac{2a^2(1-x)}{a^4-1}=\frac{2x-1}{1-a^4}-\frac{1-x}{1+a}.$

53. $\frac{x-2a}{b+c-a}+\frac{x-2b}{a+c-b}+\frac{x-2c}{a+b-c}-\frac{2x}{a+b+c}=1.$

54. $\frac{a^2+x}{b^2-x}-\frac{a^2-x}{b^2+x}=\frac{4abx+2a^2-2b^2}{b^4-x^2}.$

55. $\frac{a^2+ax+x^2}{a^3+a^2x+ax^2+x^3}-\frac{a^3-a^2x+ax^2}{a^4+2a^2x^2+x^4}=\frac{1}{a+x}.$

56. $\frac{a^2-x}{x-2a}-\frac{2a+x}{a^2-x}=\frac{a^4}{a^2x+2ax-2a^3-x^2}.$

57. $\frac{a}{10a-5b}-\frac{2ax-bx}{20ax+50a^2}=\frac{2a^2-b(b-x)}{20a^2+8ax-10ab-4bx}.$

58. $\frac{2(x-a)}{a^2-c^2-2ax+x^2}+\frac{c-x}{a^2-ac+cx-2ax+x^2}=\frac{1}{x-(a+c)}.$

59. $\frac{an}{a-x}+\frac{(a+n)(anx+nx^2+x^3)}{x^3+nx^2-a^2x-a^2n}=\frac{ax}{n+x}+\frac{nx^2}{x^2-a^2}.$

60. $1-\dfrac{1-\dfrac{1}{a^2}}{\dfrac{a}{x}\left(1-\dfrac{1}{a}\right)}=-\dfrac{1}{a^3}.$ 61. $\dfrac{a+x-\dfrac{a^2}{a+x}}{a+x}=1-\dfrac{2ax}{(a+x)^2}.$

General Problems.

7. A problem in which the given numbers have particular values can be *generalized* by assuming literal numbers for the given numbers. A problem so stated is a *general problem*, and its solution is called a *general solution*.

Some of the advantages of the general solution of a problem were pointed out in Ch. I., § 1, Art. 17. A *particular* solution can always be obtained from the general solution by assigning to the literal numbers in the latter, particular numerical values.

We will now obtain general solutions of some of the problems already solved in Ch. V. The student should compare the solutions here given with the corresponding solutions in that chapter.

Ch. V., Pr. 1. The greater of two numbers is m times the less, and their sum is s. What are the numbers?

Let x stand for the less required number. Then mx stands for the greater.

By the condition of the problem, we have

$$x+mx=s;$$

whence, $x=\dfrac{s}{1+m}$, the less number,

and $mx=\dfrac{ms}{1+m}$, the greater.

If $m=3$ and $s=84$ (as in the particular problem), we have

$$x=\frac{84}{1+3}=21,$$

and $mx=3\times 21=63.$

When the numbers are equal, $m = 1$, and we obtain

$$x = \frac{s}{2} \text{ and } mx = \frac{s}{2},$$

for all values of s; that is, either of the two numbers is half their sum.

If the greater be twice the less, $m = 2$, and we obtain

$$x = \frac{s}{3} \text{ and } mx = \frac{2s}{3},$$

for all values of s; that is, the less number is one-third their sum, and the greater is two-thirds their sum.

Ch. V., Pr. 2. Find two consecutive integers whose sum is n.

Let x stand for the less number; then $x + 1$ stands for the greater. By the condition of the problem, we have

$$x + (x + 1) = n;$$

whence $$x = \frac{n-1}{2}, \text{ the less number,}$$

and $$x + 1 = \frac{n+1}{2}, \text{ the greater.}$$

When $n = 163$, as in the particular problem, we have

$$x = \tfrac{162}{2} = 81, \text{ and } x + 1 = \tfrac{164}{2}, = 82.$$

Notice that n must be an odd number in order that $n - 1$ and $n + 1$ may be exactly divisible by 2; that is, that the required numbers may be integers.

Ch. V., Pr. 3. A is a years old, and B is b years old. After how many years will A be n times as old as B?

Let x stand for the required number of years.

Then after x years, A will be $a + x$ years old, and B will be $b + x$ years old. By the condition of the problem, we have

$$a + x = n(b + x);$$

whence $$x = \frac{nb - a}{1 - n}.$$

If, as in the particular problem, $a = 40$, $b = 10$, and $n = 3$, we obtain

$$x = \frac{3 \times 10 - 40}{1 - 3} = 5.$$

Ch. V., Pr. **10**. A carriage, starting from a point A, travels m miles daily; a second carriage, starting from a point B, p miles behind A, travels in the same direction n miles daily. After how many days will the second carriage overtake the first, and at what distance from B will the meeting take place?

Let x stand for the number of days after which the carriages meet. Then the number of miles traveled by the first carriage will be mx; the number of miles traveled by the second will be nx.

Therefore, by the condition of the problem,

$$mx = nx - p;$$

whence $x = \frac{p}{n-m}$, the number of days

after which the carriages meet.

The distance traveled by the first carriage is $\frac{mp}{n-m}$ miles, and the distance traveled by the second carriage is $\frac{np}{n-m}$ miles.

They, therefore, meet $\frac{np}{n-m}$ miles from B.

If, as in the particular problem, $m = 35$, $p = 84$, and $n = 49$, we obtain

$$\frac{p}{n-m} = \frac{84}{49-35} = 6, \quad \frac{mp}{n-m} = \frac{35 \times 84}{49-35} = 210,$$

and

$$\frac{np}{n-m} = \frac{49 \times 84}{49-35} = 294.$$

Ch. V., Pr. **11**. One man asked another what time it was, and received the answer: "It is between n and $n+1$ o'clock, and the hour-hand is directly over the minute-hand." What time was it?

At n o'clock the minute-hand points to 12 and the hour-hand to n. The hour-hand is therefore $5n$ minute-divisions in advance of the minute-hand.

Let x stand for the number of minute-divisions passed over by the minute-hand from n o'clock until it is directly over the hour-hand between n and $n+1$ o'clock.

Then the number of minute-divisions passed over by the hour-hand is equal to the number of minute-divisions passed over by the minute-hand, minus $5n$; that is, to $x - 5n$.

But since the minute-hand moves 12 times as fast as the hour-hand, we have

$$x = 12(x - 5n);$$

whence

$$x = \frac{60n}{11}.$$

Consequently, the time was $\frac{60n}{11}$ minutes past n.

If $n = 1$, it was $\frac{60}{11}$, or $5\frac{5}{11}$ minutes past 1.

If $n = 2$, it was $\frac{120}{11}$, or $10\frac{10}{11}$ minutes past 2.

If $n = 3$, it was $\frac{180}{11}$, or $16\frac{4}{11}$ minutes past 3.

Etc.

If $n = 11$, it was $\frac{660}{11}$, or 60 minutes past 11; *i.e.*, 12 o'clock.

If $n = 12$, it was $\frac{720}{11}$, or $65\frac{5}{11}$ minutes past 12; *i.e.*, $5\frac{5}{11}$ minutes past 1.

Notice that the two hands coincide at 12 o'clock, but not between 12 and 1.

EXERCISES II.

Find the general solution of each of the following problems, and from this solution obtain the particular solution for the numerical values assigned to the literal numbers in the problem.

1. Find a number, such that the result of adding it to n shall be equal to n times the number.

Let $n = 2$; 5.

2. Divide a into two parts, such that $\frac{1}{m}$ of the first, plus $\frac{1}{n}$ of the second, shall be equal to b.

Let $a = 100$, $b = 30$, $m = 3$, $n = 5$.

3. Find a number, such that the sum of the results of subtracting it from a and from b shall be equal to c.

Let $a = 3$, $b = 6$, $c = 5$.

4. Find a number, such that n times the number exceeds the sum of q and m times the number by p.

Let $n = 10$, $m = 8$, $p = 15$, $q = 7$.

5. One boy said to another: "Think of some number, add 3 to it, multiply the sum by 2, add 4 to the product, divide the result by 2, subtract 1 from the quotient, multiply the difference by 4, add 4 to the product, divide the result by 4, tell me the result, and I will tell you the number you have in mind." If the result be d, what number does the boy think of?

6. A sum of d dollars is divided between A and B. B receives b dollars as often as A receives a dollars. How much does each receive?

Let $d = 7000$, $a = 3$, $b = 2$.

7. A father's age exceeds his son's age by m years, and the sum of their ages is n times the son's age. What are their ages?

Let $m = 20$, $n = 4$; $m = 25$, $n = 7$.

8. If two trains start together and run in the same direction, one at the rate of m_1 miles an hour, and the other at the rate of m_2 miles an hour, after how many hours will they be d miles apart?

Let $d = 200$, $m_1 = 35$, $m_2 = 30$.

9. A farmer can plow a field in a days, and his son in b days; in how many days can they plow the field, working together?

Let $a = 10$, $b = 15$.

10. A father was n years ago m times as old as his son, and he is now p times as old. What is the age of the son?

Let $n = 2$, $m = 15$, $p = 8$.

11. How many grains of gold c_1 carats fine must be combined with m grains of gold c_2 carats fine to give a mixture c_3 carats fine?

Let $c_1 = 18$, $c_2 = 12$, $c_3 = 16$, $m = 40$.

12. A pupil was told to add m to a certain number, and to divide the sum by n. But he misunderstood the problem, and subtracted n from the number and multiplied the remainder by m. Nevertheless he obtained the correct result. What was the number?

Let $m = 12$, $n = 13$.

13. What time is it, if the number of hours which have elapsed since noon is m times the number of hours to midnight?

Let $m = \frac{1}{2}$.

14. A starts from P and walks to Q, a distance of d miles. At the same time B starts from Q and walks to P. If A walk at the rate of m miles a day and B at the rate of n miles a day, at what distance from P do they meet, and how many days after they start?

Let $m = 20$, $n = 30$, $d = 600$.

15. Two friends, A and B, each intending to visit the other, start from their houses at the same time. A could reach B's house in m minutes, and B could reach A's house in n minutes. After how many minutes do they meet?

Let $m = 12\frac{1}{4}$, $n = 10\frac{1}{2}$.

16. A farmer wishes to receive a certain sum for his eggs and intends to sell them at a cents a dozen. But he breaks b eggs, and in order to receive the desired sum he then sells the unbroken ones at c cents a dozen. How many eggs had he originally?

Let $a = 25$, $b = 24$, $c = 30$.

17. Two couriers start at the same time and move in the same direction, the first from a place d miles ahead of the second. The first courier travels at the rate of m_1 miles an hour, and the second at the rate of m_2 miles an hour. After how many hours will the second courier overtake the first?

Let $d = 15$, $m_1 = 17$, $m_2 = 20$.

From the result of the preceding example find the results of Exx. 18–20.

18. At what rate must the second courier travel in order to overtake the first after h hours?

Let $d = 18$, $m_1 = 15$, $h = 3$.

19. At what rate must the first courier travel in order that the second may overtake him after h hours?

Let $d = 12$, $m_2 = 22$, $h = 3$.

20. How many miles behind the first courier must the second start in order to overtake the first after h hours?

Let $m_1 = 18$, $m_2 = 21$, $h = 4$.

21. In a company are a men and b women; and to every m unmarried men there are n unmarried women. How many married couples are in the company?

Let $a = 13$, $b = 17$, $m = 3$, $n = 5$.

22. An officer, who has saved each year a dollars of his salary is transferred to a new post. In the latter place he finds the cost of living $\frac{1}{n}$ dearer, and in consequence spends b dollars more than his salary. What is his salary?

Let $a = 423$, $n = 8$, $b = 120$.

23. The annual dues of a certain club are at first a dollars. Subsequently the yearly expenses increased by d dollars, while the number of members decreased by n. In consequence the annual dues were increased by b dollars. How many members were originally in the club?

Let $a = 25$, $d = 315$, $n = 7$, and $b = 2$.

24. A merchant sells $\frac{1}{a}$ of his oranges plus $\frac{1}{a}$ of an orange; then $\frac{1}{a}$ of the oranges remaining plus $\frac{1}{a}$ of an orange; and so on. After he has sold oranges n times in this way he has left m oranges. How many oranges had he at first?

Let $a = 2$, $n = 4$, $m = 25$.

25. A father divided his property equally among his sons. To the oldest he gave d dollars and $\frac{1}{n}$ of what remained; to the second son he gave $2d$ dollars and $\frac{1}{n}$ of what was then left; to the third son he gave $3d$ dollars and $\frac{1}{n}$ of the remainder; and so on. What was the amount of his property?

Let $d = 1500$, $n = 11$; $d = 2000$, $n = 6$.

26. Three brothers wish to divide a number of apples among themselves. Since the number of apples is not divisible by 3, one of them proposes that they give a apples to their sister, so that an equal division will be possible. But the older brother proposes that, after giving her a apples, they each give her $\frac{1}{n}$ of his share, so that she will have as many apples as each of them has. How many apples are there?

Let $a = 5$, $n = 9$.

27. Two couriers start from the same place and move in the same direction, one h hours after the other. The first one travels at the rate of m_1 miles an hour, and the second at the rate of m_2 miles an hour. After how many hours will the second courier overtake the first?

Let $h = 2$, $m_1 = 15$, $m_2 = 20$.

From the result of the preceding example, find the results of Exx. 28–30.

28. At what rate must the second courier travel in order to overtake the first after H hours?

Let $H = 6$, $h = 2$, $m_1 = 12$.

29. At what rate must the first courier travel in order that the second may overtake him after H hours?

Let $H = 4$, $h = 1$, $m_2 = 20$.

30. How many hours after the first courier starts must the second start in order to overtake the first after H hours?

Let $H = 6$, $m_1 = 14$, $m_2 = 22$.

31. Two boys run a race from A to B, a distance of d yards. The first runs a yards a second; after reaching B, he turns and runs back at the same rate to meet the other boy, who runs b yards a second. How many seconds after they start does the faster runner meet the other?

Let $d = 253$, $a = 2.5$, $b = 2.1$.

32. An accommodation train leaves A every h hours, and runs to B at the rate of m miles an hour. At the same time an express train leaves B and runs to A at the rate of n miles an hour. What time elapses after an express train meets an accommodation train until it meets the next accommodation train?

Let $h = 3$, $m = 20$, $n = 40$.

33. At what time between n and $n + 1$ o'clock will the hands of a clock be in a straight line?

Let $n = 1;\ 2;\ 3;\cdots$ to 12.

34. At what time between n and $n + 1$ o'clock are the minute-hand and the hour-hand of a clock at right angles to each other?

Let $n = 1;\ 2;\ 3;\cdots$ to 12.

35. At what time between n and $n + 1$ o'clock will the minute-hand be 20 minute-divisions in advance of the hour-hand?

Let $n = 1;\ 2;\ 3;\cdots$ to 12.

36. At what time between n and $n + 1$ o'clock will the hour-hand be d minute-divisions in advance of the minute-hand?

Let $n = 1;\ 2;\ 3;\cdots$ to 12, and with each value of n let $d = 5;\ 8;\ 20$.

37. At what time between n and $n + 1$ o'clock does the second-hand bisect the angle between the hour-hand and the minute-hand?

Let $n = 1;\ 2;\ 3;\cdots$ to 12.

38. A ship, having on board p persons, is provisioned for d days. After sailing d_1 days, p_1 persons were landed, and in consequence the allowance of food for each person was increased by $\frac{1}{n}$ of a pound. After sailing d_2 days longer, p_2 more persons were landed. It was then found that the journey could be completed d_3 days sooner than was expected, and therefore the allowance for each person was again increased by $\frac{1}{n}$ of a pound. What was the original allowance of food for each person?

Let $p = 340$, $d = 80$, $d_1 = 30$, $p_1 = 20$, $n = 2$, $d_2 = 20$, $p_2 = 20$, $d_3 = 10$.

CHAPTER XII.

INTERPRETATION OF THE SOLUTIONS OF PROBLEMS.

1. In solving equations we do not concern ourselves with the meaning of the results. When, however, an equation has arisen in connection with a problem, the interpretation of the result becomes important.

In this chapter we shall interpret the solutions of some linear equations in connection with the problems from which they arise.

Positive Solutions.

2. Pr. A company of 20 people, men and women, proposed to arrange a fair for the benefit of a poor family. Each man contributed $3, and each woman $1. If $55 were contributed, how many men and how many women were in the company?

Let x stand for the number of men; then the number of women was $20 - x$. The amount contributed by the men was $3x$ dollars, that by the women $20 - x$ dollars. By the condition of the problem, we have

$$3x + (20 - x) = 55;$$

whence $$x = 17\tfrac{1}{2}.$$

The result, $17\frac{1}{2}$, satisfies the equation, but not the problem. For the number of men, which x represents, must be an *integer*. This implied condition could not be introduced into the equation.

The interpretation of the result is that the conditions stated in the problem are impossible, since they are inconsistent with the implied condition that the number of men must be an integer.

If the problem be generalized, its solution will show how the given data in the particular problem can be modified so that all the conditions, expressed and implied, shall be consistent.

The generalized problem may be stated thus:

A company of m people, men and women, proposed to arrange a fair for the benefit of a poor family. Each man contributed a dollars, and each woman b dollars. If n dollars were contributed, how many men and how many women were in the company?

The solution of the equation of this problem is

$$x = \frac{n - bm}{a - b}.$$

In order that x may be an integer, $n - bm$ must be exactly divisible by $a - b$. Thus, if, in the particular problem, the number of people were 21 instead of 20, the other data being the same, we should have

$$x = \frac{55 - 1 \times 21}{3 - 1} = \frac{34}{2} = 17.$$

Let the student obtain consistent results by changing the other data.

3. Pr. What time is it, if 4 times the number of hours which have elapsed since noon exceeds 5 by as much as 6 times the number of hours still remaining till midnight is less than 7?

Let x stand for the number of hours which have elapsed since noon; then $12 - x$ is the number of hours still remaining till midnight. From the condition of the problem, we have

$$4x - 5 = 7 - 6(12 - x);$$

whence $$x = 30.$$

This result is a positive integer, yet it cannot be taken as the solution of the problem, since the number of hours between noon and midnight cannot exceed 12. The conditions expressed in the problem are therefore impossible, since they are inconsistent with an implied condition.

Now, let the data of the problem remain the same, except that 6 times the number of hours still remaining till midnight shall be less than n.

We then have

$$x = \frac{67 - n}{2}.$$

In order that x may be less than 12, $67 - n$ must be less than 24. Thus, if $n = 48$, we have

$$x = \frac{67 - 48}{2} = 9\frac{1}{2}.$$

That is, it is half-past 9 o'clock.

Observe that this problem was generalized only in part.

4. When a positive solution is inconsistent with a condition expressed or implied in the problem, it shows that the conditions of the problem are impossible, as in Arts. 2 and 3.

But if all the conditions of a problem, expressed and implied, be consistent with one another, a positive solution will satisfy these conditions and therefore give the solution of the problem.

Negative Solutions.

5. Pr. A father is 40 years old, and his son 10 years old. After how many years will the father be seven times as old as his son?

Let x stand for the required number of years. Then after x years the father will be $40 + x$ years old, and the son $10 + x$ years old. By the condition of the problem, we have

$$40 + x = 7(10 + x), \tag{1}$$

whence $$x = -5.$$

This result satisfies the equation, but not the condition of the problem. For since the question of the problem is "*after* how many years?" the result, if added to the number of years in the ages of father and son, should increase them, and therefore be *positive*. Consequently, at no time in the future will the father be seven times as old as his son. But since to add

-5 is equivalent to subtracting 5, we conclude that the question of the problem should have been, "How many years ago?"

The equation of the problem, with this modified question, is obtained as follows:

x years ago the number of years in the father's age was $40 - x$, and in the son's age $10 - x$. By the condition of the modified problem, we have

$$40 - x = 7(10 - x); \qquad (2)$$

whence $$x = 5.$$

Notice that equation (2) could have been obtained from equation (1) by changing x into $-x$.

6. The interpretation of a negative result in a given problem is often facilitated by the following principle:

If $-x$ be substituted for x in an equation which has a negative root, the resulting equation will have a positive root of the same absolute value; and vice versa.

E.g., the equation $$x + 1 = -x - 3$$

has the negative root -2; while the equation

$$-x + 1 = x - 3$$

has the positive root 2.

In general, the equation $$ax = b \qquad (1)$$

has the root $\frac{b}{a}$. And the equation

$$-ax = b \qquad (2)$$

has the root $-\frac{b}{a}$. If the root $\frac{b}{a}$ be negative, then the root $-\frac{b}{a}$ is positive; and *vice versa*.

If a negative result be obtained, we change the sign of x in the equation of the problem, and thus obtain a new equation which has a positive root. This new equation, it is true, is not the equation of the proposed problem; but the problem can be modified so that its conditions will be satisfied by the positive root, as in Art. 5.

7. Pr. 1. Two pocket-books contain together \$ 100. If one-half of the contents of one pocket-book, and one-third of the contents of the other be removed, the amount of money left in both will be \$ 70. How many dollars does each pocket-book contain?

Let x stand for the number of dollars contained in the first pocket-book; then the number of dollars contained in the second is $100 - x$. When one-half of the contents of the first, and one-third of the contents of the second are removed, the number of dollars remaining in the first is $\frac{x}{2}$, and in the second $\frac{2}{3}(100 - x)$.

Therefore, by the condition of the problem, we have

$$\tfrac{1}{2}x + \tfrac{2}{3}(100 - x) = 70,$$

or

$$3x + 400 - 4x = 420;$$

whence

$$x = -20.$$

Substituting $-x$ for x in the given equation, we obtain

$$-\tfrac{1}{2}x + \tfrac{2}{3}(100 + x) = 70,$$

or

$$\tfrac{2}{3}(100 + x) - \tfrac{1}{2}x = 70.$$

This equation corresponds to the following conditions:

If x stand for the number of dollars in one pocket-book, then $100 + x$ stands for the number of dollars in the other; that is, one pocket-book contains 100 dollars more than the other. The second condition of the problem, obtained from the equation, is: two-thirds of the contents of one pocket-book exceeds one-half of the contents of the other by \$ 70. Therefore the modified problem reads as follows:

Two pocket-books contain a certain amount of money, and one contains 100 dollars more than the other. If one-third of the contents be removed from the first pocket-book, and one-half of the contents from the second, the first will then contain \$ 70 more than the second. How much money is contained in each pocket-book?

Pr. 2. Two couriers are traveling along a road in the direction from M to N; one courier is traveling at the rate of 15 miles an hour, the other at the rate of 12 miles an hour. The former is seen at the station A at noon, and the other is seen two hours later at the station B, which is 25 miles distant from A in the direction in which the couriers are traveling. Where do the couriers meet?

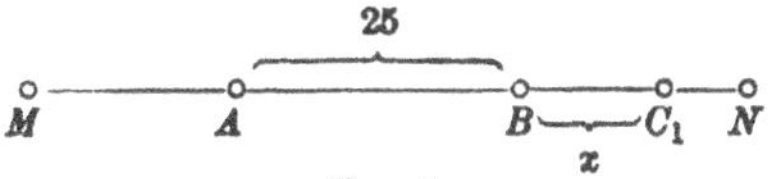

FIG. 8.

Let us assume that the meeting takes place to the right of B, at some point C_1; and let x stand for the number of miles from B to C_1 (Fig. 8).

The first courier, in moving from A to C_1, travels $25+x$ miles, in $\frac{25+x}{15}$ hours. The second courier, in moving from B to C_1, travels x miles in $\frac{x}{12}$ hours.

From the condition of the problem it is evident that, if the place of meeting be to the right of B, the number of hours it takes the first courier to travel from A to C_1 exceeds by 2 the number of hours it takes the second courier to travel from B to C_1; hence

$$\frac{25+x}{15}-\frac{x}{12}=2, \tag{1}$$

or

$$100+4x-5x=120;$$

whence

$$x=-20.$$

To understand the meaning of this negative result, we substitute $-x$ for x in equation (1). We thus obtain

$$\frac{25-x}{15}+\frac{x}{12}=2. \tag{2}$$

The resulting equation has the positive solution 20.

From equation (2) we see that the number of miles traveled by the first courier from A to their point of meeting is $25-x$;

that is, is less than 25 (since $x = 20$). Consequently they must meet to the left of B, at some point C_2 (Fig. 9).

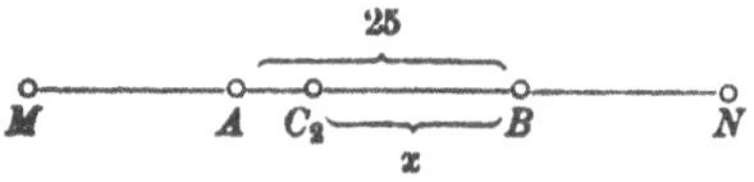

FIG. 9.

Then the first courier travels from A to C_2, a distance of $25 - x$ miles, in $\frac{25 - x}{15}$ hours; and since the second courier is seen at B after the meeting, he travels from C_2 to B, a distance of x miles, in $\frac{x}{12}$ hours. Since the second courier reaches B two hours after the first leaves A, we have

$$\frac{25 - x}{15} + \frac{x}{12} = 2.$$

This equation is identical with equation (2). The answer to the problem is: The couriers meet 20 miles to the left of the station B.

It is also evident from the conditions of the problem that two hours after the first courier was seen at A he must have already passed the station B, and consequently must have passed the other courier before the latter was seen at B.

8. The problems of Arts. 5 and 7 show that the required modification of an assumption, question, or condition of a problem which has led to a negative result, consists in making the assumption, question, or condition the opposite of what it originally was.

Thus, if a positive result signify the length of time after a certain event, a negative result will signify the length of time before that event, and *vice versa;* if a positive result signify a distance toward the right from a certain point, a negative result will signify a distance toward the left from the same point; and *vice versa.*

Zero Solutions.

9. A zero result gives in some cases the answer to the question; in other cases it proves its impossibility.

Pr. **1.** A father is 40 years old, and his son is 10 years old. After how many years will the father be four times as old as his son?

Let x stand for the required number of years. Then

$$40 + x = 4(10 + x);$$

whence $$3x = 0, \text{ or } x = \tfrac{0}{3} = 0.$$

This result is the direct answer to the question of the problem. At the present time the father is four times as old as his son.

Pr. **2.** A merchant has two kinds of wine, one worth \$ 7.25 a gallon, and the other \$ 5.50 a gallon. How many gallons of each kind must be taken to make a mixture of 16 gallons worth \$ 88?

Let x stand for the number of gallons of the first kind; then $16 - x$ will stand for the number of gallons of the second kind.

Therefore, by the condition of the problem, we have

$$7.25x + 5.5(16 - x) = 88;$$

whence $$x = 0.$$

That is, no mixture which contains the first kind of wine can be made to satisfy the condition. In fact, 16 gallons of the second kind are worth \$ 88.

Pr. **3.** The denominator of a fraction is three times its numerator. If 5 be added to the numerator and 10 to the denominator, the resulting fraction will be equal to $\frac{1}{2}$. What is the fraction?

Let x stand for the required numerator; then $3x$ will stand for the denominator. Therefore, by the condition of the problem, we have

$$\frac{x + 5}{3x + 10} = \frac{1}{2},$$

or $$2x + 10 = 3x + 10;$$

whence $$x = 0.$$

That is, there is no fraction which will satisfy the given condition.

Indeterminate Solutions.

10. Pr. **1.** A father is 40 years old, and his son 10 years old. After how many years will the father be 30 years older than his son?

Let x stand for the required number of years. Then, by the condition of the problem, we have

$$40 + x = 10 + x + 30, \text{ or } 40 + x = 40 + x.$$

Both members of the equation are identical, and therefore we can assign to x any value whatever, *i.e.*, the problem is indeterminate. Solving this equation as a linear equation, we obtain

$$x - x = 40 - 40, \text{ or } (1 - 1)x = 40 - 40;$$

whence

$$x = \frac{40 - 40}{1 - 1} = \frac{0}{0}.$$

The indeterminate result, $\frac{0}{0}$, therefore means that any finite number satisfies the condition of the problem. It is evident from the problem that the father will be at any time 30 years older than his son.

Pr. **2.** A merchant buys 4 pieces of goods. In the second there are 3 yards less than in the first, in the third 7 yards less than in the first, and in the fourth 10 yards less than in the first. The number of yards in the first and fourth is equal to the number of yards in the second and third. How many yards are there in the first piece?

Let $x =$ the number of yards in the first piece,

then $x - 3 =$ the number of yards in the second piece,

$x - 7 =$ the number of yards in the third piece,

$x - 10 =$ the number of yards in the fourth piece.

Therefore, by the condition of the problem, we have

$$x + (x - 10) = (x - 3) + (x - 7),$$

or

$$2x - 10 = 2x - 10.$$

This equation is an identity, and is therefore satisfied by any finite value of x.

If it be solved in the usual way, we obtain

$$(2-2)x = 10 - 10,$$

or

$$x = \frac{10-10}{2-2} = \frac{0}{0}.$$

That is, the conditions of the problem will be satisfied by any number of yards in the first piece.

Infinite Solutions.

11. Pr. **1**. What number must be added to the numerator and denominator of $\frac{2}{5}$ to give 1?

Let x stand for the required number.

Then, by the condition of the problem, we have

$$\frac{2+x}{5+x} = 1,$$

or

$$2 + x = 5 + x.$$

Evidently no finite value of x will satisfy this equation, since 2 plus any finite number cannot be equal to 5 plus the same finite number.

But if the equation be solved as a linear equation, we obtain

$$x - x = 5 - 2,$$

$$(1-1)x = 5 - 2;$$

whence

$$x = \frac{5-2}{1-1} = \frac{3}{0} = \infty.$$

The meaning of this result is, that the greater the number which is added to the terms of the given fraction, the more nearly does the value of the resulting fraction approach 1.

The impossibility of satisfying the equation by a finite value of x means, of course, the impossibility of the problem from whose conditions the equation was obtained.

Pr. **2**. A cistern has three pipes. Through the first it can be filled in 24 minutes; through the second in 36 minutes; through the third it can be emptied in $14\frac{2}{5}$ minutes. In what time will the cistern be filled if all the pipes be opened at the same time?

Let x stand for the number of minutes after which the cistern will be filled. In one minute $\frac{1}{24}$ of its capacity enters through the first pipe, and hence in x minutes $\frac{x}{24}$ of its capacity enters. For a similar reason, $\frac{x}{36}$ of its capacity enters through the second pipe in x minutes; and in the same time $\frac{5x}{72}$ of its capacity is discharged through the third pipe.

Therefore, after x minutes there is in the cistern

$$\frac{x}{24}+\frac{x}{36}-\frac{5x}{72}, \quad =(\tfrac{1}{24}+\tfrac{1}{36}-\tfrac{5}{72})x,$$

of its capacity. But by the condition of the problem, that the cistern is then filled, we have

$$(\tfrac{1}{24}+\tfrac{1}{36}-\tfrac{5}{72})x=1;$$

whence
$$x=\frac{1}{\frac{1}{24}+\frac{1}{36}-\frac{5}{72}}=\frac{1}{0}=\infty.$$

This result means that the cistern will never be filled. This is also evident from the data of the problem, since the third pipe in a given time discharges from the cistern exactly as much as enters through the other pipes.

The Problem of the Couriers.

12. We will next generalize the problem of the couriers which was solved in Art. 7, and interpret the general solution.

Two couriers are traveling along a road in the direction from M to N; one courier is traveling at the rate of m_1 miles an hour, the other at the rate of m_2 miles an hour. The former is seen at the station A at noon, and the other is seen h hours later at the station B, which is d miles from A in the direction in which the couriers are traveling.

Where do the couriers meet?

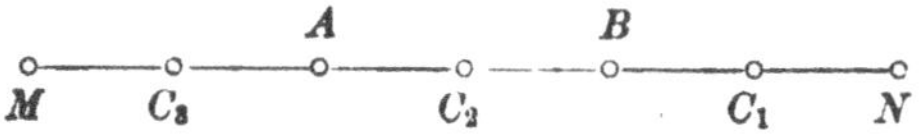

FIG. 10.

Let us assume that the couriers meet to the right of B at a point C_1, and let x stand for the number of miles from B to the place of meeting C_1 (Fig. 10).

The first courier, moving at the rate of m_1 miles an hour, travels $d+x$ miles, from A to C_1, in $\frac{d+x}{m_1}$ hours; the second courier, moving at the rate of m_2 miles an hour, travels x miles, from B to C_1 in $\frac{x}{m_2}$ hours. By the condition of the problem it is evident that, if the place of meeting be to the right of B, the number of hours it takes the first courier to travel from A to C_1 exceeds by h the number of hours it takes the second courier to travel from B to C_1. We therefore have

$$\frac{d+x}{m_1} - \frac{x}{m_2} = h, \qquad (1)$$

or $$dm_2 + xm_2 - xm_1 = hm_1m_2;$$

whence $$x = \frac{hm_1m_2 - dm_2}{m_2 - m_1} = \frac{m_2(hm_1 - d)}{m_2 - m_1}.$$

(i.) **A Positive Result.**—The result will be positive either when $hm_1 > d$ and $m_2 > m_1$, or when $hm_1 < d$ and $m_2 < m_1$. A positive result means that the problem is possible with the assumption made; *i.e.*, that the couriers meet at a point to the right of B. That under these conditions only this assumption is possible is evident from the following considerations: The product hm_1 denotes the distance over which the first courier passes in h hours, and hence shows how far he has moved from A at the moment that the second courier is seen at B. If $hm_1 > d$, the first courier has passed B when the second courier is seen at that station; that is, the second courier is behind the first at that time. And since also $m_2 > m_1$, the second courier is traveling the faster, and will, therefore, overtake the first, and at a point beyond the station B.

On the other hand, if $hm_1 < d$, the first courier has not yet reached B when the second courier is seen at that station; that is, the first courier is behind the second at that time. Moreover, since $m_2 < m_1$, the first courier is traveling the faster, and will, therefore, overtake the second, and at a point to the right at B.

(ii.) **A Negative Result.**—The result will be negative either when $hm_1 > d$ and $m_2 < m_1$, or when $hm_1 < d$ and $m_2 > m_1$.

Such a result shows that the assumption that the couriers meet to the right of B is untenable, since, as we have seen, in that case the result is positive.

If we substitute $-x$ for x in equation (1), we obtain

$$\frac{d-x}{m_1} + \frac{x}{m_2} = h. \qquad (2)$$

This equation is satisfied by a positive root having the same absolute value as the root of equation (1).

It is readily seen that the new equation corresponds to the assumption that the meeting of the couriers takes place to the left of B, either at the point C_2 or at the point C_3 (Fig. 10).

That this assumption is in accord with either of the two sets of conditions, namely, that $hm_1 > d$ and $m_2 < m_1$, or that $hm_1 < d$ and $m_2 > m_1$, will be evident from the following considerations:

If we suppose that the meeting takes place at C_2, the first courier in passing from A to C_2 travels $d - x$ miles in $\frac{d-x}{m_1}$ hours; the second courier in passing from C_2 to B travels x miles in $\frac{x}{m_2}$ hours; the sum of the numbers of hours, according to the condition of the problem, must be equal to h. This condition is, in fact, expressed by equation (2).

If, on the other hand, the couriers meet at C_3, the distance from C_3 to A being $x - d$ miles, it takes the first courier $\frac{x-d}{m_1}$ hours to travel from C_3 to A, and the second courier $\frac{x}{m_2}$ hours to travel from C_3 to B. From the condition of the problem it follows that $\frac{x}{m_2}$ must exceed $\frac{x-d}{m_1}$ by h. We therefore have

$$\frac{x}{m_2} - \frac{x-d}{m_1} = h, \text{ or } \frac{x}{m_2} + \frac{d-x}{m_1} = h.$$

This equation is identical with equation (2).

That under the assumed conditions the couriers can meet only at some point to the left of B can also be inferred from the following considerations, which are independent of the negative result: If $hm_1 > d$, the first courier has passed B when the second courier is seen at that station; that is, the second courier is behind the first at that time. And since also $m_2 < m_1$, the first courier is traveling the faster, and must therefore have overtaken the second, and at some point to the left of B.

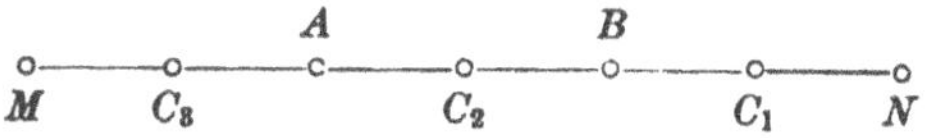

FIG. 10.

On the other hand, if $hm_1 < d$, the first courier has not yet reached B when the second is seen at that station; that is, the first courier is behind the second at that time. And since also $m_2 > m_1$, the second courier is traveling the faster, and must therefore have overtaken the first, and at some point to the left of B.

(iii.) **A Zero Result.**—A zero result is obtained when $hm_1 = d$, and $m_2 \neq m_1$; that is, the meeting takes place at B. This is also evident from the assumed conditions. For the first courier evidently reaches B h hours after he was seen at A; and since the second courier is seen at B h hours after the first was seen at A, the meeting must take place at B.

(iv.) **Indeterminate Result.**—An indeterminate result is obtained if $hm_1 = d$, and $m_2 = m_1$. In this case every point of the road can be regarded as their place of meeting. For the first courier evidently reaches B at the time at which the second courier is seen at that station; and since they are traveling at the same rate, they must be together all the time; in other words, the problem under these conditions becomes indeterminate.

(v.) **An Infinite Result.** — An infinite result is obtained when $hm_1 \neq d$, and $m_2 = m_1$. In this case a meeting of the couriers is impossible, since both travel at the same rate, and when the second is seen at B the first either has not yet reached B or has already passed that station.

An infinite result also means that the more nearly equal m_1 and m_2 are, the further removed is the place of meeting.

EXERCISES.

Solve the following problems, and interpret the results. Modify those problems which have negative solutions so that they will be satisfied by positive solutions.

1. A and B together have $100. If A spend one-third of his share, and B spend one-fourth of his share, they will then have $80 left. What are their respective shares?

2. In a number of two digits, the digit in the tens' place exceeds the digit in the units' place by 5. If the digits be interchanged, the resulting number will be less than the original number by 45. What is the number?

3. A father is 40 years old, and his son is 13 years old; after how many years will the father be four times as old as his son?

4. The sum of the first and third of three consecutive even numbers is equal to twice the second. What are the numbers?

5. In a number of two digits, the tens' digit is two-thirds of the units' digit. If the digits be interchanged, the resulting number will exceed the original number by 36. What is the number?

6. A father is 26 years older than his son, and the sum of their ages is 26 years less than twice the father's age. How old is the son?

7. A teacher proposes 30 problems to a pupil. The latter is to receive 8 marks in his favor for each problem solved, and 12 marks against him for each problem not solved. If the number of marks against him exceed those in his favor by 420, how many problems will he have solved?

8. A cistern has two pipes. To fill the cistern, it takes the smaller pipe 5 hours longer than the larger one. If both pipes be open, the cistern will be filled in 6 hours. In what time will each pipe fill the cistern?

9. In a number of two digits the tens' digit is twice the units' digit. If the digits be interchanged, the resulting number will exceed the original number by 18. What is the number?

10. In a number of two digits, the digit in the units' place is two-thirds of the digit in the tens' place. If the digits be interchanged and the resulting number be divided by the original number, the quotient will be equal to $\frac{23}{32}$. What is the number?

11. On a building are at work 6 more masons than carpenters. Each mason receives $2.50 a day, and each carpenter $2 a day. The amount earned by the masons exceeds the amount earned by the carpenters in the same time by one-fifth of the amount earned by the carpenters. How many masons and how many carpenters are at work?

12. In a number of two digits, the digit in the units' place exceeds the digit in the tens' place by 4. If the sum of the digits be divided by 2, the quotient will be less than the first digit by 2. What is the number?

13. A has $100, and B has $30. A spends twice as much money as B, and then has left three times as much as B. How much does each one spend?

Discuss the solutions of the following general problems. State under what conditions each solution is positive, negative, zero, indeterminate, or infinite. Also, in each problem, assign a set of particular values to the general numbers which will give an admissible solution.

14. In a number of two digits, the tens' digit is m times the units' digit. If the digits be interchanged, the resulting number will exceed the original number by n. What is the number?

15. A father is a years old, and his son is b years old. After how many years will the father be n times as old as his son?

16. What number, added to the denominators of the fractions $\frac{a}{b}$ and $\frac{c}{d}$, will make the resulting fractions equal?

17. The number of years in A's age is a, and in B's age is b. After how many years will the number of years in A's age divided by the number in B's age be equal to $\frac{m}{n}$?

18. Having two kinds of wine worth a and b dollars a gallon, respectively, how many gallons of each kind must be taken to make a mixture of n gallons worth c dollars a gallon?

19. Two couriers, A and B, are traveling in the same direction over the same route, A at the rate of m_1 miles an hour, and B at the rate of m_2 miles an hour. When A is at the station P, B is at the station Q, which is d miles from P, in the direction in which the couriers are traveling. When and where do the couriers meet?

20. Two couriers, A and B, start at the same time from two stations, distant d miles from each other, and travel in the same direction. A travels n times as fast as B. Where will A overtake B?

CHAPTER XIII.

SIMULTANEOUS LINEAR EQUATIONS.

§ 1. SYSTEMS OF EQUATIONS.

1. If the linear equation in two unknown numbers

$$x + y = 5 \qquad (1)$$

be solved for y, we obtain

$$y = 5 - x.$$

This value of y contains the unknown number x, and is therefore not definitely determined. We may substitute in it any particular numerical value for x, and obtain a corresponding value for y.

Thus, when $x = 1$, $y = 4$;
when $x = 2$, $y = 3$;
when $x = 3$, $y = 2$; etc.

In like manner the equation could have been solved for x in terms of y, and corresponding sets of values obtained.

Any set of corresponding values of x and y satisfies the given equation, and is therefore a solution.

An equation which, like the above, has an indefinite number of solutions, is called an **Indeterminate Equation.**

2. The equation

$$y - x = 1 \qquad (2)$$

also has an unlimited number of solutions.

Solving this equation for y, we have

$$y = 1 + x.$$

Then, when $x = 1$, $y = 2$;
when $x = 2$, $y = 3$;
when $x = 3$, $y = 4$; etc.

Now, observe that equations (1) and (2) have one common solution, namely $x = 2$, $y = 3$. It seems evident, and we shall later prove, that these equations have only this solution in common.

Equations (1) and (2) express different relations between the unknown numbers, and are called **Independent Equations.**

3. But the equation

$$x + y = 5 \qquad (1)$$

and the equivalent equation

$$2x + 2y = 10 \qquad (3)$$

give the same value for y in terms of x, namely,

$$y = 5 - x.$$

Consequently these equations are satisfied by an unlimited number of common sets of values of x and y, and not by one definite set of values.

In fact, equations (1) and (3) express the same relation between the unknown numbers, and are therefore not independent.

4. The equations $\quad x + y = 5 \qquad (1)$

and $\quad 3x + 3y = 16 \qquad (4)$

are not satisfied by any common set of values of x and y.

For any set of values which reduces $x + y$ to 5 must reduce $3x + 3y$, or $3(x + y)$, to 15, and not to 16. These two equations express inconsistent relations between the unknown numbers, and are called **Inconsistent Equations.**

5. The three equations

$$x + y = 5, \qquad (1)$$

$$y - x = 1, \qquad (2)$$

$$2x + y = 9, \qquad (5)$$

are not satisfied by any common set of values of x and y. For, by Art. 2, equations (1) and (2) are satisfied by the values $x = 2$, $y = 3$. But equation (5) is evidently not satisfied by this set of values. The three equations express three independent relations between x and y.

6. A System of Simultaneous Equations is a group of equations which are to be satisfied by the same set, or sets, of values of the unknown numbers.

A **Solution** of a system of simultaneous equations is a set of values of the unknown numbers which converts all of the equations into identities, that is, which satisfies all of the equations.

The examples of Arts. 1–5 are illustrations of the following general principles, which will be proved later:

A system of equations has a definite number of solutions,

(i.) *When the number of equations is the same as the number of unknown numbers.*

(ii.) *And when the equations are all independent and consistent.*

EXERCISES I.

Of which of the following systems are the equations inconsistent? Of which are the equations not independent? Of which are the equations consistent and independent?

1. $\begin{cases} 3x + 5y = 11, \\ 4x + 7y = 15. \end{cases}$

2. $\begin{cases} 2x + 3y = 4, \\ 4x - 6y = 8. \end{cases}$

3. $\begin{cases} 3x + 10y = 42, \\ 6x + 20y = 84. \end{cases}$

4. $\begin{cases} 6x - 9y = 4, \\ 4x - 6y = 9. \end{cases}$

5. $\begin{cases} 18x - 15y = 51, \\ 6x - 5y = 17. \end{cases}$

6. $\begin{cases} 5x + 4y = 6, \\ 7x + 6y = 10. \end{cases}$

7. $\begin{cases} 8x + 2y = 5, \\ 12x + 3y = 7. \end{cases}$

8. $\begin{cases} 18x - 15y = 50, \\ 6x - 5y = 17. \end{cases}$

9. $\begin{cases} 3x + 5y = 30, \\ 7x + 2y = 70. \end{cases}$

10. $\begin{cases} 7x - 2y = 10, \\ 8x - 5y = 25. \end{cases}$

11. $\begin{cases} ax - by = c, \\ nax - nby = c. \end{cases}$

12. $\begin{cases} mx + ny = p, \\ m(x + y) = mp. \end{cases}$

§ 2. EQUIVALENT SYSTEMS.

1. *Two systems of equations are equivalent when every solution of either system is a solution of the other.*

E.g., the systems (I.) and (II.):

$$\left.\begin{aligned} 3x + 2y &= 8, \\ x - y &= 1, \end{aligned}\right\} \quad \text{(I.)} \qquad \left.\begin{aligned} 3x + 2y &= 8, \\ 2x - 2y &= 2, \end{aligned}\right\} \quad \text{(II.)}$$

are equivalent. For they are both satisfied by the same solution, $x = 2$, $y = 1$, and, as we shall see later, by no other solution.

2. The principles of the equivalence of equations given in Ch. IV. were there proved for equations which contain one or more unknown numbers. They can therefore be applied to any equation of a system of equations.

The solution of a system of two or more equations depends also upon the following principles of the equivalence of systems:

(i.) *If any equation of a system be replaced by an equivalent equation, the resulting system will be equivalent to the given one.*

E.g., the system

$$\left.\begin{aligned} 3x + 2y &= 8, \\ x - y &= 1, \end{aligned}\right\} \quad \text{(I.)}$$

is equivalent to the system

$$\left.\begin{aligned} 3x + 2y &= 8, \\ 2x - 2y &= 2 \end{aligned}\right\} \quad \text{(II.)}$$

in which the equation $x - y = 1$ of the given system is replaced by the equivalent equation $2x - 2y = 2$.

It is evident that if each equation of a system be replaced by an equivalent equation, the resulting system will be equivalent to the given one.

(ii.) *If any equation of a system be replaced by an equation obtained by adding or subtracting corresponding members of two or more of the equations of the system, the resulting system will be equivalent to the given one.*

E.g., the system

$$\left.\begin{aligned} 3x + 2y &= 8, \\ 2x - 2y &= 2, \end{aligned}\right\} \qquad \text{(II.)}$$

is equivalent to the system

$$\left.\begin{aligned} 3x + 2y &= 8, \\ (3x + 2y) + (2x - 2y) &= 10, \end{aligned}\right\} \text{ or } \left.\begin{aligned} 3x + 2y &= 8, \\ 5x &= 10. \end{aligned}\right\} \qquad \text{(III.)}$$

(iii.) *If one equation of a system be solved for one of the unknown numbers, and the resulting value be substituted for this unknown number in each of the other equations, the derived system will be equivalent to the given one.*

E.g., the system

$$\left.\begin{aligned} x - y &= 2, \\ 5x - 3y &= 12, \end{aligned}\right\} \qquad \text{(IV.)}$$

is equivalent to the system

$$\left.\begin{aligned} x &= 2 + y, \\ 5(2 + y) - 3y &= 12. \end{aligned}\right\} \qquad \text{(V.)}$$

The proofs of the principles enunciated are as follows:

(i.) Let

$$\left.\begin{aligned} A &= B, \\ C &= D, \end{aligned}\right\} \qquad \text{(I.)}$$

be two equations in two unknown numbers, say x and y; and let $C' = D'$ be equivalent to $C = D$.

Then the system

$$\left.\begin{aligned} A &= B, \\ C' &= D', \end{aligned}\right\} \qquad \text{(II.)}$$

is equivalent to the system (I.). For, by definition of equivalent equations, the same sets of values which satisfy $C = D$ also satisfy $C' = D'$, and *vice versa*. Therefore any one of these sets of values which also satisfies $A = B$ is a solution of both systems. Consequently, every solution of either system is a solution of the other.

In like manner, the principle can be proved for a system of any number of equations.

(ii.) Let

$$\left.\begin{aligned} A &= B, \\ C &= D, \end{aligned}\right\} \qquad \text{(I.)}$$

be two equations in two unknown numbers, say x and y. Then the systems

$$\left.\begin{aligned} A &= B, \\ A + C &= B + D, \end{aligned}\right\} \quad \text{(II.)} \quad \text{and} \quad \left.\begin{aligned} A &= B, \\ A - C &= B - D, \end{aligned}\right\} \quad \text{(III.)}$$

are each equivalent to the system (I.).

For any set of values which makes A and B equal, and C and D equal, makes $A + C$ and $B + D$ equal, and $A - C$ and $B - D$ equal. Therefore every solution of (I.) is a solution of (II.) and of (III.). Likewise, any set of values which makes A and B equal and $A + C$ and $B + D$ equal, or $A - C$ and $B - D$ equal, makes C and D equal. Therefore every solution of (II.) and of (III.) is a solution of (I.).

In like manner, the principle can be proved for a system of any number of equations.

(iii.) Let

$$\left.\begin{array}{ll} A = B, & (1) \\ C = D, & (2) \end{array}\right\} \qquad (\text{I.})$$

be two equations in two unknown numbers, say x and y; and let $x = P$ be the equation derived by solving (1) for x, and $C' = D'$ be the equation obtained by substituting P for x in (2).

Then the system

$$\left.\begin{array}{ll} x = P, & (3) \\ C' = D', & (4) \end{array}\right\} \qquad (\text{II.})$$

is equivalent to the system (I.).

Since equation (3) is equivalent to equation (1), any solution of the system (1) must satisfy equation (3); that is, must give to x and P one and the same value. But (4) differs from (2) only in having P where (2) has x. Therefore, since x and P have the same value, any value of x, with the corresponding value of y, which makes C and D equal must make C' and D' equal. Therefore every solution of the system (I.) is a solution of the system (II.).

Since equation (1) is equivalent to equation (3), any solution of the system (II.) must satisfy equation (1); that is, must make A and B equal. But (2) differs from (4) only in having x where (4) has P. Therefore, since any solution of (II.) makes x and P equal and C' and D' equal, it must also make C and D equal. Therefore every solution of the system (II.) is a solution of the system (I.).

Consequently, the two systems are equivalent.

In like manner, the principle can be proved for a system of any number of equations.

3. Elimination is the process of deriving from two or more equations of a system an equation with one less unknown number than the equations from which it is derived. The unknown number which does not appear in the derived equation is said to have been *eliminated*.

E.g., if the equations $x + y = 7,$ (1)

$x - y = 1$ (2)

be added, we obtain $2x = 8,$ (3)

in which the unknown number y does not appear. We say that y has been eliminated from the given equations.

In this chapter we shall apply these principles to the solutions of systems of linear equations.

§ 3. SYSTEMS OF LINEAR EQUATIONS.

Linear Equations in Two Unknown Numbers.

1. There are several methods for solving two simultaneous equations in two unknown numbers. The object in all of them is to obtain from the given system an equivalent system of which one equation contains only one of the unknown numbers.

Elimination by Addition and Subtraction.

2. Ex. 1. Solve the system

$$\left.\begin{aligned} 3x + 4y &= 24, \quad (1) \\ 5x - 6y &= 2. \quad (2) \end{aligned}\right\} \quad \text{(I.)}$$

To eliminate x, we multiply both members of equation (1) by 5, and both members of equation (2) by 3, thereby making the coefficients of x in the two equations equal. We then have

$$\left.\begin{aligned} 15x + 20y &= 120, \quad (3) \\ 15x - 18y &= 6. \quad (4) \end{aligned}\right\} \quad \text{(II.)}$$

The system (II.) is equivalent to the system (I.), by § 2, Art. 2 (i.). The system (II.) is, by § 2, Art. 2 (i.) and (ii.), equivalent to the system

$$\left.\begin{aligned} 3x + 4y &= 24, \quad (1) \\ (15x + 20y) - (15x - 18y) &= 120 - 6; \quad (5) \end{aligned}\right\} \quad \text{(III.)}$$

or, performing the indicated operations, to

$$\left.\begin{aligned} 3x + 4y &= 24, \quad (1) \\ 38y &= 114; \quad (6) \end{aligned}\right\} \quad \text{(IV.)}$$

or, to

$$\left.\begin{aligned} 3x + 4y &= 24, \quad (1) \\ y &= 3. \quad (7) \end{aligned}\right\} \quad \text{(V.)}$$

The system (V.) gives the required solution, since equation (7) gives the value of y, and equation (1) the corresponding value of x, by § 2, Art. 2 (iii.). Substituting 3 for y in (1), we obtain

$$3x + 12 = 24; \qquad (8)$$

whence $$x = 4. \qquad (9)$$

Consequently the required solution is $x = 4$, $y = 3$.

This solution may be written 4, 3, it being understood that the first number is the value of x, and the second the value of y. This way of representing the solution will be used in subsequent work.

The work above has been given in full in order to emphasize that by each step one system has been replaced by an equivalent system. In practice the work may be contracted as follows:

Multiplying (1) by 5, $15x + 20y = 120.$ (3)

Multiplying (2) by 3, $15x - 18y = 6.$ (4)

Subtracting (4) from (3), $38y = 114;$ (6)

whence $y = 3.$ (7)

Substituting 3 for y in (1), $3x + 12 = 24;$ (8)

whence $x = 4.$ (9)

If we wish first to eliminate y, we multiply both members of (1) by 3 only, and both members of (2) by 2 only, since the coefficients of y in the equations have the common factor 2. The work then proceeds as follows:

Multiplying (1) by 3, $9x + 12y = 72.$ (10)

Multiplying (2) by 2, $10x - 12y = 4.$ (11)

Adding (10) and (11), $19x = 76;$ (12)

whence $x = 4.$ (13)

Substituting 4 for x in (1), $12 + 4y = 24;$ (14)

whence $y = 3.$ (15)

Ex. 2. Solve the system

$$\frac{7+x}{5}-\frac{2x-y}{4}=3y-5, \qquad (1)$$

$$\frac{4x-3}{6}+\frac{5y-7}{2}=18-5x. \qquad (2)$$

Clearing (1) of fractions,

$$28+4x-10x+5y=60y-100. \qquad (3)$$

Clearing (2) of fractions,

$$4x-3+15y-21=108-30x. \qquad (4)$$

Transferring and uniting terms,

$$6x+55y=128, \qquad (5)$$

$$34x+15y=132. \qquad (6)$$

Multiplying (5) by 3, $18x+165y=384.$ (7)

Multiplying (6) by 11, $374x+165y=1452.$ (8)

Subtracting (7) from (8), $356x=1068;$ (9)

whence $x=3.$ (10)

Substituting 3 for x in (5), $18+55y=128;$ (11)

whence $y=2.$ (12)

Consequently, the required solution is 3, 2.

Notice that (1) and (2), (3) and (4), (5) and (6), (7) and (8), and (10) and any preceding equation except (9), form equivalent systems. In forming with (10) an equivalent system, we naturally take the simplest of the preceding equations, in this case (5).

Ex. 3. Solve the system

$$\frac{2x-b}{a}-\frac{2y-a}{b}=2, \qquad (1)$$

$$\frac{2ax-a^2}{b}+\frac{2by+b^2}{a}=a+b. \qquad (2)$$

Clearing (1) and (2) of fractions,

$$2bx-b^2-2ay+a^2=2ab, \qquad (3)$$

$$2a^2x-a^3+2b^2y+b^3=a^2b+ab^2. \qquad (4)$$

Transferring terms,

$$2\,bx - 2\,ay = b^2 + 2\,ab - a^2, \qquad (5)$$

$$2\,a^2x + 2\,b^2y = a^3 + a^2b + ab^2 - b^3. \qquad (6)$$

Multiplying (5) by a^2,

$$2\,a^2bx - 2\,a^3y = a^2b^2 + 2\,a^3b - a^4. \qquad (7)$$

Multiplying (6) by b,

$$2\,a^2bx + 2\,b^3y = a^3b + a^2b^2 + ab^3 - b^4. \qquad (8)$$

Subtracting (7) from (8),

$$2\,(a^3 + b^3)\,y = a^4 - a^3b + ab^3 - b^4 \qquad (9)$$

$$= a^3(a - b) + b^3(a - b)$$

$$= (a^3 + b^3)(a - b);$$

whence $$y = \frac{a - b}{2}. \qquad (10)$$

Substituting $\frac{a - b}{2}$ for y in (5),

$$2\,bx - a(a - b) = b^2 + 2\,ab - a^2;$$

whence $$x = \frac{a + b}{2}.$$

Consequently, the required solution is $\frac{a + b}{2}$, $\frac{a - b}{2}$.

3. The examples of the preceding article illustrate the following method of elimination by addition and subtraction.

Simplify the given equations, if necessary, and transfer the terms in x and y to the first members, and the terms free from x and y to the second members.

Determine the L. C. M. of the coefficients of the unknown number to be eliminated, and multiply both members of each equation by the quotient of the L. C. M. divided by the coefficient of that unknown number in the equation.

The coefficients of the unknown number to be eliminated being now equal, or equal and opposite, in the two equations, subtract, or add, corresponding members, and equate the results. A final equation in one unknown number will thus be derived.

The solution of the given system is then obtained by solving this derived equation, and substituting the value of the unknown number thus obtained in the simplest of the preceding equations.

EXERCISES II.

Solve the following systems of equations by the method of addition and subtraction:

1. $\begin{cases} x + y = 17, \\ x - y = 7. \end{cases}$

2. $\begin{cases} x + y = a, \\ x - y = b. \end{cases}$

3. $\begin{cases} 7x + 11y = 2, \\ 7x - 11y = 0. \end{cases}$

4. $\begin{cases} 3x + ay = 5a^2, \\ 3x - ay = a^2. \end{cases}$

5. $\begin{cases} x - 12y = 3, \\ x + 4y = 19. \end{cases}$

6. $\begin{cases} 3x + y = 31, \\ 5x - 2y = 15. \end{cases}$

7. $\begin{cases} 4x - 7y = 19, \\ x + 9y = 37. \end{cases}$

8. $\begin{cases} x + 5y = 14, \\ 3x - 4y = 4. \end{cases}$

9. $\begin{cases} 10x - 3y = 25, \\ 5x - 9y = -25. \end{cases}$

10. $\begin{cases} 3x + 10y = 12, \\ 12x - 5y = 3. \end{cases}$

11. $\begin{cases} nx - ay = 0, \\ n^2x - ay = an. \end{cases}$

12. $\begin{cases} 6x + y = 6, \\ 4x + 3y = 11. \end{cases}$

13. $\begin{cases} 12x + 15y = 8, \\ 16x + 9y = 7. \end{cases}$

14. $\begin{cases} 5x + 4y = 49\frac{1}{2}, \\ 2x + 7y = 63. \end{cases}$

15. $\begin{cases} 5x - 3y = 12, \\ 19x - 5y = 73\frac{1}{3}. \end{cases}$

16. $\begin{cases} 3x + 16y = 5, \\ -5x + 28y = 19. \end{cases}$

17. $\begin{cases} 24x + 7y = 27, \\ 8x - 33y = 115. \end{cases}$

18. $\begin{cases} 16x + 17y = 274, \\ 24x - 105y = 150. \end{cases}$

19. $\begin{cases} 21x + 8y = -66, \\ 28x - 23y = 13. \end{cases}$

20. $\begin{cases} 18x - 20y = 1, \\ 15x + 16y = 9. \end{cases}$

21. $\begin{cases} 12x - 14y = -4, \\ 8x - 21y = -8.5. \end{cases}$

22. $\begin{cases} 33x + 54y = -9, \\ 44x - 81y = 294. \end{cases}$

23. $\begin{cases} 96x - 38y = 18, \\ 84x - 57y = 87. \end{cases}$

24. $\begin{cases} 3x + \dfrac{7y}{2} = 22, \\ 11y - \dfrac{2x}{5} = 20. \end{cases}$

25. $\begin{cases} \dfrac{2x-y}{2} + 14 = 18, \\[2ex] \dfrac{2y+x}{3} + 16 = 19. \end{cases}$

26. $\begin{cases} \dfrac{\frac{2x}{3} - \frac{5x}{12}}{\frac{7}{4}} - \dfrac{\frac{3x}{2} - \frac{y}{3}}{\frac{23}{2}} = 2, \\[2ex] x - y = \frac{1}{6}(x+y). \end{cases}$

27. $\begin{cases} 4x + \dfrac{15-x}{4} = 2y + 5 + \dfrac{7x+11}{16}, \\[2ex] 3y - \dfrac{2x+y}{5} = 2x + \dfrac{2y+4}{3}. \end{cases}$

28. $\begin{cases} \dfrac{x}{m-a} + \dfrac{y}{m-b} = 1, \\[2ex] \dfrac{x}{n-a} + \dfrac{y}{n-b} = 1. \end{cases}$

29. $\begin{cases} ax - by = a^2 + b^2, \\ bx + ay = a^2 + b^2. \end{cases}$

Elimination by Substitution.

4. Ex. 1. Solve the system

$$\left.\begin{aligned} 5x - 2y &= 1, & (1) \\ 4x + 5y &= 47. & (2) \end{aligned}\right\} \quad \text{(I.)}$$

If we wish to eliminate x, we proceed as follows:

Solving (1) for x, $\quad x = \dfrac{1+2y}{5}$. (3)

Substituting $\dfrac{1+2y}{5}$ for x in (2),

$$4\left(\frac{1+2y}{5}\right) + 5y = 47. \quad (4)$$

(3), (4): (II.)

The system (II.) is equivalent to the system (I.), by § 2, Art. 2 (iii.).

Solving (4) for y, $\quad y = 7$. (5)

Substituting 7 for y in (3), $\quad x = 3$. (6)

(5), (6): (III.)

The system (III.) is, by § 2, Art. 2 (iii.), equivalent to the system (II.), and hence to the given system. Therefore the required solution is 3, 7.

If we wish first to eliminate y, we proceed as follows:

Solving (1) for y, $\quad y = \dfrac{5x-1}{2}$. (7)

Substituting $\frac{5x-1}{2}$ for y in (2), $4x+5\left(\frac{5x-1}{2}\right)=47.$ (8)

Solving (8) for x, $x=3.$ (9)

Substituting (3) for x in (7), $y=7.$ (10)

Ex. 2. Solve the system $bx+ay=2a,$ (1)

$$b(x-1)-a(y-1)=a+b. \quad (2)$$

Solving (1) for x, $x=\frac{2a-ay}{b}.$ (3)

Substituting $\frac{2a-ay}{b}$ for x in (2),

$$b\left(\frac{2a-ay}{b}-1\right)-a(y-1)=a+b; \quad (4)$$

whence $y=\frac{a-b}{a}.$ (5)

Substituting $\frac{a-b}{a}$ for y in (3),

$$x=\frac{2a-a+b}{b}=\frac{a+b}{b}. \quad (6)$$

Hence the required solution is $\frac{a+b}{b}, \frac{a-b}{a}.$

This system could have been solved more easily by the method of addition and subtraction.

Notice that (1) and (2), (3) and (4), and (3) and (5) form equivalent systems.

5. The examples of the preceding article illustrate the following method of elimination by substitution:

Solve the simpler equation for the unknown number to be eliminated in terms of the other, and substitute the value thus obtained in the other equation. The derived equation will contain but one unknown number.

The solution of the given system is then obtained by solving the derived equation, and substituting the value of the unknown number thus obtained in the expression for the other unknown number.

EXERCISES III.

Solve the following systems of equations by the method of substitution:

1. $\begin{cases} x = 2y - 3, \\ y = 2x - 15. \end{cases}$ 2. $\begin{cases} x = 6 - y, \\ y = 3x - 4. \end{cases}$ 3. $\begin{cases} x = 3y - 7, \\ y = 3x - 19. \end{cases}$

4. $\begin{cases} x - y = 18, \\ x = 4y. \end{cases}$ 5. $\begin{cases} x = \frac{5}{8}y, \\ x-4=\frac{6}{5}(y+6). \end{cases}$ 6. $\begin{cases} \frac{1}{2}y - 3x = 2, \\ y = 14x. \end{cases}$

7. $\begin{cases} x + y = a, \\ x = ny. \end{cases}$ 8. $\begin{cases} 7x = 5y, \\ 15y=28x-70. \end{cases}$ 9. $\begin{cases} 5x = 8y - 11, \\ 6y = 7x - 21. \end{cases}$

10. $\begin{cases} 7x - 3 = 5y, \\ 7y - 3 = 8x. \end{cases}$ 11. $\begin{cases} ay = bx, \\ a + y = b + x. \end{cases}$ 12. $\begin{cases} x - 3y = 0, \\ 25x+48y=287. \end{cases}$

13. $\begin{cases} 3x + 4y = 2, \\ 9x+20y=8. \end{cases}$ 14. $\begin{cases} 5x+7y=49, \\ 7x+5y=47. \end{cases}$ 15. $\begin{cases} \frac{2}{3}x = 10 - \frac{1}{2}y, \\ 4\frac{3}{4}y = 5x - 7. \end{cases}$

16. $\begin{cases} \frac{1}{2}x + \frac{1}{3}y = 7, \\ 2x + 3y = 43. \end{cases}$ 17. $\begin{cases} \frac{1}{3}x - \frac{1}{2}y = 2, \\ 2x + 3y = 60. \end{cases}$ 18. $\begin{cases} 3\frac{1}{3}x = 8y, \\ 2\frac{1}{2}x = 3(y + 5). \end{cases}$

19. $\begin{cases} \dfrac{7+x}{5} - \dfrac{2x-y}{4} = 3y - 5, \\ \dfrac{5y-7}{2} + \dfrac{4x-3}{6} = 18 - 5x. \end{cases}$ 20. $\begin{cases} \dfrac{7x+3y}{4} - \dfrac{9x-3y}{2} = 3, \\ \dfrac{11x+7y}{2} - \dfrac{15x+9y}{5} = 16. \end{cases}$

21. $\begin{cases} \dfrac{x+y}{8a} + \dfrac{x-y}{12b} = 1, \\ \dfrac{x}{4a+6b} + \dfrac{y}{4a-6b} = 1. \end{cases}$ 22. $\begin{cases} (x-a)(a+b)=(a-b)(y-a), \\ \dfrac{x}{a^3-b^3} = \dfrac{y}{a^3+b^3}. \end{cases}$

23. $\begin{cases} \left(\dfrac{1}{a} + \dfrac{1}{b}\right)x - \left(\dfrac{1}{a} - \dfrac{1}{b}\right)y = 4, \\ \dfrac{x}{a+b} + \dfrac{y}{a-b} = 2. \end{cases}$

Elimination by Comparison.

6. Ex. 1. Solve the system

$$\left.\begin{aligned} 7x + 2y &= 20, \quad (1) \\ 13x - 3y &= 17. \quad (2) \end{aligned}\right\} \qquad \text{(I.)}$$

To eliminate y, we proceed as follows:

$$\left.\begin{aligned} &\text{Solving (1) for } y, \quad y = \frac{20 - 7x}{2}. \quad (3) \\ &\text{Solving (2) for } y, \quad y = \frac{13x - 17}{3}. \quad (4) \end{aligned}\right\} \qquad \text{(II.)}$$

The system (II.) is equivalent to the system (I.), by § 2, Art. 2 (i.).

Substituting in (4) for y, its value given in (3),

$$\frac{20 - 7x}{2} = \frac{13x - 17}{3}. \qquad (5)$$

The last equation and equation (3) form a system which is equivalent to the system (II.), and hence to the given system.

$$\left.\begin{aligned} &\text{Solving (5) for } x, \quad x = 2. \quad (6) \\ &\text{Substituting 2 for } x \text{ in (3)}, \quad y = 3. \quad (7) \end{aligned}\right\} \qquad \text{(III.)}$$

The system (III.) is equivalent to the system formed by equations (3) and (5), and therefore to the given system. Consequently the required solution is 2, 3.

To eliminate x, we proceed as follows:

Solving (1) and (2) for x, $$x = \frac{20 - 2y}{7}, \qquad (8)$$

$$x = \frac{17 + 3y}{13}. \qquad (9)$$

Equating these expressions for x,

$$\frac{20 - 2y}{7} = \frac{17 + 3y}{13}; \qquad (10)$$

whence $y = 3$.

Substituting 3 for y in (8), $x = 2$.

Ex. 2. Solve the system

$$\frac{3x+y}{9}-\frac{5y-2x}{7}=\frac{3y+1}{14}, \tag{1}$$

$$(x+20)^2-(x+y)(x-y)=(y+22)^2. \tag{2}$$

Clearing (1) of fractions,

$$42x+14y-90y+36x=27y+9. \tag{3}$$

Removing parentheses in (2),

$$x^2+40x+400-x^2+y^2=y^2+44y+484. \tag{4}$$

Simplifying (3) and (4),

$$78x-103y=9, \tag{5}$$

$$10x-11y=21. \tag{6}$$

Solving (5) for x, $$x=\frac{9+103y}{78}. \tag{7}$$

Solving (6) for x, $$x=\frac{21+11y}{10}. \tag{8}$$

Equating these expressions for x,

$$\frac{9+103y}{78}=\frac{21+11y}{10}; \tag{9}$$

whence $$y=9. \tag{10}$$

Substituting 9 for y in (8), $x=12$. (11)

Consequently the required solution is 12, 9.

Notice that (1) and (2), (3) and (4), (5) and (6), (7) and (8), and (10) and any preceding equation except (9) form equivalent systems. In forming with (10) an equivalent system, we naturally take the simplest of the preceding equations, in this case (8).

Ex. 3. Solve the system

$$\frac{ay+bx}{ab}-ab=\frac{x+y}{a+b}=\frac{a^2x-b^2y}{a^3-b^3}.$$

Notice that this continued equation is equivalent to two independent equations, which can be chosen in two ways:

$$\left.\begin{aligned}\frac{ay+bx}{ab}-ab&=\frac{x+y}{a+b}, &(1)\\ \frac{ay+bx}{ab}-ab&=\frac{a^2x-b^2y}{a^3-b^3}. &(2)\end{aligned}\right\}\quad\text{(I.)}$$

$$\left.\begin{aligned}\frac{ay+bx}{ab}-ab&=\frac{x+y}{a+b}, &(1)\\ \frac{x+y}{a+b}&=\frac{a^2x-b^2y}{a^3-b^3}. &(3)\end{aligned}\right\}\quad\text{(II.)}$$

Both systems have the same solution. We will solve the system (II.), which appears to be the simpler.

Clearing of fractions,

$$a^2y+abx+aby+b^2x-a^3b^2-a^2b^3=abx+aby, \qquad (4)$$

$$a^3x+a^3y-b^3x-b^3y=a^3x-ab^2y+a^2bx-b^3y. \qquad (5)$$

Transferring and uniting terms,

$$a^2y+b^2x=a^2b^2(a+b), \qquad (6)$$

$$ay-bx=0. \qquad (7)$$

Solving (6) for y, $\quad y=\dfrac{a^2b^2(a+b)-b^2x}{a^2}. \qquad (8)$

Solving (7) for y, $\quad y=\dfrac{bx}{a}. \qquad (9)$

Equating these expressions for y,

$$\frac{a^2b^2(a+b)-b^2x}{a^2}=\frac{bx}{a}; \qquad (10)$$

whence $\quad x=a^2b.$

Substituting a^2b for x in (9),

$$y=ab^2.$$

Consequently the required solution is a^2b, ab^2.

Notice that the equations (6) and (7) could have been more easily solved by the method of addition and subtraction, or by the method of substitution.

7. The examples of the preceding article illustrate the following method of elimination by comparison:

Solve the given equations for the unknown number to be eliminated, and equate the expressions thus obtained. The derived equation will contain but one unknown number.

The solution of the given system is then obtained by solving this derived equation, and substituting the value of the unknown number thus obtained in the simplest of the preceding equations.

EXERCISES IV.

Solve the following systems of equations by the method of comparison:

1. $\begin{cases} x = 3y - 2, \\ x = 5y - 12. \end{cases}$

2. $\begin{cases} y = 3x - 17, \\ y = 2x - 10. \end{cases}$

3. $\begin{cases} 5y = 2x + 1, \\ 8y = 5x - 11. \end{cases}$

4. $\begin{cases} 5x = 7y - .1, \\ 7x = 9y + 1.7. \end{cases}$

5. $\begin{cases} \frac{1}{5}x = \frac{1}{8}y - 1, \\ \frac{1}{5}y = \frac{1}{4}x - 2. \end{cases}$

6. $\begin{cases} 2\frac{1}{2}x - 3\frac{1}{8}y = 10, \\ 7\frac{1}{8}x - 5\frac{1}{2}y = 55. \end{cases}$

7. $\begin{cases} 1\frac{1}{2}x = 7y - 38, \\ 1\frac{1}{2}y = 7x - 72. \end{cases}$

8. $\begin{cases} 5x + 9y = 28, \\ 7x + 3y = 20. \end{cases}$

9. $\begin{cases} 21x - 23y = 2, \\ 7x - 19y = 12. \end{cases}$

10. $\begin{cases} \frac{1}{7}x + 7y = 99, \\ \frac{1}{7}y + 7x = 51. \end{cases}$

11. $\begin{cases} \frac{x}{2} + \frac{y}{3} - 7 = 0, \\ \frac{x}{3} + \frac{y}{2} - 8 = 0. \end{cases}$

12. $\begin{cases} \frac{x}{2} + \frac{y}{6} = 11, \\ \frac{x}{5} + \frac{y}{24} = \frac{7}{2}. \end{cases}$

13. $\begin{cases} 8x + 9y = 26, \\ 32x - 3y = 26. \end{cases}$

14. $\begin{cases} 63x - 46y = 29, \\ 42x - 69y = 96. \end{cases}$

15. $\begin{cases} \frac{x+2}{3} + 8y = 31, \\ \frac{y+5}{4} + 10x = 192. \end{cases}$

16. $\begin{cases} 5x + 4y = 9a - b, \\ 7x - 6y = a - 13b. \end{cases}$

17. $\begin{cases} x + ay + 1 = 0, \\ y + c(x+1) = 0. \end{cases}$

18. $\begin{cases} ax = c - bk(y-1), \\ by = b - \frac{1}{a}(c - ax). \end{cases}$

19. $\begin{cases} ax - by = a^2 + b^2, \\ (a-b)x + (a+b)y = 2(a^2 - b^2). \end{cases}$

20. $\begin{cases} \frac{7x-21}{6} + \frac{3y-x}{3} = 4 + \frac{3x-19}{2}, \\ \frac{2x+y}{2} - \frac{9x-7}{8} = \frac{3y+9}{4} - \frac{4x+5y}{16}. \end{cases}$

The General Solution of a System of Two Linear Equations in Two Unknown Numbers.

8. Any linear equation in two unknown numbers can evidently be brought to the form

$$ax + by = c,$$

in which ax stands for the algebraic sum of all the terms in x, by for the algebraic sum of all the terms in y, and c for the algebraic sum of all the terms free from x and y.

9. Let
$$a_1x + b_1y = c_1, \quad (1)$$
$$a_2x + b_2y = c_2, \quad (2)$$
be any two linear equations.

To eliminate y we must make the coefficients of y equal in the two equations.

Multiplying (1) by b_2, $a_1b_2x + b_1b_2y = b_2c_1$. (3)

Multiplying (2) by b_1, $a_2b_1x + b_1b_2y = b_1c_2$. (4)

Subtracting (4) from (3), $(a_1b_2 - a_2b_1)x = b_2c_1 - b_1c_2$; (5)

whence, if $a_1b_2 - a_2b_1 \neq 0$, $x = \frac{b_2c_1 - b_1c_2}{a_1b_2 - a_2b_1}$. (6)

To eliminate x we must make the coefficients of x equal in the two equations.

Multiplying (1) by a_2, $a_1a_2x + a_2b_1y = a_2c_1$. (7)

Multiplying (2) by a_1, $a_1a_2x + a_1b_2y = a_1c_2$. (8)

Subtracting (7) from (8), $(a_1b_2 - a_2b_1)y = a_1c_2 - a_2c_1$; (9)

whence, if $a_1b_2 - a_2b_1 \neq 0$, $y = \frac{a_1c_2 - a_2c_1}{a_1b_2 - a_2b_1}$ (10)

But if $a_1b_2 - a_2b_1 = 0$, we have no authority for dividing both members of equations (5) and (9) by $a_1b_2 - a_2b_1$.

Consequently, the *general solution* of two linear equations in two unknown numbers is

$$x = \frac{b_2c_1 - b_1c_2}{a_1b_2 - a_2b_1}, \qquad y = \frac{a_1c_2 - a_2c_1}{a_1b_2 - a_2b_1},$$

when $a_1b_2 - a_2b_1 \neq 0$.

10. *Every system of two independent and consistent equations of the first degree in two unknown numbers has one, and only one, solution.*

For the system

$$a_1x + b_1y = c_1, \tag{1}$$

$$a_2x + b_2y = c_2, \tag{2}$$

is, by Art. 9, equivalent to the system

$$(a_1b_2 - a_2b_1)\,x = b_2c_1 - b_1c_2, \tag{3}$$

$$(a_1b_2 - a_2b_1)\,y = a_1c_2 - a_2c_1. \tag{4}$$

But equations (3) and (4) are each linear in one unknown number, and each, therefore, has one, and only one, solution. Consequently, the given system has one, and only one, solution.

11. *Three independent linear equations in two unknown numbers cannot be satisfied by any common set of values of the unknown numbers.*

If the values of x and y which constitute the solution of equations (1) and (2), Art. 9, satisfy a third equation,

$$a_3x + b_3y = c_3, \tag{3}$$

we have
$$a_3 \times \frac{b_2c_1 - b_1c_2}{a_1b_2 - a_2b_1} + b_3 \times \frac{a_1c_2 - a_2c_1}{a_1b_2 - a_2b_1} = c_3.$$

From this relation, we obtain

$$c_3 = c_1 \times \frac{a_3b_2 - a_2b_3}{a_1b_2 - a_2b_1} + c_2 \times \frac{a_1b_3 - a_3b_1}{a_1b_2 - a_2b_1}$$
$$= lc_1 + mc_2, \tag{i.}$$

wherein
$$l = \frac{a_3b_2 - a_2b_3}{a_1b_2 - a_2b_1}, \text{ and } m = \frac{a_1b_3 - a_3b_1}{a_1b_2 - a_2b_1}.$$

We also have

$$b_1 \times \frac{a_3b_2 - a_2b_3}{a_1b_2 - a_2b_1} + b_2 \times \frac{a_1b_3 - a_3b_1}{a_1b_2 - a_2b_1} = \frac{b_3(a_1b_2 - a_2b_1)}{a_1b_2 - a_2b_1}$$
$$= b_3.$$

That is,
$$b_3 = lb_1 + mb_2. \tag{ii.}$$

In like manner, it can be shown that

$$a_3 = la_1 + ma_2. \tag{iii.}$$

Observe that (ii.) and (iii.) are identities, and hold for all values of the a's and b's which do not reduce $a_1b_2 - a_2b_1$ to 0; while (i.) is not an identity, but imposes a condition upon the values of the known numbers in the three equations.

When this condition is satisfied, c_3 is obtained from c_1 and c_2, just as a_3 is obtained from a_1 and a_2, and b_3 from b_1 and b_2. That is, when the solution of (1) and (2) is also a solution of (3), the last equation is not independent of the other two.

Discussion of the Solution of a System of Two Linear Equations in Two Unknown Numbers.

12. The denominator $a_1b_2 - a_2b_1$ not equal to Zero. — In this case the system always has a definite solution. The value of either unknown number is positive, negative, or 0, according as its numerator has the same sign as its denominator, the opposite sign, or is 0. It is important to notice that the known numbers a_1, b_1, c_1, a_2, b_2, c_2, may have any values whatever, including 0, provided only $a_1b_2 - a_2b_1 \neq 0$.

13. The Denominator $a_1b_2 - a_2b_1$ equal to Zero. — As long as the denominator is not equal to 0, however near its value may be to 0, the values of x and y in Art. 9 constitute the solution of the system. If we assume that these values still give a solution when $a_1b_2 - a_2b_1 = 0$, we must determine the nature of the solution, and consider it in connection with the equations of the system.

The denominator $a_1b_2 - a_2b_1$ will reduce to 0 when a_1, a_2, b_1, b_2 are all different from 0, if $a_1b_2 = a_2b_1$. This denominator will also reduce to 0 in the following cases, among others:

$$a_1 = 0,\ a_2 = 0,\ b_1 \neq 0,\ b_2 \neq 0; \quad b_1 = 0,\ b_2 = 0,\ a_1 \neq 0,\ a_2 \neq 0;$$

$$a_1 = 0,\ b_1 = 0,\ a_2 \neq 0,\ b_2 \neq 0; \quad a_2 = 0,\ b_2 = 0,\ a_1 \neq 0,\ b_1 \neq 0.$$

We will now discuss these cases in the above order. It is to be understood in the following discussion that each of the numbers a_1, a_2, b_1, b_2 is different from 0, unless the contrary is stated.

14. $a_1b_2 = a_2b_1$. — We have to consider two cases, according as the numerators in the expressions for x and y are equal to 0, or are different from 0.

(i.) If $b_2c_1 - b_1c_2 \neq 0$, then $x = \infty$.

But in this case also $y = \infty$.

For from $a_1b_2 = a_2b_1$, we derive

$$a_1 = \frac{a_2b_1}{b_2}, \text{ since } b_2 \neq 0.$$

Substituting this value of a_1 in $a_1c_2 - a_2c_1$, the numerator of the expression for y, we obtain

$$\frac{a_2b_1}{b_2} \times c_2 - a_2c_1, \quad = \frac{-a_2(b_2c_1 - b_1c_2)}{b_2}$$
$$\neq 0, \text{ since } a_2 \neq 0,\ b_2 \neq 0,\ b_2c_1 - b_1c_2 \neq 0.$$

Therefore, $a_1c_2 - a_2c_1 \neq 0$, and hence $y = \infty$.

Consequently, in this case the system does not have a finite solution.

That the equations of the system are inconsistent can be shown as follows:

From $a_1b_2 = a_2b_1$, we obtain

$$\frac{a_2}{a_1} = \frac{b_2}{b_1} = k, \text{ say};$$

whence $\quad a_2 = ka_1$, and $b_2 = kb_1$.

But since $\quad b_2c_1 - b_1c_2 \neq 0,\ b_2c_1 \neq b_1c_2$;

therefore, $\quad \frac{c_2}{c_1} \neq \frac{b_2}{b_1}$, and hence $\frac{c_2}{c_1} \neq k$, or $c_2 \neq kc_1$.

The given system is therefore equivalent to

$$a_1x + b_1y = c_1, \qquad (1)$$

$$k(a_1x + b_1y) = c_2 \neq kc_1. \qquad (2)$$

But any set of finite values of x and y which reduces

$$a_1x + b_1y \text{ to } c_1$$

must reduce $k(a_1x + b_1y)$ to kc_1, and not to c_2.

See the particular example given in § 1, Art. 4.

(ii.) If $b_2c_1 - b_1c_2 = 0$, then $x = \frac{0}{0}$.

But in this case also $y = \frac{0}{0}$.

For from $a_1b_2 = a_2b_1$, we obtain

$$a_1 = \frac{a_2b_1}{b_2}, \text{ since } b_2 \neq 0.$$

Substituting this value of a_1 in $a_1c_2 - a_2c_1$, the numerator of the expression for y, we obtain

$$\frac{a_2b_1}{b_2} \times c_2 - a_2c_1 = \frac{-a_2(b_2c_1 - b_1c_2)}{b_2}$$
$$= 0, \text{ since } b_2c_1 - b_1c_2 = 0, \text{ and } b_2 \neq 0.$$

Therefore, $a_1c_2 - a_2c_1 = 0$, and hence $y = \frac{0}{0}$.

Consequently, in this case the equations of the system do not have a determinate solution.

That the equations are equivalent can be shown as follows:

From $a_1b_2 - a_2b_1 = 0$, we obtain $\frac{a_2}{a_1} = \frac{b_2}{b_1}$;

from $a_1c_2 - a_2c_1 = 0$, we obtain $\frac{a_2}{a_1} = \frac{c_2}{c_1}$.

Therefore,
$$\frac{a_2}{a_1} = \frac{b_2}{b_1} = \frac{c_2}{c_1} = k, \text{ say};$$

whence
$$a_2 = ka_1,\ b_2 = kb_1,\ c_2 = kc_1.$$

The given system is, therefore, equivalent to

$$a_1x + b_1y = c_1. \quad (1)$$
$$k(a_1x + b_1y) = kc_1. \quad (2)$$

That is, equation (2) is equivalent to (1), and the system has an unlimited number of finite solutions.

See the particular example given in § 1, Art. 3.

15. $a_1 = 0$ and $a_2 = 0$. — We have to consider two cases, according as the numerators of the expressions for x and y are, or are not, equal to 0.

(i.) If $b_2c_1 - b_1c_2 \neq 0$, then $x = \infty$.

And since $a_1 = 0$ and $a_2 = 0$, therefore $a_1c_2 - a_2c_1 = 0$; consequently, $y = \frac{0}{0}$.

(ii.) If $b_2c_1 - b_1c_2 = 0$, then $x = \frac{0}{0}$. And, as in (i.), $y = \frac{0}{0}$.

But in this case the value of y is not truly indeterminate.

For from $b_2c_1 - b_1c_2 = 0$, we obtain

$$c_1 = \frac{b_1c_2}{b_2}, \text{ since } b_2 \neq 0.$$

Substituting this value of c_1 in the expression for y, we obtain

$$y = \frac{a_1c_2 - a_2 \times \frac{b_1c_2}{b_2}}{a_1b_2 - a_2b_1} = \frac{c_2(a_1b_2 - a_2b_1)}{b_2(a_1b_2 - a_2b_1)}$$
$$= \frac{c_2}{b_2}.$$

16. $b_1 = 0$ and $b_2 = 0$. This case is similar to that in Art. 15.

(i.) If $a_1c_2 - a_2c_1 \neq 0$, then $y = \infty$, and $x = \frac{0}{0}$.

(ii.) If $a_1c_2 - a_2c_1 = 0$, then $y = \frac{0}{0}$, and $x = \frac{c_2}{a_2}$.

17. $a_1 = 0$ and $b_1 = 0$. In this case $x = \infty$, and $y = \infty$.

18. $a_2 = 0$ and $b_2 = 0$. In this case $x = \infty$, and $y = \infty$.

Linear Equations in Three or More Unknown Numbers.

19. The following examples will illustrate the methods for solving systems of three linear equations in three unknown numbers:

Ex. 1. Solve the system

$$\left.\begin{array}{lr} x+y-z=6, & (1)\\ x+y+z=12, & (2)\\ x-3y-z=10. & (3)\end{array}\right\}$$

To eliminate z we proceed as follows:

Adding (1) and (2), and dividing by 2, $x+y=9$. (4)

Adding (2) and (3), and dividing by 2, $x-y=11$. (5)

We thus obtain a system of two equations in the same two unknown numbers. From equations (4) and (5), we have

$$x=10, \quad (6)$$

$$y=-1. \quad (7)$$

Substituting these values for x and y in equation (1),

$$10-1-z=6, \text{ or } z=3.$$

Notice that (1), (2), (3); (1), (4), (5); and (1), (6), (7) form equivalent systems. Therefore, the required solution is 10, -1, 3.

Ex. 2. Solve the system

$$2x-3y+5z=11, \quad (1)$$

$$5x+4y-6z=-5, \quad (2)$$

$$-4x+7y-8z=-14. \quad (3)$$

To eliminate x, we proceed as follows:

Multiplying (1) by 5, $10x-15y+25z=55$. (4)

Multiplying (2) by 2, $10x+8y-12z=-10$. (5)

Subtracting (4) from (5), $23y-37z=-65$. (6)

Multiplying (1) by 2, $4x-6y+10z=22$. (7)

Adding (3) and (7), $y+2z=8$. (8)

Solving (6) and (8), $y=2$,

$z=3$.

Substituting 2 for y and 3 for z in (1), $x=1$.

Notice that (1), (2), (3); and (1), (6), (8) form equivalent systems. Consequently the required solution is 1, 2, 3.

Ex. 3. Solve the system

$$ay - cz = 0, \tag{1}$$

$$z - x = -b, \tag{2}$$

$$ax + by = a^2 + b(a + c). \tag{3}$$

Notice that by eliminating z from (1) and (2) we obtain an equation in x and y, which with equation (3) gives a system of two equations in the same two unknown numbers.

Solving (2) for z, $\quad z = x - b. \quad (4)$

Substituting $x - b$ for z in (1),

$$ay - cx + cb = 0. \tag{5}$$

Multiplying (3) by a, $\quad a^2x + aby = a^3 + a^2b + abc. \quad (6)$

Multiplying (5) by b, $\quad -bcx + aby = -b^2c. \quad (7)$

Subtracting (7) from (6), $\quad (a^2 + bc)x = a^3 + a^2b + abc + b^2c$

$$= a^2(a + b) + bc(a + b)$$

$$= (a^2 + bc)(a + b); \quad (8)$$

whence $\quad x = a + b.$

Substituting $a + b$ for x in (4), $\quad z = a.$

Substituting a for z in (1), $\quad y = c.$

To solve three simultaneous equations in three unknown numbers, eliminate one of the unknown numbers from any two of the equations; next eliminate the same unknown number from the third equation and either of the other two. Two equations in the same two unknown numbers are thus derived.

Solve these equations for the two unknown numbers, and substitute the values thus obtained in the simplest equation which contains the third unknown number.

20. From four equations in four unknown numbers, we can by eliminating one of the unknown numbers obtain three equations in three unknown numbers. We then solve these equations for the three unknown numbers and substitute the values thus obtained in the simplest equation which contains the fourth unknown number.

Ex. 1. Solve the system $4x - 3y + 2z - u = 40,$ (1)

$$5x + 4y - 3z - 2u = 76, \quad (2)$$

$$6x + y - 5z - u = 9, \quad (3)$$

$$x + y - z - u = 42. \quad (4)$$

Since the coefficients of u in the equations are the simplest, we first eliminate u.

Multiplying (1) by 2, $8x - 6y + 4z - 2u = 80.$ (5)

Subtracting (2) from (5), $3x - 10y + 7z = 4.$ (6)

Subtracting (1) from (3), $2x + 4y - 7z = -31.$ (7)

Subtracting (3) from (4), $-5x + 4z = 33.$ (8)

We have thus obtained three equations in the three unknown numbers, x, y, z. Notice that we could have obtained a third equation [instead of (8)] by subtracting (1) from (4); but by subtracting (3) from (4), we eliminated both y and u, and thus obtained a somewhat simpler system in x, y, z. We have now only to eliminate y from (6) and (7) to obtain a system of two equations in x and z.

Multiplying (6) by 2, $6x - 20y + 14z = 8.$ (9)

Multiplying (7) by 5, $10x + 20y - 35z = -155.$ (10)

Adding (9) and (10), $16x - 21z = -147.$ (11)

We now have the system (8) and (11) in x and z.

From (8) and (11), $x = -\frac{105}{41},$ (12)

$$z = \frac{207}{41}. \quad (13)$$

From (6), $y = \frac{97}{41}.$ (14)

From (4), $u = -\frac{1937}{41}.$ (15)

It is frequently possible to shorten the work by employing some simple device.

Ex. 2. Solve the system $x + y + z = a,$ (1)

$$x + y + u = b, \quad (2)$$

$$x + z + u = c, \quad (3)$$

$$y + z + u = d. \quad (4)$$

Adding all four equations, we obtain

$$x+y+z+u=\frac{a+b+c+d}{3}. \tag{5}$$

Subtracting from (5), the given equations in turn, we have

from (1) and (5), $u=\frac{b+c+d-2a}{3}$;

from (2) and (5), $z=\frac{a-2b+c+d}{3}$;

from (3) and (5), $y=\frac{a+b-2c+d}{3}$;

from (4) and (5), $x=\frac{a+b+c-2d}{3}$.

Number of Solutions of a System of Linear Equations.

21. The examples of the preceding articles illustrate the following principles:

(i.) *A system of n independent and consistent linear equations, in n unknown numbers, has one, and only one, determinate solution.*

From the given system a system of $n-1$ equations in $n-1$ unknown numbers can be derived by eliminating one of the unknown numbers. By eliminating from the latter system another unknown number, a second system of $n-2$ equations in $n-2$ unknown numbers is derived; and so on.

Finally, a single equation in one unknown number is obtained.

By the principles of equivalent equations the given system is equivalent to a second system which contains the following equations: any one of the given equations in n unknown numbers, any one of the $n-1$ derived equations in $n-1$ unknown numbers, and so on, to any one of the three derived equations in three unknown numbers, either of the two derived equations in two unknown numbers, and the last derived equation in one unknown number.

The last equation in one unknown number has one, and only one, definite solution. If the value of this unknown number be substituted in the next to the last equation of the second system described above, one and only one definite value for a second unknown number is obtained. If the values of these two unknown numbers be substituted in the equation in three unknown numbers, one, and only one, definite value of a third unknown number is obtained; and so on.

Consequently the given system is satisfied by one, and only one, definite set of values of the unknown numbers.

(ii.) *A system of n independent linear equations, in more than n unknown numbers, has an indefinite number of solutions.*

For, by each elimination of an unknown number, we derive a set of equations, one less in number, and containing one less unknown number.

Finally, as in (i.), we obtain a single equation. But since the original system contained more unknown numbers than equations, the last derived equation will contain more than one unknown number. Since this equation, therefore, has an indefinite number of solutions, we conclude that the given system has likewise an indefinite number of solutions.

(iii.) *A system of n independent linear equations, in less than n unknown numbers, does not have a determinate finite solution.*

For, if we take from the given system as many equations as there are unknown numbers, the system formed by these equations will have by (i.) one, and only one, definite solution.

But since the other equations of the given system are independent of the equations selected, that is, express independent relations between the unknown numbers, they cannot be satisfied by this solution.

Therefore, the given system cannot be satisfied by any one definite set of values of the unknown numbers.

EXERCISES V.

Solve the following systems of equations:

1. $\begin{cases} x+y=28, \\ x+z=30, \\ y+z=32. \end{cases}$

2. $\begin{cases} x+y=2c, \\ x+z=2b, \\ y+z=2a. \end{cases}$

3. $\begin{cases} x-y=2, \\ y-z=3, \\ x+z=9. \end{cases}$

4. $\begin{cases} 3x-y=7, \\ 3y-z=5, \\ 3z-x=0. \end{cases}$

5. $\begin{cases} 3x+5y=35, \\ 3y+5z=27, \\ 3z+5x=34. \end{cases}$

6. $\begin{cases} x+y+z=50, \\ y=3x-21, \\ z=4x-33. \end{cases}$

7. $\begin{cases} 3x+2y-4z=15, \\ 5x-3y+2z=28, \\ 3y+4z-x=24. \end{cases}$

8. $\begin{cases} x+y-z=1, \\ 8x+3y-6z=1, \\ 3z-4x-y=1. \end{cases}$

9. $\begin{cases} x+y-z=c, \\ x+z-y=b, \\ y+z-x=a. \end{cases}$

10. $\begin{cases} 4x-3y+2z=9, \\ 2x+5y-3z=4, \\ 5x+6y-2z=18. \end{cases}$

11. $\begin{cases} 2x - 4y + 9z = 28, \\ 7x + 3y - 5z = 3, \\ 9x + 10y - 11z = 4. \end{cases}$

12. $\begin{cases} x - 2y + 3z = 6, \\ 2x + 3y - 4z = 20, \\ 3x - 2y + 5z = 26. \end{cases}$

13. $\begin{cases} \dfrac{x}{6} + \dfrac{y}{9} + \dfrac{z}{10} = 9, \\ \dfrac{x}{3} + \dfrac{y}{2} - \dfrac{z}{25} = 11, \\ \dfrac{x}{2} - \dfrac{y}{18} + \dfrac{z}{10} = 10. \end{cases}$

14. $\begin{cases} \dfrac{x}{a+b} + \dfrac{y}{b+c} = b - a, \\ \dfrac{y}{c-a} + \dfrac{z}{c+a} = c + a, \\ \dfrac{x}{b-c} - \dfrac{z}{a-b} = b - c. \end{cases}$

15. $\begin{cases} x+y+z=a+b+c, \\ bx+cy+az=a^2+b^2+c^2, \\ cx+ay+bz=a^2+b^2+c^2. \end{cases}$

16. $\begin{cases} x + y + z = A, \\ ax + by + cz = 0, \\ a^2x + b^2y + c^2z = 0. \end{cases}$

17. $\begin{cases} (c+a)x - (c-a)y = 2bc, \\ (a+b)y - (a-b)z = 2ac, \\ (b+c)z - (b-c)x = 2ab. \end{cases}$

18. $\begin{cases} x + y + z = (a+b+c)^2, \\ ay + bz + cx = 3(ab^2 + bc^2 + ca^2), \\ ax + by + cz = a^3 + b^3 + c^3 + 6abc. \end{cases}$

19. $\begin{cases} x + y + z = 6, \\ x + y + u = 7, \\ x + z + u = 8, \\ y + z + u = 9. \end{cases}$

20. $\begin{cases} x + y + z - u = 11, \\ x + y - z + u = 17, \\ x - y + z + u = 9, \\ -x + y + z + u = 12. \end{cases}$

21. $\begin{cases} 7x - 2z + 3u = 17, \\ 4y - 2z + t = 15, \\ 5y - 3x - 2u = 8, \\ 4y - 3u + 2t = 17, \\ 3z + 8u = 33. \end{cases}$

22. $\begin{cases} 3x-4y+3z+3v-6u=11, \\ 3x-5y+2z-4u=11, \\ 10y-3z+3u-2v=2, \\ 5z+4u+2v-2x=3, \\ 6u-3v+4x-2y=6. \end{cases}$

§ 4. SYSTEMS OF FRACTIONAL EQUATIONS.

1. If some or all of the equations of a system be fractional, and lead, when cleared of fractions, to linear equations, the solution of the system can be obtained by the methods of the preceding paragraph.

Any solution of the linear system which is derived by clearing of fractions, is a solution of the given system, unless it is a solution of the L. C. D. (equated to 0) of one or more of the fractional equations. (See Ch. X., Art. 5.)

Ex. Solve the system $6x - 5y = 0,$ (1)

$$\frac{6x+1}{4y+5} = \frac{13}{11}. \qquad (2)$$

Clearing (2) of fractions, $66x + 11 = 52y + 65.$ (3)

Transferring and uniting terms in (3), and dividing by 2,

$$33x - 26y = 27. \qquad (4)$$

The solution of (1) and (4) is 15, 18.

In clearing (2) of fractions, we multiplied by $11(4y+5)$.

Since $x = 15$, $y = 18$ is not a solution of $4y + 5 = 0$, it is a solution of the given system.

2. When the equations of a system contain only the reciprocals of the unknown numbers, they can be solved directly for these reciprocals.

Ex. 1. Solve the system $\frac{2}{x} + \frac{3}{y} = 2,$ (1)

$$\frac{4}{x} - \frac{5}{z} = \frac{3}{4}, \qquad (2)$$

$$\frac{5}{y} + \frac{3}{z} = \frac{29}{12}. \qquad (3)$$

We will solve the system for $\frac{1}{x}, \frac{1}{y}, \frac{1}{z}$.

Multiplying (1) by 2, $\frac{4}{x} + \frac{6}{y} = 4.$ (4)

Subtracting (2) from (4), $\frac{6}{y}+\frac{5}{z}=\frac{13}{4}$. (5)

We now solve (3) and (5) for $\frac{1}{y}$ and $\frac{1}{z}$.

Multiplying (3) by 5, $\frac{25}{y}+\frac{15}{z}=\frac{145}{12}$. (6)

Multiplying (5) by 3, $\frac{18}{y}+\frac{15}{z}=\frac{39}{4}$. (7)

Subtracting (7) from (6), $\frac{7}{y}=\frac{28}{12}$; (8)

whence $\frac{1}{y}=\frac{1}{3}$.

Substituting $\frac{1}{3}$ for $\frac{1}{y}$ in (1), $\frac{2}{x}+1=2$;

whence $\frac{1}{x}=\frac{1}{2}$.

Substituting $\frac{1}{2}$ for $\frac{1}{x}$ in (2), $2-\frac{5}{z}=\frac{3}{4}$;

whence $\frac{1}{z}=\frac{1}{4}$

The required solution is, therefore, 2, 3, 4.

Ex. 2. Solve the system

$$\left.\begin{aligned} x+y&=xy, & (1)\\ 2x+2z&=xz, & (2)\\ 3y+3z&=yz. & (3)\end{aligned}\right\} \quad \text{(I.)}$$

Observe that the given equations are neither linear nor fractional. Yet they can be transformed so that they will contain only the reciprocals of x, y, and z.

$$\left.\begin{aligned} &\text{Dividing (1) by } xy, & \frac{1}{y}+\frac{1}{x}&=1. & (4)\\ &\text{Dividing (2) by } xz, & \frac{2}{z}+\frac{2}{x}&=1. & (5)\\ &\text{Dividing (3) by } yz, & \frac{3}{z}+\frac{3}{y}&=1. & (6)\end{aligned}\right\} \quad \text{(II.)}$$

Multiplying (4) by 2, $\frac{2}{y}+\frac{2}{x}=2.$ (7)

Subtracting (5) from (7), $\frac{2}{y}-\frac{2}{z}=1.$ (8)

Solving (6) and (8) for $\frac{1}{y}$ and $\frac{1}{z}$, $\frac{1}{y}=\frac{5}{12}$,

$$\frac{1}{z}=-\frac{1}{12}.$$

Substituting $\frac{5}{12}$ for $\frac{1}{y}$ in (4), $\frac{1}{x}=\frac{7}{12}.$

Consequently, a solution of the given system is $\frac{12}{7}$, $\frac{12}{5}$, -12.

It is important to notice that we cannot assume that the system (II.) is equivalent to the system (I.), since the equations of (II.) are derived from the equations of (I.) by dividing by expressions which contain the unknown numbers.

But if any solution of (I.) be lost by this transformation, it must be a solution of the expressions (equated to 0) by which the equations of (I.) were divided; that is, of

$$\left.\begin{aligned} xy&=0,\\ xz&=0,\\ yz&=0.\end{aligned}\right\} \qquad \text{(III.)}$$

The system (III.) has the solution 0, 0, 0, and this solution evidently satisfies the system (I.).

We therefore conclude that the given system has the two solutions $\frac{12}{7}$, $\frac{12}{5}$, -12, and 0, 0, 0.

Ex. 3. Solve the system

$$\frac{3}{x+y+z}+\frac{6}{2x-y}+\frac{1}{y-3z}=1, \qquad (1)$$

$$\frac{6}{x+y+z}+\frac{4}{2x-y}-\frac{1}{y-3z}=3, \qquad (2)$$

$$\frac{15}{x+y+z}-\frac{2}{2x-y}-\frac{3}{y-3z}=5. \qquad (3)$$

This system can be readily solved by making the following substitutions:

Let $$\frac{3}{x+y+z}=u,\ \frac{2}{2x-y}=v,\ \frac{1}{y-3z}=w. \qquad \text{(I.)}$$

Then the given system becomes

$$u+3v+w=1, \qquad (4)$$

$$2u+2v-w=3, \qquad (5)$$

$$5u-v-3w=5. \qquad (6)$$

Solving equations (4), (5), and (6), we obtain

$$u=\tfrac{1}{2},\ v=\tfrac{1}{2},\ w=-1.$$

Substituting these values in the system (I.), we have

$$x+y+z=6, \qquad (7)$$

$$2x-y=4, \qquad (8)$$

$$y-3z=-1. \qquad (9)$$

Solving equations (7), (8), and (9), we obtain

$$x=3,\ y=2,\ z=1.$$

3. As in a system of integral equations, so in a system of fractional equations, the equations must be consistent and independent.

Ex. Solve the system

$$3x+4y=11, \qquad (1)$$

$$\frac{1}{x-1}+\frac{1}{y-2}=0. \qquad (2)$$

Clearing (2) of fractions and uniting terms,

$$x+y=3. \qquad (3)$$

Solving (1) and (3), $x=1,\ y=2.$

These values constitute a solution of equations (1) and (3), but not of (1) and (2).

For they form a solution of the L. C. D. (equated to 0) of the fractions in (2); that is, of

$$(x-1)(y-2)=0. \qquad (4)$$

We conclude, therefore, that equations (1) and (2) do not have any solution.

It can, in fact, be shown that they are inconsistent. For (1) is equivalent to

$$3x - 3 + 4y - 8 = 0,$$

or to

$$3(x-1) + 4(y-2) = 0.$$

Dividing both members of the last equation by $(x-1)(y-2)$, we obtain

$$\frac{3}{y-2} + \frac{4}{x-1} = 0. \qquad (5)$$

Equation (5) is evidently inconsistent with (2).

It should be noticed that in clearing (2) of fractions no unnecessary factor was used. The explanation of the apparent contradiction of the principle proved in Ch. X., Art. 4, is that this principle holds only when the fractional equation contains but one unknown number.

EXERCISES VI.

Solve the following systems of equations:

1. $\begin{cases} \dfrac{1}{x} + \dfrac{1}{y} = \dfrac{1}{2}, \\ \dfrac{1}{x} - \dfrac{1}{y} = \dfrac{1}{6}. \end{cases}$

2. $\begin{cases} \dfrac{3}{x} + \dfrac{5}{y} = 2, \\ \dfrac{9}{x} - \dfrac{10}{y} = 1. \end{cases}$

3. $\begin{cases} 7x - \dfrac{3}{y} = 16, \\ 3x - \dfrac{2}{y} = 4. \end{cases}$

4. $\begin{cases} \dfrac{4}{x} - 3y = 8, \\ \dfrac{5}{x} - 6y = 1. \end{cases}$

5. $\begin{cases} \dfrac{14}{x} + \dfrac{25}{y} = 7, \\ \dfrac{49}{x} - \dfrac{50}{y} = -3. \end{cases}$

6. $\begin{cases} \dfrac{1}{x} + \dfrac{1}{y} = a, \\ \dfrac{1}{x} - \dfrac{1}{y} = b. \end{cases}$

7. $\begin{cases} \dfrac{p}{x} + \dfrac{q}{y} = a, \\ \dfrac{q}{x} + \dfrac{p}{y} = b. \end{cases}$

8. $\begin{cases} 6y - 6x = xy, \\ 10y + 3x = 6xy. \end{cases}$

9. $\begin{cases} \dfrac{3}{x-4} + \dfrac{4}{y-1} = 3, \\ \dfrac{9}{x-4} - \dfrac{2}{y-1} = 2. \end{cases}$

10. $\begin{cases} \dfrac{3}{2x-3y}+\dfrac{5}{y-2}=8, \\ \dfrac{7}{2x-3y}+\dfrac{3}{y-x}=10. \end{cases}$

11. $\begin{cases} \dfrac{3x+7}{10y+3}=1, \\ \dfrac{12x+5}{7y+1}=2. \end{cases}$

12. $\begin{cases} \dfrac{7}{2x+3y}=\dfrac{11}{2x-3y}, \\ \dfrac{x}{10y-7}=\dfrac{9}{10}. \end{cases}$

13. $\begin{cases} \dfrac{x+y-1}{x-y+1}=a, \\ \dfrac{y-x+1}{x-y+1}=ab. \end{cases}$

14. $\begin{cases} \dfrac{x+a-b}{y-a-b}=\dfrac{x-b}{y-a}, \\ \dfrac{b}{x-a}=\dfrac{a}{y+b}. \end{cases}$

15. $\begin{cases} x-\dfrac{2y-x}{23-x}=20-\dfrac{59-2x}{2}, \\ y+\dfrac{y-3}{x-18}=30-\dfrac{73-3y}{3}. \end{cases}$

16. $\begin{cases} \dfrac{2n}{x+ny}-\dfrac{1}{n-ny}=1, \\ \dfrac{10n}{x+ny}+\dfrac{3}{n-ny}=1. \end{cases}$

17. $\begin{cases} \dfrac{a}{2x+ay}+\dfrac{a+1}{ax-2y}=\dfrac{2a+3}{5a(a-1)}, \\ \dfrac{a^2}{2x+ay}-\dfrac{a(a-1)}{ax-2y}=\dfrac{2a-3}{5(a+1)}. \end{cases}$

18. $\begin{cases} \dfrac{c}{x}+\dfrac{n}{y}=\dfrac{cn}{xy}, \\ \dfrac{n}{x-c}+\dfrac{c}{y-n}=\dfrac{cn}{xy-nx-cy+cn}. \end{cases}$

19. $\begin{cases} xy=3\frac{1}{3}(y+2x), \\ yz=2\frac{1}{2}(3z+4y), \\ xz=1\frac{5}{7}(5z+6x). \end{cases}$

20. $\begin{cases} \dfrac{1}{x}+\dfrac{1}{y}=5, \\ \dfrac{1}{x}+\dfrac{1}{z}=6, \\ \dfrac{1}{y}+\dfrac{1}{z}=7. \end{cases}$

21. $\begin{cases} \dfrac{1}{z}+\dfrac{1}{y}=a, \\ \dfrac{1}{z}+\dfrac{1}{x}=b, \\ \dfrac{1}{x}+\dfrac{1}{y}=c. \end{cases}$

22. $\begin{cases} \dfrac{1}{x}+\dfrac{2}{y}+\dfrac{2}{z}=16, \\ \dfrac{1}{y}+\dfrac{2}{z}+\dfrac{2}{x}=15, \\ \dfrac{1}{z}+\dfrac{2}{x}+\dfrac{2}{y}=14. \end{cases}$

EXERCISES VII.

Solve the following systems of equations by the methods given in this chapter:

1. $\begin{cases} 65x+68y=-3, \\ 39x-119y=158. \end{cases}$

2. $\begin{cases} 125x+252y=53, \\ 85x+243y=44. \end{cases}$

3. $\begin{cases} 22x-46y=126, \\ 25x+69y=391. \end{cases}$

4. $\begin{cases} a(x+y)-b(x-y)=2a^2, \\ (a^2-b^2)(x-y)=4a^2b. \end{cases}$

5. $\begin{cases} \dfrac{x}{n+1} + \dfrac{y}{n-1} = \dfrac{1}{n-1}, \\[2ex] \dfrac{x}{n-1} + \dfrac{y}{n+1} = \dfrac{1}{n^2-1}. \end{cases}$

6. $\begin{cases} \dfrac{x-1}{a} = \dfrac{1-y}{b}, \\[2ex] \dfrac{x+y}{a^2+b^2} = \dfrac{x-y}{a^2-b^2}. \end{cases}$

7. $\begin{cases} \dfrac{2z-y}{3z+y} = \dfrac{5}{11}, \\[2ex] \dfrac{5z-3y}{y+z} = \dfrac{9}{5}. \end{cases}$

8. $\begin{cases} \dfrac{x-1}{y-1} = \dfrac{3}{4}, \\[2ex] \dfrac{x+3}{y+3} = \dfrac{10}{13}. \end{cases}$

9. $\begin{cases} 5\frac{1}{4}x - 9\frac{1}{3}y = 21, \\ 5\frac{3}{5}x - 4\frac{2}{3}y = 70. \end{cases}$

10. $\begin{cases} \dfrac{1}{2}(x+y) = 1 + \dfrac{x-y}{2a}, \\[2ex] \dfrac{a}{2}(x-y) = 1 + \dfrac{x-y}{2a}. \end{cases}$

11. $\begin{cases} \dfrac{ax+by}{2} + x = \dfrac{a+1}{a}, \\[2ex] \dfrac{ax+by}{2} + y = \dfrac{b+1}{b}. \end{cases}$

12. $\begin{cases} ax - by = \dfrac{a^2+b^2}{2}, \\[2ex] (a-b)x - (a+b)y = 0. \end{cases}$

13. $\begin{cases} a^2x - b^2y = 0, \\ (a^2+b^2)x + (a^2-b^2)y = a^4 + b^4. \end{cases}$

14. $\begin{cases} (a+b)x + (a-b)y = a^2 + b^2, \\ (a-b)x + (a+b)y = a^2 - b^2. \end{cases}$

15. $\begin{cases} x + y = z + 10, \\ y = 2x - 13, \\ z = 2y - 11. \end{cases}$

16. $\begin{cases} yz = 2(y+z), \\ xz = 3(x+z), \\ xy = 4(x+y). \end{cases}$

17. $\begin{cases} x - \dfrac{3x+5y}{17} + 17 = 5y + \dfrac{4x+7}{3}, \\[2ex] \dfrac{22-6y}{3} - \dfrac{5x-7}{11} = \dfrac{x+1}{6} - \dfrac{8y+5}{18}. \end{cases}$

18. $\begin{cases} x - \dfrac{2x+7y}{15} + 17 = 5y + \dfrac{4x+7}{3}, \\[2ex] \dfrac{22-6y}{3} - \dfrac{5x-28}{4} = \dfrac{x+1}{6} - \dfrac{8y+5}{18}. \end{cases}$

19. $\begin{cases} \frac{1}{2}(a+b-c)x + \frac{1}{2}(a-b+c)y = a^2 + (b-c)^2, \\ \frac{1}{2}(a-b+c)x + \frac{1}{2}(a+b+c)y = a^2 - (b-c)^2. \end{cases}$

20. $\begin{cases} \dfrac{x}{n^2-1} - \dfrac{y}{a^2-1} = a^2 - n^2, \\ \dfrac{x}{a^2+1} - \dfrac{y}{n^2+1} = a^2 + n^2 - 2. \end{cases}$

21. $\begin{cases} \dfrac{a-1}{a^2y - 2ay} - \dfrac{x+y}{2y} = \dfrac{1}{a}, \\ \dfrac{x}{2a} + \dfrac{y}{2a-4} = \dfrac{a+1}{a^3-4a}. \end{cases}$

22. $\begin{cases} \dfrac{1-9x}{1-y} = -4, \\ \dfrac{x}{2x-10} + \dfrac{5y}{30-6x} = \dfrac{1}{20-4x}. \end{cases}$

23. $\begin{cases} \dfrac{y-6}{x-4} - \dfrac{10}{16-x^2} = \dfrac{y+6}{x+4}, \\ \dfrac{5}{x^2-3x} + \dfrac{3}{3y-xy} = -\dfrac{10}{xy}. \end{cases}$

24. $\begin{cases} x + \dfrac{y}{2} = 1, \\ y + \dfrac{z}{3} = 1, \\ z + \dfrac{x}{4} = 1. \end{cases}$

25. $\begin{cases} \dfrac{1}{x} + \dfrac{1}{y} - \dfrac{1}{z} = a, \\ \dfrac{1}{x} - \dfrac{1}{y} + \dfrac{1}{z} = b, \\ \dfrac{1}{x} - \dfrac{1}{y} - \dfrac{1}{z} = -c. \end{cases}$

26. $\begin{cases} (x-1)(4y+3) = (4x-8)(y+2), \\ (x-2)(3z+1) = (3x-8)(z+1), \\ (y+1)(2z+3) = (2y+1)(z+2). \end{cases}$

27. $\begin{cases} (2x-9)(3y-7) = (3x-11)(2y-6), \\ (3x-1)(4z-1) = (4x+2)(3z-2), \\ (4y-5)(5z-4) = (5y-9)(4z-1). \end{cases}$

28. $\begin{cases} \dfrac{xy}{x+y}=a, \\ \dfrac{xz}{x+z}=b, \\ \dfrac{yz}{y+z}=c. \end{cases}$

29. $\begin{cases} \dfrac{xy}{ay+bx}=1, \\ \dfrac{yz}{cz+dy}=1, \\ \dfrac{xz}{ez+fx}=1. \end{cases}$

30. $xyz=a(yz-zx-xy)=b(zx-xy-yz)=c(xy-yz-zx).$

31. $\begin{cases} x+ay+a^2z+a^3=0, \\ x+by+b^2z+b^3=0, \\ x+cy+c^2z+c^3=0. \end{cases}$

32. $\begin{cases} ax+by+cz=A, \\ a^2x+b^2y+c^2z=A^2, \\ a^3x+b^3y+c^3z=A^3. \end{cases}$

33. $\begin{cases} x+y+z=a+b+c, \\ bx+cy+az=cx+ay+bz=0. \end{cases}$

34. $\begin{cases} \dfrac{bx+ay}{c}=\dfrac{a-b}{(a-c)(b-c)}, \\ \dfrac{cy+bz}{a}=\dfrac{b-c}{(b-a)(c-a)}, \\ \dfrac{az+cx}{b}=\dfrac{c-a}{(a-b)(c-b)}. \end{cases}$

35. $\begin{cases} \dfrac{x}{a+b}+\dfrac{y}{b-c}+\dfrac{z}{c+a}=2c, \\ \dfrac{x}{a-b}-\dfrac{y}{b-c}+\dfrac{z}{c-a}=2a, \\ \dfrac{x}{a-b}-\dfrac{y}{b-c}-\dfrac{z}{c+a}=2a-2c. \end{cases}$

36. $\begin{cases} x+y+z=a+b+c, \\ c(x-y)+a(y-z)+b(z-x)=0, \\ b(x+y-c-a)+c(y+z-a-b)+a(z+x-b-c)=0. \end{cases}$

37. $\begin{cases} (4-x)(244-y)=z, \\ (7-x)(124-y)=z, \\ (13-x)(64-y)=z. \end{cases}$

38. $\begin{cases} xy-yz+xz=4xyz, \\ 3xy+2yz-6xz=-xyz, \\ 4xy+3yz+2xz=19xyz. \end{cases}$

39. $\begin{cases} x+y+z=A, \\ (b+c)x+(c+a)y+(a+b)z=0, \\ bcx+cay+abz=0. \end{cases}$

40. $\begin{cases} (z+x)a-(z-x)b=2yz, \\ (x+y)b-(x-y)c=2xz, \\ (y+z)c-(y-z)a=2xy. \end{cases}$

41. $\begin{cases} x+y+z=0, \\ (b+c)x+(c+a)y+(a+b)z=0, \\ bcx+cay+abz=1. \end{cases}$

42. $\begin{cases} \dfrac{x}{a^2-2a}+\dfrac{y}{a^2+ab+2a}=\dfrac{a^2+2(a+b)}{a^2-2b+ab-4}, \\ \dfrac{1}{bx^2-x}=\dfrac{b(1-ay)+\dfrac{1}{x}}{b^3x^3-1}. \end{cases}$

43. $\begin{cases} \dfrac{n^2}{ax+a}=\dfrac{1}{xy-2ax-2a+y}+\dfrac{1}{2a-y}, \\ a\left[\dfrac{a}{n}(x-n^2)+1\right]=y. \end{cases}$

44. $\begin{cases} \dfrac{n}{x-n^2y+1}=\dfrac{x-n^2y+n}{n^3y-(n-1)(x+1)-n^2y}, \\ \dfrac{xy}{2n^2x-n^2x^2-n^2}=-\dfrac{1}{1-\dfrac{1}{x}}. \end{cases}$

45. $\begin{cases} \dfrac{2x-1}{2x+1}+\dfrac{8x^2+3y}{8x^3+1}=1, \\ \dfrac{18}{x^3+2x^2+x}-\dfrac{1}{xy-x^2}=\dfrac{x}{x^3-(y-2)x^2-(2y-1)x-y}. \end{cases}$

46. $\begin{cases} \dfrac{1}{x+\dfrac{1}{y-\dfrac{a}{x}}}=\dfrac{1}{x-\dfrac{1}{y-\dfrac{b}{x}}}. \\ \dfrac{1}{y}\left(1-\dfrac{1}{x}\right)=1. \end{cases}$

47. $\begin{cases} \dfrac{1}{1-x+y}-\dfrac{1}{x+y-1}=\dfrac{2}{3}, \\ \dfrac{1}{\dfrac{1}{1-x+y}-\dfrac{1}{1-x-y}}=\dfrac{3}{4}. \end{cases}$

48. $\begin{cases} \dfrac{8}{2x-3y+17}+5x-8y+39=0, \\ \dfrac{5}{2x-3y+17}+16y-10x=88\frac{1}{2}. \end{cases}$

49. $\begin{cases} (a-b)x+(b-c)y+(c-a)z=2(a^2+b^2+c^2-ab-ac-bc), \\ (a-b)y+(b-c)z+(c-a)x=ab+ac+bc-a^2-b^2-c^2, \\ x+y+z=0. \end{cases}$

50. $\begin{cases} yz+xz+xy=xyz, \\ yu+xu+xy=xyu, \\ zu+xu+xz=xuz, \\ zu+yu+yz=uyz. \end{cases}$

51. $\begin{cases} a^4+a^3x+a^2y+az+u=0, \\ b^4+b^3x+b^2y+bz+u=0, \\ c^4+c^3x+c^2y+cz+u=0, \\ d^4+d^3x+d^2y+dz+u=0. \end{cases}$

52. $\begin{cases} x+y+z+u=16, \\ x+y+z+v=18, \\ x+y+u+v=20, \\ x+z+u+v=22, \\ y+z+u+v=24. \end{cases}$

53. $\begin{cases} 2u-3v=2a-7b+2c, \\ v+2z=7b, \\ 3z+y=3a+6b, \\ 4y-2x=8a, \\ 3x-5u=a-5b-5c. \end{cases}$

54. $\begin{cases} x+y+z+u+v=25, \\ 2x+3y+4z-5u-6v=14, \\ 3x-5y+6z-2u+3v=22, \\ 4x+5y+6z+2u+3v=105, \\ 5x+6y+2z+3u+4v=105. \end{cases}$

CHAPTER XIV.

PROBLEMS.

1. As was stated in Ch. V., Art. 2 (iii.), every problem which can be solved must contain as many conditions, expressed or implied, as there are required numbers. In solving a problem by means of one equation in one unknown number, one of the required numbers was usually, though not always, taken as the unknown number of the equation. All but one of the conditions of the problem were used to express the other required numbers in terms of the one selected as the unknown number. The remaining condition then furnished the equation of the problem.

But a problem which contains more than one condition can be solved by means of a system of equations in which the unknown numbers are usually, though not always, the required numbers of the problem. Each condition then furnishes an equation. The solution of the system of equations thus obtained gives the solution of the problem, if the conditions of the latter be consistent.

2. We will first solve by means of a system of two equations one of the problems which was solved in Ch. V. by means of one equation in one unknown number.

Pr. **1.** (Pr. 5, Ch. V.) At an election at which 943 votes were cast, A and B were candidates. A received a majority of 65 votes. How many votes were cast for each candidate?

Let x stand for the number of votes cast for A,
and y for the number of votes cast for B.

Then, by the first condition

$$x + y = 943; \tag{1}$$

and by the second condition

$$x - y = 65. \tag{2}$$

Solving (1) and (2), we obtain

$x = 504$, the number of votes cast for A,

$y = 439$, the number of votes cast for B.

Notice that if we had substituted the value of y obtained from (1), namely $943 - x$, for y in (2), we should have obtained the equation of the solution in Ch. V.,

$$x - (943 - x) = 65.$$

Pr. 2. A, B, and C compared fortunes. A said to B: "Give me \$700, and I shall have twice as much as you will have left." B said to C: "Give me \$1400, and I shall have three times as much as you will have left." C said to A: "Give me \$420, and I shall have five times as much as you will have left."

Let x stand for number of dollars in A's fortune,

y for the number of dollars in B's,

and z for the number of dollars in C's.

We then have, from A's statement to B,

$$x + 700 = 2(y - 700); \qquad (1)$$

from B's statement to C,

$$y + 1400 = 3(z - 1400); \qquad (2)$$

from C's statement to A,

$$z + 420 = 5(x - 420). \qquad (3)$$

Solving equations (1), (2), and (3), we obtain

$$x = 980, \ y = 1540, \ z = 2380.$$

Pr. 3. A tank can be filled by two pipes. If the first be opened 6 minutes, and the second 7 minutes, the tank will be filled; or if the first be opened 3 minutes, and the second 12 minutes, the tank will be filled. In what time can each pipe fill the tank?

Let x stand for the number of minutes it takes the first pipe to fill the tank, and y for the number of minutes it takes the second pipe. Let the capacity of the tank be represented by 1.

Then in 1 minute the first pipe fills $\frac{1}{x}$ of the tank, and in 6 minutes $\frac{6}{x}$ of the tank. In like manner, the second pipe fills $\frac{7}{y}$ of the tank in 7 minutes.

Therefore, by the first condition,

$$\frac{6}{x}+\frac{7}{y}=1; \qquad (1)$$

and by the second condition

$$\frac{3}{x}+\frac{12}{y}=1. \qquad (2)$$

Solving equations (1) and (2), we obtain

$$x=10\tfrac{1}{5},\ y=17.$$

Pr. 4. A teacher required each of three pupils to multiply two given numbers. The first pupil, in adding the partial products, neglected to carry 1 from a certain column. To check his work he divided his product by the less number, and obtained a quotient 971 and a remainder 214. The second pupil neglected to carry 2 from the next column (to the left), and obtained by dividing his product by the less number a quotient 965 and a remainder 198. The third pupil neglected to carry 1 from the next column (always to the left), and obtained by his division a quotient 940 and a remainder 48. What were the two numbers, and from what column did each pupil neglect to carry?

The required numbers of the problem are the two numbers given by the teacher and the numbers of the columns from which the pupils neglected to carry. But the three last numbers will be known, if the error made by the first pupil (*i.e.*, the difference between his product and the correct product) in neglecting to carry 1 is known. Thus, if the first pupil made an error of 10, he must have neglected to carry 1 from the first column to the second; evidently in that case, the second and third pupils neglected to carry from the second and third columns, respectively.

Let x stand for the less number given by the teacher,
y for the greater,
and z for the error made by the first pupil.

Then the correct product is xy; the product obtained by the first pupil is $971x + 214$; that obtained by the second pupil is $965x + 198$; and that obtained by the third pupil is $940x + 48$.

The second pupil's error is $2 \times 10 \times z, = 20z$, since he neglected to carry 2 from the next column; and the third pupil's error is $1 \times 100 \times z, = 100z$, since he neglected to carry 1 from the next succeeding column.

The conditions of the problem are:

(1) *first pupil's product + his error = exact product;*

(2) *second pupil's product + his error = exact product;*

(3) *third pupil's product + his error = exact product.*

These three conditions give the following equations, respectively:

$$971x + 214 + z = xy, \quad (1)$$

$$965x + 198 + 20z = xy, \quad (2)$$

$$940x + 48 + 100z = xy. \quad (3)$$

Solving equations (1)–(3), we obtain

$$x = 314, \ y = 972, \ z = 100.$$

Therefore, the two numbers given by the teacher were 314 and 972; and the first pupil neglected to carry from the second column, the second pupil from the third column, and the third pupil from the fourth column.

EXERCISES I.

1. Find two numbers whose sum is 19 and whose difference is 7.

2. If one number be multiplied by 3 and another by 7, the sum of the products will be 58; if the first be multiplied by 7 and the second by 3, the sum will be 42. What are the numbers?

3. In a meeting of 48 persons, a motion was carried by a majority of 18. How many persons voted for the motion and how many against it?

4. If one of two numbers be divided by 6 and the other by 5, the sum of the quotients will be 52; if the first be divided by 8 and the second by 12, the sum of the quotients will be 31. What are the numbers?

5. Find two numbers, such that if 1 be subtracted from the first and added to the second the results will be equal; while if 5 be subtracted from the first and the second be subtracted from 5, these results will also be equal.

6. If 45 be subtracted from a number, the remainder will be a certain multiple of 5; but if the number be subtracted from 135, the remainder will be the same multiple of 10. What is the number, and what multiple of 5 is the first remainder?

7. If 1 be added to the numerator of a fraction, the resulting fraction will be equal to $\frac{1}{4}$; but if 1 be added to the denominator, the resulting fraction will be equal to $\frac{1}{5}$. What is the fraction?

8. If 1 be subtracted from the numerator and denominator of a certain fraction, the resulting fraction will be equal to $\frac{1}{3}$; but if 1 be added to the numerator and denominator of the same fraction, the resulting fraction will be equal to $\frac{1}{2}$. What is the fraction?

9. A said to B: "Give me three-fourths of your marbles and I shall have 100 marbles." B said to A: "Give me one-half of your marbles and I shall have 100 marbles." How many marbles had A and B?

10. A bag contains white and black balls. One-half of the number of white balls is equal to one-third of the number of black balls, and twice the number of white balls is 6 less than the total number of balls. How many balls of each color are there?

11. The sum of two numbers is 47. If the greater be divided by the less, the quotient and the remainder will each be 5. What are the numbers?

12. A father said to his son: "After 3 years I shall be three times as old as you will be, and 7 years ago I was seven times as old as you then were." What were the ages of father and son?

13. A merchant received from one customer $26 for 10 yards of silk and 4 yards of cloth; and from another customer $23 for 7 yards of silk and 6 yards of cloth at the same prices. What was the price of the silk and of the cloth?

14. A merchant has two kinds of wine. If he mix 9 gallons of the poorer with 7 gallons of the better, the mixture will be worth $1.37½ a gallon; but if he mix 3 gallons of the poorer with 5 gallons of the better, the mixture will be worth $1.45 a gallon. What is the price of each kind of wine?

15. A man has a gold watch, a silver watch, and a chain. The gold watch and the chain cost seven times as much as the silver watch; the cost of the chain and half the cost of the silver watch is equal to three-tenths of the cost of the gold watch. If the chain cost $40, what was the cost of each watch?

16. A and B make a purchase for $48. A gives all of his money, and B three-fourths of his. If A had given three-fourths of his money and B all of his, they would have paid $1.50 less. How much money had A and B?

17. A mechanic and an apprentice together receive $40. The mechanic works 7 days and the apprentice 12 days; and the mechanic earns in 3 days $7 more than the apprentice earns in 5 days. What wages does each receive?

18. I have 7 silver balls equal in weight and 12 gold balls equal in weight. If I place 3 silver balls in one pan of a balance and 5 gold balls in the other, I must add to the gold balls 7 ounces to maintain equilibrium. If I place in one pan

4 silver balls and in the other 7 gold balls, the balance is in equilibrium. What is the weight of each gold and of each silver ball?

19. A tank has two pumps. If the first be worked 2 hours and the second 3 hours, 1020 cubic feet of water will be discharged. But if the first be worked 1 hour and the second $2\frac{1}{2}$ hours, 690 cubic feet of water will be discharged. How many cubic feet of water can each pump discharge in 1 hour?

20. It was intended to distribute $25 among a certain number of the poor, each adult to receive $2.50 and each child 75 cents. But it was found that there were 3 more adults and 5 more children than was at first supposed. Each adult was therefore given $1.75 and each child 50 cents. How many adults and how many children were there?

21. A man ordered a wine-merchant to fill two casks of different sizes with wine, one at $1.20 and the other at $1.50 a quart, paying $88.50 for both casks of wine. By mistake the casks were interchanged, so that the purchaser received more of the cheaper wine and less of the dearer. The merchant therefore returned to him $1.50. How many quarts did each cask hold?

22. A and B jointly contribute $10,000 to a business. A leaves his money in the business 1 year and 3 months, and B his money 2 years and 11 months. If their profits be equal, how much does each contribute?

23. A merchant sold 12 gallons from each of two full casks of wine, and then found that the larger contained twice as much as the smaller. After he had sold more wine from both casks, he found that each one contained one-third of its original capacity. If he had then added 4 gallons of wine to each cask, the contents of the smaller would have been three-fourths of the contents of the larger. What was the capacity of each cask?

24. One boy said to another: "Give me 5 of your nuts, and I shall have three times as many as you will have left."

"No," said the other, "give me 2 of your nuts, and I shall have five times as many as you will have left." How many nuts had each boy?

25. A father has two sons, one 4 years older than the other. After 2 years the father's age will be twice the joint ages of his sons; and 6 years ago his age was six times the joint ages of his sons. How old is the father and each of his sons?

26. If a number of two digits be divided by the sum of the digits, the quotient will be 7. If the digits be interchanged, the resulting number will be less than the original number by 27. What is the number?

27. A man walks 26 miles, first at the rate of 3 miles an hour, and later at the rate of 4 miles an hour. If he had walked 4 miles an hour when he walked 3, and 3 miles an hour when he walked 4, he would have gone 4 miles further. How far would he have gone, if he had walked 4 miles an hour the whole time?

28. Two trains leave different cities, which are 650 miles apart, and run toward each other. If they start at the same time, they will meet after 10 hours; but if the first start $4\frac{1}{3}$ hours earlier than the second, they will meet 8 hours after the second train starts. What is the speed of each train?

29. If the base of a rectangle be increased by 2 feet, and the altitude be diminished by 3 feet, the area will be diminished by 48 square feet. But if the base be increased by 3 feet, and the altitude be diminished by 2 feet, the area will be increased by 6 square feet. Find the base and the altitude of the rectangle?

30. A number of three digits is in value between 400 and 500, and the sum of its digits is 9. If the digits be reversed, the resulting number will be $\frac{36}{47}$ of the original number. What is the number?

31. The report of a cannon travels with the wind 344.42 yards a second, and against the wind 335.94 yards a second. What is the velocity of the report in still air, and what is the velocity of the wind?

32. Two messengers, A and B, travel toward each other, starting from two cities which are 805 miles distant from each other. If A start $5\frac{3}{4}$ hours earlier than B, they will meet $6\frac{1}{8}$ hours after B starts. But if B start $5\frac{3}{4}$ hours earlier than A, they will meet $5\frac{5}{8}$ hours after A starts. At what rates do A and B travel?

33. Each of two servants was to receive $160, a dress, and a pair of shoes for one year's services. One servant left after 8 months, and received the dress and $106; the other servant left after $9\frac{1}{2}$ months, and received a pair of shoes and $142. What was the value of the dress, and of the pair of shoes?

34. On the eve of a battle, one army had 5 men to every 6 men in the other. The first army lost 14,000 men, and the second lost 6000 men. The first army then had 2 men to every 3 men in the other. How many men were there originally in each army?

35. If the sum of two numbers, each of three digits, be increased by 1, the result will be 1000. If the greater be placed on the left of the less, and a decimal point be placed between them, the resulting number will be six times the number obtained by placing the smaller number on the left of the greater, with a decimal point between them. What are the numbers?

36. A vessel sails 110 miles with the current and 70 miles against the current in 10 hours. On a second trip, it sails 88 miles with the current and 84 miles against the current in the same time. How many miles can the vessel sail in still water in one hour, and what is the speed of the current?

37. A and B run a race of 400 yards. In the first heat A gives B a start of 20 seconds, and wins by 50 yards. In the second heat A gives B a start of 125 yards, and wins by 5 seconds. What is the speed of each runner?

38. A merchant had two casks containing different quantities of wine. He poured from the first cask into the second as much wine as was in the second; next he poured from the second cask into the first as much wine as was left in the first; finally he poured from the first cask into the second as much wine as was left in the second. Each cask then contained 80 quarts. How many gallons did each cask originally contain?

39. A and B both wished to buy a horse. A said to B: "Give me half of your money, and I can buy the horse." B said to A: "Give me a third of your money, and I can buy the horse." But neither was willing to lend to the other. When a second horse, worth $36 less than the first, was to be sold, A said to B: "Give me three-eighths of your money, and I can buy this horse." But B said to A: "Give me one-sixth of your money, and I can buy the horse." How much money had A and B, and what was the price of the first horse?

40. A battle between two armies, A and B, continued three days. The number of killed and wounded in each army during each day's engagement was the following fraction of the number of uninjured at the beginning of the day: The first day A lost $\frac{1}{10}$, and B lost $\frac{1}{9}$; the second day A lost $\frac{1}{9}$, and B lost $\frac{1}{16}$; the third day A lost $\frac{1}{4}$, and B lost $\frac{1}{15}$. If the total loss in the army A was twice that in the army B, and if 6000 more men survived uninjured in B than in A, how many men were there in each army before the battle?

41. A and B formed a partnership. A invested $20,000 of his own money and $5000 which he borrowed; B invested $22,000 of his own money and $8000 which he borrowed at the same rate of interest as was paid by A. At the end of a year, A's share in the profits amounted to $1750 more than the interest on his $5000, and B's share to $2000 more than the interest on his $8000. What rate per cent interest did they pay, and what rate per cent did they realize on their investments?

42. Two bodies move along the circumference of a circle in the same direction from two different points, the shorter distance between which, measured along the circumference, is 160 feet. One body will overtake the other in 32 seconds, if they move in one direction; or in 40 seconds, if they move in the opposite direction. While the one goes once around the circumference, the distance passed over by the other exceeds the circumference by 45 feet. What is the circumference of the circle, and at what rates do the bodies move?

43. A number of workmen, who receive the same wages, earn together a certain sum. Had there been 7 more workmen, and had each one received 25 cents more, their joint earnings would have increased by $18.65. Had there been 4 fewer workmen, and had each one received 15 cents less, their joint earnings would have decreased by $9.20. How many workmen are there, and how much does each one receive?

44. A courier rode from *A* toward *B*, which is 64 miles distant from *A*. Five hours after his departure, a second courier started from *B* and rode toward *A*. The couriers met 7 hours after the second courier started. If the second courier had started from *B* 2 hours before the first started from *A*, they would have met 8 hours after the second courier started. At what rate did each courier ride?

45. A farmer has enough feed for his oxen to last a certain number of days. If he were to sell 75 oxen, his feed would last 20 days longer. If, however, he were to buy 100 oxen, his feed would last 15 days less. How many oxen has he, and for how many days has he enough feed?

46. An alloy of tin and lead, weighing 40 pounds, loses 4 pounds in weight when immersed in water. Find the amount of tin and lead in the alloy, if 10 pounds of tin lose $1\frac{3}{8}$ pounds when immersed in water, and 5 pounds of lead lose .375 of a pound.

47. Two men were to receive $96 for a certain piece of work, which they could do together in 30 days. After half of the work was done, one of them stopped for 8 days, and then the other stopped for 4 days. They finally completed the work in $35\frac{1}{2}$ days. How many dollars should each one receive, and in what time could each one have done the work alone?

48. Two boys, A and B, run a race from P to Q and return. A, the faster runner, on his return meets B 90 feet from Q, and reaches P 3 minutes ahead of B. If he had run again to Q, he would have met B at a distance from P equal to one-sixth of the distance from P to Q. How far is Q from P, and how long did it take B to run from P to Q and return?

49. It took a certain number of workmen 6 hours to carry a pile of stones from one place to another. Had there been 2 more workmen, and had each one carried 4 pounds more at each trip, it would have taken them 1 hour less to complete the work. Had there been 3 fewer workmen, and had each one carried 5 pounds less at each trip, it would have taken them 2 hours longer to complete the work. How many workmen were there, and how many pounds did each one carry at every trip?

50. Three carriages travel from A to B. The second carriage travels every 4 hours 1 mile less than the first, and is 4 hours longer in making the journey. The third carriage travels every 3 hours $1\frac{3}{4}$ miles more than the second, and is 7 hours less in making the journey. How far is B from A, and how many hours does it take each carriage to make the journey?

51. Water enters a basin through one pipe and is discharged through another. Through the first pipe four more gallons enter the basin every minute than is discharged through the second. When the basin is empty, both pipes are opened, the first one hour earlier than the second, and after a certain time the basin contains 1760 gallons. The pipe through which water enters is then closed, and after one hour is again opened. If both pipes be then left open for as long a time as they were open together in the former case, the basin will contain 880 gallons. In what time can the one pipe fill the basin and the other empty it?

52. A body moves with a uniform velocity from a point A to a point B, which is 323 feet distant from A, and without stopping returns at the same rate from B to A. A second body leaves B 13 seconds after the first leaves A, and moves toward A with a uniform but less velocity than the velocity of the first. The first body meets the second 10 seconds after the latter starts, and on returning to A overtakes the second body 45 seconds after the latter starts. What is the velocity of each body?

53. A fox pursued by a dog is 60 of her own leaps ahead of the dog. The fox makes 9 leaps while the dog makes 6, but the dog goes as far in 3 leaps as the fox goes in 7. How many leaps does each make before the dog catches the fox?

54. The sum of the three digits of a number is 14; the sum of the first and the third digit is equal to the second; and if the digits in the units' and in the tens' place be interchanged, the resulting number will be less than the original number by 18. What is the number?

55. The sum of the ages of A, B, and C is 69 years. Two years ago B's age was equal to one-half of the sum of the ages of A and C, and 10 years hence the sum of the ages of B and C will exceed A's age by 31 years. What are the present ages of A, B, and C?

56. The total capacity of three casks is 1440 quarts. Two of them are full and one is empty. To fill the empty cask it takes all the contents of the first and one-fifth of the contents of the second, or the contents of the second and one-third of the contents of the first. What is the capacity of each cask?

57. Three brothers wished to buy a house worth $70,000, but none of them had enough money. If the oldest brother had given the second brother one-third of his money, or the youngest brother one-fourth of his money, each of the latter would then have had enough money to buy the house. But the oldest brother borrowed one-half of the money of the youngest and bought the house. How much money had each brother?

58. The sum of the three digits of a number is 9. The digit in the hundreds' place is equal to one-eighth of the number

composed of the two other digits, and the digit in the units' place is equal to one-eighth of the number composed of the two other digits. What is the number?

59. Find the contents of three vessels from the following data: If the first be filled with water and the second be filled from it, the first will then contain two-thirds of its original contents; if from the first, when full, the third be filled, the first will then contain five-ninths of its original contents; finally, if from the first, when full, the second and third be filled, the first will then contain 8 gallons.

60. Three boys were playing marbles. A said to B: "Give me 5 marbles, and I shall have twice as many as you will have left." B said to C: "Give me 13 marbles, and I shall have three times as many as you will have left." C said to A: "Give me 3 marbles, and I shall have six times as many as you will have left." How many marbles did each boy have?

61. Three cities, A, B, and C, are situated at the vertices of a triangle. The distance from A to C by way of B is 82 miles, from B to A by way of C is 97 miles, and from C to B by way of A is 89 miles. How far are A, B, and C from one another?

62. A father's age is twenty-one times the difference between the ages of his two sons. Six years ago his age was six times the sum of his sons' ages, and two years hence it will be twice the sum of their ages. Find the ages of the father and his two sons.

63. A regiment of 600 soldiers is quartered in a four-story building. On the first floor are twice as many men as are on the fourth; on the second and third are as many men as are on the first and fourth; and to every 7 men on the second there are 5 on the third. How many men are quartered on each floor?

64. The sum of the three digits of a number is 9. If 198 be added to the number, the digits of the resulting number are those of the given number written in reverse order. Two-thirds of the digit in the tens' place is equal to the difference between the digits in the units' and in the hundreds' place. What is the number?

65. Four men are to do a piece of work. A and B can do the work in 10 days, A and C in 12 days, A and D in 20 days, and B, C, and D in $7\frac{1}{2}$ days. In how many days can each man do the work, and in how many days can they all together do the work?

66. The sum of the four digits of a number is 11. The tens' digit is equal to the sum of the hundreds' and the thousands' digit, and the thousands' digit is equal to the sum of the hundreds' and the units' digit. If the digits be written in reverse order, the resulting number will be less than the original number by 1728. What is the number?

67. Three cities, A, B, and C, are situated at the vertices of a triangle. The distance from A to B by way of C is 4 times the direct distance. The distance from A to C by way of B is 5 miles greater than the direct distance; and the distance from B to C by way of A is 85 miles. Find the direct distances between the cities.

68. The year in which printing was invented is expressed by a figure of four digits, whose sum is 14. The tens' digit is one-half of the units' digit, and the hundreds' digit is equal to the sum of the thousands' and the units' digit. If the digits be reversed, the resulting number will be equal to the original number increased by 4905. In what year was printing invented?

69. Water enters a tank through two pipes, A and B, and is discharged through a third pipe, C. The capacity of the tank is 200 gallons. If, when the tank is empty, A be left open 7 minutes, and be then closed, and B and C be next left open 5 minutes, and then B be closed, the tank will be again empty 13 minutes after B is closed. If all the pipes be left open 5 minutes, and then A be closed, the tank will be filled 25 minutes after A is closed. If, when the tank is full, C be left open 12 minutes, and then B and A be left open 3 minutes,

the tank will contain 119 gallons. If the three pipes be left open, how long will it take to fill the empty tank?

70. A merchant paid $13,600 for a number of chests of tea of two kinds. He calculated that if he were to gain $20 on each chest of the first kind, and $15 on each chest of the second kind, he would realize $15,300. But he sold to one customer 25 chests of the first kind, and 15 chests of the second kind, gaining thereby $750. The gain on the first kind was 12%, and on the second kind 15%. The remainder of the tea he then sold to another man for $9000, at $180 a chest for the first kind, and $140 a chest for the second. How many chests of each kind did he buy, and at what price?

71. A body moves at a uniform rate from a point A toward a point B, which is 490 feet distant from A. After 5 minutes it is followed by a second body which moves at the same rate. When the first body leaves A, a third body moves from B toward A at a uniform rate, and meets the first body after 19.6 minutes, and the second 2.4 minutes later. At what rate does each body move?

Discussion of Solutions.

3. Pr. 1. A merchant has two kinds of tea; the first is worth a cents a pound, and the second b cents a pound. How much of each kind must be taken to make a mixture of one pound worth c cents?

Let x stand for the part of a pound of the first kind, and y for the part of a pound of the second kind.

Then, by the first condition,

$$x + y = 1; \quad (1)$$

and by the second condition,

$$ax + by = c. \quad (2)$$

Solving (1) and (2), we obtain

$$x = \frac{c-b}{a-b}, \quad y = \frac{a-c}{a-b}.$$

(i.) If $a > c > b$, the values of x and y are both positive, and the solution satisfies the conditions of the problem. Thus, if $a = 100$, $b = 75$, and $c = 85$, we have

$$x = \frac{2}{5}, \quad y = \frac{3}{5}.$$

If $a < c < b$, the values of x and y are both positive, and satisfy the conditions of the problem.

That is, if the value of the pound of mixture be intermediate between the values of a pound of each of the two kinds, a definite solution is always possible.

(ii.) If $c > a > b$, then x will be *positive*, and y *negative*. Therefore the solution does not satisfy the conditions of the problem. Thus, if $a = 100$, $b = 75$, $c = 110$, we obtain

$$x = \frac{7}{5}, \quad y = -\frac{2}{5}.$$

It is evident that a one-pound mixture of two kinds of tea which is worth more than either kind cannot be made.

(iii.) If $a = b = c$, then $x = \frac{0}{0}$, $y = \frac{0}{0}$. This solution indicates that the conditions of the problem may be satisfied in an indefinite number of ways. It is evident that a one-pound mixture of two kinds of tea, that are the same in price, can be made in any number of ways, if the mixture is to be the same in price.

(iv.) If $a = b$, and $a \neq c$, then $x = \infty$ and $y = \infty$.

This solution does not satisfy the conditions of the problem, since x and y must be finite proper fractions. It is also evident that a one-pound mixture of two kinds of tea which are the same in price cannot be made, if the mixture is to be of a different price.

Pr. 2. [See Ch. XII., Art. 7, Pr. 2.] Two couriers are traveling along a road in the direction from M to N; one at the rate of m_1 miles an hour, and the other at the rate of m_2 miles an hour. The former was seen at noon at the station A, and the latter h hours later at the station B, which is d miles from A in the direction in which the couriers are traveling. At what place do the couriers meet, and how many hours after the first was seen at A?

The meeting can take place either at a point to the left of A, or at a point intermediate between A and B, or at a point to the right of B.

Let us assume that the couriers meet to the right of B, at a point C_1, say (Fig. 11).

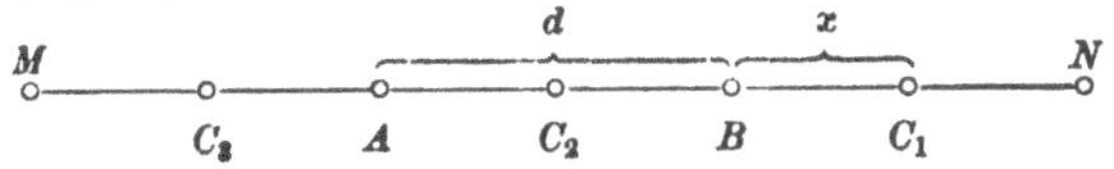

FIG. 11.

Let x stand for the number of miles from B to the place of meeting C_1; and let y stand for the number of hours from the time the first courier was seen at A until the couriers meet.

Then the first courier travels $d + x$ miles in y hours; but since he travels at the rate of m_1 miles an hour, in y hours he travels m_1y miles. Therefore

$$d + x = m_1y. \qquad (1)$$

The second courier travels x miles in $y - h$ hours; but since he travels at the rate of m_2 miles an hour, in $y - h$ hours he travels $m_2(y - h)$ miles. Therefore

$$x = m_2(y - h). \qquad (2)$$

Solving equations (1) and (2), we obtain

$$x = \frac{m_2(hm_1 - d)}{m_2 - m_1}, \quad y = \frac{hm_2 - d}{m_2 - m_1}.$$

(i.) **A Positive Result.** When $hm_1 > d$ and $m_2 > m_1$, then also $hm_2 > d$; and when $hm_1 < d$ and $m_2 < m_1$, then also $hm_2 < d$. Consequently, in either of these cases, the values of x and y are *both positive*. We therefore conclude that the couriers meet to the *right* of B, some time *after* the first courier was seen at A.

(ii.) **A Negative Result.** When $hm_2 > d$ and $m_1 > m_2$, then also $hm_1 > d$; and when $hm_2 < d$ and $m_1 < m_2$, then also $hm_1 < d$. Consequently, in either of these cases, the values of x and y are *both negative*. We therefore conclude that the couriers meet to the *left* of B, some time *before* the first courier was seen at A. Hence they must have met not only to the left of B, but to the left of A, at some point C_3.

When $hm_1 > d$, $hm_2 < d$, and $m_1 > m_2$, then the value of x is *negative*, and the value of y is *positive*. We therefore conclude that the couriers meet to the *left* of B, some time *after* the first courier was seen at A. Consequently, they meet at some point between A and B, at C_2, say.

(iii.) **A Zero Result.** When $hm_1 = d$, and $m_2 \neq m_1$, then $x = 0$, and y is *positive*. These results mean that the couriers meet at B, some time *after* the first courier was seen at A.

(iv.) **An Indeterminate Result.** When $m_1 = m_2$ and $hm_1 = d$, then $hm_2 = d$. In this case

$$x = \frac{0}{0},\ y = \frac{0}{0}.$$

Since the couriers reach B at the same time, and are now traveling at the same rate, they are always together. That is, every point on the road can be regarded as their place of meeting.

(v.) **An Infinite Result.** If $m_1 = m_2$, and $hm_1 \neq d$, then $hm_2 \neq d$. In this case $x = \infty$, and $y = \infty$. We therefore conclude that the couriers can never meet. In fact, since both couriers now travel at the same rate, and do not reach B at the same time, they can never meet.

(vi.) **One Infinite and One Indeterminate Result.**

If $m_1 = 0$, $m_2 = 0$, $d \neq 0$, then $x = \frac{0}{0}$, and $y = \infty$.

Since the couriers are at rest, the first at A and the second at B, they will never meet.

(vii.) **One Determinate and One Indeterminate Result.**

If $m_1 = 0$, $m_2 = 0$, and $d = m_2h$, then $x = \frac{0}{0}$, and $y = \frac{0}{0}$.

But in this case,
$$x = \frac{m_2(hm_1 - hm_2)}{m_1 - m_2}$$
$$= hm_2 = 0, \text{ since } m_2 = 0.$$

The meaning of these results is that the couriers are at rest at A, and hence will be always together. In fact, $d = m_2h$, $= 0$, since $m_2 = 0$, and the points A and B coincide.

EXERCISES II.

Solve the following problems, and discuss the results:

1. If an alloy of two kinds of silver be made, and a ounces of the first be taken with b ounces of the second, the mixture will be worth m dollars an ounce. If b ounces of the first be taken with a ounces of the second, the mixture will be worth n dollars an ounce. How much is an ounce of each kind of silver worth?

2. Two bodies are separated by a distance of d yards. If they move toward each other with different velocities, they will meet after m seconds; but if they both move in the same direction, the one will overtake the other after n seconds. With what velocities do the bodies move?

3. Two bodies, A and B, start from two points, which are d feet distant from each other, and move toward each other. If A start s seconds earlier than B, they will meet m seconds after B starts; but if B start t seconds earlier than A, they will meet n seconds after A starts. With what velocities do the bodies move?

CHAPTER XV.

EVOLUTION.

§ 1. DEFINITIONS AND PRINCIPLES.

1. A **qth Root** of a number or an expression is a number or an expression whose qth power is equal to the given number or expression.

E.g., since $(+5)^2 = 25$ and $(-5)^2 = 25$, therefore $+5$ and -5 are *second* roots of 25. The statement, $+5$ and -5, is usually written ± 5.

Since $(a+b)^3 = a^3 + 3a^2b + 3ab^2 + b^3$, therefore $a+b$ is a *third* root of $a^3 + 3a^2b + 3ab^2 + b^3$.

A *second* root of a number is usually called a *square* root; and a *third* root a *cube* root.

2. It follows from the definition of a root that a qth root of a number is one of q equal factors of the number.

Thus, $a+b$ is one of three equal factors of

$$a^3 + 3a^2b + 3ab^2 + b^3;$$

either $+3$ or -3 is one of two equal factors of 9.

3. The **Radical Sign**, $\sqrt{\ }$, is used to denote a root, and is placed before the number or expression whose root is to be found.

The **Radicand** is the number or expression whose root is required.

The **Index** of a root is the number which indicates what root of the radicand is to be found, and is written over the radical sign. The index 2 is usually omitted.

E.g., $\sqrt[2]{9}$, or $\sqrt{9}$, denotes a second, or square root of 9; the radicand is 9, and the index is 2. In general, $\sqrt[q]{a}$ denotes a qth root of the radicand a; the index is q.

4. A vinculum is often used in connection with the radical sign to indicate what part of an expression following the sign is affected by it.

E.g., $\sqrt{9}+16$ means the sum of $\sqrt{9}$ and 16, while $\sqrt{9+16}$ means a square root of the sum $9+16$. Likewise, $\sqrt[3]{a^3}\times b^6$ means the product of $\sqrt[3]{a^3}$ and b^6, while $\sqrt[3]{a^3\times b^6}$ means a cube root of a^3b^6.

Parentheses may be used instead of the vinculum in connection with the radical sign; as $\sqrt{}(9+16)$ for $\sqrt{9+16}$.

5. Like and Unlike Roots. — Two roots are said to be *like* or *unlike* according as their indices are equal or unequal, whether or not their radicands are equal. Thus,

$\sqrt[3]{a}$, $\sqrt[3]{b}$ are *like* roots; $\sqrt{a}$, $\sqrt[3]{a}$ are *unlike* roots.

6. In this chapter we shall consider only roots of numbers which are powers with exponents equal to or multiples of the indices of the required roots; as $\sqrt{16}, =\sqrt{4^2}$, $\sqrt[3]{a^3}$, $\sqrt[5]{a^{10}}$, $\sqrt[q]{a^{kq}}$.

An **Even Root** of a number is one whose *index* is *even;* as $\sqrt{a^2}$, $\sqrt[4]{a^4}$, $\sqrt[2q]{a^{2q}}$.

An **Odd Root** of a number is one whose *index* is *odd;* as $\sqrt[3]{8}$, $\sqrt[5]{a^{10}}$, $\sqrt[2q+1]{a^{2q+1}}$.

7. (i.) *A positive number has at least two even roots, equal and opposite; i.e., one positive and one negative.*

E.g., since $(\pm 4)^2=16$, $\sqrt{16}=\pm 4$; since $(\pm a)^4=a^4$, $\sqrt[4]{a^4}=\pm a$. In general, since $(\pm a)^{2q}=a^{2q}$, $\sqrt[2q]{a^{2q}}=\pm a$.

(ii.) *A positive or a negative number has at least one odd root of the same sign as the number itself.*

E.g., since $(-3)^3=-27$, $\sqrt[3]{-27}=-3$; since $2^5=32$, $\sqrt[5]{32}=2$; since $(-a)^7=-a^7$, $\sqrt[7]{-a^7}=-a$.

In general, since $(+a)^{2q+1}=+a^{2q+1}$, $\sqrt[2q+1]{+a^{2q+1}}=+a$; since $(-a)^{2q+1}=-a^{2q+1}$, $\sqrt[2q+1]{-a^{2q+1}}=-a$.

The principle enunciated in (ii.), when the radicand is negative, may also be stated as follows:

(iii.) *An odd root of a negative number is equal and opposite to a like root of a positive number which has the same absolute value.*

E.g., since $\sqrt[3]{-8}=-2$ and $-\sqrt[3]{8}=-2$,

therefore $\sqrt[3]{-8}=-\sqrt[3]{8}$;

since $\sqrt[2q+1]{-a^{2q+1}}=-a$ and $-\sqrt[2q+1]{a^{2q+1}}=-a$,

therefore $\sqrt[2q+1]{-a^{2q+1}}=-\sqrt[2q+1]{a^{2q+1}}$.

Consequently, to find an odd root of a negative number, find a like root of the positive number which has the same absolute value, and prefix the negative sign to this root.

(iv.) Since $0^2=0$, therefore $\sqrt{0}=0$. In general, since $0^n=0$, therefore $\sqrt[n]{0}=0$.

(v.) Since $(+4)^2=+16$ and $(-4)^2=+16$, there is no number, *with which we are as yet familiar*, whose square is -16. Consequently $\sqrt{-16}$ cannot be expressed in terms of the numbers as yet used in this book.

In general, since $(\pm a)^{2n}=+a^{2n}$, we cannot express $\sqrt[2n]{-a^{2n}}$ in terms of numbers hitherto used.

The roots of numbers which are not powers with exponents equal to or multiples of the indices of the required roots, and *even* roots of *negative* numbers, will be considered in the Chapters XVII.–XIX.

8. It was shown in Art. 7 that a positive number, which is the qth power of a number, has *at least one qth root*, and when q is even *at least two;* also that any negative number, which is an odd power of a negative number, has *at least one odd root.*

It will be proved in Chapters XX. and XXI. that any number has *two square* roots, *three cube* roots, *four fourth* roots, and *five fifth* roots; and in Part II. that, in general, any number has q qth roots.

Principal Roots.

9. The **Principal Root** of a *positive* number is its one *positive* root.

E.g., 3 is the principal square root of 9; 5 is the principal cube root of 125.

The **Principal Odd Root** of a *negative* number is its one *negative* root.

E.g., -2 is the principal cube root of -8.

10. It is important to notice that the relation

$$\sqrt[q]{a^q} = (\sqrt[q]{a})^q = a$$

is true only for the principal qth root.

For, by Art. 8, the $\sqrt[q]{a^q}$ has q values, the principal value being a. But, by definition of a root, the $(\sqrt[q]{a})^q = a$, for every qth root of a.

E.g., $\sqrt{4^2} = \sqrt{16} = \pm 4$, if other than principal roots be admitted; and $(\sqrt{4})^2 = (\pm 2)^2 = 4$.

Therefore, $\sqrt{4^2} = (\sqrt{4})^2$, only for the principal square root.

In subsequent work the radical sign will be understood to denote only the principal root, *unless the contrary is expressly stated.*

E.g., $\sqrt{9} = 3$, $-\sqrt{16} = -4$, $\sqrt[3]{-27} = -3$.

EXERCISES I.

Write

1. Two square roots of 49. **2.** Two fourth roots of 81.

3. Two sixth roots of 64. **4.** Two square roots of 5^{2n}.

Write one cube root of

5. 64. **6.** -125. **7.** 1000. **8.** $-a^{3n}$.

Find the value of the indicated principal root of each of the following numbers:

9. $\sqrt{256}$. **10.** $\sqrt[3]{216}$. **11.** $\sqrt[3]{-512}$. **12.** $\sqrt{625}$.

13. $\sqrt[3]{-729}$. **14.** $\sqrt[4]{1296}$. **15.** $\sqrt[5]{243}$. **16.** $\sqrt[6]{64}$.

From the definition of a root, express a as a root of the second member of each of the following equations:

17. $a^3 = b$. **18.** $a^4 = b^3$. **19.** $a^5 = b^7$. **20.** $a^n = b^m$.

21–24. In Exx. 17–20, express b as a root of the first member of each of the equations.

Evolution.

11. Evolution is the process of finding any required root of a given number or expression.

Since even roots of negative numbers are not considered in this chapter, and since, by Art. 7 (iii.), an odd root of a negative number can be found from the like root of a positive number, we shall give now only methods for finding principal roots of positive numbers and expressions.

It is to be kept in mind therefore that *in the following principles the radicands are limited to positive values, and the roots to principal roots.*

12. *The like principal roots of equal numbers or expressions are equal.*

If $a = b$, then $\sqrt[q]{a} = \sqrt[q]{b}$.

For $\sqrt[q]{a} = \sqrt[q]{a}$, by Axiom (i.). (1)

Since $b = a$, we can, by Axiom (iii.), substitute b for a in the second member of (1). We thus obtain

$$\sqrt[q]{a} = \sqrt[q]{b}.$$

13. *The principal root of a power of a number is equal to the same power of the like principal root of the number, and conversely;* or, stated symbolically,

$$\sqrt[q]{a^p} = (\sqrt[q]{a})^p.$$

In particular, $\sqrt[q]{a^q} = (\sqrt[q]{a})^q$.

E.g., $\sqrt[3]{8^5} = (\sqrt[3]{8})^5 = 2^5 = 32$; $\sqrt[5]{32^5} = (\sqrt[5]{32})^5 = 32$.

Let the qth root of a be denoted by R,

or $\sqrt[q]{a} = R$.

Then $(\sqrt[q]{a})^q = R^q$, by Ch. III., § 3, Art. 5,

or $a = R^q$, since $(\sqrt[q]{a})^q = a$, by definition of a root;

and $a^p = (R^q)^p$, by Ch. III., § 3, Art. 5,

or $a^p = (R^p)^q$, since $(R^q)^p = (R^p)^q$.

Whence $\sqrt[q]{a^p} = \sqrt[q]{(R^p)^q}$, by Art. 12,

$= R^p$, by Art. 10.

Substituting $\sqrt[q]{a}$ for R in the last equation, we have

$$\sqrt[q]{a^p} = (\sqrt[q]{a})^p.$$

14. *The principal root of a power is obtained by dividing the exponent of the power by the index of the root;* or, stated symbolically,

$$\sqrt[q]{a^{kq}} = a^{\frac{kq}{q}} = a^k.$$

E.g., $$\sqrt[3]{a^6} = a^{\frac{6}{3}} = a^2.$$

Let the q root of a^{kq} be denoted by R,

or $$R = \sqrt[q]{a^{kq}}.$$

Then $R^q = (\sqrt[q]{a^{kq}})^q$, by Ch. III., § 3, Art. 5,

$= a^{kq}$, by definition of a root,

$= (a^k)^q.$

Whence $\sqrt[q]{R^q} = \sqrt[q]{(a^k)^q}$, by Art. 12,

or $R = a^k$, by Art. 10.

Substituting $\sqrt[q]{a^{kq}}$ for R in the last equation, we have

$$\sqrt[q]{a^{kq}} = a^k = a^{\frac{kq}{q}}.$$

15. *The principal root of a product of two or more factors is equal to the product of the like principal roots of the factors, and conversely;* or, stated symbolically,

$$\sqrt[q]{(ab)} = \sqrt[q]{a} \times \sqrt[q]{b}, \text{ and } \sqrt[q]{a} \times \sqrt[q]{b} = \sqrt[q]{(ab)}.$$

E.g., $$\sqrt{(16 \times 25)} = \sqrt{16} \times \sqrt{25} = 4 \times 5 = 20;$$

$$\sqrt[3]{(8a^3b^6)} = \sqrt[3]{8} \times \sqrt[3]{a^3} \times \sqrt[3]{b^6} = 2 \times a \times b^2 = 2ab^2.$$

Let $\sqrt[q]{a} = R$, and $\sqrt[q]{b} = R_1$.

Then $(\sqrt[q]{a})^q = R^q$, and $(\sqrt[q]{b})^q = R_1{}^q$, by Ch. III., § 3, Art. 5,

or $a = R^q$, and $b = R_1{}^q$, by definition of a root.

Therefore $ab = R^qR_1{}^q = (RR_1)^q$.

Whence $\sqrt[q]{(ab)} = RR_1$, by Arts. 10 and 12.

Substituting $\sqrt[q]{a}$ for R, and $\sqrt[q]{b}$ for R_1 in the last equation, we have

$$\sqrt[q]{(ab)} = \sqrt[q]{a}\sqrt[q]{b}.$$

In like manner, the principle can be proved for the qth root of a product of any number of factors.

16. *The principal root of a quotient of two numbers is equal to the quotient of the like principal roots of the numbers, and conversely;* or, stated symbolically,

$$\sqrt[q]{\frac{a}{b}}=\frac{\sqrt[q]{a}}{\sqrt[q]{b}}, \text{ and } \frac{\sqrt[q]{a}}{\sqrt[q]{b}}=\sqrt[q]{\frac{a}{b}}.$$

E.g., $\sqrt{\frac{25}{16}}=\frac{\sqrt{25}}{\sqrt{16}}=\frac{5}{4}$; $\sqrt[3]{\frac{27\,a^3}{b^6}}=\frac{\sqrt[3]{(27\,a^3)}}{\sqrt[3]{b^6}}=\frac{3\,a}{b^2}$.

Let $\sqrt[q]{a}=R$ and $\sqrt[q]{b}=R_1$.

Then $a=R^q$ and $b=R_1{}^q$.

And $\frac{a}{b}=\frac{R^q}{R_1{}^q}=\left(\frac{R}{R_1}\right)^q$.

Whence $\sqrt[q]{\frac{a}{b}}=\frac{R}{R_1}$, by Arts. 10 and 12.

Substituting $\sqrt[q]{a}$ for R and $\sqrt[q]{b}$ for R_1 in the last equation, we have

$$\sqrt[q]{\frac{a}{b}}=\frac{\sqrt[q]{a}}{\sqrt[q]{b}}.$$

17. *The principal root of the principal root of a number is equal to that principal root of the number whose index is equal to the product of the indices of the given roots, and conversely;* or, stated symbolically,

$$\sqrt[q]{\sqrt[r]{a}}=\sqrt[qr]{a}, \text{ and } \sqrt[qr]{a}=\sqrt[q]{\sqrt[r]{a}}=\sqrt[r]{\sqrt[q]{a}}.$$

E.g., $\sqrt[3]{\sqrt{64}}=\sqrt[6]{64}=2$; $\sqrt[4]{256}=\sqrt{\sqrt{256}}=\sqrt{16}=4$.

Let $R=\sqrt[q]{\sqrt[r]{a}}$.

Then $R^q=\sqrt[r]{a}$ and $(R^q)^r=a$;

or $R^{qr}=a$.

Therefore $R=\sqrt[qr]{a}$.

Substituting $\sqrt[q]{\sqrt[r]{a}}$ for R in the last equation, we have

$$\sqrt[q]{\sqrt[r]{a}}=\sqrt[qr]{a}.$$

The principle and its converse can be easily extended.

E.g., $\sqrt[8]{256}=\sqrt{\sqrt[4]{256}}=\sqrt{\sqrt{\sqrt{256}}}=\sqrt{\sqrt{16}}=\sqrt{4}=2$.

18. In the principles of the preceding articles the roots were limited to principal values, because with the exception of that in Art. 17, they do not always hold for other than these values.

E.g., the identity $\sqrt[4]{2^8} = 2^2$, by Art. 14, is true only for the *positive* fourth root of 2^8. For $\sqrt[4]{2^8} = \sqrt[4]{256} = \pm 4$, two of the four fourth roots, if the root be not limited to its principal value; while $2^2 = 4$.

The identity $\sqrt{4^2} = (\sqrt{4})^2$, by Art. 13, is true only for the positive square roots. For $\sqrt{4^2} = \sqrt{16} = \pm 4$, if the root be not limited to its principal value; while $(\sqrt{4})^2 = (\pm 2)^2 = 4$. But $\sqrt{4^3} = (\sqrt{4})^3$, by Art. 13, is true for both the positive and the negative square roots. For

$$\sqrt{4^3} = \sqrt{64} = \pm 8, \text{ and } (\sqrt{4})^3 = (\pm 2)^3 = \pm 8.$$

Observe, also, that the principles of Arts. 15 and 16, with certain restrictions, hold for other than the principal values.

E.g., the identities $\sqrt{(16 \times 9)} = \sqrt{16} \times \sqrt{9}$ and $\sqrt{\frac{16}{9}} = \frac{\sqrt{16}}{\sqrt{9}}$ hold in the following cases:

$$+12 = (+4)(+3), \qquad +\frac{4}{3} = \frac{+4}{+3},$$

$$+12 = (-4)(-3), \qquad +\frac{4}{3} = \frac{-4}{-3},$$

$$-12 = (+4)(-3), \qquad -\frac{4}{3} = \frac{+4}{-3},$$

$$-12 = (-4)(+3), \qquad -\frac{4}{3} = \frac{-4}{+3}.$$

But evidently the *positive* square root of 16×9 cannot be equal to the product of the *positive* square root of 16 and the *negative* square root of 9; and so on.

The principle of Art. 17, however, is true for all values of the roots.

E.g., the identity $\sqrt[6]{64} = \sqrt[3]{\sqrt{64}}$ is true for both the positive and the negative values of the roots. For $\sqrt[6]{64} = \pm 2$, two of the six sixth roots, and $\sqrt[3]{\sqrt{64}} = \sqrt[3]{\pm 8} = \pm 2$, one of the three cube roots of 8 and one of the three cube roots of -8. It will be proved in Part II. that this principle holds for all six sixth roots.

§ 2. ROOTS OF MONOMIALS.

1. The principal (*positive*) root of a positive number or expression can be found by applying the principles of Arts. 13–17.

The *negative even* root of a positive number or expression is found by prefixing the negative sign to its principal root.

The *negative odd* root of a negative number or expression is found by prefixing the negative sign to the principal root of the radicand taken positively.

Ex. 1. $$\sqrt{(16\,a^2b^4)} = \sqrt{16} \times \sqrt{a^2} \times \sqrt{b^4} \quad (1)$$
$$= 4\,ab^2, \text{ the principal square root.}$$

Therefore $\pm\sqrt{(16\,a^2b^4)} = \pm\,4\,ab^2.$

In the following examples we shall give only the *positive* even roots.

Ex. 2. $$\sqrt[3]{(-27\,x^3y^6z^9)} = \sqrt[3]{-27} \times \sqrt[3]{x^3} \times \sqrt[3]{y^6} \times \sqrt[3]{z^9} \quad (1)$$
$$= -3\,xy^2z^3.$$

Ex. 3. $$\sqrt[4]{\frac{16\,a^8b^{12}}{625\,c^{16}}} = \frac{\sqrt[4]{(16\,a^8b^{12})}}{\sqrt[4]{(625\,c^{16})}} = \frac{\sqrt[4]{16} \times \sqrt[4]{a^8} \times \sqrt[4]{b^{12}}}{\sqrt[4]{625} \times \sqrt[4]{c^{16}}} \quad (1)$$
$$= \frac{2\,a^2b^3}{5\,c^4}.$$

Ex. 4. $$\sqrt[3]{(27\,a^3)^2} = (\sqrt[3]{27\,a^3})^2 = (3\,a)^2 = 9\,a^2.$$

Notice that the work of the last example is much simplified by taking the power of the root instead of the root of the power.

It is frequently necessary to modify the form of the radicand before finding the indicated root.

Ex. 5. $$\sqrt{(35 \times 7 \times 20)} = \sqrt{(7 \times 5 \times 7 \times 4 \times 5)}$$
$$= \sqrt{(7^2 \times 5^2 \times 4)} = 7 \times 5 \times 2 = 70.$$

Ex. 6. $$\sqrt{(8\,a^3b \times 2\,ab^5)} = \sqrt{(16\,a^4b^6)} = 4\,a^2b^3.$$

Ex. 7.
$$\sqrt{[(a+b)^3(a^2-b^2)(a-b)]} = \sqrt{[(a+b)^3(a+b)(a-b)(a-b)]}$$
$$= \sqrt{[(a+b)^4(a-b)^2]}$$
$$= (a+b)^2(a-b).$$

Ex. 8. $\sqrt[2n+1]{(-a^{4n+2}b^{4n^2-1})} = -\sqrt[2n+1]{a^{4n+2}} \times \sqrt[2n+1]{b^{4n^2-1}} = -a^2b^{2n-1}.$

The reduction indicated by (1) above in Exx. 1, 2, and 3, and similar reductions in the succeeding examples, should be performed mentally.

EXERCISES II.

Simplify each of the following expressions:

1. $\sqrt{(16\,a^2b^8)}$.
2. $\sqrt{(36\,a^4b^{10}c^6)}$.
3. $\sqrt[3]{(8\,a^3b^6)}$.
4. $\sqrt[3]{(-64\,a^9b^{12}x^{15})}$.
5. $\sqrt[4]{(8\,a^{10}b^4 \times 2\,a^2b^4)}$.
6. $\sqrt[5]{(-a^{20}x^{35})}$.
7. $\sqrt{(3\,ax^{3n} \times 27\,a^3x^{3n})}$.
8. $\sqrt[3]{(9\,a^4x^{14}y^{2n} \times 3\,a^2x^{10}y^n)}$.
9. $\sqrt[4]{(5\tfrac{1}{16}\,x^{4n}y^{8n-12})}$.
10. $\sqrt[6]{(64\,x^{10}y^{11}z^2 \times x^2y^7z^4)}$.
11. $\sqrt{[81\,a^4(a^2+x^2)^6]}$.
12. $\sqrt{(6\tfrac{1}{4}\,a^6b^4c^{4p-6})}$.
13. $\sqrt[2n+1]{(-a^{2n+1}b^{6n+3})}$.
14. $\sqrt[3]{[3\tfrac{3}{8}\,x^{3n-9}(x-1)^9]}$.
15. $\sqrt[5]{(.00032\,a^{5n+15}b^{x(5n-10)})}$.
16. $\sqrt{\dfrac{a^2b^4}{100\,n^2x^4}}$.
17. $\sqrt{\dfrac{49\,a^{10}}{b^4c^6}}$.
18. $\sqrt[3]{-\dfrac{a^{21}x^{15}}{343}}$.
19. $\sqrt[3]{\dfrac{27\,a^3b^6}{64\,x^9y^{12}}}$.
20. $\sqrt{\dfrac{9\,a^6b^{4m}}{c^{10}d^{2n}}}$.
21. $\sqrt[3]{\dfrac{216\,a^{18}}{b^6x^3}}$.
22. $\sqrt[4]{\dfrac{625\,x^4y^{12}}{a^8b^{16}}}$.
23. $\sqrt[3]{\dfrac{.064\,a^{12}}{b^3x^{15n}}}$.
24. $\sqrt[5]{\dfrac{a^5x^{20}}{(a-x)^{10}}}$.
25. $\sqrt[6]{\dfrac{a^{12}b^{24}c^6}{64\,x^{18}y^{42}}}$.
26. $\sqrt[3]{\dfrac{a^9(x^3+1)^{6n}}{216\,x^{3n}y^{12}}}$.
27. $\sqrt[5]{-\dfrac{243\,a^{15}x^{10}}{b^5y^{20}}}$.
28. $\sqrt[n]{\dfrac{x^{mn}}{a^nb^{2n}}}$.
29. $\sqrt{\dfrac{a^{14}b^{21}(c+d)^7}{c^{49}d^{28}}}$.
30. $\sqrt[x]{\dfrac{3^xa^{(n-1)x}}{2^{2x}b^{2(nx-x)}}}$.
31. $\sqrt{\sqrt{(625\,a^4b^8c^{12})}}$.
32. $\sqrt{\sqrt[3]{(49^3 \times 64^3)}}$.
33. $\sqrt[3]{\sqrt{(5^6x^{12}y^6)}}$.
34. $\sqrt[3]{\sqrt{(256\,a^{12}b^{18})}}$.
35. $\sqrt{\sqrt{(16\,a^{4n-4})}}$.
36. $\sqrt[3]{\sqrt[n]{(27^na^{3n}b^{6n})^2}}$.
37. $\sqrt[5]{\sqrt[3]{(-3^{15}a^{30}b^{15n}}}$
38. $\sqrt[m]{\sqrt[n]{(a^{mn}b^{m^2n^2-m^3n^3})}}$.
39. $\sqrt{(100\,a^{2n}b^4)^3}$.
40. $\sqrt[3]{(\tfrac{27}{64})^2}$.
41. $\sqrt[3]{\left(\dfrac{125\,a^3}{216\,b^6}\right)^4}$.
42. $\sqrt[2n+3]{\left(\dfrac{a^{4n+6}}{b^{6n+9}}\right)^9}$.
43. $\sqrt[3]{(8\,a^6b^9x^{12})^5}$.
44. $\sqrt[n]{(a^{2n}b^{3n}x^{n^2-n})^{10}}$.
45. $\sqrt[p]{(a^{p^3}b^{pq}c^{mnp})^r}$.

§ 3. SQUARE ROOTS OF MULTINOMIALS.

1. A **Rational Number** is a number which can be expressed as an integer or as a fraction; as $2, \frac{2x}{3y}, \sqrt{(27a^6)}$.

A **Rational Expression** is an expression which involves only rational numbers; as $\frac{2}{3}a + \frac{1}{2}b,\ ab + \sqrt{a^2}$.

All numbers and expressions hitherto used in this book are rational.

2. Since the square of a rational monomial is a monomial, and the square of a rational binomial is a trinomial, therefore a binomial cannot be the square of a rational expression.

E.g., $a^2 + b^2$ is not the square of any rational expression.

Hence we need consider the square roots of only multinomials of three or more terms.

3. The square root of a trinomial which is the square of a binomial and the square roots of certain multinomials can be found by inspection (Ch. VIII., § 1, Arts. 10 and 19), and will not be further considered in this chapter.

4. The process of finding the square root of a multinomial is the inverse of the process of squaring the multinomial. This process will be first illustrated by finding the square root of a trinomial.

Since $$(a+b)^2 = a^2 + 2ab + b^2,$$

we have $$\sqrt{(a^2 + 2ab + b^2)} = a + b. \qquad (1)$$

From the identity (1) we infer:

(i.) *The first term of the root is the square root of the first term of the trinomial;* i.e.,

$$a = \sqrt{a^2}.$$

(ii.) *If the square of the first term of the root be subtracted from the trinomial, the remainder will be*

$$2ab + b^2, \ = (2a + b)b.$$

Twice the first term of the root, $2a$, is called the trial divisor.

(iii.) *The second term of the root is obtained by dividing the first term of the remainder by the trial divisor;* i.e.,

$$b = \frac{2ab}{2a}.$$

(iv.) *If twice the first term of the root plus the second term, $2a + b$ (the complete divisor), be multiplied by the second term, b, and the product be subtracted from the first remainder, the second remainder will be* 0.

The work may be arranged as follows:

$$\begin{array}{r|lll}
a^2 + 2ab + b^2 & \underline{a + b} & & \\
\underline{a^2 \qquad\qquad} & 2a & \text{trial divisor} & (1) \\
2ab \qquad & 2ab \div 2a = b, & \text{second term of root} & (2) \\
 & 2a + b & \text{complete divisor} & (3) \\
\underline{2ab + b^2} & = (2a + b)b & & (4)
\end{array}$$

Ex. Find the square root of $4x^4 - 12x^2y + 9y^2$.

The work, arranged as above, is:

$$\begin{array}{r|lll}
4x^4 - 12x^2y + 9y^2 & \underline{2x^2 - 3y} & & \\
\underline{4x^4 \qquad\qquad\qquad} & 4x^2 & \text{trial divisor} & (1) \\
-12x^2y \qquad & -12x^2y \div 4x^2 = -3y, & \text{second term of root} & (2) \\
 & 4x^2 - 3y, & \text{complete divisor} & (3) \\
\underline{-12x^2y + 9y^2} & = (4x^2 - 3y)(-3y) & & (4)
\end{array}$$

Steps (2) and (4) should be performed mentally.

The work may then be arranged as follows:

$$\begin{array}{r|l}
4x^4 - 12x^2y + 9y^2 & \underline{2x^2 - 3y} \\
\underline{4x^4 \qquad\qquad\qquad} & 4x^2 \\
-12x^2y \qquad & \\
\underline{-12x^2y + 9y^2} & \underline{4x^2 - 3y}
\end{array}$$

5. When the multinomial is the square of a trinomial, the process of finding the root is an extension of the method of Art. 4.

The multinomial whose root is required should be arranged to powers of a letter of arrangement.

Since $(a+b+c)^2 = (a+b)^2 + 2(a+b)c + c^2$

$$= (a^2 + 2ab + b^2) + 2ac + 2bc + c^2, \quad (1)$$

we have $\sqrt{[(a^2 + 2ab + b^2) + 2ac + 2bc + c^2]} = a + b + c. \quad (2)$

The first two terms of the root are found by inspection, or by the method of Art. 4. The work may be arranged as follows:

$a^2+2ab+b^2+2ac+2bc+c^2$	$a+b+c$	required root
a^2	$2a$	trial divisor
$2ab$	$2ab \div 2a = b$,	second term of root
$2ab+b^2$	$2a+b$	complete divisor of first stage
$2ac$	$2ac \div 2a = c$,	third term of root
	$2a+2b+c$,	complete divisor of second stage
$2ac+2bc+c^2$		

Observe that $2ac$ is the first term of the remainder after subtracting $(a+b)^2 = a^2 + 2ab + b^2$. For, in finding the first two terms of the root we first subtracted a^2 and then $2ab + b^2$.

Ex. 1. Find the square root of

$$4x^4 - 12x^3 + 29x^2 - 30x + 25.$$

The work follows:

$4x^4 - 12x^3 + 29x^2 - 30x + 25$	$2x^2 - 3x + 5$
$4x^4$	$4x^2$
$-12x^3$	
$-12x^3 + 9x^2$	$4x^2 - 3x$
$20x^2$	
$20x^2 - 30x + 25$	$4x^2 - 6x + 5$

Observe that only the trial divisor and the complete divisor of each stage are written, the other steps being performed mentally.

Notice, also, that the first three terms are not the square of the sum of the first two terms of the root. But if $29x^2$ had been separated into $9x^2 + 20x^2$, such would have been the case.

6. The method of the preceding articles can be extended to find square roots which are multinomials of any number of terms.

The work consists of successive repetitions of the following steps:

After one or more terms of the root have been found, obtain each succeeding term, by dividing the first term of the remainder at that stage by twice the first term of the root.

Find the next remainder by subtracting from the last remainder the expression

$$(2a + b)b,$$

wherein a stands for the part of the root already found, and b for the term last found.

Ex. 1.

$$\begin{array}{l|l}
36-24a+40a^2-24a^3+13a^4-6a^5+a^6 & 6-2a+3a^2-a^3 \\ \hline
36 & 12 \\ \hline
\quad -24a & \\
\quad -24a+4a^2 & 12-2a \\ \hline
\qquad\quad 36a^2 & \\
\qquad\quad 36a^2-12a^3+9a^4 & 12-4a+3a^2 \\ \hline
\qquad\qquad -12a^3+4a^4 & \\
\qquad\qquad -12a^3+4a^4-6a^5+a^6 & 12-4a+6a^2-a^3 \\ \hline
\end{array}$$

If the polynomial whose root is required contains fractional terms, the root can be found by the preceding method.

Ex. 2.

$$\begin{array}{l|l}
\frac{4x^2}{25a^2}-\frac{4bx}{5ac}+\frac{b^2}{c^2}-\frac{12dx}{5af}+\frac{6bd}{cf}+\frac{9d^2}{f^2} & \frac{2x}{5a}-\frac{b}{c}-\frac{3d}{f} \\ \hline
\frac{4x^2}{25a^2} & \frac{4x}{5a} \\ \hline
\quad -\frac{4bx}{5ac} & \\
\quad -\frac{4bx}{5ac}+\frac{b^2}{c^2} & \frac{4x}{5a}-\frac{b}{c} \\ \hline
\qquad\quad -\frac{12dx}{5af} & \\
\qquad\quad -\frac{12dx}{5af}+\frac{6bd}{cf}+\frac{9d^2}{f^2} & \frac{4x}{5a}-\frac{2b}{c}-\frac{3d}{f} \\ \hline
\end{array}$$

EXERCISES III.

Find the square root of each of the following expressions:

1. $x^4 - 4x^3 + 8x + 4.$
2. $4m^4 - 4m^3 + 5m^2 - 2m + 1.$
3. $x^4 - 2x^3 + 3x^2 - 2x + 1.$
4. $4x^4 + 12x^3 + 5x^2 - 6x + 1.$
5. $9x^4 + 12x^3 - 26x^2 - 20x + 25.$
6. $4x^4 - 28x^3 + 51x^2 - 7x + \frac{1}{4}.$
7. $x^4y^4 - 4x^3y^3 + 6x^2y^2 - 4xy + 1.$
8. $\frac{1}{9}x^4 + \frac{4}{3}x^3y + 2x^2y^2 - 12xy^3 + 9y^4.$
9. $x^4 - 6ax^3 + 13a^2x^2 - 12a^3x + 4a^4.$
10. $4a^2 + 9b^2 + 16c^2 - 12ab + 16ac - 24bc.$
11. $49x^8 + 42x^6 - 19x^4 - 12x^2 + 4.$
12. $25a^4b^2 - 30a^3b^3 + 29a^2b^4 - 12ab^5 + 4b^6.$
13. $25x^4 - 30ax^3 + 49a^2x^2 - 24a^3x + 16a^4.$
14. $a^4 + 4a^3 + 4a^2 + 2a + 4 + \frac{1}{a^2}.$
15. $\frac{4}{x^4} - \frac{12}{x^3} + \frac{25}{x^2} - \frac{24}{x} + 16.$
16. $\frac{x^4}{y^4} - 4x^3 + 4x^2y^2 + 6xy - 12y^3 + 9\frac{y^4}{x^2}.$
17. $x^4 + \frac{2x^3}{a} + \frac{x^2}{a^2} + 2ax + 2 + \frac{a^2}{x^2}.$
18. $\frac{y^4}{16} - \frac{y^3}{2x} + \frac{5y^2}{2x^2} - \frac{6y}{x^3} + \frac{9}{x^4}.$
19. $1 + 2x - x^2 + 3x^4 - 2x^5 + x^6.$
20. $x^6 + 4x^5 + 10x^4 + 20x^3 + 25x^2 + 24x + 16.$
21. $x^6 - 6ax^5 + 15a^2x^4 - 20a^3x^3 + 15a^4x^2 - 6a^5x + a^6.$
22. $x^6 - 4x^5 + 10x^4 - 20x^3 + 25x^2 - 24x + 16.$
23. $1 - 4a + 64a^6 - 64a^5 - 32a^3 + 48a^4 + 12a^2.$
24. $4a^6 + 17a^2 - 22a^3 + 13a^4 - 24a - 4a^5 + 16.$
25. $25x^8 + 70x^6 - 44x^7 + 73x^4 - 76x^5 + 36x^2 - 60x^3 + 4x^{10} - 12x^9.$
26. $9x^6 + 6x^5y + 43x^4y^2 + 2x^3y^3 + 45x^2y^4 - 28xy^5 + 4y^6.$

27. $\frac{x^4}{y^4} - 4\frac{x^3}{y^3} + 4\frac{x^2}{y^2} + 2\frac{x}{y} - 14 + 20\frac{y}{x} + \frac{y^2}{x^2} - 10\frac{y^3}{x^3} + 25\frac{y^4}{x^4}.$

28. $x^4 + 4x^3 + 6x^2 + 5x + 5 + \frac{5}{x} + \frac{9}{4x^2} + \frac{1}{x^3} + \frac{1}{x^4}.$

29. $a^{2m}x^{2n} + 10a^{2m-2}x^{2n+1} - 6a^{m+1}x^{n+1} + 25a^{2m-4}x^{2n+2} - 30a^{m-1}x^{n+2}$
$+ 9a^2x^2.$

§ 4. CUBE ROOTS OF MULTINOMIALS.

1. Since the cube of a rational monomial is a monomial, and the cube of a rational binomial is a multinomial of four terms, a binomial or a trinomial cannot be the cube of a rational expression.

E.g., $a^3 + b^3$ is not the cube of any rational expression.

Hence we need consider only the cube roots of multinomials of four or more terms.

2. The cube root of a multinomial of four terms, which is the cube of a binomial, can be found by inspection (Ch. VIII., § 1, Art. 20), and will not be further considered in this chapter.

3. The process of finding the cube root of a multinomial is the inverse of the process of cubing the multinomial. This process will be first illustrated by finding the cube root of a multinomial of four terms.

Since $$(a+b)^3 = a^3 + 3a^2b + 3ab^2 + b^3$$
$$= a^3 + (3a^2 + 3ab + b^2)b, \qquad (1)$$

we have $$\sqrt[3]{(a^3 + 3a^2b + 3ab^2 + b^3)} = a + b. \qquad (2)$$

From the identity (2) we infer:

(i.) *The first term of the root is the cube root of the first term of the multinomial;* i.e., $a = \sqrt[3]{a^3}$.

(ii.) *If the cube of the first term of the root be subtracted from the multinomial, the remainder will be*

$$3a^2b + 3ab^2 + b^3 = (3a^2 + 3ab + b^2)b.$$

Three times the square of the first term of the root, $3a^2$, is called the trial divisor.

(iii.) *The second term of the root is obtained by dividing the first term of the remainder by the trial divisor;* i.e.,

$$b=\frac{3a^2b}{3a^2}.$$

(iv.) *If the sum* $3a^2+3ab+b^2$, *the complete divisor, be multiplied by the second term of the root, and this product be subtracted from the first remainder, the second remainder will be* 0.

The work may be arranged as follows:

$a^3+3a^2b+3ab^2+b^3$	$a+b$		
a^3	$3a^2$	trial divisor	(1)
$3a^2b$	$3a^2b \div 3a^2=b,$	second term of root	(2)
	$3a^2+3ab+b^2,$	complete divisor	(3)
$3a^2b+3ab^2+b^3$	$=(3a^2+3ab+b^2)\times b$		(4)

Ex. 1. Find the cube root of $27x^3+54x^2y+36xy^2+8y^3$.

The work, arranged as above, is:

$27x^3+54x^2y+36xy^2+8y^3$	$3x+2y$		
$27x^3$	$3(3x)^2=27x^2,$	trial divisor	(1)
$54x^2y$	$54x^2y \div 27x^2=2y,$	second term of root	(2)
	$3(3x)^2+3(3x)(2y)+(2y)^2=27x^2+18xy+4y^2,$	complete divisor	(3)
$54x^2y+36xy^2+8y^3$	$=(27x^2+18xy+4y^2)(2y)$		(4)

Steps (2) and (4) should be performed mentally.

4. The method of the preceding article can be extended to find cube roots which are multinomials of any number of terms, as the method of finding square roots was extended.

The work consists of successive repetitions of the following steps:

After one or more terms of the root have been found, obtain each succeeding term by dividing the first term of the remainder at that stage by three times the square of the first term of the root.

Find the next remainder by subtracting from the last remainder the expression

$$(3\,a^2 + 3\,ab + b^2)\,b,$$

wherein a stands for the part of the root already found, and b for the term last found.

The given multinomial should be arranged to powers of a letter of arrangement.

Ex.

$$\begin{array}{l|l}
27-27\,x+90\,x^2-55\,x^3+90\,x^4-27\,x^5+27\,x^6 & 3-x+3\,x^2 \\ \cline{2-2}
\underline{27} & 3(3)^2=27 \\
\quad -27\,x & \\
\quad \underline{-27\,x+\;9x^2-\quad x^3} & 3(3)^2+3(3)(-x)+(-x)^2=27-9\,x+x^2 \\ \cline{2-2}
\qquad\qquad 81\,x^2-54\,x^3 & 3(3-x)^2+3(3-x)(3\,x^2)+(3\,x^2)^2 \\
\qquad\qquad \underline{81\,x^2-54\,x^3+90\,x^4-27\,x^5+27\,x^6} & \underline{=27-18\,x+30\,x^2-9\,x^3+9\,x^4}
\end{array}$$

EXERCISES IV.

Find the cube root of each of the following expressions:

1. $x^6 - 6\,x^5 + 15\,x^4 - 20\,x^3 + 15\,x^2 - 6\,x + 1.$
2. $8\,x^6 - 36\,x^5 + 66\,x^4 - 63\,x^3 + 33\,x^2 - 9\,x + 1.$
3. $156\,a^4 - 144\,a^5 - 99\,a^3 + 64\,a^6 + 39\,a^2 - 9\,a + 1.$
4. $a^3 - 6\,a^2b + 3\,a^2c + 12\,ab^2 - 12\,abc + 3\,ac^2 - 8\,b^3 + 12\,b^2c - 6\,bc^2 + c^3.$
5. $1 + 3\,x + 6\,x^2 + 7\,x^3 + 6\,x^4 + 3\,x^5 + x^6.$
6. $1 - 6\,x + 9\,x^2 + 4\,x^3 - 9\,x^4 - 6\,x^5 - x^6.$
7. $8\,x^3 - 12\,x^2 + 12\,x - 7 + \dfrac{3}{x} - \dfrac{3}{4\,x^2} + \dfrac{1}{8\,x^3}.$
8. $27\,a^6x^6 + 54\,a^5x^5 + 9\,a^4x^4 - 28\,a^3x^3 - 3\,a^2x^2 + 6\,ax - 1.$
9. $8\,x^6 - 36\,x^5y + 42\,x^4y^2 + 9\,x^3y^3 - 21\,x^2y^4 - 9\,xy^5 - y^6.$
10. $8\,a^6 + 48\,a^5b + 60\,a^4b^2 - 80\,a^3b^3 - 90\,a^2b^4 + 108\,ab^5 - 27\,b^6.$
11. $8\,x^6 - 36\,ax^5 + 102\,a^2x^4 - 171\,a^3x^3 + 204\,a^4x^2 - 144\,a^5x + 64\,a^6.$
12. $x^3 + 3\,x^7 - 9\,x^{11} - 27\,x^{15} - 6\,x^5 - 54\,x^{13} + 28\,x^9.$
13. $108\,a^5 - 48\,a^4 + 8\,a^3 + 54\,a^7 - 12\,a^8 + a^9 - 112\,a^6.$
14. $8\,a^6 - 48\,a^5x + 60\,a^4x^2 - 27\,x^6 - 108\,ax^5 - 90\,a^2x^4 + 80\,a^3x^3.$
15. $1 + 3\,x - 8\,x^3 - 6\,x^4 + 6\,x^5 + 8\,x^6 - 3\,x^8 - x^9.$

16. $\frac{125y^6}{x^6} - \frac{150y^5}{x^5} - \frac{165y^4}{x^4} + \frac{172y^3}{x^3} + \frac{99y^2}{x^2} - \frac{54y}{x} - 27.$

17. $a^{3m} - 6a^{2m+1}x^n + 12a^{m+2}x^{2n} - 8a^3x^{3n}.$

18. $64x^{3n} - 144x^{3n-1} + 12x^{3n-2} + 117x^{3n-3} - 6x^{3n-4} - 36x^{3n-5}$
$+ 8x^{3n-6}.$

§ 5. HIGHER ROOTS.

1. The process of finding the nth root of a multinomial is the inverse of raising a multinomial (the required root) to the nth power.

The methods may be derived from the following identities:

$$(a+b)^4 = a^4 + 4a^3b + 6a^2b^2 + 4ab^3 + b^4,$$
$$= a^4 + (4a^3 + 6a^2b + 4ab^2 + b^3)b. \qquad (1)$$
$$(a+b)^5 = a^5 + 5a^4b + 10a^3b^2 + 10a^2b^3 + 5ab^4 + b^5,$$
$$= a^5 + (5a^4 + 10a^3b + 10a^2b^2 + 5ab^3 + b^4)b. \qquad (2)$$

etc.

The work for the fifth root consists of successive repetitions of the following steps (the first term of the root being the fifth root of the first term of the given multinomial):

After one or more terms have been found, obtain each succeeding term by dividing the first term of the remainder at that stage by five times the fourth power of the first term of the root.

Find the next remainder by subtracting from the last remainder the expression

$$(5a^4 + 10a^3b + 10a^2b^2 + 5ab^3 + b^4) \times b,$$

wherein a stands for the part of the root already found, and b for the term last found.

2. Since $\sqrt[4]{N} = \sqrt{\sqrt{N}}$, wherein N stands for any multinomial, the fourth root is most easily found as the square root of the square root of the given multinomial.

In like manner, since $\sqrt[6]{N} = \sqrt[3]{\sqrt{N}}$, the sixth root can be most easily found as the cube root of the square root of the given multinomial. And so on for any root whose index can be factored.

EXERCISES V.

Find the fourth root of each of the following expressions:

1. $x^8 + 4x^6 + 6x^4 + 4x^2 + 1$.

2. $a^8 + 4a^7b + 10a^6b^2 + 16a^5b^3 + 19a^4b^4 + 16a^3b^5 + 10a^2b^6 + 4ab^7 + b^8$.

3. $16x^8 - 160x^7 + 408x^6 + 440x^5 - 2111x^4 - 1320x^3 + 3672x^2 + 4320x + 1296$.

4. $625x^8 + 5500x^7 + 17150x^6 + 20020x^5 + 721x^4 - 8008x^3 + 2744x^2 - 352x + 16$.

5. $x^{4n} + 8x^{4n-1} + 24x^{4n-2} + 32x^{4n-3} + 16x^{4n-4} - 12x^{3n} - 72x^{3n-1} - 144x^{3n-2} - 96x^{3n-3} + 54x^{2n} + 216x^{2n-1} + 216x^{2n-2} - 108x^n - 216x^{n-1} + 81$.

Find the fifth root of each of the following expressions:

6. $x^{10} - 10x^8 + 40x^6 - 80x^4 + 80x^2 - 32$.

7. $a^{10} + 5a^9b + 15a^8b^2 + 30a^7b^3 + 45a^6b^4 + 51a^5b^5 + 45a^4b^6 + 30a^3b^7 + 15a^2b^8 + 5ab^9 + b^{10}$.

8. $x^{10} - 35x^9 + 505x^8 - 3850x^7 + 16505x^6 - 39277x^5 + 49515x^4 - 34650x^3 + 13635x^2 - 2835x + 243$.

Find the sixth roots of each of the following expressions:

9. $64x^{12} - 192x^{10} + 240x^8 - 160x^6 + 60x^4 - 12x^2 + 1$.

10. $a^{12} + 6a^{11}b + 21a^{10}b^2 + 50a^9b^3 + 90a^8b^4 + 126a^7b^5 + 141a^6b^6 + 126a^5b^7 + 90a^4b^8 + 50a^3b^9 + 21a^2b^{10} + 6ab^{11} + b^{12}$.

§ 6. ROOTS OF ARITHMETICAL NUMBERS.

Square Roots.

1. Since the squares of the numbers 1, 2, 3, ⋯, 9, 10, are 1, 4, 9, ⋯, 81, 100, respectively, the square root of any integer of *one or two* digits is a number of *one* digit.

Since the squares of the numbers 10, 11, ⋯, 100, are 100, 121, ⋯, 10000, the square root of any integer of *three or four* digits is a number of *two* digits.

In general, the square root of any integer of $2n-1$ or $2n$ digits is a number of n digits.

Therefore, to find the number of digits in the square root of a given integer, we first mark off the digits from right to left in groups of two. The number of digits in the square root will be equal to the number of groups, counting any one digit remaining on the left as a group.

2. The method of finding square roots of numbers is then derived from the identity

$$(a+b)^2 = a^2 + (2a+b)b, \qquad (1)$$

wherein a denotes *tens*, and n denotes *units*, if the square root be a number of two digits.

Ex. 1. Find the square root of 1296.

We see that the root is a number of *two* digits, since the given number divides into *two* groups. The digit in the *tens'* place is 3, the square root of 9, the square next less than 12. Therefore, in the identity (1), a denotes 3 *tens*, or 30.

The work then proceeds as follows:

	$a+b$		
12'96	$30+6=36$		
9 00	$2a=60,$	trial divisor	(1)
3 96	$(2ab+b^2)\div 2a = 396 \div 60 = 6+$		(2)
3 96	$=(2a+b)\times b = (60+6)\times 6$		(3)

Observe that the first remainder, 396, is equal to $2ab+b^2$, and that we cannot separate it into the sum of two terms, one of which is $2ab$. We cannot, therefore, determine b by dividing $2ab$ by $2a$, as in finding square roots of algebraic expressions.

Consequently step (2) *suggests* the value of b but does not definitely determine it. As a rule we take the integral part of the quotient, 6 in the above example, and test that value by step (3).

The preceding method may be extended to find roots which contain any number of digits.

At any stage of the work a stands for the part of the root already found, and b for the digit to be found.

Ex. 2. Find the square root of 51529.

The root is a number of *three* digits, since the given number divides into *three* groups. The digit in the *hundreds'* place is 2, the square root of 4, the square next less than 5. Therefore in the identity (1), a denotes 2 *hundreds*, or 200, in the first stage of the work.

The work then proceeds as follows:

5′ 15′ 29	$200 + 20 + 7 = 227$	
4 00 00	$2a = 400$, trial divisor	(1)
1 15 29	$(2ab + b^2) \div 2a = 11529 \div 400 = 20+$	(2)
84 00	$= (2a + b)b = (400 + 20) \times 20$	(3)
31 29	$(2ab + b^2) \div 2a = 3129 \div 440 = 7+$	(4)
31 29	$= (2a + b)b = (440 + 7) \times 7$	(5)

Observe that in the second stage of the work, a stands for the part of the root already found, 220, and b for the next figure of the root.

In practice the work may be arranged more compactly, omitting unnecessary ciphers, and in each remainder writing only the next group of figures. Thus:

5′ 15′ 29	227	
4		
1 15	$11 \div 4 = 2+$	(2)
84	42	
31 29	$312 \div 44 = 7+$	(4)
31 29	447	

Observe that the trial divisor at any stage is twice the part of the root already found, as in (2) and (4).

The abbreviated work in the last example illustrates the following method:

After one or more figures of the root have been found, obtain the next figure of the root by dividing the remainder at that stage (omitting the last figure) by the trial divisor at that stage.

See lines (2) and (4).

Annex this quotient to the part of the root already found, and also to the trial divisor to form the complete divisor.

Find the next remainder by subtracting from the last remainder the product of the complete divisor and the figure of the root last found.

3. Since the number of decimal places in the square of a decimal fraction is twice the number of decimal places in the fraction, the number of decimal places in the square root of a decimal fraction is one half the number of decimal places in the fraction.

Consequently, in finding the square root of a decimal fraction, the decimal places are divided into groups of two from the decimal point to the right, and the integral places from the decimal point to the left as before.

Ex.

14′ 46.28′ 09	38.03
9	
5 46	
5 44	68
2.28 09	
2.28 09	76.03

Observe that in finding the second figure of the root, we have $\frac{54}{6} = 9$; but $69 \times 9 = 621$, which is greater than the number 546 from which it is to be subtracted. Hence we take the next less figure 8.

Cube Roots.

4. Since the cubes of the numbers 1, 2, 3, $\cdots$ 9, 10 are 1, 8, 27, $\cdots$ 729, 1000, respectively, the cube root of any integer of *one*, *two*, or *three* digits is a number of *one* digit. *The cube roots of such numbers can be found only by inspection.*

Since the cubes of 10, 11, $\cdots$ 100 are 1000, 1331, $\cdots$, 970294, 1000000, respectively, the cube root of any integer of *four*, *five*, or *six* digits is a number of *two* digits.

In general, the cube root of any integer of $3n - 2$, $3n - 1$, or $3n$ digits is a number of n digits.

Therefore, to find the number of digits in the cube root of a given integer, we first mark off the digits from right to left in groups of *three*. The number of digits in the cube root will be equal to the number of groups, counting one or two digits remaining on the left as a group.

5. The method of finding cube roots of numbers is derived from the identity

$$(a+b)^3 = a^3 + (3a^2 + 3ab + b^2)b,$$

wherein a denotes *tens*, and b denotes *units*, if the cube root is a number of two digits.

Ex. Find the cube root of 59319.

The digits in the *tens'* place of the root is 3, the cube root of 27, the cube next less than 59. Therefore in identity (1), a denotes 3 *tens* or 30.

The work may then be arranged as follows:

$$\begin{array}{r|lr}
59'\,319 & a+b & \\
\underline{27\ \ 000} & \underline{30+9} & \\
32\ \ 319 & 3a^2 = 3(30)^2 = 2700 & (1) \\
 & (3a^2b + 3ab^2 + b^3) \div 3a^2 = 32319 \div 2700 = 9+ & (2) \\
 & \qquad 3a^2 \;= 3(30)^2 \;= 2700 & \\
 & \qquad 3ab = 3(30)9 \;= \;\;810 & \\
 & \qquad b^2 = \qquad 9^2 = \;\;\;\underline{81} & \\
\underline{32\ \ 319} & \underline{= (3a^2 + 3ab + b^2) \times b = 3591 \times 9} & (3)
\end{array}$$

As in finding square roots of numbers, step (2) ***suggests*** the value of b, but does not definitely determine it. If the value of b makes $(3a^2 + 3ab + b^2) \times b$ greater than the number from which it is to be subtracted, we must try the next less number.

In practice the work may be arranged more compactly, omitting unnecessary ciphers, and in each remainder writing only the next group of figures.

The work of Ex. 1 may be arranged thus:

$$\begin{array}{r|rr}
59'\,319 & \underline{39} & \\
\underline{27} & & \\
32\ \ 319 & 2700 & (1) \\
 & 810 & (2) \\
 & \underline{81} & (3) \\
\underline{32\ \ 319} & \underline{3591} &
\end{array}$$

6. The preceding method may be extended to find roots that contain any number of digits.

At any stage of the work a stands for the part of the root already found, and b for the digit to be found.

In general, the method consists of a repetition of the following steps:

The trial divisor at any stage is three times the square of the part of the root already found; as 27 in the preceding example.

After one or more figures of the root have been found obtain the next figure of the root by dividing the remainder at that stage (omitting the last two figures) by the trial divisor. In the last example, $9+ = 323 \div 27$.

Annex this quotient to the part of the root already found.

Add to the trial divisor (with two ciphers annexed) three times the product of the part of the root already found (with one cipher annexed) and the figure of the root just found, and also the square of the figure of the root just found. The sum is called the complete divisor.

Find the next remainder by subtracting from the last remainder the product of the complete divisor and the figure of the root last found.

Evidently, in finding the cube root of a decimal fraction the decimal places are divided into groups of *three* figures from the decimal point to the right, and the integral places from the decimal point to the left as before.

Ex. Find the cube root of 11089.567.

The work proceeds as follows:

11′ 089.567	22.3
8	1200
3 089	120
	4
2 648	1324
441.567	1452.00
	19.80
	.09
441.567	1471.89

EXERCISES VI.

Find the square root of each of the following numbers:

1. 196. **2.** 841. **3.** 1296. **4.** 65.61.
5. 7396. **6.** 3481. **7.** 667489. **8.** 5.1529.
9. 170569. **10.** 1664.64. **11.** 582169. **12.** 57198969.
13. 1.737124. **14.** 556.0164. **15.** .00099225.

Find the cube root of each of the following numbers:

16. 2744. **17.** 39304. **18.** 110.592. **19.** 148877.
20. 328509. **21.** 2460375. **22.** 1.191016. **23.** 64481201.
24. 74088000. **25.** 340068392. **26.** 426.957777.
27. 584067.412279. **28.** 375601280.458951. **29.** .041063625.
30. .000071939391679.

Find the value of each of the following indicated roots:

31. $\sqrt[4]{279841}$. **32.** $\sqrt[4]{9904930.7841}$. **33.** $\sqrt[4]{164204746.7776}$.
34. $\sqrt[6]{3010936384}$. **35.** $\sqrt[6]{42611.309937355041}$.

CHAPTER XVI.

INEQUALITIES.

1. One number is greater or less than a second number according as the remainder of subtracting the second number from the first is positive or negative. Thus,

$a > b$, when $a - b$ is *positive*, *i.e.*, when $a - b > 0$.

$a < b$, when $a - b$ is *negative*, *i.e.*, when $a - b < 0$.

This statement is in accordance with Ch. II., § 2, Art. 18.

2. An **Inequality** is a statement that two numbers or expressions are unequal; as $a^2 + b^2 > a^2$.

The members or sides of an inequality are the numbers or expressions which are connected by one of the signs of inequality, $>$ or $<$.

3. Two inequalities are of the **Same** or **Opposite Species**, or are said to subsist in the *same* or *opposite sense*, according as they have the *same* or *opposite* sign of inequality.

E.g., $8 > 3$ and $-5 > -7$ are inequalities of the same species; $0 > -1$ and $0 < 1$ are inequalities of opposite species.

4. Observe that a relation of inequality between two numbers can be stated in two ways; as $7 > 3$, or $3 < 7$.

That is, *if the members of an inequality be interchanged, the sign of inequality must be reversed.*

Principles of Inequalities.

5. *If one number be greater than a second, and this second number be greater than a third, then the first number is greater than the third;* that is,

If $a > b$ and $b > c$, then $a > c$.

In like manner, if $a < b$ and $b < c$, then $a < c$.

E.g., $3 > 2$, $2 > 1$, and $3 > 1$; $-3 < -2$, $-2 < 0$, and $-3 < 0$.

6. Addition and Subtraction. — The following principles of inequalities involve the operations of addition and subtraction:

(i.) *If the same number, or equal numbers, be added to or subtracted from both members of an inequality, the resulting inequality will be of the same species;* that is,

If $a > b$, then $a \pm m > b \pm m$.

E.g., $3 > 2$, and $3 + 1 > 2 + 1$, and $3 - 1 > 2 - 1$.

(ii.) *If the corresponding members of two or more inequalities of the same species be added, the resulting inequality will be of the same species;* that is,

If $a_1 > b_1$, $a_2 > b_2$, $a_3 > b_3$, $\cdots$, then $a_1 + a_2 + a_3 + \cdots > b_1 + b_2 + b_3 + \cdots$.

E.g., $-5 > -7$, $3 > 2$, $0 > -4$, and $-5 + 3 + 0 > -7 + 2 - 4$; *i.e.*, $-2 > -9$.

(iii.) *If the members of one inequality be subtracted from the corresponding members of another inequality of the same species, the resulting inequality will not necessarily be of the same species;* that is,

If $a_1 > b_1$ and $a_2 > b_2$, then $a_1 - a_2$ may or may not $> b_1 - b_2$.

E.g., $11 > 6$, $4 > 3$, and $11 - 4 > 6 - 3$; $5 > 4$, $3 > 1$, but $5 - 3 < 4 - 1$.

(iv.) *If the members of an inequality be added to the corresponding members of an equality, the resulting inequality will be of the same species;* that is,

If $a = b$, and $c > d$, then $a + c > b + d$.

E.g., $2 \times 3 = 6$, $5 > 2$, and $2 \times 3 + 5 > 6 + 2$.

(v.) *If the members of an inequality be subtracted from the corresponding members of an equality, the resulting inequality will be of the opposite species;* that is,

If $a = b$, and $c > d$, then $a - c < b - d$.

E.g., $4 = 4$, $3 > -2$, and $4 - 3 < 4 - (-2)$, or $1 < 6$.

The proofs of the principles enunciated in (i.) and (ii.) follow; the other principles are easily proved in a similar manner.

(i.) If $a > b$, then $a - b$ is positive; and $a - b \pm m \mp m$ is positive.

Therefore $(a \pm m) - (b \pm m)$ is positive; and hence $a \pm m > b \pm m$.

(ii.) If $a_1 > b_1$, $a_2 > b_2$, $a_3 > b_3$, $\cdots$,

then $a_1 - b_1$, $a_2 - b_2$, $a_3 - b_3$, $\cdots$, are positive.

Therefore, $(a_1 - b_1) + (a_2 - b_2) + (a_3 - b_3) + \cdots$ is positive,

or $(a_1 + a_2 + a_3 + \cdots) - (b_1 + b_2 + b_3 + \cdots)$ is positive.

Consequently, $a_1 + a_2 + a_3 + \cdots > b_1 + b_2 + b_3 + \cdots$.

7. Multiplication and Division. — The following principles of inequalities involve the operations of multiplication and division:

(i.) *If both members of an inequality be multiplied or divided by the same positive number, or by equal positive numbers, the resulting inequality will be of the same species;* that is, if

$$a > b, \text{ then } an > bn, \text{ and } \frac{a}{n} > \frac{b}{n},$$

wherein n is a positive number.

E.g., $\quad -3 > -5$, and $-15 > -25$, and $-1 > -\frac{5}{3}$.

(ii.) *If both members of an inequality be multiplied or divided by the same negative number, or by equal negative numbers, the resulting inequality will be of the opposite species;* that is, if

$$a > b, \text{ then } a(-n) < b(-n), \text{ and } \frac{a}{-n} < \frac{b}{-n},$$

wherein $-n$ is a negative number.

E.g., $\quad 2 > -1$, and $2(-3) < (-1)(-3)$, or $-6 < 3$;

$$\text{and } \frac{2}{-2} < \frac{-1}{-2}, \text{ or } -1 < \frac{1}{2}.$$

(iii.) *If all the members of two or more inequalities of the same species be positive, and if the corresponding members be multiplied together, the resulting inequality will be of the same species;* that is, if

$$a_1 > b_1, \ a_2 > b_2, \ a_3 > b_3, \text{ then } a_1a_2a_3 > b_1b_2b_3,$$

wherein a_1, b_1, a_2, b_2, a_3, b_3 are all positive.

E.g., $\quad 12 > 4$, $3 > 2$, and $12 \times 3 > 4 \times 2$, or $36 > 8$.

(iv.) *If all the members of an odd number of inequalities of the same species be negative, and if the corresponding members be multiplied together, the resulting inequality will be of the same species;* that is, if

$$-a_1 > -b_1, \ -a_2 > -b_2, \ -a_3 > -b_3,$$

$$\text{then } (-a_1)(-a_2)(-a_3) > (-b_1)(-b_2)(-b_3).$$

E.g., $\quad -3 < -2$, $-5 < -4$, $-4 < -3$,

and $(-3)(-5)(-4) < (-2)(-4)(-3)$, or $-60 < -24$.

(v.) *If all the members of an even number of inequalities of the same species be negative, and if the corresponding members be multiplied together, the resulting inequality will be of the opposite species;* that is, if

$$-a_1 > -b_1, \ -a_2 > -b_2, \text{ then } (-a_1)(-a_2) < (-b_1)(-b_2).$$

E.g., $\quad -2 > -3$, $-4 > -5$, and $(-2)(-4) < (-3)(-5)$, or $8 < 15$.

(vi.) *If the members of one inequality be divided by the corresponding members of another inequality of the same species, the resulting inequality will not necessarily be of the same species;* that is, if

$$a > b \text{ and } c > d, \text{ then } \frac{a}{c} \text{ may or may not } > \frac{b}{d}.$$

E.g., $9 > 4$ and $3 > 2$, and $\frac{9}{3} > \frac{4}{2}$; $9 > 8$ and $3 > 2$, but $\frac{9}{3} < \frac{8}{2}$.

(vii.) *If the members of an equation which are both positive be divided by the corresponding members of an inequality which are either both positive or both negative, the resulting inequality will be of the opposite species;* that is, if

$$a = b \text{ and } \pm c > \pm d, \text{ then } \frac{a}{\pm c} < \frac{b}{\pm d},$$

wherein a and b are positive.

E.g., $3 = 3$, $12 > 5$, and $\frac{3}{12} < \frac{3}{5}$; $5 = 5$, $-2 > -5$, and $\frac{5}{-2} < -1$.

The proofs of the principles enunciated in (ii.) and (v.) follow; the other principles can be easily proved in a similar way.

(ii.) If $a > b$, then $a - b$ is positive. Let $-m$ be any negative number. Then

$$-m(a-b), = -ma - (-mb), \text{ and } \frac{(a-b)}{-m}, = \frac{a}{-m} - \frac{b}{-m}$$

are negative. Therefore

$$-ma < -mb, \text{ and } \frac{a}{-m} < \frac{b}{-m}.$$

(v.) Let $-a_1 > -b_1$, and $-a_2 > -b_2$.

Then $(-a_1)(-a_2) < (-b_1)(-a_2)$, by (ii.).

Also $(-b_1)(-a_2) < (-b_1)(-b_2)$, by (ii.).

Therefore $(-a_1)(-a_2) < (-b_1)(-b_2)$, by Art. 5.

The principle can be easily extended to include any even number of inequalities.

8. Powers and Roots. — The following principles follow directly from those of the preceding article:

(i.) *If both members of an inequality be positive, and be raised to the same positive integral power, the resulting inequality will be of the same species;* that is, if

$$a > b, \text{ then } a^n > b^n,$$

wherein a and b are positive, and n is a positive integer.

E.g., $9 > 4$, and $81 > 16$.

(ii) *If both members of an inequality be negative, and be raised to the same positive* **odd** *power, the resulting inequality will be of the same species;* that is, if

$$-a > -b, \text{ then } (-a)^{2n+1} > (-b)^{2n+1}.$$

E.g., $-3 < -2$, and $(-3)^3 < (-2)^3$, or $-27 < -8$.

(iii.) *If both members of an inequality be negative, and be raised to the same positive* **even** *power, the resulting inequality will be of the opposite species;* that is, if

$$-a > -b, \text{ then } (-a)^{2n} < (-b)^{2n}.$$

E.g., $-2 > -5$, and $(-2)^2 < (-5)^2$, or $4 < 25$.

(iv.) *If the same principal root of both members of an inequality be taken, the resulting inequality will be of the same species;* that is, if

$$a > b, \ \sqrt[n]{a} > \sqrt{b}.$$

E.g., $9 > 4$, and $3 > 2$; $-27 < -8$, and $-3 < -2$.

(v.) *If the same negative even root of both members of an inequality be taken, the resulting inequality will be of the opposite species;* that is, if

$$a > b, \ -\sqrt[2n]{a} < -\sqrt[2n]{b}.$$

E.g., $9 > 4$, and $-3 < -2$.

9. Transformation of Inequalities. — The preceding principles enable us to make the following transformations of inequalities:

(i.) *Any term may be transferred from one member of an inequality to the other, if its sign be reversed.*

E.g., if $a - b > c$, then $a > b + c$.

(ii.) *If the signs of both members of an inequality be reversed from + to −, or from − to +, the sign of inequality must be reversed.*

E.g., $-3 < 5$, and $3 > -5$.

(iii.) *An inequality may be cleared of fractions by multiplying both members by the L. C. D., taken positively.*

E.g., if $\frac{a}{-3} - \frac{b}{5} < \frac{c}{6}$, then $-10a - 6b < 5c$.

Notice that the L. C. D. must be positive. If it be negative, the inequality must be reversed by (ii.).

E.g., if $$\frac{x}{b-c}-\frac{y}{b+c}>\frac{z}{b^2-c^2},$$

then $x(b+c)-y(b-c)>z$, if b^2-c^2 be positive, *i.e.*, if $b>c$

while $x(b+c)-y(b-c)<z$, if b^2-c^2 be negative, *i.e.*, if $b<c$.

(iv.) *Common positive factors can be canceled from both members of an inequality.*

E.g., $$8>-12, \text{ and } 2>-3.$$

If $$x(a^2-b^2)<(a+b)^2,$$

then $x(a-b)<(a+b)$, when $a+b$ is positive;

but $x(a-b)>(a+b)$, when $a+b$ is negative.

(v.) *If the reciprocals of the members of an inequality, which are either both positive or both negative, be taken, the resulting inequality will be of the opposite species.*

E.g., $3>2$, and $\frac{1}{3}<\frac{1}{2}$; $-5<-2$, and $-\frac{1}{5}>-\frac{1}{2}$.

10. An **Absolute Inequality** is one which holds for all values of the literal numbers involved; as $a^2+b^2>a^2$.

Such inequalities are analogous to identical equations.

A **Conditional Inequality** is one which holds only for values of the literal numbers lying between certain limits.

E.g., $x+1>2$, only for values of x greater than 1; that is, for values of x between 1 and $+\infty$.

$x^2+1>2$, only for values of x greater than 1 and less than -1; that is, for values of x between 1 and $+\infty$, and between -1 and $-\infty$.

Absolute Inequalities.

11. Ex. **1.** Prove that if $a\neq b$, then $a^2+b^2>2ab$.

We have $$(a-b)^2>0, \quad (1)$$

since the square of any positive or negative number is positive, and therefore greater than 0.

From (1), $$a^2-2ab+b^2>0;$$

whence $$a^2+b^2>2ab, \text{ by Art. 9 (i.).}$$

Ex. **2.** The sum of any positive fraction and its reciprocal is greater than 2; that is, $\frac{a}{b}+\frac{b}{a}>2$, wherein $a\neq b$.

Dividing both members of the inequality

$$a^2 + b^2 > 2ab$$

by the *positive* number ab, we have

$$\frac{a}{b} + \frac{b}{a} > 2.$$

Ex. 3. Which is greater, $\frac{a+4b}{a+5b}$ or $\frac{a+2b}{a+3b}$, in which a and b are *positive?*

We can determine which fraction is greater by finding their difference.

$$\frac{a+4b}{a+5b} - \frac{a+2b}{a+3b} = \frac{(a+4b)(a+3b)-(a+2b)(a+5b)}{(a+5b)(a+3b)}$$

$$= \frac{12b^2 - 10b^2}{(a+5b)(a+3b)} = \frac{2b^2}{(a+5b)(a+3b)}.$$

Since this remainder is *positive*, we have

$$\frac{a+4b}{a+5b} > \frac{a+2b}{a+3b}.$$

Conditional Inequalities.

12. Ex. 1. Between what limits must x lie to satisfy the inequality

$$x > 5x - 10?$$

Transferring terms, we have

$$-4x > -10;$$

whence $x < \frac{5}{2}$, by Art. 7 (ii.).

That is, the given inequality will be satisfied by all values of x between $\frac{5}{2}$ and $-\infty$.

Ex. 2. What values of x satisfy the inequality

$$x^2 > 9?$$

From Art. 8 (iv.), $x > 3$;

from Art. 8 (v.), $x < -3$.

Therefore the given inequality is satisfied by all values of x between 3 and $+\infty$, and between -3 and $-\infty$.

Ex. 3. What values of x satisfy the inequality

$$x^2 + 5x > -6?$$

Transferring -6 to the first member, $x^2 + 5x + 6 > 0$.

or $(x+2)(x+3) > 0$.

In order that the product $(x+2)(x+3)$ may be greater than 0, *i.e.*, *positive*, the two factors must be either *both positive* or *both negative*.

The factors $x+2$ and $x+3$ will be both positive, when $x>-2$.

Thus, if $x=-1$, then $(x+2)(x+3)=(-1+2)(-1+3)=2$.

The factors will be both negative, when $x<-3$.

Thus if $x=-4$, then $(x+2)(x+3)=(-4+2)(-4+3)=$ 2.

Therefore the given inequality will be satisfied by all values of x between -2 and $+\infty$, and between -3 and $-\infty$.

Ex. 4. What values of x satisfy the inequality

$$\frac{x}{x-a}-\frac{2a}{x+a}>\frac{8a^2}{x^2-a^2},$$

wherein a is positive?

Notice that we cannot clear this inequality of fractions at once, since we do not know whether, for the value of x which satisfies the inequality, x^2-a^2 will be positive or negative; that is, whether the sign of inequality must be kept the same or reversed.

Transferring $\frac{8a^2}{x^2-a^2}$ to the first member, and adding fractions,

$$\frac{x^2-ax-6a^2}{x^2-a^2}>0,$$

or

$$\frac{(x+2a)(x-3a)}{x^2-a^2}>0.$$

Multiplying numerator and denominator by x^2-a^2,

$$\frac{(x+2a)(x-3a)(x^2-a^2)}{(x^2-a^2)^2}>0.$$

Since the denominator $(x^2-a^2)^2$ will be positive for all values of x, we may now clear of fractions. We then have

$$(x-3a)(x-a)(x+a)(x+2a)>0.$$

This inequality will be satisfied when all the factors of the first member are positive, or all negative, or when two of them are positive and the other two are negative.

When $x>3a$, the factors will all be positive. Therefore the inequality is satisfied by all values of x between $3a$ and $+\infty$.

When $x<3a$ but $>a$, the first factor is negative, and the three other factors are positive. Therefore no value of x between $3a$ and a satisfies the inequality.

When $x<a$ but $>-a$, the first two factors are negative, and the two other factors are positive. Therefore the inequality is satisfied by all values of x between a and $-a$.

When $x < -a$ but $> -2a$, the first three factors are negative, and the last factor is positive. Therefore no value of x between $-a$ and $-2a$ satisfies the inequality.

When $x < -2a$, all four factors are negative. Therefore the inequality is satisfied by all values of x between $-2a$ and $-\infty$.

Consequently the values of x between the following limits satisfy the given inequality,

$$+\infty > x > 3a; \quad +a > x > -a; \quad -2a > x > -\infty.$$

Ex. 5. What values of x satisfy the inequalities

$$2x + 1 < 3x, \qquad (1)$$

$$5(x+1) > 6x? \qquad (2)$$

From (1), $\quad -x < -1$, or $x > 1$.

From (2), $\quad -x > -5$, or $x < 5$.

Hence the values of x lie between 1 and 5.

The only integral values of x which satisfy the inequalities are 2, 3, 4.

Ex. 6. What values of x and y satisfy the inequality

$$5x + 3y > 11, \qquad (1)$$

and the equality $\quad 3x + 5y = 13? \qquad (2)$

Multiplying (1) by 3, $\quad 15x + 9y > 33. \qquad (3)$

Multiplying (2) by 5, $\quad 15x + 25y = 65. \qquad (4)$

Subtracting (4) from (3), $\quad -16y > -32$, or $y < 2$.

Multiplying (1) by 5, $\quad 25x + 15y > 55. \qquad (5)$

Multiplying (2) by 3, $\quad 9x + 15y = 39. \qquad (6)$

Subtracting (6) from (5), $\quad 16x > 16$, or $x > 1$.

Notice that not *any* value of x greater than 1 taken with *any* value of y less than 2, will satisfy both (1) and (2). But such values of x and y as satisfy (1) and (2) simultaneously, must be greater than 1 for x, and less than 2 for y. If we assign to x any value greater than 1, we can determine from (2) the corresponding value of y, which will always be less than 2; these corresponding values of x and y will then satisfy (1).

E.g., let $x = \frac{4}{3}$; then from (2), $y = \frac{9}{5}$, < 2; these values of x and y satisfy (1).

EXERCISES I.

Prove the following inequalities, in which the literal numbers are all positive and unequal:

1. $a^2 + b^2 + c^2 > ab + ac + bc$. **2.** $a^2b^2 + b^2c^2 + a^2c^2 > abc(a + b + c)$.

3. $ab(a + b) + bc(b + c) + ac(a + c) > 6abc$.

4. If $l^2+m^2+n^2=1$, and $l_1^2+m_1^2+n_1^2=1$, then $ll_1+mm_1+nn_1<1$.

5. $a^3+b^3>a^2b+ab^2$.

6. $a^4+b^4>a^3b+ab^3$.

7. $(a+b)(b+c)(c+a)>8\,abc$.

8. $3(a^2+b^2+c^2)>(a+b+c)^2$.

9. $a^3-b^3>3\,a^2b-3\,ab^2$, if $a>b$;
$<3\,a^2b-3\,ab^2$, if $a<b$.

10. $(ab+xy)(ax+by)>4\,abxy$.

11. $3(1+a^2+a^4)>(1+a+a^2)^2$.

12. $2(a^3+b^3+c^3)>ab(a+b)+ac(a+c)+bc(b+c)$.

13. $a^3+b^3+c^3>3\,abc$.

14. $a^4+b^4+c^4>abc(a+b+c)$.

15. $(a+b+c)^3>3(a+b)(a+c)(b+c)$.

16. $(a+b+c)(a^2+b^2+c^2)>9\,abc$.

17. $9(a^3+b^3+c^3)>(a+b+c)^3$.

18. $\frac{a_1}{a_2}+\frac{a_2}{a_3}+\frac{a_3}{a_1}>3$.

19. $\frac{2a}{b+c}+\frac{2b}{a+c}+\frac{2c}{a+b}>3$.

20. $(n-1)(a_1^2+a_2^2+a_3^2+\cdots+a_n^2)$
$>2(a_1a_2+a_1a_3+\cdots+a_2a_3+a_2a_4+\cdots+a_{n-1}a_n)$.

21. $n(a_1^2+a_2^2+a_3^2+\cdots+a_n^2)>(a_1+a_2+a_3+\cdots+a_n)^2$.

If x be positive, which fraction is the greater:

22. $\frac{x+4}{x+3}$ or $\frac{x+2}{x+1}$?

23. $\frac{x+5}{x-6}$ or $\frac{x+3}{x-4}$, if $x>6$?

Determine the limits between which the values of x must lie to satisfy each of the following inequalities:

24. $x-8>4$.

25. $1-x<2$.

26. $x+2>8-x$.

27. $-3(x+10)>-20$.

28. $\left(\frac{4x+1}{4}\right)^2-\left(\frac{4x-3}{4}\right)^2<15$.

29. $x-2\left(1-\frac{1}{a}\right)<\frac{2(x+1)}{3a}$.

30. $\frac{3x-8}{4}-x<\frac{37-2x}{3}+9$.

31. $\frac{8x+1}{3}-10x>\frac{7-6x}{2}+12$.

32. $x+\frac{x}{n}+\frac{x+1}{n^2}>n$.

33. $\frac{11a-x}{4a+b}>\frac{a-x}{b-a}$.

34. $x-\frac{a}{1-a}<1-\frac{x-1}{a-1}$.

35. $\frac{x}{a+b}+\frac{x}{a-b}<2a$.

36. $x^2-3x+2>0$.

37. $x^2-x-6>0$.

38. $\frac{x+1}{x-2}>0$.

39. $\frac{x^2-a^2}{x+2a}>0$.

40. $\frac{x^2-3x+2}{x^2+3x+2}>0$.

41. $\frac{6x^2-7x+2}{2x^2-5x-3}<0$.

Determine the limits between which the values of x must lie to satisfy simultaneously each of the following systems of inequalities:

42. $\begin{cases} 6x + 1 > 0, \\ 25 - 4x > 0. \end{cases}$

43. $\begin{cases} 5x - 8 < x, \\ 3 - x > 4x - 17. \end{cases}$

44. $\begin{cases} 17 - 3x < 12x - 133, \\ 17 + 3x < 30. \end{cases}$

45. $\begin{cases} \frac{1}{3}x - \frac{1}{4}x + \frac{1}{2}x > x + 5, \\ \frac{1}{3}(x + 2) > -\frac{1}{7}(x - 2). \end{cases}$

46. $\begin{cases} \dfrac{x-7}{9} + \dfrac{x-4}{5} > 9, \\ \frac{1}{2}x + 3 > \frac{2}{3}x + 4. \end{cases}$

47. $\begin{cases} \dfrac{x-7}{9} + \dfrac{x-4}{5} < 9, \\ x - \frac{5}{8}x > 15. \end{cases}$

48. $\begin{cases} x^2 - 12x + 32 > 0, \\ x^2 - 13x + 22 > 0. \end{cases}$

49. $\begin{cases} x^2 - 3x - 4 > 0, \\ x^2 - x - 6 > 0. \end{cases}$

What value of x satisfies each of the following systems:

50. $\begin{cases} 2x^2 - 5x + 2 = 0, \\ x^2 - 1 > 0. \end{cases}$

51. $\begin{cases} x^2 + x - 6 = 0, \\ x^2 + 3x - 4 > 0. \end{cases}$

Determine the limits between which the values of x and y must lie to satisfy the following systems:

52. $\begin{cases} 2x + 3y = -4, \\ x - y > 2. \end{cases}$

53. $\begin{cases} 7x + y = 15, \\ 3x - 2y > 14. \end{cases}$

Determine the limits between which the values of a and b must lie to make each of the following values of x positive:

54. $x = \dfrac{a}{11 - 2a}.$

55. $x = \dfrac{8 - 3a}{15 - 4a}.$

56. $x = \dfrac{4a - 1}{9 - 2a}.$

57. $x = \dfrac{3b - 4}{2 - 3a}.$

58. $x = \dfrac{3a - 2b - 8}{3b - 5}.$

Problems.

13. **Pr. 1.** Divide 80 into two parts, such that the greater part shall exceed twice the sum of 4 and the less part.

Let x stand for the greater part; then $80 - x$ will stand for the less.

By the given condition,

$$x > 2(80 - x + 4), \text{ or } 3x > 168;$$

whence $$x > 56.$$

Therefore any number greater than 56 (and less than 80) will satisfy the condition of the problem.

E.g., if $x = 60$, the greater part, then $80 - x = 20$, the less part; and $60 > 2 \times 24$.

Pr. 2. In moving a certain distance the front wheel of a carriage makes 250 revolutions more than the hind wheel; the circumference of the front wheel is 5 feet, and the circumference of the hind wheel is between 6 and 7 feet. What is the distance the carriage moves?

Let x stand for the number of feet the carriage moves.

Then $\frac{x}{5}$ will stand for the number of revolutions made by front wheel.

Also $\frac{x}{6}>$ number of revolutions made by hind wheel, and $\frac{x}{7}<$ number of revolutions made by hind wheel.

Then, by the given condition, $\frac{x}{5}-250$ is the number of revolutions made by hind wheel.

Therefore $\frac{x}{5}-250<\frac{x}{6}$, whence $x<7500$;

and $\frac{x}{5}-250>\frac{x}{7}$, whence $x>4375$.

Any assumed distance between 4375 feet and 7500 feet will make the circumference of the hind wheel between 6 and 7 feet.

E.g., if $x=6000$, then $\frac{6000}{5}=1200$, the number of revolutions made by front wheel; and $1200-250=950$, number of revolutions made by hind wheel.

Therefore, $\frac{6000}{950}=6\frac{6}{19}$, the number of feet in circumference of hind wheel.

Pr. 3. A man receives from an investment an integral number of dollars a day. He calculates that if he were to receive $6 more a day his investment would yield over $270 a week; but that, if he were to receive $14 less a day, his investment would not yield as much as $270 in two weeks. How much does he receive a day from his investment?

Let x stand for the number of dollars which he receives a day.

Then, by the first condition,

$$7(x+6)>270;\ \text{whence } x>32\tfrac{4}{7}.$$

And, by the second condition,

$$14(x-14)<270;\ \text{whence } x<33\tfrac{2}{7}.$$

Therefore he receives $33 a day from his investment.

EXERCISES II.

1. What integers have each the property that one-half of the integer, increased by 5, is greater than four-thirds of it, diminished by 3?

2. What integers have each the property that, if one-half of the integer be added to itself, the sum will be greater than one-third of the integer, increased by 2?

3. What integers have each the property that, if 9 be subtracted from three times the integer, the remainder will be less than twice the integer, increased by 12?

4. What fractions have each the property that, if 1 be subtracted from the denominator of the fraction, the resulting fraction will be equal to $\frac{1}{2}$; but if 20 be added to the numerator, the resulting fraction will lie between 2 and 3?

5. A has three times as much money as B. If B gives A $10, then A will have more than seven times as much as B will have left. What are the possible amounts of money which A and B have?

6. What integers have each the property that, the sum of three-tenths of the integer and 5 is less than one-half of the integer, while five times the integer is less than it double, increased by 360?

7. Find a multiple of 25, such that three-fourths of it is greater than one-half of it, increased by 15, while five times the number is less than three times the number, increased by 200.

8. What positive numbers have each the property that, if the number be subtracted from a and be added to b, the product of the resulting numbers will be greater than the product of the given numbers?

9. In a class-room can be placed 6 benches, but it contains fewer. If 5 pupils be seated on each bench, then 4 pupils will be without seats. But if 6 pupils be seated on each bench, some seats will be unoccupied. How many benches are in the room?

10. A traveler had covered more than two-thirds of his journey. After traveling 2 miles further, he found that he had covered seven-tenths of his journey. What is the least possible number of miles in his journey?

11. A man wishes to make a purchase for $14, but has not enough money. If he borrows one-third as much money as he now has, he will be able not only to make the purchase, but will have left more money than he now lacks? How much money has he?

12. A man bought 2 horses at the same price. The amount paid for both horses exceeds seven-tenths of the price of one horse by more than $259; and one-half the price of a horse is less than one-fourth of the price, increased by $50.25. If the purchaser paid the amount in five-dollar bills, how much did he pay for each horse?

13. The daily pay roll of a contractor, who engages masons at $4.40 a day and carpenters at $3.60 a day, amounts to more than $104. If the wages of each man were increased by a small amount, the pay roll would amount to $112. If there are 3 fewer carpenters than masons, how many carpenters and how many masons are at work?

A Property of Fractions.

14. *If the denominators of the fractions* $\frac{n_1}{d_1}, \frac{n_2}{d_2}, \frac{n_3}{d_3}, \cdots$ *be all positive, then the fraction* $\frac{n_1+n_2+n_3+\cdots}{d_1+d_2+d_3+\cdots}$ *is greater than the least, and less than the greatest, of the given fractions.*

Let $\frac{n_1}{d_1}$ be the greatest of the given fractions, and let

$$\frac{n_1}{d_1} = x, \text{ or } n_1 = d_1x. \qquad (1)$$

Then

$$\frac{n_2}{d_2} < x, \text{ or } n_2 < d_2x; \qquad (2)$$

$$\frac{n_3}{d_3} < x, \text{ or } n_3 < d_3x; \qquad (3)$$

etc.

From the equation (1), and the inequalities (2), (3), etc., we have

$$n_1 + n_2 + n_3 + \cdots < d_1x + d_2x + d_3x + \cdots$$
$$< (d_1 + d_2 + d_3 + \cdots)x.$$

Therefore

$$\frac{n_1+n_2+n_3+\cdots}{d_1+d_2+d_3+\cdots} < x,$$
$$< \frac{n_1}{d_1}.$$

In like manner it can be proved that $\frac{n_1+n_2+n_3+\cdots}{d_1+d_2+d_3+\cdots}$ is greater than the least of the given fractions.

Properties of Powers.

15. *The value of a positive integral power increases as its exponent increases, if its base be greater than* 1; *and decreases as its exponent increases, if its base be positive and less than* 1; *i.e., if its base be a positive proper fraction.*

Thus, the powers

$$2^2 = 4, \; 2^3 = 8, \; 2^4 = 16, \text{ etc.,}$$

and their exponents increase together; while the powers

$$\left(\frac{1}{2}\right)^2 = \frac{1}{4}, \; \left(\frac{1}{2}\right)^3 = \frac{1}{8}, \; \left(\frac{1}{2}\right)^4 = \frac{1}{16}, \text{ etc.,}$$

decrease as their exponents increase.

In general,

$$\left(\frac{a}{b}\right)^{n+1} > \left(\frac{a}{b}\right)^{n}, \text{ when } a > b;$$

$$\left(\frac{a}{b}\right)^{n+1} < \left(\frac{a}{b}\right)^{n}, \text{ when } a < b;$$

wherein a, b, and n are positive integers.

For $$\left(\frac{a}{b}\right)^{n} = \left(\frac{a}{b}\right)^{n};$$

and $$\frac{a}{b} > 1, \text{ when } a > b.$$

Therefore $$\left(\frac{a}{b}\right)^{n+1} > \left(\frac{a}{b}\right)^{n}, \text{ by Art. 7 (i.).}$$

Also, $$\frac{a}{b} < 1, \text{ when } a < b.$$

Therefore $$\left(\frac{a}{b}\right)^{n+1} < \left(\frac{a}{b}\right)^{n}, \text{ by Art. 7 (i.).}$$

We therefore conclude that a positive integral power whose base is greater than 1 increases without limit, *i.e.*, becomes greater than any assigned positive number, however great, when its exponent increases without limit; and that a positive integral power whose base is less than 1 and positive decreases without limit, *i.e.*, becomes less than any assigned positive number, however small, when its exponent increases without limit.

These conclusions may be stated symbolically thus:

$$\left(\frac{a}{b}\right)^{\infty} = \infty, \text{ when } a > b;$$

$$\left(\frac{a}{b}\right)^{\infty} = 0, \text{ when } a < b.$$

16. *The value of a positive integral power of a positive base increases as the base increases, the exponent remaining constant.*

E.g., $2^3 = 8$, $3^3 = 27$, $4^3 = 64$, and $64 > 27 > 8$.

For, if $a > b$, then by Art. 8 (i.) $a^n > b^n$.

17. If d be a positive number and n a positive integer, then

$$(1 + d)^n > 1 + nd.$$

We have $a^n - b^n = (a - b)(a^{n-1} + a^{n-2}b + \cdots + ab^{n-2} + b^{n-1})$,

or $$a^n = b^n + (a - b)(a^{n-1} + a^{n-2}b + \cdots + ab^{n-2} + b^{n-1}).$$

Let $$a > b.$$

Then
$$a^{n-1} > b^{n-1}$$
$$a^{n-2}b > b^{n-1}$$

$$ab^{n-2} > b^{n-1}$$
$$b^{n-1} = b^{n-1}$$

Therefore $a^{n-1} + a^{n-2}b + \cdots + ab^{n-2} + b^{n-1} > b^{n-1} + b^{n-1} + \cdots n$ terms
$$> nb^{n-1}.$$

Consequently, since $a - b$ is positive,
$$a^n > b^n + n(a-b)b^{n-1}.$$

Now let $a = 1 + d$, and $b = 1$.

Then $$(1+d)^n > 1 + nd.$$

CHAPTER XVII.

IRRATIONAL NUMBERS.

1. If a be the qth power of a number, say b, then $\sqrt[q]{a}, = \sqrt[q]{b^q}$, has, as we have seen in Ch. XV., a definite value; as $\sqrt[4]{16} = 2$.

In this chapter we shall consider roots of positive numbers which are not powers with exponents equal to or multiples of the indices of the required roots.

2. *The qth root of a positive fraction, whose terms (either or both) are not qth powers of positive integers, cannot be expressed either as an integer or as a fraction.*

In the above enunciation the words *positive fraction* are to be understood as meaning also a positive integer, that is, a fraction whose denominator is 1.

The proof of the general case will be first illustrated by the particular example $\sqrt{2}$.

The $\sqrt{2}$ must be a number whose square is 2. But since $1^2 = 1$ and $2^2 = 4$, the $\sqrt{2}$ cannot be an integer.

Let us assume that $\sqrt{2}$ can be expressed as a fraction, $\frac{n}{d}$, reduced to its lowest terms.

Then from
$$\sqrt{2} = \frac{n}{d}, \tag{1}$$
we have
$$2 = \frac{n^2}{d^2}. \tag{2}$$

Since $\frac{n}{d}$ is in its lowest terms, $\frac{n^2}{d^2}$ is in its lowest terms [Ch. IX., Art. 34 (iii.)]. Consequently, by Ch. IX., Art. 34 (v.),
$$n^2 = 2, \text{ and } d^2 = 1. \tag{3}$$
But since 2 is not the square of an integer, the first of equations (3), and therefore also (1), is untenable.

Consequently $\sqrt{2}$ cannot be expressed as a fraction.

In general, $\sqrt[q]{\frac{N}{D}}$, wherein N and D (either or both) are not qth powers of positive integers, cannot be expressed as a fraction $\frac{n}{d}$.

Since the value of a fraction is unchanged by dividing both of its terms by the same number, we can without loss of generality assume that both $\frac{N}{D}$ and $\frac{n}{d}$ are in their lowest terms.

Let us assume $$\sqrt[q]{\frac{N}{D}}=\frac{n}{d}. \qquad (1)$$

Then $$\frac{N}{D}=\frac{n^q}{d^q}. \qquad (2)$$

Since $\frac{n}{d}$ is in its lowest terms, $\frac{n^q}{d^q}$ is in its lowest terms [Ch. IX., Art. 34 (iii.)]. Consequently, by Ch. IX., Art. 34 (v.),

$$n^q = N, \text{ and } d^q = D. \qquad (3)$$

But since, by hypothesis, N and D (either or both) are not qth powers of positive integers, equations (3) (one or both), and therefore also (1), are untenable. Consequently, $\sqrt[q]{\frac{N}{D}}$ cannot be expressed as a fraction.

Observe that, if d be assumed equal to 1, the preceding proof shows that $\sqrt[q]{\frac{N}{D}}$ cannot be expressed as a positive integer; also, if D be assumed equal to 1, that the $\sqrt[q]{N}$ cannot be expressed as a positive integer, or as a positive fraction.

3. We have now proved that $\sqrt{2}$, and in general, $\sqrt[q]{\frac{N}{D}}$, wherein N and D (either or both) are not qth powers of positive integers, cannot be expressed in terms of the numbers as yet comprised in our number system.

It is therefore necessary to exclude such roots from our consideration or to enlarge our idea of number. The latter alternative is in accordance with the generalizing spirit of Algebra.

We therefore assume that $\sqrt{2}$, and in general, $\sqrt[q]{\frac{N}{D}}$, is a number, and include it in our number system.

The properties of these new numbers must be consistent with the definition of a root; that is, with the relations,

$$(\sqrt{2})^2 = 2, \text{ and } \left(\sqrt[q]{\frac{N}{D}}\right)^q=\frac{N}{D}.$$

4. Before operating with or upon the numbers thus introduced into the number system, we must prove that they obey the fundamental laws of Algebra, which were proved in Chs. II. and III. only for integers and fractions.

The following property will lead to another definition of the $\sqrt{2}$, and in general of the $\sqrt[q]{\frac{N}{D}}$, which is consistent with that given in Art. 3, and from which the fundamental laws can be easily deduced.

If $\frac{N}{D}$ be a fraction whose terms (either or both) are not qth powers of positive integers, numbers can always be found, both greater and less than $\sqrt[q]{\frac{N}{D}}$, which differ from $\sqrt[q]{\frac{N}{D}}$ by as little as we please; that is, by less than any assigned number, however small.

The proof of the general case will first be illustrated by the particular example $\sqrt{2}$.

Since 2 lies between 1^2 and 2^2, the $\sqrt{2}$ lies between 1 and 2, *i.e.*, $1 < \sqrt{2} < 2$.

The interval between 1 and 2 we now divide into ten equal parts, and form the series of powers

$$1^2,\ 1.1^2,\ 1.2^2,\ 1.3^2,\ 1.4^2,\ 1.5^2,\ \cdots,\ 1.9^2,\ 2^2.$$

Then 2, which lies between 1^2 and 2^2, must lie between two consecutive powers of this series, or between two consecutive numbers of the equivalent series

$$1,\ 1.21,\ 1.44,\ 1.69,\ 1.96,\ 2.25,\ \cdots,\ 3.61,\ 4.$$

Since 2 lies between 1.96 and 2.25 (that is, between 1.4^2 and 1.5^2), the $\sqrt{2}$ must lie between 1.4 and 1.5, *i.e.*, $1.4 < \sqrt{2} < 1.5$.

The interval between 1.4 and 1.5, = .1, we next divide into ten equal parts, and form the series of powers

$$1.4^2,\ 1.41^2,\ 1.42^2,\ \cdots,\ 1.49^2,\ 1.5^2.$$

Then 2, which lies between 1.4^2 and 1.5^2, must lie between two consecutive powers of this series, or between two consecutive numbers of the equivalent series

$$1.9600,\ 1.9881,\ 2.0164,\ \cdots,\ 2.2201,\ 2.2500.$$

Since 2 lies between 1.9881 and 2.0164 (that is, between 1.41^2 and 1.42^2), the $\sqrt{2}$ lies between 1.41 and 1.42, *i.e.*, $1.41 < \sqrt{2} < 1.42$.

This method of procedure can be continued indefinitely. These results may be summarized as follows:

(a)		(b)	
$1 < \sqrt{2} < 2$	and	$2 - 1$	$= 1$
$1.4 < \sqrt{2} < 1.5$	and	$1.5 - 1.4$	$= .1$
$1.41 < \sqrt{2} < 1.42$	and	$1.42 - 1.41$	$= .01$
$1.414 < \sqrt{2} < 1.415$	and	$1.415 - 1.414$	$= .001$
$1.4142 < \sqrt{2} < 1.4143$	and	$1.4143 - 1.4142$	$= .0001$
etc.		etc.	

It follows from tables (*a*) and (*b*) that there can be found two numbers, one greater and the other less than $\sqrt{2}$, which differ from each other by as little as we please, and which therefore differ from $\sqrt{2}$, which lies between them, by as little as we please.

Observe that the numbers of the one series, which are always less than $\sqrt{2}$, continually *increase* toward $\sqrt{2}$, while the numbers of the other series, which are always greater than $\sqrt{2}$, continually *decrease* toward $\sqrt{2}$.

Either of these two values is an approximation to $\sqrt{2}$.

E.g., 1.41 is an approximation to $\sqrt{2}$, and the error is less than .01; 1.4142 is a closer approximation to $\sqrt{2}$, since the error is less than .0001.

In the proof of the general case, which now follows, it is necessary to represent the two values between which the required root lies at any stage of the work in terms of common fractions instead of decimal fractions.

Thus, $$1.4142 < \sqrt{2} < 1.4143$$
could have been written

$$\frac{14}{10}+\frac{1}{100}+\frac{4}{1000}+\frac{2}{10000}<\sqrt{2}<\frac{14}{10}+\frac{1}{100}+\frac{4}{1000}+\frac{3}{10000},$$

or $$\frac{14}{10}+\frac{1}{10^2}+\frac{4}{10^3}+\frac{2}{10^4}<\sqrt{2}<\frac{14}{10}+\frac{1}{10^2}+\frac{4}{10^3}+\frac{3}{10^4}.$$

Let $\frac{N}{D}$ be a fraction, in which N and D (either or both) are not qth powers of positive integers. By Ch. XVI., Art. 16, the powers

$$0^q,\ \left(\frac{1}{10}\right)^q,\ \left(\frac{2}{10}\right)^q,\ \left(\frac{3}{10}\right)^q,\ \cdots,\ \left(\frac{n}{10}\right)^q,\ \cdots$$

increase without limit. Therefore, whatever positive value $\frac{N}{D}$ may have, there will always be two consecutive powers of the above series between which $\frac{N}{D}$ lies.

Let $\left(\frac{k_1}{10}\right)^q$ and $\left(\frac{k_1+1}{10}\right)^q$ be the two powers between which $\frac{N}{D}$ is found to lie, wherein k_1 is 0 or any *positive integer*.

Then since $\frac{N}{D}$ lies between $\left(\frac{k_1}{10}\right)^q$ and $\left(\frac{k_1+1}{10}\right)^q$, the $\sqrt[q]{\frac{N}{D}}$ lies between $\frac{k_1}{10}$ and $\frac{k_1+1}{10}$; *i.e.*,

$$\frac{k_1}{10}<\sqrt[q]{\frac{N}{D}}<\frac{k_1+1}{10}. \qquad \text{(I.)}$$

Compare the corresponding step in the example $\sqrt{2}$, whereby we found that $\sqrt{2}$ lies between 1.4 and 1.5; *i.e.*, between $\frac{14}{10}$ and $\frac{15}{10}$. In this case $k_1 = 14$ and $k_1 + 1 = 15$.

The interval between $\frac{k_1+1}{10}$ and $\frac{k_1}{10}$, $=\frac{1}{10}$, we now divide into ten equal parts, and form the series of powers

$$\left(\frac{k_1}{10}\right)^q, \left(\frac{k_1}{10}+\frac{1}{10^2}\right)^q, \left(\frac{k_1}{10}+\frac{2}{10^2}\right)^q, \cdots, \left(\frac{k_1}{10}+\frac{9}{10^2}\right)^q, \left(\frac{k_1+1}{10}\right)^q.$$

Then $\frac{N}{D}$, which lies between $\left(\frac{k_1}{10}\right)^q$ and $\left(\frac{k_1+1}{10}\right)^q$, must lie between two consecutive powers of this series.

Let $\left(\frac{k_1}{10}+\frac{k_2}{10^2}\right)^q$ and $\left(\frac{k_1}{10}+\frac{k_2+1}{10^2}\right)^q$, wherein k_2 is one of the numbers, 0, 1, 2, ⋯, 8, 9, be the two powers between which $\frac{N}{D}$ is found to lie.

Then $\sqrt[q]{\frac{N}{D}}$ lies between $\frac{k_1}{10}+\frac{k_2}{10^2}$ and $\frac{k_1}{10}+\frac{k_2+1}{10^2}$; *i.e.*,

$$\frac{k_1}{10}+\frac{k_2}{10^2}<\sqrt[q]{\frac{N}{D}}<\frac{k_1}{10}+\frac{k_2+1}{10^2}. \qquad \text{(II.)}$$

The method can evidently be carried on indefinitely; that is, we can find two powers

$$\left(\frac{k_1}{10}+\frac{k_2}{10^2}+\cdots+\frac{k_p}{10^p}\right)^q \text{ and } \left(\frac{k_1}{10}+\frac{k_2}{10^2}+\cdots+\frac{k_p+1}{10^p}\right)^q$$

between which $\frac{N}{D}$ lies. We therefore have

$$\frac{k_1}{10}+\frac{k_2}{10^2}+\cdots+\frac{k_p}{10^p}<\sqrt[q]{\frac{N}{D}}<\frac{k_1}{10}+\frac{k_2}{10^2}+\cdots+\frac{k_p+1}{10^p}, \qquad \text{(III.)}$$

wherein p is any positive integer from 1 to $+\infty$.

The two numbers between which $\sqrt[q]{\frac{N}{D}}$ is found to lie at any stage of the work evidently differ by $\frac{1}{10^p}$.

As p increases without limit, 10^p also increases without limit (Ch. XVI., Art. 15), and hence $\frac{1}{10^p}$ decreases without limit (Ch. III., § 4, Art. 20). Since, therefore, these two numbers can be made to differ from each other by less than any assigned number, however small, the $\sqrt[q]{\frac{N}{D}}$, which lies between them, will differ from either of them by less than any assigned number, however small.

Either of these numbers is an approximate value of $\sqrt[q]{\frac{N}{D}}$.

5. It is important to keep clearly in mind that, although approximate values have been obtained for $\sqrt{2}$, and in general for $\sqrt[q]{\frac{N}{D}}$, these numbers have as exact values as have integers and fractions.

Consider the following examples:

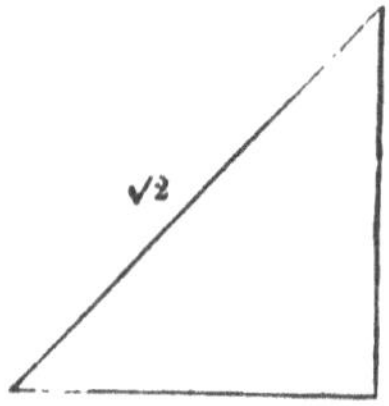

Fig. 12.

It is proved in Geometry that the hypothenuse of a right triangle is equal to the square root of the sum of the squares of its sides. If, therefore, the sides of a right triangle be equal, and each be 1 inch say, the hypothenuse will be $\sqrt{2}$ inches (Fig. 12). Now the length of the hypothenuse is just as exact and definite as the length of either side, and yet it can be expressed only by $\sqrt{2}$ (1 inch being taken as the unit of length). Therefore $\sqrt{2}$ is just as definite and exact as is 1.

Again, $\sqrt{2} \times \sqrt{2} = 2$, by the definition of a root. Now no approximate value of $\sqrt{2}$ multiplied by itself will give exactly 2. Therefore the number which multiplied by itself gives 2 must have an exact value. This exact value, to be sure, cannot be expressed in terms of integers and fractions. But when fractions were introduced it was equally impossible to express their values in terms of numbers previously comprised in the number system; that is, in terms of integers. Yet they were added to the number system.

6. In Art 4 we found that the $\sqrt[q]{\frac{N}{D}}$ lies between the two series of numbers:

$$\frac{k_1}{10},\ \frac{k_1}{10}+\frac{k_2}{10^2},\ \cdots,\ \frac{k_1}{10}+\frac{k_2}{10^2}+\cdots+\frac{k_p}{10^p}, \qquad (1)$$

$$\frac{k_1+1}{10},\ \frac{k_1}{10}+\frac{k_2+1}{10^2},\ \cdots,\ \frac{k_1}{10}+\frac{k_2}{10^2}+\cdots+\frac{k_p+1}{10^p}, \qquad (2)$$

wherein $p = 1, 2, 3, \cdots, \infty$.

These two series have the following properties:

(i.) *The numbers*

$$\frac{k_1}{10}+\frac{k_2}{10^2}+\cdots+\frac{k_p}{10^p},\ \textit{wherein } p = 1, 2, 3, \cdots, \infty,$$

of the first series increase as p increases, but remain always less than the numbers

$$\frac{k_1}{10}+\frac{k_2}{10^2}+\cdots+\frac{k_p+1}{10^p},\ \textit{wherein } p = 1,\ 2,\ 3,\ \cdots\ \infty,$$

of the second series; and the numbers of the second series decrease as p increases, but remain always greater than the numbers of the first series. That is, the numbers of the one series more and more nearly approach the numbers of the other series, but never meet them.

(ii.) *The difference between a number of the one series and the corresponding number of the other series can be made less than any assigned number, however small, by taking p sufficiently great.*

7. Two series of numbers which possess the properties (i.) and (ii.), Art. 6, are said to have a common limit, which lies between them. Two such series therefore define the number which is their common limit. This number is approached by both series and not reached by either.

8. From the nature of the process by which the series (1) and (2), Art. 6, were derived, it follows that $\sqrt[q]{\frac{N}{D}}$ is their common limit. These two series can therefore be used to define $\sqrt[q]{\frac{N}{D}}$, and the relation

$$\frac{k_1}{10} + \frac{k_2}{10^2} + \cdots + \frac{k_p}{10^p} < \sqrt[q]{\frac{N}{D}} < \frac{k_1}{10} + \frac{k_2}{10^2} + \cdots + \frac{k_p + 1}{10^p}$$

is in subsequent theory to be understood as expressing this definition.

The two numbers between which the $\sqrt[q]{\frac{N}{D}}$ lies can be reduced to a common denominator 10^p. Let us designate 10^p by n. Then since these two numbers differ by $\frac{1}{10^p}$, $=\frac{1}{n}$, they may be represented by $\frac{m}{n}$ and $\frac{m+1}{n}$ respectively.

In the theory which follows, we shall let

$$\frac{k_1}{10} + \frac{k_2}{10^2} + \cdots + \frac{k_p}{10^p} = \frac{m}{n}, \qquad (1)$$

$$\frac{k_1}{10} + \frac{k_2}{10^2} + \cdots + \frac{k_p + 1}{10^p} = \frac{m+1}{n}. \qquad (2)$$

The above definition may now be expressed thus:

$$\frac{m}{n} < \sqrt[q]{\frac{N}{D}} < \frac{m+1}{n}. \qquad (I.)$$

Irrational Numbers.

9. An **Irrational Number** is a number which cannot be expressed either as an integer or as a fraction, but which can be inclosed between two series of fractions ultimately differing from each other, and therefore from the inclosed number, by less than any assigned number however small.

An irrational number, I, is therefore defined by the relation

$$\frac{m}{n} < I < \frac{m+1}{n},$$

wherein $\frac{m}{n}$ and $\frac{m+1}{n}$ have the properties (i.) and (ii.), Art. 6.

E.g., $\sqrt{2}$, $\sqrt{\frac{2}{3}}$, $\sqrt[q]{a}$, wherein a is a positive integer which is not the qth power of an integer.

The existence of such a number is based upon the knowledge of how the fractions $\frac{m}{n}$ and $\frac{m+1}{n}$ are obtained; that is, upon knowing the values of the k's in the expressions for which $\frac{m}{n}$ and $\frac{m+1}{n}$ stand.

10. Whatever value p and therefore $\frac{m}{n}$ and $\frac{m+1}{n}$, may have, there will always be numbers, integers or fractions, lying between any two numbers of the series $\frac{m}{n}$ and $\frac{m+1}{n}$. But no such number can be selected which will not be passed by numbers of one or the other series, if p be sufficiently increased. Therefore there is no number in the system defined so as to include only integers and fractions, which is greater than every number of the series (1.) and less than every number of the series (2.); that is, which is approached by both series and not reached by either. Since, however, these series cannot meet, we conclude that there was a gap between them which could not be filled by any integer or fraction.

Consequently by including irrational numbers in the number system, continuity has been introduced where before it was lacking.

Negative Irrational Numbers.

11. If the fractions of the series which define an irrational number be negative, the number thus defined is called a **Negative Irrational Number.**

Therefore a negative irrational number is defined by the relation

$$-\frac{m+1}{n} < -I < -\frac{m}{n},$$

wherein the two series of fractions, $-\frac{m+1}{n}$ and $-\frac{m}{n}$, have the properties (i.) and (ii.), Art. 6.

12. If I be a positive irrational number defined by the relation

$$\frac{m}{n} < I < \frac{m+1}{n},$$

the negative irrational number defined by the relation

$$-\frac{m+1}{n} < -I < -\frac{m}{n}$$

is called its *equal and opposite.*

The absolute value of an irrational number is its value without regard to quality.

The Fundamental Operations with Irrational Numbers.

13. Addition. — Let I_1 and I_2 be two positive irrational numbers defined by the relations

$$\frac{m_1}{n_1} < I_1 < \frac{m_1+1}{n_1}, \tag{1}$$

$$\frac{m_2}{n_2} < I_2 < \frac{m_2+1}{n_2}. \tag{2}$$

If the corresponding *rational* numbers of the series which define I_1 and I_2 be added, we obtain the two series of rational numbers

$$\frac{m_1}{n_1} + \frac{m_2}{n_2} \text{ and } \frac{m_1+1}{n_1} + \frac{m_2+1}{n_2}.$$

The numbers of these series have the properties (i.) and (ii.), Art. 6.

For, since $\frac{m_1}{n_1}$ increases as n_1 increases, and $\frac{m_2}{n_2}$ increases as n_2 increases, therefore $\frac{m_1}{n_1} + \frac{m_2}{n_2}$ increases as n_1 and n_2 increase. For a similar reason, $\frac{m_1+1}{n_1} + \frac{m_2+1}{n_2}$ decreases as n_1 and n_2 increase. And since

$$\frac{m_1}{n_1} < \frac{m_1+1}{n_1} \text{ and } \frac{m_2}{n_2} < \frac{m_2+1}{n_2},$$

therefore $$\frac{m_1}{n_1} + \frac{m_2}{n_2} < \frac{m_1+1}{n_1} + \frac{m_2+1}{n_2}.$$

The difference

$$\left(\frac{m_1+1}{n_1} + \frac{m_2+1}{n_2}\right) - \left(\frac{m_1}{n_1} + \frac{m_2}{n_2}\right), \ = \frac{1}{n_1} + \frac{1}{n_2},$$

can be made less than any assigned number, however small. For $\frac{1}{n_1}$ can be made less than any assigned number, say $\frac{1}{2}d$; and $\frac{1}{n_2}$ can be made less than any assigned number, say $\frac{1}{2}d$. Therefore, $\frac{1}{n_1} + \frac{1}{n_2}$ can be made less than $\frac{1}{2}d + \frac{1}{2}d, \ = d$.

Therefore, the two series of numbers

$$\frac{m_1}{n_1} + \frac{m_2}{n_2} \text{ and } \frac{m_1+1}{n_1} + \frac{m_2+1}{n_2}$$

define a positive number which lies between them. This number is defined as the sum $I_1 + I_2$. That is,

$$\frac{m_1}{n_1} + \frac{m_2}{n_2} < I_1 + I_2 < \frac{m_1+1}{n_1} + \frac{m_2+1}{n_2}.$$

Exactly similar reasoning will apply if either or both of the irrational numbers be negative.

Let I_1 be a positive irrational number defined by the relation

$$\frac{m_1}{n_1} < I_1 < \frac{m_1 + 1}{n_1},$$

and $-I_2$ be a negative irrational number defined by the relation

$$-\frac{m_2 + 1}{n_2} < -I_2 < -\frac{m_2}{n_2}.$$

It can be shown as above that the rational numbers of the series

$$\frac{m_1}{n_1} + \left(-\frac{m_2 + 1}{n_2}\right) \text{ and } \frac{m_1 + 1}{n_1} + \left(-\frac{m_2}{n_2}\right)$$

have the properties (i.) and (ii.), Art. 6.

They, therefore, define a positive or negative number which lies between them. This number is defined as the sum $I_1 + (-I_2)$. That is,

$$\frac{m_1}{n_1} + \left(-\frac{m_2 + 1}{n_2}\right) < I_1 + (-I_2) < \frac{m_1 + 1}{n_1} + \left(-\frac{m_2}{n_2}\right).$$

The method can evidently be extended to the sum of three or more irrational numbers.

Similar reasoning will apply if one or more of the numbers to be added be rational.

14. Subtraction.—The following definition of Subtraction of irrational numbers is a natural extension of the principle of subtraction for rational numbers.

To subtract an irrational number from a rational or irrational number is equivalent to adding an equal and opposite irrational number.

The Associative and Commutative Laws for Addition and Subtraction of Irrational Numbers.

15. These fundamental laws hold also for irrational numbers; that is,

$$I_1 + I_2 = I_2 + I_1,$$

$$I_1 + I_2 + I_3, \ = I_1 + (I_2 + I_3) = I_1 + (I_3 + I_2) = \text{etc.}$$

For, by the definition of $I_1 + I_2$,

$$\frac{m_1}{n_1} + \frac{m_2}{n_2} < I_1 + I_2 < \frac{m_1 + 1}{n_1} + \frac{m_2 + 1}{n_2};$$

and, by the definition of $I_2 + I_1$,

$$\frac{m_2}{n_2} + \frac{m_1}{n_1} < I_2 + I_1 < \frac{m_2 + 1}{n_2} + \frac{m_1 + 1}{n_1}.$$

But since $$\frac{m_2}{n_2}+\frac{m_1}{n_1}=\frac{m_1}{n_1}+\frac{m_2}{n_2},$$

and $$\frac{m_2+1}{n_2}+\frac{m_1+1}{n_1}=\frac{m_1+1}{n_1}+\frac{m_2+1}{n_2},$$

therefore $I_1+I_2=I_2+I_1$.

The Associative Law can be proved in a similar manner.

E.g., $\sqrt{2}+\sqrt{3}=\sqrt{3}+\sqrt{2}$, $\sqrt{2}+\sqrt{3}+(-\sqrt{5})=\sqrt{2}+(-\sqrt{5})+\sqrt{3}$.

16. Multiplication. — Let I_1 and I_2 be two positive irrational numbers defined by the relations

$$\frac{m_1}{n_1}<I_1<\frac{m_1+1}{n_1}, \tag{1}$$

$$\frac{m_2}{n_2}<I_2<\frac{m_2+1}{n_2}. \tag{2}$$

If the corresponding *rational* numbers of the series which define I_1 and I_2 be multiplied, we obtain the two series of rational numbers,

$$\frac{m_1}{n_1}\cdot\frac{m_2}{n_2} \text{ and } \frac{m_1+1}{n_1}\cdot\frac{m_2+1}{n_2}.$$

The numbers of these series have the properties (i.) and (ii.), Art. 6.

For, since $\frac{m_1}{n_1}$ increases as n_1 increases, and $\frac{m_2}{n_2}$ increases as n_2 increases, therefore $\frac{m_1}{n_1}\cdot\frac{m_2}{n_2}$ increases as n_1 and n_2 increase. For a similar reason $\frac{m_1+1}{n_1}\cdot\frac{m_2+1}{n_2}$ decreases as n_1 and n_2 increase. And since

$$\frac{m_1}{n_1}<\frac{m_1+1}{n_1} \text{ and } \frac{m_2}{n_2}<\frac{m_2+1}{n_2},$$

therefore $$\frac{m_1}{n_1}\cdot\frac{m_2}{n_2}<\frac{m_1+1}{n_1}\cdot\frac{m_2+1}{n_2}.$$

The difference

$$\frac{m_1+1}{n_1}\cdot\frac{m_2+1}{n_2}-\frac{m_1}{n_1}\cdot\frac{m_2}{n_2}, =\frac{m_1+1}{n_1}\cdot\frac{m_2+1}{n_2}-\frac{m_1+1}{n_1}\cdot\frac{m_2}{n_2}+\frac{m_1+1}{n_1}\cdot\frac{m_2}{n_2}$$
$$-\frac{m_1}{n_1}\cdot\frac{m_2}{n_2}=\frac{m_1+1}{n_1}\left(\frac{m_2+1}{n_2}-\frac{m_2}{n_2}\right)+\frac{m_2}{n_2}\left(\frac{m_1+1}{n_1}-\frac{m_1}{n_1}\right),$$

can be made less than any assigned number, however small.

For, since $\frac{m_1+1}{n_1}$ decreases, it is always less than some positive finite rational number, say R; and since $\frac{m_2}{n_2}<\frac{m_2+1}{n_2}$, it is also less than some positive finite rational number, say R_1. Moreover, $\frac{m_2+1}{n_2}-\frac{m_2}{n_2}$ and $\frac{m_1+1}{n_1}-\frac{m_1}{n_1}$ can each be made less than any assigned number, say d.

Therefore the given difference can be made less than $dR + dR_1$, $= d(R + R_1)$.

But $d(R + R_1)$ can be made less than any assigned number, say δ, by taking d less than $\dfrac{\delta}{R + R_1}$.

Therefore the two series

$$\frac{m_1}{n_1} \cdot \frac{m_2}{n_2} \text{ and } \frac{m_1 + 1}{n_1} \cdot \frac{m_2 + 1}{n_2},$$

define a positive number which lies between them. This number is defined as the product $I_1 \cdot I_2$. That is,

$$\frac{m_1}{n_1} \cdot \frac{m_2}{n_2} < I_1 \cdot I_2 < \frac{m_1 + 1}{n_1} \cdot \frac{m_2 + 1}{n_2}.$$

The following definition of multiplication of irrational numbers is consistent with the preceding result.

The product of two irrational numbers is the product of their absolute values, with a sign determined by the laws of signs for the product of two rational numbers.

The preceding method can evidently be extended to the product of three or more irrational numbers.

Similar reasoning will apply if one or more of the numbers to be multiplied be rational.

The Associative, Commutative, and Distributive Laws for Multiplication of Irrational Numbers.

17. These fundamental laws hold also for irrational numbers. That is,

$$I_1 I_2 = I_2 I_1;$$

$$I_1 I_2 I_3 = I_1(I_2 I_3) = I_3(I_1 I_2) = \text{etc.};$$

$$(I_1 \pm I_2) I_3 = I_1 I_3 \pm I_2 I_3.$$

For, by definition of $I_1 I_2$,

$$\frac{m_1}{n_1} \cdot \frac{m_2}{n_2} < I_1 I_2 < \frac{m_1 + 1}{n_1} \cdot \frac{m_2 + 1}{n_2},$$

and by definition of $I_2 I_1$,

$$\frac{m_2}{n_2} \cdot \frac{m_1}{n_1} < I_2 I_1 < \frac{m_2 + 1}{n_2} \cdot \frac{m_1 + 1}{n_1}.$$

But since $\dfrac{m_1}{n_1} \cdot \dfrac{m_2}{n_2} = \dfrac{m_2}{n_2} \cdot \dfrac{m_1}{n_1}$, and $\dfrac{m_1 + 1}{n_1} \cdot \dfrac{m_2 + 1}{n_2} = \dfrac{m_2 + 1}{n_2} \cdot \dfrac{m_1 + 1}{n_1}$, therefore $I_1 I_2 = I_2 I_1$.

The other laws can be proved in a similar manner.

18. Reciprocal of an Irrational Number. — The reciprocal of the numbers of the series which define I have the properties (i.) and (ii.), Art. 6.

For, since, as n increases, $\frac{m}{n}$ increases and $\frac{m+1}{n}$ decreases, therefore $\frac{1}{\frac{m}{n}}$ decreases and $\frac{1}{\frac{m+1}{n}}$ increases.

And since $\frac{m}{n} < \frac{m+1}{n}$, therefore $\frac{1}{\frac{m+1}{n}} < \frac{1}{\frac{m}{n}}$.

The difference

$$\frac{1}{\frac{m}{n}} - \frac{1}{\frac{m+1}{n}}, = \frac{\frac{m+1}{n} - \frac{m}{n}}{\frac{m}{n} \cdot \frac{m+1}{n}}$$

can be made less than any assigned number, however small, by taking n sufficiently great.

For, since $\frac{m}{n}$ increases, it is always greater than some positive finite rational number, say R. And since $\frac{m+1}{n} > \frac{m}{n}$, $\frac{m+1}{n}$ is always greater than the same positive finite rational number.

Therefore, the difference

$$\frac{\frac{m+1}{n} - \frac{m}{n}}{\frac{m}{n} \cdot \frac{m+1}{n}} < \frac{\frac{m+1}{n} - \frac{m}{n}}{R^2}.$$

Moreover, $\frac{m+1}{n} - \frac{m}{n}$ can be made less than any assigned number, say d. Therefore the given difference can be made less than $\frac{d}{R^2}$. But $\frac{d}{R^2}$ can be made less than any assigned number, say δ, by taking d less than $R^2\delta$.

Therefore the two series of numbers, $\frac{1}{\frac{m+1}{n}}$ and $\frac{1}{\frac{m}{n}}$, define a positive number which lies between them. This number is defined as the reciprocal of I. That is, $\frac{1}{\frac{m+1}{n}} < \frac{1}{I} < \frac{1}{\frac{m}{n}}$.

19. Division. — Division by an irrational number may be defined as follows:

To divide any number by an irrational number, not 0, is equivalent to multiplying by the reciprocal of the irrational number.

From this definition it follows that the fundamental laws hold also for division of irrational numbers.

20. It follows from the preceding theory that the laws governing the fundamental operations with irrational numbers are the same as those governing these operations with rational numbers.

E.g.,
$$\sqrt{2}-(\sqrt{3}-\sqrt{5})=\sqrt{2}-\sqrt{3}+\sqrt{5};$$
$$\sqrt{2}\div(\sqrt{3}\div\sqrt{5})=\sqrt{2}\div\sqrt{3}\times\sqrt{5};$$
$$(\sqrt{2}\sqrt{3})^3=(\sqrt{2})^3(\sqrt{3})^3;$$
$$(\sqrt{2})^3(\sqrt{2})^4=(\sqrt{2})^7;\ \text{etc.}$$

CHAPTER XVIII.

SURDS.

1. In Ch. XV. we considered only roots whose radicands are powers with exponents equal to or multiples of the indices of the roots.

In Ch. XVII. we assumed the existence of roots of numbers which are not powers with exponents equal to or multiples of the indices of the required roots, and proved that such roots obey the fundamental laws of Algebra; as $\sqrt{2} \times \sqrt{3} = \sqrt{3} \times \sqrt{2}$, etc.

Such roots were called **Irrational Numbers**; as $\sqrt{2}$, $\sqrt[3]{5}$.

2. A **Radical** is an indicated root of a number or expression; as $\sqrt{7}$, $\sqrt{9}$, $\sqrt[3]{(a+b)}$.

A **Radical Expression** is an expression which contains radicals; as $2\sqrt{7}$, $\sqrt{x} + \sqrt{y}$, $\sqrt{(a + \sqrt{b})}$.

A **Surd** is an irrational root of a rational number; as $\sqrt{7}$, $\sqrt{a}$.

Observe that $\sqrt{(1+\sqrt{7})}$ is not a surd, since $1 + \sqrt{7}$ is not a rational number.

It is important to notice the difference between arithmetical and algebraical irrationality. Thus, $\sqrt{a}$ is algebraically irrational; but if $a = 4$, then $\sqrt{a}, = \sqrt{4}, = 2$ is arithmetically rational.

Classification of Surds.

3. A **Quadratic Surd**, or a **Surd of the Second Order**, is one with index 2; as $\sqrt{3}$, $\sqrt{a}$.

A **Cubic Surd**, or a **Surd of the Third Order**, is one with index 3; as $\sqrt[3]{(a+b)}$, $\sqrt[3]{7}$.

A **Biquadratic Surd**, or a **Surd of the Fourth Order**, is one with index 4; as $\sqrt[4]{(ab)}$, $\sqrt[4]{5}$.

A Simple Monomial Surd Number is a single surd number, or a rational multiple of a single surd number; as $\sqrt{3}$, $2\sqrt{5}$.

A Simple Binomial Surd Number is the sum of two simple surd numbers, or of a rational number and a simple surd number; as $\sqrt{2}+\sqrt[3]{3}$, $3+\sqrt{6}$.

4. The principles enunciated in Ch. XV., and their proofs, are such that they hold also for irrational roots. Each principle will be restated as occasion for its use arises in this chapter.

As in Ch. XV., we shall limit the radicands to positive values, and the roots to principal roots.

Reduction of Surds.

5. A surd is in its *simplest form* when the radicand is integral, and does not contain a factor with an exponent equal to or a multiple of the index of the root; as $\sqrt{2}$, $\sqrt[3]{(a^2b)}$, $\sqrt[n]{a^m}$.

A surd can be reduced to its simplest form by applying one or more of the following principles:

(i.) $$\sqrt[q]{a^{kq}} = a^{\frac{kq}{k}} = a^q \qquad \text{(Ch. XV., § 1, Art. 14.)}$$

(ii.) $$\sqrt[q]{(ab)} = \sqrt[q]{a} \times \sqrt[q]{b} \qquad \text{(Ch. XV., § 1, Art. 15.)}$$

(iii.) $$\sqrt[q]{\frac{a}{b}} = \frac{\sqrt[q]{a}}{\sqrt[q]{b}} \qquad \text{(Ch. XV., § 1, Art. 16.)}$$

(iv.) *In a root of a power (or a power of a root) the index of the root and the exponent of the power may both be multiplied or divided by one and the same number;* or, stated symbolically,

$$\sqrt[q]{a^p} = \sqrt[kq]{a^{kp}}, \text{ and } \sqrt[q]{a^p} = \sqrt[\frac{q}{k}]{a^{\frac{p}{k}}}.$$

E.g., $$\sqrt[3]{a^2} = \sqrt[6]{a^4};\quad \sqrt[6]{a^9} = \sqrt{a^3}.$$

Let $$R = \sqrt[q]{a^p}.$$

Then $$R^q = (\sqrt[q]{a^p})^q, \text{ by Ch. III., § 3, Art. 5,}$$
$$= a^p, \text{ by definition of a root.}$$

Therefore $R^{kq} = a^{kp}$, by Ch. III., § 3, Art. 5.

Whence $R = \sqrt[kq]{a^{kp}}$, by Ch. XV., § 1, Art. 12.

Substituting $\sqrt[q]{a^p}$ for R in the last equation, we have $\sqrt[q]{a^p} = \sqrt[kq]{a^{kp}}$.

In a similar manner the second part of the principle can be proved.

The following examples will illustrate the methods of reducing surds to their simplest forms:

Ex. 1. $\sqrt{(a^5b^2)} = \sqrt{(a^4b^2)} \times \sqrt{a} = a^2b\sqrt{a}$.

Ex. 2. $\sqrt{18} = \sqrt{9} \times \sqrt{2} = 3\sqrt{2}$.

Ex. 3. $\sqrt[n]{(a^{n+1}b^{2n+2})} = \sqrt[n]{(a^nb^{2n})} \times \sqrt[n]{(ab^2)} = ab^2\sqrt[n]{(ab^2)}$.

Observe that the radicand is separated into two factors, one of which is a power with the highest exponent which is equal to or a multiple of the index of the required root. The result is then obtained by multiplying the rational root of this factor by the irrational root of the second factor.

Ex. 4. $\sqrt{\dfrac{3a^2}{4b^2}} = \dfrac{\sqrt{(3a^2)}}{\sqrt{(4b^2)}} = \dfrac{\sqrt{a^2} \times \sqrt{3}}{\sqrt{(4b^2)}} = \dfrac{a\sqrt{3}}{2b}$.

When the required root of the denominator of a fraction cannot be expressed rationally, multiply both terms of the fraction by the expression of lowest degree which will make the denominator a power with an exponent equal to the index of the root.

Ex. 5. $\sqrt{\dfrac{2}{3}} = \sqrt{\dfrac{6}{9}} = \dfrac{\sqrt{6}}{3}$.

Ex. 6. $\sqrt[3]{\dfrac{3a}{4b^2}} = \sqrt[3]{\dfrac{3a \times 2b}{4b^2 \times 2b}} = \dfrac{\sqrt[3]{(6ab)}}{2b}$.

Ex. 7. $\sqrt[n]{\dfrac{a}{b^5}} = \sqrt[n]{\dfrac{ab^{n-5}}{b^n}} = \dfrac{\sqrt[n]{(ab^{n-5})}}{b}$.

By (iv.), a given surd can frequently be reduced to an equivalent surd of a lower order.

Ex. 8. $\sqrt[6]{(27a^3b^6)} = \sqrt[6]{b^6} \times \sqrt[6]{(3a)^3} = b\sqrt{(3a)}$.

EXERCISES I.

Reduce each of the following surds to its simplest form:

1. $\sqrt{32}$. 2. $\sqrt{75}$. 3. $\sqrt{108}$. 4. $\sqrt{x^3}$.

5. $\sqrt{(a^2b)}$. 6. $\sqrt{(a^4b^5)}$. 7. $\sqrt{(4a^7x^{11})}$. 8. $\sqrt{(50ax^2y^3)}$.

9. $\sqrt{(5b^{3n}x^{2n})}$. 10. $\sqrt{(a^{2n}y^{4n+1}z)}$. 11. $\sqrt{(a^{3n+4}b^{9n-2})}$.

12. $\sqrt{(a^2b^3+a^2c^3)}$. 13. $\sqrt{(ab^3c^4-b^2c^6)}$.

14. $\sqrt{(b-c)(b^3-c^3)}$. 15. $\sqrt{(a^2-1)(1+a)}$.

16. $\sqrt{(9x^3-18x^2+9x)}$. 17. $\sqrt{(4a^3b-8a^2b^2+4ab^3)}$.

18. $\sqrt[3]{192}$. 19. $\sqrt[3]{(-10\frac{1}{8})}$. 20. $\sqrt[3]{(-a^{10})}$. 21. $\sqrt[3]{(a^3b^5)}$.

22. $\sqrt[3]{(16a^5x^9)}$. 23. $\sqrt[3]{(a^6b^3c^4)}$. 24. $\sqrt[3]{(-a^{7x}b^{3x})}$.

25. $\sqrt[3]{(32a^{m+6n})}$. 26. $\sqrt[3]{(-54x^{4n+11}y^{14})}$. 27. $\sqrt[3]{(a^6-a^3x^2)}$.

28. $\sqrt{(a^{6n}b^n-a^{7n})}$. 29. $\sqrt[5]{729}$.

30. $\sqrt[4]{(4a^6x^8)}$. 31. $\sqrt[4]{(a^{x+11}b^{4x+1})}$.

32. $\sqrt[6]{(a^{18}x^7)}$. 33. $\sqrt[5]{(a^{7n-5}b^{11n+28}c^{10})}$.

34. $\sqrt[9]{(-a^{15m}b^{10m})}$. 35. $\sqrt[n]{(a^{2n+1}b)}$.

36. $\sqrt[n]{(a^{nx+m}b^{3n}c^3)}$. 37. $\sqrt[m]{(3a^{4m}b^{m+3n}m^n)}$.

38. $\sqrt[x-1]{(a^{x^2-3x+2}b^{x^2-1})}$. 39. $\sqrt[n]{(a^{n+1}b^n-a^nb^{n+1})}$.

40. $\sqrt[8]{(a^3)^4+a^{3^4}}$. 41. $\sqrt[6]{[(-x)^2]^3-(-x^2)^3-x^{2^3}}$.

42. $\sqrt{\frac{9}{10}}$. 43. $\sqrt{\frac{8}{27}}$. 44. $\sqrt{\frac{x^3}{a}}$. 45. $\sqrt{\frac{3a^2}{4}}$.

46. $\sqrt{\frac{64a}{81b}}$. 47. $\sqrt{\frac{18a^2x^3}{125b^5}}$. 48. $\sqrt{\frac{4a^2x}{9b^4y^5}}$. 49. $\sqrt{\frac{162}{a^7}}$.

50. $\sqrt{\frac{16a^8}{45b^3x^5}}$. 51. $\sqrt{\frac{a^2-b^2c^4}{a^2b^2c^4}}$. 52. $\sqrt{\frac{98a^2(x^2-1)^{2n+3}}{b^6x^{4n+3}}}$.

53. $\sqrt[3]{\frac{8}{9}}$. 54. $\sqrt[3]{\frac{a}{b^3}}$. 55. $\sqrt[3]{\frac{a}{27b}}$. 56. $\sqrt[3]{\frac{3a^2x^3}{4b^3y^4}}$.

57. $\sqrt[3]{\frac{ax^{4n}}{8b^2}}$. 58. $\sqrt[3]{\frac{128a^7x^3}{b^6y^{13}}}$. 59. $\sqrt[3]{\frac{ax^8-a^2x^7}{64b^{12}}}$.

60. $\sqrt[3]{\left(1-\frac{1}{a^3}\right)}$. 61. $\sqrt{\left(\frac{a^3}{x^4}-\frac{a^5}{x^6}\right)}$. 62. $\sqrt[4]{\frac{16a^5x^{16}}{b^3y^{11}}}$.

63. $\sqrt[5]{\frac{a^6b^8}{x^5}}$. 64. $\sqrt[6]{\frac{a^6}{6\,b^7x^{23m}}}$. 65. $\sqrt[2n]{\frac{an^2}{x^{6n}}}$. 66. $\sqrt[4]{25}$.

67. $\sqrt[4]{(81\,a^2)}$. 68. $\sqrt[6]{a^3}$. 69. $\sqrt[14]{(4\,a^{12})}$. 70. $\sqrt[4]{\frac{9}{36}}$.

71. $\sqrt[4]{\frac{a^2}{b^2}}$. 72. $\sqrt[10]{\frac{32}{x^{15}y^{20}}}$. 73. $\sqrt[mx]{\frac{1}{a^{nx}}}$.

74. $\sqrt[6]{(8\,a^9b^{15})}$. 75. $\sqrt[12]{(64\,a^8x^{10})}$. 76. $\sqrt[15x]{(a^{30x}b^{20})}$.

77. $\sqrt[12n]{(81\,a^8b^{60})}$. 78. $\sqrt[3n]{(a^{2n+n^2}b^n)}$. 79. $\sqrt[10]{(a^2-2ax+x^2)^3}$.

Addition and Subtraction of Surds.

6. Similar or Like Surds are rational multiples of one and the same simple monomial surd, as $\sqrt{12}, = 2\sqrt{3}$, and $5\sqrt{3}$.

7. The addition and subtraction of *unlike* surds can only be indicated, as $\sqrt{2} + \sqrt[3]{5}$, $\sqrt{4} - 2\sqrt[5]{7}$.

But *like* surds, or such surds as can be reduced to *like* surds, can be united by algebraic addition into a single like surd.

Ex. 1. $\sqrt{12}+2\sqrt{27}-9\sqrt{48}=2\sqrt{3}+6\sqrt{3}-36\sqrt{3}=-28\sqrt{3}$.

Ex. 2. $8\sqrt[3]{40} + 3\sqrt[3]{135} - 2\sqrt[3]{625} = 16\sqrt[3]{5} + 9\sqrt[3]{5} - 10\sqrt[3]{5}$.
$= 15\sqrt[3]{5}$.

Ex. 3. $\sqrt{2} - \sqrt{\tfrac{1}{2}} + \sqrt{.02} = \sqrt{2} - \tfrac{1}{2}\sqrt{2} + \tfrac{1}{10}\sqrt{2} = \tfrac{3}{5}\sqrt{2}$.

Ex. 4. $\sqrt{(a^5b)} + 2\sqrt{(a^3b^3)} + \sqrt{(ab^5)}$
$= a^2\sqrt{(ab)} + 2\,ab\sqrt{(ab)} + b^2\sqrt{(ab)} = (a+b)^2\sqrt{(ab)}$.

EXERCISES II.

Simplify each of the following expressions:

1. $\sqrt{24} - \sqrt{6} + \sqrt{150}$.
2. $2\sqrt{8} + 5\sqrt{72} - 7\sqrt{18}$.
3. $\sqrt{54} + 2\sqrt{24} - 9\sqrt{96}$.
4. $5\sqrt{3} - 2\sqrt{48} + 5\sqrt{108}$.
5. $3\sqrt{75} + 4\tfrac{1}{2}\sqrt{192} - 2\tfrac{3}{4}\sqrt{12}$.
6. $\sqrt{2\tfrac{9}{20}} + \sqrt{5\tfrac{41}{80}} - \sqrt{\tfrac{2}{5}}$.
7. $4\sqrt{\tfrac{3}{4}} - \tfrac{2}{7}\sqrt{\tfrac{3}{16}} - 2\sqrt{27}$.
8. $2\sqrt{\tfrac{5}{3}} + \sqrt{60} - \sqrt{15} + \sqrt{\tfrac{3}{5}}$.
9. $\sqrt{2\tfrac{3}{11}} + \sqrt{5\tfrac{9}{11}} - \sqrt{\tfrac{9}{11}}$.
10. $\sqrt{1\tfrac{1}{3}}+3\sqrt{5\tfrac{1}{3}}-2\sqrt{8\tfrac{1}{3}}+3\sqrt{16\tfrac{1}{3}}$.

11. $8\sqrt[3]{48}+3\sqrt[3]{162}-2\sqrt[3]{384}.$ **12.** $5\sqrt[3]{54}+9\sqrt[3]{250}-\sqrt[3]{686}.$

13. $2\frac{3}{5}\sqrt[3]{500}+\frac{3}{4}\sqrt[3]{256}-3\frac{1}{2}\sqrt[3]{32}-\frac{2}{3}\sqrt[3]{108}.$

14. $1\frac{1}{3}\sqrt[3]{\frac{135}{8}}+3\sqrt[3]{135}-2\sqrt[3]{625}-\sqrt[3]{5}+8\sqrt[3]{40}.$

15. $1.5\sqrt[3]{1\frac{3}{5}}-2\frac{1}{4}\sqrt[3]{12.8}-3\frac{2}{3}\sqrt[3]{5\frac{2}{5}}+4.6\sqrt[3]{43.2}.$

16. $\sqrt[3]{40}-5\sqrt[3]{\frac{1}{25}}+4\sqrt[3]{(-.625)}-\frac{2}{3}\sqrt[3]{16\frac{7}{8}}.$

17. $2\sqrt{3}-\sqrt{12}+\sqrt[4]{9}.$ **18.** $\sqrt[3]{24}+3\sqrt[6]{9}-5\sqrt[3]{192}.$

19. $\sqrt{(4a^3)}+\sqrt{(9a^3)}+\sqrt{(25a^3)}-\sqrt{(81a^3)}.$

20. $\sqrt{(12a^2b)}+\sqrt{(75a^2b)}-\sqrt{(27a^2b)}.$

21. $\sqrt[3]{(64a^8b^5)}+\sqrt[3]{(125a^8b^5)}-\sqrt[3]{(a^8b^5)}.$

22. $a\sqrt{(a^3b^7)}+b^2\sqrt{(a^5b^3)}-2ab^2\sqrt{(a^3b^3)}+\sqrt{(a^{21}b^{25})}.$

23. $5x\sqrt{(12a^3)}-2a\sqrt{(27ax^2)}-2x\sqrt{(75a^5)}+ax\sqrt{(48a)}$
$+4a\sqrt{(108a^3x^2)}.$

24. $3z\sqrt[3]{(250x^4z^3)}-5x\sqrt[3]{(128xz^5)}+3xz\sqrt[3]{(16xz^2)}.$

25. $5ab\sqrt[3]{(243a^2b)}+3\sqrt[3]{(72a^5b^4)}-2b\sqrt[3]{(1125a^5b)}.$

26. $3a^2b\sqrt[3]{(32a^2b)}+5\sqrt[3]{(108a^8b^4)}-ab\sqrt[3]{(500a^5b)}.$

27. $\sqrt[4]{(9a^2b^2)}+\sqrt{(27a^3b)}+5\sqrt[4]{(729a^6b^2)}.$

28. $2\sqrt[3]{(3x^2y)}-\sqrt[6]{(9x^4y^2)}+\sqrt[3]{(125x^4y)}-\sqrt[6]{(x^8y^2)}.$

29. $\sqrt{(9a+27)}+3\sqrt{(4a+12)}.$

30. $\sqrt{(4a^3+4a^2b)}+\sqrt{(4ab^2+4b^3)}.$

31. $7x\sqrt{(25a+75)}-5\sqrt{(9x^2a+27x^2)}.$

32. $2\sqrt{(2x^3)}-\sqrt{(8x)}-\sqrt{(2x^3-4x^2+2x)}.$

33. $\sqrt[5]{(a^3b^5+3b^6)}+a\sqrt[5]{(32a^3+96b)}-\sqrt[5]{(a^8+3a^5b)}.$

34. $\sqrt{(a^2-b^2)(a+b)}+\sqrt{(a-b)^3}.$

35. $\sqrt{a^3-a^2b}-\sqrt{ab^2-b^3}-\sqrt{(a+b)(a^2-b^2)}.$

36. $3a\sqrt{\frac{a-x}{a+x}}-3x\sqrt{\frac{a-x}{a+x}}-2a\sqrt{\frac{a-x}{a+x}}+4x\sqrt{\frac{a-x}{a+x}}.$

Reduction of Surds of Different Orders to Equivalent Surds of the Same Order.

8. Surds of different orders with unequal indices can be reduced to equivalent surds of the same order with equal indices by the principle

$$\sqrt[q]{a^q} = \sqrt[kq]{a^{kq}} \qquad \text{[Art. 5 (iv.)]}$$

Ex. Reduce $\sqrt{3}$, $\sqrt[4]{2a}$, and $\sqrt[6]{5b}$ to equivalent surds of the same order.

We have

$$\sqrt{3} = \sqrt[12]{(3)^6} = \sqrt[12]{729};$$

$$\sqrt[4]{(2a)} = \sqrt[12]{(2a)^3} = \sqrt[12]{(8a^3)};$$

$$\sqrt[6]{(5b)} = \sqrt[12]{(5b)^2} = \sqrt[12]{(25b^2)}.$$

Observe that the L.C.M. of the given indices is taken as the common index of the equivalent surds, and that each radicand is raised to a power whose exponent is equal to the quotient of this L.C.M. divided by the index of the given root.

9. Any rational number can be expressed in the form of a surd by writing under the radical sign a power of the number whose exponent is equal to the index.

E.g., $2 = \sqrt{4} = \sqrt[3]{8} = \cdots = \sqrt[r]{2^r}.$

10. Two surds, or a surd and a rational number, can be compared by first reducing them to equivalent surds of the same order, and then comparing the resulting radicands.

Ex. Which is greater, $\sqrt{2}$ or $\sqrt[3]{3}$?

We have $\sqrt{2} = \sqrt[6]{8}$, and $\sqrt[3]{3} = \sqrt[6]{9}$.

Since $9 > 8$, therefore $\sqrt[6]{9} > \sqrt[6]{8}$, or $\sqrt[3]{3} > \sqrt{2}$.

EXERCISES III.

Reduce to equivalent surds of the same order:

1. $\sqrt[3]{2}$, $\sqrt[4]{5}$.
2. $\sqrt{3}$, $\sqrt[4]{6}$.
3. $\sqrt[3]{7}$, $\sqrt[5]{10}$.
4. $\sqrt{\frac{1}{2}}$, $\sqrt[3]{\frac{1}{4}}$.
5. 5, $\sqrt[4]{10}$.
6. 6, $\sqrt[3]{4}$.
7. $\sqrt[6]{2}$, $\sqrt[8]{3}$.
8. $\sqrt[10]{15}$, $\sqrt[15]{10}$.
9. $\sqrt[n]{(3a^2)}$, b^4.

10. $\sqrt[n]{a^3}$, $\sqrt[m]{b^5}$. **11.** $\sqrt[n+1]{(x^3y)}$, $\sqrt[n-1]{(xy^3)}$. **12.** $\sqrt{5}$, $\sqrt[3]{10}$, $\sqrt[4]{15}$.

13. $\sqrt[4]{2}$, 3, $\sqrt[6]{5}$. **14.** $\sqrt[m]{a^2}$, b^3, $\sqrt[p]{c^4}$. **15.** $\sqrt[2m]{a^3}$, $\sqrt[4m]{b^3}$, $\sqrt[6m]{c^5}$.

16. $\sqrt[3]{(a+b)}$, $\sqrt[4]{(a-b)}$. **17.** $\sqrt[4]{(x-y^2)}$, $\sqrt[6]{(x^2-y)}$.

Which is the greater,

18. $2\sqrt{3}$ or $3\sqrt{2}$? **19.** $\sqrt{5}$ or $\sqrt[3]{10}$? **20.** $\frac{1}{2}\sqrt[3]{25}$ or $\frac{1}{3}\sqrt{11}$?

21. $\sqrt[3]{a^2}$ or $\sqrt[2]{a}$, when $a<1$? **22.** $\sqrt[4]{x^3}$ or $\sqrt[5]{x^4}$, when $x>1$?

Which is the greatest,

23. $\sqrt{3}$, $\sqrt[3]{5}$, or $\sqrt[4]{10}$? **24.** $\sqrt[2]{\frac{2}{3}}$, $\sqrt[3]{\frac{3}{2}}$, or $\sqrt[4]{\frac{7}{4}}$?

Multiplication of Surds.

11. Multiplication of Monomial Surds. — The product of two or more monomial surds is found by applying the principle

$$\sqrt[q]{a} \times \sqrt[q]{b} = \sqrt[q]{(ab)}.$$

Ex. 1. $5\sqrt[3]{4} \times 2\sqrt[3]{6} = 10\sqrt[3]{24} = 20\sqrt[3]{3}$.

Ex. 2. $\sqrt[5]{(2a^4)} \times \sqrt[5]{(5a^3b^2)} \times \sqrt[5]{(7ab^3)} = \sqrt[5]{(70a^8b^5)} = ab\sqrt[5]{(70a^3)}$.

If the surds are of different orders, they should first be reduced to equivalent surds of the same order.

Ex. 3. $\sqrt{a} \times \sqrt[3]{a^2} = \sqrt[6]{a^3} \times \sqrt[6]{a^4} = \sqrt[6]{a^7} = a\sqrt[6]{a}$.

Ex. 4.

$$\begin{aligned}\sqrt[3]{(a^2b)} \times \sqrt[4]{(a^3b^2)} \times \sqrt[6]{(a^4b^5)} &= \sqrt[12]{(a^8b^4)} \times \sqrt[12]{(a^9b^6)} \times \sqrt[12]{(a^8b^{10})}\\ &= \sqrt[12]{(a^{25}b^{20})} = a^2b\sqrt[12]{(ab^8)}.\end{aligned}$$

Ex. 5.

$$\begin{aligned}2\sqrt[3]{5} \times 3\sqrt[4]{20} \times \sqrt{10} &= 6\sqrt[3]{5} \times \sqrt[4]{(2^2 \times 5)} \times \sqrt{(2 \times 5)}\\ &= 6\sqrt[12]{5^4} \times \sqrt[12]{(2^6 \times 5^3)} \times \sqrt[12]{(2^6 \times 5^6)}\\ &= 6\sqrt[12]{(5^{13} \times 2^{12})} = 60\sqrt[12]{5}.\end{aligned}$$

When the radicands contain numerical factors it is frequently advisable to express them as powers of the smallest possible bases, as in Ex. 5. In all cases the results should be reduced to equivalent surds in their simplest forms.

It is frequently desirable to introduce the coefficient of a surd under the radical sign.

Ex. 6. $2\sqrt{3} = \sqrt{4} \times \sqrt{3} = \sqrt{12}$.

Ex. 7. $3a\sqrt[3]{(2ab)} = \sqrt[3]{(27a^3)} \times \sqrt[3]{(2ab)} = \sqrt[3]{(54a^4b)}$.

12. Multiplication of Multinomial Surd Numbers. — The work may be arranged as in multiplication of rational multinomials.

Ex. Multiply $2 - 3\sqrt{2} + 5\sqrt{6}$ by $\sqrt{2} - \sqrt{3}$.

We have

$$\begin{array}{rrrr} & 2 & -3\sqrt{2} & +5\sqrt{6} \\ & \sqrt{2} & -\sqrt{3} & \\ \hline 2\sqrt{2} & -6 & +10\sqrt{3} & \\ -15\sqrt{2} & & -2\sqrt{3} & +3\sqrt{6} \\ \hline -13\sqrt{2} & -6 & +8\sqrt{3} & +3\sqrt{6} \end{array}$$

Observe that the terms of each partial product are simplified, and that similar surds are then written in the same column.

13. Conjugate Surds. — Two binomial quadratic surds which differ only in the sign of a surd term are called **Conjugate Surds**.

E.g., $\sqrt{3} + \sqrt{2}$ and $-\sqrt{3} + \sqrt{2}$; $1 - \sqrt{5}$ and $1 + \sqrt{5}$.

Either of two conjugate surds is called the conjugate of the other.

The product of two conjugate surds is a rational number.

For, $(\sqrt{a} + \sqrt{b})(\sqrt{a} - \sqrt{b}) = (\sqrt{a})^2 - (\sqrt{b})^2 = a - b$.

14. Type-Forms. — Many products are more easily obtained by using the type-forms given in Ch. VI., § 1.

Ex.
$$\begin{aligned}(\sqrt{2} + \sqrt{3})^2 &= (\sqrt{2})^2 + 2\sqrt{2} \times \sqrt{3} + (\sqrt{3})^2 \\ &= 2 + 2\sqrt{6} + 3 = 5 + 2\sqrt{6}.\end{aligned}$$

EXERCISES IV.

Simplify each of the following expressions:

1. $\sqrt{3} \times \sqrt{6}$.
2. $\sqrt{27} \times 3\sqrt{18}$.
3. $-2\sqrt{45} \times 4\sqrt{15}$.
4. $4\sqrt{\frac{3}{2}} \times 5\sqrt{\frac{3}{8}}$.
5. $\sqrt{\frac{5}{7}} \times \sqrt{\frac{7}{125}}$.
6. $\frac{1}{2}\sqrt{\frac{6}{11}} \times 3\sqrt{\frac{22}{3}}$.
7. $\sqrt[3]{4} \times \sqrt[3]{2}$.
8. $2\sqrt[3]{4} \times 3\sqrt[3]{16}$.
9. $2\sqrt[3]{\frac{25}{16}} \times \frac{1}{5}\sqrt[3]{\frac{2}{5}}$.
10. $\sqrt[3]{(a^2b)} \times \sqrt[3]{(ab^2)}$.
11. $\sqrt[4]{8} \times \sqrt[4]{10}$.
12. $9\sqrt[4]{54} \times (-3\sqrt[4]{24})$.
13. $\sqrt[6]{(2a^4x)} \times \sqrt[6]{(4ax)} \times \sqrt[6]{(8a^2x^5)}$.

14. $\sqrt{2} \times 2\sqrt[4]{4}$. **15.** $\sqrt{\tfrac{1}{2}} \times \sqrt[6]{12}$. **16.** $\sqrt[4]{54} \times \sqrt[6]{486}$.

17. $\sqrt{\tfrac{10}{11}} \times \sqrt[3]{\tfrac{11}{20}}$. **18.** $\sqrt[3]{12} \times \sqrt[4]{6}$. **19.** $\sqrt[3]{\tfrac{5}{6}} \times \sqrt[4]{\tfrac{18}{25}}$.

20. $\sqrt[4]{\tfrac{27}{64}} \times \sqrt[3]{\tfrac{81}{16}}$. **21.** $\sqrt[5]{\tfrac{7}{15}} \times \sqrt[6]{\tfrac{15}{14}}$. **22.** $\sqrt[3]{2} \times \sqrt[6]{\tfrac{1}{3}} \times \sqrt[3]{3}$.

23. $\sqrt[5]{54} \times 3\sqrt{6} \times 5\sqrt[3]{2}$. **24.** $\sqrt{2} \times \sqrt[3]{3} \times \sqrt[4]{6} \times \sqrt[6]{1\tfrac{1}{3}}$.

25. $\sqrt{10} \times \sqrt[3]{100} \times \sqrt[4]{500}$. **26.** $\sqrt[3]{12} \times \sqrt[4]{108} \times \sqrt[6]{486}$.

27. $\sqrt[4]{3} \times \sqrt[3]{\tfrac{1}{3}} \times \sqrt[6]{\tfrac{1}{2}} \times \sqrt[12]{12}$. **28.** $12\sqrt[4]{14} \times \sqrt{2\tfrac{1}{7}} \times \sqrt[8]{\tfrac{49}{800}}$.

29. $1\tfrac{1}{2}\sqrt{\tfrac{1}{2}} \times \tfrac{1}{3}\sqrt[4]{2} \times 4\sqrt{6} \times 2\sqrt{3}$. **30.** $x^2\sqrt[6]{(3x^5)} \times 3x^3\sqrt[4]{(2x^3)}$.

31. $3a^2\sqrt{(a-2)} \times 2a^3\sqrt[4]{(a+2)}$.

32. $a^3\sqrt[10]{a^4(x-a)^7} \times a^{12}\sqrt[5]{a^3(x-a)^2}$.

33. $\sqrt{(x^2+x)} \times \sqrt{(ax+a)}$. **34.** $\sqrt{(12x^3-12x)} \times \sqrt{(3x^2-3)}$.

35. $\sqrt{(x^2-x)} \times \sqrt{(x^4+x^3)}$. **36.** $\sqrt{(ax+a)} \times \sqrt{(bx+b)}$.

37. $\sqrt{(a^2-b^2)} \times \sqrt{\dfrac{a+b}{a-b}}$. **38.** $\sqrt{(6x^2-6)} \times \sqrt{\dfrac{3x-3}{2x+2}}$.

39. $\dfrac{a-2x}{a+3x}\sqrt{\dfrac{a^2-2ax}{ax-3x^2}} \times \dfrac{a^2-9x^2}{a^2-4ax+4x^2}\sqrt{\dfrac{ax-2x^2}{a^2-3ax}}$.

40. $b\sqrt[5]{(8b^3-12b^2y+6by^2-y^3)} \times \sqrt[5]{(4b^3-4b^2y+by^2)}$.

41. $\dfrac{x^3-8z^3}{\sqrt{(x^3+2x^2z+4xz^2)}} \times \dfrac{x^2}{x-2z}\sqrt{\dfrac{xz}{x^2+2xz+4z^2}}$.

42. $\dfrac{a^2-25m^2}{4a^2-9m^2}\sqrt[3]{\dfrac{4a^2-12am+9m^2}{a^2+10am+25m^2}} \times \sqrt[3]{\dfrac{4a^2-12am+9m^2}{a^2+5am}}$.

43. $\dfrac{a^2-4ay+4y^2}{a^3-27y^3}\sqrt{\dfrac{a^3+3a^2y+9ay^2}{ay-2y^2}} \times \dfrac{a-3y}{a-2y}\sqrt{\dfrac{a^2y+3ay^2+9y^3}{a^2-2ay}}$.

44. $(\sqrt{2}-\sqrt{3}+\sqrt{18})\sqrt{2}$. **45.** $(2\sqrt{\tfrac{1}{2}}-2\sqrt{\tfrac{1}{3}}+5\sqrt{1\tfrac{1}{3}})\sqrt{3}$.

46. $(4\sqrt[3]{9}-2\tfrac{3}{4}\sqrt[3]{36}+5\tfrac{3}{5}\sqrt[3]{225}) \times 2\sqrt[3]{30}$.

47. $(\sqrt{2}+\sqrt[3]{2}+\sqrt[4]{2})\sqrt[6]{\tfrac{1}{2}}$. **48.** $\left(\sqrt{\dfrac{a}{x}}+\sqrt{\dfrac{ax}{y}}\right)\sqrt{(axy)}$.

49. $(3+\sqrt{5})(2-\sqrt{5})$. **50.** $(9-7\sqrt{13})(5-6\sqrt{13})$.

51. $(13-\sqrt{5})(7+3\sqrt{5})$. **52.** $(4+\sqrt{6}+\sqrt{22})(3+2\sqrt{6}-\sqrt{33})$.

53. $(\sqrt{3\tfrac{1}{3}}+\sqrt{2}+1\tfrac{1}{3}\sqrt{3})(\sqrt{2}+\sqrt{1\tfrac{1}{5}}-4\sqrt{\tfrac{1}{5}})$.

54. $(5+\sqrt[3]{4}-2\sqrt[4]{5})(\sqrt{6}+\sqrt{5})$. **55.** $(2\sqrt{3}+\sqrt[3]{2})(2\sqrt{3}-\sqrt[3]{4})$.

56. $(\sqrt{3} - 5\sqrt[3]{6})(\sqrt[3]{9^2} - 2\sqrt{27})$.

57. $(\frac{1}{4}\sqrt[4]{8} + \sqrt[3]{4} - 2\sqrt{2})(2\sqrt[3]{4} - 10\sqrt{2})$.

58. $(\sqrt[6]{2} - 1 + \frac{1}{2}\sqrt[6]{32} - \frac{1}{2}\sqrt[3]{4} + \frac{1}{2}\sqrt{2} - \frac{1}{2}\sqrt[3]{2})(\sqrt[6]{32} + \sqrt[3]{4})$.

Find the value of each of the following powers:

59. $(\sqrt{7})^3$.
60. $(2\sqrt{3})^4$.
61. $(\sqrt{x})^3$.
62. $(\sqrt[3]{ab})^2$.
63. $(5\sqrt[3]{4})^5$.
64. $(\sqrt[4]{xy})^2$.
65. $(\frac{1}{2}\sqrt[2]{6ab})^3$.
66. $(2a\sqrt[3]{3b})^6$.
67. $(\sqrt[6]{a^2b^2})^3$.
68. $(\sqrt[6]{3ab})^3$.
69. $(\sqrt[3]{n} \times \sqrt[q]{x})^{3q}$.
70. $\left[\frac{\sqrt[4]{(a^3x^2)}}{\sqrt[3]{(ax^2)}}\right]^2$.
71. $\left[\frac{\sqrt[5]{(a^4b)}}{\sqrt[4]{(a^3b)}}\right]^{10}$.
72. $\left(\sqrt[4]{\frac{3ax + 3x^2}{n^2}} \times \sqrt[3]{\frac{n}{6a + 6x}}\right)^4$.

Find the value of each of the following expressions, without performing the actual multiplication:

73. $(\sqrt{5} - \sqrt{10})^2$.
74. $(\frac{1}{2} + 2\sqrt{2})^2$.
75. $(\sqrt[4]{8} - \sqrt[4]{2})^2$.
76. $(\sqrt{6} - \sqrt[4]{40})^2$.
77. $(\sqrt{3} - \sqrt{6})^3$.
78. $(3\sqrt{2} + 4\sqrt{3})^3$.
79. $(\sqrt{6} - 2\sqrt[3]{2})^3$.
80. $(1 + \sqrt{2} - \sqrt{3})^2$.
81. $(\frac{1}{2}\sqrt{3} + \frac{3}{4}\sqrt{5} - \sqrt{10})^2$.
82. $(\sqrt{2} + \sqrt{3} + 1)^3$.
83. $\sqrt[4]{(5+2\sqrt{6})} \times \sqrt{(3-\sqrt{6})}$.
84. $(8 - 3\sqrt{7})(8 + 3\sqrt{7})$.
85. $\sqrt[3]{(2+\sqrt{12})}\sqrt[3]{(2-\sqrt{12})}$.
86. $\sqrt[4]{(\sqrt{23}-\sqrt{7})}\sqrt[4]{(\sqrt{23}+\sqrt{7})}$.
87. $(\sqrt{3 - \sqrt{5}} + \sqrt{3 + \sqrt{5}})^2$.
88. $(\sqrt{8 + \sqrt{39}} - \sqrt{8 - \sqrt{39}})^2$.
89. $(\sqrt{2ab} + \sqrt{3ab})^2$.
90. $(n - \sqrt{1 - n^2})^2$.
91. $(\sqrt{a + x} + \sqrt{a - x})^2$.
92. $(\sqrt[3]{2a^2} + \sqrt[3]{2a})^3$.
93. $(x + 2\sqrt{x^2 - 1})^3$.
94. $(2\sqrt[3]{ab^2} - 3\sqrt[6]{a^2b^3})^3$.

95. $(a\sqrt{ax} + 2a\sqrt{2ax})^3$.

96. $(\sqrt{a + b} + \sqrt{a} - \sqrt{b})(\sqrt{a + b} - \sqrt{a} + \sqrt{b})$.

97. $(a^2 + x^2 + x\sqrt{a^2 - x^2})(a^2 + x^2 - x\sqrt{a^2 - x^2})$.

98. $[\sqrt{\sqrt{(a + b)} + \sqrt{(a - b)}} - \sqrt{\sqrt{(a + b)} - \sqrt{(a - b)}}]^2$.

99. $\left(x + \frac{p}{2} + \sqrt{\frac{p^2}{4} - q}\right)\left(x + \frac{p}{2} - \sqrt{\frac{p^2}{4} - q}\right)$.

In each of the following expressions introduce the coefficient under the radical sign:

100. $7\sqrt{3}$. **101.** $\frac{1}{2}\sqrt{2}$. **102.** $\frac{1}{2}\sqrt[3]{4}$. **103.** $\frac{3}{4}\sqrt[3]{\frac{16}{9}}$.

104. $2a\sqrt{a}$. **105.** $5x^2\sqrt{(3xy)}$. **106.** $ab\sqrt{\frac{1}{ab}}$. **107.** $4a^2b\sqrt[3]{(2a)}$.

108. $a\sqrt[n]{a}$. **109.** $a^2b\sqrt[n-1]{(ab)}$. **110.** $a^{n+1}\sqrt{a^{n-2}}$. **111.** $x^ny^m\sqrt[3]{(x^my^n)}$.

112. $\frac{1}{8}(\sqrt{5}-1)\sqrt{(6+2\sqrt{5})}$. **113.** $(a+b)\sqrt{\frac{ab}{a^2+2ab+b^2}}$.

114. $(m-n)\sqrt{\frac{m+n}{m-n}}$. **115.** $(m+n)\sqrt{\frac{m^4-m^3n+m^2n^2-mn^3+n^4}{m+n}}$.

Division of Surds.

15. Division of Monomial Surds. — The quotient of one monomial surd divided by another is obtained by applying the principle

$$\frac{\sqrt[q]{a}}{\sqrt[q]{b}}=\sqrt[q]{\frac{a}{b}}.$$

Ex. 1. $\frac{\sqrt{8}}{\sqrt{2}}=\sqrt{\frac{8}{2}}=\sqrt{4}=2.$

If the surds be of different orders, they should first be reduced to equivalent surds of the same order.

Ex. 2. $\frac{\sqrt[3]{(4a^2)}}{\sqrt[2]{(3a)}}=\frac{\sqrt[6]{(16a^4)}}{\sqrt[6]{(27a^3)}}=\sqrt[6]{\frac{16a}{27}}=\frac{1}{3}\sqrt[6]{(432a)}.$

16. Division of Multinomial Surd Numbers. — It is better to write the quotient of one multinomial surd number by another as a fraction, and then to simplify this fraction by the method to be given in Art. 25. But if the divisor is a monomial, the work proceeds as follows:

$$(\sqrt{72}+\sqrt{32}-4)\div 2\sqrt{2}=\frac{\sqrt{36}}{2}+\frac{\sqrt{16}}{2}-\frac{2}{\sqrt{2}}.$$
$$=3+2-\sqrt{2}=5-\sqrt{2}.$$

17. Type-Forms. — Many quotients are more easily obtained by using the type-forms given in Ch. VI., § 2.

Ex. $(\sqrt[3]{a^2}-\sqrt[3]{b^2})\div(\sqrt[3]{a}-\sqrt[3]{b})=[(\sqrt[3]{a})^2-(\sqrt[3]{b})^2]\div(\sqrt[3]{a}-\sqrt[3]{b})$
$=\sqrt[3]{a}+\sqrt[3]{b}.$

EXERCISES V.

Simplify each of the following expressions:

1. $3\sqrt{2} \div 2\sqrt{3}$.
2. $\sqrt{60} \div \sqrt{5}$.
3. $\sqrt{99} \div \sqrt{22}$.
4. $\sqrt{15} \div \sqrt{\tfrac{3}{5}}$.
5. $\sqrt{\tfrac{5}{3}} \div \sqrt{15}$.
6. $\sqrt{21} \div \sqrt{\tfrac{7}{3}}$.
7. $\sqrt{30} \div \sqrt{\tfrac{5}{6}}$.
8. $5 \div \sqrt{5}$.
9. $6 \div \sqrt{3}$.
10. $8 \div \sqrt{2}$.
11. $20 \div 3\sqrt{10}$.
12. $18 \div 7\sqrt{6}$.
13. $10 \div \sqrt[3]{5}$.
14. $15 \div \sqrt[4]{3}$.
15. $9\sqrt[3]{7} \div 2\sqrt[3]{21}$.
16. $2\sqrt[3]{6} \div \sqrt[6]{2}$.
17. $6\sqrt{2} \div \sqrt[3]{9}$.
18. $\sqrt[3]{20} \div 3\sqrt[9]{16}$.
19. $(4 + \sqrt{6} - 5\sqrt{14}) \div 2\sqrt{2}$.
20. $(3\sqrt{10} - 4\sqrt{15} + 5) \div \sqrt{5}$.
21. $(\sqrt{2} - 3\sqrt[4]{4}) \div \sqrt[4]{2}$.
22. $(\sqrt[3]{3} - 3\sqrt[6]{6}) \div \sqrt[6]{3}$.
23. $\sqrt{x} \div \sqrt[3]{x}$.
24. $\sqrt{x^3} \div \sqrt[3]{x^2}$.
25. $\sqrt{x} \div \sqrt[4]{x}$.
26. $\sqrt[3]{x^2} \div \sqrt[4]{x^3}$.
27. $\sqrt{(a^2x)} \div \sqrt{x}$.
28. $\sqrt[3]{x^2} \div n\sqrt{x}$.
29. $\sqrt{(14\,ab)} \div \sqrt[3]{(28\,a^2b^2)}$.
30. $\sqrt[3]{(15\,x^2y)} \div \sqrt[4]{(25\,xy^3)}$.
31. $2\,a^2\sqrt{n} \div 5\sqrt[3]{(4\,n)}$.
32. $x\sqrt[2n]{x^{n-8}} \div n\sqrt[n]{x^{n-4}}$.
33. $(a - \sqrt{ab}) \div \sqrt{a}$.
34. $(5\sqrt{\tfrac{3}{5}\,x} + 3\,x\sqrt{x}) \div x\sqrt{(15\,x)}$.

Simplify each of the following expressions, without performing the actual division:

35. $(1 - x) \div (1 - \sqrt{x})$.
36. $(ax - bx) \div (\sqrt{a} - \sqrt{b})$.
37. $\left(1 - \dfrac{1}{x}\right) \div (1 + \sqrt{x})$.
38. $\left(\dfrac{a}{b} - \dfrac{x}{y}\right) \div \left(\sqrt{\dfrac{a}{x}} + \sqrt{\dfrac{b}{y}}\right)$.
39. $(1 - x\sqrt{x}) \div (1 - \sqrt{x})$.
40. $(a\sqrt{a} + b\sqrt{b}) \div (\sqrt{a} + \sqrt{b})$.
41. $(x\sqrt[3]{x} - y\sqrt[3]{y}) \div (\sqrt[3]{x} - \sqrt[3]{y})$.
42. $(x\sqrt[3]{x} + y\sqrt[3]{y}) \div (\sqrt[3]{x} + \sqrt[3]{y})$.

Surd Factors.

18. The principles of factoring given in Ch. VIII. can be applied to obtain irrational factors.

Ex. 1. $a + \sqrt{a} = \sqrt{a}(\sqrt{a} + 1)$.

Ex. 2. $a - b = (\sqrt{a})^2 - (\sqrt{b})^2 = (\sqrt{a} + \sqrt{b})(\sqrt{a} - \sqrt{b})$.

Ex. 3. $\sqrt[6]{a} - \sqrt[4]{b} = (\sqrt[12]{a})^2 - (\sqrt[8]{b})^2 = (\sqrt[12]{a} + \sqrt[8]{b})(\sqrt[12]{a} - \sqrt[8]{b})$.

Ex. 4. $a - b = (\sqrt[3]{a})^3 - (\sqrt[3]{b})^3$

$$= (\sqrt[3]{a} - \sqrt[3]{b})[(\sqrt[3]{a})^2 + \sqrt[3]{a} \times \sqrt[3]{b} + (\sqrt[3]{b})^2]$$

$$= (\sqrt[3]{a} - \sqrt[3]{b})[\sqrt[3]{a^2} + \sqrt[3]{(ab)} + \sqrt[3]{b^2}].$$

In general, the difference between any two numbers can be expressed as the difference of two squares, or of two cubes, or of two fourth powers, etc., and be factored accordingly.

Ex. 5. $x^2 + 1 = x^2 + 2x + 1 - 2x$

$$= (x+1)^2 - (\sqrt{2x})^2 = (x+1+\sqrt{2x})(x+1-\sqrt{2x}).$$

19. An expression of the second degree in a letter of arrangement, say x, can be transformed into the difference of two squares, and hence be factored. This transformation is effected by means of the principle which follows:

From the identity

$$(mx + n)^2 = m^2x^2 + 2mnx + n^2$$

we infer:

If a trinomial, arranged to descending powers of a letter, say x, be the square of a binomial, the third term is equal to the square of the quotient obtained by dividing the coefficient of x by twice the square root of the coefficient of x^2; that is,

$$n^2 = \left(\frac{2mn}{2m}\right)^2.$$

Consequently, if to any binomial of the form $m^2x^2 + 2mnx$ the term $\left(\frac{2mn}{2m}\right)^2, = n^2$, be added, the resulting trinomial will be the square of a binomial.

This step is called *completing the square.*

E.g., if to $9x^2 + 5x$ we add

$$\left(\frac{5}{2 \times 3}\right)^2, = \frac{25}{36},$$

we have $$9x^2 + 5x + \frac{25}{36}, = \left(3x + \frac{5}{6}\right)^2.$$

Ex. Factor $25x^2+13x+1$.

To transform $25x^2+13x+1$ into the difference of two squares, we first complete $25x^2+13x$ to the square of a binomial by adding $\left(\frac{13}{2\times 5}\right)^2, =\frac{169}{100}$; and, in order that the value of the given expression may remain unchanged, we also subtract $\frac{169}{100}$ from it. We then have

$$\begin{aligned}25x^2+13x+1 &= 25x^2+13x+\frac{169}{100}-\frac{169}{100}+1\\ &=\left(5x+\frac{13}{10}\right)^2-\left(\frac{\sqrt{69}}{10}\right)^2\\ &=\left(5x+\frac{13}{10}+\frac{1}{10}\sqrt{69}\right)\left(5x+\frac{13}{10}-\frac{1}{10}\sqrt{69}\right).\end{aligned}$$

If the coefficient of x^2 in the expression to be factored be 1, the term to be added to complete the square is evidently the square of half the coefficient of x.

Ex. 1. Factor x^2-5x-1.

We have
$$\begin{aligned}x^2-5x-1 &= x^2-5x+(\tfrac{5}{2})^2-(\tfrac{5}{2})^2-1\\ &=\left(x-\frac{5}{2}\right)^2-\left(\frac{\sqrt{29}}{2}\right)^2\\ &=(x-\tfrac{5}{2}+\tfrac{1}{2}\sqrt{29})(x-\tfrac{5}{2}-\tfrac{1}{2}\sqrt{29}).\end{aligned}$$

Ex. 2. Factor $-3x^2+4xy+2y^2$.

Since the coefficient of x^2 is not the square of a rational number, the work is simplified by first taking out the factor -3. We then have

$$-3x^2+4xy+2y^2=-3(x^2-\tfrac{4}{3}xy-\tfrac{2}{3}y^2).$$

Completing $x^2-\frac{4}{3}xy$ to the square of a binomial by adding $(\frac{2}{3}y)^2, =\frac{4}{9}y^2$, to the expression *within the parentheses*, and also subtracting $\frac{4}{9}y^2$ from it, we obtain

$$\begin{aligned}-3x^2+4xy+2y^2 &= -3(x^2-\tfrac{4}{3}xy+\tfrac{4}{9}y^2-\tfrac{10}{9}y^2)\\ &=-3\left[\left(x-\frac{2}{3}y\right)^2-\left(\frac{\sqrt{10}}{3}y\right)^2\right]\\ &=-3\left(x-\frac{2}{3}y+\frac{\sqrt{10}}{3}y\right)\left(x-\frac{2}{3}y-\frac{\sqrt{10}}{3}y\right).\end{aligned}$$

This method can of course be applied when the factors are rational, but the methods given in Ch. VIII., § 1, Arts. 10–14, are, as a rule, to be preferred.

EXERCISES VI.

Factor each of the following expressions:

1. $\sqrt{(ax)}+\sqrt{(bx)}$. **2.** $a-\sqrt{(ab)}$.

3. $n\sqrt{x}+x\sqrt{n}$. **4.** $an-\sqrt{(an^2)}$.

5. $a-\sqrt[n]{(ax)}$. **6.** $\sqrt{(ac)}-\sqrt{(bc)}-\sqrt{(ad)}+\sqrt{(bd)}$.

7. $\sqrt{a^3}-\sqrt{b^3}$. **8.** $a\sqrt{a}+b\sqrt{b}$.

9. $x\sqrt[3]{x}-16$. **10.** $a^{10}+x\sqrt[3]{x^2}$.

11. $a+b-2\sqrt{(ab)}$. **12.** $a+\sqrt{(ax^2)}+\dfrac{x^2}{4}$.

13. $x^2-2\sqrt{(ax^2)}+a$. **14.** $a^2+a(b-1)\sqrt{n}-bn$.

15. $4x^2+9$. **16.** $25a^2b^2+1$. **17.** $36a^2b^4+25x^4y^2$.

Resolve $x-1$ into two factors, one of which is

18. $\sqrt{x}+1$. **19.** $\sqrt[3]{x}-1$. **20.** $\sqrt[4]{x}+1$. **21.** $\sqrt[5]{x}-1$.

Resolve $ab+c$ into two factors, one of which is

22. $\sqrt[3]{(ab)}+\sqrt[3]{c}$. **23.** $\sqrt[5]{(ab)}+\sqrt[5]{c}$.

Factor each of the following expressions:

24. $x^2-2x-11$. **25.** $166+6x-x^2$.

26. $x^2+14x+29$. **27.** $4x^2-4xy-17y^2$.

28. $5x^2-11xy-4y^2$. **29.** $3+2x-11x^2$.

30. $x^2-2mx-1$. **31.** $x^2-2ax+a^2-b^2$.

32. $ax^2+bxy+cy^2$. **33.** $m^2x^2-4mx+4-nm^2$.

Rationalization.

20. To *rationalize* a surd expression is to free it from irrational numbers.

Thus, $\sqrt[3]{4}$ is rationalized by multiplying it by $\sqrt[3]{2}$, since $\sqrt[3]{4}\times\sqrt[3]{2}=\sqrt[3]{8}=2$.

A **Rationalizing Factor** for an irrational expression is an expression which, multiplying the irrational expression, gives a rational product.

E.g., $\sqrt[3]{2}$ is a rationalizing factor for $\sqrt[3]{4}$, and *vice versa.*

21. A rationalizing factor for a monomial surd number is easily determined by inspection.

Ex. 1. A rationalizing factor for $\sqrt[q]{a^p}$ is $\sqrt[q]{a^{q-p}}$, and *vice versa.*

Ex. 2. A rationalizing factor for $\sqrt[5]{(a^3b)}$ is $\sqrt[5]{(a^2b^4)}$, and *vice versa.*

22. A rationalizing factor for a binomial quadratic surd is its conjugate (Art. 13).

Ex. 1. $(\sqrt{2}-\sqrt{3})(\sqrt{2}+\sqrt{3})=2-3=-1.$

Either of the given binomial surds is a rationalizing factor for the other.

23. A rationalizing factor for any irrational binomial can be found by reference to the following identities:

$$(x-y)(x^{m-1}+x^{m-2}y+\cdots+xy^{m-2}+y^{m-1})=x^m-y^m, \qquad \text{(i.)}$$

for all positive integral values of m;

$$(x+y)(x^{m-1}-x^{m-2}y+\cdots-xy^{m-1}+y^{m-2})=x^m+y^m, \qquad \text{(ii.)}$$

for odd positive values of m;

$$(x+y)(x^{m-1}-x^{m-2}y+\cdots+xy^{m-1}-y^{m-2})=x^m-y^m, \qquad \text{(iii.)}$$

for even positive values of m.

Ex. Rationalize $\sqrt[5]{2}+2\sqrt[4]{3}$.

Let $\sqrt[5]{2}+2\sqrt[4]{3}=x+y$, in (iii.). Then we are to find a factor of the form

$$(\sqrt[5]{2})^{m-1}-(\sqrt[5]{2})^{m-2}(2\sqrt[4]{3})+\cdots, \qquad (2)$$

such that its product by $\sqrt[5]{2}+2\sqrt[4]{3}$ shall be rational; that is, such that $(\sqrt[5]{2})^m-(2\sqrt[4]{3})^m$ shall be rational. Therefore m must be 20, the L.C.M. of the two indices 5 and 4.

Notice that type-form (iii.) is used instead of (ii.), since m, in order to be divisible by 4, must be *even*.

Substituting 20 for m in (2), we obtain the rationalizing factor

$$(\sqrt[5]{2})^{19}-(\sqrt[5]{2})^{18}(2\sqrt[4]{3})+\cdots.$$

The product of this factor by $\sqrt[5]{2}+2\sqrt[4]{3}$ is

$$(\sqrt[5]{2})^{20}-(2\sqrt[4]{3})^{20}=2^4-2^{20}\times 3^5.$$

24. Rationalizing factors for a trinomial quadratic surd can be found by reference to the following identity:

$$(\sqrt{a}+\sqrt{b}+\sqrt{c})(\sqrt{a}+\sqrt{b}-\sqrt{c})(\sqrt{a}-\sqrt{b}+\sqrt{c})$$
$$(\sqrt{a}-\sqrt{b}-\sqrt{c})=a^2+b^2+c^2-2ab-2ac-2bc.$$

The rationalizing factors for any one of the four factors on the left are the other three.

E.g., $\sqrt{3}-\sqrt{5}+\sqrt{7}$ has the rationalizing factors

$$(\sqrt{3}+\sqrt{5}+\sqrt{7})(\sqrt{3}+\sqrt{5}-\sqrt{7})(\sqrt{3}-\sqrt{5}-\sqrt{7}).$$

In rationalizing numerical examples, the work is simplified if, after multiplying by one of the three rationalizing factors, the product, which is a binomial, be multiplied by its conjugate. Thus,

$$(\sqrt{3}-\sqrt{5}+\sqrt{7})(\sqrt{3}-\sqrt{5}-\sqrt{7})=3+5-2\sqrt{15}-7$$
$$=1-2\sqrt{15}.$$

Therefore the second rationalizing factor is $1+2\sqrt{15}$.

EXERCISES VII.

Find the expressions which will rationalize the following:

1. $\sqrt{3}$. **2.** $\sqrt[3]{4}$. **3.** $\sqrt[3]{7}$. **4.** $\sqrt[4]{12}$. **5.** $\sqrt[5]{16}$.

6. $\sqrt{a}$. **7.** $\sqrt[3]{(ab^2)}$. **8.** $\sqrt[4]{(a^3x)}$. **9.** $\sqrt[11]{(a^9b^2)}$. **10.** $\sqrt[n]{(a^2b^{n-2})}$.

11. $\sqrt[m-n]{(a^mb^n)}$. **12.** $3+\sqrt{2}$. **13.** $2\sqrt{14}-\sqrt{3}$.

14. $\frac{1}{3}\sqrt{3}-\frac{1}{2}\sqrt{2}$. **15.** $1-\sqrt{3}+2\sqrt{5}$. **16.** $\sqrt{5}+\sqrt{3}-\sqrt{2}$.

17. $\sqrt{(a^2-1)}-\sqrt{(a^2+1)}$. **18.** $\sqrt{(x^2+y^2)}-\sqrt{(2xy)}$.

19. $\sqrt[3]{7}-\sqrt[3]{5}$. **20.** $2\sqrt[3]{9}+\sqrt[3]{6}$. **21.** $\sqrt{2}-\sqrt[3]{3}$.

22. $\sqrt{3}+\sqrt[4]{5}$. **23.** $\sqrt[3]{2}-\sqrt[4]{5}$. **24.** $\sqrt[5]{2}-\sqrt[4]{7}$.

Reduction of a Fraction with an Irrational Denominator to an Equivalent Fraction with a Rational Denominator.

25. Multiply both numerator and denominator of the given fraction by the rationalizing factor for the denominator.

Ex. 1. $$\frac{\sqrt{2}}{\sqrt{3}}=\frac{\sqrt{2}}{\sqrt{3}}\times\frac{\sqrt{3}}{\sqrt{3}}=\frac{\sqrt{6}}{3}.$$

Ex. 2. $$\frac{1}{1+\sqrt{2}}=\frac{1}{1+\sqrt{2}}\times\frac{1-\sqrt{2}}{1-\sqrt{2}}=\frac{1-\sqrt{2}}{1-2}=-1+\sqrt{2}.$$

Ex. 3. $$\frac{3}{2-\sqrt[3]{5}}=\frac{3}{\sqrt[3]{8}-\sqrt[3]{5}}\times\frac{\sqrt[3]{8^2}+\sqrt[3]{8}\times\sqrt[3]{5}+\sqrt[3]{5^2}}{\sqrt[3]{8^2}+\sqrt[3]{8}\times\sqrt[3]{5}+\sqrt[3]{5^2}}$$

$$=\frac{3\sqrt[3]{64}+3\sqrt[3]{40}+3\sqrt[3]{25}}{(\sqrt[3]{8})^3-(\sqrt[3]{5})^3}$$

$$=\frac{12+6\sqrt[3]{5}+3\sqrt[3]{25}}{3}=4+2\sqrt[3]{5}+\sqrt[3]{25}.$$

Ex. 4.

$$\frac{\sqrt{(1+x)}+\sqrt{(1-x)}}{\sqrt{(1+x)}-\sqrt{(1-x)}}=\frac{\sqrt{(1+x)}+\sqrt{(1-x)}}{\sqrt{(1+x)}-\sqrt{(1-x)}}\times\frac{\sqrt{(1+x)}+\sqrt{(1-x)}}{\sqrt{(1+x)}+\sqrt{(1-x)}}$$

$$=\frac{1+x+2\sqrt{(1-x^2)}+1-x}{(1+x)-(1-x)}=\frac{1+\sqrt{(1-x^2)}}{x}.$$

Ex. 5.

$$\frac{3\sqrt{2}-1}{2\sqrt{2}-\sqrt{3}+\sqrt{6}}=\frac{3\sqrt{2}-1}{2\sqrt{2}-\sqrt{3}+\sqrt{6}}\times\frac{2\sqrt{2}-\sqrt{3}-\sqrt{6}}{2\sqrt{2}-\sqrt{3}-\sqrt{6}}$$

$$=\frac{12-2\sqrt{6}-5\sqrt{3}-2\sqrt{2}}{5-4\sqrt{6}}\times\frac{5+4\sqrt{6}}{5+4\sqrt{6}}$$

$$=\frac{12+38\sqrt{6}-41\sqrt{3}-70\sqrt{2}}{-71}.$$

EXERCISES VIII.

Change each of the following fractions into an equivalent fraction with a rational denominator:

1. $\frac{1}{\sqrt{2}}$. 2. $\frac{6}{\sqrt{3}}$. 3. $\frac{12}{5\sqrt{3}}$. 4. $\frac{20}{3\sqrt{10}}$.

5. $\dfrac{6}{\sqrt[3]{3}}$. 6. $\dfrac{8}{3\sqrt[3]{4}}$. 7. $\dfrac{10}{3\sqrt[4]{25}}$. 8. $\dfrac{14}{2\sqrt[5]{16}}$.

9. $\dfrac{x}{\sqrt{x}}$. 10. $\dfrac{x}{\sqrt[3]{x^2}}$. 11. $\dfrac{xy}{2\sqrt{(2x)}}$. 12. $\dfrac{a}{\sqrt[n]{a^4}}$.

13. $\dfrac{1}{2-\sqrt{3}}$. 14. $\dfrac{12}{5-\sqrt{21}}$. 15. $\dfrac{\sqrt{(2m)}+3n}{\sqrt{(2m)}-3n}$. 16. $\dfrac{1+\sqrt{2}}{2-\sqrt{2}}$.

17. $\dfrac{3\sqrt{5}-2\sqrt{2}}{2\sqrt{5}-\sqrt{18}}$. 18. $\dfrac{a\sqrt{b}+b\sqrt{a}}{\sqrt{a}+\sqrt{b}}$. 19. $\dfrac{3a^2-2a\sqrt{(ab)}}{3a+2\sqrt{(ab)}}$.

20. $\dfrac{1}{(\sqrt{10}-\sqrt{2}-\sqrt{3})}$. 21. $\dfrac{12}{\sqrt{6}-\sqrt{10}+2}$.

22. $\dfrac{2-\sqrt{3}}{1+\sqrt{2}+\sqrt{3}}$. 23. $\dfrac{3+4\sqrt{3}}{\sqrt{6}+\sqrt{2}-\sqrt{5}}$.

24. $\dfrac{\sqrt{(x^2+x)}+\sqrt{(x^2-x)}}{\sqrt{(x^2+x)}-\sqrt{(x^2-x)}}$. 25. $\dfrac{(a^2+b)-a\sqrt{(a^2+b)}}{a+\sqrt{(a^2+b)}}$.

26. $\dfrac{(n+1)+\sqrt{(n^2-1)}}{(n+1)-\sqrt{(n^2-1)}}$. 27. $\dfrac{a+\sqrt{(a^2-4a)}}{a-2+\sqrt{(a^2-4a)}}$.

28. $\dfrac{\sqrt{a}+\sqrt{(a+x)}}{\sqrt{a}+\sqrt{x}+\sqrt{(a+x)}}$. 29. $\dfrac{x^2+px+q}{x+\frac{p}{2}\pm\sqrt{\left(\frac{p^2}{4}-q\right)}}$.

30. $\dfrac{5-\sqrt[3]{2}}{\sqrt[3]{4}+2}$. 31. $\dfrac{1+\sqrt[4]{2}}{\sqrt[4]{5}-\sqrt[3]{2}}$. 32. $\dfrac{\sqrt[3]{(a^2b^2)}}{\sqrt[3]{a}+\sqrt[3]{b}}$.

33. $\dfrac{x(x-1)}{\sqrt[3]{x^2}-\sqrt{x}}$. 34. $\dfrac{a-x}{\sqrt[4]{a}+\sqrt[4]{x}}$. 35. $\dfrac{a-1}{\sqrt{a}-\sqrt[4]{a}}$.

Properties of Quadratic Surds.

26. *The product and the quotient of two like quadratic surds are rational.*

For $m\sqrt{x}\times n\sqrt{x}=mn\sqrt{x^2}=mnx$, and $\dfrac{m\sqrt{x}}{n\sqrt{x}}=\dfrac{m}{n}$.

E.g., $3\sqrt{2}\times 5\sqrt{2}=15\sqrt{4}=30$; $\dfrac{4\sqrt{3}}{2\sqrt{3}}=2$.

27. *If the product, or the quotient, of two quadratic surds be rational, they must be like surds.*

Let $\sqrt{x}\sqrt{y} = R$, a rational number.

Then $$\sqrt{x} = \frac{R}{\sqrt{y}} = \frac{R}{y}\sqrt{y}.$$

Since $\sqrt{x}$ is a rational multiple of $\sqrt{y}$, therefore $\sqrt{x}$ and $\sqrt{y}$ must be like surds.

In like manner the principle can be proved for the quotient.

28. *The product and the quotient of two unlike quadratic surds is irrational.*

For, by Art. 27, whenever the product of two quadratic surds is rational, they must be like surds.

E.g., $$\sqrt{2}\sqrt{3} = \sqrt{6}; \quad \frac{\sqrt{2}}{\sqrt{3}} = \sqrt{\frac{2}{3}}.$$

29. *A quadratic surd cannot be equal to the sum of a rational number and another quadratic surd;* or

$$\sqrt{a} \neq b + \sqrt{c},$$

wherein $\sqrt{a}$ and $\sqrt{c}$ are surds, and b is rational.

For if $\sqrt{a} = b + \sqrt{c}$, then $a = b^2 + c + 2b\sqrt{c}$.

Solving the last equation for $\sqrt{c}$, we obtain

$$\sqrt{c} = \frac{a - b^2 - c}{2b}.$$

This equation asserts that $\sqrt{c}$, an irrational number, is equal to $\frac{a - b^2 - c}{2b}$, a rational number. But that is a contradiction of terms, and therefore the hypothesis $\sqrt{a} = b + \sqrt{c}$ is untenable.

30. *If* $$a + \sqrt{b} = x + \sqrt{y}, \tag{1}$$

wherein $\sqrt{b}$ and $\sqrt{y}$ are surds, and a and x are rational, then $a = x$ and $b = y$.

For if $a \neq x$, let $a = x + m$, wherein $m \neq 0$.

Then (1) becomes $x + m + \sqrt{b} = x + \sqrt{y}$, or $m + \sqrt{b} = \sqrt{y}$. (2)

But by the preceding article (2) is untenable, unless $m = 0$.

Therefore $a = x$, and hence $\sqrt{b} = \sqrt{y}$, or $b = y$.

31. *If* $\sqrt{(a+\sqrt{b})} = \sqrt{x} + \sqrt{y}$, *then* $\sqrt{(a-\sqrt{b})} = \sqrt{x} - \sqrt{y}$.

From $$\sqrt{(a+\sqrt{b})} = \sqrt{x} + \sqrt{y},$$

we obtain $$a + \sqrt{b} = x + y + 2\sqrt{(xy)}.$$

Whence, by Art. 30, $$a = x + y, \qquad (1)$$

and $$\sqrt{b} = 2\sqrt{(xy)}. \qquad (2)$$

Subtracting (2) from (1),

$$a - \sqrt{b} = x + y - 2\sqrt{(xy)}. \qquad (3)$$

Therefore $$\sqrt{(a-\sqrt{b})} = \sqrt{x} - \sqrt{y}.$$

Evolution of Surd Expressions.

32. A root of a monomial surd number is found by applying the principle

$$\sqrt[q]{\sqrt[r]{a}} = \sqrt[qr]{a}. \quad \text{(Ch. XV., § 1, Art. 17.)}$$

Ex. 1. $$\sqrt[4]{\sqrt[3]{5}} = \sqrt[12]{5}.$$

It is important to notice that $\sqrt[q]{\sqrt[r]{a}} = \sqrt[r]{\sqrt[q]{a}}$.

Ex. 2. $$\sqrt[3]{\sqrt[5]{(8x^3)}} = \sqrt[5]{\sqrt[3]{(8x^3)}} = \sqrt[5]{(2x)}.$$

EXERCISES IX.

Simplify each of the following expressions:

1. $\sqrt[4]{\sqrt[3]{a^8}}$.
2. $\sqrt[6]{\sqrt[3]{a^2}}$.
3. $\sqrt[3]{\sqrt[5]{(-x^3)}}$.
4. $\sqrt[3]{\sqrt[4]{(a^9x^{12})}}$.
5. $\sqrt[9]{\sqrt{(a^{27}b^{45})}}$.
6. $\sqrt[4]{(2a\sqrt[3]{a^2})}$.
7. $\sqrt[3]{(a\sqrt{a})}$.
8. $\sqrt[m]{\sqrt[n]{a^m}}$.
9. $\sqrt[2]{\sqrt[3]{(\frac{25}{49}a^2b^6c^8)}}$.
10. $\sqrt[5]{(a^2\sqrt{a})}$.
11. $\sqrt[4]{\dfrac{a^2b^4}{\sqrt[3]{(a^2b^2)}}}$.
12. $\sqrt{\dfrac{2}{\sqrt[3]{2}}}$.
13. $\sqrt[n-1]{\dfrac{a}{\sqrt[n]{a}}}$.
14. $\sqrt[3]{\dfrac{a^2}{\sqrt{a}}}$.
15. $\sqrt{\sqrt{\sqrt{x^{12}}}}$.
16. $2\sqrt{\{2\sqrt{[2\sqrt{(2\sqrt{2})}]}\}}$.
17. $a\sqrt{a\sqrt{\{a\sqrt{[a\sqrt{(a\sqrt{a})}]}\}}}$.
18. $2\sqrt{\sqrt[6]{7}} + 3\sqrt{\sqrt[3]{\sqrt{7}}} - 3\sqrt[4]{\sqrt[3]{7}} - \sqrt[3]{\sqrt{\sqrt{7}}}$.
19. $\sqrt[2m]{\sqrt[3n]{a^5}} \times \sqrt[6m]{\sqrt[n]{a^8}} \times \sqrt[m]{\sqrt[6n]{a^9}} \times \sqrt[6m]{\sqrt[n]{a}}$.

Square Roots of Simple Binomial Surds.

33. Ex. **1.** Find a square root of $3 + 2\sqrt{2}$.

Let $\sqrt{(3 + 2\sqrt{2})} = \sqrt{x} + \sqrt{y}$. (1)

Then, by Art. 31, $\sqrt{(3 - 2\sqrt{2})} = \sqrt{x} - \sqrt{y}$. (2)

Multiplying (1) by (2), $\sqrt{(9 - 8)} = x - y$,

or $x - y = 1$. (3)

Squaring (1), $3 + 2\sqrt{2} = x + y + 2\sqrt{(xy)}$;

whence, by Art. 30, $x + y = 3$. (4)

Solving (4) and (5), we have $x = 2$, $y = 1$.

Therefore $\sqrt{(3 + 2\sqrt{2})} = \sqrt{2} + \sqrt{1} = \sqrt{2} + 1$.

This example could have been solved by inspection. We change $3 + 2\sqrt{2}$ into the form

$$m + 2\sqrt{(mn)} + n = (\sqrt{m} + \sqrt{n})^2.$$

We then have

$$\sqrt{(3 + 2\sqrt{2})} = \sqrt{(2 + 2\sqrt{2} + 1)} = \sqrt{(\sqrt{2} + 1)^2} = \sqrt{2} + 1.$$

Ex. **2.** Solve, by inspection, $\sqrt{(21 - 3\sqrt{24})}$.

We have
$$\begin{aligned}\sqrt{(21 - 3\sqrt{24})} &= \sqrt{(21 - 2\sqrt{54})}\\ &= \sqrt{(18 - 2\sqrt{54} + 3)}\\ &= \sqrt{(\sqrt{18} - \sqrt{3})^2}\\ &= \sqrt{18} - \sqrt{3} = 3\sqrt{2} - \sqrt{3}.\end{aligned}$$

In solving by inspection, first write the surd term of the given binomial surd in the form $2\sqrt{mn}$, as $3\sqrt{24} = 2\sqrt{54}$.

Then find by inspection two numbers whose sum is equal to the rational term of the given binomial surd, and whose product is equal to mn.

EXERCISES X.

Find a square root of each of the following expressions:

1. $7 + \sqrt{48}$. **2.** $5 - \sqrt{24}$. **3.** $2 + \sqrt{3}$.

4. $1\frac{1}{2} + \sqrt{2}$. **5.** $3 - \sqrt{5}$. **6.** $6 + \sqrt{11}$.

7. $9+\sqrt{32}$.	**8.** $8-\sqrt{28}$.	**9.** $6+4\sqrt{2}$.
10. $7+2\sqrt{10}$.	**11.** $8-2\sqrt{15}$.	**12.** $11-6\sqrt{2}$.
13. $11+4\sqrt{7}$.	**14.** $30-10\sqrt{5}$.	**15.** $\frac{5}{7}+\frac{1}{7}\sqrt{21}$.
16. $\frac{9}{11}-\frac{4}{11}\sqrt{2}$.	**17.** $\frac{11}{12}-\frac{1}{6}\sqrt{10}$.	**18.** $\frac{21}{22}-\frac{2}{11}\sqrt{5}$.

19. $4a+2\sqrt{(4a^2-b^2)}$. **20.** $n-2\sqrt{(n-1)}$.

21. $1+2a\sqrt{(1-a^2)}$. **22.** $2a-2\sqrt{(2an-n^2)}$.

23. $10n^2+1-6n\sqrt{(n^2+1)}$. **24.** $a-x-2\sqrt{(a-x-1)}$.

Roots of Multinomial Surds.

34. The square and cube roots of multinomial surds can be found by the methods given in Ch. XV.

Ex. 1. Find the square root of

$$(5-4\sqrt{5})a^2+10a^4-2\sqrt{10}\,a+10\sqrt{2}\,a^3+2.$$

Arranging the given expression to descending powers of a, we have

$$\begin{array}{l|l}
10a^4+10\sqrt{2}\,a^3+(5-4\sqrt{5})a^2-2\sqrt{10}\,a+2 & \sqrt{10}\,a^2+\sqrt{5}\,a-\sqrt{2} \\
\underline{10a^4} & 2\sqrt{10}\,a^2 \\
\quad 10\sqrt{2}\,a^3 & \\
\quad \underline{10\sqrt{2}\,a^3+5a^2} & 2\sqrt{10}\,a^2+\sqrt{5}\,a \\
\qquad -4\sqrt{5}\,a^2 & \\
\qquad \underline{-4\sqrt{5}\,a^2-2\sqrt{10}\,a+2} & 2\sqrt{10}\,a^2+2\sqrt{5}\,a-\sqrt{2}
\end{array}$$

If the required root be a binomial, it can be found by inspection.

Ex. 2. Find the cube root of

$$8a+150\sqrt[3]{(ax^2)}-60\sqrt[3]{(a^2x)}-125x.$$

We have

$$\sqrt[3]{8a-60\sqrt[3]{(a^2x)}+150\sqrt[3]{(ax^2)}-125x}$$

$$=\sqrt[3]{(2\sqrt[3]{a})^3-3(2\sqrt[3]{a})^2(5\sqrt[3]{x})+3(2\sqrt[3]{a})(5\sqrt[3]{x})^2-(5\sqrt[3]{x})^3}$$

$$=\sqrt[3]{(2\sqrt[3]{a}-5\sqrt[3]{x})^3}=2\sqrt[3]{a}-5\sqrt[3]{x}.$$

Approximate Values of Surd Numbers.

35. The method for finding an approximate value of a surd number given in Ch. XVII. would be too laborious for practical use, and was given for theoretical purposes.

An approximate value of a surd number can, however, be found to any degree of accuracy by the methods given in Ch. XV.

Ex. 1. Find an approximate value of $\sqrt{2}$ correct to three decimal places.

The work proceeds as follows:

2.00'00'00'00	1.4142
1	2
1 00	
96	24
4 00	
2 81	281
1 19 00	
1 12 96	2824
6 04 00	2828

The work is simplified by neglecting the decimal point, writing it only in the result. Observe that it is necessary to find the root to four decimal places in order to determine whether to take the figure found in the third place or the next greater figure, according to the well-known principle of Arithmetic.

Ex. 2. Find the value of $\sqrt[3]{(1-x)}$ to three terms.

The work proceeds as follows:

$1-x$	$1-\frac{1}{3}x-\frac{1}{9}x^2$
1	$3\times 1^2=3$
$-x$	$3\times 1^2+3\times 1\times(-\frac{1}{3}x)+(-\frac{1}{3}x)^2=3-x+\frac{1}{9}x^2$
$-x+\frac{1}{3}x^2-\frac{1}{27}x^3$	
$-\frac{1}{3}x^2+\frac{1}{27}x^3$	

An approximate value of a fractional surd is obtained most simply by rationalizing its denominator, then finding the required root of the numerator of the resulting fraction, and dividing this value by the denominator.

Ex. 3. Find an approximate value of $\sqrt{\frac{1}{2}}$, correct to three decimal places.

We have $\sqrt{\frac{1}{2}} = \frac{1}{2}\sqrt{2}$, and $\sqrt{2} = 1.4142 + \cdots$.

Therefore $\sqrt{\frac{1}{2}} = .707$, correct to three places of decimals.

EXERCISES XI.

Find by inspection the square root of each of the following expressions:

1. $a^2 + 2a\sqrt{b} + b$.
2. $4a + 9x - 12\sqrt{(ax)}$.
3. $9 + 6\sqrt[3]{3} + \sqrt[3]{9}$.
4. $\sqrt[3]{5} + 2\sqrt[3]{2} + 2\sqrt[6]{80}$.
5. $x^2 - 4x\sqrt{a} + 2x + 4a - 4\sqrt{a} + 1$.
6. $a - 2a\sqrt{x} + x + 2\sqrt{(ax)} - 2x\sqrt{a} + ax$.

Find by inspection the cube root of each of the following expressions:

7. $x\sqrt{x} + 3\sqrt{x} - 3x - 1$.
8. $8a + 150\sqrt[3]{(ax^2)} - 60\sqrt[3]{(a^2x)} - 125x$.
9. $144\sqrt{a} - 108a + 27a\sqrt{a} - 64$.
10. $4n + 12n\sqrt[3]{n^2} + 12n\sqrt[3]{n} + 4n^2$.
11. $8x^3 + 66x^2 + 33x - 36x^2\sqrt{x} - 63x\sqrt{x} - 9\sqrt{x} + 1$.

Find an approximate value of each of the following expressions, correct to four figures:

12. $\sqrt{8}$.
13. $\frac{1}{2}\sqrt{2.5}$.
14. $\sqrt{2}$.
15. $\frac{2}{3}\sqrt{1.25}$.
16. $\sqrt{345.06}$.
17. $\sqrt{10862.321}$.
18. $\sqrt{54.0001}$.
19. $\dfrac{2}{\sqrt{5}}$.
20. $\dfrac{3}{\sqrt{8}}$.
21. $\dfrac{1}{2\sqrt{4}}$.
22. $\dfrac{5}{\sqrt{75}}$.
23. $\dfrac{1+\sqrt{3}}{1-\sqrt{3}}$.
24. $\dfrac{3+2\sqrt{7}}{5-4\sqrt{11}}$.
25. $\dfrac{\sqrt{17}}{\sqrt{2.5}+\sqrt{6}}$.

Find an approximate value of each of the following expressions, to include four terms:

26. $\sqrt{(1-x)}$.
27. $\sqrt{(a^2+b^2)}$.
28. $\sqrt{(x^2 - xy + y^2)}$.
29. $\sqrt[3]{(1+x^3)}$.
30. $\sqrt[3]{(a^3-b^3)}$.
31. $\sqrt[3]{(x^3+x^2y+xy^2+y^3)}$.

CHAPTER XIX.

IMAGINARY AND COMPLEX NUMBERS.

1. Since even powers of both positive and negative numbers are *positive*, even roots of negative numbers cannot be expressed in terms of numbers as yet comprised in the number system.

E.g., since $(\pm 4)^2 = 16$, the $\sqrt{-16}$ cannot be expressed as a positive or as a negative number.

It is therefore necessary either to exclude such roots from our consideration or to again enlarge our ideas of number. The latter alternative is in accordance with the generalizing spirit of Algebra.

We therefore assume that $\sqrt{-1}$ *and in general* $\sqrt[2n]{-a}$, *are numbers, and include them in the number system.*

2. These new numbers are defined by the relations

$$(\sqrt{-1})^2 = -1, \text{ and in general } (\sqrt[2n]{-a})^{2n} = -a.$$

Imaginary Numbers.

3. The square root of a negative number is called an **Imaginary Number**; as $\sqrt{-3}$, $\sqrt{-8}$.

The study of these numbers is simplified by first considering the properties of $\sqrt{-1}$, which is taken as the **Imaginary Unit.*** This new unit is commonly designated by the letter i, and its opposite by $-i$.

* The designation, *imaginary*, is unfortunate, since, as will be shown in Part II., such numbers are no more imaginary (in the ordinary meaning of the word) than common fractions or negative numbers. Dr. George Bruce Halsted, Professor of Mathematics in the University of Texas, has suggested **Neomon** for the *imaginary unit*, and **Neomonic** for *imaginary*.

We then have by definition

$$(\pm i)^2 = -1.$$

For the sake of distinction all numbers, rational and irrational, which have been used hitherto in this book are called **Real Numbers.**

4. Multiples and Fractional Parts of the Imaginary Unit. — Just as multiples and fractional parts of the real units 1 and -1 are numbers, so we assume that multiples and fractional parts of the new unit i, and of its opposite $-i$, are numbers.

E.g., just as $$3 = 1 + 1 + 1,$$
so $$3\sqrt{-1} = \sqrt{-1} + \sqrt{-1} + \sqrt{-1}, \text{ or } 3i = i + i + i;$$
just as $$-3 = -1 - 1 - 1,$$
so $$-3\sqrt{-1} = -\sqrt{-1} - \sqrt{-1} - \sqrt{-1}, \text{ or } -3i = -i - i - i;$$
just as $$\tfrac{2}{3} = \tfrac{1}{3} + \tfrac{1}{3},$$
$$\text{so } \tfrac{2}{3}\sqrt{-1} = \frac{\sqrt{-1}}{3} + \frac{\sqrt{-1}}{3}, \text{ or } \tfrac{2}{3}i = \frac{i}{3} + \frac{i}{3}.$$

5. Two or more multiples or fractions of the imaginary unit can be united by addition or subtraction into a single multiple or fraction of that unit.

E.g., $$3\sqrt{-1} + 2\sqrt{-1} = 5\sqrt{-1}, \text{ or } 3i + 2i = 5i;$$
$$6\sqrt{-1} - 8\sqrt{-1} = -2\sqrt{-1}, \text{ or } 6i - 8i = -2i;$$
$$\tfrac{1}{2}\sqrt{-1} + \tfrac{3}{4}\sqrt{-1} = 1\tfrac{1}{4}\sqrt{-1}, \text{ or } \tfrac{1}{2}i + \tfrac{3}{4}i = 1\tfrac{1}{4}i;$$
$$a\sqrt{-1} + b\sqrt{-1} = (a+b)\sqrt{-1}, \text{ or } ai + bi = (a+b)i.$$

6. Multiplication by *i*. — It is necessary to extend the meaning of multiplication to include the case in which the multiplier is the imaginary unit. We therefore define multiplication, when the multiplier is the imaginary unit, by assuming that the Commutative Law holds, that is, by the relation

$$\boldsymbol{\sqrt{-1} \times a = a\sqrt{-1}}, \text{ or } \boldsymbol{ia = ai}.$$

E.g., $$i2 = 2i = i + i.$$

That is, i is used like a real factor.

Observe that the multiplier is here written on the left of the multiplicand.

7. The following particular cases of Art. 6 deserve special mention:

$$i \cdot 1 = 1 \cdot i = i;$$
$$i \cdot 0 = 0 \cdot i = 0.$$

8. It follows directly from Arts. 5 and 6 that the Distributive and Associative Laws hold when the imaginary unit is a factor of the product.

E.g., $(a \pm b)i = ai \pm bi$; $aibi = abii = abi^2$.

9. Division by *i*. — It follows from the definition of division that $\frac{ai}{i}$ is a number which multiplied by i gives ai.

But $$a \times i = ai.$$

Therefore $$\frac{ai}{i} = a.$$

Observe again that i is used like a real factor.

10. We now have, in addition to the double series of real numbers, the double series of imaginary numbers:

$$\cdots -3i,\ -2i,\ -i,\ 0,\ i,\ 2i,\ 3i,\ \cdots.$$

Between any two consecutive numbers of this series there are fractional and irrational multiples of i. Thus, between i and $2i$ lie $\frac{3}{2}i$, $\sqrt{2}\,i$, etc.

11. Powers of *i*. — The following values of the positive integral powers of $\sqrt{-1}$ or i follow directly from the definition of i and Art. 8:

$$\begin{aligned}
&\sqrt{-1} = \sqrt{-1}, && \text{or } i = i,\\
&(\sqrt{-1})^2 = -1, && i^2 = -1,\\
&(\sqrt{-1})^3 = (\sqrt{-1})^2(\sqrt{-1}) = -\sqrt{-1}, && i^3 = i^2 \cdot i = -i,\\
&(\sqrt{-1})^4 = (\sqrt{-1})^2(\sqrt{-1})^2 = +1, && i^4 = i^2 \cdot i^2 = +1,\\
&(\sqrt{-1})^5 = (\sqrt{-1})^4(\sqrt{-1}) = +\sqrt{-1}, && i^5 = i^4 \cdot i = +i,\\
&(\sqrt{-1})^6 = (\sqrt{-1})^4(\sqrt{-1})^2 = -1, && i^6 = i^4 \cdot i^2 = -1,\\
&(\sqrt{-1})^7 = (\sqrt{-1})^6(\sqrt{-1}) = -\sqrt{-1}, && i^7 = i^6 \cdot i = -i,\\
&(\sqrt{-1})^8 = (\sqrt{-1})^4(\sqrt{-1})^4 = +1, && i^8 = i^4 \cdot i^4 = +1,\\
&\qquad\text{etc.} && \qquad\text{etc.}
\end{aligned}$$

The preceding results give the following properties of powers of i:

(*a*) *All even powers of i are real.*

(1) *If the exponent of the power of i be exactly divisible by 4, the power is equal to* $+1$.

(2) *If the exponent of the power of i be exactly divisible by 2 only, and not by 4, the power is equal to* -1.

(*b*) *All odd powers of i are imaginary.*

(1) *If the exponent of the power be one greater than a multiple of 4, the power is equal to* $+i$.

(2) *If the exponent of the power be one less than a multiple of 4, the power is equal to* $-i$.

That is, if n be any positive integer,

$$i^{4n}=+1,\quad i^{4n+2}=-1,\quad i^{4n+1}=+i,\quad i^{4n-1}=-i.$$

12. Since

$$(\sqrt{-a})^2=-a, \text{ and } (\sqrt{a}\times\sqrt{-1})^2=(\sqrt{a})^2(\sqrt{-1})^2=-a,$$

we have $$(\sqrt{-a})^2=(\sqrt{a}\times\sqrt{-1})^2.$$

Whence $$\sqrt{-a}=\sqrt{a}\times\sqrt{-1}.$$

E.g., $$\sqrt{-9}=\sqrt{9}\times\sqrt{-1}=3\sqrt{-1}.$$

13. Addition of Imaginary Numbers. — Imaginary numbers are united by addition and subtraction just as real numbers are united.

Ex. 1. $\sqrt{-9}+\sqrt{-16}=3\sqrt{-1}+4\sqrt{-1}=7\sqrt{-1}=7i.$

Ex. 2. $4\sqrt{-5}-10\sqrt{-5}+3\sqrt{-5}=-3\sqrt{-5}$
$=-3\sqrt{5}\sqrt{-1}=3\sqrt{5}\,i.$

Ex. 3. $\sqrt{(-1-2x-x^2)}-\sqrt{-x^2}=(1+x)\sqrt{-1}-x\sqrt{-1}$
$=\sqrt{-1}=i.$

Ex. 4. $i^{13}+i^{15}=i+(-i)=0.$

14. Multiplication of Imaginary Numbers. — The following principles enable us to simplify a product of imaginary factors:

$$\sqrt{-a}\times\sqrt{b}=\sqrt{(ab)}\times\sqrt{-1}=\sqrt{(-ab)};$$
$$\sqrt{-a}\times\sqrt{-b}=-\sqrt{(ab)}.$$

For $\sqrt{-a}\times\sqrt{b}=\sqrt{a}\sqrt{-1}\sqrt{b}=\sqrt{a}\sqrt{b}\sqrt{-1}$
$=\sqrt{(ab)}\sqrt{-1}=\sqrt{(-ab)}$;

and $\sqrt{-a}\times\sqrt{-b}=\sqrt{a}\sqrt{-1}\times\sqrt{b}\sqrt{-1}=\sqrt{a}\sqrt{b}(\sqrt{-1})^2$
$=-\sqrt{(ab)}$.

Ex. 1. $\sqrt{-9}\times\sqrt{16}=3\sqrt{-1}\times 4=12\sqrt{-1}=12\,i$.

Ex. 2. $\sqrt{-2}\times\sqrt{-8}=-\sqrt{16}=-4$.

Ex. 3. $\sqrt{-5}\times\sqrt{-10}\times\sqrt{-15}=\sqrt{5}\times\sqrt{10}\times\sqrt{15}\times(\sqrt{-1})^3$
$=-5\sqrt{30}\sqrt{-1}=-5\sqrt{30}\,i$.

15. Division of Imaginary Numbers. — The following principles enable us to simplify quotients which involve imaginary numbers:

$$\frac{\sqrt{-a}}{\sqrt{b}}=\sqrt{\frac{a}{b}}\times\sqrt{-1};\quad \frac{\sqrt{a}}{\sqrt{-b}}=-\sqrt{\frac{a}{b}}\times\sqrt{-1};\quad \frac{\sqrt{-a}}{\sqrt{-b}}=\sqrt{\frac{a}{b}}.$$

For $$\frac{\sqrt{-a}}{\sqrt{b}}=\frac{\sqrt{a}\sqrt{-1}}{\sqrt{b}}=\sqrt{\frac{a}{b}}\times\sqrt{-1}.$$

$$\frac{\sqrt{a}}{\sqrt{-b}}=\frac{\sqrt{a}}{\sqrt{b}\times\sqrt{-1}}=\frac{\sqrt{a}\times\sqrt{-1}}{\sqrt{b}\times(\sqrt{-1})^2}=-\sqrt{\frac{a}{b}}\times\sqrt{-1};$$

and $$\frac{\sqrt{-a}}{\sqrt{-b}}=\frac{\sqrt{a}\sqrt{-1}}{\sqrt{b}\sqrt{-1}}=\frac{\sqrt{a}}{\sqrt{b}}=\sqrt{\frac{a}{b}}.$$

Ex. 1. $$\frac{\sqrt{-8}}{\sqrt{2}}=\sqrt{\frac{8}{2}}\times\sqrt{-1}=2\sqrt{-1}=2\,i.$$

Ex. 2. $$\frac{\sqrt{-25}}{\sqrt{-15}}=\sqrt{\tfrac{5}{3}}=\tfrac{1}{3}\sqrt{15}.$$

Ex. 3. $$\frac{1}{\sqrt{-1}}=\frac{\sqrt{-1}}{(\sqrt{-1})^2}=\frac{\sqrt{-1}}{-1}=-\sqrt{-1}=-i.$$

EXERCISES I.

Simplify each of the following expressions:

1. $2\sqrt{-9}-3\sqrt{-25}$.
2. $7\sqrt{-81}+5\sqrt{-144}$.
3. $-6\sqrt{-4}+5\sqrt{-36}$.
4. $2\sqrt{-12}-3\sqrt{-27}$.
5. $8\sqrt{-75}+\sqrt{-147}$.
6. $-5\sqrt{-8}-3\sqrt{-32}$.
7. $5\sqrt{-20}-3\sqrt{-45}+2\sqrt{-180}$.

8. $\sqrt{-7} - 3\sqrt{-28} - 5\sqrt{-175}$.

9. $6\sqrt{-\frac{2}{3}} - 2\sqrt{-\frac{3}{2}} + \sqrt{-24} - \sqrt{-96} - \sqrt{-4\frac{1}{6}} + \sqrt{-8\frac{1}{6}}$.

10. $2\sqrt{-a^2} + 5\sqrt{(-9a^2)} - 3\sqrt{(-16a^2)}$.

11. $2\sqrt{(-a^4b)} - 4\sqrt{(-a^2b^3)} + 2\sqrt{-b^3}$.

12. $\sqrt{(-a^2b^2)} + \sqrt{(-a^2 - b^2 - 2ab)}$. **13.** $\sqrt{-(x-y)^2} + \sqrt{-y^2}$.

14. $\sqrt{[-(1+x)^2y]} + \sqrt{[-(1-x)^2y]} - \sqrt{-(x-y)^2} - \sqrt{(-4y)}$
$+ \sqrt{-(x+y)^2}$.

15. $\sqrt{3} \cdot \sqrt{-3}$. **16.** $\sqrt{3} \cdot \sqrt{-12}$. **17.** $\sqrt{-18} \cdot \sqrt{2}$.

18. $\sqrt{32} \cdot \sqrt{-2}$. **19.** $\sqrt{-50} \cdot \sqrt{-2}$. **20.** $\sqrt{-3} \cdot \sqrt{-6}$.

21. $\sqrt{-5} \cdot \sqrt{-15}$. **22.** $\sqrt{-7} \cdot \sqrt{-14}$.

23. $\sqrt{-2} \cdot \sqrt{-6} \cdot \sqrt{-24}$. **24.** $\sqrt{-3} \cdot \sqrt{-27} \cdot \sqrt{-18}$.

25. $\sqrt{-5} \cdot \sqrt{-20} \cdot \sqrt{8}$. **26.** $\sqrt{6} \cdot \sqrt{-12} \cdot \sqrt{-3}$.

27. $\sqrt{-15} \cdot \sqrt{10} \cdot \sqrt{2}$. **28.** $\sqrt{-7} \cdot \sqrt{-14} \cdot \sqrt{-42}$.

29. $\sqrt{-a} \cdot \sqrt{-a}$. **30.** $\sqrt{-x^2} \cdot \sqrt{-y^2}$. **31.** $\sqrt{-x^3} \cdot \sqrt{-x^4}$.

32. $\sqrt{-a} \cdot \sqrt{(-ab)}$. **33.** $\sqrt{(3x^2)} \cdot \sqrt{-3}$.

34. $\sqrt{(-9a)} \cdot \sqrt{(ab)}$.

35. $\sqrt{(-a^2b)} \cdot \sqrt{(-ab^3)} \cdot \sqrt{(-ab^2)}$.

36. $\sqrt{(-x^4y^3)} \cdot \sqrt{(-xy)} \cdot \sqrt{x^3}$.

37. $\sqrt{(-m^4n^2)} \cdot \sqrt{(-mn^3)} \cdot \sqrt{(-m^3n^7)} \cdot \sqrt{(-m^2n)}$.

38. $(\sqrt{-3} + \sqrt{-2})(\sqrt{-3} + \sqrt{-5})$.

39. $(\sqrt{-3} + \sqrt{-8})(\sqrt{-6} - \sqrt{-1})$.

40. $(\sqrt{-7} + 2\sqrt{-9})(\sqrt{-7} - 2\sqrt{-9})$.

41. $(3\sqrt{-5} + 4\sqrt{-6})(2\sqrt{-5} - 3\sqrt{-6})$.

42. $(\sqrt{-a} + \sqrt{-b})(\sqrt{-a} - \sqrt{-b})$.

43. $[\sqrt{-(a+b)} + \sqrt{-b}][\sqrt{-(a+b)} - \sqrt{-b}]$.

44. $[\sqrt{(-a^3b^5)} + \sqrt{(-a^7b^9)}][\sqrt{(-a^5b^7)} - \sqrt{(-a^9b^{11})}]$.

45. $\sqrt{-x^2}$. **46.** $(\sqrt{-x})^2$. **47.** $\sqrt{-a^4}$. **48.** $(\sqrt{-a})^4$.

49. $\sqrt{(1-x)} \cdot \sqrt{(x-1)}$. **50.** $\sqrt{(a-b)} \cdot \sqrt{(b-a)}$. **51.** i^{20}.

52. i^{41}. **53.** i^{54}. **54.** i^{101}. **55.** $i^4 + i^{34}$.

56. $i^{24} - i^{42}$. **57.** $i^{35} - i^{91}$. **58.** $(\sqrt{-a})^{29}$. **59.** $(\sqrt{-a})^{170}$.

60. $(-a\sqrt{-a})^{42}$. **61.** $\sqrt{-27} \div \sqrt{-3}$. **62.** $\sqrt{-8} \div \sqrt{-2}$.

63. $\sqrt{-6} \div \sqrt{-3}$. **64.** $5\sqrt{-35} \div 2\sqrt{7}$. **65.** $6\sqrt{-18} \div 3\sqrt{9}$.

66. $\sqrt{3} \div \frac{1}{2}\sqrt{-5}$. 67. $\sqrt{-a} \div \sqrt{-a^2}$. 68. $\sqrt{(-ab)} \div \sqrt{-b}$.

69. $\sqrt{a^3} \div \sqrt{-a}$. 70. $(\sqrt{-6} + \sqrt{-8}) \div \sqrt{-2}$.

71. $(\sqrt{-12} - \sqrt{18}) \div \sqrt{-3}$.

72. $\frac{1}{i}$. 73. $\frac{1}{i^7}$. 74. $\frac{1}{i^8}$. 75. $\frac{1}{i^{33}}$.

76. $\frac{1}{i + i^5}$. 77. $\frac{1}{i^2 + i^6}$. 78. $\frac{1}{i^4 + i^8}$.

Complex Numbers.

16. A **Complex Number** is the algebraic sum of a real and an imaginary number; as, $3 \pm 2i$.

The general form of a complex number is evidently $a + bi$, wherein a and b are real numbers.

When $b = 0$, we have any real number.

When $a = 0$, we have any imaginary number.

17. Two complex numbers which differ only in the sign of their imaginary terms are called **Conjugate Complex Numbers**; as, $2 - 3i$ and $2 + 3i$.

18. *Two complex numbers are said to be equal when the real term of one is equal to the real term of the other, and the imaginary term of one is equal to the imaginary term of the other;* as, $2 + 3i = 2 + 3i$.

That is, if $$a + bi = c + di,$$

then $$a = c, \text{ and } bi = di, \text{ or } b = d.$$

Observe that the preceding statement is a definition of the meaning of the sign of equality between two complex numbers.

19. From the preceding article it follows that, if $a + bi = 0 = 0 + 0i$, then $a = 0$, $b = 0$.

20. Addition and Subtraction of Complex Numbers. — The following definition of Addition and Subtraction of Complex Numbers is a natural extension of the definition of these operations for real numbers:

Two complex numbers are added or subtracted by adding or subtracting the real parts by themselves and the imaginary parts by themselves.

E.g., $(2+3i)+(-5+6i)=(2-5)+(3i+6i)$
$$=-3+9i.$$

In general $(a+bi)\pm(c+di)=(a\pm c)+(b\pm d)i.$

21. The Commutative and Associative Laws hold for algebraic addition of complex numbers.

This principle follows immediately from the definition of addition and subtraction.

That is, $(a+bi)+(c+di)=(c+di)+(a+bi)$;

$$(a+bi)+(c+di)+(e+fi)=(a+bi)+[(e+fi)+(c+di)]$$
$$=\text{etc.}$$

22. The sum or difference of two complex numbers is, in general, a complex number.

E.g., $(2+3i)+(-4+2i)=(2-4)+(3+2)i$
$$=-2+5i.$$

But the sum of two conjugate complex numbers is real.

E.g., $(2+3i)+(2-3i)=4.$

23. Multiplication of Complex Numbers. — We define multiplication by a complex number by assuming that the Distributive Law holds; that is, by the relation

$$(a+bi)(c+di)=ac+bci+adi+bdi^2$$
$$=(ac-bd)+(bc+ad)i.$$

24. The Commutative, Associative, and Distributive Laws hold for multiplication of complex numbers.

This principle follows immediately from the definition of multiplication.

That is, $(a+bi)(c+di)=(c+di)(a+bi)$;

$$(a+bi)(c+di)(e+fi)=(a+bi)[(e+fi)(c+di)]=\text{etc.}$$

25. The product of two complex numbers is, in general, a complex number.

E.g., $(2+3i)(-4+2i) = -8+4i-12i-6$
$= -14-8i.$

But the product of two conjugate complex numbers is *real and positive.*

E.g., $(-2+3i)(-2-3i) = (-2)^2-(3i)^2$
$= 4+9 = 13.$

In general, $(a+bi)(a-bi) = a^2-(bi)^2$
$= a^2+b^2.$

26. The square of a complex number is a complex number.

E.g.,
$$(2+3\sqrt{-1})^2 = 4+12\sqrt{-1}+(3\sqrt{-1})^2$$
$$= 4+12\sqrt{-1}-9$$
$$= -5+12\sqrt{-1} = -5+12i.$$

But the cube of a complex number is sometimes real.

E.g.,
$$(-\tfrac{1}{2}\pm\tfrac{1}{2}\sqrt{-3})^3 = -\tfrac{1}{8}\pm\tfrac{3}{8}\sqrt{-3}+\tfrac{9}{8}\mp\tfrac{3}{8}\sqrt{-3}$$
$$= 1.$$

From these results we see that $\sqrt[3]{1}$ has three values,
$$1,\ -\tfrac{1}{2}+\tfrac{1}{2}\sqrt{-3},\ -\tfrac{1}{2}-\tfrac{1}{2}\sqrt{-3}.$$

27. *If the sum of the squares of two real numbers be* 0, *each of the numbers is* 0; that is, if $a^2+b^2=0$, then $a=0$, $b=0$, wherein a and b are real numbers.

For, from $a^2+b^2=0$,

we have $a^2=-b^2$,

or $a=b\sqrt{-1}$.

But by hypothesis a is real, and therefore cannot be equal to the imaginary number $b\sqrt{-1}$, unless a and b be both 0.

28. *A product of two complex numbers is* 0 *when, and only when, a factor is* 0.

From $(a+bi)(c+di)=0$,

we have $(ac-bd)+(bc+ad)i = 0 = 0+0i$.

Therefore, by Art. 19,
$$ac-bd=0, \text{ and } bc+ad=0.$$

Whence $$(ac-bd)^2+(bc+ad)^2=0,$$

or $$(a^2+b^2)(c^2+d^2)=0.$$

Consequently, by Ch. III., § 3, Art. 20, either

$$a^2+b^2=0, \text{ or } c^2+d^2=0.$$

But if $a^2+b^2=0$, then by the preceding article $a=0, b=0$, and therefore $a+bi=0$. And if $c^2+d^2=0$, then $c=0, d=0$, and hence $c+di=0$.

29. Division of Complex Numbers. — The quotient of one complex number by another is a complex number.

For $$\frac{a+bi}{c+di}=\frac{(a+bi)(c-di)}{(c+di)(c-di)}$$

$$=\frac{(ac+bd)+(bc-ad)i}{c^2+d^2}$$

$$=\frac{ac+bd}{c^2+d^2}+\frac{bc-ad}{c^2+d^2}\cdot i.$$

30. It follows from the preceding article that a fraction whose denominator is a complex number can be expressed as a complex number by multiplying both numerator and denominator by the conjugate of the denominator.

Ex. 1.

$$\frac{1+\sqrt{-2}}{2\sqrt{-3}}=\frac{(1+\sqrt{-2})(-2\sqrt{-3})}{(2\sqrt{-3})(-2\sqrt{-3})}=\frac{-2\sqrt{-3}+2\sqrt{6}}{12}$$

$$=\tfrac{1}{6}\sqrt{6}-\tfrac{1}{6}\sqrt{-3}.$$

Notice that it was necessary in the last example to multiply numerator and denominator only by $-\sqrt{-3}$.

$$\frac{1+\sqrt{-2}}{2\sqrt{-3}}=\frac{(1+\sqrt{-2})(-\sqrt{-3})}{(2\sqrt{-3})(-\sqrt{-3})}=\frac{-\sqrt{-3}+\sqrt{6}}{6}$$

$$=\tfrac{1}{6}\sqrt{6}-\tfrac{1}{6}\sqrt{-3}.$$

Ex. 2.

$$\frac{1}{2+\sqrt{-3}}=\frac{2-\sqrt{-3}}{(2+\sqrt{-3})(2-\sqrt{-3})}=\frac{2-\sqrt{-3}}{7}=\tfrac{2}{7}-\tfrac{1}{7}\sqrt{-3}.$$

Ex. 3. $\dfrac{2-\sqrt{-5}}{\sqrt{2}-\sqrt{-3}}=\dfrac{(2-\sqrt{-5})(\sqrt{2}+\sqrt{-3})}{(\sqrt{2}-\sqrt{-3})(\sqrt{2}+\sqrt{-3})}$

$$=\frac{2\sqrt{2}+\sqrt{15}-\sqrt{-10}+2\sqrt{-3}}{2+3}$$

$$=\frac{2\sqrt{2}+\sqrt{15}}{5}+\frac{-\sqrt{10}+2\sqrt{3}}{5}\sqrt{-1}.$$

Complex Factors.

31. Any quadratic expression which is the product of two complex factors can be resolved into these factors by the method used to resolve a quadratic expression into irrational factors.

Ex. Factor x^2-2x+3.

Completing x^2-2x to the square of a binomial in x, we have

$$\begin{aligned}x^2-2x+3&=x^2-2x+1-1+3\\&=(x-1)^2-(\sqrt{-2})^2\\&=(x-1+\sqrt{-2})(x-1-\sqrt{-2}).\end{aligned}$$

Square Root of a Complex Number.

32. If $\sqrt{(a+bi)}=\sqrt{x}+i\sqrt{y}$,

then $\sqrt{(a-bi)}=\sqrt{x}-i\sqrt{y}$.

For, from $\sqrt{(a+bi)}=\sqrt{x}+i\sqrt{y}$,

we have $a+bi=x-y+2\sqrt{(xy)}\cdot i$.

Therefore, by Art. 18, $a=x-y$,

and $b=2\sqrt{(xy)}$.

Consequently, $a-bi=x-y-2\sqrt{(xy)}\cdot i$

$$=(\sqrt{x}-i\sqrt{y})^2.$$

Whence $\sqrt{(a-bi)}=\sqrt{x}-i\sqrt{y}$.

33. The square root of a complex number can be expressed as a complex number.

Assuming $\sqrt{(13-20\sqrt{-3})}=\sqrt{x}-i\sqrt{y}$, (1)

we have $\sqrt{(13+20\sqrt{-3})}=\sqrt{x}+i\sqrt{y}$. (2)

47. $\dfrac{3}{1+\sqrt{-2}}$.
48. $\dfrac{7}{2-\sqrt{-3}}$.
49. $\dfrac{14}{3+2\sqrt{5}\cdot i}$.

50. $\dfrac{13}{2+3i}$.
51. $\dfrac{17}{3-5\sqrt{-1}}$.
52. $\dfrac{6\sqrt{2}}{3-5\sqrt{-8}}$.

53. $\dfrac{3+2\sqrt{-1}}{2-3\sqrt{-1}}$.
54. $\dfrac{3+4\sqrt{-5}}{4-3\sqrt{-5}}$.
55. $\dfrac{\sqrt{5}-3\sqrt{-7}}{\sqrt{7}+3\sqrt{-5}}$.

56. $\dfrac{2-\frac{1}{3}\sqrt{6}\cdot i}{2+\frac{1}{3}\sqrt{6}\cdot i}$.
57. $\dfrac{1+i}{(1+i)^3}$.
58. $\dfrac{x+iy}{ix-y}$.

59. $\dfrac{2}{\sqrt{2}-\sqrt{-2}-2\sqrt{-1}}$.
60. $\dfrac{10+\sqrt{-5}}{1-\sqrt{-3}+\sqrt{-5}}$.

Factor each of the following expressions:

61. $a+b$.
62. x^2+y^2.
63. $x+1$.
64. $x^2-6x+29$.
65. $x^2+4x+67$.
66. $x^2-14x+61$.
67. $x^2+10x+97$.
68. $3x^2+5x+28$.
69. $5x^2-6x+2$.
70. $4x^2+4xy+3y^2$.
71. $16x^2-8xy+5y^2$.

Find the square root of each of the following expressions:

72. $1+\sqrt{-3}$.
73. $5-\sqrt{-11}$.
74. $3+4i$.
75. $-3+4i$.
76. $8-\sqrt{-17}$.
77. $-2+4\sqrt{-6}$.
78. $-15+3\sqrt{-11}$.
79. $2a+a\sqrt{-5}$.
80. $3n-n\sqrt{-7}$.

Make the indicated substitution in each of the following expressions, and simplify the results:

81. In $x^2-6x+14$, let $x=3+\sqrt{-5}$.

82. In $x^2-16x+92$, let $x=8-2\sqrt{-7}$.

83. In $3x^2-5x+7$, let $x=2-3\sqrt{-2}$.

84. In $5x^3+2x^2-3x-1$, let $x=1-2i$.

85. In $-\sqrt{(4x^2+4x+9)}$, let $x=-\frac{1}{6}(1\pm4\sqrt{-5})$.

86. Show that $8x^3+12x^2+18x+27=[\frac{2}{3}(-2\pm\sqrt{-5})]^3$, when

$$x=-\tfrac{1}{6}(1\pm4\sqrt{-5}).$$

Simplify each of the following expressions:

87. $\dfrac{5}{4-\sqrt{-14}}-\dfrac{3}{2-\sqrt{-14}}.$

88. $\dfrac{\sqrt{-2}}{6+\sqrt{-6}}+\dfrac{\sqrt{-\frac{1}{2}}}{3-\sqrt{-\frac{3}{2}}}.$

89. $\dfrac{3}{1+i}-\dfrac{5}{4-2i}+\dfrac{4}{1-i}.$

90. $\dfrac{a+bi}{a-bi}\pm\dfrac{a-bi}{a+bi}.$

91. $\dfrac{ai-b}{a+bi}\pm\dfrac{ai+b}{a-bi}.$

92. $\left[\sqrt{\dfrac{a^2-x^2}{x}}+\sqrt{\dfrac{x^2-a^2}{a}}\right]\times\left[\sqrt{(-x^2-6ax-9a^2)}+\dfrac{x+3a}{5x}\right].$

CHAPTER XX.

QUADRATIC EQUATIONS.

1. A **Quadratic Equation** is an equation of the second degree in the unknown number or numbers.

E.g., $x^2 = 25$, $x^2 - 5x + 6 = 0$, $x^2 + 2xy = 7$.

A **Complete Quadratic Equation**, in one unknown number, is one which contains a term (or terms) in x^2, a term (or terms) in x, and a term (or terms) free from x, as $x^2 - 2ax + b = cx - d$.

A **Pure Quadratic Equation** is an incomplete quadratic equation which has no term in x, as $x^2 - 9 = 0$.

In this chapter we shall consider quadratic equations in only one unknown number.

2. The following example illustrates a principle of the equivalence of a quadratic equation to two derived linear equations.

The equation
$$x^2 + 6x + 9 = 16, \qquad (1)$$
or
$$(x + 3)^2 = 16, \qquad (2)$$
is equivalent to the two equations
$$x + 3 = 4, \qquad (3)$$
$$x + 3 = -4, \qquad (4)$$
obtained by equating the positive square root of the first member in turn to the positive and to the negative square root of the second member.

For (1), or its equivalent (2), is equivalent to
$$(x + 3)^2 - 16 = 0. \qquad (5)$$

This equation is equivalent to

$$x+3-4=0 \text{ and } x+3+4=0$$

jointly. But the latter equations are equivalent to

$$x+3=4, \qquad (3)$$

$$x+3=-4. \qquad (4)$$

Equations (3) and (4) are usually written

$$x+3=\pm 4.$$

In general, *if the positive square root of the first member of an equation be equated in turn to the positive and to the negative square root of the second member, these two derived equations are jointly equivalent to the given equation.*

For, the equation $M=N$ (1)

is equivalent to $M-N=0$; (2)

that is, to $(\sqrt{M}+\sqrt{N})(\sqrt{M}-\sqrt{N})=0.$ (3)

The last equation is equivalent to

$\sqrt{M}-\sqrt{N}=0$ and $\sqrt{M}+\sqrt{N}=0$ (4)

jointly; that is, to $\sqrt{M}=\pm\sqrt{N}.$ (5)

Pure Quadratic Equations.

3. Any pure quadratic equation can be reduced to the form $x^2=m$. From this equation we obtain $x=\pm\sqrt{m}$, by Art. 2.

Ex. Solve the equation $(2x-5)(2x+5)=11$.

Simplifying, $x^2=9$;

whence $x=\pm 3.$

EXERCISES I.

Solve each of the following equations:

1. $x^2=289.$
2. $x^2=2809.$
3. $x^2=3.61.$
4. $x^2=53.29.$
5. $\frac{2}{3}x^2=1536.$
6. $\frac{4}{5}x^2=1479.2.$
7. $9x^2-36=5x^2.$
8. $7x^2-8=9x^2-10.$
9. $(3x-4)(3x+4)=65.$
10. $(7+x)^2+(7-x)^2=130.$
11. $(3x-5)^2+(3x+5)^2=122.$

12. $(2x-3)(3x-4)-(x-13)(x-4)=40.$

13. $(5x-7)(3x+8)-(x-10)(9-x)=1634.$

14. $(4+x)(3-x)(2-x)-(x+2)(x+3)(x-4)=30.$

15. $(5-x)(3-x)(1+x)+(5+x)(3+x)(1-x)=16.$

16. $8(2-x)^2=2(8-x)^2.$ 17. $(3-x)^2=3(1-x)^2.$

18. $ax^2+b=bx^2+a.$ 19. $a(x^2+b)=b(x^2+a).$

Solution by Factoring.

4. The principle on which the solution of an equation by factoring depends was proved in Ch. VIII., § 4, Art. 1. The methods given in Ch. VIII., § 1, Arts. 10–14, Ch. XVIII., Art. 19, and Ch. XIX., Art. 31, enable us to factor any quadratic expression. The roots of the given quadratic equation are the roots of the equations obtained by equating to 0 each of its factors.

Ex. 1. Solve the equation $4(x-\frac{3}{2})^2=6x+20.$

Reducing the first member,

$$4x^2-12x+9=6x+20.$$

Transferring and uniting terms,

$$4x^2-18x-11=0.$$

Factoring first member,

$$4(x-\tfrac{9}{4}+\tfrac{5}{4}\sqrt{5})(x-\tfrac{9}{4}-\tfrac{5}{4}\sqrt{5})=0.$$

Equating each factor to 0, $x-\frac{9}{4}+\frac{5}{4}\sqrt{5}=0,$

$$x-\tfrac{9}{4}-\tfrac{5}{4}\sqrt{5}=0.$$

Whence $x=\frac{9}{4}-\frac{5}{4}\sqrt{5}$, and $x=\frac{9}{4}+\frac{5}{4}\sqrt{5}.$

Ex. 2. Solve the equation $4m^2x^2+4m^2n+1=4mx.$

Transferring terms, $4m^2x^2-4mx+1+4m^2n=0,$

or $$(2mx-1)^2-(2m\sqrt{-n})^2=0.$$

Equating to 0 the factors of the first member,

$$2mx-1+2m\sqrt{-n}=0,$$

$$2mx-1-2m\sqrt{-n}=0.$$

Whence $x=\dfrac{1}{2m}-\sqrt{-n}$, and $x=\dfrac{1}{2m}+\sqrt{-n}.$

EXERCISES II.

Solve each of the following equations:

1. $x^2 - 7x = 4x$.
2. $x^2 - 2x - 17 = 0$.
3. $x^2 - 6x + 8 = 0$.
4. $x^2 - 4x + 8 = 0$.
5. $x^2 - 4x - 71 = 0$.
6. $x^2 + 10x + 24 = 0$.
7. $13x - 6 - 6x^2 = 0$.
8. $(x + 10)^2 = 28$.
9. $6x - x^2 = 18$.
10. $7x^2 - 3x = 160$.
11. $(2x - 1)^2 = 2$.
12. $x(5x - 2) = -6$.
13. $x^2 - 2\sqrt{2}\,x - 1 = 0$.
14. $36x^2 - 36\sqrt{5}\,x + 17 = 0$.
15. $(x + 8)(x + 3) = x - 6$.
16. $(x + 7)(x - 7) = 2(x + 50)$.
17. $(2x+1)(x+2) = 3x^2 - 4$.
18. $(x - 1)(2x + 3) = 4x^2 - 22$.
19. $x^2 - 3 = \frac{1}{6}(x - 3)$.
20. $x(x + 5) = 5(40 - x) + 27$.
21. $7x(x - 1) = 7 - 4(x - 1)$.
22. $(2x + 1)^2 = x(x + 2)$.
23. $(x-2)^2 = (1+x)(4-x) - 8$.
24. $7 + x(5x + 8) = 10(x + 1)$.
25. $(3x + 2)(2x - 1) + 4(x + 2) = -(5x + 2)(2x - 2)$.
26. $(3x - 5)(2x - 5) = (x + 3)(x - 1)$.
27. $(x + 1)(2x + 3) = (x + 1)(5x - 3)$.
28. $\frac{5}{21}x(x + 1) - \frac{1}{7}(2x^2 + x - 1) = \frac{4}{35}(x + 1)$.
29. $x^2 - 2ax + a^2 = b^2$.
30. $x^2 - 2mx - 1 = 0$.
31. $x^2 - 2ax + a^2 + b^2 = 0$.
32. $x^2 - 4b^2 = a(2x - 5a)$.
33. $n^2x^2 + 2mnx + 2m^2 = 0$.
34. $x^2 - (a + b)x + ab = 0$.
35. $(a^2 + b^2)x - abx^2 - ab = 0$.
36. $x^2 + 2mx + m^2 = n$.
37. $x^2 + 2a + 1 = 2(x - a)$.
38. $a^2x^2 - 2ax + 1 = a^2$.
39. $4x^2 - 12ax + 9a^2 = 4b^2$.
40. $(x + a)^2 = 5ax - (x - a)^2$.
41. $x^2 - 4(a + b) + 1 = 2x$.
42. $x^2 = 2(a + b)x + 2(a^2 + b^2)$.
43. $(m^2 - 1)x^2 - 2(m^2 + 1)x + m^2 - 1 = 0$.
44. $a^2(a^2 + b^2x^2) + b(b^3 - a^3x) = a^3x$.
45. $mnx^2 - (m + n)(mn + 1)x + (m^2 + 1)(n^2 + 1) = 0$.
46. $(3a^2 + b^2)(x^2 - x + 1) = (3b^2 + a^2)(x^2 + x + 1)$.
47. $(m - n)^2(m + n)x^2 + 2(m - n)(m + n)^2x + 6m^2n + 2n^3 = 0$.
48. $(m^2 - n^2)^2x^2 + 2(m - n)(m + n)^3x + 2(m^2 + n^2)^2 + 8m^2n^2 = 0$.

Solution by Completing the Square.

5. The following examples illustrate the solution of a quadratic equation by the method commonly called *Completing the Square.*

Ex. 1. Solve the equation $x^2 - 5x + 6 = 0$.

Transferring 6, $x^2 - 5x = -6$.

To complete the square in the first member, we add $(-\frac{5}{2})^2$, $= \frac{25}{4}$, to this member, and therefore also to the second. We then have

$$x^2 - 5x + \tfrac{25}{4} = \tfrac{25}{4} - 6 = \tfrac{1}{4}.$$

Equating square roots, $x - \frac{5}{2} = \pm \frac{1}{2}$, by Art. 2.

Whence $x = \frac{5}{2} \pm \frac{1}{2}$.

Therefore the required roots are 3 and 2.

Ex. 2. Solve the equation

$$7x^2 + 5x + 1 = 0.$$

Transferring 1, $7x^2 + 5x = -1$.

Dividing by 7, $x^2 + \frac{5}{7}x = -\frac{1}{7}$.

Adding $\left(\dfrac{5}{2 \times 7}\right)^2 = \dfrac{25}{196}$, $x^2 + \dfrac{5}{7}x + \dfrac{25}{196} = \dfrac{25}{196} - \dfrac{1}{7} = \dfrac{-3}{196}$.

Equating square roots, $x + \dfrac{5}{14} = \pm \dfrac{\sqrt{-3}}{14}$.

Whence $x = -\dfrac{5}{14} \pm \dfrac{\sqrt{-3}}{14}$.

Therefore the required roots are

$$-\tfrac{5}{14} + \tfrac{1}{14}\sqrt{-3} \text{ and } -\tfrac{5}{14} - \tfrac{1}{14}\sqrt{-3}.$$

Ex. 3. Solve the equation

$$(a^2 - b^2)x^2 - 2a^2x + a^2 = 0.$$

Transferring a^2, $(a^2 - b^2)x^2 - 2a^2x = -a^2$.

Dividing by $a^2 - b^2$, $x^2 - \dfrac{2a^2x}{a^2 - b^2} = \dfrac{-a^2}{a^2 - b^2}$.

Adding $\left(-\frac{a^2}{a^2-b^2}\right)^2, = \frac{a^4}{(a^2-b^2)^2}$, to both members,

$$x^2 - \frac{2a^2x}{a^2-b^2} + \frac{a^4}{(a^2-b^2)^2} = -\frac{a^2}{a^2-b^2} + \frac{a^4}{(a^2-b^2)^2} = \frac{a^2b^2}{(a^2-b^2)^2}.$$

Equating square roots, $\quad x - \frac{a^2}{a^2-b^2} = \pm \frac{ab}{a^2-b^2}.$

Whence $\quad x = \frac{a^2 \pm ab}{a^2-b^2}.$

Therefore the required roots are $\frac{a}{a-b}$ and $\frac{a}{a+b}$.

The preceding examples illustrate the following method of procedure:

Bring the terms in x and x^2 to the first member, and the terms free from x to the second member, uniting like terms.

If the resulting coefficient of x^2 be not $+1$, divide both members by this coefficient.

Complete the square by adding to both members the square of half the coefficient of x.

Equate the positive square root of the first member to the positive and negative square roots of the second member.

Solve the resulting equations.

6. Equal Roots. — From the equation

$$x^2 - 6x + 9 = 0 \tag{1}$$

we obtain, by factoring the first member,

$$(x-3)(x-3) = 0. \tag{2}$$

The roots of (1) are therefore the roots of the linear equations

$$x - 3 = 0,$$
$$x - 3 = 0.$$

Consequently (1) has the root 3 twice.

From the same equation, we obtain, by transferring 9 and completing the square,

$$x^2 - 6x + 9 = -9 + 9 = 0,$$

or $\quad (x-3)^2 = 0.$

Whence $x - 3 = \pm 0,$

or $x = 3 + 0 = 3,$ and $x = 3 - 0 = 3.$

In general, if the first member of an equation whose second member is zero be the square of a linear expression in x, the two roots of the equation are equal.

EXERCISES III.

Solve each of the following equations:

1. $x^2 - 4x + 3 = 0.$
2. $x^2 - 5x = -4.$
3. $x^2 + 2x + 1 = 0.$
4. $2x^2 - 7x + 3 = 0.$
5. $3x^2 - 53x + 34 = 0.$
6. $14x - 49x^2 - 1 = 0.$
7. $x^2 - 4x + 7 = 0.$
8. $(2x - 1)(x - 2) = (x + 1)^2.$
9. $x^2 - 2x + 6 = 0.$
10. $x^2 - 1 + x(x - 1) = x^2.$
11. $(3x - 2)(x - 1) = 14.$
12. $110x^2 - 21x + 1 = 0.$
13. $780x^2 - 73x + 1 = 0.$
14. $(2x - 3)^2 = 8x.$
15. $(3x - 2)(x - 1) = 14.$
16. $(5x - 3)^2 - 7 = 40x - 47.$
17. $(2x + 1)(x + 2) = 3x^2 - 4.$
18. $(x + 1)(2x + 3) = 4x^2 - 22.$
19. $(x - 7)(x - 4) + (2x - 3)(x - 5) = 103.$
20. $10(2x + 3)(x - 3) + (7x + 3)^2 = 20(x + 3)(x - 1).$
21. $(x - 1)(x - 3) + (x - 3)(x - 5) = 32.$
22. $(x - 1)(x - 2) + (x - 3)(x - 4) = (x - 1)^2 - 2.$
23. $4x^2 - 4bx - a^2 + b^2 = 0.$
24. $(ax - b)(bx - a) = c^2.$
25. $(m - n)x^2 - (m + n)x + 2n = 0.$
26. $4x^2 - 4(3a + 2b)x + 24ab = 0.$
27. $(a + b)^2x^2 + 2abx = -\dfrac{a^2b^2}{(a + b)^2}.$
28. $x^2 - 2(a + b)x + (a + b + c)(a + b - c) = 0.$
29. $x^2 - (a - 1)x + a = 0.$
30. $x^2 - 2cx + ac + bc - ab = 0.$
31. $(a^2 - b^2)x^2 - 2(a^2 + b^2)x + a^2 - b^2 = 0.$
32. $x^2 - 4mnx = (m^2 - n^2)^2.$
33. $d^2x^2 - 4abdx + 4a^2b^2 - 9c^2 = 0.$
34. $abcx^2 - (a^2b^2 + c^2)x + abc = 0.$
35. $x^2\sqrt{6} - (\sqrt{2} + \sqrt{3})x + 1 = 0.$

36. $x^2 - 2x\sqrt{(a+b)} + 2b = 0.$

37. $\dfrac{(x-4a)(x+2b-2)}{(a+b-1)(a-b+1)} = -3.$

38. $\dfrac{x^2}{a^2+ab+ac} = \dfrac{b+c-x}{a+b+c} - \dfrac{(b+c)x^2}{a^3+a^2b+a^2c}.$

General Solution.

7. The most general form of the quadratic equation in one unknown number is evidently

$$ax^2 + bx + c = 0,$$

in which ax^2 is the algebraic sum of all the terms in x^2, bx is the algebraic sum of all the terms in x, and c is the algebraic sum of all the terms free from x.

The coefficient a is assumed to be *positive* and not 0, but b and c may either or both be positive or negative, or 0.

Dividing by a, $$x^2 + \frac{b}{a}x + \frac{c}{a} = 0. \quad (1)$$

Transferring $\frac{c}{a}$, $$x^2 + \frac{b}{a}x = -\frac{c}{a}. \quad (2)$$

Adding $\left(\frac{b}{2a}\right)^2, = \frac{b^2}{4a^2}$, $$x^2 + \frac{b}{a}x + \frac{b^2}{4a^2} = \frac{b^2}{4a^2} - \frac{c}{a}$$

$$= \frac{b^2-4ac}{4a^2}. \quad (3)$$

Equating square roots, $$x + \frac{b}{2a} = \pm\frac{\sqrt{(b^2-4ac)}}{2a}. \quad (4)$$

Equation (4) is equivalent to

$$x + \frac{b}{2a} = \frac{\sqrt{(b^2-4ac)}}{2a}, \quad (5)$$

$$x + \frac{b}{2a} = -\frac{\sqrt{(b^2-4ac)}}{2a}. \quad (6)$$

From (5) and (6) we obtain

$$x = -\frac{b}{2a} + \frac{\sqrt{(b^2-4ac)}}{2a},$$

and $$x = -\frac{b}{2a} - \frac{\sqrt{(b^2-4ac)}}{2a}.$$

8. The roots of any quadratic equation can be obtained by substituting in the general solution the particular values of the coefficients a, b, and c.

Ex. 1. Solve the equation $3x^2 + 7x - 10 = 0$.

We have $a = 3,\ b = 7,\ c = -10$.

Substituting these values in the general solution, we obtain

$$x = -\frac{7}{6} + \frac{\sqrt{[49 - 4 \times 3(-10)]}}{6} = 1,$$

and $$x = -\frac{7}{6} - \frac{\sqrt{[49 - 4 \times 3(-10)]}}{6} = -\frac{10}{3}.$$

EXERCISES IV.

Solve each of the following equations:

1. $2x^2 = 3x + 2$.
2. $5x^2 - 6x + 1 = 0$.
3. $9x(x + 1) = 28$.
4. $x^2 - b^2 = 2ax - a^2$.
5. $x^2 + 6ax + 1 = 0$.
6. $x^2 + 1 = 2\frac{1}{8}x$.
7. $(x - 5)^2 + (x - 10)^2 = 37$.
8. $2x(3n - 4x) = n^2$.
9. $n^2(x^2 + 1) = a^2 + 2n^2x$.
10. $x^2 + (x + a)^2 = a^2$.

Fractional Equations which lead to Quadratic Equations.

9. The principles given in Ch. X. for solving fractional equations which lead to linear equations hold also for fractional equations which lead to quadratic equations.

Ex. 1. Solve the equation $\dfrac{4}{x - 1} = \dfrac{3x}{x^2 - 1} + 2.$ (1)

Multiplying by $x^2 - 1$, $4(x + 1) = 3x + 2(x^2 - 1)$. (2)

Transferring and uniting terms, $2x^2 - x = 6$. (3)

Dividing by 2, $x^2 - \frac{1}{2}x = 3$. (4)

The roots of equation (4) are $2, -\frac{3}{2}$. Since neither is a root of the L.C.D. (equated to 0) of the fractions in the given equation, *i.e.*, of $x^2 - 1 = 0$, they are the roots of that equation.

Ex. 2. Solve the equation

$$\frac{1}{1-x}+\frac{1}{x+1}=\frac{2}{2x-1}-\frac{1}{x+2}. \qquad (1)$$

Uniting fractions in each member,

$$\frac{2}{1-x^2}=\frac{5}{(2x-1)(x+2)}. \qquad (2)$$

Clearing of fractions,

$$2(2x-1)(x+2)=5-5x^2. \qquad (3)$$

Performing indicated operations, transferring and uniting terms,

$$9x^2+6x=9. \qquad (4)$$

The roots of this equation are $\frac{-1+\sqrt{10}}{3}$ and $\frac{-1-\sqrt{10}}{3}$.

Since neither is a root of the L.C.D. equated to 0, that is, of

$$(1-x^2)(2x-1)(x+2)=0,$$

they are the roots of the given equation.

Ex. 3. Solve the equation

$$\frac{1}{n+x}-\frac{1}{n-x}=\frac{x^2-2n-n^2}{x^2-n^2}. \qquad (1)$$

Uniting the fractions in the first member,

$$\frac{-2x}{n^2-x^2}=\frac{x^2-2n-n^2}{x^2-n^2}. \qquad (2)$$

Clearing of fractions,

$$2x=x^2-2n-n^2. \qquad (3)$$

Transferring and uniting terms,

$$x^2-2x=n^2+2n. \qquad (4)$$

Completing the square,

$$x^2-2x+1=n^2+2n+1. \qquad (5)$$

Equating square roots,

$$x-1=\pm(n+1).$$

Therefore the roots of (5) are

$$1+(n+1)=2+n,$$

and

$$1-(n+1)=-n.$$

The number $2+n$ is not a root of the L.C.D. equated to 0, that is, of

$$x^2-n^2=0.$$

Therefore $2+n$ is a root of the given equation.

But $-n$ is a root of $x^2-n^2=0$, or of $(x-n)(x+n)=0$, and is therefore not a root of the given equation. It can easily be seen that this root was introduced by multiplying the given equation by the factor $x+n$ which was not necessary to clear it of fractions.

For, transferring and uniting the fractions in equation (2), we obtain

$$\frac{x^2-2x-2n-n^2}{x^2-n^2}=0.$$

Factoring the numerator,

$$\frac{(x+n)(x-n-2)}{x^2-n^2}=0.$$

Reducing to lowest terms,

$$\frac{x-n-2}{x-n}=0.$$

The numerator equated to 0 gives $x-n-2=0$, whence $x=2+n$, as above.

10. The work of solving an equation can sometimes be simplified by a simple substitution.

Ex. Solve the equation $\dfrac{x+5}{x+2}-\dfrac{x+2}{x+5}=\dfrac{3}{2}.$

If we let $\dfrac{x+5}{x+2}=y$, the given equation becomes

$$y-\frac{1}{y}=\frac{3}{2}.$$

The roots of this equation are $2, -\frac{1}{2}$.

We now have to solve the two equations

$$\frac{x+5}{x+2}=2, \text{ whence } x=1;$$

and

$$\frac{x+5}{x+2}=-\tfrac{1}{2}, \text{ whence } x=-4.$$

This method can be used when the fractional equation contains only two expressions in the unknown number, one of which is the reciprocal of the other.

EXERCISES V.

Solve each of the following equations:

1. $15x + \frac{2}{x} = 11.$

2. $x + \frac{1}{3} = \frac{1}{x} + 3.$

3. $x + \frac{1}{x} = 7 + \frac{1}{7}.$

4. $x - \frac{3}{5} = \frac{5}{3} - \frac{1}{x}.$

5. $\frac{x-6}{x+30} = \frac{3}{x-6}.$

6. $\frac{x}{x+120} = \frac{14}{3x-10}.$

7. $\frac{x}{x-6} + 1 = \frac{60}{x+4}.$

8. $\frac{3}{1+x} + \frac{3}{1-x} = 8.$

9. $\frac{x-1}{x} - \frac{3x}{x-1} = 2.$

10. $\frac{x+4}{x-4} + \frac{x-4}{x+4} = \frac{10}{3}.$

11. $\frac{7}{x} - \frac{15}{x+2} + \frac{5}{x-8} = 0.$

12. $\frac{3}{x^2-1} - \frac{1}{4} = \frac{1}{2-2x}.$

13. $\frac{x^2-40}{x+8} = \frac{3}{5}(8-x).$

14. $5 - \frac{x+605}{x^2} = \frac{605-x}{x^2} - 5.$

15. $\frac{3}{2(x^2-1)} - \frac{1}{4(x+1)} = \frac{1}{8}.$

16. $\frac{x+1}{x-1} + \frac{x+3}{x-3} = \frac{11}{2}.$

17. $\frac{1}{x-1} + \frac{2(x-1)}{x+1} = \frac{5}{x^2-1}.$

18. $\frac{1}{x-1} + \frac{1}{x-2} + \frac{1}{x-3} = 0.$

19. $\frac{9x+1}{9x-3x^2} = \frac{x}{21-7x} - \frac{x+3}{21x}.$

20. $\frac{x+3}{x+2} + \frac{x-3}{x-2} = \frac{2x-3}{x-1}.$

21. $\frac{x-2}{x+2} + \frac{x+2}{x-2} = \frac{2(x+3)}{x-3}.$

22. $\frac{5x-1}{x+3} + \frac{7x^2-106}{8x^2-72} = -\frac{1}{8}.$

23. $\frac{4x+67}{40x^2-36} + \frac{x}{30x^2-27} = \frac{2}{3}.$

24. $\frac{x+24}{5x^2-5} = \frac{x-7}{x+1} - \frac{1}{2x-2}.$

25. $\frac{4x+4}{5x-5} = \frac{(x+1)^2}{x^2+1}.$

26. $\frac{2x-2}{5x+5} = \frac{x^2-1}{x^2+1}.$

27. $\frac{x^3+3x^2+3x+1}{x^3+2x^2+2x+1} = \frac{9}{7}.$

28. $\frac{3x-16}{2x-12} + \frac{2x-12}{3x-16} = \frac{5}{2}.$

29. $\frac{3x-15}{5x-9} + \frac{5x-9}{3x-15} = \frac{10}{3}.$

30. $1 + 2x[1 + x(1+x)] + \frac{2x^4 - x}{1-x} = x - 1.$

31. $x - \frac{a}{b} = \frac{b}{a} - \frac{1}{x}$.

32. $\frac{m(m-1)}{x-1} = x + 1$.

33. $\frac{x-1}{a} = \frac{b}{x+1}$.

34. $\frac{x-a}{x+a} = \frac{b-x}{b+x}$.

35. $\frac{x}{x-1} - \frac{x}{x+1} = m$.

36. $\frac{a}{x-b} + \frac{b}{x-a} = 2$.

37. $\frac{(a+x)(a-x)}{(x+1)(x-1)} = -a^4$.

38. $\frac{an}{x+4n} - \frac{an}{x-4n} = 2$.

39. $x = \frac{3}{(a-b)^2 x} - \frac{2}{a-b}$.

40. $\frac{1}{(a^2-x)^2} = \frac{4x}{4x^3 + 5a^6}$.

41. $\frac{x^2+1}{n^2 x - 2n} - \frac{1}{2 - nx} = \frac{x}{n}$.

42. $\frac{1}{n+x} - \frac{1}{n-x} = \frac{x^2 - 2n - n^2}{x^2 - n^2}$.

43. $\frac{a-2b}{8x^2 - 2b^2} = \frac{1}{2x+b} - \frac{1}{2a}$.

44. $1 - \frac{2c}{x-a} = \frac{a^2+b^2}{a^2 + x^2 - 2ax}$.

45. $\frac{a}{nx - x} - \frac{a-1}{x^2 - 2nx^2 + n^2x^2} = 1$.

46. $\frac{5x^2+4}{n^2 - 4x^2} = \frac{1}{n-2x} - \frac{1}{n+2x}$.

47. $\frac{n+x}{n-x} + \frac{n-x}{n+x} = \frac{n^2}{n^2 - x^2}$.

48. $\frac{x-a+b}{x+a-b} = \frac{a-b-x}{a+b+x}$.

49. $\frac{ax}{ax+1} = \frac{1-a}{a^2x^2 - a - a^2x + ax}$.

50. $\left(\frac{a+x}{a-x}\right)^2 + \frac{7}{2} \cdot \frac{a+x}{a-x} + 3 = 0$.

51. $\frac{11a - x}{3a - x} = \frac{121a^2 - x^2}{9a^2 - x^2}$.

52. $\frac{x+1}{x-1} + \frac{a-b}{a+b} = \frac{x-1}{x+1} + \frac{a+b}{a-b}$.

53. $\frac{x}{(n+1)(n-1)^2} = \frac{n+2}{n^2-1} - \frac{1}{x}$.

54. $\frac{mx+a}{mx-a} - \frac{mx-a}{mx+a} = \frac{mx+3a}{mx+a}$.

55. $\frac{n-p+1}{nx+px} = \frac{1}{(n+p)^2} + \frac{n-p}{x^2}$.

56. $\frac{1}{a+b+x} = \frac{1}{a} + \frac{1}{b} + \frac{1}{x}$.

57. $\frac{x+a}{x-a} + \frac{x+b}{x-b} = \frac{a}{b} + \frac{b}{a}$.

58. $(a+b)^2 x - \frac{b}{x}[a - (a+b)x] = a(a+b)$.

59. $\frac{(a-x)^3 + (x-b)^3}{(a-x)^2 + (x-b)^2} = a - b$.

60. $\frac{(a-x)^3 + (x-b)^3}{(a-x)^2 + (x-b)^2} = \frac{a^3 - b^3}{a^2 + b^2}$.

61. $\dfrac{(a-x)^3+(x-b)^3}{(a+b-2x)^2}=\dfrac{a^3-b^3}{(a+b)^2}.$

62. $\dfrac{x^2-2nx+2ax-n^2}{x^3-a^3}+\dfrac{x+2n}{x^2+ax+a^2}=\dfrac{1}{x-a}.$

63. $\dfrac{x}{a^3-1}\left[1-\dfrac{3(a^2+1)}{x}\right]+\dfrac{1}{a-1}\left[1-\dfrac{2a-(a^2+1)}{x}\right]=0.$

64. $\dfrac{\frac{a}{b}+1.}{ax^3-ax^2-bx+b}-\dfrac{\frac{a}{b}-1.}{bx^3-bx^2+ax-a}=\dfrac{2b(x+1)}{abx^4-b^2x^2+a^2x^2-ab}.$

Theory of Quadratic Equations.

11. *A quadratic equation has two, and only two, roots.*

For, by Art. 7, the equations

$$x+\frac{b}{2a}=\frac{\sqrt{(b^2-4ac)}}{2a} \qquad (1)$$

and

$$x+\frac{b}{2a}=\frac{-\sqrt{(b^2-4ac)}}{2a} \qquad (2)$$

are jointly equivalent to the equation

$$ax^2+bx+c=0. \qquad (3)$$

But equations (1) and (2) have each one, and only one, root. Therefore ax^2+bx+c has two, and only two, roots.

12. Relations between the roots of a quadratic equation and the coefficients of its terms.—If the roots of the quadratic equation

$$ax^2+bx+c=0, \text{ or } x^2+\frac{b}{a}x+\frac{c}{a}=0$$

be designated by r_1 and r_2, we have

$$r_1=-\frac{b}{2a}+\frac{\sqrt{(b^2-4ac)}}{2a},$$

$$r_2=-\frac{b}{2a}-\frac{\sqrt{(b^2-4ac)}}{2a}.$$

The sum of the roots is $\quad r_1+r_2=-\dfrac{b}{a} \qquad (1)$

The product of the roots is

$$r_1r_2=\left[-\frac{b}{2a}+\frac{\sqrt{(b^2-4ac)}}{2a}\right]\times\left[-\frac{b}{2a}-\frac{\sqrt{(b^2-4ac)}}{2a}\right]$$

$$=\left[-\frac{b}{2a}\right]^2-\left[\frac{\sqrt{(b^2-4ac)}}{2a}\right]^2=\frac{b^2}{4a^2}-\frac{b^2-4ac}{4a^2}$$

$$=\frac{c}{a}. \qquad (2)$$

The relations (1) and (2) may be expressed thus:

(i.) *The sum of the roots of a quadratic equation is equal to the quotient obtained by dividing the coefficient of the first power of the unknown number, with sign reversed, by the coefficient of the second power of the unknown number.*

In particular, if the coefficient of the second power of the unknown number be 1, the sum of the roots is equal to the coefficient of the first power of the unknown number, with sign reversed.

(ii.) *The product of the roots of a quadratic equation is equal to the quotient obtained by dividing the term free from the unknown number by the coefficient of the second power of the unknown number.*

In particular, if the coefficient of the second power of the unknown number be 1, the product of the roots is equal to the term free from the unknown number.

E g., the roots of the equation $6x^2 - x - 2 = 0$ are $\frac{2}{3}$ and $-\frac{1}{2}$; their sum is $\frac{1}{6}$ (the coefficient of x, with sign reversed, divided by the coefficient of x^2), and their product is $-\frac{1}{3}$ (the term free from x divided by the coefficient of x^2).

The roots of the equation $x^2 - 5x + 6 = 0$ are 2 and 3; their sum is 5 (the coefficient of x, with sign reversed), and their product is 6 (the term free from x).

13. Formation of an Equation from its Roots.—The relations of the last article enable us to form an equation if its roots be given. We may always assume that the coefficient of the second power of the unknown number is 1.

Ex. 1. Form the equation whose roots are $-1, 2$.

We have $r_1 + r_2 = -1 + 2 = 1$, the coefficient of x, with sign reversed; and $r_1 r_2 = -1 \times 2 = -2$, the term free from x.

Therefore the required equation is

$$x^2 - x - 2 = 0.$$

Ex. 2. Form the equation whose roots are $1 + 2\sqrt{3}$, $1 - 2\sqrt{3}$.

We have $\quad r_1 + r_2 = (1 + 2\sqrt{3}) + (1 - 2\sqrt{3}) = 2;$

and $\quad r_1 r_2 = (1 + 2\sqrt{3})(1 - 2\sqrt{3}) = 1 - 12 = -11.$

Therefore the required equation is

$$x^2 - 2x - 11 = 0.$$

Ex. 3. Form the equation whose roots are $2 + 5\sqrt{-3}$, $2 - 5\sqrt{-3}$.

We have $\quad r_1 + r_2 = (2 + 5\sqrt{-3}) + (2 - 5\sqrt{-3}) = 4;$

and $\quad r_1 r_2 = (2 + 5\sqrt{-3})(2 - 5\sqrt{-3}) = 4 + 75 = 79.$

Therefore the required equation is $x^2 - 4x + 79 = 0$.

14. The roots of a quadratic equation, all of whose terms are in the first member, are the roots of the two linear factors into which this member can be resolved. Consequently a quadratic equation whose roots are given can be formed by multiplying together the two linear factors which (equated to 0) have as roots the given roots.

Ex. **1.** Form the equation whose roots are $-1, 2$.

Since -1 is the root of $\quad x+1=0,$

and 2 is the root of $\quad x-2=0,$

the required quadratic is $(x+1)(x-2)=0,$

or $\quad x^2-x-2=0.$

Ex. **2.** Form the equation whose roots are $1+2\sqrt{3}$, $1-2\sqrt{3}$.

Since $1+2\sqrt{3}$ is the root of $x-(1+2\sqrt{3})=0$, or $x-1-2\sqrt{3}=0$,

and $1-2\sqrt{3}$ is the root of $x-(1-2\sqrt{3})=0$, or $x-1+2\sqrt{3}=0$,

the required quadratic is

$$(x-1-2\sqrt{3})(x-1+2\sqrt{3})=0,$$

or

$$x^2-2x-11=0.$$

When the roots are irrational or imaginary, the method of the preceding article is to be preferred.

15. When one root of a given quadratic equation is known, the other root can be found without solving the equation.

Ex. **1.** One root of the equation

$$x^2-5x+6=0$$

is 3; what is the other root?

Since 5 is the sum of the roots, the required root is $5-3, =2$.

Or, since x^2-5x+6 is the product of two linear factors, one of which is $x-3$, the other factor is

$$\frac{x^2-5x+6}{x-3}, =x-2.$$

The required root is therefore the root of

$$x-2=0, \text{ or } x=2.$$

Ex. **2.** One root of the equation

$$x^2+2x-1=0$$

is $-1+\sqrt{2}$; what is the other root?

The required root is $-2-(-1+\sqrt{2}), =-1-\sqrt{2}$.

16. The value of an expression which is symmetrical in the roots of a given quadratic equation can be found without solving the equation.

Ex. 1. If r_1 and r_2 be the roots of $x^2 + px + q = 0$, find the value of $r_1^2 + r_2^2$.

We have
$$r_1^2 + r_2^2 = (r_1 + r_2)^2 - 2r_1r_2$$
$$= (-p)^2 - 2q = p^2 - 2q.$$

17. We can sometimes form an equation whose roots have definite relations to the roots of a given quadratic equation without solving the latter.

Ex. 1. Form an equation whose roots are the reciprocals of the roots of the equation
$$x^2 - 8x + 15 = 0.$$

Let r_1 and r_2 be the roots of the given equation.

Then $\frac{1}{r_1}$ and $\frac{1}{r_2}$ are the roots of the required equation.

We have $\frac{1}{r_1} + \frac{1}{r_2} = \frac{r_1 + r_2}{r_1r_2} = \frac{8}{15}$, the coefficient of x (with sign reversed) in the required equation; and $\frac{1}{r_1} \times \frac{1}{r_2} = \frac{1}{r_1r_2} = \frac{1}{15}$, the term free from x in the required equation.

Consequently the required equation is
$$x^2 - \tfrac{8}{15}x + \tfrac{1}{15} = 0,$$
or
$$15x^2 - 8x + 1 = 0.$$

Nature of the Roots of a Quadratic Equation.

18. In many applications it is important to know, without having to solve an equation, the nature of its roots; *i.e.*, whether they are both *real and unequal*, whether they are both *real and equal*, whether they are *imaginary*, etc.

In the general solution
$$r_1 = -\frac{b}{2a} + \frac{\sqrt{(b^2 - 4ac)}}{2a},$$
$$r_2 = -\frac{b}{2a} - \frac{\sqrt{(b^2 - 4ac)}}{2a},$$
of the equation
$$ax^2 + bx + c = 0,$$
a, b, and c are limited to real, rational values.

(i.) *The two roots are real and unequal when $b^2 - 4ac$ is positive*; i.e., *when* $b^2 - 4ac > 0$.

E.g., in
$$x^2 + 4x - 12 = 0,$$
$$a = 1,\ b = 4,\ c = -12;\ \text{and since } b^2 - 4ac, = 16 + 48,$$
is *positive*, the roots of this equation are real and unequal.

(ii.) *The two roots are real and equal when $b^2 - 4ac$ is equal to 0*; i.e., *when $b^2 = 4ac$.*

E.g., in

$$x^2 - 4x + 4 = 0,$$

$$a = 1,\ b = -4,\ c = 4;\ \text{and since } b^2 = 4ac,$$

the roots of this equation are real and equal.

(iii.) *The two roots are conjugate complex numbers when $b^2 - 4ac$ is negative*; i.e., *when $b^2 - 4ac < 0$.*

E.g., in

$$x^2 - 2x + 3 = 0$$

$$a = 1,\ b = -2,\ c = 3, \text{ and since } b^2 - 4ac,\ = 4 - 12,\ = -8,$$

is *negative*, the roots of this equation are imaginary.

(iv.) *The two roots are real and rational when $b^2 - 4ac$ is positive and the square of a rational number.*

E.g., in

$$x^2 - 5x + 4 = 0$$

$$a = 1,\ b = -5,\ c = 4, \text{ and since } b^2 - 4ac,\ = 25 - 16,\ = 9,$$

is the square of a rational number, the roots of this equation are real and rational.

(v.) *The two roots are real, but conjugate irrationals, when $b^2 - 4ac$ is positive and not the square of a rational number.*

E.g., in

$$x^2 - 5x + 2 = 0$$

$$a = 1,\ b = -5,\ c = 2, \text{ and since } b^2 - 4ac,\ = 25 - 8,\ = 17,$$

is positive and not the square of a rational number, the roots of this equation are conjugate irrational numbers.

(vi.) *The two roots are equal and opposite when $b = 0$.*

We then have $r_1 = \dfrac{\sqrt{(-4ac)}}{2a}$, and $r_2 = -\dfrac{\sqrt{(-4ac)}}{2a}$.

Notice that in this case the roots are real or imaginary according as ac is negative or positive.

E.g., from $2x^2 + 5 = 0$ we have $x = \sqrt{-\frac{5}{2}} = \pm \frac{1}{2}\sqrt{-10}$.

(vii.) *One root is zero when $c = 0$.*

We then have $r_1 = -\dfrac{b}{2a} + \dfrac{\sqrt{b^2}}{2a} = -\dfrac{b}{2a} + \dfrac{b}{2a} = 0$;

$$r_2 = -\frac{b}{2a} - \frac{\sqrt{b^2}}{2a} = -\frac{b}{2a} - \frac{b}{2a} = -\frac{b}{a}.$$

E.g., from $2x^2 - 3x = 0$ we have $x(2x - 3) = 0$;

whence $x = 0$, and $x = \frac{3}{2}$.

(viii.) *Both roots are zero when* $b = 0$ *and* $c = 0$.

We then have $$r_1 = 0 + 0,\ r_2 = 0 - 0.$$

E.g., from $3x^2 = 0$ we obtain $x = 0$ and $x = 0$.

19. As long as a is not equal to 0, however near its value may be to 0, the values of r_1 and r_2 given above constitute the solution of the equation. If, then, we assume that these values still give the roots when $a = 0$, we must determine the nature of the roots under this assumption.

(i.) *One root is infinite and the other* $-\frac{c}{b}$, *when* $a = 0$, $b \neq 0$, $c \neq 0$.

We then have $$r_1 = -\frac{b}{0} + \frac{\sqrt{b^2}}{0} = \frac{-b+b}{0} = \frac{0}{0}.$$

This indeterminate result can be evaluated.

For $$r_1 = \frac{-b+\sqrt{(b^2-4ac)}}{2a} = \frac{[-b+\sqrt{(b^2-4ac)}][-b-\sqrt{(b^2-4ac)}]}{2a[-b-\sqrt{(b^2-4ac)}]}$$
$$= \frac{b^2-(b^2-4ac)}{2a[-b-\sqrt{(b^2-4ac)}]} = \frac{2c}{-b-\sqrt{(b^2-4ac)}}.$$

If in $\frac{2c}{-b-\sqrt{(b^2-4ac)}}$ we let $a = 0$, we obtain
$$r_1 = \frac{2c}{-b-b} = -\frac{c}{b}.$$

Also, $$r_2 = \frac{-b-\sqrt{b^2}}{0} = \frac{-2b}{0} = \infty.$$

In this case we are apparently dealing with the linear equation
$$bx + c = 0,$$
and not with a quadratic. But in applications of Algebra it is frequently necessary to consider the coefficient of x^2 as growing smaller and smaller without limit, *i.e.*, as approaching 0.

The meaning of the results given above are that as a grows smaller and smaller without limit, one root grows larger and larger without limit, and the other root becomes more and more nearly equal to $-\frac{c}{b}$.

E.g., the equation $0 \cdot x^2 + 2x + 3 = 0$ has one root ∞ and one root $-\frac{3}{2}$.

(ii.) *Both roots are infinite when* $a = 0$, $b = 0$, $c \neq 0$.

We have $$r_1 = \frac{-b+\sqrt{(b^2-4ac)}}{2a} = \frac{b^2-(b^2-4ac)}{2a[-b-\sqrt{(b^2-4ac)}]}$$
$$= \frac{2c}{-b-\sqrt{(b^2-4ac)}}.$$

And $$r_2 = \frac{-b-\sqrt{(b^2-4ac)}}{2a} = \frac{b^2-(b^2-4ac)}{2a[-b+\sqrt{(b^2-4ac)}]}$$
$$= \frac{2c}{-b+\sqrt{(b^2-4ac)}}.$$

If in the expressions thus obtained for the roots we let $a = 0$, $b = 0$, $c \neq 0$, we obtain

$$r_1 = \frac{2c}{0} = \infty, \text{ and } r_2 = \frac{2c}{0} = \infty.$$

Attention is called to the remarks at the end of (i.).

E.g., the equation $\quad 0 \cdot x^2 + 0 \cdot x + 2 = 0$

has two infinite roots.

20. If, in simplifying a quadratic equation, the terms of the second degree in the unknown number be canceled, an infinite root is lost.

E.g., solve the equation $\quad (1 + 2x)(2 - 3x) = 5 - 6x^2$.

Performing indicated operations, $2 + x - 6x^2 = 5 - 6x^2$.

Transferring and uniting terms, $\quad x = 3$.

In canceling $-6x^2$ an infinite root was lost.

It was for this reason that in the principle for adding or subtracting the same number or expression to or from both members of an equation the roots were limited to finite values.

EXERCISES VI.

Form the equations whose roots are:

1. $8, 2$. **2.** $-5, -3$. **3.** $10, 10$. **4.** $7, -3$.

5. $4, -10$. **6.** $-9, -9$. **7.** $2\frac{1}{2}, 1\frac{3}{5}$. **8.** $-\frac{2}{3}, -1\frac{1}{2}$.

9. $-\frac{1}{4}, 8$. **10.** $\frac{3}{4}, -\frac{5}{8}$. **11.** $-\frac{1}{2}, -\frac{1}{3}$. **12.** $2, 0$.

13. a, b. **14.** $-a, -1$. **15.** $a^2, -4a^2$. **16.** $2a^3, \frac{1}{2}a^3$.

17. $\frac{a}{b}, \frac{b}{a}$. **18.** $-\frac{a}{2n}, -\frac{a}{2n}$. **19.** $a - \frac{1}{2}, a - \frac{1}{3}$.

20. $(a - b)a, (a + b)b$. **21.** $(a+b)^2, -(a-b)^2$.

22. $\frac{a+b}{a-b}, 1$. **23.** $\frac{a}{2a-2b}, \frac{b}{2b-2a}$. **24.** $\sqrt{2}, -\sqrt{2}$.

25. $5\sqrt{7}, -5\sqrt{7}$. **26.** $\sqrt{-6}, -\sqrt{-6}$. **27.** $\frac{1}{2}\sqrt{-3}, -\frac{1}{2}\sqrt{-3}$.

28. $1 + \sqrt{7}, 1 - \sqrt{7}$. **29.** $-2 + 3\sqrt{5}, -2 - 3\sqrt{5}$.

30. $\frac{1}{2} - \frac{1}{3}\sqrt{11}, \frac{1}{2} + \frac{1}{3}\sqrt{11}$. **31.** $3 - \sqrt{-5}, 3 + \sqrt{-5}$.

32. $-2 - 5\sqrt{-3}, -2 + 5\sqrt{-3}$. **33.** $\frac{2}{3} - \frac{1}{3}\sqrt{-1}, \frac{2}{3} + \frac{1}{3}\sqrt{-1}$.

Find the second root of each of the following equations, without solving the equation:

34. $x^2 - 16 = 0$, when $r_1 = 4$.

35. $x^2 - 3x = 0$, when $r_1 = 3$.

36. $x^2 + 2x = 0$, when $r_1 = -2$.

37. $3x^2 - 4x = 0$, when $r_1 = 0$

38. $x^2 - 9x + 20 = 0$, when $r_1 = 4$.

39. $x^2 - 20x + 100 = 0$, when $r_1 = 10$.

40. $x^2 - 4x - 12 = 0$, when $r_1 = 6$.

41. $6x^2 - x - 1 = 0$, when $r_1 = -\frac{1}{3}$.

42. $x^2 - (a + b)^2 = 0$, when $r_1 = a + b$.

43. $x^2 - (a^2 - b^2)x = 0$, when $r_1 = 0$.

44. $b^2x^2 + 2abx + a^2 = 0$, when $r_1 = -\frac{a}{b}$.

45. $x^2 - (a^2 + b^2)x + (a^2 - b^2)ab = 0$, when $r_1 = (a + b)b$.

46. $(a^2 - b^2)x^2 + 4abx - a^2 + b^2 = 0$, when $r_1 = \frac{a - b}{a + b}$.

47. $x^2 - 12 = 0$, when $r_1 = -2\sqrt{3}$.

48. $x^2 - x\sqrt{10} = 0$, when $r_1 = \sqrt{10}$.

49. $x^2 - (2 - \sqrt{5})x = 0$, when $r_1 = 0$.

50. $x^2 - 2x - 1 = 0$, when $r_1 = 1 - \sqrt{2}$.

51. $x^2 + x\sqrt{5} + 1 = 0$, when $r_1 = -\frac{1}{2}(\sqrt{5} + 1)$.

52. $x^2 + 25 = 0$, when $r_1 = 5\sqrt{-1}$.

53. $x^2 - 6x + 13 = 0$, when $r_1 = 3 + 2\sqrt{-1}$.

54. $3x^2 + 2x + 27 = 0$, when $r_1 = \frac{1}{3}(-1 - 4\sqrt{-5})$.

55. $(a + b)^2x^2 - (a + b)cx - ac = 0$, when $r_1 = \frac{c - \sqrt{(c^2 + 4ac)}}{2(a + b)}$.

56. $abx^2 - (a^2 + b^2)x + ab - c^2 = 0$, when $r_1 = \frac{a^2 + b^2 - \sqrt{[(a^2 - b^2)^2 + 4abc^2]}}{2ab}$.

If r_1 and r_2 stand for the roots of the equation $x^2 + px + q = 0$, express each of the following symmetrical expressions in terms of p and q:

57. $r_1^3 + r_2^3$.

58. $r_1^4 + r_2^4$.

59. $r_1^5 + r_2^5$.

60. $\frac{1}{r_1} + \frac{1}{r_2}$.

61. $\frac{1}{r_1^2} + \frac{1}{r_2^2}$.

62. $\frac{1}{r_1^3} + \frac{1}{r_2^3}$.

63. $\frac{r_1}{r_2} + \frac{r_2}{r_1}$.

64. $\frac{r_1^2}{r_2^2} + \frac{r_2^2}{r_1^2}$.

65. $\frac{r_1 + r_2}{r_1} + \frac{r_1 + r_2}{r_2}$.

66–74. Express each of the relations given in Exx. 57–65 in terms of the roots of the equation $x^2 + x - 6 = 0$.

Without solving the equations $x^2 + px + q$ and $x^2 - 10x + 40 = 0$, form the equations whose roots are

75. The opposites of the roots of the given equations.

76. The reciprocals of the roots of the given equations.

77. Twice the roots of the given equations.

78. n times the roots of the given equations.

79. One-third of the roots of the given equations.

80. One-nth of the roots of the given equations.

81. The roots of the given equations increased by 2.

82. The roots of the given equations diminished by 5.

83. The squares of the roots of the given equations.

84. The cubes of the roots of the given equations.

85. The product of the roots and the reciprocal of the product of the roots of the given equations.

86. The sum of the roots and the reciprocal of the sum of the roots of the given equations.

Without solving the following equations, determine the nature of the roots of each one:

87. $x^2 + 17x + 70 = 0.$

88. $x^2 - 6x = 27.$

89. $x^2 + 12x = -40.$

90. $x^2 - 6x + 9 = 0.$

91. $x^2 + 5x - 14 = 0.$

92. $x^2 + 20x = -100.$

93. $x^2 - x = 12.$

94. $x^2 - 8x + 25 = 0.$

95. $x^2 - 13x + 22 = 0.$

96. $x^2 - 8x = 16.$

97. $4x^2 - 12x = -9.$

98. $9x^2 + 12x = -5.$

99. $9x^2 - 12x + 4 = 0.$

100. $8x^2 - 2x - 25 = 0.$

101. $9x^2 - 6x = -82.$

102. $12x^2 + 7x = 12.$

103. $16x^2 + 8x + 49 = 0.$

104. $10x^2 - 21x - 10 = 0.$

105. $6x^2 - 5x = -1.$

106. $16x^2 + 24x = -9.$

107. $16x^2 + 40x + 25 = 0.$

108. $25x^2 + 4x - 77 = 0.$

109. $20x^2 + 19x = -3.$

110. $4x^2 + 52x = 87.$

111. $25x^2 + 80x - 64 = 0.$

112. $16x^2 - 24x + 13 = 0.$

For what values of m are the roots of each of the following equations equal? For what values of m are the roots irrational? And for what values of m are the roots complex numbers?

113. $mx^2 + 4x + 1 = 0.$

114. $2x^2 + mx + 1 = 0.$

115. $3x^2 + 6x + m = 0.$

116. $mx^2 + mx + 1 = 0.$

Problems.

21. Pr. **1.** The sum of two numbers is 15 and their product is 56; what are the numbers?

Let x stand for one of the numbers; then, by the first condition, $15 - x$ stands for the other number. By the second condition

$$x(15 - x) = 56.$$

The roots of this equation are 7 and 8.

Therefore $x = 7$, one of the numbers, and $15 - x = 8$, the other number.

Observe that if we take $x = 8$, then $15 - x = 7$. That is, the two required numbers are the two roots of the quadratic equation.

Pr. **2.** A number is composed of two digits whose product is 30; if the digits be interchanged, the resulting number will exceed the original number by 9. What is the number?

Let x stand for the digit in the tens' place.

Then, by the first condition, $\frac{30}{x}$ stands for the digit in the units' place.

The required number is $10x + \frac{30}{x}$, and the number obtained by interchanging the digits is $10 \times \frac{30}{x} + x$.

Finally, by the second condition,

$$\left(10x + \frac{30}{x}\right) + 9 = 10 \times \frac{30}{x} + x.$$

The roots of this equation are 5 and -6. The problem therefore admits of two solutions.

Taking 5 for the digit in the tens' place, we have $\frac{30}{5} = 6$, the digit in the units' place. The corresponding number is 56.

Taking -6 for the digit in the tens' place, we have $\frac{30}{-6} = -5$, the digit in the units' place. The corresponding number therefore is -65.

Pr. 3. The sum of the ages of a father and son is 65 years. The product of the number of years in the father's age by the number of years in the son's age exceeds 600 by as much as five times the number of years in the father's age exceeds 100. What are the ages of father and son?

Let x stand for the number of years in the father's age; then $65 - x$ stands for the number of years in the son's age.

By the second condition

$$x(65 - x) - 600 = 5x - 100.$$

The roots of this equation are 50 and 10.

Only the solution 50 is consistent with the implied condition that a father must be older than his son.

Therefore $\qquad x = 50,$ the father's age;

and $\qquad 65 - x = 15,$ the son's age.

Had the problem referred to the ages of two persons, with no implied condition as to which one is the older, both solutions would have been admissible.

Pr. 4. Divide 100 into two parts whose product is 2600.

Let x stand for the less part; then $100 - x$ stands for the greater part.

By the second condition,

$$x(100 - x) = 2600.$$

The roots of this equation are $50 + 10\sqrt{-1}$ and $50 - 10\sqrt{-1}$.

An imaginary result always indicates inconsistent conditions in the problem. The inconsistency of these conditions may be shown as follows:

Let d stand for the difference between the two parts of 100. Then $50 + \frac{1}{2}d$ stands for the greater part, and $50 - \frac{1}{2}d$ stands for the less part.

The product of the two parts is

$$(50 + \tfrac{1}{2}d)(50 - \tfrac{1}{2}d), = 2500 - (\tfrac{1}{2}d)^2 = 2500 - \tfrac{1}{4}d^2.$$

Since d^2 is always positive for all *real* values of d, the product $2500 - \frac{1}{4}d^2$ must be less than 2500. Consequently 100 cannot be divided into two parts whose product is greater than 2500.

Pr. 5. Two men, A and B, start from P and Q respectively, 21 miles apart, and walk A to Q and B to P. They meet after walking 3 hours, and A, the faster walker, reaches Q $1\frac{3}{4}$ hours before B reaches P. At what point do they meet, and what are their respective rates?

Let x stand for number of miles A walks until they meet, and therefore $21-x$ for number of miles B walks.

Then $\frac{x}{3}$ is the number of miles A walks in one hour, and $\frac{21-x}{3}$ is the number of miles B walks in one hour.

Evidently x is the number of miles B walks after they meet, and $21-x$ the number of miles A walks after they meet. Consequently, since the number of miles divided by rate gives time, $\frac{x}{\frac{21-x}{3}}$ is the number of hours it takes B to walk to P after they meet, and $\frac{21-x}{\frac{x}{3}}$ is the number of hours it takes A to walk to Q.

We then have

$$\frac{21-x}{\frac{x}{3}}+\frac{7}{4}=\frac{x}{\frac{21-x}{3}}.$$

The roots of this equation are 12 and -63.

Taking the first root, we have:

distance A walks from P to place of meeting, M say, is 12 miles (Fig. 13);

distance B walks from Q to place of meeting is $21-x$, $=9$ miles.

The rate at which A walks is $\frac{12}{3}$, $=4$ miles an hour;

the rate at which B walks is $\frac{9}{3}$, $=3$ miles an hour.

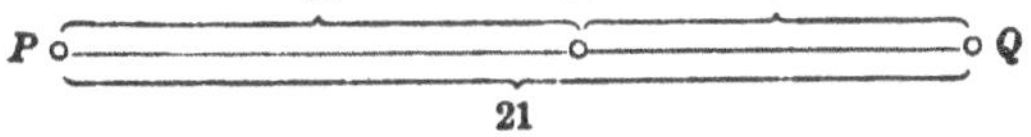

Fig. 13.

The second solution is not consistent apparently with the wording of the problem, yet it admits of a perfectly intelligible interpretation. For, if we assume that a negative distance is measured in an opposite direction to that which is consistent with a positive result, we have:

distance A walks from P to place of meeting, M, is -63 miles;
distance B walks from Q to place of meeting is $21-x$, $=+84$ miles.

This means that while B walks from Q toward and beyond P, a distance of 84 miles, A walks from P, not toward Q, as the problem states and as the first solution gives, but away from Q, a distance of 63 miles.

The following figure represents the state of affairs:

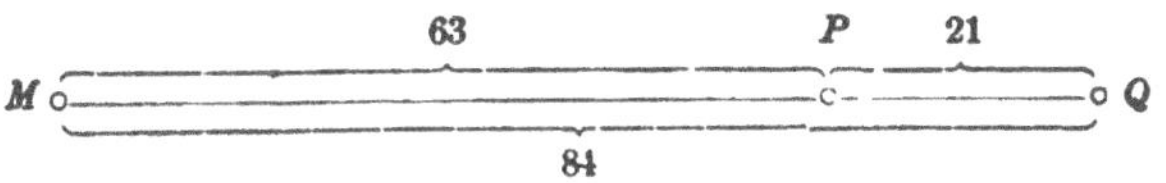

FIG. 14.

Now B's rate is $\frac{21-x}{3}$, 28 miles an hour; and A's rate is $\frac{x}{3}$, $=-21$ miles an hour; *i.e.*, 21 miles an hour in a direction away from Q.

Observe that the positive direction is not the same for A and B.

22. When the solution of a problem leads to a quadratic equation, it is necessary to determine whether either or both of the roots of the equation satisfy the conditions expressed and implied in the problem.

Positive results, in general, satisfy all the conditions of the problem.

A *negative result*, as a rule, satisfies the conditions of the problem, when they refer to abstract numbers. When the required numbers refer to quantities which can be understood in opposite senses, as opposite directions, gain and loss, etc., an intelligible meaning can usually be given to a negative result.

An *imaginary result* always implies inconsistent conditions.

23. The interpretation of a negative result is often facilitated by the following principle:

If a given quadratic equation have a negative root, then the equation obtained from the given one by changing the sign of x has a positive root of the same absolute value.

Let $-r$ be a root of

$$ax^2 + bx + c = 0. \qquad (1)$$

Then, since $-r$ must satisfy the equation, we have

$$a(-r)^2 + b(-r) + c = 0,$$

or

$$ar^2 - br + c = 0. \qquad (2)$$

But equation (2) shows that r satisfies the equation

$$ax^2 - bx + c = 0,$$

which is obtained from (1) by changing the sign of x.

Pr. A man bought muslin for $3.00. If he had bought three yards more for the same money, each yard would have cost him 5 cents less. How many yards did he buy?

Let x stand for the number of yards the man bought. Then 1 yard cost $\frac{300}{x}$ cents.

If he had bought $x+3$ yards for the same money, each yard would have cost $\frac{300}{x+3}$ cents.

Therefore

$$\frac{300}{x} - \frac{300}{x+3} = 5;$$

whence $x = 12$ and -15.

The root 12 satisfies the equation and also the conditions of the problem; the root -15 has no meaning.

But if x be replaced by $-x$ in the equation, we obtain a new equation

$$\frac{300}{-x} - \frac{300}{-x+3} = 5,$$

or

$$\frac{300}{x-3} - \frac{300}{x} = 5, \qquad (2)$$

whose roots are -12 and $+15$.

Equation (2) evidently corresponds to the problem: A man bought muslin for \$3.00. If he had bought 3 yards less for the same money, each yard would have cost him 5 cents more.

Notice, however, that the intelligible result, 12, of the first statement has become -12 and is meaningless in the second statement.

Attention is called to the remarks in Ch. XII., Arts. 6 and 8.

24. The following problem illustrates the interpretation of all kinds of roots, except imaginary.

Pr. Find a point on a line joining two sources of light, A and B, which is equally illuminated by both.

The further an object is removed from a source of light, the feebler is the illumination of the object. It has been found by experiment that when the object is removed to a point 2, 3, 4, ... times its original distance from the source of light, its illumination is 2^2, 3^2, 4^2, ... times as feeble. That is, if the object be twice its original distance from the source of light, its illumination is one-fourth of its original illumination.

The position of the point will evidently depend upon the intensities of the two sources of light, A and B.

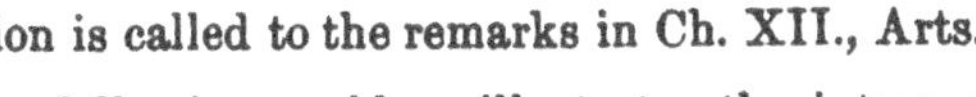

Fig. 15.

Let the distance between A and B be a units;
and let m stand for the intensity of the source of light A,
and n for the intensity of the source of light B.

Let us assume that the point of equal illumination is between A and B, at C say, a distance x units from A, and $a-x$ units from B.

Then the intensity of the illumination at C, due to the source of light A, is $\frac{m}{x^2}$; that due to the source of light B is $\frac{n}{(a-x)^2}$.

Consequently, by the condition of equal illumination, we have

$$\frac{m}{x^2}=\frac{n}{(a-x)^2}.$$

The roots of this equation are

$$\frac{\sqrt{m}}{\sqrt{m}+\sqrt{n}}a \text{ and } \frac{\sqrt{m}}{\sqrt{m}-\sqrt{n}}a.$$

Since m and n are positive, both roots are real.

(i.) If $m > n$, *i.e.*, if the light at A be more intense than that at B, both roots are positive.

There are therefore two equally illuminated points, distant $\frac{\sqrt{m}}{\sqrt{m}+\sqrt{n}}a$ and $\frac{\sqrt{m}}{\sqrt{m}-\sqrt{n}}a$ units, respectively, from A. But since $\frac{\sqrt{m}}{\sqrt{m}+\sqrt{n}}$ is a proper fraction, it follows that

$$\frac{\sqrt{m}}{\sqrt{m}+\sqrt{n}}a < a;$$

therefore the first point of equal illumination lies between A and B according to the hypothesis. Moreover, since $m > n$, $\sqrt{m}+\sqrt{n} < 2\sqrt{m}$.

Consequently $\frac{\sqrt{m}}{\sqrt{m}+\sqrt{n}}a > \frac{\sqrt{m}}{2\sqrt{m}}a, = \frac{1}{2}a.$

Therefore the first point of equal illumination is nearer to B than to A, at C_1 (Fig. 15).

Again, since $\sqrt{m}-\sqrt{n} < \sqrt{m}$, therefore $\frac{\sqrt{m}}{\sqrt{m}-\sqrt{n}} > 1$, and $\frac{\sqrt{m}}{\sqrt{m}-\sqrt{n}}a > a$, that is, the distance of the second point of equal illumination from A is greater than the distance between the two points A and B.

This result contradicts the hypothesis that the point of equal illumination lies between A and B. But it is consistent with the wording of the problem, which does not require that the point of equal illumination shall lie between A and B.

The second point of equal illumination, C_2 say, is further from A than B is from A, and since it must be nearer the feebler light B, it must be to the right of both. Notice that the less the difference between the intensities of the two lights, the further this *point of equal illumination is removed.*

(ii.) If $m < n$, *i.e.*, if the light at B be more intense than that at A, we have:

$$\frac{\sqrt{m}}{\sqrt{m}+\sqrt{n}}a \text{ is positive, and } \frac{\sqrt{m}}{\sqrt{m}-\sqrt{n}}a \text{ is negative.}$$

Since $\frac{\sqrt{m}}{\sqrt{m}+\sqrt{n}}$ is a proper fraction, we have

$$\frac{\sqrt{m}}{\sqrt{m}+\sqrt{n}}a < a,$$

and the corresponding point of equal illumination must lie between A and B, according to the hypothesis. Moreover, since $m < n$, $\sqrt{m}+\sqrt{n} > 2\sqrt{m}$;

consequently $$\frac{\sqrt{m}}{\sqrt{m}+\sqrt{n}}a < \frac{\sqrt{m}}{2\sqrt{m}}a, = \tfrac{1}{2}a.$$

Therefore this point of equal illumination is nearer to A than it is to B, at C_3 (Fig. 15).

The negative result $\frac{\sqrt{m}}{\sqrt{m}-\sqrt{n}}a$ can also be shown to be consistent with the wording of the problem. The distance $a-x$ of the point of equal illumination from B is in this case

$$a - \frac{\sqrt{m}}{\sqrt{m}-\sqrt{n}}a, = \frac{-\sqrt{n}}{\sqrt{m}-\sqrt{n}}a.$$

Moreover, since $m < n$, the distance $\frac{\sqrt{m}}{\sqrt{m}-\sqrt{n}}a$ of this point of equal illumination, C_4 say, from A is numerically less than its distance $\frac{-\sqrt{n}}{\sqrt{m}-\sqrt{n}}a$ from B; and since the absolute value of $\frac{-\sqrt{n}}{\sqrt{m}-\sqrt{n}}$ is greater than 1, the absolute value of $\frac{-\sqrt{n}}{\sqrt{m}-\sqrt{n}}a$ is greater than a. Consequently, this point of equal illumination must lie to the left of A.

Notice again that the less the difference between the intensities of the two lights, the further this point of equal illumination is removed.

(iii.) If $a = 0$ and $m \neq n$, then

$$\frac{\sqrt{m}}{\sqrt{m}+\sqrt{n}}a = 0 \text{ and } \frac{\sqrt{m}}{\sqrt{m}-\sqrt{n}}a = 0.$$

Both results show that if two lights of unequal intensity be placed at the same point, that point is the only point of equal illumination.

(iv.) If $a = 0$ and $m = n$, the two roots are

$$\frac{a\sqrt{m}}{\sqrt{m}+\sqrt{n}} = 0 \text{ and } \frac{a\sqrt{m}}{\sqrt{m}-\sqrt{n}} = \frac{0}{0}.$$

The first result shows that if two lights of equal intensity are placed at the same point, this point is a point of equal illumination.

The second result shows that a point at any distance from the two lights is equally illuminated by both of them.

(v.) If $m = n$, *i.e.*, if the lights be of equal intensity, the two roots are

$$\frac{a\sqrt{m}}{\sqrt{m}+\sqrt{m}} = \frac{a}{2} \text{ and } \frac{a\sqrt{m}}{\sqrt{m}-\sqrt{m}}, = \frac{a\sqrt{m}}{0}, = \infty.$$

The first root shows that, if the two lights be of equal intensity, a point of equal illumination is midway between them. This result seems self-evident.

The second root shows that as the intensity of the two lights approaches equality, *i.e.*, as their difference approaches zero, the second point of equal illumination is further and further removed.

EXERCISES VII.

1. If 1 be added to the square of a number, the sum will be 50. What is the number?

2. If 5 be subtracted from a number, and 1 be added to the square of the remainder, the sum will be 10. What is the number?

3. One of two numbers exceeds 50 by as much as the other is less than 50, and their product is 2400. What are the numbers?

4. The product of two consecutive integers exceeds the smaller by 17,424. What are the numbers?

5. If 27 be divided by a certain number, and the same number be divided by 3, the results will be equal. What is the number?

6. What number, added to its reciprocal, gives 2.9?

7. What number, subtracted from its reciprocal, gives n? Let $n = 6.09$.

8. If n be divided by a certain number, the result will be the same as if the number were subtracted from n. What is the number? Let $n = 4$.

9. If the product of two numbers be 176, and their difference be 5, what are the numbers?

10. If a certain number be divided by 8, the result will be the same as if 16 were divided by the number and $3\frac{1}{2}$ were added to the quotient. What is the number?

11. A certain number was to be added to $\frac{1}{2}$, but by mistake $\frac{1}{2}$ was divided by the number. Nevertheless the correct result was obtained. What was the number?

12. The sum of the two digits of a number is 9. If the digits be interchanged and the original number be divided by the resulting number, the quotient will be one-fourth of the latter. What is the number?

13. If 100 marbles be so divided among a certain number of boys that each boy shall receive four times as many marbles as there are boys, how many boys are there?

14. The area of a rectangle, one of whose sides is 7 inches longer than the other, is 494 square inches. How long is each side?

15. The difference between the squares of two consecutive numbers is equal to three times the square of the less number. What are the numbers?

16. A merchant received $48 for a number of yards of cloth. If the number of dollars a yard be equal to three-sixteenths of the number of yards, how many yards did he sell?

17. In a company of 14 persons, men and women, the men spent $24 and the women $24. If each man spent $1 more than each woman, how many men and how many women were in the company?

18. A pupil was to add a certain number to 4, then to subtract the same number from 9, and finally to multiply the results. But he added the number to 9, then subtracted 4 from the number, and multiplied these results. Nevertheless he obtained the correct product. What was the number?

19. If the first factor of the product 21×84 be increased by a certain number, and the second factor be diminished by the same number, the result will be the same as if the first factor were diminished by this number, the second factor increased by the same number, and the sign of the product reversed. What is the number?

20. A man paid \$80 for wine. If he had received 4 gallons less for the same money, he would have paid \$1 more a gallon. How many gallons did he buy?

21. A man left \$31,500 to be divided equally among his children. But since 3 of the children died, each remaining child received \$3375 more. How many children survived?

22. Two bodies move from the vertex of a right angle along its sides at the rate of 12 feet and 16 feet a second respectively. After how many seconds will they be 90 feet apart?

23. A tank can be filled by two pipes, by the one in two hours less than by the other. If both pipes be open $1\frac{7}{8}$ hours, the tank will be filled. How long does it take each pipe to fill the tank?

24. From a thread, whose length is equal to the perimeter of a square, 36 inches are cut off, and the remainder is equal in length to the perimeter of another square whose area is four-ninths of that of the first. What is the length of the thread?

25. A number of coins can be arranged in a square, each side containing 51 coins. If the same number of coins be arranged in two squares, the side of one square will contain 21 more coins than the side of the other. How many coins does the side of each of the latter squares contain?

26. A farmer wished to receive \$2.88 for a certain number of eggs. But he broke 6 eggs, and in order to receive the desired amount he increased the price of the remaining eggs by $2\frac{2}{5}$ cents a dozen. How many eggs had he originally?

27. A man paid \$33 for a number of bottles of wine of one kind, and \$27 for a number of bottles of another kind, buying altogether 33 bottles. Each bottle of the first kind cost 70 cents more than each bottle of the second kind. How many bottles of each kind did he buy?

28. Two bodies move toward each other from A and B respectively, and meet after 35 seconds. If it takes the one 24 seconds longer than the other to move from A to B, how long does it take each one to move that distance?

29. It takes a boat's crew 4 hours and 12 minutes to row 12 miles down a river with the current, and back again against the current. If the speed of the current be 3 miles an hour, at what rate can the crew row in still water?

30. A man paid \$300 for a drove of sheep. By selling all but 10 of them at a profit of \$2.50 each, he received the amount he paid for all the sheep. How many sheep did he buy?

31. A manufacturer paid \$640 to 36 employés, men and women. Each man received as many dollars as there were women, and each woman received as many dollars as there were men. How many women and how many men were there?

32. Two men start at the same time to go from A to B, a distance of 36 miles. One goes 3 miles more an hour than the other, and arrives at B 1 hour earlier. At what rate does each man travel?

33. It took a number of men as many days to dig a ditch as there were men. If there had been 6 more men, the work would have been done in 8 days. How many men were there?

34. The front wheel of a carriage makes 6 revolutions more than the hind wheel in running 36 yards; if the circumference of each wheel were 1 yard longer, the front wheel would make but 3 revolutions more than the hind wheel in running the same distance. What is the circumference of each wheel?

35. A and B receive different wages. A receives \$ 48 for working a certain number of days, and B receives \$ 27 for working 6 days less. If A had worked as many days as B, and B as many days as A, they would have received equal amounts. How many days does each work?

36. Two men formed a partnership with a joint capital of \$ 500. The first left his money in the business 5 months, and the second his money 2 months. Each realized \$ 450, including invested capital. How much did each invest?

37. Two trains run toward each other from A and B respectively, and meet at a point which is 15 miles further from A than it is from B. After the trains meet, it takes the first train $2\frac{2}{3}$ hours to run to B, and the second train $3\frac{3}{8}$ hours to run to A. How far is A from B?

38. The perimeter of a rectangular lawn having around it a path of uniform width is 420 feet. The area of the lawn and path together exceeds twice the difference of their areas by 1200 square yards, and the width of the path is one-sixth of the shorter side of the lawn. Find the dimensions of lawn and path.

39. Water enters a forty-gallon cask through one pipe and is discharged through another. In 4 minutes one gallon more is discharged through the second pipe than enters through the first. The first pipe can fill the cask in 3 minutes less time than it takes the second to discharge 66 gallons. How long does it take the first pipe to fill the cask?

40. In a rectangle, whose sides are a and b inches respectively, a second rectangle is constructed. The sides of the inner rectangle are equally distant from the sides of the outer, and the area of the inner rectangle is one-nth of the remaining part of the outer. What are the lengths of the sides of the inner rectangle?

Let $a = 70$, $b = 52\frac{1}{2}$, $n = 1$.

41. A lamp and a candle are 4 feet apart. At what point on the straight line joining the two will the illumination from the candle be equal to that from the lamp, if the light of the lamp be 9 times as intense as that of the candle?

CHAPTER XXI.

EQUATIONS OF A HIGHER DEGREE THAN THE SECOND.

1. The solution of equations of higher degree than the second is, in general, beyond the scope of this book. We shall consider in this chapter a few higher equations which can be solved by means of quadratic equations.

2. A **Binomial Equation** is an equation of the form $x^n = a$, wherein n is a positive integer.

Certain binomial equations can be factored into linear and quadratic factors or factors which can be brought to quadratic form by proper substitutions.

Ex. **1.** Solve the equation $\quad x^3 - 1 = 0. \qquad (1)$

Factoring, $\quad (x-1)(x^2+x+1) = 0. \qquad (2)$

This equation is equivalent to the two equations

$$x - 1 = 0, \text{ whence } x = 1;$$

and $$x^2 + x + 1 = 0, \text{ whence } x = -\tfrac{1}{2} \pm \tfrac{1}{2}\sqrt{-3}.$$

Notice that this example gives the three cube roots of 1, since $x^3 - 1 = 0$ is equivalent to

$$x^3 = 1, \text{ or } x = \sqrt[3]{1}.$$

Therefore the three cube roots of 1 are

$$1,\ -\tfrac{1}{2} + \tfrac{1}{2}\sqrt{-3},\ -\tfrac{1}{2} - \tfrac{1}{2}\sqrt{-3}.$$

Compare Ch. XIX., Art. 26.

In general, the three cube roots of any number can be found by multiplying the principal cube root of the number in turn by the three algebraic cube roots of 1.

E.g., $$\sqrt[3]{8} = 2\sqrt[3]{1} = 2,\ -1 \pm \sqrt{-3};$$

the three cube roots of a are $\sqrt[3]{a}$, $\sqrt[3]{a}(-\tfrac{1}{2} \pm \tfrac{1}{2}\sqrt{-3})$, wherein $\sqrt[3]{a}$ denotes the principal cube root of a.

Ex. **2.** Solve the equation $x^4 + 1 = 0$.

Factoring, $\quad (x^2 + 1 + x\sqrt{2})(x^2 + 1 - x\sqrt{2}) = 0.$

This equation is equivalent to the two equations

$$x^2 + 1 + x\sqrt{2} = 0, \text{ whence } x = \tfrac{1}{2}\sqrt{2}(-1 \pm \sqrt{-1});$$

and $$x^2 + 1 - x\sqrt{2} = 0, \text{ whence } x = \tfrac{1}{2}\sqrt{2}(1 \pm \sqrt{-1}).$$

Since the given equation is equivalent to

$$x^4 = -1, \text{ or } x = \sqrt[4]{-1},$$

we conclude that the four fourth roots of -1 are

$$\tfrac{1}{2}\sqrt{2}(-1 \pm \sqrt{-1}),\ \tfrac{1}{2}\sqrt{2}(1 \pm \sqrt{-1}).$$

Compare Ch. XIX., Exercises II., Exx. 38 and 39.

The four fourth roots of any negative number can be found by multiplying the principal fourth root of the radicand *taken positively* in turn by the four fourth roots of -1.

E.g., $\sqrt[4]{-16} = 2\sqrt[4]{-1} = \sqrt{2}(-1 \pm \sqrt{-1}),\ \sqrt{2}(1 \pm \sqrt{-1})$.

Ex. 3. Solve the equation $x^5 - 2 = 0$.

The roots of this equation can be obtained by multiplying the principal fifth root of 2 by the five (as will appear) fifth roots of 1.

We have therefore to solve $x^5 - 1 = 0$.

Factoring, $$(x - 1)(x^4 + x^3 + x^2 + x + 1) = 0. \quad (1)$$

This equation is equivalent to the two equations

$$x - 1 = 0, \text{ whence } x = 1;$$

and $$x^4 + x^3 + x^2 + x + 1 = 0.$$

The last equation belongs to a class which will be considered in Art. 4, to which the student is referred for its solution. The roots are found to be

$$-\tfrac{1}{4}[1 - \sqrt{5} \pm \sqrt{(-10 - 2\sqrt{5})}],\ -\tfrac{1}{4}[1 + \sqrt{5} \pm \sqrt{(-10 + 2\sqrt{5})}].$$

Consequently the five fifth roots of 2 are $\sqrt[5]{2}$,

$$-\tfrac{1}{4}\sqrt[5]{2}[1 - \sqrt{5} \pm \sqrt{(-10-2\sqrt{5})}],\ -\tfrac{1}{4}\sqrt[5]{2}[1 + \sqrt{5} \pm \sqrt{(-10+2\sqrt{5})}],$$

wherein $\sqrt[5]{2}$ denotes the principal fifth root of 2.

3. Ex. 1. Solve the equation $x^4 - 9 = 2x^2 - 1$.

Since $x^4 = (x^2)^2$, we may take x^2 as the unknown number and solve this equation as a quadratic in x^2.

We then have $$(x^2)^2 - 2x^2 - 8, = 0.$$

Factoring, $$(x^2 - 4)(x^2 + 2) = 0.$$

Whence $$x^2 - 4 = 0, \text{ or } x = \pm 2;$$

and $$x^2 + 2 = 0, \text{ or } x = \pm\sqrt{-2}.$$

In general, any equation containing only two powers of the unknown number, *one of which is the square of the other*, can be solved as a quadratic equation.

Ex. 2. Solve the equation $x^6 - 3x^3 = 40$.

Since $x^6 = (x^3)^2$, we take x^3 as the unknown number.

We then have $(x^3)^2 - 3x^3 = 40$.

Solving this equation for x^3, we obtain

$$x^3 = 8, \text{ whence } x = \sqrt[3]{8};$$

and

$$x^3 = -5, \text{ whence } x = -\sqrt[3]{5}.$$

Therefore, by Art. 2, Ex. 1, the six roots of the given equation are

$$2,\ -1 \pm \sqrt{-3},\ -\sqrt[3]{5},\ \tfrac{1}{2}\sqrt[3]{5}(1 \mp \sqrt{-3}),$$

wherein $\sqrt[3]{5}$ denotes the principal cube root of 5.

In like manner any equation containing two powers of a quadratic expression, one of which is the square of the other, can be solved as a quadratic equation.

Ex. 3. Solve the equation $(x^2 - 3x + 1)^2 = 6 + 5(x^2 - 3x + 1)$.

In this example $x^2 - 3x + 1$ is regarded as the unknown number, and may temporarily be represented by the letter y. The equation then becomes

$$y^2 = 6 + 5y.$$

The roots of this equation are 6 and -1.

In all cases, after having solved the quadratic equation in the assumed unknown number, it is necessary to solve the two equations obtained by equating the expression which was regarded as the unknown number to each of the two roots of the quadratic equation.

We therefore have the two equations

$$x^2 - 3x + 1 = 6, \text{ whence } x = \tfrac{3}{2} \pm \tfrac{1}{2}\sqrt{29};$$

$$x^2 - 3x + 1 = -1, \text{ whence } x = 2,\ x = 1.$$

Therefore the roots of the given equation are

$$\tfrac{3}{2} \pm \tfrac{1}{2}\sqrt{29},\ 2,\ 1.$$

The attention of the student is called to the fact that, in each example, we have obtained as many roots as there are units in the degree of the equation.

Frequently equations which do not at first appear to come under this case can, by a proper arrangement of terms, be made to do so.

Ex. 4. Solve the equation $x^4 + 2x^3 - 7x^2 - 8x + 12 = 0$.

The given equation can be written

$$x^4 + 2x^3 + x^2 - 8x^2 - 8x + 12 = 0,$$

or
$$(x^2 + x)^2 - 8(x^2 + x) + 12 = 0.$$

If we now let $x^2 + x = y$, we have

$$y^2 - 8y + 12 = 0.$$

The roots of this equation are 2, 6.

We then have to solve the equations

$$x^2 + x = 2, \qquad (1)$$

and
$$x^2 + x = 6. \qquad (2)$$

The roots of (1) are 1, -2; and the roots of (2) are 2, -3.

4. Ex. Solve the equation $x^4 + x^3 + x^2 + x + 1 = 0$. (1)

Dividing by x^2, $$x^2 + x + 1 + \frac{1}{x} + \frac{1}{x^2} = 0. \qquad (2)$$

Rearranging terms, $$x^2 + 1 + \frac{1}{x^2} + x + \frac{1}{x} = 0. \qquad (3)$$

Since $\left(x + \frac{1}{x}\right)^2 = x^2 + 2 + \frac{1}{x^2}$, if 1 be added to the second term of the first member, the first member can be arranged as a quadratic expression in $x + \frac{1}{x}$.

We then have $$\left(x + \frac{1}{x}\right)^2 + \left(x + \frac{1}{x}\right) = 1. \qquad (4)$$

From equation (4) we obtain

$$x + \frac{1}{x} = -\frac{1}{2} \pm \frac{1}{2}\sqrt{5}. \qquad (5)$$

We now have to solve the two equations

$$x + \frac{1}{x} = -\frac{1}{2} + \frac{1}{2}\sqrt{5}, \qquad (6)$$

and
$$x + \frac{1}{x} = -\frac{1}{2} - \frac{1}{2}\sqrt{5}. \qquad (7)$$

The work of solving these equations can be simplified by representing $-\frac{1}{2} + \frac{1}{2}\sqrt{5}$ by m, and $-\frac{1}{2} - \frac{1}{2}\sqrt{5}$ by n.

From (6) we then have

$$x + \frac{1}{x} = m, \qquad (8)$$

or
$$x^2 - mx = -1. \qquad (9)$$

The roots of (9) are

$$x = \frac{m}{2} \pm \frac{1}{2}\sqrt{(-4 + m^2)}.$$

If we now substitute $-\frac{1}{2}+\frac{1}{2}\sqrt{5}$ for m in these values of x, we obtain

$$\begin{aligned} x &= -\tfrac{1}{4}+\tfrac{1}{4}\sqrt{5} \pm \tfrac{1}{2}\sqrt{[-4+(-\tfrac{1}{2}+\tfrac{1}{2}\sqrt{5})^2]} \\ &= -\tfrac{1}{4}[1-\sqrt{5} \mp \sqrt{(-10-2\sqrt{5})}], \text{ the roots of equation (6).} \end{aligned}$$

The solution of equation (7) is left as an exercise for the student. The roots of (7) are

$$x = -\tfrac{1}{4}[1+\sqrt{5} \mp \sqrt{(-10+2\sqrt{5})}].$$

Observe that an equation can be solved by this method only when its degree is even and the coefficients of terms equally distant from the beginning and end of the expression in the first member (the second member being 0) are equal.

5. Ex. Solve the equation $\dfrac{x^2+x+1}{x^2-x+2}+\dfrac{x^2-x+2}{x^2+x+1}=2\tfrac{1}{6}$.

If we let $\dfrac{x^2+x+1}{x^2-x+2}=y$, the given equation becomes $y+\dfrac{1}{y}=2\tfrac{1}{6}$.

The roots of this equation are $\tfrac{3}{2}$, $\tfrac{2}{3}$.

We now have to solve the two equations

$$\frac{x^2+x+1}{x^2-x+2}=\frac{3}{2}, \qquad (1)$$

and

$$\frac{x^2+x+1}{x^2-x+2}=\frac{2}{3}. \qquad (2)$$

The roots of (1) are found to be 4, 1; and the roots of (2) are found to be $-\frac{5}{2}\pm\frac{1}{2}\sqrt{29}$.

An equation can be solved by this method when it contains only two expressions in the unknown number, one of which is the reciprocal of the other, and when the numerators and denominators of these expressions are of degree not higher than the second.

EXERCISES.

Solve each of the following equations:

1. $x^3+1=0$.

2. $(x-1)^3=8$.

3. $(x+2)^3+4=0$.

4. $(x+1)^3=(3-x)^3$.

5. $x^3=(2a-x)^3$.

6. $x^4-1=0$.

7. $(x+1)^4=16$.

8. $x^4+625(x+1)^4=0$.

9. $x^5+1=0$.

10. $x^5=(1-x)^5$.

11. $x^6+1=0$.

12. $x^6-1=0$.

13. $x^6=64(1-x)^6$.

14. $x^8-256=0$.

15. $x^4+9=10x^2$.

16. $x^4-5x^2+4=0$.

17. $x^4-6x^2=-1$.

18. $x^4-20x^2=-16$.

19. $(x^2-9)(x^2-16)=15x^2$.

20. $(x^2-10)(x^2-18)=13x^2$.

21. $(x^2-5)^2+(x^2-8)^2=17.$ **22.** $6x^6-12x^3=288.$

23. $x^6-65x^3=-64.$ **24.** $x^8+5x^4=6.$

25. $x^{10}+3x^5=32.$ **26.** $(x^2+5)^2-4x^2=160.$

27. $(x-2)^6-19(x-2)^3=216.$ **28.** $a^2+x^4=2a(n^2+x^2)-n^2(n^2-2x^2).$

29. $\dfrac{(a+x)^4+(a-x)^4}{(a+x)^3+(a-x)^3}=2a.$ **30.** $\dfrac{(3+x)^3+(3-x)^3}{(3+x)^4+(3-x)^4}=\dfrac{3}{9-x^2}.$

31. $\dfrac{(2+x)^5+(2-x)^5}{(2+x)^4+(2-x)^4}=2.$ **32.** $\dfrac{(a+x)^5+(a-x)^5}{(a+x)^4+(a-x)^4}=2a.$

33. $\dfrac{x^4+10x^2+1}{x^4-10x^2+1}=\dfrac{a}{a-1}.$ **34.** $\dfrac{x^4+6x^2+1}{x^4-6x^2+1}=\dfrac{3}{2}.$

35. $(3x^2-5x+1)^2-9x^2+15x=7.$

36. $15x^2-35x-3(7x-3x^2+8)^2+310=0.$

37. $1-2(5x^2+3x+2)^2=6x+10x^2-215.$

38. $\left(x+\dfrac{8}{x}\right)^2+x=42-\dfrac{8}{x}.$ **39.** $\dfrac{x^2-a^2}{x^2+a^2}+\dfrac{x^2+a^2}{x^2-a^2}=\dfrac{34}{15}.$

40. $\dfrac{x^2-5x+3}{x^2+5x-3}-\dfrac{x^2+5x-3}{x^2-5x+3}=\dfrac{8}{3}.$

41. $\dfrac{4x^2-x+1}{x^2-4x-1}+\dfrac{3x^2-12x-3}{4x^2-x+1}+4=0.$

42. $x^3+x^2-x-1=0.$ **43.** $x^3-2x^2+2x-1=0.$

44. $3x^3-13x^2+13x-3=0.$ **45.** $4x^3-21x^2+21x-4=0.$

46. $x^4-4x^3+3x^2+2x-6=0.$

47. $16x^4-32x^3+12x^2+4x-552=0.$

48. $16x^4-96x^3+236x^2-276x+120=0.$

49. $x^4-14x^3+71x^2-154x+120=0.$

50. $x^4-x^3-1\frac{3}{4}x^2+x+1=0.$

51. $x^4-4x^3+8x=165.$ **52.** $(x^2-x+1)^2=3x(x-1)+1.$

53. $(x^2-7x)(x^2-3x)+2(7x-6)(x-2)=10x.$

54. $x^4+(x-1)^4=97.$ **55.** $(3-x)^4+(2-x)^4=17.$

56. $\dfrac{x^4+(2-x)^4}{x^3+(2-x)^3}=2.$ **57.** $\dfrac{(3-x)^5+x^5}{(3-x)^2+x^2}=27.$

58. $x^4-2x^3+2x^2-2x+1=0.$ **59.** $2x^4-5x^3+4x^2-5x+2=0.$

60. $6x^4-35x^3+62x^2-35x+6=0.$ **61.** $2x^5+3x^4-5x^3-5x^2+3x+2=0.$

62. $\dfrac{(x+1)^4}{x^4+1}=8.$ **63.** $\dfrac{x^4+1}{(x-1)^4}=17.$

64. $\dfrac{(x+1)(x^3+1)}{(x-1)(x^3-1)}=\dfrac{28}{13}.$ **65.** $\dfrac{(x^2+1)(x^3+1)}{(x^2-1)(x^3-1)}=\dfrac{221}{189}.$

66. $\dfrac{(x+1)^5}{(x^2+1)(x^3+1)}=8.$

CHAPTER XXII.

IRRATIONAL EQUATIONS.

1. An **Irrational Equation** is an equation whose members are irrational in the unknown number or numbers; as, $\sqrt{(x+1)} = 3$.

Notice that we cannot speak of the *degree* of an irrational equation.

2. The solution of an irrational equation depends upon the following principle:

If both members of an equation be raised to the same positive integral power, the resulting equation will have as roots the roots of the given equation, and, in general, additional roots.

Let $$M = N$$
be the given equation.

Squaring both members, $$M^2 = N^2.$$

Whence $$M^2 - N^2 = 0,$$

or $$(M - N)(M + N) = 0.$$

This equation is equivalent to the two equations

$M - N = 0$, or $M = N$, the given equation;

and $M + N = 0$, or $M = -N$, an additional equation.

That is, the equation obtained by squaring both members of the given equation is equivalent to the given equation and an additional equation which differs from the given one in the sign of one of its members.

In like manner the principle can be proved for any positive integral powers of the members of the given equation.

E.g., if both members of the equation

$$x + 1 = 2$$

be squared, we have $$(x+1)^2 = 4.$$

The roots of this equation are 1 and -3.

The root 1 satisfies the given equation; the root -3 is a root of the equation

$$x + 1 = -2,$$

which was introduced by squaring, and does not satisfy the given equation.

3. To solve an irrational equation, we must first derive from it a rational, integral equation. This step, which is usually effected by raising both members of the equation to the same positive integral power one or more times, is called *rationalizing the equation.*

In the following examples the indicated roots will be limited to *principal* values:

Ex. 1. Solve the equation $x + \sqrt{(25 - x^2)} = 7$.

Before squaring, it is better to have the radical by itself in one member.

Transferring x, $\qquad \sqrt{(25 - x^2)} = 7 - x. \qquad (1)$

Squaring, $\qquad 25 - x^2 = 49 - 14x + x^2. \qquad (2)$

The roots of this equation are 3, 4.

Both roots of (2) satisfy the given equation, since $3 + \sqrt{(25 - 9)} = 7$, and $4 + \sqrt{(25 - 16)} = 7$.

Therefore no root was introduced by squaring both members of the given equation.

This is also evident from the following considerations:

Any root of the additional equation,

$$\sqrt{(25 - x^2)} = -(7 - x), \text{ or } -\sqrt{(25 - x^2)} = 7 - x, \qquad (3)$$

obtained by changing the sign of one of the members of the given equation when prepared for squaring, must be a root of the rational integral equation (2).

But both roots of this equation, 3 and 4, make the first member of (3) negative, and the second member positive.

That is, equation (3) is an impossible equation.

Ex. 2. Solve the equation $x - \sqrt{(25 - x^2)} = 1$.

Transferring x, $\qquad -\sqrt{(25 - x^2)} = 1 - x. \qquad (1)$

Squaring, $\qquad 25 - x^2 = 1 - 2x + x^2. \qquad (2)$

The roots of this equation are 4 and -3.

The number 4 is a root of the given equation, since

$$4 - \sqrt{(25 - 16)} = 1;$$

but the number -3 is not a root of the given equation, since $-3 - \sqrt{(25 - 9)} = -7$, not 1.

Therefore, the root -3 is a root of the additional equation

$$-\sqrt{(25 - x^2)} = -(1 - x), \text{ or } \sqrt{(25 - x^2)} = 1 - x,$$

introduced by squaring.

That -3 is not a root of the given equation is also evident from the form of the equation. For any real value of x which makes $x - \sqrt{(25 - x^2)}$ equal to 1 must be greater than 1, and therefore cannot be equal to -3.

Ex. 3. Solve the equation $\sqrt{(x+1)}+\sqrt{(2x+3)}=5$.

This equation can be rationalized more easily when only one radical is in each member.

Transferring $\sqrt{(x+1)}$, $\sqrt{(2x+3)}=5-\sqrt{(x+1)}$. (1)

Squaring, $2x+3=25-10\sqrt{(x+1)}+x+1$. (2)

Transferring and uniting terms, $x-23=-10\sqrt{(x+1)}$. (3)

Squaring, $x^2-46x+529=100x+100$. (4)

The roots of this equation are 143 and 3.

The number 3 is a root of the given equation, since

$$\sqrt{(3+1)}+\sqrt{(6+3)}=5.$$

But the number 143 is not a root, since

$$\sqrt{(143+1)}+\sqrt{(286+3)}=29, \text{ and not } 5.$$

Therefore the number 143 is a root of one of the additional equations,

$\sqrt{(2x+3)}=-[5-\sqrt{(x+1)}]$, or $-\sqrt{(2x+3)}=5-\sqrt{(x+1)}$, (5)

introduced by the first squaring, and

$$x-23=10\sqrt{(x+1)}, \quad (6)$$

introduced by the second squaring. We find 143 to be a root of (6).

That a root of the given equation, if *real*, must be less than 143 is also evident from the form of the equation.

Ex. 4. Solve the equation $\sqrt{(x-2)}-\sqrt{(6x-11)}+\sqrt{(x+3)}=0$.

Transferring $\sqrt{(x+3)}$, $\sqrt{(x-2)}-\sqrt{(6x-11)}=-\sqrt{(x+3)}$. (1)

Squaring, $x-2-2\sqrt{[(x-2)(6x-11)]}+6x-11=x+3$. (2)

Transferring and uniting terms, and dividing by -2,

$$\sqrt{(6x^2-23x+22)}=3x-8. \quad (3)$$

Squaring, $6x^2-23x+22=9x^2-48x+64$. (4)

The roots of this equation are 6 and $\frac{7}{3}$.

The number 6 is a root of the given equation, since

$$\sqrt{(6-2)}-\sqrt{(36-11)}+\sqrt{(6+3)}=0.$$

The number $\frac{7}{3}$ is not a root, since

$$\sqrt{(\tfrac{7}{3}-2)}-\sqrt{(14-11)}+\sqrt{(\tfrac{7}{3}+3)}=\tfrac{2}{3}\sqrt{3}, \text{ and not } 0.$$

Consequently the root $\frac{7}{3}$ was introduced at some stage of the work by squaring.

We find that $\frac{7}{3}$ is a root of the additional equation

$$-\sqrt{(6x^2-23x+22)}=3x-8,$$

introduced by the second squaring.

Ex. 5. Solve the equation $\dfrac{21}{\sqrt{(2x+1)}}-2\sqrt{x}-\sqrt{(2x+1)}=0$.

Clearing of fractions, $21 - 2\sqrt{(2x^2 + x)} - 2x - 1 = 0.$ (1)

Transferring and uniting terms, and dividing by 2,

$$-\sqrt{(2x^2 + x)} = x - 10. \quad (2)$$

Squaring, $2x^2 + x = x^2 - 20x + 100.$

The roots of this equation are 4 and -25.

The number 4 is a root of the given equation, since

$$\frac{21}{\sqrt{(8+1)}} - 2\sqrt{4} - \sqrt{(8+1)} = 0.$$

The number -25 is not a root of the given equation, since

$$\frac{21}{\sqrt{(-50+1)}} - 2\sqrt{-25} - \sqrt{(-50+1)} = \frac{21}{7\sqrt{-1}} - 10\sqrt{-1} - 7\sqrt{-1}$$

$$= -3\sqrt{-1} - 10\sqrt{-1} - 7\sqrt{-1} \neq 0.$$

But -25 is a root of the equation (2), obtained by clearing the given equation of fractions, since $-\sqrt{(2 \times 625 - 25)} = -25 - 10$. At the same time it is not a root of the L.C.D. equated to 0, *i.e.*, of

$$\sqrt{(2x+1)} = 0.$$

The explanation of this apparent contradiction of the principle that if a root be introduced, it is a root of the L.C.D. equated to 0, is found in a property of imaginary numbers. For, when $x = -25$,

$$\sqrt{(2x+1)} = \sqrt{-49},$$

and $\sqrt{x}, = \sqrt{-25}$, are imaginary; therefore

$$\sqrt{(2x+1)} \times \sqrt{x} = -\sqrt{(2x^2 + x)},$$

and not $\sqrt{(2x^2 + x)}$. But $\sqrt{(2x+1)} \times \sqrt{(2x+1)}, = [\sqrt{(2x+1)}]^2$, $= 2x + 1$, whether $2x + 1$ be positive or negative.

Consequently, when $x = -25$, the correct equation obtained by clearing the given equation of fractions is

$$\sqrt{(2x^2 + x)} = x - 10, \quad (3)$$

and not (2) as above. Equation (3) is evidently not satisfied by -25. This root was therefore introduced in rationalizing equation (2), which leads to the same rational equation as does (3).

The preceding examples illustrate the following method of solving irrational equations:

Transform the given equation so that one radical stands by itself in one member of the equation.

Equate equal powers of the two members when so transformed.

Repeat this process until a rational equation is obtained.

4. In the preceding article the indicated roots in the equations were limited to principal values.

At the same time an irrational equation, if written arbitrarily, may be inconsistent with the laws governing the relations between numbers. In such a case the equation is *impossible*, that is, it cannot be satisfied by either real or imaginary values of the unknown numbers.

E.g., $\sqrt{(x+6)} + \sqrt{(x+1)} = 1$ is an impossible equation.

For it cannot be satisfied by any complex value of x, since by Ch. XIX., Arts. 33 and 22, $\sqrt{(x+6)} + \sqrt{(x+1)}$ must be complex if x be complex, and hence cannot be equal to 1.

It cannot be satisfied by any *real positive* value of x, since, in that case, either $\sqrt{(x+1)}$ or $\sqrt{(x+6)}$ is greater than 1.

It cannot be satisfied by any *real negative* value of x, since, if x be negative and its absolute value be less than 1, $\sqrt{(x+6)}$ will be greater than 1, and if x be negative and its absolute value be greater than 1, $\sqrt{(x+1)}$ will be imaginary.

5. But if the restriction to *principal* roots be removed, any irrational equation contains in itself the statements of two or more equations.

E.g., if both *positive* and *negative* square roots be admitted, the equation

$$\sqrt{(x+6)} + \sqrt{(x+1)} = 1$$

is equivalent to the four equations

$$\sqrt{(x+6)} + \sqrt{(x+1)} = 1, \tag{1}$$

$$\sqrt{(x+6)} - \sqrt{(x+1)} = 1, \tag{2}$$

$$-\sqrt{(x+6)} + \sqrt{(x+1)} = 1, \tag{3}$$

$$-\sqrt{(x+6)} - \sqrt{(x+1)} = 1, \tag{4}$$

in which the roots are limited to *principal* values.

The same rational integral equation will evidently be derived by rationalizing any one of these four equations. Therefore the roots of this rational equation must comprise the roots of these four irrational equations. Consequently, in solving an irrational equation, we must expect to obtain not only its roots but also the roots of the other three equations obtained by changing the signs of the radicals in all possible ways. Some of these equations can be rejected at once as impossible. The roots of the other irrational equations will be the roots of the rational equation. Thus, of the above equations, (1), (3), and (4) can be rejected at once as impossible.

The rational equation derived from any one of the four equations is

$$x + 1 = 4\,; \text{ whence } x = 3.$$

The number 3 is a root of the one equation not rejected, since

$$\sqrt{(3+6)} - \sqrt{(3+1)} = 1.$$

The same conclusions could have been reached by substituting the roots of the integral equation successively in the irrational equations, rejecting those which are not satisfied by any root.

Special Devices.

6. Ex. **1.** Solve the equation

$$\sqrt{(3x^2-2x+4)}-3x^2+2x=-16.$$

Since $$-3x^2+2x=-(3x^2-2x+4)+4,$$

we may take $\sqrt{(3x^2-2x+4)}$ as the unknown number, replacing it temporarily by y. We then have

$$y-y^2+4=-16.$$

The roots of this equation are 5, and -4.

Equating $\sqrt{(3x^2-2x+4)}$ to each of these roots, we have

$$\sqrt{(3x^2-2x+4)}=5, \text{ whence } x=3,\ -\tfrac{7}{3}.$$

$$\sqrt{(3x^2-2x+4)}=-4, \text{ whence } x=\tfrac{1}{3}(1\pm\sqrt{37}).$$

The numbers 3, $-\frac{7}{3}$ satisfy the given equation, and are therefore roots of that equation. The numbers $\frac{1}{3}\sqrt{(1\pm\sqrt{37})}$ do not satisfy the given equation.

But if the value of the radical be not restricted to the principal root, the given equation comprises the two equations

$$\sqrt{(3x^2-2x+4)}-3x^2+2x=-16, \qquad (1)$$

$$-\sqrt{(3x^2-2x+4)}-3x^2+2x=-16. \qquad (2)$$

Then $\frac{1}{3}(1\pm\sqrt{37})$ are roots of (2).

Ex. **2**. Solve the equation $\sqrt[4]{(3x^2+13)}+\sqrt{(3x^2+13)}=6.$

Assuming $\sqrt[4]{(3x^2+13)}$ as the unknown number, and representing it by y, we have $$y+y^2=6.$$

The roots of this equation are 2 and -3.

Equating $\sqrt[4]{(3x^2+13)}$ to each of these roots, we have

$$\sqrt[4]{(3x^2+13)}=2, \text{ whence } x=\pm 1,$$

$$\sqrt[4]{(3x^2+13)}=-3, \text{ whence } x=\pm\sqrt{\tfrac{68}{3}}=\pm\tfrac{2}{3}\sqrt{51}.$$

The numbers ± 1 are roots of the given equation, since $\sqrt[4]{16}+\sqrt{16}=6$.

The numbers $\pm\frac{2}{3}\sqrt{51}$ are evidently not roots of the given equation, but are found to be roots of the equation

$$-\sqrt[4]{(3x^2+13)}+\sqrt{(3x^2+13)}=6.$$

The preceding examples illustrate the following principle:

If a radical equation contain one radical, and an expression which is equal to the radicand or which can be made to differ from the radicand (or a multiple of the radicand) by a constant term, it can be solved as a quadratic equation. The same is true if the equation contain two radicals, one the square of the other, and in addition only constant terms. In both cases, the radicand must, in general, be a linear or a quadratic expression.

7. Ex. Solve the equation $9x^2 - (7x + 18\sqrt{x}) = 80$.

This equation can be written in the form

$$9x^2 - 6x + 1 = x + 18\sqrt{x} + 81,$$

or

$$(3x - 1)^2 = (\sqrt{x} + 9)^2.$$

Equating square roots, $3x - 1 = \pm(\sqrt{x} + 9)$. (1)

If we let $\sqrt{x} = y$ in the first of these equations, we have

$$3y^2 - 1 = y + 9.$$

The roots of this equation are 2, $-\frac{5}{3}$, and the corresponding values of x are 4, $\frac{25}{9}$.

Solving the second of equations (1) in like manner, we obtain

$$x = -\tfrac{1}{18}(47 \pm \sqrt{-95}).$$

The numbers 4 and $-\frac{1}{18}(47 \pm \sqrt{-95})$ satisfy the given equation and are therefore roots of that equation.

The number $\frac{25}{9}$ is found to satisfy the equation

$$9x^2 - (7x - 18\sqrt{x}) = 80.$$

Observe that, if the given equation had been at once rationalized, it would have led to an equation of the fourth degree.

8. Ex. Solve the equation

$$\sqrt{(x^2 - 3x + 4)} - \sqrt{(x^2 - 5x + 7)} = 1. \quad (1)$$

Let us assume $x^2 - 3x + 4 = A$ and $x^2 - 5x + 7 = B$. (2)

Then $A - B = (x^2 - 3x + 4) - (x^2 - 5x + 7) = 2x - 3$. (3)

Since

$$\sqrt{A} + \sqrt{B} = \frac{A - B}{\sqrt{A} - \sqrt{B}},$$

we have

$$\sqrt{(x^2 - 3x + 4)} + \sqrt{(x^2 - 5x + 7)} = \frac{2x - 3}{\sqrt{(x^2 - 3x + 4)} - \sqrt{(x^2 - 5x + 7)}}, \quad (4)$$

or

$$\sqrt{(x^2 - 3x + 4)} + \sqrt{(x^2 - 5x + 7)} = \frac{2x - 3}{1} = 2x - 3, \quad (5)$$

since

$$\sqrt{(x^2 - 3x - 4)} - \sqrt{(x^2 - 5x + 7)} = 1.$$

Adding corresponding members of (1) and (5),

$$2\sqrt{(x^2 - 3x + 4)} = 2x - 2. \quad (6)$$

Dividing by 2 and rationalizing,

$$x^2 - 3x + 4 = x^2 - 2x + 1.$$

Whence $x = 3$.

The number 3 is found to be a root of the given equation.

The purpose of the method is to obtain the sum of two radicals when their difference is given, or to obtain their difference when their sum is given.

In practice the steps indicated by (2) and (4) can be omitted.

9. Irrational equations containing cube and higher roots in general lead to rational, integral equations of a higher degree than the second, and therefore cannot be solved by means of quadratic equations. But in some cases their solutions can be effected by special devices.

Ex. 1. Solve the equation $\sqrt[3]{(8x+4)} - \sqrt[3]{(8x-4)} = 2$.

Cubing,

$$8x+4-3[\sqrt[3]{(8x+4)}]^2\sqrt[3]{(8x-4)}+3\sqrt[3]{(8x+4)}[\sqrt[3]{(8x-4)}]^2 - 8x+4 = 8. \qquad (1)$$

Transferring and uniting terms, and dividing by -3,

$$[\sqrt[3]{(8x+4)}]^2\sqrt[3]{(8x-4)} - \sqrt[3]{(8x+4)}[\sqrt[3]{(8x-4)}]^2 = 0. \qquad (2)$$

Factoring, $\sqrt[3]{(8x+4)}\sqrt[3]{(8x-4)}[\sqrt[3]{(8x+4)} - \sqrt[3]{(8x-4)}] = 0. \qquad (3)$

This equation is equivalent to the three equations

$$\sqrt[3]{(8x+4)} = 0, \text{ whence } x = -\tfrac{1}{2}; \qquad (4)$$

$$\sqrt[3]{(8x-4)} = 0, \text{ whence } x = \tfrac{1}{2}; \qquad (5)$$

and $\sqrt[3]{(8x+4)} - \sqrt[3]{(8x-4)} = 0,$

whence $8x+4 = 8x-4. \qquad (6)$

Equation (6) is not satisfied by any finite value of x.

The numbers $-\frac{1}{2}$ and $\frac{1}{2}$ are found to satisfy the given equation.

Ex. 2. Solve the equation

$$\sqrt[3]{(8x^3+12x^2+18x+27)} = \sqrt{(4x^2+4x+9)}.$$

The terms $8x^3$ and 27 in the first member suggest the cube of $2x+3$, and the terms $4x^2$ and 9 in the second member the square of $2x+3$. Modifying the radicands so that $(2x+3)^3$ and $(2x+3)^2$ shall appear in the first and second members respectively, we have

$$\sqrt[3]{(8x^3+36x^2+54x+27-24x^2-36x)} = \sqrt{(4x^2+12x+9-8x)},$$

or $\sqrt[3]{[(2x+3)^3 - 12x(2x+3)]} = \sqrt{[(2x+3)^2 - 8x]}.$

Equating *sixth* powers, we obtain

$$(2x+3)^6 - 24x(2x+3)^4 + 144x^2(2x+3)^2$$
$$= (2x+3)^6 - 24x(2x+3)^4 + 192x^2(2x+3)^2 - 512x^3.$$

Uniting like terms, $48x^2(2x+3)^2 - 512x^3 = 0,$

or $x^2[6(2x+3)^2 - 64x] = 0,$

or $x^2(24x^2+8x+54) = 0.$

The roots of the last equation are

$$0, 0, -\frac{1}{6}(1 \pm 4\sqrt{-5}).$$

The numbers 0, 0 are evidently roots of the given equation.

The numbers $-\frac{1}{6}(1 \pm 4\sqrt{-5})$ are found to satisfy the equation

$$\sqrt[3]{(8x^3+12x^2+18x+27)} = -\sqrt{(4x^2+4x+9)}.$$

Compare Ch. XIX., Exercises II., Exx. 85 and 86.

EXERCISES

Solve each of the following equations, and check the results. If a result does not satisfy an equation as written, determine what signs the radical terms must have in order that the result may satisfy the equation.

1. $\sqrt{(3x+4)} - 4 = 0.$

2. $\sqrt{(16+x)} = 2\sqrt{(x+6)}.$

3. $\sqrt{[5+\sqrt{(x-4)}]} = 3.$

4. $\sqrt[3]{(10x+35)} - 1 = 4.$

5. $\sqrt{(x^2-9)} = 4.$

6. $4x = 3\sqrt{(2x^2-4)}.$

7. $3 - \sqrt{(3x^2-4x+9)} = 0.$

8. $\sqrt{(x+9)} = 2\sqrt{(x-3)}.$

9. $2 - \sqrt{(3x^2-11x)} = 0.$

10. $5x = 2\sqrt{(3x^2-x+15)}.$

11. $\sqrt[8]{x} - \sqrt[4]{3} = 0.$

12. $\sqrt[6]{(1\frac{1}{3}x+8)} - \sqrt[3]{4} = 0.$

13. $\sqrt{\dfrac{x}{x-3}} - \sqrt{2} = 0.$

14. $\sqrt[3]{} - 2 + \sqrt[12]{\dfrac{24}{x-2}} = 0.$

15. $\sqrt{(4x+9)} - 2\sqrt{x} = 1.$

16. $\sqrt{[(x-5)-7+\sqrt{(x-12)}]} = 0.$

17. $\sqrt{(x^4+15)} - x^2 = \sqrt{5}.$

18. $\sqrt{[12+\sqrt{\{4+\sqrt{(x^2+23)}\}}]} = 4.$

19. $\dfrac{x-1}{\sqrt{x}+1} = 4 + \dfrac{\sqrt{x}-1}{2}.$

20. $\dfrac{x+\sqrt{(x^2+7)}}{28} = \dfrac{1}{\sqrt{(x^2+7)}}.$

21. $\sqrt{(3x+2)} - \sqrt{(2x^2+12x+10)} = 0.$

22. $\sqrt{(x+2\sqrt{5})} = \sqrt{(x+\sqrt{5})} + \sqrt{(-x+\sqrt{5})}.$

23. $\dfrac{2x+\sqrt{(4x^2-1)}}{2x-\sqrt{(4x^2-1)}} = 4.$

24. $\sqrt{\left(\dfrac{4}{x^2}+5\right)} - \sqrt{\left(\dfrac{4}{x^2}-5\right)} = 2.$

25. $\sqrt{[4x^2-\sqrt[3]{(3x-5)}]} = 2x.$

26. $\sqrt{[4x-\sqrt{(2x+3)}]} = 3.$

27. $\dfrac{\sqrt{x}-2}{\sqrt{x}+3} = \dfrac{\sqrt{x}+1}{\sqrt{x}+21}.$

28. $\dfrac{\sqrt{(3x+1)}+\sqrt{(3x)}}{\sqrt{(3x+1)}-\sqrt{(3x)}} = 2.$

29. $\sqrt{(2+x)} + \sqrt{x} = \dfrac{4}{\sqrt{(2+x)}}.$

30. $\dfrac{x-\sqrt{(x+1)}}{x+\sqrt{(x+1)}} = \dfrac{5}{11}.$

31. $3x - 2\sqrt{x} - 1 = 0.$

32. $x + \sqrt{(10x+6)} = 9.$

33. $\sqrt{(x+2)} - \sqrt{(x^2+2x)} = 0.$

34. $7\sqrt{x} = 3\sqrt{(x^2+3x-59)}.$

35. $x + 5 - \sqrt{(x+5)} = 6.$

36. $x - 6\sqrt{x} + 5 = 0.$

37. $(5-\sqrt{x})^2 = 2(7+\sqrt{x}).$

38. $x - 7\sqrt{(51-x)} = 33.$

39. $4\sqrt{(75-x)} = x - 54.$

40. $(\sqrt[4]{x}-3)^2 + (\sqrt[4]{x}-2)^2 = 1.$

41. $\sqrt{(x-2)} + 2\sqrt{(x+3)} - 2\sqrt{(3x-2)} = 0.$

42. $\sqrt{(x+3)} + \sqrt{(x+8)} = 5\sqrt{x}.$

43. $\sqrt{(2x+9)} + \sqrt{(3x-15)} = \sqrt{(7x+8)}.$

44. $x^2 + \sqrt{[4x^2+\sqrt{(16x^2+12x)}]} = (x+1)^2.$

45. $\sqrt{\dfrac{3x-4}{x-5}} + \sqrt{\dfrac{x-5}{3x-4}} = \dfrac{5}{2}.$

46. $\sqrt{\dfrac{3x+6}{7x-3}} + \sqrt{\dfrac{7x-3}{3x+6}} = \dfrac{13}{6}.$

47. $\frac{1}{\sqrt{(x+2)}}+\frac{1}{\sqrt{(3x-2)}}=\frac{4}{\sqrt{(3x^2+4x-4)}}.$

48. $\frac{1}{x-\sqrt{(2-x^2)}}+\frac{1}{x+\sqrt{(2-x^2)}}=1.$

49. $\frac{1}{1-\sqrt{(1-x^2)}}-\frac{1}{1+\sqrt{(1-x^2)}}=\frac{\sqrt{3}}{x^2}.$

50. $3x-x\sqrt{x}=2\sqrt{x}.$

51. $\sqrt{x}+\sqrt[4]{x^3}=2\sqrt[4]{x}.$

52. $x\sqrt{x}-x=12\sqrt{x}.$

53. $x-\sqrt[4]{x^3}=6\sqrt{x}.$

54. $x^2-x+2\sqrt{(x^2-x-11)}=14.$

55. $x^2+24=2x+6\sqrt{(2x^2-4x+16)}.$

56. $\sqrt{(2x^2-3x+5)}+2x^2-3x=1.$

57. $\sqrt{(7x^2+8x-19)}-7x^2-8x=-39.$

58. $\sqrt{(5x^2-2x-3)}-5x^2+2x=-33.$

59. $x^2-5x-2\sqrt{x}=-3.$

60. $8\sqrt{x}=x(x-7)-7.$

61. $2x\sqrt{(4x^2-27x)}=-5x^2+27x+9.$

62. $\sqrt{(3x^2+7x-1)}-\sqrt{(3x^2-4x+5)}=8.$

63. $\sqrt{(2x^2-7x+7)}+\sqrt{(2x^2+9x-1)}=6.$

64. $\sqrt{(5x^2-2x+6)}-\sqrt{(5x^2+2x-3)}=1.$

65. $\sqrt{(x+\sqrt{x})}-\sqrt{(x-\sqrt{x})}=\frac{3}{2}\sqrt{\frac{x}{x+\sqrt{x}}}.$

66. $\frac{x^2-1}{-4+4\sqrt{x}}=1.$

67. $\sqrt[3]{(1+\sqrt{x})}=2-\sqrt[3]{(1-\sqrt{x})}.$

68. $\sqrt[3]{x}+\sqrt[3]{(28-x)}=4.$

69. $\sqrt[3]{(37+x)}-\sqrt[3]{x}=1.$

70. $\sqrt[3]{(36+x)}+\sqrt[3]{(36-x)}=6.$

71. $\sqrt[3]{(14+x)}+\sqrt[3]{(14-x)}=4.$

72. $\sqrt[3]{(50+x)}+\sqrt[3]{(22-x)}=6.$

73. $x-\sqrt[4]{(97-x^4)}=1.$

74. $x+\sqrt[4]{(337-x^4)}=7.$

75. $x-\sqrt[5]{(x^5-242)}=2.$

76. $\sqrt[4]{(41+x)}+\sqrt[4]{(41-x)}=4.$

77. $\sqrt[4]{(a+x)}+\sqrt[4]{(a-x)}=\sqrt[4]{(2a)}.$

78. $\sqrt[5]{(20-x)}+\sqrt[5]{(13+x)}=3.$

79. $\sqrt[5]{(24+x)}+\sqrt[5]{(40-x)}=4.$

80. $\sqrt[3]{(x^3-12x^2+12x+1)}=\sqrt{(x^2-8x+1)}.$

81. $\sqrt[3]{(27x^3+72x^2+48x+8)}=\sqrt{(9x^2+16x+4)}.$

82. $\sqrt[3]{(x^3-9x^2+9x-1)}=\sqrt{(x^2-6x+1)}.$

83. $\sqrt[3]{(x^3+3x^2+12x+64)}=\sqrt{(x^2+2x+16)}.$

84. $\sqrt{(x^2+b^2)}+a=x.$

85. $\sqrt{[\sqrt{(cx+a^2)}-a]}=c.$

86. $\sqrt{\frac{a^2+x^2}{a^2-x^2}}=\frac{a}{b}$.

87. $\frac{\sqrt{x}+\sqrt{b}}{\sqrt{x}-\sqrt{b}}=\frac{a}{b}$.

88. $\sqrt{\left(\frac{a^2}{x}+b\right)}-\sqrt{\left(\frac{a^2}{x}-b\right)}=c$.

89. $\frac{1+\sqrt{(1-x)}}{1-\sqrt{(1-x)}}=n$.

90. $\sqrt{(a+x)}+\sqrt{(a-x)}=\frac{b}{\sqrt{(a+x)}}$.

91. $\frac{1}{x}+\frac{1}{a}=\sqrt{\left[\frac{1}{a^2}-\sqrt{\left(\frac{1}{a^2x^2}+\frac{1}{x^4}\right)}\right]}$.

92. $\sqrt{(a+x)}+\sqrt{(b+x)}+\sqrt{(c+x)}=0$.

93. $\frac{x^2}{a-\sqrt{(a^2-x^2)}}-\frac{x^2}{a+\sqrt{(a^2-x^2)}}=a$.

94. $\sqrt{(1-x+x^2)}+\sqrt{(1+x+x^2)}=m$.

95. $\frac{a\sqrt{(x-b)}+b\sqrt{(a-x)}}{\sqrt{(a-x)}+\sqrt{(x-b)}}=x$.

96. $\frac{b\sqrt{(a-x)}-a\sqrt{(b-x)}}{\sqrt{(a-x)}-\sqrt{(b-x)}}=x$.

97. $\frac{\sqrt{(1-x^3)}}{\sqrt{(1-x)}}+\frac{\sqrt{(1+x^3)}}{\sqrt{(1+x)}}=mx$.

98. $\frac{a+x+\sqrt{(a^2-x^2)}}{a+x-\sqrt{(a^2-x^2)}}=\frac{b}{x}$.

99. $\frac{x}{\sqrt{x}+\sqrt{(a-x)}}+\frac{x}{\sqrt{x}-\sqrt{(a-x)}}=\frac{b}{\sqrt{x}}$.

100. $\frac{a+x}{\sqrt{x}+\sqrt{(a+x)}}-\frac{a+x}{\sqrt{a}-\sqrt{(a+x)}}=0$.

101. $\frac{3\sqrt[3]{(x-5)}+5\sqrt[3]{(3-x)}}{\sqrt[3]{(3-x)}+\sqrt[3]{(x-5)}}=x$.

102. $\frac{a\sqrt[3]{(b-x)}+b\sqrt[3]{(a-x)}}{\sqrt[3]{(a-x)}+\sqrt[3]{(b-x)}}=x$.

103. $\frac{\sqrt{a}-\sqrt{[a-\sqrt{(a^2-ax)}]}}{\sqrt{a}+\sqrt{[a-\sqrt{(a^2-ax)}]}}=b$.

104. $\frac{\sqrt[3]{(a-x)}+\sqrt[3]{(x-b)}}{\sqrt[3]{(a-b)}}=1$.

105. $\sqrt{[a+\sqrt{(a^2-x^2)}]}+\sqrt{[a-\sqrt{(a^2-x^2)}]}=n\sqrt{\frac{a+x}{a+\sqrt{(a^2-x^2)}}}$.

106. $\frac{(a+x)\sqrt[4]{(a-x)}+(a-x)\sqrt[4]{(a+x)}}{\sqrt[4]{(a+x)}+\sqrt[4]{(a-x)}}=a$.

CHAPTER XXIII

SIMULTANEOUS QUADRATIC AND HIGHER EQUATIONS.

To obtain a definite solution of a system of two or more quadratic or higher equations, as many equations must be given as there are unknown numbers. And, as in systems of linear equations, the given equations must be consistent and independent.

The solution of a system of quadratic or higher equations in general involves the solution of an equation of higher degree than the second, and therefore cannot be effected by the methods for solving quadratic equations. But there are many special systems whose solutions can be made to depend upon the solutions of quadratic equations.

§ 1. SIMULTANEOUS QUADRATIC EQUATIONS.

1. Elimination by Substitution. — When one equation of a system of two equations is of the first degree, the solution can be obtained by the method of substitution.

Ex. Solve the system $y + 2x = 5,$ (1)

$$x^2 - y^2 = -8. \quad (2)$$

Solving (1) for y, $y = 5 - 2x.$ (3)

Substituting $5 - 2x$ for y in (2),

$$x^2 - 25 + 20x - 4x^2 = -8. \quad (4)$$

From this equation we obtain $x = 1,$ (5)

$$x = 5\tfrac{2}{3}. \quad (6)$$

Substituting 1 for x in (3), $y = 3.$

Substituting $5\frac{2}{3}$ for x in (3), $y = -6\frac{1}{3}.$

Notice that the system (1), (2) is equivalent to the system (3), (4), which is equivalent to the two systems (3), (5) and (3), (6).

Therefore the solutions of the given system are $x = 1$, $y = 3$; $x = 5\frac{2}{3}$, $y = -6\frac{1}{3}$.

These solutions may be written 1, 3; $5\frac{2}{3}$, $-6\frac{1}{3}$, if the first number of each pair be understood to be the value of x, and the second the corresponding value of y.

Observe that, if we had substituted the value 1 for x in (2), we should have obtained $y = \pm 3$.

But the values $x = 1$ and $y = -3$ do not satisfy equation (1).

By the principle of equivalent equations, proved in Ch. XIII., § 2, Art. 2 (iii.); equation (3), obtained from (1) by solving for y, and equation (4), obtained by substituting this value for y in (2), form a system equivalent to the given system. This principle does not, however, prove that (2) and (4) are necessarily equivalent to the given system. In this example, since (2) and (4) give more solutions than (1) and (4), the system formed by (2) and (4) cannot be equivalent to the given system. Therefore, having obtained the values of one of the unknown numbers, we should obtain the values of the other by substituting in the equation of the first degree.

This advice was unnecessary in solving systems of linear simultaneous equations, since then both equations were of the first degree.

EXERCISES I.

Solve each of the following systems:

1. $\begin{cases} xy = 54, \\ 3x = 2y. \end{cases}$

2. $\begin{cases} 4x - 3y = 24, \\ xy = 96. \end{cases}$

3. $\begin{cases} 2x^2 - 3y^2 = 24, \\ 2x = 3y. \end{cases}$

4. $\begin{cases} 3x - 2y = 1, \\ x^2 + y^2 = 74. \end{cases}$

5. $\begin{cases} 3x - y = 5, \\ x^3 + y^3 = 1825. \end{cases}$

6. $\begin{cases} 2x + 3y = 10, \\ x(x + y) = 25. \end{cases}$

7. $\begin{cases} 3x^2 - 7y^2 = 84, \\ 3x + 7y = 42. \end{cases}$

8. $\begin{cases} 3x - 2y = 1, \\ x + 3y = 4xy. \end{cases}$

9. $\begin{cases} 4x^2 - xy = 0, \\ 2x - 3y = 6. \end{cases}$

10. $\begin{cases} x^2 + xy + y^2 = 343, \\ 2x - y = 21. \end{cases}$

11. $\begin{cases} 4x - 6y = -18, \\ x^2 + 6xy - 5y^2 = -26. \end{cases}$

12. $\begin{cases} 2x^2 - 3xy + y^2 = 14, \\ 2x - y = 7. \end{cases}$

13. $\begin{cases} (x-7)(y+3) = 48, \\ x + y = 18. \end{cases}$

14. $\begin{cases} (x-3)(y-4) = -6, \\ 4x + 3y = 10. \end{cases}$

15. $\begin{cases} 4x - 3y = 5, \\ 2x^2 + 5y^2 - 6y = 7. \end{cases}$

16. $\begin{cases} 2x + 3y = -13, \\ x^2 - 4xy + 3y^2 = 96. \end{cases}$

17. $\begin{cases} 2x - 3y = 11, \\ \frac{4}{x} - \frac{3}{y} = -\frac{17}{7}. \end{cases}$

18. $\begin{cases} x + 2y = 1, \\ \frac{x}{y} + \frac{y}{x} + 3\frac{1}{8} = 0. \end{cases}$

19. $\begin{cases} 7x + y = 11, \\ \frac{5}{x} - \frac{3}{y} = 4\frac{1}{4}. \end{cases}$

20. $\begin{cases} x - y = 11, \\ \frac{2x}{y} - \frac{y}{2x} = \frac{169}{60}. \end{cases}$

21. $\begin{cases} \frac{x+5}{y-7} = \frac{y+2}{x-4}, \\ 3x + 1 = 2y. \end{cases}$

22. $\begin{cases} \frac{3x-5}{2y-7} - \frac{y-2}{2x-5} = 2, \\ 5x = 4y - 1. \end{cases}$

23. $\begin{cases} \frac{x-1}{y-1} = 2, \\ \frac{x^2+x+1}{y^2+y+1} = \frac{19}{7}. \end{cases}$

24. $\begin{cases} \frac{x+1}{y+1} = \frac{3}{2}, \\ \frac{x^2+y}{x+y^2} = \frac{32}{15}. \end{cases}$

25. $\begin{cases} 5x - 7y + 6 = 0, \\ x^2 + y^2 = 18. \end{cases}$

26. $\begin{cases} 7x + 3y = 10, \\ x^2 + x + y = 3. \end{cases}$

27. $\begin{cases} 5xy + 2y^2 + 4x + 16 = 0, \\ 11x + 5y = 4. \end{cases}$

28. $\begin{cases} x^2 + xy + 5x + 10y = 29, \\ x + 2y = 3. \end{cases}$

2. Elimination by Addition and Subtraction. — When both equations of a system of two quadratic equations contain only the squares of the unknown numbers, the solution can be obtained by the method of addition and subtraction.

Ex. 1. Solve the system

$$9x^2 - 8y^2 = 28, \quad (1)$$

$$7x^2 + 3y^2 = 31. \quad (2)$$

We will first eliminate y^2.

Multiplying (1) by 3, $27x^2 - 24y^2 = 84.$ (3)

Multiplying (2) by 8, $56x^2 + 24y^2 = 248.$ (4)

Adding (3) and (4), $83x^2 = 332.$ (5)

Whence $x = 2,$ (6)

and $x = -2.$ (7)

Substituting 2 for x in (1), $y = \pm 1.$ (8)

Substituting -2 for x in (1), $y = \pm 1.$ (9)

The given system is equivalent to the system (3), (4), which is equivalent to the system (5), (1); this last system is equivalent to the two systems (6), (1) and (7), (1).

The solutions of the system (6), (1) are 2, 1; 2, −1.

The solutions of the system (7), (1) are −2, 1; −2, −1.

Therefore, the given system has the four solutions, 2, 1; 2, −1; −2, 1; −2, −1.

Notice that the number of solutions is equal to the product of the degrees of the equations of the system.

Many other examples are most easily solved by this method.

Ex. 2. Solve the system $x^2 + 3y = 18,$ (1)

$2x^2 - 5y = 3.$ (2)

We will first eliminate y.

Multiplying (1) by 5, $5x^2 + 15y = 90.$ (3)

Multiplying (2) by 3, $6x^2 - 15y = 9.$ (4)

Adding (3) and (4), $11x^2 = 99.$ (5)

Whence, $x = 3,$ (6)

and $x = -3.$ (7)

Substituting 3 for x in (1), $y = 3.$ (8)

Substituting -3 for x in (1), $y = 3.$ (9)

Therefore, the given system has the two solutions 3, 3; −3, 3.

Notice that this example could also have been solved by the method of substitution.

Ex. 3. Solve the system $x^2 + xy = 6,$ (1)

$x + y = 3.$ (2)

Multiplying (2) by x, $x^2 + xy = 3x.$ (3)

Subtracting (3) from (1), $0 = 6 - 3x.$ (4)

Whence, $x = 2.$ (5)

Substituting 2 for x in (2), $y = 1.$

Notice that in multiplying equation (2) by x, the root $x = 0$ was introduced into equation (3). But, since equation (1) is not satisfied by $x = 0$, the given system is equivalent to the system (1), (3). The latter system is equivalent to the system (4), (1).

EXERCISES II.

Solve each of the following systems:

1. $\begin{cases} x^2 + y^2 = 13, \\ x^2 - y^2 = 5. \end{cases}$

2. $\begin{cases} x^2 + y^2 = a, \\ x^2 - y^2 = b. \end{cases}$

3. $\begin{cases} 2x^2 - 3y = 20, \\ x^2 + 5y = 36. \end{cases}$

4. $\begin{cases} 2x^2 + 7y = 32, \\ 3x^2 - 5y = 17. \end{cases}$

5. $\begin{cases} 7x + xy = 20, \\ 2xy + 5x = 22. \end{cases}$

6. $\begin{cases} 4x + 3y^2 = 34, \\ 7y^2 = 2x. \end{cases}$

7. $\begin{cases} x^2 + xy = a, \\ x + y = b. \end{cases}$

8. $\begin{cases} 2x^2 + 3y^2 = 77, \\ 2x^2 - 3y^2 = 23. \end{cases}$

9. $\begin{cases} 5xy + 3x^2 = 132, \\ 5xy - 3x^2 = 78. \end{cases}$

10. $\begin{cases} 5x^2 + 3y^2 = 107, \\ 4x^2 - 3y^2 = 37. \end{cases}$

11. $\begin{cases} 9x^2 + 5y^2 = 29, \\ 5x^2 - 3y^2 = -7. \end{cases}$

12. $\begin{cases} ax^2 + by^2 = c, \\ a_1x^2 + b_1y^2 = c_1. \end{cases}$

13. $\begin{cases} x^2 + 4x + y^2 + 3y = 30, \\ x^2 + 4x + 6 = y^2 + 3y. \end{cases}$

14. $\begin{cases} x^2 + 5xy + y^2 = 43, \\ x^2 + 5xy - y^2 = 25. \end{cases}$

15. $\begin{cases} 4x = xy + 5, \\ 7y = xy + 6. \end{cases}$

16. $\begin{cases} 3x = x^2 + y^2 - 1, \\ 3y = x^2 + y^2 - 7. \end{cases}$

17. $\begin{cases} 2x^2 - 3y^2 + 5y = 16, \\ 3y^2 - 2x^2 + 5x = 9. \end{cases}$

18. $\begin{cases} 3xy = 5x - 7y - 1, \\ 2xy = 3x + 5y - 9. \end{cases}$

19. $\begin{cases} x + y = 7xy, \\ x - y = 3xy. \end{cases}$

20. $\begin{cases} x + y = 3xy - 121, \\ x - y = 2xy - 86. \end{cases}$

21. $\begin{cases} x = \dfrac{4}{y} + 3, \\ y = \dfrac{5}{x} + 1. \end{cases}$

22. $\begin{cases} \dfrac{1}{x} = 9y + 2, \\ \dfrac{1}{y} = 5x + 2. \end{cases}$

23. $\begin{cases} 3 - \dfrac{x}{y} = 12x, \\ 5 + \dfrac{y}{x} = 20y. \end{cases}$

24. $\begin{cases} \dfrac{x}{y} = 10x - 3, \\ \dfrac{y}{x} = 9y - 4. \end{cases}$

25. $\begin{cases} \dfrac{x + y}{x - y} + 3x = 2\frac{3}{8}, \\ 5\dfrac{x + y}{x - y} - 7x = -8\frac{3}{8}. \end{cases}$

26. $\begin{cases} 3x + \sqrt{\dfrac{x}{y}} = 30, \\ 5x - 2\sqrt{\dfrac{x}{y}} = 39. \end{cases}$

3. Method of Factoring. — The method of this article depends upon the following principle:

A system of two integral equations

$$\left.\begin{aligned} P \times Q = 0, \\ R \times S = 0, \end{aligned}\right\} \text{(I.)}$$

whose first members (when all terms are brought to these members) can be resolved into factors, is equivalent to the four systems,

$$\left.\begin{aligned} P = 0, \\ R = 0, \end{aligned}\right\} (a), \quad \left.\begin{aligned} P = 0, \\ S = 0, \end{aligned}\right\} (b), \quad \left.\begin{aligned} Q = 0, \\ R = 0, \end{aligned}\right\} (c), \quad \left.\begin{aligned} Q = 0, \\ S = 0, \end{aligned}\right\} (d),$$

obtained by taking each factor of one equation with each factor of the other.

For, every solution of the given system must reduce either P or Q, or both P and Q, to 0, and at the same time must reduce either R or S, or both R and S, to 0.

Now any solution of (I.) which reduces P to 0 and R to 0, is a solution of (a); any solution which reduces P to 0 and S to 0, is a solution of (b); and so on. Therefore, every solution of the given system is a solution of at least one of the derived systems.

And any solution of (a) reduces P to 0 and R to 0, and therefore reduces $P \times Q$ to 0 and $R \times S$ to 0. Therefore, every solution of (a) is a solution of (I.).

In like manner it can be shown that every solution of the three other derived systems is a solution of the given system.

Ex. 1. Solve the system $(x - 2y)(x - 3y) = 0,$

$$(x + y - 4)(x - y + 2) = 0.$$

The given system is equivalent to the four systems

$$\left.\begin{aligned} x - 2y &= 0, \\ x + y - 4 &= 0, \end{aligned}\right\} (a), \qquad \left.\begin{aligned} x - 2y &= 0, \\ x - y + 2 &= 0, \end{aligned}\right\} (b),$$

$$\left.\begin{aligned} x - 3y &= 0, \\ x + y - 4 &= 0, \end{aligned}\right\} (c), \qquad \left.\begin{aligned} x - 3y &= 0, \\ x - y + 2 &= 0, \end{aligned}\right\} (d).$$

The solution of (a) is $\frac{8}{3}$, $\frac{4}{3}$; the solution of (b) is -4, -2; the solution of (c) is 3, 1; the solution of (d) is -3, -1.

These are therefore the solutions of the given equations.

Ex. 2. Solve the system $2x^2 - 7xy + 6y^2 = 0, \qquad (1)$

$$x^2 + y^2 = 13. \qquad (2)$$

The first member of (1) is $(x - 2y)(2x - 3y)$, and the first member of (2), when 13 is transferred to that member, cannot be resolved into rational factors. The given system is therefore equivalent to the two systems

$$\left.\begin{aligned} x - 2y &= 0, \\ x^2 + y^2 &= 13, \end{aligned}\right\} (a), \qquad \left.\begin{aligned} 2x - 3y &= 0, \\ x^2 + y^2 &= 13, \end{aligned}\right\} (b).$$

The solutions of (a) and (b), and therefore of the given system, are respectively

$$2\sqrt{\tfrac{13}{5}},\ \sqrt{\tfrac{13}{5}};\ -2\sqrt{\tfrac{13}{5}},\ -\sqrt{\tfrac{13}{5}};\ 3,\ 2;\ -3,\ -2.$$

Ex. 3. Solve the system $(x - 7)(y - 3) = 0,$

$$(x - 5)(y - 9) = 0.$$

If a solution is to contain simultaneous values of x and y, $x - 7 = 0$ and $x - 5 = 0$ cannot be taken as a system. Therefore the given system is equivalent only to the two systems

$$\left.\begin{array}{r} x-7=0, \\ y-9=0, \end{array}\right\} (a), \qquad \left.\begin{array}{r} x-5=0, \\ y-3=0, \end{array}\right\} (b).$$

The solutions are evidently 7, 9, and 5, 3.

The solutions of such systems can be written at sight.

4. When all the terms which contain the unknown numbers in both equations of the system are of the second degree, a system can always be derived whose solution is obtained by the method of the preceding article.

Ex. Solve the system $\qquad x^2 + xy + 2y^2 = 74, \qquad (1)$

$$2x^2 + 2xy + y^2 = 73. \qquad (2)$$

Multiplying (1) by 73, $\; 73x^2 + 73xy + 146y^2 = 74 \times 73. \qquad (3)$

Multiplying (2) by 74, $\; 148x^2 + 148xy + 74y^2 = 74 \times 73. \qquad (4)$

Subtracting (3) from (4), $\; 75x^2 + 75xy - 72y^2 = 0,$

or $\qquad 25x^2 + 25xy - 24y^2 = 0,$

or $\qquad (5x - 3y)(5x + 8y) = 0.$

Therefore the given system is equivalent to

$$\left.\begin{array}{r} 5x-3y=0, \\ x^2+xy+2y^2=74, \end{array}\right\} (a), \qquad \left.\begin{array}{r} 5x+8y=0, \\ x^2+xy+2y^2=74, \end{array}\right\} (b).$$

The solutions of these systems, and hence of the given system, are respectively

$$3, 5; \ -3, -5; \ 8, -5; \ -8, 5.$$

In applying this method to such systems, we must first derive from the given equations a homogeneous equation in which there is no term free from the unknown numbers. The factors of the first member of this equation can always be obtained. See Ch. VIII., § 1, Art. 12; Ch. XVIII., Art. 19; and Ch. XIX., Art. 31.

5. A system of two quadratic equations, whose terms in the unknown numbers are all of the second degree, can also be solved by a special device.

Ex. 1. Solve the system $\; x^2 + 4y^2 = 13, \qquad (1)$

$$xy + 2y^2 = 5. \qquad (2)$$

In both equations, let $\qquad y = tx. \qquad (3)$

Then from (1), $x^2 + 4x^2t^2 = 13$, whence $x^2 = \dfrac{13}{1+4t^2}$; (4)

and from (2), $x^2t + 2x^2t^2 = 5$, whence $x^2 = \dfrac{5}{t+2t^2}$. (5)

Equating values of x^2, $\dfrac{13}{1+4t^2} = \dfrac{5}{t+2t^2}$. (6)

From (6), $t = \frac{1}{3}$, (7)

and $t = -\frac{5}{2}$. (8)

When $t = \frac{1}{3}$, $x^2 = \dfrac{13}{1+4t^2} = 9$, whence $x = \pm 3$.

When $t = -\frac{5}{2}$, $x^2 = \frac{1}{2}$, whence $x = \pm\sqrt{\frac{1}{2}}$.

When $x = \pm 3$, $y = tx = \frac{1}{3}(\pm 3) = \pm 1$.

When $x = \pm\sqrt{\frac{1}{2}}$, $y = -\frac{5}{2}(\pm\sqrt{\frac{1}{2}}) = \mp\frac{5}{2}\sqrt{\frac{1}{2}}$.

It is to be noted that after assuming $y = tx$, we have a system of three equations in three unknown numbers x, y, and t. Then the system (1), (2), (3) is equivalent to (3), (4), (5), which is in turn equivalent to (3), (4), and (6). From (6) we obtain the values of t, from (4) the corresponding values of x, and from (3) the corresponding values of y.

The solutions of the given system therefore are

$$3, 1;\ -3, -1;\ \sqrt{\tfrac{1}{2}}, -\tfrac{5}{2}\sqrt{\tfrac{1}{2}};\ -\sqrt{\tfrac{1}{2}}, \tfrac{5}{2}\sqrt{\tfrac{1}{2}}.$$

EXERCISES III.

Solve each of the following systems:

1. $\begin{cases}(x-8)(y-6)=0,\\ x+y=13.\end{cases}$

2. $\begin{cases}(x-3)(y-4)=0,\\ 4x+3y=36.\end{cases}$

3. $\begin{cases}(x-1)(y-2)=0,\\ (x+1)(y+2)=12.\end{cases}$

4. $\begin{cases}(x-5)(y-3)=0,\\ (x-4)(y-7)=0.\end{cases}$

5. $\begin{cases}x^2+xy=78,\\ y^2-xy=7.\end{cases}$

6. $\begin{cases}x^2=28xy-675,\\ y^2=225-8xy.\end{cases}$

7. $\begin{cases}x^2+4y^2=13,\\ xy+2y^2=5.\end{cases}$

8. $\begin{cases}x^2+3xy=54,\\ xy+4y^2=115.\end{cases}$

9. $\begin{cases} x^2 + xy + y^2 = 52, \\ xy - x^2 = 8. \end{cases}$

10. $\begin{cases} x^2 - xy + y^2 = 21, \\ y^2 - 2xy + 15 = 0. \end{cases}$

11. $\begin{cases} x^2 + xy + 2y^2 = 74, \\ 2x^2 + 2xy + y^2 = 73. \end{cases}$

12. $\begin{cases} x^2 + xy + 4y^2 = 6, \\ 3x^2 + 8y^2 = 14. \end{cases}$

13. $\begin{cases} 2x^2 + 3xy + y^2 = 15, \\ 5x^2 + 4y^2 = 24. \end{cases}$

14. $\begin{cases} x^2 - 2xy + 3y^2 = 9, \\ x^2 - 4xy + 5y^2 = 5. \end{cases}$

15. $\begin{cases} x^2 + xy + y^2 = 57, \\ x^2 - xy + y^2 = 43. \end{cases}$

16. $\begin{cases} x^2 + xy + y^2 = 13x, \\ x^2 - xy + y^2 = 7x. \end{cases}$

17. $\begin{cases} x^2 + y^2 = 61 - 3xy, \\ x^2 - y^2 = 31 - 2xy. \end{cases}$

18. $\begin{cases} x^2 + 5xy + y^2 = 79, \\ x^2 + 3xy + y^2 = 59. \end{cases}$

19. $\begin{cases} 2x^2 - 5xy + 3y^2 = 6, \\ 3x^2 - 8xy + 2y^2 = 5. \end{cases}$

20. $\begin{cases} x + 4\sqrt{(xy)} + 4y + \sqrt{x} + 2\sqrt{y} = 0, \\ \sqrt{x} + 2\sqrt{y} = 3. \end{cases}$

21. $\begin{cases} (x - 15)(y - 7) = 0, \\ \dfrac{1}{x} + \dfrac{1}{y} = \dfrac{1}{6}. \end{cases}$

22. $\begin{cases} (x - 3)(y - 2) = 0, \\ \dfrac{7}{x} + \dfrac{3}{y} = 2. \end{cases}$

23. $\begin{cases} \left(\dfrac{x}{y} - 2\right)\left(\dfrac{y}{x} + 3\right) = 0, \\ x^2 - 2xy + 3y^2 = 12. \end{cases}$

24. $\begin{cases} \dfrac{x^2}{y^2} - \dfrac{y^2}{x^2} = 0, \\ x^2 + 2xy = 12. \end{cases}$

25. $\begin{cases} x - 2\sqrt{(xy)} + y - \sqrt{x} + \sqrt{y} = 0, \\ \sqrt{x} + \sqrt{y} = 5. \end{cases}$

6. *If the members of one equation of a system of two equations*

$$\left.\begin{aligned} PR &= QS, \quad (1) \\ P &= Q, \quad (2) \end{aligned}\right\} \qquad \text{(I.)}$$

contain as factors the corresponding members of the second equation, then the system is equivalent to the following two systems:

$$\left.\begin{aligned} R &= S, \\ P &= Q, \end{aligned}\right\} (a), \quad \text{and} \quad \left.\begin{aligned} P &= 0, \\ Q &= 0, \end{aligned}\right\} (b).$$

The first equation of the system (a) is obtained by dividing the members of equation (1) of the given system by the corresponding members of equation (2), and the second equation is equation (2) of the given system.

The system (b) is obtained by equating to 0 the two members of equation (2) of the given system.

The given system is equivalent to the system

$$\left.\begin{aligned} PR - QS &= 0, \\ P - Q &= 0, \end{aligned}\right\}, \qquad \text{(II.)}$$

or, replacing Q by P, to the system

$$\left.\begin{aligned} P(R - S) &= 0, \\ P - Q &= 0, \end{aligned}\right\}. \qquad \text{(III.)}$$

The system (III.) is, by Art. 3, equivalent to the two systems

$$\left.\begin{aligned} R - S &= 0, \\ P - Q &= 0, \end{aligned}\right\}, \quad \text{and} \quad \left.\begin{aligned} P &= 0, \\ P - Q &= 0, \end{aligned}\right\}.$$

That is, to

$$\left.\begin{aligned} R &= S, \\ P &= Q, \end{aligned}\right\} (a), \quad \text{and} \quad \left.\begin{aligned} P &= 0, \\ Q &= 0, \end{aligned}\right\} (b).$$

Ex. 1. Solve the system $(x-1)(x-y+2)=(y+1)(x+y)$,
$x - y + 2 = x + y$.

By the above principle, the given system is equivalent to the following two systems:

$$\left.\begin{aligned} x - 1 &= y + 1, \\ x - y + 2 &= x + y, \end{aligned}\right\} (a), \quad \text{and} \quad \left.\begin{aligned} x - y + 2 &= 0, \\ x + y &= 0, \end{aligned}\right\} (b).$$

The solution of (a) is 3, 1; the solution of (b) is -1, 1.

Ex. 2. Solve the system $x^2 - y^2 = 8$,
$x + y = 3$.

The given system is equivalent to the two systems

$$\left.\begin{aligned} x - y &= \tfrac{8}{3}, \\ x + y &= 3, \end{aligned}\right\} (a), \quad \text{and} \quad \left.\begin{aligned} x + y &= 0, \\ 3 &= 0, \end{aligned}\right\} (b).$$

The solution of system (a) is $\frac{17}{6}$, $\frac{1}{6}$. Since the equation $3 = 0$ is impossible for finite values of x, the system (b) is impossible.

EXERCISES IV.

Solve each of the following systems:

1. $\begin{cases} x^2 - 4y^2 = 21, \\ x - 2y = 3. \end{cases}$

2. $\begin{cases} 2xy - x^2 - y^2 - 16 = 0, \\ x - y = 4. \end{cases}$

3. $\begin{cases} 15x^2 + 2xy - y^2 = 15, \\ 5x - y = 3. \end{cases}$

4. $\begin{cases} (x^2-1)(y^2-1) = 2800, \\ (x-1)(y-1) = 40. \end{cases}$

5. $\begin{cases} (3x - 2y)(2x - 3y) = \frac{50}{3}(x - y), \\ (3x - 2y) = \frac{10}{3}(x - y). \end{cases}$

6. $\begin{cases} 2x^2 - xy - 3y^2 = -12, \\ x + y = 3. \end{cases}$

7. $\begin{cases} 3x^2 - 2xy - y^2 = 3, \\ 3x + y = 1. \end{cases}$

8. $\begin{cases} x - y = 2, \\ 3(x^2 - y^2) = 8x + 6y. \end{cases}$

9. $\begin{cases} 3x^2 - 4xy + y^2 = 39, \\ x - y = 3. \end{cases}$

10. $\begin{cases} x^2 - y^2 + (x + y)^2 = 24, \\ x + y = 4. \end{cases}$

11. $\begin{cases} (2x + 3y)^2 - (x - 5y)^2 = -209, \\ x + 8y = 11. \end{cases}$

12. $\begin{cases} (3x - 1)^2 - (4y + 2)^2 = 60, \\ 3x + 4y = 5. \end{cases}$

13. $\begin{cases} \frac{1}{x} - \frac{1}{y} = 2, \\ \frac{1}{x^2} - \frac{1}{y^2} = 16. \end{cases}$

14. $\begin{cases} \frac{x^2}{y^2} - \frac{y^2}{x^2} = 3\frac{3}{4}, \\ \frac{x}{y} + \frac{y}{x} = 2\frac{1}{2}. \end{cases}$

15. $\begin{cases} \sqrt{x} - \sqrt{y} = 1, \\ x - y = 9. \end{cases}$

16. $\begin{cases} \sqrt{(x-1)} - \sqrt{(y+9)} = 1, \\ x - y = 13. \end{cases}$

7. Symmetrical Equations. — A **Symmetrical Equation** is one which remains the same when the unknown numbers are interchanged.

A system of two symmetrical equations can be solved by first finding the values of $x + y$ *and* $x - y$.

Ex. 1. Solve the system

$$x + y = 3, \quad (1)$$

$$xy = 2. \quad (2)$$

Squaring (1), $x^2 + 2xy + y^2 = 9.$ (3)

Multiplying (2) by 4, $4xy = 8.$ (4)

Subtracting (4) from (3), $x^2 - 2xy + y^2 = 1.$ (5)

Equating square roots, $x - y = 1,$ (6)

and $x - y = -1.$ (7)

From (1) and (6), $x = 2,\ y = 1;$

from (1) and (7), $x = 1,\ y = 2.$

We find, by substitution, that the two sets of values 2, 1 and 1, 2 satisfy the given system.

Notice that these solutions differ only in having the values of x and y interchanged. This we should expect from the definition of symmetrical equations.

Observe that the system (3), (4), and therefore the equivalent system (3), (5), is equivalent to the two systems:

$$\left.\begin{matrix} x + y = 3 \\ xy = 2 \end{matrix}\right\} \text{the given system, and } \left\{\begin{matrix} x + y = -3, \\ xy = 2. \end{matrix}\right.$$

Consequently, if in the system (3), (5), equation (3) be replaced by (1), the resulting system

$$x + y = 3, \quad (1)$$

$$x^2 - 2xy + y^2 = 1, \quad (5)$$

is equivalent to the given system.

Ex. 2. Solve the system $x^2 + y^2 = 13,$ (1)

$xy = 6.$ (2)

Multiplying (2) by 2, $2xy = 12.$ (3)

Adding (3) to (1), $x^2 + 2xy + y^2 = 25.$ (4)

Subtracting (3) from (1), $x^2 - 2xy + y^2 = 1.$ (5)

Equating square roots of (4), $x + y = \pm 5.$ (6)

Equating square roots of (5), $x - y = \pm 1.$ (7)

The given system is equivalent to the system (4), (5), which is equivalent to the systems (6), (7).

The latter systems are

$$\left.\begin{matrix} x+y=5, \\ x-y=1, \end{matrix}\right\}(a), \quad \left.\begin{matrix} x+y=5, \\ x-y=-1, \end{matrix}\right\}(b), \quad \left.\begin{matrix} x+y=-5, \\ x-y=+1, \end{matrix}\right\}(c), \quad \left.\begin{matrix} x+y=-5, \\ x-y=-1, \end{matrix}\right\}(d).$$

The solutions of these four systems are respectively 3, 2; 2, 3; $-2, -3$; $-3, -2$.

The solutions of (6) and (7) should be obtained mentally, without writing the equivalent systems (*a*), (*b*), (*c*), (*d*).

In thus solving mentally, each sign of the second member of (6) should be taken in turn with each sign of the second member of (7).

Ex. 3. Solve the system
$$\frac{1}{x}+\frac{1}{y}=\frac{1}{x+y}, \qquad (1)$$
$$\frac{1}{x^2}+\frac{1}{y^2}=\frac{1}{a^2}. \qquad (2)$$

Clearing (1) of fractions,
$$(x+y)^2=xy. \qquad (3)$$

Clearing (2) of fractions,
$$x^2+y^2=\frac{x^2y^2}{a^2}. \qquad (4)$$

Subtracting (4) from (3),
$$2xy=xy-\frac{x^2y^2}{a^2}. \qquad (5)$$

Transferring terms and factoring,
$$xy\left(\frac{xy}{a^2}+1\right)=0. \qquad (6)$$

Whence
$$xy=0, \qquad (7)$$
and
$$xy=-a^2. \qquad (8)$$

From (3) and (7) we have $x=0, y=0$. This solution does not satisfy the given equations, and was therefore introduced by clearing of fractions.

From (3) and (8) we have
$$(x+y)^2=-a^2.$$

Whence
$$x+y=\pm a\sqrt{-1}. \qquad (9)$$

Subtracting $2xy$, $=-2a^2$, from (4), $(x-y)^2=3a^2$.

Whence
$$x-y=\pm a\sqrt{3}. \qquad (10)$$

From (9) and (10) we obtain

$$\frac{a(\sqrt{-1}+\sqrt{3})}{2}, \quad \frac{a(\sqrt{-1}-\sqrt{3})}{2}; \quad \frac{a(\sqrt{-1}-\sqrt{3})}{2},$$
$$\frac{a(\sqrt{-1}+\sqrt{3})}{2}; \quad \frac{a(\sqrt{3}-\sqrt{-1})}{2}, \quad \frac{-a(\sqrt{3}+\sqrt{-1})}{2};$$
$$\frac{-a(\sqrt{-1}+\sqrt{3})}{2}, \quad \frac{-a(\sqrt{-1}-\sqrt{3})}{2}.$$

We find that these sets of values satisfy the given equations.

When the equations are symmetrical, except for sign, the solution can be obtained by a similar method.

Ex. 4. Solve the system $x - y = 1,$ (1)

$xy = 2.$ (2)

Squaring (1), $x^2 - 2xy + y^2 = 1.$ (3)

Adding four times (2) to (3), $x^2 + 2xy + y^2 = 9.$ (4)

Equating square roots of (4), $x + y = \pm 3.$ (5)

The solutions of (5) and (1) are 2, 1, and -1, -2.

Notice that the solutions in this case differ not only in having the values of x and y interchanged, but also in sign. This we should have expected.

8. Many systems which are not symmetrical can be solved by the method of the preceding article.

Ex. 1. Solve the system $2x + 3y = 8,$ (1)

$xy = 2.$ (2)

We should first obtain the value of $2x - 3y$.

Squaring (1), $4x^2 + 12xy + 9y^2 = 64.$ (3)

Subtracting 24 times (2) from (3), $4x^2 - 12xy + 9y^2 = 16.$ (4)

Equating square roots of (4), $2x - 3y = \pm 4.$ (5)

The solutions of (1) and (5) are 3, $\frac{2}{3}$; 1, 2.

9. A system of simultaneous quadratic equations can frequently be solved by some special device.

Ex. 1. Solve the system $x^2 + y = 7,$ (1)

$x + y^2 = 11.$ (2)

Subtracting (1) from the product of (2) by x,

$$xy^2 - y = 11x - 7. \quad (3)$$

Adding (3) to twice (2),

$$xy^2 - y + 2x + 2y^2 = 11x + 15,$$

or

$$y^2(x + 2) - y = 9x + 15. \quad (4)$$

Solving (4) for y,

$$y=\frac{1}{2(x+2)}\pm\sqrt{\left(\frac{9x+15}{x+2}+\frac{1}{4(x+2)^2}\right)}$$

$$=\frac{1}{2(x+2)}\pm\frac{6x+11}{2(x+2)}. \qquad (5)$$

From (5), taking the sign + between the fractions, we have

$$y=\frac{3x+6}{x+2}=3;$$

taking the sign − between the fractions,

$$y=-\frac{6x+10}{2(x+2)}=-\frac{3x+5}{x+2}.$$

Substituting 3 for y in (2), $x=2$.

Therefore one solution is 2, 3.

If we substitute $-\dfrac{3x+5}{x+2}$ for y in either (1) or (2), we are led to the cubic equation

$$x^3+2x^2-10x=19.$$

Methods for solving cubic equations will be given in Part II.

Ex. 2. Solve the system

$$\frac{3}{4}\sqrt{(x-y)}=1+\frac{1}{\sqrt{(x-y)}}, \qquad (1)$$

$$\sqrt{(x+y)}+\sqrt{(x-y)}=5. \qquad (2)$$

Clearing (1) of fractions,

$$3(x-y)=4\sqrt{(x-y)}+4. \qquad (3)$$

Solving (3) as a quadratic in $\sqrt{(x-y)}$, we obtain

$$\sqrt{(x-y)}=2, \text{ and } -\tfrac{2}{3},$$

whence $$x-y=4, \qquad (4)$$

and $$x-y=\tfrac{4}{9}. \qquad (5)$$

Substituting 2 for $\sqrt{(x-y)}$ in (2), $\sqrt{(x+y)}+2=5$,

whence $$x+y=9. \qquad (6)$$

Substituting $-\tfrac{2}{3}$ for $\sqrt{(x-y)}$ in (2), $\sqrt{(x+y)}-\tfrac{2}{3}=5$;

whence $$x+y=\tfrac{289}{9}. \qquad (7)$$

From (4) and (6), $x = 6\frac{1}{2},\ y = 2\frac{1}{2}$;

from (5) and (7), $x = 16\frac{5}{18},\ y = 15\frac{5}{6}$.

The solution $6\frac{1}{2}, 2\frac{1}{2}$ is found to satisfy the given equations; the solution $16\frac{5}{18}, 15\frac{5}{6}$ is found to satisfy the system obtained by changing the sign of the radical $\sqrt{(x-y)}$.

EXERCISES V.

Solve each of the following systems:

1. $\begin{cases} x + y = 12, \\ xy = 32. \end{cases}$

2. $\begin{cases} x + y = 33, \\ xy = 272. \end{cases}$

3. $\begin{cases} x + y = a, \\ xy = b. \end{cases}$

4. $\begin{cases} 2x + 3y = 5, \\ 8xy = 3. \end{cases}$

5. $\begin{cases} \frac{1}{2}x + 5y = 37, \\ xy = 28. \end{cases}$

6. $\begin{cases} x + 2y = 10, \\ xy = 12. \end{cases}$

7. $\begin{cases} x - y = 2, \\ xy = 48. \end{cases}$

8. $\begin{cases} x - y = 8, \\ xy = -15. \end{cases}$

9. $\begin{cases} x - y = m, \\ xy = n. \end{cases}$

10. $\begin{cases} 3x - 5y = -13, \\ 2xy = -4. \end{cases}$

11. $\begin{cases} 6x - 7y = 58, \\ 3xy = -60. \end{cases}$

12. $\begin{cases} x^2 + y^2 = 40, \\ xy = 12. \end{cases}$

13. $\begin{cases} x^2 + y^2 = 181, \\ xy = -90. \end{cases}$

14. $\begin{cases} x^2 + y^2 = 52, \\ xy = 24. \end{cases}$

15. $\begin{cases} 16x^2 + y^2 = 145, \\ xy = 3. \end{cases}$

16. $\begin{cases} 25x^2 + 9y^2 = 148, \\ 5xy = 8. \end{cases}$

17. $\begin{cases} 9x^2 + 25y^2 = 29b^2, \\ xy = \frac{3}{8}b^2. \end{cases}$

18. $\begin{cases} 9x^2 + y^2 = 37a^2, \\ xy = -2a^2. \end{cases}$

19. $\begin{cases} 5x^2 + 2y^2 = 5a^2 + 8b^2, \\ xy = 2ab. \end{cases}$

20. $\begin{cases} x^2 + y^2 = 137, \\ x + y = 15. \end{cases}$

21. $\begin{cases} x^2 + y^2 = 61, \\ x + y = 11. \end{cases}$

22. $\begin{cases} 5x + 3y = 11, \\ 25x^2 + 9y^2 = 73. \end{cases}$

23. $\begin{cases} 3x + 7y = -4, \\ 9x^2 + 49y^2 = 58. \end{cases}$

24. $\begin{cases} x^2 - y^2 = 68, \\ xy = 288. \end{cases}$

25. $\begin{cases} x^2 - y^2 = 28, \\ xy = 48. \end{cases}$

26. $\begin{cases} x^2 - 4y^2 = -3, \\ xy = -1. \end{cases}$

27. $\begin{cases} 9x^2 - 25y^2 = -64, \\ xy = -4. \end{cases}$

28. $\begin{cases} x^2 + y^2 = 53, \\ x - y = 5. \end{cases}$

29. $\begin{cases} x^2 + y^2 = 74, \\ x - y = 2. \end{cases}$

30. $\begin{cases} x^2 + y^2 = 61, \\ x - y = 1. \end{cases}$

31. $\begin{cases} 9x^2 + y^2 = 82, \\ 3x - y = 10. \end{cases}$

32. $\begin{cases} 16x^2 + 49y^2 = 113, \\ 4x + 7y = 1. \end{cases}$

33. $\begin{cases} \dfrac{1}{x} + \dfrac{1}{y} = 3, \\ \dfrac{1}{xy} = 2. \end{cases}$

34. $\begin{cases} \dfrac{1}{x} + \dfrac{1}{y} = \dfrac{1}{2}, \\ \dfrac{2}{xy} = \dfrac{1}{9}. \end{cases}$

35. $\begin{cases} xy = 80, \\ \dfrac{1}{x} - \dfrac{1}{y} = \dfrac{1}{5}. \end{cases}$

36. $\begin{cases} \dfrac{1}{x} + \dfrac{1}{y} = 1, \\ x + y = 4. \end{cases}$

37. $\begin{cases} x + y = 16, \\ \dfrac{1}{x} + \dfrac{1}{y} = \dfrac{1}{3}. \end{cases}$

38. $\begin{cases} \dfrac{x}{y} - \dfrac{y}{x} = \dfrac{16}{15}, \\ x - y = 2. \end{cases}$

39. $\begin{cases} \dfrac{x}{y} + \dfrac{y}{x} = 3\frac{1}{3}, \\ \dfrac{1}{x} + \dfrac{1}{y} = 1\frac{1}{3}. \end{cases}$

40. $\begin{cases} \dfrac{1}{x} + \dfrac{1}{y} = 10, \\ \dfrac{1}{x^2} + \dfrac{1}{y^2} = 58. \end{cases}$

41. $\begin{cases} x^2 - y^2 = \frac{5}{6}xy, \\ \dfrac{1}{x} + \dfrac{1}{y} = \dfrac{5}{6}. \end{cases}$

42. $\begin{cases} x^2 + y^2 = 45, \\ \dfrac{1}{x} + \dfrac{1}{y} = \dfrac{1}{2}. \end{cases}$

43. $\begin{cases} x^2 + y^2 = 2\frac{1}{2}xy, \\ \dfrac{1}{x} + \dfrac{1}{y} = 1\frac{1}{2}. \end{cases}$

44. $\begin{cases} \dfrac{1}{x} + \dfrac{1}{y} = \dfrac{1}{a} + \dfrac{1}{b}, \\ x^2 + y^2 = a^2 + b^2. \end{cases}$

45. $\begin{cases} \sqrt{x} + \sqrt{y} = 5, \\ xy = 36. \end{cases}$

46. $\begin{cases} \sqrt{x} + \sqrt{y} = 7, \\ x + y = 37. \end{cases}$

47. $\begin{cases} x^2 + x + y = 18 - y^2, \\ xy = 6. \end{cases}$

48. $\begin{cases} x^2 + mxy + y^2 = a^2, \\ nxy = b. \end{cases}$

49. $\begin{cases} x + xy + y = 29, \\ x^2 + xy + y^2 = 61. \end{cases}$

50. $\begin{cases} x^2 + y^2 = 45, \\ xy = 2(x + y). \end{cases}$

51. $\begin{cases} x^2 + y^2 + 7\,xy = 171, \\ xy = 2(x + y). \end{cases}$

52. $\begin{cases} x^2 + y^2 - (x - y) = 20, \\ xy + x - y = 1. \end{cases}$

53. $\begin{cases} x^2 + y^2 - x - y = 22, \\ x + y + xy = -1. \end{cases}$

54. $\begin{cases} x^2 + y^2 + x + y = m, \\ x + y = xy. \end{cases}$

55. $\begin{cases} x^2 + y^2 + x - y = a, \\ xy + x - y = b. \end{cases}$

56. $\begin{cases} x + y = 2, \\ x^2 + y^2 + xy = 3. \end{cases}$

57. $\begin{cases} x + y = 4\frac{3}{4}, \\ x^2 + y^2 + xy = 19\frac{9}{16}. \end{cases}$

58. $\begin{cases} x^2 + xy + y^2 = 397, \\ x + y = 23. \end{cases}$

59. $\begin{cases} x + y = 9, \\ x^2 + y^2 - xy = 21. \end{cases}$

60. $\begin{cases} x^2 + xy + y^2 = 2\,m, \\ x^2 - xy + y^2 = 2\,n. \end{cases}$

61. $\begin{cases} 4\,xy = 96 - x^2y^2, \\ x + y = 6. \end{cases}$

62. $\begin{cases} x^2 + xy + y^2 = \dfrac{3\,a^2 + b^2}{(a^2 - b^2)^2}, \\ (a^2 - b^2)xy = 1. \end{cases}$

63. $\begin{cases} \sqrt{[(2 + x)(1 + y)]} = 2, \\ \sqrt{(2+x)} - \sqrt{(1+y)} = \frac{1}{8}. \end{cases}$

64. $\begin{cases} x + \sqrt{x} + y + \sqrt{y} = 18, \\ (x + \sqrt{x})(y + \sqrt{y}) = 72. \end{cases}$

65. $\begin{cases} x + y + \sqrt{(x + y)} = 6, \\ 2(x^2 - y^2) = 3. \end{cases}$

66. $\begin{cases} 2\,x + 5\sqrt{(2\,x + y + 2)} = 22 - y, \\ 4\,x^2 + 2\,x + y = 34 - x - y^2. \end{cases}$

Simultaneous Quadratic Equations in Three Unknown Numbers.

10. No definite methods can be given for solving simultaneous quadratic equations in three unknown numbers.

Ex. 1. Solve the system

$$xy = 2, \quad (1)$$

$$xz = 3, \quad (2)$$

$$yz = 6. \quad (3)$$

Multiplying corresponding members of (1), (2), and (3),

$$x^2y^2z^2 = 36. \qquad (4)$$

Equating square roots of (4), $xyz = \pm 6.$ (5)

Dividing (5) in turn by (1), (2), and (3),

$$z = \pm 3, \quad y = \pm 2, \quad x = \pm 1.$$

The required solutions are therefore, 1, 2, 3, and −1, −2, −3.

Ex. 2. Solve the system

$$x(y + z) = 5, \qquad (1)$$

$$y(x + z) = 8, \qquad (2)$$

$$z(x + y) = 9. \qquad (3)$$

Adding corresponding members of (1), (2), and (3), and dividing by 2,

$$xy + xz + yz = 11. \qquad (4)$$

Subtracting in turn (1), (2), and (3) from (4),

$$yz = 6, \qquad (5)$$

$$xz = 3, \qquad (6)$$

$$xy = 2. \qquad (7)$$

The solutions of equations (5), (6), and (7) are 1, 2, 3, and − 1, − 2, − 3.

Ex. 3. Solve the system

$$x^2 + y^2 + z^2 = 169, \qquad (1)$$

$$x + y + z = 19, \qquad (2)$$

$$x(y + z) = 48. \qquad (3)$$

Multiplying (2) by x, $x^2 + x(y + z) = 19x.$ (4)

Subtracting (3) from (4), $x^2 = 19x - 48.$ (5)

Whence $x = 3$ and 16.

Substituting 3 for x in (1) and (2), we obtain

from (1), $y^2 + z^2 = 160,$ (6)

from (2), $y + z = 16.$ (7)

Substituting 16 for x in (1) and (2), we obtain

from (1), $y^2 + z^2 = -87,$ (8)

from (2), $y + z = 3.$ (9)

Solving (6) and (7), we have $y = 4, \quad z = 12;$

and $y = 12, \quad z = 4.$

Solving (8) and (9), we have

$$y = \frac{3 \pm \sqrt{-183}}{2}, \quad z = \frac{3 \mp \sqrt{-183}}{2}.$$

The solutions of the given system are therefore

$$3,\ 4,\ 12; \quad 3,\ 12,\ 4; \quad 16,\ \frac{3 \pm \sqrt{-183}}{2},\ \frac{3 \mp \sqrt{-183}}{2};$$

Ex. 4. Solve the system $x^2 + xy + y^2 = 37,$ (1)

$x^2 + xz + z^2 = 28,$ (2)

$y^2 + yz + z^2 = 19.$ (3)

Subtracting (2) from (1), $x(y - z) + y^2 - z^2 = 9,$ (4)

or $(y - z)(x + y + z) = 9.$ (5)

Subtracting (3) from (2), $(x - y)(x + y + z) = 9.$ (6)

From (5) and (6) we obtain $y - z = x - y,$ (7)

or $x = 2y - z.$ (8)

Substituting $2y - z$ for x in (1),

$(2y - z)^2 + (2y - z)y + y^2 = 37,$ (9)

or $7y^2 - 5yz + z^2 = 37.$ (10)

Multiplying (3) by 37, $37y^2 + 37yz + 37z^2 = 37 \times 19.$ (11)

Multiplying (10) by 19, $133y^2 - 95yz + 19z^2 = 37 \times 19.$ (12)

Subtracting (11) from (12),

$96y^2 - 132yz - 18z^2 = 0,$ (13)

or $48y^2 - 66yz - 9z^2 = 0,$ (14)

or $(6y - 9z)(8y + z) = 0.$ (15)

Now the given system is equivalent to the system (1), (5), (6), which is equivalent to the system (1), (3), (8). The latter system is equivalent to the system (1), (3), (10), which is equivalent to the system (8), (3), (15). The latter system is equivalent to the two systems

$$\left.\begin{aligned} x &= 2y - z, \\ 6y - 9z &= 0, \\ y^2 + yz + z^2 &= 19, \end{aligned}\right\} (a), \qquad \left.\begin{aligned} x &= 2y - z, \\ 8y + z &= 0, \\ y^2 + yz + z^2 &= 19, \end{aligned}\right\} (b).$$

In solving systems (a) and (b), we first solve the last two equations for y and z.

The solutions of (a) and (b), and therefore of the given system, are 4, 3, 2; -4, -3, -2; $10\sqrt{\frac{1}{3}}$, $\sqrt{\frac{1}{3}}$, $-8\sqrt{\frac{1}{3}}$; $-10\sqrt{\frac{1}{3}}$, $-\sqrt{\frac{1}{3}}$, $8\sqrt{\frac{1}{3}}$.

EXERCISES VI.

Solve each of the following systems:

1. $\begin{cases} xy = 30, \\ yz = -60, \\ xz = -50. \end{cases}$

2. $\begin{cases} xy = a^2, \\ yz = b^2, \\ xz = c^2. \end{cases}$

3. $\begin{cases} x\sqrt{y} = a, \\ x\sqrt{z} = b, \\ y\sqrt{z} = c. \end{cases}$

4. $\begin{cases} x^2 + y^2 = 13, \\ x^2 + z^2 = 34, \\ y^2 + z^2 = 29. \end{cases}$

5. $\begin{cases} x^2 + yz = 5, \\ y^2 + xz = 5, \\ z^2 + xy = 5. \end{cases}$

6. $\begin{cases} x^2 + y^2 = a^2, \\ y^2 + z^2 = b^2, \\ z^2 + x^2 = c^2. \end{cases}$

7. $\begin{cases} x(y + z) = 6, \\ y(x + z) = 23\frac{1}{3}, \\ z(x + y) = 22\frac{2}{3}. \end{cases}$

8. $\begin{cases} x(y + z) = a, \\ y(x + z) = b, \\ z(x + y) = c. \end{cases}$

9. $\begin{cases} x(x + y + z) = 6, \\ y(x + y + z) = 12, \\ z(x + y + z) = 18. \end{cases}$

10. $\begin{cases} \dfrac{xyz}{x + y} = 2, \\ \dfrac{xyz}{x + z} = \dfrac{3}{2}, \\ \dfrac{xyz}{y + z} = \dfrac{6}{5}. \end{cases}$

11. $\begin{cases} \dfrac{x + y}{xyz} = -(a - b)^2, \\ \dfrac{x + z}{xyz} = -(a - c)^2, \\ \dfrac{y + z}{xyz} = -(b - c)^2. \end{cases}$

12. $\begin{cases} x^2 + xy + y^2 = 61, \\ x^2 + xz + z^2 = 21, \\ y^2 + yz + z^2 = 13. \end{cases}$

13. $\begin{cases} x^2 + xy + y^2 = a^2, \\ x^2 + xz + z^2 = b^2, \\ y^2 + yz + z^2 = c^2. \end{cases}$

14. $\begin{cases} 3x = 5y, \\ x(z + 2) = yz + 32, \\ x(z - 1) = (y + 1)z - 1. \end{cases}$

15. $\begin{cases} x + y + z = a, \\ x(x + y) = b^2, \\ z(z + y) = c^2. \end{cases}$

16. $\begin{cases} xz = y^2, \\ x + y + z = 19, \\ x^2 + y^2 + z^2 = 133. \end{cases}$

17. $\begin{cases} \sqrt{(x^2 + y^2 + z^2)} = 13, \\ x + y + z = 19, \\ x(y + z) = 48. \end{cases}$

18. $\begin{cases} xz = 360, \\ y(z - 10) = 40, \\ x(z + 8) = 400 + y(z - 2). \end{cases}$

19. $\begin{cases} y = \frac{1}{2}(x + z), \\ x^2 + y^2 = 458, \\ y^2 + z^2 = 730. \end{cases}$

20. $\begin{cases} (x + y)(z + x) = a, \\ (y + z)(x + y) = b, \\ (z + x)(y + z) = c. \end{cases}$

21. $\begin{cases} (y + z)(x + y + z) = 6, \\ (z + x)(x + y + z) = 8, \\ (x + y)(x + y + z) = -6. \end{cases}$

22. $\begin{cases} x^2 + y^2 + z^2 = 29, \\ xy + xz + yz = -10, \\ x + y - z = -5. \end{cases}$

23. $\begin{cases} (x + y + z)(x + y - z) = 24\frac{3}{4}, \\ (x - y + z)(y + z - x) = 2\frac{1}{4}, \\ (x^2 + y^2) = 2z^2. \end{cases}$

§ 2. SIMULTANEOUS HIGHER EQUATIONS.

1. The solutions of certain equations of higher degree than the second can be made to depend upon the solutions of quadratic equations.

Ex. 1. Solve the system $\quad x^3 + y^3 = 9, \quad (1)$

$$x + y = 3. \quad (2)$$

Dividing (1) by (2), $\quad x^2 - xy + y^2 = 3. \quad (3)$

Subtracting (3) from the square of (2),

$$3xy = 6, \text{ or } xy = 2. \quad (4)$$

The solutions of (2) and (4), and therefore of the given systems, are 1, 2, and 2, 1.

Ex. 2. Solve the system $\quad x^4 + x^2y^2 + y^4 = 21, \quad (1)$

$$x^2 + xy + y^2 = 7. \quad (2)$$

Dividing (1) by (2), $\quad x^2 - xy + y^2 = 3. \quad (3)$

Subtracting (3) from (2), $\quad 2xy = 4, \quad (4)$

or $\quad xy = 2. \quad (5)$

Adding (5) to (2), $\quad x^2 + 2xy + y^2 = 9. \quad (6)$

Subtracting (5) from (3), $\quad x^2 - 2xy + y^2 = 1. \quad (7)$

Equating square roots, from (6), $x + y = \pm 3,$ (8)

from (7), $x - y = \pm 1.$ (9)

The given system is equivalent to the system (2), (3), which is equivalent to (2), (5), or (3), (5).

Therefore the given system is equivalent to (6), (7), or (8), (9).

The solutions of (8) and (9), and hence of the given system, are 2, 1; 1, 2; $-1, -2$; $-2, -1$.

Ex. 3. Solve the system $x^4 + y^4 = 17,$ (1)

$$x + y = 3. \quad (2)$$

We first find the value of xy.

Let $xy = z.$ (3)

Squaring (2), $x^2 + 2xy + y^2 = 9,$ (4)

or $x^2 + y^2 = 9 - 2z,$ (5)

Squaring (5), $x^4 + 2x^2y^2 + y^4 = 81 - 36z + 4z^2,$ (6)

or $x^4 + y^4 = 81 - 36z + 2z^2.$ (7)

Since $x^4 + y^4 = 17$, we have from (7)

$$2z^2 - 36z + 81 = 17. \quad (8)$$

Whence $z = 16$, and 2. (9)

Therefore, from (3) and (9), $xy = 16,$ (10)

and $xy = 2.$ (11)

The solutions of (2) and (10) and of (2) and (11) are readily found, and should be checked by substitution.

Ex. 4. Solve the system

$$x^5 + y^5 = 33, \quad (1)$$

$$x + y = 3. \quad (2)$$

Subtracting (1) from the fifth power of (2),

$$5x^4y + 10x^3y^2 + 10x^2y^3 + 5xy^4 = 210,$$

or $xy(x^3 + 2x^2y + 2xy^2 + y^3) = 42.$ (3)

Adding $x^2y + xy^2$ to the expression within the parentheses, and subtracting the same, we obtain

$$xy(x^3 + 3x^2y + 3xy^2 + y^3 - x^2y - xy^2) = 42,$$

or $xy(x + y)^3 - x^2y^2(x + y) = 42.$ (4)

Substituting 3 for $x+y$ in (4),

$$27\,xy - 3\,x^2y^2 = +\,42,$$

or $$x^2y^2 - 9\,xy = -\,14. \qquad (5)$$

Solving (5) as a quadratic in xy,

$$xy = 7, \qquad (6)$$

and $$xy = 2. \qquad (7)$$

By solving (2) and (6), and (2) and (7), we obtain the values of x and y. The results should be checked by substitution.

Ex. 5. Solve the system

$$(x^2+y^2)(x^3+y^3) = 45, \qquad (1)$$

$$x+y = 3. \qquad (2)$$

From (2) $$x^2+y^2 = 9 - 2\,xy, \qquad (3)$$

and $$x^3+y^3 = 27 - 3\,xy(x+y)$$

$$= 27 - 9\,xy, \text{ since } x+y=3. \qquad (4)$$

Substituting in (1) for x^2+y^2 and x^3+y^3 their values from (3) and (4),

$$(9-2\,xy)(27-9\,xy) = 45,$$

or $$2\,x^2y^2 - 15\,xy = -\,22. \qquad (5)$$

Equation (5) can be solved as a quadratic in xy, and the results combined with equation (2). The results should be checked by substitution.

EXERCISES VII.

Solve each of the following systems:

1. $\begin{cases} x+y=5, \\ x^3+y^3=35. \end{cases}$

2. $\begin{cases} x-y=1, \\ x^3-y^3=7. \end{cases}$

3. $\begin{cases} 2(x+y)=5, \\ 32(x^3+y^3)=2285. \end{cases}$

4. $\begin{cases} (x-1)^3+(y-2)^3=28, \\ x+y=7. \end{cases}$

5. $\begin{cases} (x-7)^3+(5-y)^3=9, \\ x-y=5. \end{cases}$

6. $\begin{cases} x^4-y^4=554, \\ x^2+y^2=34. \end{cases}$

7. $\begin{cases} x^4+y^4=82, \\ xy=3. \end{cases}$

8. $\begin{cases} x^4+y^4=97, \\ xy=6. \end{cases}$

9. $\begin{cases} x^4 + y^4 = 17, \\ xy = 2. \end{cases}$

10. $\begin{cases} x^4 + y^4 = 97, \\ x + y = 5. \end{cases}$

11. $\begin{cases} x^4 + y^4 = 257, \\ x - y = 3. \end{cases}$

12. $\begin{cases} (x-3)^4 + (y-2)^4 = 82, \\ x + y = 9. \end{cases}$

13. $\begin{cases} (x-7)^4 + (y-3)^4 = 257, \\ x - y + 1 = 0. \end{cases}$

14. $\begin{cases} (x^2 + y^2)(x - y) = 200, \\ xy(x - y) = 96. \end{cases}$

15. $\begin{cases} (x^2 - y^2)(x + y) = 9, \\ xy(x + y) = 6. \end{cases}$

16. $\begin{cases} (x + y)(x^2 + y^2) = 175, \\ (x - y)(x^2 - y^2) = 7. \end{cases}$

17. $\begin{cases} (x + y)(x^2 + y^2) = 585, \\ (x - y)(x^2 - y^2) = 441. \end{cases}$

18. $\begin{cases} x^5 + y^5 = 1056, \\ x + y = 6. \end{cases}$

19. $\begin{cases} x^5 - y^5 = 211, \\ x - y = 1. \end{cases}$

20. $\begin{cases} x - y = 342, \\ \sqrt[3]{x} - \sqrt[3]{y} = 6. \end{cases}$

21. $\begin{cases} x + y = 30, \\ \sqrt[3]{(x+7)} + \sqrt[3]{(y-9)} = 4. \end{cases}$

22. $\begin{cases} x - y = 68, \\ \sqrt[3]{(x-5)} + \sqrt[3]{(9-y)} = 6. \end{cases}$

23. $\begin{cases} x + y + \sqrt{(x + y)} = 12, \\ x^3 + y^3 = 189. \end{cases}$

24. $\begin{cases} \dfrac{x^2 + y^2}{x^2 - y^2} = \dfrac{17}{8}, \\ xy^2 = 45. \end{cases}$

25. $\begin{cases} \sqrt[4]{(x+7)} + \sqrt[4]{(y-5)} = 3, \\ x + y = 15. \end{cases}$

26. $\begin{cases} \sqrt[4]{(90-x)} - \sqrt[4]{(9-y)} = 2, \\ x + y = 17. \end{cases}$

27. $\begin{cases} x^4 + y^4 = 4097, \\ x + y + \sqrt{(x + y)} = 12. \end{cases}$

28. $\begin{cases} \sqrt[5]{(250-x)} + \sqrt[5]{(2-y)} = 2, \\ x + y = 10. \end{cases}$

29. $\begin{cases} \sqrt[5]{(x-8)} - \sqrt[5]{(y-6)} = 2, \\ x - y = 4. \end{cases}$

EXERCISES VIII.

MISCELLANEOUS EXAMPLES.

Solve each of the following systems by the methods given in this chapter, or by special devices:

1. $\begin{cases} x + y = 100, \\ xy = 2400. \end{cases}$

2. $\begin{cases} x - 2y = 2, \\ xy = 12. \end{cases}$

3. $\begin{cases} x(7x - 8y) = 159, \\ 5x + 2y = 7. \end{cases}$

4. $\begin{cases} x + y = x^2, \\ 3y - x = y^2. \end{cases}$

5. $\begin{cases} x^2 - y^2 = 8(x - y), \\ x^2 + y^2 = 50. \end{cases}$

6. $\begin{cases} x^2 - y^2 = 2(x + y), \\ x^2 + y^2 = 100. \end{cases}$

7. $\begin{cases} \dfrac{x}{y} = \dfrac{5}{36}(x + y), \\ \dfrac{y}{x} = \dfrac{4}{45}(x + y). \end{cases}$

8. $\begin{cases} x\left(1 + \dfrac{x}{y}\right) = 90, \\ y\left(1 + \dfrac{y}{x}\right) = 40. \end{cases}$

9. $\begin{cases} \dfrac{x + y}{x - y} + \dfrac{x - y}{x + y} = 5\frac{1}{5}, \\ 2x^2 - 3y^2 = 24. \end{cases}$

10. $\begin{cases} \dfrac{x^2 - 2xy + 3y^2}{3x^2 - 2xy + y^2} = \dfrac{1}{3}, \\ x^2 - 3y = 1. \end{cases}$

11. $\begin{cases} (x - 7)(y - 3) = 0, \\ (x - 5)(y - 9) = 0. \end{cases}$

12. $\begin{cases} (5x - 3)(3y + 2) = 0, \\ (4x + 5)(2y - 3) = 0. \end{cases}$

13. $\begin{cases} (2x - 5)(3y + 7) = 0, \\ x^2 + 3xy + 5x - 2y = 19\frac{3}{8}. \end{cases}$

14. $\begin{cases} (2x + 3y - 7)(x - 4y + 2) = 0, \\ x^2 + y^2 + 2x - 7y = 2. \end{cases}$

15. $\begin{cases} x - y = 5(x^2 - y^2), \\ 2x^2 + 3xy + 4x + 5y = 0. \end{cases}$

16. $\begin{cases} x^2 + y^2 + 5x - 9y = 84, \\ x^2 - y^2 + 5x + 9y = 84. \end{cases}$

17. $\begin{cases} 3x^2 + 5y^2 + 4x + 3y = 9, \\ 3x^2 + 5y^2 + 2x - 4y = 14. \end{cases}$

18. $\begin{cases} x^2 + y^2 = 485, \\ x^2y^2 = 57834 - 5xy. \end{cases}$

19. $\begin{cases} x^2 + y^2 + x - y = 62, \\ (x^2 + y^2)(x - y) = 61. \end{cases}$

20. $\begin{cases} 3(x+y)^2 = \frac{8}{7}(x+y) + \frac{1280}{7}, \\ \dfrac{x^2y^2}{5} - \dfrac{2xy}{15} = 43. \end{cases}$

21. $\begin{cases} \sqrt{(3x + 3y - 5)} + 16 - 6y = 6x, \\ 2x^2y^2 + 2 = 5xy. \end{cases}$

22. $\begin{cases} 2x + 3y + 6xy = 11, \\ 4x^2 + 9y^2 + 12xy = x^2y^2 - 11. \end{cases}$

23. $\begin{cases} 8\left(x - \dfrac{1}{x}\right)\left(y - \dfrac{1}{y}\right) = 3, \\ \left(x - \dfrac{1}{x}\right)^2 + \left(y - \dfrac{1}{y}\right)^2 = \dfrac{3229}{3600}. \end{cases}$

24. $\begin{cases} x^2 + y^2 = \dfrac{40}{x+y}, \\ xy = \dfrac{12}{x+y}. \end{cases}$

25. $\begin{cases} \dfrac{x^3 - y^3}{x^2y - xy^2} = \dfrac{7}{2}, \\ x + y = 6. \end{cases}$

26. $\begin{cases} x^3 - y^3 = 26, \\ x - y = 2. \end{cases}$

27. $\begin{cases} x + y = 18, \\ x^3 + y^3 = 4914. \end{cases}$

28. $\begin{cases} x^3y^2 - x^2y^3 = 1152, \\ x^2y - xy^2 = 48. \end{cases}$

29. $\begin{cases} x + y = 19, \\ \sqrt[3]{x} + \sqrt[3]{y} = 4. \end{cases}$

30. $\begin{cases} x - y - \sqrt{(x - y)} = 2, \\ x^3 - y^3 = 2044. \end{cases}$

31. $\begin{cases} x^4 + y^4 = 641, \\ xy\,(x^2 + y^2) = 290. \end{cases}$

32. $\begin{cases} 2\,(ab + xy) + (a + b)\,(x + y) = 0, \\ 2\,(cd + xy) + (c + d)\,(x + y) = 0. \end{cases}$

33. $\begin{cases} \left(\dfrac{x+y}{x-y}\right)^2 = 64, \\ xy = 63. \end{cases}$

34. $\begin{cases} \dfrac{x^2}{y^2} + \dfrac{4x}{y} = \dfrac{85}{9}, \\ x - y = 2. \end{cases}$

35. $\begin{cases} \dfrac{xy}{x+y} = a, \\ \dfrac{x^2y^2}{x^2+y^2} = b^2. \end{cases}$

36. $\begin{cases} \dfrac{1}{x} + \dfrac{1}{y} = \dfrac{1}{m}, \\ \dfrac{1}{x^3} + \dfrac{1}{y^3} = \dfrac{1}{n^3}. \end{cases}$

37. $\begin{cases} x^n + y^n = a^n, \\ xy = b. \end{cases}$

38. $\begin{cases} x^3 + xy^2 = y, \\ y^3 - x = x^2y. \end{cases}$

39. $\begin{cases} xy + \dfrac{x}{y} = \dfrac{5}{3}, \\ \dfrac{1}{xy} + \dfrac{y}{x} = \dfrac{20}{3}. \end{cases}$

40. $\begin{cases} \dfrac{x^3}{y} - \dfrac{y^3}{x} = \dfrac{15}{2}, \\ \dfrac{x}{y} - \dfrac{y}{x} = \dfrac{3}{2}. \end{cases}$

41. $\begin{cases} \dfrac{x^2}{y^2} + \dfrac{y^2}{x^2} + \dfrac{x}{y} + \dfrac{y}{x} = \dfrac{27}{4}, \\ x - y = 2. \end{cases}$

42. $\begin{cases} x^4 + y^4 = 14\,x^2y^2, \\ x + y = a. \end{cases}$

43. $\begin{cases} 3y(x^2+y^2)=10x, \\ 2x(x^2-y^2)=48y. \end{cases}$

44. $\begin{cases} (x^2+y^2)(x^3+y^3)=455, \\ x+y=5. \end{cases}$

45. $\begin{cases} x^4+x^2y^2+y^4=133, \\ x^2-xy+y^2=7. \end{cases}$

46. $\begin{cases} x^4+x^2y^2+y^4=84x^2, \\ x^2+xy+y^2=14x. \end{cases}$

47. $\begin{cases} x^4-x^2+y^4-y^2=84, \\ x^2+xy+y^2=19. \end{cases}$

48. $\begin{cases} x^2-x^2y^2+y^2+23=0, \\ x-xy+y+1=0. \end{cases}$

49. $\begin{cases} x^3+y^3+xy(x+y)=13, \\ x^2y^2(x^2+y^2)=468. \end{cases}$

50. $\begin{cases} (x+y)(xy+1)=18xy, \\ (x^2+y^2)(x^2y^2+1)=320x^2y^2. \end{cases}$

51. $\begin{cases} \dfrac{(ab+1)(x^2+1)}{x+1}=\dfrac{(a^2+1)(xy+1)}{y+1}, \\ \dfrac{(ab+1)(y^2+1)}{y+1}=\dfrac{(b^2+1)(xy+1)}{x+1}. \end{cases}$

52. $\begin{cases} x+y+\sqrt{(xy)}=14, \\ x^2+y^2+xy=84. \end{cases}$

53. $\begin{cases} x+y-\sqrt{(xy)}=7, \\ x^2+y^2+xy=133. \end{cases}$

54. $\begin{cases} \dfrac{x-\sqrt{(x^2-y^2)}}{x+\sqrt{(x^2-y^2)}}=x, \\ \dfrac{x}{y}=\sqrt{\left(\dfrac{1+x}{1-y}\right)}. \end{cases}$

55. $\begin{cases} \dfrac{x-\sqrt{(x^2-y^2)}}{x+\sqrt{(x^2-y^2)}}=x, \\ \dfrac{x}{y}=\sqrt{\left(\dfrac{5}{2}\cdot\dfrac{1-x^2}{1-y^2}\right)}. \end{cases}$

56. $\begin{cases} \sqrt{(x^2+y^2)}+\sqrt{(x^2-y^2)}=29, \\ x^4-y^4=a^4. \end{cases}$

57. $\begin{cases} \sqrt{(x+y)}-\sqrt{(x-y)}=a, \\ \sqrt[4]{(x+y)}+\sqrt[4]{(x-y)}=b. \end{cases}$

58. $\begin{cases} x+y-\sqrt{\dfrac{x+y}{x-y}}=\dfrac{6}{x-y}, \\ x^2+y^2=41. \end{cases}$

59. $\begin{cases} \frac{3}{4}\sqrt{(x-y)}=1+\dfrac{1}{\sqrt{(x-y)}}, \\ \sqrt{(x+y)}+\sqrt{(x-y)}=5. \end{cases}$

60. $\begin{cases} y^2=xz, \\ x+y+z=28, \\ xyz=512. \end{cases}$

61. $\begin{cases} x^3+y^3=z^3, \\ x+y=z+a, \\ xy=b. \end{cases}$

62. $\begin{cases} x+y+z=5, \\ x^2+y^2=z^2, \\ x^3+y^3+z^3=8. \end{cases}$

63. $\begin{cases} x+y+z=4, \\ xy+xz+yz=-4, \\ x-y+z=8. \end{cases}$

64. $\begin{cases} (x+y)^2+(y+z)^2+(z+x)^2=44, \\ (x-y)^2+(y-z)^2+(z-x)^2=96, \\ x^2+y^2-z^2=-15. \end{cases}$

65. $\begin{cases} xu=yz, \\ x+u=18, \\ y+z=14, \\ x^2+y^2+z^2+u^2=340. \end{cases}$ 66. $\begin{cases} xu=yz, \\ x+u=7, \\ y+z=5, \\ x^4+y^4+z^4+u^4=1394. \end{cases}$

§ 3. APPLICATIONS OF THE CRITERIA OF THE NATURE OF THE ROOTS OF QUADRATIC EQUATIONS TO SIMULTANEOUS QUADRATIC EQUATIONS.

Ex. 1. Given the system $y = x + c,$ (1)

$$x^2 + y^2 = 25, \quad (2)$$

find a value of c which will make the two solutions equal, *i.e.*, the two values of x equal to each other, and the two values of y equal to each other.

Substituting in (2) the value of y given in (1), we obtain, on rearranging terms,

$$2x^2 + 2cx + c^2 - 25 = 0. \quad (3)$$

In order that the two solutions shall be equal, the two values of x must be equal, *i.e.*, equation (3) must have equal roots. But, by Ch. XX., Art. 18 (ii.), that equation will have equal roots when

$$(2c)^2 = 4 \times 2(c^2 - 25),$$

or when $4c^2 = 8c^2 - 200,$

or when $c^2 = 50.$

Whence $c = \pm 5\sqrt{2}.$

Therefore, the solutions of the system

$$y = x + 5\sqrt{2},$$
$$x^2 + y^2 = 25$$

will be equal; and the solutions of the system

$$y = x - 5\sqrt{2},$$
$$x^2 + y^2 = 25$$

will be equal.

Ex. 2. Given the system $y = mx + c$, (1)

$$b^2x^2 + a^2y^2 = a^2b^2, \quad (2)$$

find a value of c which will make the two solutions equal. In this example we regard the values of a, b, and m as fixed, and the required value of c will be expressed in terms of these letters.

Substituting in (2) the value of y from (1), we obtain, on rearranging terms,

$$(b^2 + a^2m^2)x^2 + 2a^2mcx + a^2c^2 - a^2b^2 = 0. \quad (3)$$

In order that the two roots of equation (3) shall be equal, we must have

$$(2a^2mc)^2 = 4(b^2 + a^2m^2)(a^2c^2 - a^2b^2). \quad (4)$$

In equation (4), c is the unknown number whose value must be found in terms of a, b, m.

We obtain $c = \pm\sqrt{(a^2m^2 + b^2)}$.

Consequently the two solutions of the system

$$y = mx + \sqrt{(a^2m^2 + b^2)},$$
$$b^2x^2 + a^2y^2 = a^2b^2$$

will be equal, and the two solutions of the system

$$y = mx - \sqrt{(a^2m^2 + b^2)},$$
$$b^2x^2 + a^2y^2 = a^2b^2$$

will be equal.

Ex. 3. Given the system $y = mx + c$, (1)

$$b^2x^2 - a^2y^2 = a^2b^2, \quad (2)$$

find values of m and c, such that the two solutions shall be infinite, *i.e.*, such that both values of x and y shall be ∞.

Substituting in (2) the value of y from (1), we obtain

$$(b^2 - a^2m^2)x^2 - 2a^2mcx - a^2c^2 - a^2b^2 = 0. \quad (3)$$

In order that the two solutions shall be infinite, the two roots of equation (3) must be infinite. We therefore have, by Ch. XX., Art. 19 (ii.),

$$b^2 - a^2m^2 = 0, \quad (4)$$

and

$$2a^2mc = 0. \quad (5)$$

From (4) we obtain $m = \pm\dfrac{b}{a}$;

and, since $a \neq 0$, and $m \neq 0$, we have from (5)

$$c = 0.$$

Therefore the two solutions of the system

$$y = \frac{b}{a}x,$$

$$b^2x^2 - a^2y^2 = a^2b^2,$$

and of the system

$$y = -\frac{b}{a}x,$$

$$b^2x^2 - a^2y^2 = a^2b^2$$

are infinite.

EXERCISES IX.

In each of the following systems determine a value of c such that the two solutions shall be equal:

1. $\begin{cases} y - 3x = c, \\ x^2 + y^2 = 16. \end{cases}$
2. $\begin{cases} y + 2x = c, \\ x^2 + y^2 = a^2. \end{cases}$
3. $\begin{cases} 2x - 3y = c, \\ 9x^2 + 25y^2 = 225. \end{cases}$
4. $\begin{cases} 5x - y = c, \\ y^2 = 4x + 1. \end{cases}$
5. $\begin{cases} y = mx + c, \\ y^2 = 4px. \end{cases}$
6. $\begin{cases} y = mx + c, \\ b^2x^2 - a^2y^2 = a^2b^2. \end{cases}$

In each of the following systems determine values of m and c such that both solutions shall be infinite:

7. $\begin{cases} y = mx + c, \\ 9x^2 - 16y^2 = 144. \end{cases}$
8. $\begin{cases} y = mx + c, \\ 25y^2 - x^2 = 25. \end{cases}$

§ 4. PROBLEMS.

PR. 1. Find a number of two digits such that, if it be divided by the product of the digits the quotient will be 2, and if 27 be added to the number, the sum will be equal to the number obtained by interchanging the digits.

Let x stand for the digit in the units' place, and y for the digit in the tens' place.

Then, by the first condition,

$$\frac{10y + x}{xy} = 2; \tag{1}$$

and by the second condition

$$10y + x + 27 = 10x + y. \tag{2}$$

Solving equations (1) and (2), we obtain $x = 6$, and $\frac{3}{2}$.

Since a digit must be an integer, we reject the solution $x = \frac{5}{2}$. When $x = 6$, $y = 3$. The required number is therefore 36.

PR. 2. The front wheel of a carriage makes 6 more revolutions than the hind wheel in traveling 360 feet. But if the circumference of each wheel were 3 feet greater, the front wheel would make only 4 revolutions more than the hind wheel in traveling the same distance as before. What are the circumferences of the two wheels?

Let x stand for the number of feet in the circumference of front wheel, and y for the number of feet in the circumference of hind wheel. Then in traveling 360 feet the front wheel makes $\frac{360}{x}$ revolutions, and the hind wheel makes $\frac{360}{y}$ revolutions.

Therefore, by the first condition,

$$\frac{360}{x} = \frac{360}{y} + 6. \tag{1}$$

If 3 feet were added to the circumference of each wheel, the front wheel would make $\frac{360}{x+3}$ revolutions, and the hind wheel $\frac{360}{y+3}$ revolutions.

Therefore, by the second condition,

$$\frac{360}{x+3} = \frac{360}{y+3} + 4. \tag{2}$$

Solving equations (1) and (2), we obtain, as the only admissible solution, $x = 12$, $y = 15$.

Therefore, the circumference of the front wheel is 12 feet, and the circumference of the hind wheel is 15 feet.

EXERCISES X.

1. The square of one number increased by ten times a second number is 84, and is equal to the square of the second number increased by ten times the first.

2. The sum of two numbers is 20, and the sum of the square of the one diminished by 13 and the square of the other increased by 13 is 272. What are the numbers?

3. Find two numbers such that their difference added to the difference of their squares shall be 150, and their sum added to the sum of their squares shall be 330.

4. Find two numbers whose sum is equal to their product, and also to the difference of their squares.

5. The sum of the fourth powers of two numbers is 1921, and the sum of their squares is 61. What are the numbers?

6. If a number of two digits be multiplied by its tens' digit, the product will be 390. If the digits be interchanged and the resulting number be multiplied by its tens' digit, the product will be 280. What is the number?

7. If a number of two digits be divided by the product of its digits, the quotient will be 2. If 27 be added to the number, the sum will be equal to the number obtained by interchanging the digits. What is the number?

8. The product of the two digits of a number is equal to one-half of the number. If the number be subtracted from the number obtained by interchanging the digits, the remainder will be equal to three-halves of the product of the digits of the number. What is the number?

9. If the difference of the squares of two numbers be divided by the first number, the quotient and the remainder will each be 5. If the difference of the squares be divided by the second number, the quotient will be 13 and the remainder 1. What are the numbers?

10. The sum of the three digits of a number is 9. If the digits be written in reverse order, the resulting number will exceed the original number by 396. The square of the middle digit exceeds the product of the first and the third digit by 4. What is the number?

11. A rectangular field is 119 yards long and 19 yards wide. How many yards must be added to its width and how many yards must be taken from its length, in order that its area may remain the same while its perimeter is increased by 24 yards?

12. Two points move with different velocities along the sides of a right angle toward the vertex, the one from a distance of 50 feet and the other from a distance of $136\frac{1}{2}$ feet. After 7 seconds the distance between the points is 85 feet, and after 9 seconds the distance is 68 feet. What is the velocity of each point?

13. The floor of a room contains $30\frac{1}{3}$ square yards, one wall contains 21 square yards, and an adjacent wall contains 13 square yards. What are the dimensions of the room?

14. If the numerator of a fraction be increased by 2 and the denominator be diminished by 2, the resulting fraction will be the reciprocal of the original fraction. And if the numerator be diminished by 2 and the denominator be increased by 2, the resulting fraction, increased by $1\frac{1}{15}$, will be equal to the reciprocal of the original fraction? What is the fraction?

15. A merchant bought a number of pieces of cloth of two different kinds. He bought of each kind as many pieces and paid for each yard half as many dollars as that kind contained yards. He bought altogether 19 pieces and paid for them $921.50. How many pieces of each kind did he buy?

16. The diagonal of a rectangle is $20\frac{3}{5}$ feet. If the length of one side be increased by 14 feet and the length of the other side be diminished by $2\frac{3}{5}$ feet, the diagonal will be increased by $12\frac{3}{5}$ feet. What are the lengths of the sides of the rectangle?

17. A certain number of coins can be arranged in the form of one square, and also in the form of two squares. In the first arrangement each side of the square contains 29 coins, and in the second arrangement one square contains 41 more coins than the other. How many coins are there in a side of each square of the second arrangement?

18. A piece of cloth after being wet shrinks in length by one-eighth and in breadth by one-sixteenth. The piece contains after shrinking 3.68 fewer square yards than before shrinking, and the length and breadth together shrink 1.7 yards. What was the length and breadth of the piece?

19. A merchant paid $125 for two kinds of goods. He sold the one kind for $91 and the other for $36. He thereby gained as much per cent on the first kind as he lost on the second. How much did he pay for each kind?

20. Two workmen can do a piece of work in 6 days. How long will it take each of them to do the work, if it takes one 5 days longer than the other?

21. Two men, A and B, receive different wages. A earns $42, and B $40. If A had received B's wages a day, and B had received A's wages, they would have earned together $4 more. How many days does each work, if A works 8 days more than B, and what wages does each receive?

22. A man has two square boards of the same size. If he cover one with coins of one kind and the other with coins of another kind, he will need in all 340 coins. And 6 coins of the one kind placed side by side cover the same distance as 7 coins of the other kind. How many coins of each kind can be placed on a board?

23. A number of workmen remove a pile of stone from one place to another. Had there been 8 more workmen, and had each one carried 5 pounds less at each trip, they would have completed the work in 7 hours. Had there been 8 fewer workmen and had each one carried 11 pounds more at each trip, they would have completed the work in 9 hours. How many workmen were there and how many pounds did each one carry at every trip?

24. A man has two square fields in which he wishes to plant trees, the outer rows to be on the edges of the fields. If he plants the trees in the first field $2\frac{1}{2}$ yards apart, and in the second field $2\frac{1}{4}$ yards apart, he will need 11,113 trees. But if he plants the trees in the first field $2\frac{3}{4}$ yards apart, and in the second field 3 yards apart, he will need only 7816 trees. How long is a side of each of the fields?

25. A tank can be filled by one pipe and emptied by another. If, when the tank is half full of water, both pipes be left open

12 hours, the tank will be emptied. If the pipes be made smaller, so that it will take the one pipe one hour longer to fill the tank and the other one hour longer to empty it, the tank, when half full of water, will then be emptied in $15\frac{3}{4}$ hours. In what time will the empty tank be filled by the one pipe, and the full tank be emptied by the other?

26. A besieged garrison has enough provisions for 12 days. If 120 men withdraw, and if the ration of each man be diminished by five-eighths of a pound, the provisions will last 16 days. The provisions will last also 16 days, if 200 men withdraw, and if the ration of each man be diminished by three-eighths of a pound. How many men are in the garrison and what is the ration of each man?

CHAPTER XXIV.

RATIO, PROPORTION, AND VARIATION.

§ 1. RATIO.

1. The **Ratio** of one number to another is the relation between the numbers which is expressed by the quotient of the first divided by the second.

E.g., the ratio of 6 to 4 is expressed by $\frac{6}{4}$, $= \frac{3}{2}$.

The ratio of one number to another is frequently expressed by placing a colon between them; thus, the ratio of 5 to 7 is frequently written $5:7$.

The first number in a ratio is called the **First Term,** or the **Antecedent** of the ratio, and the second number the **Second Term,** or the **Consequent** of the ratio.

Thus, in the ratio $a:b$, a is called the first term, and b the second term.

2. A **Ratio of Equality** is one whose terms are equal; as $4:4$.

A **Ratio of Greater Inequality** is one whose first term is greater than its second term; as $6:4$.

A **Ratio of Less Inequality** is one whose first term is less than its second term; as $3:5$.

Inverse Ratios are two ratios in which the first term of the one is the second term of the other, and *vice versa;* as $4:7$ and $7:4$.

The **Compound Ratio** of two given ratios is a ratio whose first term is the product of the first terms of the given ratios, and whose second term is the product of the second terms.

E.g., $ac:bd$ is the compound ratio of $a:b$ and $c:d$.

The **Duplicate Ratio** of a given ratio is one whose terms are the squares of the terms of the given ratio.

E.g., $a^2 : b^2$ is the duplicate ratio of $a : b$.

The **Triplicate Ratio** of a given ratio is one whose terms are the cubes of the terms of the given ratio.

E.g., $a^3 : b^3$ is the triplicate ratio of $a : b$.

The **Sub-duplicate Ratio** of a given ratio is one whose terms are the square roots of the terms of the given ratio.

E.g., $\sqrt{a} : \sqrt{b}$ is the sub-duplicate ratio of $a : b$.

3. Since, by definition, a ratio is a fraction, all the properties of fractions are true of ratios.

E.g., $a : b = ma : mb$.

4. The definition given in Art. 1 has reference to the ratio of one *number* to another. But it is frequently necessary to compare concrete quantities, as the length of one line with the length of another line, the area of one field with the area of another field, etc.

If two concrete quantities of the same kind can be expressed by two rational numbers in terms of the same unit, then the ratio of the one quantity to the other is defined as the ratio of the one number to the other.

E.g., the ratio of $2\frac{1}{2}$ yards to $1\frac{1}{7}$ yards is $2\frac{1}{2} : 1\frac{1}{7}, = \frac{2\frac{1}{2}}{1\frac{1}{7}} = \frac{35}{16}$.

Observe that by this definition the ratio of two concrete quantities is a number.

It is important to notice that the quantities to be compared must be of the same kind. Dollars cannot be compared with pounds, nor apples with marbles.

5. If two concrete quantities cannot be expressed by two rational numbers in terms of the same unit, they are said to be **Incommensurable** one to the other.

Thus, if the lengths of the two sides of a right triangle be equal, the length of the hypothenuse cannot be expressed by a rational number in terms of a side as a unit, or any fraction of a side as a unit.

If a side be taken as the unit, the hypothenuse is expressed by $\sqrt{2}$, an irrational number. And the ratio of the hypothenuse to a side is $\sqrt{2}:1, =\sqrt{2}$, a number comprised in the number system.

In the following article we will prove that the ratio of any two incommensurable quantities can be expressed as a number comprised in the number system.

6. Let P and Q be two incommensurable quantities. We assume that the ratio $P:Q$ is greater than the ratio $P':Q$, wherein P' is less than P and is commensurable with Q, and that the ratio $P:Q$ is less than the ratio $P'':Q$, wherein P'' is greater than P and is commensurable with Q.

Let us take $\frac{1}{n}Q$ as the unit. Then we can find two consecutive integral multiples of $\frac{1}{n}Q$, which are therefore commensurable with Q, between which P lies. Let $\frac{m}{n}Q$ and $\frac{m+1}{n}Q$ be these multiples. The ratios of these multiples to Q are respectively $\frac{m}{n}$ and $\frac{m+1}{n}$. Then by the hypothesis

$$\frac{m}{n} < P:Q < \frac{m+1}{n}.$$

The two rational numbers $\frac{m}{n}$ and $\frac{m+1}{n}$, between which the ratio $P:Q$ lies, have the properties (i.) and (ii.), Art. 6, Ch. XVII. They therefore define an irrational number.

EXERCISES I.

What is the ratio of

1. $6a$ to $9b$? **2.** $\frac{3}{5}a^2b$ to $\frac{6}{11}ab^2$? **3.** $9\frac{1}{2}x^3y$ to $7\frac{3}{5}xy^3$?

4. $\frac{1}{a}$ to $\frac{1}{b}$? **5.** $\frac{a}{b}$ to $\frac{c}{d}$? **6.** $\frac{a}{x-3}$ to $\frac{1}{(x-3)^2}$?

7. a^2-b^2 to $(a-b)^2$? **8.** x^2-x-30 to $x^2-10x+24$?

9. Which is the greater ratio, $a+2b:a+b$ or $a+3b:a+2b$?

10. For what value of x will the ratio $2+x:5+x$ be equal to the ratio $2:3$?

11. For what values of x will the ratio $x^2-x+1:x^2+x+1$ be equal to the ratio $3:7$?

12. If $\frac{6x+2y}{3x-y}=10$, what is the value of the ratio $x:y$?

Find the compound ratio of

13. 3 : 4 and 5 : 6. **14.** 2 : 5, 3 : 7, and $\frac{1}{8} : \frac{1}{6}$.

Find the duplicate, the triplicate, and the sub-duplicate ratio of

15. 9 : 4. **16.** 5 : 3. **17.** $a^6 : b^3$. **18.** $x^5 : y^7$.

19. Find two numbers whose ratio is 7 : 5, and the difference of whose squares is 96.

20. A works 6 days with 2 horses, and B works 5 days with 3 horses. What is the ratio of A's work to B's work?

21. The ratio of a father's age to his son's age is 9 : 5. If the father is 28 years older than the son, how old is each?

§ 2. PROPORTION.

1. A **Proportion** is an equation whose members are two equal ratios.

E.g., 4 : 3 = 8 : 6, read *the ratio of* 4 *to* 3 *is equal to the ratio of* 8 *to* 6, or 4 *is to* 3 *as* 8 *is to* 6.

Instead of the equality sign a double colon is frequently used, as 4 : 3 : : 8 : 6.

2. Four numbers are said to be *in proportion*, or to be *proportional*, when the first is to the second as the third is to the fourth.

E.g., the numbers 4, 3, 8, 6 are proportional, since 4 : 3 = 8 : 6.

The individual numbers are called the **Proportionals**, or **Terms** of the proportion.

The **Extremes** of a proportion are its first and last terms; as 4 and 6 above.

The **Means** of a proportion are its second and third terms; as 3 and 8 above.

The **Antecedents** and **Consequents** of a proportion are the antecedents and consequents of its two ratios.

E.g., 4 and 6 are the antecedents, and 3 and 8 the consequents of the proportion 4 : 3 = 6 : 8.

3. *In any proportion the product of the extremes is equal to the product of the means.*

If $a : b = c : d$, we are to prove $ad = bc$.

By § 1, Art. 1, $\frac{a}{b} = \frac{c}{d}$.

Clearing of fractions, $ad = bc$.

4. *If the product of two numbers be equal to the product of two other numbers, the four numbers are in proportion.*

Let $ad = bc$.

Dividing by bd, $\frac{a}{b} = \frac{c}{d}$, or $a : b = c : d$; (1)

by cd, $\frac{a}{c} = \frac{b}{d}$, or $a : c = b : d$; (2)

by ab, $\frac{d}{b} = \frac{c}{a}$, or $d : b = c : a$; (3)

by ac, $\frac{d}{c} = \frac{b}{a}$, or $d : c = b : a$. (4)

Interchanging the ratios in (1), (2), (3), (4),

$c : d = a : b$; (5)

$b : d = a : c$; (6)

$c : a = d : b$; (7)

$b : a = d : c$. (8)

Notice that the two numbers of either product may be taken as the extremes, the other two as the means. In (1) to (4), a and d are the extremes, c and b the means; in (5) to (8), d and a are the means, c and b the extremes.

5. In Art. 4, we may regard the proportions (2) to (8) as being derived from (1), and thus obtain the following properties of a proportion:

(i.) *The means may be interchanged;* as in (2).

(ii.) *The extremes may be interchanged;* as in (3).

(iii.) *The means may be interchanged, and at the same time the extremes;* as in (4).

(iv.) *The means may be taken as the extremes, and the extremes as the means;* as (8) from (1), (7) from (2), (6) from (3), and (5) from (4).

6. *If any three terms of a proportion be given, the remaining term can be found.*

Ex. What is the second term of a proportion, whose first, third, and fourth terms are 10, 16, and 8 respectively?

Letting x stand for the second term, we have

$$10 : x = 16 : 8, \text{ or } 16x = 80; \text{ whence } x = 5.$$

7. *The products, or the quotients, of the corresponding terms of two proportions form again a proportion.*

If $$a : b = c : d, \text{ or } \frac{a}{b} = \frac{c}{d}, \quad (1)$$

and $$x : y = z : u, \text{ or } \frac{x}{y} = \frac{z}{u}, \quad (2)$$

we have, multiplying corresponding members of (1) and (2),

$$\frac{ax}{by} = \frac{cz}{du}.$$

Whence $$ax : by = cz : du.$$

Dividing the members of (1) by the corresponding members of (2), we have

$$\frac{\frac{a}{x}}{\frac{b}{y}} = \frac{\frac{c}{z}}{\frac{d}{u}}, \text{ or } \frac{a}{x} : \frac{b}{y} = \frac{c}{z} : \frac{d}{u}.$$

8. *In any proportion, the sum of the first two terms is to the first (or the second) term as the sum of the last two terms is to the third (or the fourth) term.*

Let $$a : b = c : d.$$

Then $$\frac{a}{b} = \frac{c}{d}.$$

Adding 1 to both members, $\frac{a}{b} + 1 = \frac{c}{d} + 1,$

or $$\frac{a + b}{b} = \frac{c + d}{d}.$$

Whence $a + b : b = c + d : d.$

In like manner it can be proved that

$$a + b : a = c + d : c.$$

These two proportions are said to be derived from the given proportion by **Composition.**

9. *In any proportion, the difference of the first two terms is to the first (or the second) term as the difference of the last two terms is to the third (or the fourth) term.*

If $$a : b = c : d,$$

then $a - b : a = c - d : c$, and $a - b : b = c - d : d.$

The proof is similar to that of Art. 8.

These two proportions are said to be derived from the given proportion by **Division.**

10. *In any proportion, the sum of the first two terms is to their difference as the sum of the last two terms is to their difference.*

Let $$a : b = c : d.$$

By Art. 8, $$a + b : b = c + d : d;$$

and by Art. 9, $$a - b : b = c - d : d.$$

Then by Art. 7, $$\frac{a+b}{a-b} : 1 = \frac{c+d}{c-d} : 1,$$

or $$\frac{a+b}{a-b} = \frac{c+d}{c-d}.$$

Whence $a + b : a - b = c + d : c - d.$

This proportion is said to be derived from the given one by **Composition and Division.**

11. A **Continued Proportion** is one in which the consequent of each ratio is the antecedent of the following ratio; as,

$$a : b = b : c = c : d = \text{etc.}$$

12. In the continued proportion

$$a : b = b : c,$$

b is called a **Mean Proportional** between a and c, and c is called the **Third Proportional** to a and b.

13. *The mean proportional between any two numbers is equal to the square root of their product.*

From $$a : b = b : c,$$

we have, by Art. 3, $$b^2 = ac.$$

Whence $$b = \sqrt{(ac)}.$$

14. The following examples are applications of the preceding theory:

Ex. 1. Find a mean proportional between 5 and 20.

Let x stand for the required proportional.

Then, by Art. 13, $x = \sqrt{(5 \times 20)} = 10.$

Ex. 2. If $$a : b = c : d,$$

then $$ab + cd : ab - cd = b^2 + d^2 : b^2 - d^2.$$

Let $$\frac{a}{b} = \frac{c}{d} = x.$$

Then $$a = bx \text{ and } c = dx.$$

Therefore $$ab + cd = b^2x + d^2x,$$

and $$ab - cd = b^2x - d^2x.$$

We then have

$$\frac{ab + cd}{ab - cd} = \frac{b^2x + d^2x}{b^2x - d^2x} = \frac{b^2 + d^2}{b^2 - d^2}.$$

Whence $$ab + cd : ab - cd = b^2 + d^2 : b^2 - d^2.$$

Ex. 3. Solve the equation

$$\frac{\sqrt{(2 + x)} + \sqrt{(2 - x)}}{\sqrt{(2 + x)} - \sqrt{(2 - x)}} = 2, = \frac{2}{1}.$$

By composition and division,

$$\frac{\sqrt{(2 + x)}}{\sqrt{(2 - x)}} = \frac{3}{1}.$$

Squaring and clearing of fractions,

$$2 + x = 18 - 9x.$$

Whence $$x = \tfrac{8}{5}.$$

EXERCISES II.

Verify each of the following proportions:

1. $2\frac{1}{2} : 1\frac{1}{3} = 1\frac{1}{2} : \frac{4}{5}$.

2. $14\frac{2}{3} : 4\frac{2}{5} = 200 : 60$.

3. $\dfrac{4\,ab}{a^2 - b^2} : \dfrac{a^2 + b^2}{a - b} = \dfrac{2\,ab}{a^4 - b^4} : \dfrac{1}{2\,a - 2\,b}$.

4. $a^2 - b^2 : a - b = a^3 + a^2b + ab^2 + b^3 : a^2 + b^2$.

Form proportions from each of the following products, in eight different ways:

5. $2\,x = 3\,y$.

6. $m^2 = n^2$.

7. $x^2 = 1$.

8. $a(1 + a) = a + b$.

9. $a^3 - b^3 = x^2 - y^2$.

Find a fourth proportional to

10. 1, 2, and 8.

11. $\frac{2}{5}$, $\frac{3}{5}$, and $\frac{4}{5}$.

12. ab, ac, and b.

Find a third proportional to

13. 2 and 6.

14. $\frac{1}{3}$ and $\frac{1}{6}$.

15. a and b.

Find a mean proportional between

16. 2 and 18.

17. $\frac{1}{3}$ and $\frac{3}{4}$.

18. a^2b and ab^2.

19. $\dfrac{a + b}{a - b}$ and $\dfrac{a^2 - b^2}{a^2b^2}$.

20. $\dfrac{a^2 + 1}{a^2 - 1}$ and $\frac{1}{4}(a^4 - 1)$.

Find a value of x to satisfy each of the following proportions:

21. $x : 2 = 12 : 3$.

22. $\frac{1}{2} + \sqrt{a} : \frac{1}{4} - a = x : \sqrt{a} - 2a$.

23. $a + b + \dfrac{2\,b^2}{a - b} : \dfrac{(a + b)^2}{2\,ab} - 1 = x : a - b$.

24. $a^2 - 4 : 4\,a^3 + 4 = \dfrac{\frac{1}{2}a + 1}{2(a^2 - a) + 4} : x$.

If $a : b = c : d$, prove that

25. $a + c : b + d = a^2d : b^2c$.

26. $\dfrac{1}{a} + \dfrac{1}{b} : \dfrac{1}{a} - \dfrac{1}{b} = \dfrac{1}{c} + \dfrac{1}{d} : \dfrac{1}{c} - \dfrac{1}{d}$.

27. $a^2 + b^2 : a^2 - b^2 = c^2 + d^2 : c^2 - d^2$.

28. $(a \pm b)^2 : ab = (c \pm d)^2 : cd$.

29. $2a + 3b : 4a + 5b = 2c + 3d : 4c + 5d$.

30. $ka + lb : ma + nb = kc + ld : mc + nd$.

31. $a + b : c + d = \sqrt{(a^2 + b^2)} : \sqrt{(c^2 + d^2)}$.

32. $\sqrt[3]{(a^3 + b^3)} : \sqrt[3]{(c^3 + d^3)} = a : c$.

33. $\sqrt{(a^2 + b^2)} : \sqrt{(c^2 + d^2)} = \sqrt[3]{(a^3 + b^3)} : \sqrt[3]{(c^3 + d^3)}$.

If b be a mean proportional between a and c, prove that

34. $a^2 \pm b^2 : b^2 \pm c^2 = a : c$.

35. $3a + 7b : 3b + 7c = 5a - 7b : 5b - 7c$.

36. $(a + 2b + c) : 1 = (b + c)^2 : c$.

37. $a^2b^2c^2\left(\frac{1}{a^3} + \frac{1}{b^3} + \frac{1}{c^3}\right) = a^3 + b^3 + c^3$.

If $a : b = b : c = c : d$, prove that

38. $(b + c)(b + d) = (a + c)(c + d)$.

39. $(a + d)(b + c) - (a + c)(b + d) = (b - c)^2$.

40. $(a^2 + b^2 + c^2)(b^2 + c^2 + d^2) = (ab + bc + cd)^2$.

41. $(b - c)^2 + (c - a)^2 + (d - b)^2 = (a - d)^2$.

If $a : b = c : d = e : f$, prove that

42. $a^2 : b^2 = ce : df$.

43. $c^2 + d^2 : e^2 + f^2 = cd : ef$.

44. $a^2 + c^2 + e^2 : b^2 + d^2 + f^2 = c^2 : d^2$.

45. $(a^2 + c^2 + e^2)(b^2 + d^2 + f^2) = (ab + cd + ef)^2$.

46. $\sqrt{(ab)} + \sqrt{(cd)} + \sqrt{(ef)} = \sqrt{[(a + c + e)(b + d + f)]}$.

47. If $(a+b+c+d)(a-b-c+d) = (a-b+c-d)(a+b-c-d)$, prove that a, b, c, d are proportional.

48. If $\frac{5a^2 - 8c^2}{3a^2 - 4c^2} = \frac{5b^2 - 8d^2}{3b^2 - 4d^2}$, then a, b, c, d are proportional.

49. If $\frac{ay - bx}{c} = \frac{cx - az}{b} = \frac{bz - cy}{a}$, then $x : a = y : b = z : c$.

Solve each of the following equations:

50. $\frac{\sqrt{(1 + x)} + \sqrt{(1 - x)}}{\sqrt{(1 + x)} - \sqrt{(1 - x)}} = 3$.

51. $\frac{\sqrt{(2x + 1)} - \sqrt{(5 - x)}}{\sqrt{(2x + 1)} + \sqrt{(5 - x)}} = \frac{1}{2}$.

52. $\frac{\sqrt{a} + \sqrt{(a - x)}}{\sqrt{a} - \sqrt{(a - x)}} = \frac{1}{a}$.

53. $\frac{\sqrt{(a+x)} - \sqrt{(a-x)}}{\sqrt{(a+x)} + \sqrt{(a-x)}} = \frac{1}{\sqrt{b}}$.

54. Find three numbers in a continued proportion whose sum is 39, and whose product is 721.

55. Find two numbers such that if 1 be added to the first and 8 to the second, the sums will be in the ratio 1 : 2, and if 1 be subtracted from each number, the remainders will be in the ratio 2 : 3.

56. What is the ratio of the numerator of a fraction to its denominator, if the fraction be unchanged when a is added to its numerator and b to its denominator?

57. The sum of the means of a proportion is 7, the sum of the extremes is 8, and the sum of the squares of all the terms is 65. What is the proportion?

§ 3. VARIATION.

1. Frequently two numbers or quantities are so related to each other that a change in the value of one produces a corresponding change in the value of the other.

Thus, the distance a train runs in one hour depends upon its speed, and increases or decreases when its speed increases or decreases.

The illumination made by a light depends upon the intensity of the light, and varies when the intensity varies.

The value of y given by the equation $y = 2x - 3$ depends upon the value of x and varies when the value of x varies.

Thus, if $x = 1$, $y = -1$; if $x = 2$, $y = 1$, etc.

We shall in this chapter consider only the simplest kinds of variation.

2. Direct Variation. —Two quantities are said to *vary directly*, one as the other, when their ratio is constant.

Thus, if x varies directly as y, then $\frac{x}{y} = k$, a constant.

For example, if a train runs at a uniform speed, the number of miles it runs varies directly as the number of hours. If it runs at the rate of 30 miles an hour, in 1 hour it will run 30 miles, in 2 hours 60 miles, in 3 hours 90 miles, and so on; and the ratios 1 : 30, 2 : 60, 3 : 90, etc., are equal.

The symbol of direct variation, $\propto$, is read *varies directly as*.

The word *directly* is frequently omitted.

If $y = 3x$, then $y \propto x$ (read y *varies as* x), since $\frac{y}{x} = 3$, a constant.

3. Inverse Variation. — One quantity is said to *vary inversely* as another when the first varies as the *reciprocal* of the second.

Thus, if x varies inversely as y, then $x \propto \frac{1}{y}$.

Therefore, $\dfrac{x}{\frac{1}{y}} = k$, a constant.

Whence, $xy = k$.

That is, if one quantity varies inversely as another, the product of the quantities is constant.

For example, if a train runs 120 miles, the number of hours it takes will vary inversely as the number of miles it runs an hour. If it runs at the rate of 30 miles an hour, it will take 4 hours; if it runs at the rate of 40 miles an hour, it will take 3 hours, and so on; and the products 30×4, 40×3, 60×2, etc., are equal.

If 6 men can do a piece of work in 12 hours, 3 men can do the same work in 24 hours, and 1 man in 72 hours, and the products 6×12, 3×24, 1×72 are equal. That is, the number of hours varies inversely as the number of men working.

If $y = \frac{3}{x}$, y varies inversely as x, since $xy = 3$.

4. Joint Variation. — One quantity is said to *vary* as two others *jointly*, when it varies as the product of the others.

Thus, if x varies as y and z jointly, then $\frac{x}{yz} = k$, a constant.

For example, the number of miles a train runs varies as the number of hours and the number of miles it runs an hour jointly. It will run 40 miles in 2 hours at a rate of 20 miles an hour, 90 miles in 3 hours at the rate of 30 miles an hour, 120 miles in 5 hours at the rate of 24 miles an hour; and

$$\frac{40}{2\times 20}=\frac{90}{3\times 30}=\frac{120}{5\times 24}.$$

5. One quantity is said to vary directly as a second and inversely as a third, when it varies as the first and the reciprocal of the third jointly.

Thus, if x varies directly as y and inversely as z, then $\dfrac{x}{y\cdot\dfrac{1}{z}}=k$, a constant; or $\dfrac{xz}{y}=k$.

6. *If x depend only on y and z, and if x varies as y when z is constant, and as z when y is constant, then x varies as y and z jointly when y and z both vary.*

Let x, y, z; x', y', z; x'', y', z' be three sets of corresponding values. Then since z has the same value for x and x', we have

$$\frac{x}{y}=\frac{x'}{y'}. \qquad (1)$$

And since y' has the same value for x' and x'', we have

$$\frac{x'}{z}=\frac{x''}{z'}. \qquad (2)$$

From (1) and (2), we have

$$\frac{xx'}{yz}=\frac{x'x''}{y'z'},\ \text{ or }\ \frac{x}{yz}=\frac{x''}{y'z'},$$

which proves the principle enunciated.

7. In all the preceding cases of variation, the constant can be determined when any set of corresponding values of the quantities is known.

Ex. 1. If $x\propto y$, and $x=3$ when $y=5$, what is the value of the constant?

We have $\dfrac{x}{y}=k$, or $x=ky$.

Therefore, when $x=3$ and $y=5$,

$3=5k$, whence $k=\frac{3}{5}$.

Consequently $x=\frac{3}{5}y$.

Ex. 2. If x varies inversely as y, and if $y = 4$ when $x = 7$, find the value of x when $y = 12$.

From $xy = k$, we obtain $k = 28$.

Therefore $xy = 28$.

Consequently, when $y = 12$, $12x = 28$; whence $x = 2\frac{1}{3}$.

Ex. 3. The area of a circle varies directly as the square of its radius. If the area be 1256.64 when the radius is 20, what is the area when the radius is 30?

Let A stand for the area, and r for the radius.

Then from $A = kr^2$, we obtain $k = \frac{1256 \cdot 64}{400} = 3.1416$.

Therefore $A = 3.1416\, r^2$.

Consequently, when $r = 30$, $A = 3.1416 \times 30^2 = 2827.44$.

Ex. 4. The volume of a gas varies inversely as the pressure when the temperature is constant. When the pressure is 15, the volume is 20; what is the volume when the pressure is 20?

Let v stand for the volume and p for the pressure.

Then from $pv = k$ we obtain $k = 300$.

Therefore $pv = 300$.

Consequently, when $p = 20$, $20v = 300$; whence $v = 15$.

EXERCISES III.

If $x \propto y$, what is the expression for x in terms of y,

1. If $x = 10$ when $y = \frac{3}{8}$? 2. If $x = 2\frac{2}{3}$ when $y = 2\frac{1}{9}$?

3. If $x = a$ when $y = 2a$? 4. If $x = 3a^2b$ when $y = 5ab^2$?

5. If $x \propto y^2$, and $x = 5$ when $y = 7$, what is the expression for x in terms of y?

6. If $x \propto \sqrt{y}$, and $x = 3(a^3 + b^3)$ when $y = 25(a^2 + 2ab + b^2)$, what is the expression for y in terms of x?

If $x \propto \frac{1}{y}$, what is the expression for x in terms of y,

7. If $x = 10$ when $y = \frac{4}{5}$? 8. If $x = 3\frac{1}{4}$ when $y = \frac{15}{26}$?

9. If $x \propto \frac{1}{y^2}$, and $x = 4\frac{1}{2}$ when $y = \frac{2}{3}$, what is the expression for y in terms of x?

10. If $x \propto \frac{1}{\sqrt{y}}$, and $x = 4$ when $y = 25$, what is the expression for x in terms of y?

11. If $x \propto y$, and $x = 10$ when $y = 5$, what is the value of x when $y = 12\frac{1}{2}$?

12. If $x \propto y$, and $x = a$ when $y = \frac{3}{4}a^2$, what is the value of y when $x = a^2b$?

13. If $x \propto y^2$, and $x = 5$ when $y = -3$, what is the value of x when $y = 15$?

14. If $x \propto \sqrt{y}$, and $x = a + m$ when $y = (a - m)^2$, what is the value of x when $y = (a + m)^4$?

15. If $x \propto \frac{1}{y}$, and $x = 3$ when $y = \frac{2}{3}$, what is the value of x when $y = 4\frac{1}{7}$?

16. If $x \propto \frac{1}{\sqrt{y}}$, and $x = 15$ when $y = 36$, what is the value of x when $y = 12\frac{1}{4}$?

17. If $x \propto -\frac{1}{y^3}$, and $x = -\frac{4a^2}{3b^3}$ when $y = \frac{9b}{2ac}$, what is the value of y when $x = -2ab$?

18. The area of a circle whose radius is 20 feet is 1257 square feet. What is the area of a circle whose radius is 30 feet, if it be known that the area varies as the square of the radius?

19. The circumference of a circle whose radius is 6 feet is 37.7 feet. What is the circumference of a circle whose radius is 9.5 feet, if it be known that the circumference varies as the radius?

20. An ox is tied by a rope 20 yards long in the center of a field, and eats all the grass within his reach in $2\frac{1}{2}$ days. How many days would it have taken the ox to eat all the grass within his reach if the rope had been 10 yards longer?

21. The volume of a sphere whose radius is 7 inches is 1437.3 cubic inches. What is the volume of a sphere whose radius is 10 inches, if it be known that the volume varies as the cube of the radius?

22. The surface of a sphere whose radius is 10 feet is 1257.14 square feet. What is the surface of a sphere whose radius is 7 feet, if it be known that the surface varies as the square of the radius?

It has been found by experiment that the distance a body falls from rest varies as the square of the time.

23. If a body falls 256 feet in 4 seconds, how far will it fall in 10 seconds?

24. From what height must a body fall to reach the earth after 15 seconds?

It has been found by experiment that the velocity acquired by a body falling from rest varies as the time.

25. If the velocity of a falling body is 160 feet after 5 seconds, what will be the velocity after 8 seconds?

26. How long must a body have been falling to have acquired a velocity of 256 feet?

27. The volume of a cylinder of revolution, the radius of whose base is 7 inches and whose altitude is 10 inches, is 1540 cubic inches. What is the volume of a cylinder of revolution, the radius of whose base is 8 inches and whose altitude is 12 inches, if it be known that the volume varies as the square of the radius of the base and the altitude jointly?

28. The lateral surface of a cylinder of revolution, the radius of whose base is 6 inches and whose altitude is 14 inches, is 528 square inches. What is the lateral surface of a cylinder of revolution, the radius of whose base is 5 inches and whose altitude is 11 inches, if it be known that the surface varies as the altitude and the radius jointly?

29. The surface of a cube whose edge is 5 inches is 150 square inches. What is the surface of a cube whose edge is

9 inches, if it be known that the surface varies as the square of its edge?

30. It has been found by experiment that the weight of a body varies inversely as the square of its distance from the center of the earth. If a body weighs 30 pounds on the surface of the earth (approximately 4000 miles from the center), what would be its weight at a distance of 24,000 miles from the surface of the earth?

It has been found by experiment that the illumination of an object varies inversely as the square of its distance from the source of light.

31. If the illumination of an object at a distance of 10 feet from a source of light is 2, what is the illumination at a distance of 40 feet?

32. To what distance must an object which is now 10 feet from a source of light be removed in order that it shall receive only one-half as much light?

33. At what distance will a light of intensity 10 give the same illumination as a light of intensity 8 gives at a distance of 50 feet?

CHAPTER XXV.

DOCTRINE OF EXPONENTS.

1. We have already abbreviated such products as

$$aa,\ aaa,\ aaaa,\ \cdots,\ aaa \cdots n \text{ factors},$$

by a^2, a^3, a^4, $\cdots$, a^n, respectively, and called them the *second, third, fourth,* $\cdots$, *nth* powers of a.

This definition of the symbol a^n requires the exponent n to be a *positive integer.*

Thus, 2^5 means the product of 5 factors, each equal to 2. But 2^0 has, as yet, no meaning, since 2 cannot be taken 0 times as a factor; likewise 2^{-5} has, as yet, no meaning, since 2 cannot be taken -5 times as a factor. For a similar reason $2^{\frac{1}{2}}$ and $2^{\sqrt{3}}$ are, as yet, meaningless.

2. Nevertheless, having introduced into Algebra the symbol a^n, it is natural to inquire what it may mean when n is 0, a *rational negative* or *fractional number,* or an *irrational number.*

We shall find that, by enlarging our conception of *powers,* quite clear and definite meanings can be given to such expressions as 2^0, 3^{-2}, $4^{\frac{1}{2}}$, $5^{\sqrt{2}}$.

3. The discussion of powers, in general, therefore naturally divides itself into five cases:

(1) Powers with *positive integral exponents.*

(2) Powers with *zero exponents.*

(3) Powers with *negative integral exponents.*

(4) Powers with *fractional* (*positive* or *negative*) *exponents.*

(5) Powers with *irrational exponents.*

The consideration of powers with imaginary exponents is reserved for Part II.

Positive Integral Powers.

4. The principles upon which operations with positive integral powers depend have been proved in the preceding chapters.

For the sake of emphasis, and for convenience of reference in enlarging our conceptions of powers, we restate them here:

(i.) $$a^m a^n = a^{m+n}.$$

(ii.) $$\frac{a^m}{a^n} = a^{m-n}, \text{ when } m > n;$$

$$\frac{a^m}{a^n} = 1, \text{ when } m = n,$$

$$\frac{a^m}{a^n} = \frac{1}{a^{n-m}}, \text{ when } m < n.$$

(iii.) $$(a^m)^n = a^{mn}.$$

(iv.) $$(ab)^m = a^m b^m.$$

(v.) $$\left(\frac{a}{b}\right)^m = \frac{a^m}{b^m}.$$

Zeroth Powers.

5. In discussing positive integral powers, the symbol a^n was introduced only for convenience, as an abbreviation for the result of a definite operation.

The operation came first, the symbol afterwards for the sake of brevity in writing.

The symbol a^0, as has been stated, has no meaning under the definition of a power already given.

Consequently we must either discard this symbol altogether, or attach to it a definite meaning which will be consistent with the laws of positive exponents.

Now the meaning of a symbol may be defined by assuming that it stands for the result of a definite operation, as was done in letting

$$a^n = a \cdot a \cdot a \cdot \cdots n \text{ factors};$$

or by enlarging the meaning of some operation or law which was previously restricted in its application.

In the latter way, negative numbers were introduced by extending the meaning of subtraction.

6. We now enlarge the meaning of powers by assuming that the principle

$$\frac{a^m}{a^n} = a^{m-n}$$

holds also when $m = n$.

We then have
$$\frac{a^m}{a^m} = a^{m-m} = a^0.$$

But since
$$\frac{a^m}{a^m} = 1,$$

it follows that
$$a^0 = 1.$$

That is, *the zeroth power of any base whatever, except* 0, *is equal to* 1.

E.g., $1^0 = 1,\ 5^0 = 1,\ 99^0 = 1,\ (a+b)^0 = 1$, etc.

7. Thus, by the assumption that the stated law holds when $m = n$, a definite value of the zeroth power of a number is obtained. Nevertheless, it will doubtless seem strange to the student that all numbers to the zeroth power have one and the same value, namely 1.

But it should be distinctly noted that a^0 is by definition a symbol for $\frac{a^m}{a^m}$; *i.e.*, for the quotient of two like powers of the same base. Thus,

$$2^0 = \frac{2^3}{2^3} = \frac{2^5}{2^5} = \frac{2^m}{2^m} = 1.$$

It is no more surprising that

$$a^0 = b^0 = c^0 = 1,$$

than that
$$\frac{x}{x} = \frac{y}{y} = \frac{a+b}{a+b} = 1,$$

or than that
$$9 - 8 = 6 - 5 = 3 - 2 = 1.$$

Thus, a^0 is a convenient symbol for the result of a perfectly clear and definite operation.

8. *The zeroth power of 0 is indeterminate.*

For $0^0 = 0^{n-n} = \frac{0^n}{0^n} = \frac{0}{0}$, an indeterminate number.

Negative Integral Powers.

9. We now still further enlarge the meaning of powers by assuming that the principle

$$\frac{a^m}{a^n} = a^{m-n}$$

holds not only when $m > n$ and $m = n$, but also when $m < n$. In this case, $m - n$ is a negative number.

Since $m < n$, we may assume $n = m + k$.

Then $$\frac{a^m}{a^n} = \frac{a^m}{a^{m+k}} = a^{m-(m+k)} = a^{-k}.$$

But $$\frac{a^m}{a^{m+k}} = \frac{1}{a^{m+k-m}} = \frac{1}{a^k}.$$

Therefore $$a^{-k} = \frac{1}{a^k}.$$

That is, *a power with a negative exponent is equal to* 1 *divided by a power of the same base with a positive exponent of the same absolute value as the given exponent.*

E.g., $$2^{-3} = \frac{1}{2^3} = \frac{1}{8}.$$

Observe that the words *negative integral* refer only to the exponent. The base may be either positive or negative, integral or fractional, and consequently a *negative integral* power may sometimes have a *positive* or *fractional* value.

E.g., $$(-3)^{-2} = \frac{1}{(-3)^2} = \frac{1}{9};\ (-\tfrac{1}{2})^{-4} = \frac{1}{(-\tfrac{1}{2})^4} = 16.$$

We are thus led to a quite definite and intelligible meaning of *negative powers* by extending still further the application of the stated law to the case in which $m < n$.

10. From the result of the preceding article we derive the following:

$$\left(\frac{a}{b}\right)^{-k} = \frac{1}{\left(\frac{a}{b}\right)^k} = \frac{1}{\frac{a^k}{b^k}} = \frac{b^k}{a^k} = \left(\frac{b}{a}\right)^k.$$

That is, a negative integral power of any base is equal to a positive power of the reciprocal of the base, the exponents of the powers having the same absolute values; and *vice versa.*

This reciprocal relation between positive and negative powers is very useful in reductions which involve negative powers.

E.g., $$(\tfrac{2}{3})^{-2} = (\tfrac{3}{2})^2 = \tfrac{9}{4}.$$

11. We also have $$\frac{1}{a^{-k}} = \frac{1}{\frac{1}{a^k}} = a^k.$$

This relation and the relation which defined a negative integral power may be stated thus:

Any power of a number may be transferred from the denominator to the numerator, or from the numerator to the denominator, of a fraction, if the sign of its exponent be reversed.

E.g., $$\frac{a^2}{a^{-3}} = a^2 \cdot a^3 = a^5; \quad \frac{(-a)^{-4}}{a} = \frac{1}{a(-a)^4} = \frac{1}{a^5}.$$

12. A negative integral power of zero is equal to infinity.

For $$0^{-n} = \frac{1}{0^n} = \frac{1}{0} = \infty.$$

13. A principle for negative integral powers corresponding to the principle proved in Ch. XVI., Art. 15, for positive integral powers is the following:

The value of a negative integral power decreases as its exponent increases in absolute value, if its base be greater than 1; *and increases as its exponent increases in absolute value, if its base be positive and less than* 1; i.e., *if its base be a positive proper fraction.*

Thus, the powers $2^{-1} = \frac{1}{2}$, $2^{-2} = \frac{1}{2^2} = \frac{1}{4}$, $2^{-3} = \frac{1}{2^3} = \frac{1}{8}$, etc;

decrease as the exponents

$$-1, -2, -3, \cdots$$

increase in absolute value; while the powers

$$\left(\frac{2}{3}\right)^{-1}=\frac{3}{2},\quad \left(\frac{2}{3}\right)^{-2}=\left(\frac{3}{2}\right)^{2}=\frac{9}{4},\quad \left(\frac{2}{3}\right)^{-3}=\left(\frac{3}{2}\right)^{3}=\frac{27}{8},\cdots,$$

increase as the exponents

$$-1,\ -2,\ -3\cdots,$$

increase in absolute value.

In general,

$$\left(\frac{a}{b}\right)^{-(n+1)}<\left(\frac{a}{b}\right)^{-n},\text{ when } a>b;$$

and

$$\left(\frac{a}{b}\right)^{-(n+1)}>\left(\frac{a}{b}\right)^{-n},\text{ when } a<b;$$

wherein a, b, and n are positive integers.

For, by Ch. XVI., Art. 15,

$$\left(\frac{a}{b}\right)^{n+1}>\left(\frac{a}{b}\right)^{n},\text{ when } a>b.$$

Therefore, by Ch. XVI., Art. 9 (v.),

$$\frac{1}{\left(\frac{a}{b}\right)^{n+1}}<\frac{1}{\left(\frac{a}{b}\right)^{n}},$$

or

$$\left(\frac{a}{b}\right)^{-(n+1)}<\left(\frac{a}{b}\right)^{-n},\text{ when } a>b.$$

In like manner the second part of the principle can be proved.

We therefore conclude that a negative integral power whose base is greater than 1 decreases without limit, *i.e.*, becomes less than any assigned positive number, however small, when the absolute value of its exponent increases without limit; and that a negative integral power whose base is less than 1 increases without limit, *i.e.*, becomes greater than any assigned positive number, however great, when the absolute value of its exponent increases without limit.

These conclusions may be stated symbolically thus:

$$\left(\frac{a}{b}\right)^{-\infty}=0,\text{ when } a>b;$$

$$\left(\frac{a}{b}\right)^{-\infty}=\infty,\text{ when } a<b.$$

EXERCISES I.

Find the value of each of the following expressions:

1. 2^{-3}.
2. 3^{-2}.
3. $(\frac{4}{3})^{-1}$.
4. $(3\frac{3}{4})^{-3}$.
5. $(\frac{1}{8})^{-3}$.
6. $\frac{1}{.25^{-4}}$.
7. $\frac{1}{.2^{-6}}$.
8. $(2^0)^{-6}$.

Change each of the following expressions into an equivalent expression in which all the exponents are positive:

9. x^3y^{-4}. **10.** $2\,c^{-4}d$. **11.** $3^{-1}a^2n^{-3}$. **12.** $5\,x^{-2}y^{-3}$.

13. $\dfrac{2\,n^{-3}}{a^{-1}b^2}$. **14.** $\dfrac{3\,b^2}{4\,a^{-5}c}$. **15.** $\dfrac{5\,ad^{-2}}{7^{-1}b^{-3}c}$. **16.** $\dfrac{3\,a^{-2}n^{-2}}{8\,b^{-4}}$.

17. $7 \times 3^{-2}ab^{-4}c^3d^{-5}$. **18.** $9 \times 10^{-2}(\frac{3}{5})^{-3}ax^{-5}$.

19. $(\frac{3}{2})^{-3}2^{-4}a^0b^{-6}x^{-2}y^3$. **20.** $2^{-3}5(a+1)(a-1)^{-2}x^{n-1}n^{-(x+1)}$.

21. $(1\frac{1}{5})^{-1}a^{n-1}c^{-n}(x+a)^n(x+b)^{-m-n}$.

In each of the following expressions transfer the factors from the denominator to the numerator:

22. $\dfrac{a}{b^2}$. **23.** $\dfrac{2\,x^2}{5\,y^{-3}}$. **24.** $\dfrac{3\,x^{-3}}{2^{-2}y}$. **25.** $\dfrac{5\,xy}{ab}$.

26. $\dfrac{3}{(a+b)}$. **27.** $\dfrac{4(x+y)^3}{(x-y)^2}$. **28.** $\dfrac{2\,a(x^2+1)}{3\,a^{-1}(x^2-1)^3}$.

29–35. In the examples 22–28 transfer the factors from the numerator to the denominator.

Fractional (Positive or Negative) Powers.

14. The meaning of a fractional power in which the exponent is the reciprocal of a positive integer will be determined first, then that of any fractional power.

Notice again that the word *fractional* refers to the exponent of the power and not to its value.

15. We will define, *i.e.*, fix the meaning of, the power $a^{\frac{1}{q}}$, in which q is a positive integer, by assuming that it must obey the first law of exponents, namely,

$$a^m \cdot a^n = a^{m+n}.$$

In other words, whatever meaning $a^{\frac{1}{q}}$ may have must be derived by an application of this law.

By this law, $a^{\frac{1}{2}} \cdot a^{\frac{1}{2}} = a^{\frac{1}{2}+\frac{1}{2}} = a^1 = a.$

But, since $a^{\frac{1}{2}} \cdot a^{\frac{1}{2}} = (a^{\frac{1}{2}})^2$, by definition of positive integral power of *any* base, we have

$$(a^{\frac{1}{2}})^2 = a.$$

That is, $a^{\frac{1}{2}}$ is a number whose *square is a*, or $a^{\frac{1}{2}} = \sqrt{a}$.

In like manner, $a^{\frac{1}{3}} \cdot a^{\frac{1}{3}} \cdot a^{\frac{1}{3}} = a^{\frac{1}{3}+\frac{1}{3}+\frac{1}{3}} = a^1 = a.$

But, since $a^{\frac{1}{3}} \cdot a^{\frac{1}{3}} \cdot a^{\frac{1}{3}} = (a^{\frac{1}{3}})^3,$

we have $(a^{\frac{1}{3}})^3 = a.$

That is, $a^{\frac{1}{3}}$ is a number whose *cube is a*, or $a^{\frac{1}{3}} = \sqrt[3]{a}$.

In general,

$$a^{\frac{1}{q}} \cdot a^{\frac{1}{q}} \cdot a^{\frac{1}{q}} \cdots q \text{ factors} = a^{\frac{1}{q}+\frac{1}{q}+\frac{1}{q}+\cdots q \text{ terms}}$$
$$= a^{q \cdot \frac{1}{q}} = a.$$

But, since $a^{\frac{1}{q}} \cdot a^{\frac{1}{q}} \cdot a^{\frac{1}{q}} \cdots q$ factors $= (a^{\frac{1}{q}})^q$, by definition of positive integral power, we have

$$(a^{\frac{1}{q}})^q = a.$$

That is, $a^{\frac{1}{q}}$ is a number whose *qth power is a*, or $a^{\frac{1}{q}} = \sqrt[q]{a}$.

We are thus led, by the definition of the fractional power, $a^{\frac{1}{q}}$, to the operation that is inverse to that of raising a number to a positive integral power, *i.e.*, to the operation of finding a root.

Thus, $9^{\frac{1}{2}}$ and $\sqrt{9}$, $(-243)^{\frac{1}{5}}$ and $\sqrt[5]{-243}$, $a^{\frac{1}{q}}$ and $\sqrt[q]{a}$, are only different ways of representing the same numbers.

Notice that the index of the root is the *denominator* of the exponent of the fractional power, and the radicand is the *base*.

16. From the definition of a fractional power we have

$$(9^{\frac{1}{2}})^2 = (\sqrt{9})^2 = 9,$$

$$[(-25)^{\frac{1}{3}}]^3 = (\sqrt[3]{-25})^3 = -25.$$

In general, $$(a^{\frac{1}{q}})^q = (\sqrt[q]{a})^q = a. \qquad (1)$$

Also, from Ch. XV., § 1, Art. 10,

$$(a^q)^{\frac{1}{q}} = \sqrt[q]{a^q} = a,$$

if only principal roots be considered.

Therefore $(a^{\frac{1}{q}})^q = (a^q)^{\frac{1}{q}}$, for the principal root.

17. Meaning of $a^{\frac{p}{q}}$, wherein $\frac{p}{q}$ is a *positive or a negative* fraction. We may always assume q to be positive and p to have the sign of the fraction.

Whatever meaning $a^{\frac{p}{q}}$ may have must be derived by an application of the law

$$a^m \cdot a^n = a^{m+n}.$$

By this law, $5^{\frac{2}{3}} \cdot 5^{\frac{2}{3}} \cdot 5^{\frac{2}{3}} = 5^{\frac{2}{3}+\frac{2}{3}+\frac{2}{3}} = 5^2.$

But, since $5^{\frac{2}{3}} \cdot 5^{\frac{2}{3}} \cdot 5^{\frac{2}{3}} = (5^{\frac{2}{3}})^3,$

we have $(5^{\frac{2}{3}})^3 = 5^2.$

That is, $5^{\frac{2}{3}}$ is a number whose *cube is* 5^2; or $5^{\frac{2}{3}} = \sqrt[3]{5^2}$.

In general,

$$a^{\frac{p}{q}} \cdot a^{\frac{p}{q}} \cdot a^{\frac{p}{q}} \cdots q \text{ factors} = a^{\frac{p}{q}+\frac{p}{q}+\frac{p}{q}+\cdots,\ q \text{ terms}}$$

$$= a^{q \cdot \frac{p}{q}},$$

$$= a^p.$$

But, since $a^{\frac{p}{q}} \cdot a^{\frac{p}{q}} \cdot a^{\frac{p}{q}} \cdots q \text{ factors} = (a^{\frac{p}{q}})^q,$

we have $(a^{\frac{p}{q}})^q = a^p.$

That is, $a^{\frac{p}{q}}$ is a number whose *qth power is* a^p; or $a^{\frac{p}{q}} = \sqrt[q]{a^p}$.

Notice that a fractional power is a root of an integral power. The numerator of the fractional exponent is the index of the root, and the denominator is the exponent of the power.

E.g., $23^{\frac{4}{5}} = \sqrt[5]{23^4}$; $(-19)^{\frac{2}{3}} = \sqrt[3]{(-19)^2}$; $2^{-\frac{2}{3}} = \sqrt[3]{2^{-2}} = \sqrt[3]{\frac{1}{4}}$.

18. Since fractional powers simply afford another way of indicating roots, all the principles relating to roots which were proved in Chapters XV. and XVII.–XIX. hold for such powers.

For the sake of emphasis, the following properties are restated:

(i.) *The fractional power* $a^{\frac{p}{q}}$, *wherein* a *is negative,* q *is an even number, and* $\frac{p}{q}$ *is in its lowest terms, is an imaginary or complex number.*

E.g., $$(-2)^{\frac{3}{4}} = \sqrt[4]{(-2)^3} = \sqrt[4]{-8};$$

but $$(-2)^{\frac{2}{4}} = \sqrt[4]{(-2)^2} = \sqrt[4]{4} = \sqrt{2}.$$

Such powers should be expressed as roots and be treated as imaginary or complex numbers, and will not be further considered in this chapter.

(ii.) *If a fractional power of a negative base have a real value, and if the meaning of the power be limited to this value, it can be expressed as a fractional power of a positive base.*

E.g., $$(-5)^{\frac{1}{3}} = \sqrt[3]{-5} = -\sqrt[3]{5} = -5^{\frac{1}{3}},$$

$$(-a)^{\frac{2}{3}} = \sqrt[3]{(-a)^2} = \sqrt[3]{a^2} = a^{\frac{2}{3}}.$$

We shall, therefore, in the theory of fractional powers, assume that the bases are positive, and limit the meaning of a fractional power to the principal value of the corresponding root.

(iii.) $a^{\frac{p}{q}} = (a^p)^{\frac{1}{q}} = (a^{\frac{1}{q}})^p.$

E.g., $$(4^4)^{\frac{1}{2}} = (4^{\frac{1}{2}})^4.$$

(iv.) $a^{\frac{p}{q}} = a^{\frac{kp}{kq}}$, and conversely, $a^{\frac{kp}{kq}} = a^{\frac{p}{q}}.$

E.g., $$8^{\frac{1}{3}} = 8^{\frac{2}{6}}.$$

Observe that these principles do not always hold if other than principal roots are considered.

Thus, $$(4^4)^{\frac{1}{2}} = \sqrt{4^4} = \sqrt{256} = \pm 16;$$

while $$(4^{\frac{1}{2}})^4 = (\sqrt{4})^4 = (\pm 2)^4 = 16.$$

Also $$8^{\frac{1}{3}} = \sqrt[3]{8} = 2;$$

while $$8^{\frac{2}{6}} = \sqrt[6]{8^2} = \sqrt[6]{64} = \pm 2.$$

19. Principles for fractional powers corresponding to the principles proved in Ch. XVI., Art. 15, and this chapter, Art. 13, for positive and negative integral powers are the following:

(i.) *A positive fractional power of a positive base which is greater than 1 increases as the exponent increases, and vice versa; while a positive fractional power of a positive base which is less than 1 decreases as the exponent increases, and vice versa.*

Let $a^{\frac{m}{n}}$ be a fractional power, in which $\frac{m}{n}$ and a are positive. And let $a^{\frac{m}{n}+\frac{p}{q}}$ be a fractional power whose exponent exceeds the exponent $\frac{m}{n}$ by $\frac{p}{q}$.

Then $a^{\frac{m}{n}+\frac{p}{q}} - a^{\frac{m}{n}} = a^{\frac{m}{n}}(a^{\frac{p}{q}} - 1)$.

But, if $a > 1$, $a^{\frac{p}{q}}$ is greater than 1, by Ch. XVI., Art. 8 (iv.) and (i.).

Therefore $a^{\frac{p}{q}} - 1$ is positive, and hence $a^{\frac{m}{n}}(a^{\frac{p}{q}} - 1)$ is also positive. Consequently $a^{\frac{m}{n}+\frac{p}{q}}$ is greater than $a^{\frac{m}{n}}$, when $a > 1$.

If $a < 1$, then $a^{\frac{p}{q}}$ is less than 1, by Ch. XVI., Art. 8 (iv.) and (i.).

Therefore $a^{\frac{p}{q}} - 1$, and hence $a^{\frac{m}{n}}(a^{\frac{p}{q}} - 1)$, is negative. Consequently $a^{\frac{m}{n}+\frac{p}{q}}$ is less than $a^{\frac{m}{n}}$, when $a < 1$.

(ii.) *A negative fractional power of a positive base which is greater than 1 decreases as the exponent increases in absolute value, and vice versa; while a negative power of a positive base which is less than 1 increases as the exponent increases in absolute value.*

We have $a^{-\frac{m}{n}} = \left(\frac{1}{a}\right)^{\frac{m}{n}}$.

When $a > 1$, then $\frac{1}{a} < 1$.

Therefore, by (i.), $\left(\frac{1}{a}\right)^{\frac{m}{n}}$, and hence $a^{-\frac{m}{n}}$, decreases as $\frac{m}{n}$ increases.

When $a < 1$, then $\frac{1}{a} > 1$.

Therefore, by (i.), $\left(\frac{1}{a}\right)^{\frac{m}{n}}$, and hence $a^{-\frac{m}{n}}$, increases as $\frac{m}{n}$ increases.

20. The following principle will be useful in subsequent work:

The fractional power $a^{\frac{1}{n}}$, wherein a is greater than 1, can be made to differ from 1 by less than any assigned number, however small, by increasing n indefinitely.

Let $$a^{\frac{1}{n}} - 1 = d; \tag{1}$$

wherein d is positive, since $a^{\frac{1}{n}} > 1$, by Ch. XVI., Art. 8 (iv.).

We are then to prove that d can be made less than any assigned number, however small, by increasing n indefinitely.

From (1) we have $$a^{\frac{1}{n}} = 1 + d,$$

or $$a = (1 + d)^n.$$

But, by Ch. XVI., Art. 17,

$$1 + nd < (1 + d)^n.$$

Therefore $$1 + nd < a,$$

or $$d < \frac{a-1}{n}.$$

Therefore as n increases indefinitely, $\frac{a-1}{n}$, and hence also d, decreases indefinitely, and can be made less than any assigned number, however small.

EXERCISES II.

Write each of the following expressions as an equivalent expression with radical signs:

1. $a^{\frac{1}{2}}$.
2. $b^{-\frac{1}{4}}$.
3. $x^{\frac{2}{3}}$.
4. $3\,y^{\frac{1}{3}}$.
5. $4\,x^{-\frac{3}{4}}y^{\frac{1}{3}}$.
6. $2\,ab^{-\frac{5}{6}}c$.
7. $2^{\frac{1}{2}}x^{\frac{3}{4}}y^{\frac{1}{3}}$.
8. $2\,a^{\frac{m}{n}}b^{-\frac{p}{q}}$.
9. $\left(\frac{a}{b}\right)^{\frac{4}{5}}$.
10. $\left(\frac{2\,x}{3\,y}\right)^{-\frac{4}{5}}$.
11. $\frac{4\,m^{\frac{3}{4}}}{3\,n^{\frac{5}{6}}}$.
12. $\frac{ab^{-\frac{m}{n}}}{xy^{-\frac{p}{q}}}$.

Find the value of each of the following expressions:

13. $4^{\frac{1}{2}}$.
14. $169^{\frac{1}{2}}$.
15. $16^{-\frac{1}{2}}$.
16. $144^{-\frac{1}{2}}$.
17. $27^{\frac{1}{3}}$.
18. $27^{-\frac{1}{3}}$.
19. $16^{\frac{1}{4}}$.
20. $81^{-\frac{1}{4}}$.
21. $49^{\frac{3}{2}}$.
22. $512^{\frac{2}{3}}$.
23. $216^{-\frac{2}{3}}$.
24. $32^{-\frac{3}{5}}$.
25. $64^{\frac{2}{3}}$.
26. $64^{\frac{5}{6}}$.
27. $.09^{\frac{3}{2}}$.
28. $(3\frac{3}{8})^{-\frac{2}{3}}$.

Write each of the following expressions as an equivalent expression with fractional exponents:

29. $\sqrt{a}$.
30. $\sqrt{a^3}$.
31. $\sqrt{(a^{-3}b^7)}$.
32. $\sqrt{(2\,xy^{-5})}$.
33. $\sqrt[3]{a^2}$.
34. $\sqrt[3]{(2\,x^{-1}y^3)}$.
35. $\sqrt[4]{(5\,x^{-2}y^5)}$.
36. $\sqrt[5]{(3\,a^{-7}b^6)}$.
37. $\sqrt[n]{(3\,a^2)}$.
38. $\sqrt[q]{(3\,a^{-n})}$.
39. $\sqrt[q]{[(a+b)^2(x-y)^{-p}]}$.

Simplify each of the following expressions:

40. $36^{\frac{1}{2}} + 8^{\frac{2}{3}} - 625^{.75}$. **41.** $.16^{.5} - (1\frac{2}{5}\frac{1}{1}\frac{7}{2})^{\frac{2}{3}} + 64^{-\frac{1}{2}}$.

42. $2\,a^{-\frac{3}{4}} - .4\,a^{\frac{1}{4}} + 2.5\,a^{.75} - 5\,a^{.25}$.

21. Having thus determined definite meanings for zeroth, negative, and fractional powers, it remains to prove that they obey all the principles of positive integral powers.

(I.) $$a^m a^n = a^{m+n},$$
for all rational values of m and n.

Ex. 1. $x^5 x^{-7} = x^{5-7} = x^{-2} = \dfrac{1}{x^2}$.

Ex. 2. $a^{\frac{1}{2}}b^{-\frac{3}{4}} \times a^{-3}b^4 = a^{\frac{1}{2}-3}b^{-\frac{3}{4}+4} = a^{-\frac{5}{2}}b^{\frac{13}{4}} = \dfrac{b^{\frac{13}{4}}}{a^{\frac{5}{2}}}$.

(i.) $m = 0$, and n a negative integer.

Let $n = -n_1$, so that n_1 is positive.

Then $a^m a^n = a^0 a^{-n_1} = a^{-n_1}$, since $a^0 = 1$.

But $a^{0+(-n_1)} = a^{-n_1}$, since $0 + (-n_1) = -n_1$.

Therefore $$a^0 a^{-n_1} = a^{0+(-n_1)},$$
or $$a^m a^n = a^{m+n},$$
when m is 0, and n is negative.

(ii.) m positive and n negative, and the absolute value of m less than the absolute value of n.

Let $n = -n_1$, so that n_1 is positive.

Then
$$\begin{aligned} a^m a^n &= a^m a^{-n_1} \\ &= \frac{a^m}{a^{n_1}} \\ &= \frac{1}{a^{n_1-m}} \\ &= \frac{1}{a^{-[m+(-n_1)]}} \\ &= a^{m+(-n_1)} = a^{m+n}. \end{aligned}$$

Therefore $$a^m a^n = a^{m+n},$$
when n is negative and the absolute value of m is less than the absolute value of n.

In a similar way the principle can be proved for other cases in which the exponents are 0 or negative.

That the principle holds when the exponents, either or both, are fractions, follows from the definition of a fractional power.

EXERCISES III.

Simplify each of the following expressions:

1. x^3x^0.
2. $x^{-3}x^3$.
3. $a^{-5}a^6$.
4. $m^{-3}m^{-5}$.
5. $a^3a^{\frac{1}{2}}$.
6. $a^{\frac{2}{3}}a^{\frac{3}{4}}$.
7. $b^{-\frac{5}{6}}b^{\frac{2}{3}}$.
8. $c^{-\frac{1}{3}}c^{-\frac{3}{5}}$.
9. $5\,a^{-3} \times 3\,a^5$.
10. $-\frac{5}{7}\,b^{-2} \times 1\frac{2}{5}\,b^{-3}$.
11. $3\frac{3}{4}\,x^3 \times 2\frac{2}{3}\,x^{-\frac{1}{2}}$.
12. $a^3b^{-2} \times a^{\frac{1}{3}}b^{\frac{2}{3}}$.
13. $3\,x^{-\frac{3}{4}}y^{\frac{2}{3}} \times 2\,x^{-5}y^{-\frac{1}{2}}$.
14. $2\,a^{-\frac{1}{3}}b^{-2}c^3 \times 5\,a^6c^{-\frac{7}{3}}$.
15. $a^mb^{-n} \times a^{-p}b^q$.
16. $xy^p \times x^{\frac{m}{n}}y^{-\frac{p}{q}}$.
17. $a^{\frac{1}{m-n}}b^{\frac{p}{q}} \times a^{\frac{1}{m+n}}b^{-\frac{p}{q}}$.
18. $\dfrac{12\,a^{-3}}{n^{-2}} \times \dfrac{a^2}{9\,n^3}$.
19. $\dfrac{7\,c^{-3}}{3\,a^3} \div \dfrac{35\,a^{-4}}{6\,c^2}$.
20. $\dfrac{a^{-m}b^{-n}}{\frac{1}{2}\,c} \div \dfrac{c^{-1}}{a^{-2m}b^{-2n}}$.
21. $(a^{\frac{1}{3}} + x^{-2})(a^{\frac{1}{3}} - x^{-2})$.
22. $(a^{\frac{1}{2}} + a^{-\frac{1}{2}})(a^{\frac{1}{2}} - a^{-\frac{1}{2}})$.
23. $(a^{\frac{2}{3}} - a^{\frac{1}{3}}b^{\frac{1}{3}} + b^{\frac{2}{3}})(a^{\frac{1}{3}} + b^{\frac{1}{3}})$.
24. $(x^2y^{-\frac{2}{3}} + xy^{-\frac{1}{3}} + 1)(xy^{-\frac{1}{3}} - 1)$.
25. $(a^{\frac{1}{2}} + a^{\frac{1}{4}}b^{\frac{1}{4}} + b^{\frac{1}{2}})(a^{\frac{1}{2}} - a^{\frac{1}{4}}b^{\frac{1}{4}} + b^{\frac{1}{2}})$.
26. $(a^{\frac{1}{2}} + b^{\frac{1}{2}} + a^{-\frac{1}{2}}b)(ab^{-\frac{1}{2}} - a^{\frac{1}{2}} + b^{\frac{1}{2}})$.
27. $(a^{-7} + a^{-5} - a^{-3})(a^7 + a^5 + a^3)$.
28. $(x^3 - x^{-3} - 2\,x^{-6} + 5)(10\,x^{-7} + x^{-1} - 5\,x^{-4})$.
29. $(\frac{1}{2}\,a^{-6}x^{-n} + \frac{3}{4}\,x^{-3n} - a^{-8}x^{-2n})(\frac{2}{3}\,a^2x^p - 4\,a^{-4}x^{p+2n} + a^{-1}x^{p+n})$.
30. $(x^{\frac{3}{2}} - xy^{\frac{1}{2}} + x^{\frac{1}{2}}y - y^{\frac{3}{2}})(x + x^{\frac{1}{2}}y^{\frac{1}{2}} + y)$.
31. $(a^{\frac{2}{3}} + a^{-\frac{2}{3}} - a^{\frac{1}{3}} - a^{-\frac{1}{3}})(a^{\frac{1}{3}} + a^{-\frac{1}{3}} + 1)$.
32. $(x^{\frac{2}{3}} + 2\,x^{.5} + 3\,x^{\frac{1}{3}} + 2\,x^{\frac{1}{6}} + 1)(x^{\frac{1}{3}} - 2\,x^{\frac{1}{6}} + 1)$.
33. $(a^{-1.5} + b^{-1.6} - a^{-.75}b^{-.8})(a^{-.75} + b^{-.8})$.
34. $(1\frac{4}{5}\,a^{.5}x^{1\frac{1}{3}} + 2\,a^{\frac{3}{2}} + \frac{2}{5}\,x^{1.8} + 6\,ax^{\frac{3}{5}})(a^{\frac{1}{2}} - 3\,x^{\frac{3}{5}} + \frac{5}{2}\,x^{1\frac{4}{5}}a^{-1})$.

(II.) $$\frac{a^m}{a^n} = a^{m-n},$$

for all rational values of m and n.

Ex. 1. $$\frac{x^0}{x^{-2}} = x^{0+2} = x^2.$$

Ex. 2. $$\frac{16^2}{16^{\frac{3}{4}}} = 16^{2-\frac{3}{4}} = 16^{\frac{5}{4}} = (\sqrt[4]{16})^5 = 32.$$

Ex. 3. $$\frac{x^2}{x^{-3}} = x^{2+3} = x^5.$$

Ex. 4. $$\frac{a^{-\frac{1}{2}}b^{\frac{2}{3}}}{a^{\frac{1}{4}}b^{-\frac{1}{2}}} = a^{-\frac{1}{2}-\frac{1}{4}}\,b^{\frac{2}{3}+\frac{1}{2}} = a^{-\frac{3}{4}}b^{1\frac{1}{6}} = \frac{b^{1\frac{1}{6}}}{a^{\frac{3}{4}}}.$$

We have $$\frac{a^m}{a^n} = a^m a^{-n} = a^{m+(-n)} = a^{m-n}.$$

EXERCISES IV.

Simplify each of the following expressions:

1. $\dfrac{a}{a^{-1}}$. 2. $\dfrac{x^0}{x^{-2}}$. 3. $\dfrac{5^{-2}}{5^{-3}}$. 4. $\dfrac{a^2}{a^{\frac{1}{2}}}$.

5. $\dfrac{x^{-2}}{x^{-5}}$. 6. $\dfrac{x^{\frac{2}{3}}}{x^{-\frac{1}{6}}}$. 7. $\dfrac{a^{-\frac{3}{4}}}{a^{-\frac{1}{4}}}$. 8. $\dfrac{a^{\frac{1}{2}}}{a^{-2}}$.

9. $\dfrac{x^n}{x^{-n}}$. 10. $\dfrac{x^{m-n}}{x^{-n}}$. 11. $\dfrac{x^{-1}}{x^{n-1}}$. 12. $\dfrac{x^{5-n}}{x^{-5}}$.

13. $1\frac{1}{2}\,b^{-3} \div 3\,b^2$. 14. $1 \div \frac{1}{2}\,ab^{-1}$. 15. $-\frac{2}{5}\,a^{-1}b^{\frac{1}{2}} \div 4\,a^{\frac{2}{3}}b^{-1}$.

16. $3\frac{1}{2}\,a^n b^{-4} \div \frac{7}{8}\,a^n b^{-3}$. 17. $2\,a^{\frac{3n}{4}}x^2 \div 1\frac{1}{4}\,a^{-\frac{n}{2}}\,x^{-\frac{1}{2}}$.

18. $a^{\frac{m}{n}}b^{-\frac{p}{q}} \div a^{-\frac{n}{m}}b^{\frac{q}{p}}$. 19. $\dfrac{2\,x^2y^{-\frac{1}{2}}}{3\,a^2b^{-4}} \times \dfrac{6\,a^{-5}b^{\frac{1}{2}}}{7\,x^2y^{\frac{1}{2}}}$.

20. $12\,a^{-1}b^{-1}x^{-\frac{3}{4}} \div \dfrac{4\,a^2b^{-\frac{2}{3}}}{x^{-\frac{2}{3}}}$. 21. $(a^{\frac{1}{2}} - b^{\frac{1}{2}}) \div (a^{\frac{1}{4}} + b^{\frac{1}{4}})$.

22. $(x - y) \div (x^{\frac{1}{3}} - y^{\frac{1}{3}})$. 23. $(x^{-1} + y^{-1}) \div (x^{-\frac{1}{3}} + y^{-\frac{1}{3}})$.

24. $(x^{-4} - y^{-4}) \div (x^{-1} + y^{-1})$.

25. $(x + x^{\frac{1}{2}}y^{\frac{1}{2}} + y) \div (x^{\frac{1}{2}} + x^{\frac{1}{4}}y^{\frac{1}{4}} + y^{\frac{1}{2}})$.

26. $[(a-1)^{-2}-1] \div [(a-1)^{-1}-1]$.

27. $(a^{\frac{3n}{2}} - a^{-\frac{3n}{2}}) \div (a^{\frac{n}{2}} - a^{-\frac{n}{2}})$.

28. $(3a^{-10} + a^6 - 4a^{-6}) \div (2a^{-2} + a^2 + 3a^{-6})$.

29. $(2x^{-3} - 3x^{-2} - 2x^{-1} + 2 - x) \div (x^{-1} + 1)$.

30. $(x^{-1} - 3x^{-\frac{1}{2}} + 3 - 3x^{\frac{1}{2}} - 2x) \div (x^{-\frac{3}{2}} - 2x^{-1} + x^{-\frac{1}{2}} - 2)$.

31. $(2a^7 - 3a^3 - 23a^{-1} + 15a^{-5} + 9a^{-9}) \div (a^4 + 2 - 3a^{-4})$.

32. $(6x^{\frac{1}{2}} + 9x^{-\frac{1}{2}} - 2x^{-1} - 13) \div (3x^{\frac{1}{2}} + 2x^{-\frac{1}{2}} - 5)$.

33. $(x^{\frac{3}{2}} - xy^{\frac{1}{2}} + x^{\frac{1}{2}}y - y^{\frac{3}{2}}) \div (x^{\frac{1}{2}} - y^{\frac{1}{2}})$.

34. $(x^{\frac{4}{3}} + a^{\frac{2}{3}}x^{\frac{2}{3}} + a^{\frac{4}{3}}) \div (x^{\frac{2}{3}} + a^{\frac{1}{3}}x^{\frac{1}{3}} + a^{\frac{2}{3}})$.

35. $(a^{\frac{5}{2}} - a^2b^{\frac{1}{2}} - a^{\frac{1}{2}}b^2 + b^{\frac{5}{2}}) \div (a^{\frac{3}{2}} - ab^{\frac{1}{2}} + a^{\frac{1}{2}}b - b^{\frac{3}{2}})$.

36. $(6x^{\frac{5}{4}} - 7x - 19x^{\frac{3}{4}} + 2x^{\frac{1}{2}} + 8x^{\frac{1}{4}}) \div (2x^{\frac{3}{4}} - 3x^{\frac{1}{2}} - 4x^{\frac{1}{4}})$.

(III.) $$(a^m)^n = a^{mn},$$

for all rational values of m and n.

Ex. 1. $(x^2)^{-3} = x^{2(-3)} = x^{-6} = \dfrac{1}{x^6}$.

Ex. 2. $(1024^{\frac{1}{2}})^{-\frac{3}{5}} = 1024^{-\frac{3}{10}} = \dfrac{1}{(\sqrt[10]{1024})^3} = \dfrac{1}{8}$.

Ex. 3. $(a^{-\frac{2}{3}}x^2y^{-4})^{-\frac{1}{2}} = (a^{-\frac{2}{3}})^{-\frac{1}{2}}(x^2)^{-\frac{1}{2}}(y^{-4})^{-\frac{1}{2}} = a^{\frac{1}{3}}x^{-1}y^2 = \dfrac{a^{\frac{1}{3}}y^2}{x}$.

(i.) m and n both negative integers.

Let $m = -m_1$ and $n = -n_1$, so that m_1 and n_1 are positive.

We have
$$\begin{aligned}(a^m)^n &= (a^{-m_1})^{-n_1}\\ &= \left(\frac{1}{a^{m_1}}\right)^{-n_1}\\ &= (a^{m_1})^{n_1}\\ &= a^{m_1n_1}\\ &= a^{(-m_1)(-n_1)}\\ &= a^{mn}.\end{aligned}$$

Therefore
$$(a^m)^n = a^{mn},$$
when m and n are both negative integers.

In a similar manner the principle can be proved for other cases in which the exponents are 0 or negative integers.

(ii.) m a fraction, and n a positive or a negative integer, or 0.

Let $m = \frac{p}{q}$, wherein q is a positive integer and p is a positive or a negative integer.

We then have $(a^m)^n = (a^{\frac{p}{q}})^n$

$$= [(a^{\frac{1}{q}})^p]^n, \text{ by Art. 18 (iii.)},$$

$$= (a^{\frac{1}{q}})^{pn}, \text{ by (i.)},$$

$$= a^{\frac{pn}{q}}, \text{ by Art. 18 (iii.)},$$

$$= a^{\frac{p}{q}n} = a^{mn}.$$

In a similar manner the principle can be proved when m is an integer and n is a fraction.

(iii.) m and n both fractions.

Let $m = \frac{p}{q}$ and $n = \frac{r}{s}$, wherein q and s are positive integers, and p and r are positive or negative integers.

If $(a^{\frac{p}{q}})^{\frac{r}{s}}$ be raised to the qsth, $= sq$th power, we have

$$[(a^{\frac{p}{q}})^{\frac{r}{s}}]^{qs} = \{[(a^{\frac{p}{q}})^{\frac{r}{s}}]^s\}^q$$

$$= [(a^{\frac{p}{q}})^r]^q, \text{ by (ii.)},$$

$$= [(a^{\frac{p}{q}})^q]^r$$

$$= (a^p)^r, \text{ by (ii.)},$$

$$= a^{pr}.$$

Consequently $(a^{\frac{p}{q}})^{\frac{r}{s}}$ is the qs root of a^{pr}; or, by definition of a fractional power,

$$(a^{\frac{p}{q}})^{\frac{r}{s}} = a^{\frac{pr}{qs}} = a^{\frac{p}{q} \cdot \frac{r}{s}}.$$

Therefore $(a^m)^n = (a^m)^n$,

when m and n are both fractions.

EXERCISES V.

Simplify each of the following expressions:

1. $(x^{\frac{2}{3}})^{-3}$.
2. $(a^3)^{\frac{1}{2}}$.
3. $[(-x)^{\frac{1}{3}}]^2$.
4. $(x^{-\frac{3}{5}})^4$.
5. $(x^{-\frac{3}{5}})^{15}$.
6. $(a^{-3})^{\frac{1}{6}}$.
7. $(b^3)^{-\frac{4}{5}}$.
8. $(x^{-\frac{2}{3}})^{-3}$.

9. $(x^{-\frac{1}{3}})^{-\frac{1}{2}}$. 10. $(a^n)^{-2}$. 11. $(a^{-m})^{-3}$. 12. $(a^{-\frac{p}{q}})^{-\frac{m}{n}}$.

13. $(\sqrt[3]{a^{-2}})^4$. 14. $(\sqrt{a})^{-\frac{2}{3}}$. 15. $(\sqrt[5]{x^{\frac{4}{3}}})^{-\frac{3}{4}}$ 16. $(\sqrt[q]{a^{-m}})^{-3}$.

17. $(a^{\frac{1}{2}} - a^{-\frac{1}{2}})^2$. 18. $(1 - x^{-\frac{1}{2}})^2$. 19. $(a^{\frac{1}{3}} + a^{-\frac{1}{3}})^3$.

(IV.) $(ab)^m = a^m b^m$, **for all rational values of** m.

Ex. 1. $(2x)^{-3} = 2^{-3}x^{-3} = \frac{1}{8x^3}$.

Ex. 2. $(3x^{-\frac{1}{2}}y^2)^{-4} = 3^{-4}x^2y^{-8} = \frac{x^2}{81y^8}$.

(i.) m a negative integer. Let $m = -m_1$, so that m_1 is positive.

Then
$$(ab)^m = (ab)^{-m_1}$$
$$= \frac{1}{(ab)^{m_1}}$$
$$= \frac{1}{a^{m_1}b^{m_1}}$$
$$= a^{-m_1}b^{-m_1} = a^m b^m.$$

(ii.) m a fraction.

Let $m = \frac{p}{q}$, wherein p is a positive or negative integer, and q is a positive integer.

If $(ab)^{\frac{p}{q}}$ be raised to the qth power, we have

$$[(ab)^{\frac{p}{q}}]^q = (ab)^p, \text{ since } q \text{ is an integer,}$$
$$= a^p b^p, \text{ by (i.).}$$

But
$$(a^{\frac{p}{q}}b^{\frac{p}{q}})^q = (a^{\frac{p}{q}})^q(b^{\frac{p}{q}})^q$$
$$= a^p b^p, \text{ by (III.).}$$

Therefore
$$[(ab)^{\frac{p}{q}}]^q = (a^{\frac{p}{q}}b^{\frac{p}{q}})^q;$$

whence
$$(ab)^{\frac{p}{q}} = a^{\frac{p}{q}}b^{\frac{p}{q}}.$$

Consequently, $(ab)^m = a^m b^m$, when m is a fraction.

(V.) $\left(\frac{a}{b}\right)^m = \frac{a^m}{b^m}$, **for all rational values of** m.

Ex. 1. $\left(\frac{a^{\frac{1}{2}}}{b^3}\right)^{-3} = \frac{a^{-\frac{3}{2}}}{b^{-9}} = \frac{b^9}{a^{\frac{3}{2}}}$. Ex. 2. $\left(\frac{4^{-3}}{x^2y^{-1}}\right)^{-\frac{1}{2}} = \frac{4^{\frac{3}{2}}}{x^{-1}y} = \frac{8x}{y^{\frac{1}{2}}}$.

We have
$$\left(\frac{a}{b}\right)^m = (ab^{-1})^m = a^m b^{-m} = \frac{a^m}{b^m}.$$

EXERCISES VI.

Simplify each of the following expressions:

1. $(a^{\frac{1}{2}}x^{-1})^{-2}$.
2. $(\frac{1}{4}a)^{-\frac{1}{2}}$.
3. $(8a^{-6})^{\frac{1}{3}}$.
4. $(a^{-1}b^{-3})^{-4}$.
5. $(2a^{\frac{3}{5}}x)^{\frac{5}{6}}$.
6. $(x^{\frac{1}{3}}a^{-\frac{1}{2}})^{-12}$.
7. $(\frac{1}{32}x^{-10})^{-\frac{1}{5}}$.
8. $(a^{-2}b^{\frac{1}{3}}c^{-\frac{3}{4}})^{-6}$.
9. $(2a^{-1}\sqrt[3]{x})^{-2}$.
10. $(5x^{-3}\sqrt[3]{a^2})^{-6}$.
11. $(3\sqrt{x}\sqrt[3]{x^{-1}})^{-6}$.
12. $[3\sqrt[n]{(a^{-m}b^p)}]^{-np}$.
13. $\left(\frac{x^{\frac{2}{3}}}{y^{-\frac{1}{4}}}\right)^{-6}$.
14. $\left(\frac{-2^3a^{-3}}{4b^3}\right)^{-2}$.
15. $\left(\frac{4x^{-\frac{1}{2}}}{y^5}\right)^{-\frac{1}{5}}$.
16. $\left(\frac{8a^2}{27a^{-3}y^{\frac{1}{3}}}\right)^{-\frac{1}{3}}$.
17. $\left(\frac{2x^{\frac{4}{5}}}{3a^{-2}b^3}\right)^{-5}$.
18. $\left(\frac{16a^{-8}}{81x^2y^{-1}}\right)^{-\frac{1}{4}}$.
19. $\left(\frac{5a^{-\frac{1}{6}}b^{\frac{1}{3}}}{6x^{-2}}\right)^3$.
20. $\left(\frac{2x^{-\frac{1}{10}}y^3}{3a^{-4}b^{-\frac{1}{5}}}\right)^5$.
21. $\left(\frac{\sqrt{a}}{\sqrt[3]{x^2}}\right)^{-6}$.
22. $\left(\frac{2\sqrt[3]{a^{-2}}}{3\sqrt{b^{-3}}}\right)^6$.
23. $\left(\frac{3\sqrt[4]{x^3}}{5\sqrt{a^{-3}}}\right)^2$.
24. $\left(\frac{7\sqrt[4]{x^{-\frac{2}{3}}}}{8\sqrt[3]{y^{-2}}}\right)^{-3}$.
25. $\left[\left(\frac{2a^{-\frac{1}{3}}}{x^{-1}y^{-\frac{1}{3}}}\right)^2\right]^{-3}$.
26. $\left[\left(\frac{4a^{-\frac{1}{4}}x^3}{9b^{-3}y^{\frac{1}{2}}}\right)^{-\frac{1}{2}}\right]^2$.
27. $\left[\left(\frac{\sqrt[4]{a^{-3}}}{2x^{-\frac{1}{2}}y^{\frac{2}{3}}}\right)^{-\frac{2}{3}}\right]^{-6}$.
28. $\left[\left(\frac{a^{-\frac{1}{10}}x^{15}}{b^3y^{-\frac{3}{5}}}\right)^{-\frac{1}{3}}\right]^{\frac{5}{3}}$.
29. $(a^{\frac{1}{2}}b^{-2}-1)^2$.
30. $(a^{-\frac{2}{3}}b^{-1}+ab^{\frac{2}{3}})^2$.
31. $(a^{-3}b^{\frac{1}{3}}-c^{-\frac{2}{3}})^3$.

Irrational Powers.

22. **An Irrational Power is a power whose exponent is an irrational number; as $x^{\sqrt{2}}$.**

23. **Let a be any real positive number greater than 1, and I be a positive irrational number defined by the relation (Ch. XVII., Art. 6):**

$$\frac{m}{n}<I<\frac{m+1}{n}.$$

Then the two *rational* powers $a^{\frac{m}{n}}$ and $a^{\frac{m+n}{n}}$ have the properties (i.) and (ii.), Art. 6, Ch. XVII.

For, since $\frac{m}{n}$ increases and $\frac{m+1}{n}$ decreases as n increases, therefore $a^{\frac{m}{n}}$ increases and $a^{\frac{m+1}{n}}$ decreases as n increases [Art. 19 (i.)], and $a^{\frac{m}{n}} < a^{\frac{m+1}{n}}$.

The difference $a^{\frac{m+1}{n}} - a^{\frac{m}{n}} = a^{\frac{m}{n}}(a^{\frac{1}{n}} - 1)$ is positive and can be made less than any assigned number, however small.

For, by Art. 20, $a^{\frac{1}{n}} - 1$ is positive and can be made less than any assigned number, say d. Moreover, since $a^{\frac{m}{n}} < a^{\frac{m+1}{n}}$, therefore $a^{\frac{m}{n}}$ is always less than some positive finite number, say R.

Therefore the given difference can be made less than Rd. But Rd can be made less than any assigned number, say δ, by taking d less than $\frac{\delta}{R}$.

Therefore the two series of powers $a^{\frac{m}{n}}$ and $a^{\frac{m+1}{n}}$ define a positive number which lies between them. This number is defined as a^I. That is,

$$a^{\frac{m}{n}} < a^I < a^{\frac{m+1}{n}}.$$

In the proofs of the principles which follow we shall assume that the base is greater than 1.

24. In like manner it can be shown that if $-I$ be an irrational number, defined by the relation

$$-\frac{m+1}{n} < -I < -\frac{m}{n},$$

then the two series of powers, $a^{-\frac{m+1}{n}}$ and $a^{-\frac{m}{n}}$, define a positive number which lies between them. This number is defined as a^{-I}. That is,

$$a^{-\frac{m+1}{n}} < a^{-I} < a^{-\frac{m}{n}}.$$

25. It follows directly from the definition of a^{-I}, that

$$a^{-I} = \frac{1}{a^I}.$$

26. It can now be proved that the principles of rational powers hold also for irrational powers.

Let a^{I_1} and a^{I_2} be two irrational powers defined by the relations

$$a^{\frac{m_1}{n_1}} < a^{I_1} < a^{\frac{m_1+1}{n_1}},$$

$$a^{\frac{m_2}{n_2}} < a^{I_2} < a^{\frac{m_2+1}{n_2}}.$$

(i.) If the corresponding powers of the series which define a^{l_1} and a^{l_2} be multiplied, we obtain the two series of numbers

$$a^{\frac{m_1}{n_1}}a^{\frac{m_2}{n_2}} \text{ and } a^{\frac{m_1+1}{n_1}}a^{\frac{m_2+1}{n_2}}.$$

The numbers of these series have the properties (i.) and (ii.), Art. 6, Ch. XVII.

For, since $a^{\frac{m_1}{n_1}}$ increases as n_1 increases, and $a^{\frac{m_2}{n_2}}$ increases as n_2 increases, therefore $a^{\frac{m_1}{n_1}}a^{\frac{m_2}{n_2}}$ increases as n_1 and n_2 increase. For a similar reason $a^{\frac{m_1+1}{n_1}}a^{\frac{m_2+1}{n_2}}$ decreases as n_1 and n_2 increase. And since

$$a^{\frac{m_1}{n_1}} < a^{\frac{m_1+1}{n_1}} \text{ and } a^{\frac{m_2}{n_2}} < a^{\frac{m_2+1}{n_2}},$$

therefore
$$a^{\frac{m_1}{n_1}}a^{\frac{m_2}{n_2}} < a^{\frac{m_1+1}{n_1}}a^{\frac{m_2+1}{n_2}}.$$

The difference

$$a^{\frac{m_1+1}{n_1}}a^{\frac{m_2+1}{n_2}} - a^{\frac{m_1}{n_1}}a^{\frac{m_2}{n_2}}$$

$$= a^{\frac{m_1+1}{n_1}}a^{\frac{m_2+1}{n_2}} - a^{\frac{m_2+1}{n_2}}a^{\frac{m_1}{n_1}} + a^{\frac{m_2+1}{n_2}}a^{\frac{m_1}{n_1}} - a^{\frac{m_1}{n_1}}a^{\frac{m_2}{n_2}}$$

$$= a^{\frac{m_2+1}{n_2}}\left(a^{\frac{m_1+1}{n_1}} - a^{\frac{m_1}{n_1}}\right) + a^{\frac{m_1}{n_1}}\left(a^{\frac{m_2+1}{n_2}} - a^{\frac{m_2}{n_2}}\right),$$

can be made less than any assigned number, however small. For $a^{\frac{m_1+1}{n_1}} - a^{\frac{m_1}{n_1}}$ and $a^{\frac{m_2+1}{n_2}} - a^{\frac{m_2}{n_2}}$ can each be made less than any assigned number, say d. Since $a^{\frac{m_2+1}{n_2}}$ decreases, it is always less than some positive finite number, say R_1; and since $a^{\frac{m_1}{n_1}} < a^{\frac{m_1+1}{n_1}}$, therefore $a^{\frac{m_1}{n_1}}$ is always less than some positive finite number, say R_2. Therefore the given difference can be made less than $dR_1 + dR_2$, $= d(R_1 + R_2)$.

But $d(R_1 + R_2)$ can be made less than any assigned number, say δ, by taking d less than $\dfrac{\delta}{R_1 + R_2}$.

Therefore the two series

$$a^{\frac{m_1}{n_1}}a^{\frac{m_2}{n_2}} \text{ and } a^{\frac{m_1+1}{n_1}}a^{\frac{m_2+1}{n_2}}$$

define a positive number which lies between them. This number is defined as the product $a^{l_1}a^{l_2}$.

That is,
$$a^{\frac{m_1}{n_1}}a^{\frac{m_2}{n_2}} < a^{l_1}a^{l_2} < a^{\frac{m_1+1}{n_1}}a^{\frac{m_2+1}{n_2}}.$$

In like manner it can be shown that the two series

$$a^{\frac{m_1}{n_1}+\frac{m_2}{n_2}} \text{ and } a^{\frac{m_1+1}{n_1}+\frac{m_2+1}{n_2}}$$

define a positive number which lies between them. This number is defined as $a^{I_1+I_2}$.

That is, $$a^{\frac{m_1}{n_1}+\frac{m_2}{n_2}} < a^{I_1+I_2} < a^{\frac{m_1+1}{n_1}+\frac{m_2+1}{n_2}}.$$

But since $$a^{\frac{m_1}{n_1}}a^{\frac{m_2}{n_2}} = a^{\frac{m_1}{n_1}+\frac{m_2}{n_2}} \text{ and } a^{\frac{m_1+1}{n_1}}a^{\frac{m_2+1}{n_2}} = a^{\frac{m_1+1}{n_1}+\frac{m_2+1}{n_2}},$$

the two numbers $a^{I_1+I_2}$ and $a^{I_1}a^{I_2}$ are determined by the same relation, and are therefore equal.

That is, $$a^{I_1}a^{I_2} = a^{I_1+I_2}.$$

In a similar manner the principle can be proved when the exponents, either or both, are negative irrational numbers.

(ii.) We have $$\frac{a^{I_1}}{a^{I_2}} = a^{I_1}a^{-I_2} = a^{I_1+(-I_2)} = a^{I_1-I_2},$$

wherein I_1 and I_2, either or both, are negative.

Principles (III.)-(V.) can be proved in a similar manner.

27. In the proofs of the preceding principles, the base was assumed to be greater than 1.

Similar reasoning will, however, apply when the base is less than 1 and positive.

For, if $a < 1$ and positive, it can be shown that the irrational power is defined by the relation

$$a^{\frac{m+1}{n}} < a^I < a^{\frac{m}{n}}.$$

It will be proved in Part II. that an irrational power of a negative base is a complex number.

EXERCISES VII.

MISCELLANEOUS EXAMPLES.

Simplify each of the following expressions:

1. $(-\sqrt[3]{x^{\frac{2}{3}}})^5 + (-2\sqrt[3]{x^{\frac{1}{2}}})^4 - x^{-1}\left(\dfrac{-3x^{\frac{2}{3}}}{\sqrt[3]{x^{-.5}}}\right)^2.$

2. $4.5(\sqrt[3]{a^{10}})^{\frac{3}{10}} + \left(\sqrt[3]{\dfrac{4}{a}}\right)^{-1\frac{1}{2}} - \left(\dfrac{3.375}{\sqrt[4]{a^{-3}}}\right)^{\frac{2}{3}}.$

3. $\dfrac{2x^2}{(1-x^2)^{\frac{3}{2}}} + \dfrac{1}{(1-x^2)^{\frac{1}{2}}}.$

4. $\dfrac{a^{1\frac{1}{2}}}{a^{1\frac{1}{2}} + x^{\frac{1}{4}}} + \dfrac{x^{\frac{1}{4}}}{a^{1\frac{1}{2}} - x^{\frac{1}{4}}}.$

5. $\dfrac{x}{(1-x)^{\frac{3}{2}}}+\dfrac{x^2}{(1-x)^{\frac{5}{2}}}$.

6. $\dfrac{a-b}{a^{\frac{1}{2}}-b^{\frac{1}{2}}}-\dfrac{a^{\frac{3}{2}}-b^{\frac{3}{2}}}{a-b}$.

7. $\dfrac{a-x}{a^{\frac{1}{3}}-x^{\frac{1}{3}}}-\dfrac{a+x}{a^{\frac{1}{3}}+x^{\frac{1}{3}}}$.

8. $\dfrac{a^{\frac{1}{2}}x^{\frac{1}{4}}+a^{\frac{1}{4}}x^{\frac{1}{2}}}{a^{\frac{1}{2}}+x^{\frac{1}{2}}}\cdot\dfrac{a-x}{a^{\frac{1}{4}}+x^{\frac{1}{4}}}$.

9. $\dfrac{x^{\frac{1}{2}}+1}{x+x^{\frac{1}{2}}+1}+\dfrac{1}{x^{1.5}-1}$.

10. $\dfrac{x^{\frac{1}{2}}-1}{x-x^{\frac{1}{2}}-2}-\dfrac{1}{x^{\frac{1}{2}}+1}$.

11. $\dfrac{ax}{\sqrt{(a+x)}}-\dfrac{2\,ax^2}{(a+x)^{\frac{3}{2}}}+\dfrac{ax^3}{(a+x)^{\frac{5}{2}}}$.

12. $\dfrac{1}{a^{\frac{1}{4}}+a^{\frac{1}{8}}+1}+\dfrac{1}{a^{\frac{1}{4}}-a^{\frac{1}{8}}+1}-\dfrac{2\,a^{\frac{1}{4}}}{a^{\frac{1}{2}}-a^{\frac{1}{4}}+1}$.

Find by inspection the square root of each of the following expressions:

13. $x^{\frac{1}{2}}+x^{-\frac{1}{2}}+2$.

14. $9\,a^{-4}-6\,a^{-\frac{5}{3}}+a^{\frac{2}{3}}$.

15. $25\,x^{-6}-30\,x^{-3}y^{\frac{1}{2}}+9\,y$.

16. $a^{-4}x+2\,a^{-\frac{3}{2}}x^{-\frac{3}{2}}+ax^{-4}$.

Find the square root of each of the following expressions:

17. $4\,x^{-4}-12\,x^{-3}+13\,x^{-2}-6\,x^{-1}+1$.

18. $9\,x^2+10\,x^{-2}-4\,x^{-4}+x^{-6}-12$.

19. $a^2-\frac{3}{2}\,a^{\frac{3}{2}}-\frac{3}{2}\,a^{\frac{1}{2}}+\frac{41}{16}\,a+1$.

20. $\frac{9}{4}\,x^3-5\,x^{\frac{5}{2}}y^{\frac{1}{2}}+\frac{179}{45}\,x^2y-\frac{4}{3}\,x^{\frac{3}{2}}y^{\frac{3}{2}}+\frac{4}{25}\,xy^2$.

21. $256\,x^{\frac{4}{3}}-512\,x+640\,x^{\frac{2}{3}}-512\,x^{\frac{1}{3}}+304-128\,x^{-\frac{1}{3}}+40\,x^{-\frac{2}{3}}$
$-8\,x^{-1}+x^{-\frac{4}{3}}$.

Find the cube root of each of the following expressions:

22. $x^{-6}-6\,x^{-5}+12\,x^{-4}-8\,x^{-3}$.

23. $8\,x-36\,x^{\frac{7}{6}}-27\,x^{\frac{3}{2}}+54\,x^{\frac{4}{3}}$.

24. $27\,x^3-54\,x+63\,x^{-1}-44\,x^{-3}+21\,x^{-5}-6\,x^{-7}+x^{-9}$.

25. $8\,x^{-3}+12\,x^{-2}-30\,x^{-1}-35+45\,x+27\,x^2-27\,x^3$.

26. $x^{\frac{3}{2}}-3\,x^{\frac{4}{3}}+3\,x^{\frac{7}{6}}+2\,x+3\,x^{\frac{5}{4}}-3\,x^{\frac{5}{6}}-6\,x^{\frac{13}{12}}+3\,x^{\frac{11}{12}}+x^{\frac{3}{4}}$.

27. $8\,x^3y^{-\frac{3}{2}}+13\,x^{\frac{3}{2}}+y^{\frac{3}{2}}+12\,x^{\frac{5}{2}}y^{-1}+18\,x^2y^{-\frac{1}{2}}+9\,xy^{\frac{1}{2}}+3\,x^{\frac{1}{2}}y$.

Solve each of the following equations:

28. $2x^{-2} - x^{-1} - 1 = 0.$

29. $x^{-2} + 5x^{-1} - \frac{11}{4} = 0.$

30. $3x^{-6} - 2x^{-3} - 1 = 0.$

31. $5x - 2x^{\frac{1}{2}} - 16 = 0.$

32. $3x^{\frac{1}{2}} - 2x^{\frac{1}{4}} - 4 = 0.$

33. $5x^{-\frac{1}{2}} - 2x^{-\frac{1}{4}} - 3 = 0.$

34. $x^{\frac{2}{3}} + 3x^{\frac{1}{3}} - 10 = 0.$

35. $7x^{-\frac{4}{3}} - 3x^{-\frac{2}{3}} - 4 = 0.$

36. $5x^{\frac{2}{5}} - 3x^{\frac{1}{5}} - 14 = 0.$

37. $2x^{-\frac{4}{5}} + 5x^{-\frac{2}{5}} - 7 = 0.$

CHAPTER XXVI.

PROGRESSIONS.

§ 1.

1. A **Series** is a succession of numbers, each formed according to some definite law. The separate numbers are called the **Terms** of the series.

The law may specify that each term shall be formed from the immediately preceding term in a prescribed way.

E.g., in the series

$$1 + 3 + 5 + 7 + 9 + \cdots \qquad (1)$$

each term after the first is formed by adding 2 to the preceding term.

In the series $$1 + 2 + 4 + 8 + \cdots \qquad (2)$$

each term after the first is formed by multiplying the preceding term by 2.

Or the law may state a definite relation between each term and the number of its place in the series.

E.g., in the series $$\tfrac{1}{1} + \tfrac{1}{2} + \tfrac{1}{3} + \tfrac{1}{4} + \cdots \qquad (3)$$

each term is the reciprocal of its number.

In the series $$1^2 + 2^2 + 3^2 + \cdots \qquad (4)$$

each term is the square of its number.

2. The number of terms in a series may be either *limited* or *unlimited.*

A **Finite** series is one of a *limited* number of terms.

An **Infinite** series is one of an *unlimited* number of terms.

In this chapter a few simple and yet very important series will be discussed.

§ 2. ARITHMETICAL PROGRESSION.

1. An **Arithmetical Series**, or as it is more commonly called an **Arithmetical Progression**, is a series in which each term, after the first, is formed by adding a constant number to the preceding term. See § 1, Art. 1, Ex. 1.

Evidently this definition is equivalent to the statement, that the difference between any two consecutive terms is constant.

E.g., in the series

$$1 + 3 + 5 + 7 + \cdots$$

we have $$3 - 1 = 5 - 3 = 7 - 5 = \cdots$$

For this reason the constant number of the first definition is called the **Common Difference** of the series.

2. Let a_1 stand for the first term of the series,
a_n for the nth (*any*) term of the series,
d for the common difference,
and S_n for the sum of n terms of the series.

The five numbers a_1, a_n, d, n, S_n are called the **Elements** of the progression.

3. The common difference may be either positive or negative.

If d be *positive*, each term is greater than the preceding, and the series is called a *rising*, or an *increasing* progression.

E.g., $1+2+3+4+ \cdots$ is an increasing progression, wherein $d = 1$.

If d be negative, each term is less than the preceding, and the series is called a *falling*, or a *decreasing* progression.

E.g., $1 - 1 - 3 - 5 - \cdots$ is a decreasing progression, wherein $d = -2$.

4. *In an arithmetical progression any term is equal to the first term plus the product of the common difference and a number one less than the number of the required term*, i.e.,

$$a_n = a_1 + (n - 1)d. \qquad \text{(I.)}$$

By the definition of an arithmetical progression

$$a_1 = a_1,$$

$$a_2 = a_1 + d,$$

$$a_3 = a_2 + d = a_1 + 2d,$$

$$a_4 = a_3 + d = a_1 + 3d, \text{ etc.}$$

The law expressed by the formulæ for these first four terms is evidently general, and since the coefficient of d in each is one less than the number of the corresponding term, we have

$$a_n = a_1 + (n-1)d.$$

Ex. 1. Find the 15th term of the progression

$$1 + 3 + 5 + 7 + \cdots$$

we have $a_1 = 1,\ d = 2,\ n = 15;$

therefore $a_{15} = 1 + (15-1)2 = 1 + 28 = 29.$

This formula may be used not only to find a_n, when a_1, d, and n are given, but also to find any one of the four numbers involved when the other three are given.

Ex. 2. If $a_5 = 3\,(n=5)$, and $a_1 = 1$, we have $3 = 1 + 4d$;

whence $d = \frac{1}{2}.$

5. *In any arithmetical progression, the sum of n terms is equal to one-half the product of the number of terms and the sum of the first and the nth term,* i.e.,

$$S_n = \frac{n}{2}(a_1 + a_n). \qquad \text{(II.)}$$

Since the successive terms in an arithmetical progression, from the first to the nth inclusive, may be obtained either by repeated additions of the common difference beginning with the first term, or by repeated subtractions of the common difference beginning with the nth term, we may express the sum of n terms in two equivalent ways:

$$S_n = a_1 + (a_1 + d) + (a_1 + 2d) + \cdots + (a_1 + \overline{n-2}\cdot d) + (a_1 + \overline{n-1}\cdot d),$$

$$S_n = a_n + (a_n - d) + (a_n - 2d) + \cdots + (a_n - \overline{n-2}\cdot d) + (a_n - \overline{n-1}\cdot d).$$

Whence, by addition,

$$2S_n = (a_1 + a_n) + (a_1 + a_n) + \cdots + (a_1 + a_n) + (a_1 + a_n),$$

wherein there are n binomials $a_1 + a_n$.

Therefore, $2S_n = n(a_1 + a_n)$, or $S_n = \frac{n}{2}(a_1 + a_n)$.

If the value of a_n, given in (I.), be substituted for a_1 in (II.), we obtain

$$S^n = \frac{n}{2}[2a_1 + (n-1)d]. \qquad \text{(III.)}$$

Formula (II.) is used when a_1, a_n, and n are given; and (III.) when a_1, d, and n are given.

Ex. 1. If $a_1 = 1$, $a_5 = 3$, then $S_5 = \frac{5}{2}(1 + 3) = 10$.

Ex. 2. If $a_1 = -4$, $d = 2$, $n = 12$,

then $$S_{12} = \tfrac{12}{2}[2(-4) + 11 \times 2] = 84.$$

Either (II.) or (III.) can be used to determine any one of the five elements a_1, a_n, d, n, S_n, when the three others involved in the formula are known.

Ex. 3. Given $a_1 = 1$, $a_n = 11$, $S_n = 36$, to find n.

From (II.), $36 = \frac{n}{2}(1 + 11)$, whence $n = 6$.

Ex. 4. Given $a_1 = -3$, $d = 2$, $S_n = 12$, to find n.

From (III.), $12 = \frac{n}{2}[-6 + 2(n-1)]$,

or $$n^2 - 4n = 12;$$

whence $$n = 6 \text{ and } -2.$$

The result 6 gives the series

$$-3 - 1 + 1 + 3 + 5 + 7, = 12.$$

Since the number of terms must be positive, the negative result, -2, is not admissible. But its meaning may be assumed to be that two terms, beginning with the last and counting toward the first, are to be taken.

6. Formulæ (I.) and (II.) may be used simultaneously to determine any two of the five numbers a_1, a_n, d, S_n, n, when the three others are given. In like manner (I.) and (III.) may be used.

Ex. 1. Given $d = \frac{1}{2}$, $n = 9$, $a_9 = 5$, to find a_1 and S_9.

From (I.), $5 = a_1 + 8 \cdot \frac{1}{2}$, whence $a_1 = 1$.

From (II.), using the value of a_1 just found,

$$S_9 = \tfrac{9}{2}(1 + 5) = 27.$$

Ex. 2. Given $a_1 = 3$, $n = 13$, $S_{13} = 13$, to find d and a_{13}.

From (II.), $13 = \frac{13}{2}(3 + a_{13})$, whence $a_{13} = -1$.

From (I.), $-1 = 3 + 12d$, whence $d = -\frac{1}{3}$.

Ex. 3. Given $d = -2$, $a_n = -16$, $S_n = -60$, to find a_1 and n.

From (I.), $$-16 = a_1 - 2(n-1), \quad (1)$$

and from (II.), $$-60 = \frac{n}{2}(a_1 - 16). \quad (2)$$

Solving (1) and (2), we obtain $n = 12$, $a_1 = 6$; and $n = 5$, $a_1 = -8$.

The two series are:

$$6 + 4 + 2 + 0 - 2 - 4 - 6 - 8 - 10 - 12 - 14 - 16,$$

and $$-8 - 10 - 12 - 14 - 16,$$

both of which have $d = -2$, $a_n = -16$, $S_n = -60$.

Notice that in this example the sum of the terms which are not common to the two series is 0.

Ex. 4. Given $a_1 = 2$, $d = 3$, $S_n = 40$, to find n and a_n.

From (III.), $40 = \frac{n}{2}[4 + 3(n-1)]$.

The roots of this equation are 5, $-5\frac{1}{3}$.

When $n = 5$, we have from (I.), $a_n = a_5 = 2 + 4 \times 3 = 14$.

The result $-5\frac{1}{3}$ is evidently not admissible, since n must always be a positive integer.

In general, since the equation which gives the value of n, in such examples as the last two, is quadratic, it may evidently have not only negative or fractional roots, but also surd or imaginary.

Moreover, even when the value of the unknown number is obtained from a linear equation, it is sometimes inadmissible.

Ex. 5. Given $a_1 = 1$, $d = 2$, $a_n = 18$, to find n.

From (I.), $18 = 1 + 2(n - 1)$, whence $n = \frac{19}{2}$.

This result is inadmissible.

A glance at the equation will reveal the meaning of this result. Since the first term is odd and the common difference even, the last term cannot be even.

This example also illustrates the fact that, although one of the formulæ determines one of the elements when the other three elements are given, yet these three elements cannot be arbitrarily assumed.

7. In many examples the elements necessary for determining the required element or elements directly from (I.)–(III.) are not given, but in their place equivalent data.

Ex. 1. Given $a_6 = 17$, $a_{11} = 32$, to find a_1 and d.

From (I.), $$17 = a_1 + 5d,$$

and $$32 = a_1 + 10d.$$

Solving these equations, $a_1 = 2$, $d = 3$.

Or, we could have regarded 17 as the first term and 32 as the last term of a progression of 6 terms. Then, by (I.), $32 = 17 + 5d$, whence $d = 3$.

By (I.) again, $17 = a_1 + 5 \times 3$; whence $a_1 = 2$, as above.

Ex 2. Given $S_8 = 80$, $S_{12} = 168$; find a_1 and d.

From (III.), $$80 = \frac{8}{2}(2a_1 + 7d),$$

and $$168 = \frac{12}{2}(2a_1 + 11d).$$

From these equations, $a_1 = 3$, $d = 2$.

EXERCISES I.

Find the last term and the sum of the terms of each of the following arithmetical progressions:

1. $2 + 6 + \cdots$ to 10 terms.
2. $1 + 6 + \cdots$ to 15 terms.
3. $3 + 1 - \cdots$ to 13 terms.
4. $4 - 1 - \cdots$ to 17 terms.
5. $-5 - 2 + \cdots$ to 21 terms.
6. $-11 - 7 - \cdots$ to 30 terms.
7. $13 + 11 + \cdots$ to 25 terms.
8. $14 + 9 + \cdots$ to 23 terms.

9. $1 + 1\frac{3}{4} + \cdots$ to 12 terms.

10. $1\frac{1}{3} + 1\frac{2}{3} + \cdots$ to 27 terms.

11. $3 + 1\frac{1}{2} + \cdots$ to 40 terms.

12. $4 + 1\frac{3}{4} - \cdots$ to 31 terms.

13. $9 + 11 + \cdots$ to n terms.

14. $1 + \frac{5}{6} + \cdots$ to n terms.

15. $n + 2n + \cdots$ to 16 terms, to m terms.

16. $a + (a + b) + \cdots$ to 20 terms, to n terms.

17. $x + (3x - 2y) + \cdots$ to 15 terms, to n terms.

18. $(m + 2) + (4m + 5) + \cdots$ to 40 terms, to n terms.

19. $\frac{a-1}{a} + \frac{a-3}{a} + \cdots$ to 30 terms, to n terms.

20. $\frac{a-b}{a+b} + \frac{3a-2b}{a+b} + \cdots$ to 32 terms, to n terms.

21. $\left(\frac{1}{a} - \frac{n}{x}\right) + \left(\frac{1}{a} - \frac{n-1}{x}\right) + \cdots$ to 18 terms, to n terms.

22. $n + \frac{n(n-3)}{n-1} + \cdots$ to 32 terms, to a terms.

23. $n - 1 + \frac{n^2+1}{n-1} + \cdots$ to 11 terms, to x terms.

24. $(a + b)^2 + (a^2 + b^2) + \cdots 20$ terms, to n terms.

25. Find the sum of the series

$$(e^{x+1} - e^x + 1) + (e^{x+1} + e^x - 1) + \cdots + (e^{x+1} + 19e^x - 19).$$

In each of the following arithmetical progressions find the values of the two elements not given:

26. $a_1 = 4,\ d = 5,\ n = 10.$

27. $a_1 = 1.2,\ d = -3,\ n = 16.$

28. $a_1 = 2\frac{1}{2},\ d = \frac{1}{3},\ n = 8.$

29. $a_n = 16,\ d = 2,\ n = 9.$

30. $a_n = -4,\ d = .3,\ n = 10.$

31. $a_n = 8\frac{1}{6},\ d = \frac{1}{4},\ n = 19.$

32. $a_1 = -5,\ n = 72,\ a_n = 37\frac{3}{5}.$

33. $a_1 = 2\frac{3}{5},\ n = 5,\ a_n = -1.9.$

34. $a_1 = 12.7,\ n = 14,\ a_n = 1.$

35. $d = 6,\ n = 10,\ S_n = 340.$

36. $d = -4.8,\ n = 3,\ S_n = 28.5.$

37. $d = 2\frac{1}{12},\ n = 9,\ S_n = 30.$

38. $a_1 = 3,\ n = 15,\ S_n = 90.$

39. $a_1 = 0,\ n = 100,\ S_n = -54450.$

40. $a_1 = 4\frac{1}{3},\ n = 10,\ S_n = 73\frac{1}{3}.$

41. $a_n = 13,\ n = 8,\ S_n = 100.$

42. $a_n = -36\frac{1}{3},\ n = 15,\ S_n = -247\frac{1}{2}.$

43. $a_n = 2\frac{1}{6},\ n = 12,\ S_n = -7.$

44. $a_1 = 9,\ d = -1,\ a_n = 6.$

45. $a_1 = 7,\ d = 5,\ a_n = 227.$

46. $a_1 = \frac{1}{12},\ d = \frac{49}{64},\ a_n = 12\frac{1}{3}.$

47. $a_1 = -7.5$, $a_n = 10.5$, $S_n = 15$.

48. $a_1 = 22\frac{1}{3}$, $a_n = -19\frac{2}{3}$, $S_n = 20$.

49. $a_1 = 3\frac{1}{4}$, $a_n = -145\frac{1}{12}$, $S_n = -12840$.

50. $a_1 = 2$, $d = 5$, $S_n = 245$.

51. $a_1 = -\frac{1}{4}$, $d = \frac{3}{8}$, $S_n = 282\frac{1}{2}$.

52. $a_1 = 7$, $d = \frac{1}{4}$, $S_n = 142$.

53. $a_n = 56$, $d = 5$, $S_n = 324$.

54. $a_n = -1\frac{7}{8}$, $d = -\frac{1}{8}$, $S_n = -13\frac{3}{4}$.

55. $a_n = 4\frac{2}{3}$, $d = -\frac{1}{9}$, $S_n = 69\frac{1}{3}$.

Arithmetical Means.

8. The **Arithmetical Mean** between two numbers is a third number, in value between the two, which forms with them an arithmetical progression.

E.g., 2 is an arithmetical mean between 1 and 3, between -3 and 7, etc.

Let A stand for the arithmetical mean between a and b; then by the definition of an arithmetical progression,

$$A - a = b - A,$$

whence

$$A = \frac{a + b}{2}.$$

That is, the arithmetical mean between two numbers is half their sum.

9. Arithmetical Means between two numbers are numbers, in value between the two, which form with them an arithmetical progression.

E.g., 2, 3, and 4 are three arithmetical means between 1 and 5; $\frac{3}{2}$, 2, $\frac{5}{2}$, 3, $\frac{7}{2}$, 4, $\frac{9}{2}$ are seven arithmetical means between 1 and 5.

If n arithmetical means be inserted between a and b, we have an arithmetical progression of $n + 2$ terms, the first term being a and the last b.

Therefore, from (I.), $b = a + (\overline{n + 2} - 1)d$,

whence

$$d = \frac{b - a}{n + 1}.$$

The resulting series is therefore

$$a,\ a+\frac{b-a}{n+1},\ a+2\frac{b-a}{n+1},\ a+3\frac{b-a}{n+1},\ \cdots,\ b;$$

or

$$a,\ \frac{na+b}{n+1},\ \frac{(n-1)a+2b}{n+1},\ \frac{(n-2)a+3b}{n+1},\ \cdots,\ b.$$

Ex. 1. Insert four arithmetical means between -2 and 9.

It is better to work the example independently of the general results given above. We have

$$n=6,\quad a_1=-2,\quad a_6=9.$$

From (I.), $9=-2+5d$, whence $d=\frac{11}{5}$.

The required means are $\frac{1}{5}$, $\frac{12}{5}$, $\frac{23}{5}$, $\frac{34}{5}$.

EXERCISES II.

Insert an arithmetical mean between

1. 45 and 31.
2. $17\frac{1}{2}$ and $14\frac{1}{2}$.
3. $2a$ and $-2b$.
4. $4a-3b$ and $b-2a$.
5. $(a+2b)^2$ and $(a-2b)^2$.
6. $\frac{a-b}{a+b}$ and $\frac{a+b}{a-b}$.
7. $\frac{x+1}{x-1}$ and $-\frac{x^3+1}{x^3-1}$.
8. Insert 6 arithmetical means between 7 and 35.
9. Insert 12 arithmetical means between 37 and -28.
10. Insert 8 arithmetical means between 7 and 13.
11. Insert 9 arithmetical means between $\frac{1}{5}$ and 12.
12. Insert 5 arithmetical means between $17\frac{5}{8}$ and $1\frac{1}{8}$.
13. Insert 20 arithmetical means between -16 and 26.
14. Insert 38 arithmetical means between 15 and 2.
15. Insert 6 arithmetical means between $a+b$ and $8a-13b$.
16. Insert 8 arithmetical means between $\frac{m-n}{m+n}$ and $\frac{m+n}{m-n}$.

Problems.

10. Pr. 1. The sum of four numbers in arithmetical progression is 16, and their product is 105. What are the numbers?

We can express the four required numbers in terms of *two unknown* numbers.

Let $x - 3d$, $x - d$, $x + d$, $x + 3d$ be the four required numbers.

Then, by the first condition,

$$(x - 3d) + (x - d) + (x + d) + (x + 3d) = 16;$$

whence $$x = 4.$$

By the second condition,

$$(x - 3d)(x - d)(x + d)(x + 3d) = 105,$$

or $$(x^2 - 9d^2)(x^2 - d^2) = 105.$$

Substituting 4 for x and reducing, $9d^4 - 160d^2 = -151$.

From this equation we obtain $d = \pm 1$, and $\pm \frac{1}{3}\sqrt{151}$.

The corresponding numbers are,

when $d = 1$: 1, 3, 5, 7; when $d = -1$: 7, 5, 3, 1;

when $d = \frac{1}{3}\sqrt{151}$:

$$4 - \sqrt{151},\ 4 - \tfrac{1}{3}\sqrt{151},\ 4 + \tfrac{1}{3}\sqrt{151},\ 4 + \sqrt{151};$$

when $d = -\frac{1}{3}\sqrt{151}$:

$$4 + \sqrt{151},\ 4 + \tfrac{1}{3}\sqrt{151},\ 4 - \tfrac{1}{3}\sqrt{151},\ 4 - \sqrt{151}.$$

Notice the advantage of assuming the required numbers as in the above example. Had we assumed x, $x + d$, $x + 2d$, $x + 3d$ as the required numbers, the solution would have involved an equation of the fourth degree which could not have been solved as a quadratic.

Pr. 2. The sum of the fifth and sixth terms of an arithmetical progression is 49, and the product of the first and eighth terms is 74. Find the series.

By the first condition, $$a_5 + a_6 = 49,$$

or $$(a_1 + 4d) + (a_1 + 5d) = 49,$$

or $$2a_1 + 9d = 49. \quad (1)$$

By the second condition, $$a_1a_8 = 74,$$

or $$a_1(a_1 + 7d) = 74,$$

or $$a_1^2 + 7a_1d = 74. \quad (2)$$

Solving (1) and (2), $a_1 = 2$, $d = 5$; and $a_1 = 66\frac{3}{5}$, $d = -9\frac{11}{45}$.

The two series are:

$$2,\ 7,\ 12,\ 17,\ 22,\ 27,\ 32,\ 37,\ \cdots;$$

$$66\tfrac{3}{5},\ 57\tfrac{11}{45},\ 47\tfrac{8}{9},\ 38\tfrac{8}{15},\ 29\tfrac{8}{45},\ 19\tfrac{37}{45},\ 10\tfrac{7}{15},\ 1\tfrac{1}{9},\ \cdots.$$

Pr. 3. Find the sum of all the numbers of three digits which are multiples of 7.

The numbers of three digits which are multiples of 7 are

$$7\times 15,\ 7\times 16,\ 7\times 17,\ \cdots,\ 7\times 142.$$

Their sum is $\quad 7(15+16+\cdots+142).$

The series within the parentheses is an arithmetical progression, in which $a_1=15$, $d=1$, $n=128$, and $a_{128}=142$.

Therefore $\quad S_{128}=10{,}048.$

The required sum is therefore $7\times 10{,}048, = 70{,}336.$

Pr. 4. Two persons starting at the same time from two places A and B, which are 170 miles apart, travel toward each other. The one starting from A travels 4 miles an hour; the one starting from B travels 2 miles the first hour, and each succeeding hour $\frac{1}{2}$ mile more than the preceding hour. At what distance from A will they meet, and after how many hours?

Let x stand for the number of miles from A to their place of meeting P, and y for the number of hours after which they meet.

Then, by the first condition,

$$x=4y. \tag{1}$$

The number of miles traveled in y hours by the man starting from B will be the sum of an arithmetical progression of which the first term is 2, the common difference is $\frac{1}{2}$, and the number of terms is y.

From (III.), this number is

$$\frac{y}{2}\left[4+(y-1)\frac{1}{2}\right],=\frac{y^2+7y}{4}.$$

But this number must be equal to $170-x$.

Therefore

$$\frac{y^2+7y}{4}=170-x. \tag{2}$$

From (1) and (2), $x=68$, $y=17$; and $x=-160$, $y=-40$.

The solution $x = 68$, $y = 17$ satisfies the condition of the problem, as shown in Fig. 16.

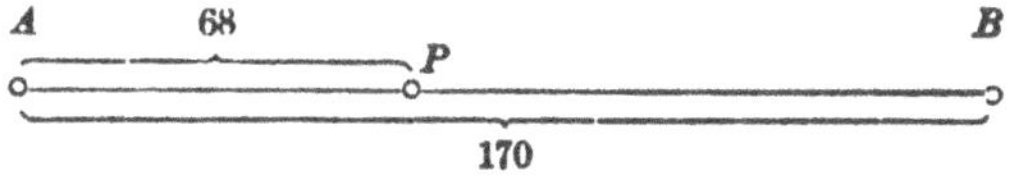

FIG. 16.

The second solution does not satisfy the conditions of the problem, but it is consistent with a modified problem. The interpretation of this solution is left as an exercise for the student.

EXERCISES III.

1. Find the sixth term and the sum of eleven terms of an A. P. whose eighth term is 11 and whose fourth term is -1.

2. The seventh term of an A. P. is 20, and the thirteenth term is 28. What is the first term, and the twentieth term?

3. The sixteenth term of an A. P. is -5, and the forty-first term is 45. What is the first term, and the sum of twenty terms?

4. The thirteenth term of an A. P. is 1, and the twentieth term is -13. What is the thirtieth term, and the sum of fifteen terms?

5. Find the sum of all the even numbers from 2 to 50 inclusive.

6. Find the sum of thirty consecutive odd numbers, of which the last is 127.

7. The sum of the eighth and fourth terms of an A. P. of twenty terms is 24, and the sum of the fifteenth and nineteenth terms is 68. What are the elements of the progression?

8. The sum of the terms of an A. P. of seven terms is 147, and the product of the first term by the last is 297. What are the elements of the progression?

9. The product of the first and fifth terms of an A. P. of ten terms is 85, and the product of the first and third terms is 55. What are the elements of the progression?

10. The first term of an A. P. of thirty terms is 100, and the sum of the first six terms is five times the sum of the six following terms. What are the elements of the progression?

11. The sum of the first and fifth terms of an A. P. of twelve terms is 6, and the sum of the squares of the second and fourth terms is 20. What are the elements of the progression?

12. The sum of the second and twentieth terms of an A. P. is 10, and their product is $23\frac{47}{64}$. What is the sum of sixteen terms?

13. The sixth term of an A. P. is 30, and the sum of the first thirteen terms is 455. What is the sum of the first thirty terms?

14. The sum of the fifth and eighth terms of an A. P. is 37, and the sum of the first ten terms is 155. What is the twentieth term?

15. What value of x will make the arithmetical mean between $x^{\frac{1}{2}}$ and $x^{\frac{1}{4}}$ equal to 6?

16. Find the sum of all even numbers of two digits.

17. How many consecutive odd numbers beginning with 7 must be taken to give a sum 775?

18. How many terms of the series 12, 10, ⋯ must be taken to give a sum 30?

19. Insert between 0 and 6 a number of arithmetical means so that the sum of the terms of the resulting A. P. shall be 39.

20. Find the number of arithmetical means between 1 and 19, if the second mean is to the last mean as 1 to 7.

21. The fifth term of a progression is three times the first term. If the thirteenth term be divided by the third, the quotient and the remainder will each be 3. What is the first term, and the common difference?

22. The sum of the terms of an A. P. of six terms is 66, and the sum of the squares of the terms is 1006. What are the elements of the progression?

23. The sum of three terms in arithmetical progression is a, and the sum of their squares is b^2. What is the first term, and the common difference?

24. The sum of the terms of an A. P. of five terms is 75, and the ratio of the first term to the last is 5. What are the elements of the progression?

25. The sum of the terms of an A. P. of twelve terms is 354, and the sum of the even terms is to the sum of the odd terms as 32 to 27. What is the common difference?

26. How many positive integers of three digits are there which are divisible by 9? Find their sum.

27. What is the sum of all positive odd numbers of four digits which are multiples of 29?

28. If the sum of m terms of an A. P. is n, and the sum of n terms is m, what is the sum of $m+n$ terms? Of $m-n$ terms?

29. Show that the sum of $2n+1$ consecutive integers is divisible by $2n+1$.

30. Show that the sum of n consecutive integers beginning with a is one-third of the sum of n consecutive integers beginning with $3a+n-1$.

31. Prove that if the same number be added to each term of an A. P., the resulting series will be an A. P.

32. Prove that if each term of an A. P. be multiplied by the same number, the resulting series will be an A. P.

33. Prove that if the corresponding terms of two arithmetical progressions be added, the resulting series will be an A. P. What is the common difference of the last series?

34. Prove that if in the equation $y = ax + b$, we substitute c, $c+d$, $c+2d$, $\cdots$, in turn for x, the resulting values of y will form an A. P.

35. Prove that if a^2, b^2, c^2 form an A. P., then

$$\frac{1}{b+c}, \quad \frac{1}{c+a}, \quad \frac{1}{a+b} \text{ form an A. P.}$$

36. If a, b, c form an A. P., then

$$\frac{2}{9}(a+b+c)^3 = a^2(b+c) + b^2(a+c) + c^2(a+b).$$

37. A man agreed to pay a debt of $19,950 by monthly installments, paying each month after the first $50 more than the preceding month, and the last month paying $2220. How much did he pay the first month, and how many months did it take him to pay the debt?

38. A laborer agreed to dig a well on the following conditions: for the first yard he was to receive $2, for the second $2.50, for the third $3, and so on. If he received $42.50 for his work, how deep was the well?

39. On a certain day the temperature rose $\frac{1}{2}°$ hourly from 5 to 11 A.M., and the average temperature for that period was 8°. What was the temperature at 8 A.M.?

40. Twenty-five trees are planted in a straight line at intervals of 5 feet. To water them, the gardener must bring water for each tree separately from a well which is 10 feet from the first tree and in line with the trees. How far has the gardener walked when he has watered all the trees?

41. Two bodies, A and B, start at the same time from two points, C and D, which are 75 feet apart, and move in the same direction. A moves 1 foot the first second, 3 feet the second, and 5 feet the third; B moves 3 feet the first second, 4 feet the second, and 5 feet the third. How long will it take A to overtake B?

42. A man bought an estate which yielded $1500 profit the first year. His personal expenses for the first year were $1250. His income from the estate increased $100 yearly, and his personal expenses increased $125 yearly. After how many years were his personal expenses equal to his income?

43. Two men started at the same time from two cities which are 1265 miles apart. After the first day, the first increased his rate 10 miles daily, and the second his rate 3 miles daily. If they met after 5 days, and if the first traveled 55 miles further than the second, what was the rate of each the first day and the last day?

44. A number of equal balls are placed in the form of a solid equilateral triangle in the following way: one ball is placed at the vertex, under this are placed two balls, under these two are placed three balls, and so on. If the number of balls is increased by 4, they can be placed in the form of a solid rectangle whose base is equal to the base of the triangle, and whose altitude is 3 balls shorter than the base. How many balls are in the triangle?

45. A tank which contains 57 gallons can be filled by four pipes. If all the pipes are opened, the tank will be emptied in 40 minutes. The number of minutes in which one gallon is discharged by the pipes separately form an A. P. whose sum is 12. How long does it take each pipe to discharge one gallon?

§ 3. GEOMETRICAL PROGRESSION.

1. A **Geometrical Series,** or as it is more commonly called, a **Geometrical Progression,** is a series in which each term after the first is formed by multiplying the preceding term by a constant number. See § 1, Art. 1, Ex. 2.

Evidently this definition is equivalent to the statement that the ratio of any term to the preceding is constant.

For this reason the constant multiplier of the first definition is called the **Ratio** of the progression.

Let a_1 stand for the first term of the series,

a_n for the nth (*any*) term,

r for the ratio,

and S_n for the sum of n terms.

The five numbers a_1, a_n, r, S_n, n, are called the **Elements** of the progression.

2. The ratio may be either larger or smaller than 1; in the former case the progression is called a *rising* or *ascending* progression; in the latter a *falling* or *descending* progression.

E.g., $1 + \frac{3}{2} + \frac{9}{4} + \frac{27}{8} + \cdots$, in which $r = \frac{3}{2}$,

and $\frac{1}{2} - 1 + 2 - 4 + 8 \cdots$, in which $r = -2$, are ascending progressions; while

$$1 + \tfrac{1}{2} + \tfrac{1}{4} + \tfrac{1}{8} + \cdots, \text{ in which } r = \tfrac{1}{2},$$

and

$$1 - \tfrac{2}{3} + \tfrac{4}{9} - \tfrac{8}{27} + \cdots, \text{ in which } r = -\tfrac{2}{3},$$

are *descending* progressions.

3. *In a geometrical progression any term is equal to the first term multiplied by a power of the ratio whose exponent is one less than the number of the required term,* i.e.,

$$a_n = a_1 r^{n-1}. \qquad \text{(I.)}$$

By the definition of a geometrical progression

$$a_1 = a_1,$$
$$a_2 = a_1 r,$$
$$a_3 = a_2 r = a_1 r^2,$$
$$a_4 = a_1 r = a_2 r^3,$$
etc.

The law expressed by the relations for these first four terms is evidently general, and since the exponent of r is one less than the number of the corresponding term, we have

$$a_n = a_1 r^{n-1}.$$

Ex. 1. If $a_1 = \frac{1}{2}$, $r = 3$, $n = 5$, then $a_5 = \frac{1}{2} \cdot 3^4 = \frac{81}{2}$.

This relation may also be used to find not only a_n when a_1, r, and n are given, but also to find the value of any one of the four numbers when the other three are given.

Ex. 2. If $a_1 = 4$, $a_6 = \frac{1}{8}$, $n = 6$, then $\frac{1}{8} = 4r^5$,

whence $r = \frac{1}{2}$.

It is important to notice that, while a_1, a_n, and r may be positive or negative, integral or fractional, n must be a positive integer. Consequently a_1, a_n, r cannot be assumed arbitrarily.

As yet the value of n can be determined from (I.) only by inspection.

4. *In a geometrical progression*

$$S_n = \frac{a_1(1-r^n)}{1-r} = \frac{a_1(r^n-1)}{r-1}, \qquad \text{(II.)}$$

or

$$S_n = \frac{a_1 - a_n r}{1-r} = \frac{a_n r - a_1}{r-1}. \qquad \text{(III.)}$$

We have $\quad S_n = a_1 + a_1 r + a_1 r^2 + \cdots + a_1 r^{n-2} + a_1 r^{n-1}, \quad (1)$

and $\quad rS_n = a_1 r + a_1 r^2 + \cdots + a_1 r^{n-2} + a_1 r^{n-1} + a_1 r^n. \quad (2)$

Consequently, subtracting (2) from (1),

$$S_n(1-r) = a_1 - a_1 r^n,$$

whence

$$S_n = \frac{a_1(1-r^n)}{1-r} = \frac{a_1(r^n-1)}{r-1}.$$

Substituting a_n for $a_1 r^{n-1}$ in (II.), we have

$$S_n = \frac{a_1 - a_n r}{1-r} = \frac{a_n r - a_1}{r-1}.$$

The first forms of (II.) and (III.) are to be used when $r < 1$, the second when $r > 1$.

Ex. 1. Given $a_1 = 3$, $r = 2$, $n = 6$, to find S_6.

Using the second form of (II.),

$$S_6 = \frac{3(2^6-1)}{2-1} = 189.$$

Formulæ (II.) and (III.) may be used not only to find S_n when a_1, r, and n, or a_1, a_n, and r are given, but also to find the value of any one of the four elements when the other three are given.

Ex. 2. Given $S_n = -63\frac{1}{2}$, $a_1 = -\frac{1}{2}$, $a_n = -32$, to find r.

By (III.), $\quad -63\frac{1}{2} = \dfrac{-\frac{1}{2} + 32r}{1-r}, \quad$ whence $r = 2$.

5. Formulæ (I.) and (II.) may be used simultaneously to determine any two of the five elements, a_1, a_n, r, S_n, n, when the three other elements are given. In like manner (I.) and (III.) may be used.

Ex. 1. Given $r = -\frac{1}{2}$, $n = 5$, $a_5 = -\frac{1}{4}$, to find a_1 and S_5.

From (I.), $-\frac{1}{4} = a_1(-\frac{1}{2})^4$, whence $a_1 = -4$;

and from (II.), using the value of a_1 just found,

$$S_5 = \frac{-4[1-(-\frac{1}{2})^5]}{1-(-\frac{1}{2})} = -\frac{11}{4}.$$

Ex. 2. Given $a_1 = 9$, $r = \frac{1}{3}$, $S_n = 13\frac{13}{27}$, to find n and a_n.

From (II.), $13\frac{13}{27} = \frac{9[1-(\frac{1}{3})^n]}{1-\frac{1}{3}}$, whence $n = 6$.

From (I.), $a_6 = 9(\frac{1}{3})^5 = \frac{1}{27}$.

Ex. 3. Given $r = 2$, $a_n = 16$, $S_n = 31\frac{1}{2}$, to find a_1 and n.

From (III.), $31\frac{1}{2} = \frac{16 \times 2 - a_1}{2-1}$, whence $a_1 = \frac{1}{2}$.

From (I.), $16 = \frac{1}{2} \cdot 2^{n-1}$, whence $n = 6$.

Ex. 4. Given $n = 7$, $a_7 = 16$, $S_7 = 31\frac{3}{4}$, to find r and a_1.

From (I.), $16 = a_1 r^6$,

and from (II.), $31\frac{3}{4} = \frac{a_1(1-r^7)}{1-r}$.

Eliminating a_1 between these two equations, we obtain

$$63r^7 - 127r^6 = -64.$$

Thus this example leads to an equation of the seventh degree, which cannot be solved.

The value of r in such equations can often be determined by inspection. Let the student verify the value $r = 2$ in the above equation. We then have $a_1 = \frac{1}{4}$.

6. In many examples the elements necessary for determining the element or elements directly from (I.)–(III.) are not given, but in their place equivalent data.

Ex. 1. Given $a_5 = 48$, $a_8 = 384$, to find a_1 and r.

From (I.), $48 = a_1 r^4$, and $384 = a_1 r^7$;

whence $r^3 = 8$, or $r = 2$. Therefore $a_1 = 3$.

Or, we could have regarded 48 as the first term and 384 as the last term of a progression of four terms. Then by (I.), $384 = 48\,r^3$, whence $r = 2$, as before.

Ex. 2. Given $S_6 = 63$, $S_9 = 511$, to find a_1 and r.

From (II.), $63 = \dfrac{a_1(r^6 - 1)}{r - 1}$, and $511 = \dfrac{a_1(r^9 - 1)}{r - 1}$; (1)

whence $$\frac{r^9 - 1}{r^6 - 1} = \frac{511}{63}, \text{ or } \frac{r^6 + r^3 + 1}{r^3 + 1} = \frac{511}{63}.$$

Therefore $$63\,r^6 - 448\,r^3 = 448;$$

whence $r^3 = 8$, and $-\frac{8}{9}$; and $r = 2$, and $-\frac{2}{3}\sqrt[3]{3}$.

We then have from (1): $a_1 = 1$, when $r = 2$; and

$$a_1 = 100\tfrac{1}{17}(2\sqrt[3]{3} + 3), \text{ when } r = -\tfrac{2}{3}\sqrt[3]{3}.$$

Such examples as the last in general lead to equations of a higher degree than the second.

EXERCISES IV.

Find the last term and the sum of the terms of each of the following geometrical progressions:

1. $3 + 6 + \cdots$ to 6 terms.
2. $5 + 2 + \cdots$ to 4 terms.
3. $2 - 4 + \cdots$ to 10 terms.
4. $1 - 3 + \cdots$ to 9 terms.
5. $32 - 16 + \cdots$ to 7 terms.
6. $3 + 2 + \cdots$ to 8 terms.
7. $1\frac{3}{5} + 2\frac{2}{5} + \cdots$ to 6 terms.
8. $\frac{2}{3} + \frac{1}{2} + \cdots$ to 5 terms.
9. $2 - 2^2 + \cdots$ to 11 terms.
10. $\frac{2}{5} - \sqrt{\frac{2}{5}} + \cdots$ to 14 terms.
11. $\dfrac{1}{\sqrt{2}} + \dfrac{1}{2} + \cdots$ to n terms.
12. $2 + \sqrt[4]{8} + \cdots$ to n terms.
13. $1 + (1 + a) + \cdots$ to 4 terms, to n terms.
14. $a^p + a^{p+q} + \cdots$ to 7 terms, to n terms.
15. $(x - y) + \left(\dfrac{y^2}{x} - \dfrac{y^3}{x^2}\right) + \cdots$ to 10 terms, to n terms.

In each of the following geometrical progressions find the values of the elements not given:

16. $a_1 = 1,\ r = 4,\ n = 5.$

17. $a_1 = 2\frac{1}{3},\ r = -2,\ n = 6.$

18. $a_n = 10,\ r = 2,\ n = 4.$

19. $a_n = 1.2,\ r = -.2,\ n = 5.$

20. $r = 2,\ n = 5,\ S_n = 62.$

21. $r = 10,\ n = 7,\ S_n = 3{,}333{,}333.$

22. $a_1 = 5,\ n = 9,\ a_n = 327{,}680.$

23. $a_1 = 74\frac{2}{3},\ n = 6,\ a_n = 2\frac{1}{3}.$

24. $a_n = 96,\ n = 4,\ S_n = 127.5.$

25. $a_n = 7,\ n = 9,\ S_n = 68{,}887.$

26. $a_1 = 1,\ r = 2,\ a_n = 64.$

27. $a_1 = 7,\ r = 10,\ a_n = 700.$

28. $a_1 = 74\frac{2}{3},\ a_n = 2\frac{1}{3},\ S_n = 147.$

29. $a_1 = 1,\ a_n = 512,\ S_n = 1023.$

30. $a_n = 44,\ r = 4,\ S_n = 55.$

31. $a_n = 3125,\ r = 5,\ S_n = 3905.$

32. $a_1 = 40,\ r = \frac{1}{2},\ S_n = 75.$

33. $a_1 = 4,\ r = 3,\ S_n = 118{,}096.$

34. $a_1 = 3,\ n = 3,\ S_n = 1953.$

35. $a_1 = 100,\ n = 3,\ S_n = 700.$

Sum of an Infinite Geometrical Progression.

7. When r is less than 1, the term a_1r^n in the formula

$$S = \frac{a_1 - a_1r^n}{1 - r}$$

decreases as n increases. As n grows indefinitely large, a_1r^n becomes indefinitely small. That is, when $n = \infty$, $a_1r^n = 0$.

We therefore have, when $r < 1$ and $n = \infty$,

$$S_\infty = \frac{a_1}{1 - r}.$$

Ex. 1. Find the sum of the infinite series

$$1 + \tfrac{1}{2} + \tfrac{1}{4} + \tfrac{1}{8} + \cdots,$$

in which $r = \frac{1}{2}.$

We have $$S_\infty = \frac{1}{1 - \frac{1}{2}} = 2.$$

The meaning of this result is that the sum of the given series approaches the finite value 2 more and more nearly as more and more terms are included in the sum, and that the sum can be made to differ from 2 by as little as we please, by taking a sufficient number of terms. The *exact* sum 2, however, can never be obtained.

This theory can be applied to find the value of a repeating (recurring) decimal.

Ex. 2. Verify that $.\dot{6} = \frac{2}{3}$.

We have $.666 \cdots = \frac{6}{10} + \frac{6}{100} + \frac{6}{1000} + \cdots,$

a geometrical progression whose first term is $\frac{6}{10}$ and whose ratio is $\frac{1}{10}$. Consequently

$$S_\infty = \frac{\frac{6}{10}}{1 - \frac{1}{10}} = \frac{6}{9} = \frac{2}{3}.$$

EXERCISES V.

Find the sum of the following infinite geometrical progressions:

1. $6 + 4 + \cdots$. **2.** $60 + 15 + \cdots$. **3.** $10 - 6 + \cdots$.

4. $\frac{1}{2} + \frac{1}{4} + \cdots$. **5.** $1 - \frac{1}{3} + \cdots$. **6.** $5 - \frac{1}{3} + \cdots$.

7. $\frac{4}{3} - \frac{2}{3} + \cdots$. **8.** $1 + \frac{1}{4} + \cdots$. **9.** $4 + \frac{12}{5} + \cdots$.

10. $\sqrt{\frac{3}{2}} + \sqrt{\frac{2}{3}} + \cdots$. **11.** $2\sqrt{\frac{1}{3}} + 1 + \cdots$. **12.** $\sqrt{.2} + \sqrt{\frac{1}{125}} + \cdots$.

13. $1 + x + x^2 + \cdots$, when $x < 1$. **14.** $1 + \frac{1}{x} + \frac{1}{x^2} + \cdots$, when $x > 1$.

15. $\frac{n}{n+1} + \frac{n}{(n+1)^2} + \cdots$. **16.** $\frac{a^2+2}{a^3-1} + \frac{a^2+2}{(a^2+a+1)^2} + \cdots$.

Find the value of each of the following repeating decimals:

17. $.44 \cdots$. **18.** $.99 \cdots$. **19.** $.2727 \cdots$. **20.** $.015015 \cdots$.

21. $.199 \cdots$. **22.** $1.0909 \cdots$. **23.** $.122323 \cdots$. **24.** $.201475475 \cdots$.

Verify each of the following identities:

25. $\sqrt{.44 \cdots} = .66 \cdots$. **26.** $\sqrt{.6944 \cdots} = .833 \cdots$.

Geometrical Means.

8. A **Geometrical Mean** between two numbers is a number, in value between the two, which forms with them a geometrical progression.

E.g., $+2$, or -2, is a geometrical mean between 1 and 4, between $\frac{1}{2}$ and 8.

Let G be the geometrical mean between a and b.

Then by definition of a geometrical progression,

$$\frac{G}{a} = \frac{b}{G}; \text{ whence } G = \sqrt{(ab)}.$$

That is, the geometrical mean between two numbers is the square root of their product.

Ex. 1. Find the geometrical mean between 1 and $\frac{4}{9}$. We have

$$G = \sqrt{(1 \times \tfrac{4}{9})} = \pm \tfrac{2}{3}.$$

9. **Geometrical Means** between two numbers are numbers, in value between the two, which form with them a geometrical progression. *E.g.*, 4 and 16 are two geometrical means between 1 and 64; and 2, 4, 8, 16, 32 are five geometrical means between 1 and 64.

If n geometrical means be inserted between a and b, we have a geometrical progression of $n + 2$ terms.

Consequently, by (I.),

$$b = ar^{n+1}, \text{ or } r = \sqrt[n+1]{\frac{b}{a}}.$$

The progression is therefore,

$$a,\ a\sqrt[n+1]{\frac{b}{a}},\ a\sqrt[n+1]{\left(\frac{b}{a}\right)^2},\ a\sqrt[n+1]{\left(\frac{b}{a}\right)^3},\ \cdots\ b,$$

or

$$a,\ \sqrt[n+1]{(a^n b)},\ \sqrt[n+1]{(a^{n-1}b^2)},\ \sqrt[n+1]{(a^{n-2}b^3)},\ \cdots\ b.$$

Ex. 2. Insert five geometrical means between 2 and 11.

We have $11 = 2r^6$, whence $r = \sqrt[6]{\tfrac{11}{2}}$.

The required means are

$$2\sqrt[6]{\tfrac{11}{2}},\ 2\sqrt[6]{(\tfrac{11}{2})^2},\ 2\sqrt[6]{(\tfrac{11}{2})^3},\ 2\sqrt[6]{(\tfrac{11}{2})^4},\ 2\sqrt[6]{(\tfrac{11}{2})^5}.$$

EXERCISES VI.

Insert a geometrical mean between

1. 2 and 8.
2. 12 and 3.
3. $\frac{1}{5}$ and $\frac{1}{125}$.
4. $\sqrt{a}$ and $\sqrt{(2a)}$.
5. $75m^3$ and $3mn^4$.
6. $\frac{p}{q}$ and $\frac{q}{p}$.
7. $(a-b)^2$ and $(a+b)^2$.
8. $(a^2+1)(a^2-1)^{-1}$ and $\frac{1}{4}(a^4-1)$.
9. Insert 5 geometrical means between 2 and 1458.
10. Insert 7 geometrical means between 2 and 512.
11. Insert 6 geometrical means between 3 and -384.
12. Insert 6 geometrical means between 5 and -640.
13. Insert 8 geometrical means between 4 and $-\frac{19683}{128}$.
14. Insert 9 geometrical means between 1 and $\frac{1024}{59049}$.

15. Insert 4 geometrical means between $\frac{3m^2}{5n^3}$ and $\frac{18.75n^2}{16m^3}$.

16. Insert 3 geometrical means between $\frac{a+b}{a-b}$ and $\left(\frac{a-b}{a+b}\right)^3$

10. **The arithmetical mean of two unequal numbers is greater than their geometrical mean.**

Let a and b be the given numbers.

We are to prove

$$A > G,$$

or $$\tfrac{1}{2}(a+b) > \sqrt{(ab)}.$$

Since $(\sqrt{a}-\sqrt{b})^2$ is always positive, if $a \neq b$,

we have $$a - 2\sqrt{(ab)} + b > 0,$$

or $$a + b > 2\sqrt{(ab)},$$

or $$\frac{a+b}{2} < \sqrt{(ab)}.$$

Consequently $A - G$ is always positive, and $A > G$.

Problems.

11. Pr. 1. In a geometrical progression of four terms, the sum of the first and fourth terms is 455, and the sum of the second and third terms is 140. What are the terms?

By the first condition, $a_1 + a_1r^3 = 455,$

or $$a_1(1 + r^3) = 455. \qquad (1)$$

By the second condition, $a_1r + a_1r^2 = 140,$

or $$a_1r(1 + r) = 140. \qquad (2)$$

Dividing (1) by (2), $$\frac{1 - r + r^2}{r} = \frac{13}{4}, \qquad (3)$$

whence $r = 4$, and $\frac{1}{4}$.

Therefore, from (2), $a_1 = 7$, and 448.

Consequently, the two series are

7, 28, 112, 448;

and 448, 112, 28, 7.

Pr. 2. The sum of the terms of a geometrical progression of five terms is 484; the sum of the second and fourth terms is 120. What is the progression?

Let $\frac{x}{r^2}$, $\frac{x}{r}$, x, xr, xr^2 be the required terms.

By the first condition, $\frac{x}{r^2}+\frac{x}{r}+x+xr+xr^2=484.$ (1)

By the second condition, $\frac{x}{r}+xr=120.$ (2)

Subtracting (2) from (1), $\frac{x}{r^2}+x+xr^2=364.$ (3)

Dividing by x, and adding 1 to both members,

$$\frac{1}{r^2}+2+r^2=\frac{364}{x}+1. \quad (4)$$

Squaring (2), $x^2\left(\frac{1}{r^2}+2+r^2\right)=14400,$ (5)

or $$\frac{1}{r^2}+2+r^2=\frac{14400}{x^2}. \quad (6)$$

Equating second members of (4) and (6),

$$\frac{364}{x}+1=\frac{14400}{x^2}, \quad (7)$$

or $$x^2+364x=14400. \quad (8)$$

From (8), $x=36$, and -400.

Substituting 36 for x in (2), we obtain

$$r=3, \text{ and } \tfrac{1}{3}.$$

The value, -400, of x gives imaginary values of r, and therefore must be rejected.

When $x=36$ and $r=3$, the series is 4, 12, 36, 108, 324; when $x=36$ and $r=\frac{1}{3}$, the series is 324, 108, 36, 12, 4.

Pr. 3. If the numbers 5, 6, 9, and 15, respectively, be added to the terms of an arithmetical progression of four terms, the resulting series will be a geometrical progression. What is the arithmetical progression?

Let $a-3d$, $a-d$, $a+d$, $a+3d$ represent the required numbers.

Then, by the condition of the problem,

$$a-3d+5,\ a-d+6,\ a+d+9,\ a+3d+15$$

is a geometrical progression.

Therefore, $$\frac{a-d+6}{a-3d+5}=\frac{a+d+9}{a-d+6},$$

or $$4d^2-2a+10d=9; \tag{1}$$

and $$\frac{a+d+9}{a-d+6}=\frac{a+3d+15}{a+d+9},$$

or $$4d^2-3a+15d=9. \tag{2}$$

Solving (1) and (2), we obtain

$$a=\tfrac{15}{2},\ d=\tfrac{3}{2};\ \text{and}\ a=-\tfrac{15}{2},\ d=-\tfrac{3}{2}.$$

Consequently, the two series are

$$3,\ 6,\ 9,\ 12;$$

and $$-3,\ -6,\ -9,\ -12.$$

EXERCISES VII.

1. The first term of a G. P. of six terms is 768, and the last term is one-sixteenth of the fourth term. What is the sum of the six terms of the progression?

2. The first term of a G. P. of ten terms is 3, and the sum of the first three terms is one-eighth of the sum of the next three terms. Find the elements of the progression.

3. The twelfth term of a G. P. is 1536, and the fourth term is 6. What is the ratio, and the sum of the first eleven terms?

4. The seventh term of a G. P. is 192, and the tenth term is -1536. What is the twelfth term, and the sum of the first five terms?

5. The tenth term of a G. P. is $-\frac{1}{512}$, and the fifth term is $\frac{243}{512}$. What is the first term, and the sum of the first fifteen terms?

6. The sum of the first and fourth terms of a G. P. of ten terms is 455, and the sum of the second and third terms is 140. What are the elements of the progression?

7. In a G. P. of eight terms, the sum of the first seven terms is $444\frac{1}{2}$, and is to the sum of the last seven terms as 1 to 2. Find the elements of the progression.

8. The sum of the first and third terms of a G. P. of four terms is a, and the sum of the second and fourth terms is b. What are the elements of the progression?

9. The sum of the first three terms of a G. P. of six terms is 117, and the sum of the last three terms is $\frac{13}{8}$. What are the elements of the progression?

10. The sum of the first six terms of a G. P. of seven terms is $157\frac{1}{2}$, and the sum of the last six terms is 315. What are the elements of the progression?

11. The sum of the first four terms of a G. P. is 15, and the sum of the terms from the second to the fifth inclusive is 30. What is the first term, and the ratio?

12. Find the elements of a G. P. of six terms whose first term is 1, and the sum of whose first six terms is twenty-eight times the sum of the first three terms.

13. The sum of the first three terms of a G. P. is 21, and the sum of their squares is 189. What is the first term?

14. The sum of the terms of a G. P. of four terms is 30, and the ratio of the last term to the sum of the second and third terms is $\frac{4}{3}$. What are the elements of the progression?

15. The product of the first and third terms of a G. P. of seven terms is 64, and the sum of the fourth and sixth terms is 6. What are the elements of the progression?

16. The product of the first three terms of a G. P. is 216, and the sum of their cubes is 1971. What is the first term, and the ratio?

17. Find the mth and the nth term of a geometrical progression whose $(m+n)$th term is k, and $(m-n)$th term is l.

18. If the numbers 1, 1, 3, 9 be added to the first four terms of an A. P., respectively, the resulting terms will be a G. P. What is the first term, and the common difference of the A. P.?

19. A G. P. and an A. P. have a common first term 3, the difference between their second terms is 6, and their third terms are equal. What is the ratio of the G. P., and the common difference of the A. P.?

20. If from each term of a G. P. of four terms the corresponding term of an A. P. of four terms be subtracted, the remainders will be 1, 5, 19, 53 respectively. What are the elements of each progression?

21. The sum of the eight terms of an A. P., whose first term is 1, is 3,294,176. The first and last terms of a G. P. of eight terms are equal to the corresponding terms of the A. P. Find the fifth term of the G. P.

22. The first and last terms of an A. P. of fifteen terms are equal to the corresponding terms of a G. P. of fifteen terms, and the ninth term of the A. P. is equal to the eighth term of the G. P. What is the ratio of the G. P.?

23. The sum of the products of the corresponding terms of two geometrical progressions having the same number of terms is 547.28. In the first progression the first term is 20 and the ratio is $3\frac{1}{2}$, and in the second progression the first term is $2\frac{1}{2}$ and the ratio is $\frac{2}{5}$. Find the last term of the first progression, and the sum of the terms of the second progression.

24. The ratios of two geometrical progressions having the same number of terms are $\frac{3}{10}$ and $\frac{1}{5}$ respectively, and their first terms are equal. The last term of the first progression is 243, and the last term of the second is 32. Find the elements of each progression.

25. Show that, if all the terms of a G. P. be multiplied by the same number, the resulting series will form a G. P.

26. Show that the series whose terms are the reciprocals of the terms of a G. P. is a G. P.

27. Show that the product of the first and last terms of a G. P. is equal to the product of any two terms which are equally distant from the first and last terms respectively.

28. If the numbers a, b, c, d form a G. P., show that

$$(a-d)^2=(b-c)^2+(c-a)^2+(d-b)^2.$$

29. If the numbers a, b, c, d form a G. P., show that

$$(a^2+b^2+c^2)(b^2+c^2+d^2)=(ab+bc+cd)^2.$$

30. If a, b, c form a G. P., show that

$$a^2b^2c^2\left(\frac{1}{a^3}+\frac{1}{b^3}+\frac{1}{c^3}\right)=a^3+b^3+c^3.$$

31. A forest now contains 14,641 trees. If the yearly increase has been 10%, how many years ago did the forest contain 10,000 trees?

32. A cask contains 150 gallons of wine. If 4 gallons of wine are drawn from the cask and replaced by water, and if the process is repeated ten times, how many gallons of pure wine will remain in the cask?

33. The number of inhabitants in a city increased in five years from 5120 to 12,500. If the rate of increase was the same from year to year, what was the increase in the population the fourth year?

34. A merchant's investment yields him each year after the first three times as much as the preceding year. If his investment paid him $9720 in four years, how much did he realize the first year and the fourth year?

35. Shah Sheran commanded the inventor of the game of chess to name a reward. The inventor asked 1 grain of wheat for the first square on the chess board, 2 grains for the second square, 4 grains for the third square, 8 grains for the fourth square, and so on. The number of squares on a chess board being 64, find the number of digits in the number of grains of wheat which the inventor of chess named as a reward, and find the last six digits on the left in the number.

36. On one of the sides of an acute angle a point is taken a feet from the vertex; from this point a perpendicular is let fall on the second side, cutting off b feet from the vertex. From the foot of this perpendicular a perpendicular is let fall on the first side, and from the foot of this perpendicular a third perpendicular is let fall on the second side, and so on indefinitely. Find the sum of all the perpendiculars.

37. Given a square whose side is $2a$. The middle points of its adjacent sides are joined by lines forming a second square inscribed in the first. In the same manner a third square is inscribed in the second, a fourth in the third, and so on indefinitely. Find the sum of the perimeters of all the squares.

§ 4. HARMONICAL PROGRESSION.

1. A **Harmonical Progression** is a series the reciprocals of whose terms form an arithmetical progression.

E.g., $$1 + \tfrac{1}{2} + \tfrac{1}{3} + \tfrac{1}{4} + \cdots$$

is a harmonical progression, since

$$1 + 2 + 3 + 4 + \cdots$$

is an arithmetical progression.

Consequently to every harmonical progression there corresponds an arithmetical progression, and *vice versa.*

2. *If three numbers be in harmonical progression, the ratio of the difference between the first and second terms to the difference between the second and third terms is equal to the ratio of the first term to the third term.*

Let the three numbers a, b, c be in harmonical progression. Then we are to prove

$$\frac{a-b}{b-c} = \frac{a}{c}.$$

By the definition of a harmonical progression

$$\frac{1}{a}, \frac{1}{b}, \frac{1}{c}$$

are in arithmetical progression. Consequently

$$\frac{1}{b} - \frac{1}{a} = \frac{1}{c} - \frac{1}{b},$$

or $$\frac{a-b}{ab} = \frac{b-c}{bc},$$

whence $$\frac{a-b}{b-c} = \frac{a}{c}.$$

3. Any term of a harmonical progression is obtained by finding the same term of the corresponding arithmetical progression and taking its reciprocal.

Ex. Find the eleventh term of the harmonical progression $4, 2, \frac{4}{3}, \cdots$.

The corresponding arithmetical progression is

$$\frac{1}{4}, \frac{1}{2}, \frac{3}{4}, \cdots,$$

and its eleventh term is $\frac{11}{4}$.

Consequently the eleventh term of the given harmonical progression is $\frac{4}{11}$.

4. No formula has been derived for the sum of n terms of a harmonical progression.

5. A Harmonical Mean between two numbers is a number, in value between them, which forms with the two numbers a harmonical progression.

E.g., $\frac{3}{2}$ is a harmonical mean between $\frac{1}{2}$ and $-\frac{3}{2}$.

Let H stand for the harmonical mean between a and b, then $\frac{1}{H}$ is an arithmetical mean between $\frac{1}{a}$ and $\frac{1}{b}$. Consequently

$$\frac{1}{H} = \frac{\frac{1}{a} + \frac{1}{b}}{2}, \text{ or } H = \frac{2ab}{a+b}.$$

Ex. Insert a harmonical mean between 2 and 5.

We have $$H = \frac{2 \times 2 \times 5}{2+5} = \frac{20}{7}.$$

6. Harmonical Means between two numbers are numbers, in value between the two, which form with them a harmonical progression.

E.g., $\frac{3}{2}, 1, \frac{3}{4}, \frac{3}{5}, \frac{1}{2}$ are five harmonical means between 3 and $\frac{3}{7}$; 1 and $\frac{3}{5}$ are two harmonical means between 3 and $\frac{3}{7}$.

To insert n harmonical means between a and b, we insert n arithmetical means between $\frac{1}{a}$ and $\frac{1}{b}$, and take their reciprocals. The n arithmetical means are

$$\frac{n\frac{1}{a}+\frac{1}{b}}{n+1},\ \frac{(n-1)\frac{1}{a}+2\frac{1}{b}}{n+1},\ \frac{(n-2)\frac{1}{a}+3\frac{1}{b}}{n+1},\ \cdots$$

or $$\frac{a+nb}{(n+1)ab},\ \frac{2a+(n-1)b}{(n+1)ab},\ \frac{3a+(n-2)b}{(n+1)ab},\ \cdots$$

Consequently the required harmonical means are

$$\frac{(n+1)ab}{a+nb},\ \frac{(n+1)ab}{2a+(n-1)b},\ \frac{(n+1)ab}{3a+(n-2)b}.$$

Ex. 2. Insert four harmonical means between 1 and 10.

We have first to insert four arithmetical means between 1 and $\frac{1}{10}$, and obtain

$$\tfrac{41}{50},\ \tfrac{32}{50},\ \tfrac{23}{50},\ \tfrac{14}{50}.$$

The required harmonical means are therefore

$$\tfrac{50}{41},\ \tfrac{50}{32},\ \tfrac{50}{23},\ \tfrac{50}{14}.$$

Relation between the Arithmetical, Geometrical, and Harmonical Means between Two Given Numbers.

7. If the two numbers be a and b, then

$$A=\frac{a+b}{2},\ G=\sqrt{(ab)},\ H=\frac{2ab}{a+b}.$$

We evidently have

$$A\cdot H=ab=G^2,$$

or $$G=\sqrt{(AH)}.$$

That is, *the geometrical mean between two numbers is also the geometrical mean between their arithmetical and harmonical means.*

Also, from $G=\sqrt{(AH)}$, it follows that G is intermediate in value between A and H.

But since $G<A$, by § 3, Art. 10, therefore $G>H$.

That is, the arithmetical, geometrical, and harmonical means are in descending order of magnitude.

Problems.

8. Pr. 1. The geometrical mean between two numbers is $\frac{1}{2}$, and the harmonical mean is $\frac{2}{5}$. What are the numbers?

Let x and y represent the two numbers.

Then $\sqrt{(xy)} = \frac{1}{2}$, or $xy = \frac{1}{4}$; (1)

and $\dfrac{2\,xy}{x+y} = \dfrac{2}{5}$, or $5\,xy = x + y$. (2)

Solving (1) and (2), we obtain $x = 1$, $y = \frac{1}{4}$, and $x = \frac{1}{4}$, $y = 1$.

Pr. **2.** If to each of three numbers in geometrical progression the second number be added, the resulting series will form a harmonical progression.

Let $\dfrac{a}{r}$, a, ar represent the three numbers.

Then we are to prove that

$$\frac{a}{r} + a,\ 2\,a,\ ar + a$$

is a harmonical progression; that is, that

$$\frac{r}{a(1+r)},\ \frac{1}{2\,a},\ \frac{1}{a(1+r)}$$

is an arithmetical progression.

We have
$$\frac{\dfrac{r}{a(1+r)} + \dfrac{1}{a(1+r)}}{2} = \frac{1}{2\,a}.$$

That is, $\dfrac{1}{2\,a}$ is an arithmetical mean between $\dfrac{r}{a(1+r)}$ and $\dfrac{1}{a(1+r)}$. Consequently, $\dfrac{a}{r} + a$, $2\,a$, $ar + a$ is a harmonical progression.

EXERCISES VIII.

Find the last term of each of the following harmonical progressions:

1. $1 + \frac{2}{3} + \frac{1}{2} + \cdots$ to 8 terms.

2. $-\frac{1}{6} - \frac{3}{16} - \frac{3}{14} - \cdots$ to 10 terms.

3. $\frac{1}{3} + \frac{1}{8} + \frac{1}{13} + \cdots$ to 15 terms.

4. $-2 + 4 + 1 + \cdots$ to 18 terms.

5. $2 - 2 - \frac{2}{3} - \cdots$ to 11 terms.

6. $8 - \frac{8}{9} - \frac{8}{17} - \cdots$ to 16 terms.

7. $\frac{1}{a}+\frac{1}{2a}+\frac{1}{3a}+\cdots$ to 25 terms.

8. $\frac{1}{\sqrt{2}}+\frac{1}{1+\sqrt{2}}+\frac{1}{2+\sqrt{2}}+\cdots$ to 30 terms.

9. $\frac{1}{\sqrt{a}+\sqrt{b}}+\frac{1}{\sqrt{a}-\sqrt{b}}+\frac{1}{\sqrt{a}-3\sqrt{b}}+\cdots$ to 19 terms.

Find the harmonical mean between

10. 2 and 4. **11.** -3 and 4. **12.** $\frac{1}{7}$ and $\frac{1}{8}$.

13. $\frac{1}{x-1}$ and $-\frac{1}{x+1}$. **14.** $\frac{a-b}{a+b}$ and $\frac{a+b}{a-b}$.

15. Insert 5 harmonical means between 5 and $\frac{1}{5}$.

16. Insert 10 harmonical means between 3 and $\frac{1}{4}$.

17. Insert 4 harmonical means between $\frac{1}{2}$ and $\frac{1}{17}$.

18. Insert 4 harmonical means between -7 and $\frac{1}{2}$.

19. Insert 5 harmonical means between $\frac{1}{a+b}$ and $\frac{1}{a-11b}$.

20. Insert 3 harmonical means between $\frac{m-n}{m+n}$ and $\frac{m+n}{m-n}$.

21. If x^2, y^2, z^2 be in A. P., prove that $y+z$, $z+x$, and $x+y$ are in H. P.

22. If y be the harmonical mean between x and z, prove that

$$\frac{1}{y-x}+\frac{1}{y-z}=\frac{1}{x}+\frac{1}{z}.$$

23. The arithmetical mean between two numbers is 6, and the harmonical mean is $\frac{35}{6}$. What are the numbers?

24. If one number exceeds another by two, and if the arithmetical mean exceeds the harmonical mean by $\frac{1}{10}$, what are the numbers?

25. The seventh term of a harmonical progression is $\frac{1}{15}$, and the twelfth term is $\frac{1}{25}$. What is the twentieth term?

26. The tenth term of a harmonical progression is $\frac{1}{5}$, and the twentieth term is $\frac{1}{10}$. What is the first term?

CHAPTER XXVII.

THE BINOMIAL THEOREM FOR POSITIVE INTEGRAL EXPONENTS.

1. The expansion of the powers of a binomial, from the first to the sixth inclusive, were given in Ch. VI., § 1, Art. 10, and the laws governing the expansion of these powers were there stated.

We will now first show that these laws hold for the expansion of the seventh power by multiplying both members of the identity

$$(a+b)^6=a^6+6\,a^5b+15\,a^4b^2+20\,a^3b^3+15\,a^2b^4+6\,ab^5+b^6$$

by $$a+b.$$

We then have

$$\begin{array}{r|r|r|r|r|r|l}
(a+b)^7=a^7+6 & a^6b+15 & a^5b^2+20 & a^4b^3+15 & a^3b^4+\ 6 & a^2b^5+1 & ab^6 \\
+1 & 6 & 15 & 20 & 15 & 6 & \quad +b^7
\end{array}$$

$$=a^7+7\,a^6b+21\,a^5b^2+35\,a^4b^3+35\,a^3b^4+21\,a^2b^5+7\,ab^6+b^7.$$

This result shows that the laws (i.)–(vi.) hold for the expansion of $(a+b)^7$. As yet, however, we cannot infer that these laws hold for the eighth power without multiplying the expansion of the seventh power by $a+b$; nor for the ninth power without next multiplying the expansion of the eighth power by $a+b$; and so on.

If, however, we prove that, provided the laws hold for any particular power, they hold for the next higher power, we can infer, without further proof, that because the laws hold for the sixth power, they hold also for the seventh; then that because they hold for the seventh, they hold also for the eighth, and so on to any higher power.

2. If the laws (i.)–(vi.) hold for the rth power, we have

$$(a+b)^r = a^r + ra^{r-1}b + \frac{r(r-1)}{1\cdot 2}a^{r-2}b^2 + \frac{r(r-1)(r-2)}{1\cdot 2\cdot 3}a^{r-3}b^3 + \cdots.$$

Notice that only the first four terms of the expansion are written. But it is often necessary to write any term (the kth, say) without having written all the preceding terms.

To derive the kth (*any*) term, observe that the following laws hold for each term of the expansion:

(i.) *The exponent of b is one less than the number of the term (counting from the left).*

Thus, in the first term we have $b^{1-1} = b^0 = 1$; in the second, $b^{2-1} = b$; in the tenth term, $b^{10-1} = b^9$; and in the kth term, b^{k-1}.

(ii.) *The exponent of a is equal to the binomial exponent less the exponent of b.*

Thus, in the first term we have $a^{r-0} = a^r$; in the second term a^{r-1}; in the tenth term a^{r-9}; and in the kth term $a^{r-(k-1)}$, $= a^{r-k+1}$.

(iii.) *The number of factors (beginning with 1 and increasing by 1) in the denominator of each coefficient, and the number of factors (beginning with r and decreasing by 1) in the numerator of each coefficient, is equal to the exponent of b in that term.*

Thus, in the coefficient of the second term the denominator is 1 and the numerator is r; in that of the second term the denominator is $1\cdot 2$ and the numerator is $r(r-1)$; in the tenth term the denominator is $1\cdot 2\cdots 9$ and the numerator is $r(r-1)\cdots(r-8)$; and in the kth term the denominator is $1\cdot 2\cdot 3\cdots(k-1)$, and the numerator is

$$r(r-1)\cdots[r-(k-2)], = r(r-1)\cdots(r-k+2).$$

Therefore the kth term in the expansion of $(a+b)^r$ is

$$\frac{r(r-1)(r-2)\cdots(r-k+2)}{1\cdot 2\cdot 3\cdots(k-1)}a^{n-k+1}b^{k-1}.$$

In like manner, any other term can be written.

Thus, the $(k-1)$th term is

$$\frac{r(r-1)(r-2)\cdots(r-k+3)}{1\cdot 2\cdot 3\cdots(k-2)}a^{r-k+2}b^{k-2}.$$

3. We can now prove that, if the laws (i.)–(vi.) hold for $(a+b)^r$, they hold also for $(a+b)^{r+1}$; that is, if they hold for any power they hold for the next higher power. Assuming, then, that the laws hold for $(a+b)^r$, we have

$$(a+b)^r = a^r + ra^{r-1}b + \frac{r(r-1)}{1\cdot 2}a^{r-2}b^2 + \cdots$$

$$+\frac{r(r-1)(r-2)\cdots(r-k+3)}{1\cdot 2\cdot 3\cdots(k-2)}a^{r-k+2}b^{k-2}$$

$$+\frac{r(r-1)(r-2)\cdots(r-k+3)(r-k+2)}{1\cdot 2\cdot 3\cdots(k-2)(k-1)}a^{r-k+1}b^{k-1}+\cdots.$$

Observe that the first three terms of the expansion are written, then all terms are omitted, except the $(k-1)$th and the kth.

Multiplying the expansion of $(a+b)^r$ by $(a+b)$, we obtain

$$(a+b)^{r+1} = a^{r+1} + ra^r b + \frac{r(r-1)}{1\cdot 2}a^{r-1}b^2 + \cdots$$

$$+\frac{r(r-1)\cdots(r-k+3)}{1\cdot 2\cdots(k-2)}a^{r-k+3}b^{k-2}$$

$$+\frac{r(r-1)\cdots(r-k+2)}{1\cdot 2\cdots(k-1)}a^{r-k+2}b^{k-1}+\cdots$$

$$+a^r b + ra^{r-1}b^2 + \cdots$$

$$+\frac{r(r-1)\cdots(r-k+3)}{1\cdot 2\cdots(k-2)}a^{r-k+2}b^{k-1}+\cdots$$

$$= a^{r+1} + (r+1)\,a^r b + \left[\frac{r(r-1)}{1\cdot 2}+r\right]a^{r-1}b^2 + \cdots$$

$$+\left[\frac{r(r-1)\cdots(r-k+2)}{1\cdot 2\cdots(k-1)}\right.$$

$$\left.+\frac{r(r-1)\cdots(r-k+3)}{1\cdot 2\cdots(k-2)}\right]a^{r-k+2}b^{k-1}+\cdots.$$

But $$\frac{r(r-1)}{1\cdot 2}+r=\frac{r^2-r+2r}{1\cdot 2}=\frac{(r+1)r}{1\cdot 2};$$

and $$\frac{r(r-1)\cdots(r-k+2)}{1\cdot 2\cdots(k-1)}+\frac{r(r-1)\cdots(r-k+3)}{1\cdot 2\cdots(k-2)}$$

$$=\frac{r(r-1)\cdots(r-k+2)+r(r-1)\cdots(r-k+3)(k-1)}{1\cdot 2\cdots(k-1)}$$

$$=\frac{r(r-1)\cdots(r-k+3)(r-k+2+k-1)}{1\cdot 2\cdots(k-1)}$$

$$=\frac{(r+1)r(r-1)\cdots(r-k+3)}{1\cdot 2\cdots(k-1)}.$$

Therefore,

$$(a+b)^{r+1}=a^{r+1}+(r+1)a^rb+\frac{(r+1)r}{1\cdot 2}a^{r-1}b^2+\cdots$$
$$+\frac{(r+1)r(r-1)\cdots(r-k+3)}{1\cdot 2\cdots(k-1)}a^{r-k+2}b^{k-1}+\cdots.$$

Observe that the laws (i.)–(vi.) hold for the above expansion of $(a+b)^{r+1}$. We therefore conclude that if the expansion holds for $(a+b)^r$, it also holds for $(a+b)^{r+1}$.

We know that the binomial formula holds for the seventh power, therefore it holds for the eighth; then, since it holds for the eighth, it holds for the ninth; and since it then holds for the ninth, it holds for the tenth; and so on for any succeeding power.

4. We may now write the expansion of $(a+b)^n$, wherein n is any positive integer:

$$(a+b)^n=a^n+na^{n-1}b+\frac{n(n-1)}{1\cdot 2}a^{n-2}b^2+\cdots$$
$$+\frac{n(n-1)\cdots(n-k+2)}{1\cdot 2\cdots(k-1)}a^{n-k+1}b^{k-1}+\cdots.$$

In particular, if $a=1$,

$$(1+b)^n=1^n+n\,1^{n-1}b+\frac{n(n-1)}{1\cdot 2}1^{n-2}b^2+\cdots$$
$$=1+nb+\frac{n(n-1)}{1\cdot 2}b^2+\cdots.$$

5. The expansion of $(a-b)^n$ can be at once written from that of $(a+b)^n$.

We have $$(a-b)^n=[a+(-b)]^n$$

$$=a^n+na^{n-1}(-b)+\frac{n(n-1)}{1\cdot 2}a^{n-2}(-b)^2+\cdots$$

$$+\frac{n(n-1)\cdots(n-k+2)}{1\cdot 2\cdots(k-1)}a^{n-k+1}(-b)^{k-1}+\cdots$$

$$=a^n-na^{n-1}b+\frac{n(n-1)}{1\cdot 2}a^{n-2}b^2+\cdots$$

$$\pm\frac{n(n-1)\cdots(n-k+2)}{1\cdot 2\cdots(k-1)}a^{n-k+1}b^{k-1}+\cdots.$$

Observe that the signs of the terms alternate, + and −, beginning with the first, or that the terms containing *even* powers of b are *positive*, and those containing *odd* powers of b are *negative*.

6. The coefficients in the expansion of $(a+b)^n$ are called **Binomial Coefficients.** They may be represented by the following abbreviations:

$$n=\frac{n}{1}=\binom{n}{1},$$

$$\frac{n(n-1)}{1\cdot 2}=\binom{n}{2},$$

$$\frac{n(n-1)(n-2)}{1\cdot 2\cdot 3}=\binom{n}{3},$$

$$\frac{n(n-1)\cdots(n-k+2)}{1\cdot 2\cdot 3\cdots(k-1)}=\binom{n}{k-1}.$$

Observe that the binomial coefficients have the same number of factors in numerator and denominator; that in the symbolic notation the upper number is the binomial exponent and the lower number is the number of factors in numerator and denominator; the numerator beginning with the binomial exponent and each succeeding factor decreasing by 1, the denominator beginning with 1 and each succeeding factor increasing by 1.

E.g.,
$$\binom{7}{4}=\frac{7\cdot 6\cdot 5\cdot 4}{1\cdot 2\cdot 3\cdot 4}.$$

The binomial expansion may now be written

$$(a+b)^n=a^n+\binom{n}{1}a^{n-1}b+\binom{n}{2}a^{n-2}b^2+\cdots$$
$$+\binom{n}{k-1}a^{n-k+1}b^{k-1}+\cdots+\binom{n}{n-1}ab^{n-1}+\binom{n}{n}b^n.$$

The terms beginning at the right-hand end of the expansion are determined from the principle that the lower number in any binomial coefficient is equal to the exponent of b in that term.

7. The binomial coefficients equally distant from the beginning and end of the expansion are equal.

The coefficient of the first term is 1, and that of the last term is

$$\binom{n}{n}=\frac{n(n-1)\cdots 1}{1\cdot 2\cdots(n-1)n}=1.$$

The coefficient of the second term is $\binom{n}{1}=\frac{n}{1}$, and that of the second from the end is

$$\binom{n}{n-1}=\frac{n(n-1)\cdots 2}{1\cdot 2\cdots(n-1)}=\frac{n}{1}.$$

The coefficient of the fifth term is

$$\binom{n}{4}=\frac{n(n-1)(n-2)(n-3)}{1\cdot 2\cdot 3\cdot 4},$$

and that of the fifth term from the end is $\binom{n}{n-4}$

$$=\frac{n(n-1)(n-2)(n-3)(n-4)(n-5)\cdots[n-(n-6)][n-(n-5)]}{1\cdot 2\cdot 3\cdot 4\cdot 5\cdots(n-5)(n-4)}$$
$$=\frac{n(n-1)(n-2)(n-3)(n-4)(n-5)\cdots 6\cdot 5}{1\cdot 2\cdot 3\cdot 4\cdot 5\cdots(n-5)(n-4)}$$
$$=\frac{n(n-1)(n-2)(n-3)}{1\cdot 2\cdot 3\cdot 4}.$$

In general, the coefficient of the kth term is

$$\binom{n}{k-1}=\frac{n(n-1)(n-2)\cdots[n-(k-3)][n-(k-2)]}{1\cdot 2\cdot 3\cdots(k-2)(k-1)}$$

$$=\frac{n(n-1)(n-2)\cdots(n-k+3)(n-k+2)}{1\cdot 2\cdot 3\cdots(k-2)(k-1)}.$$

Multiplying both numerator and denominator of the last fraction by $(n-k+1)(n-k)\cdots 3\cdot 2\cdot 1$, so that the numerator becomes the continued product of all the numbers from 1 to n inclusive, we obtain $\binom{n}{k-1}$

$$=\frac{n(n-1)\cdots(n-k+3)(n-k+2)\times(n-k+1)(n-k)\cdots 3\cdot 2\cdot 1}{1\cdot 2\cdots(k-2)(k-1)\times(n-k+1)(n-k)\cdots 3\cdot 2\cdot 1}.$$

The coefficient of the kth term from the end is

$$\binom{n}{n-k+1}=\frac{n(n-1)(n-2)\cdots[n-(n-k-1)][n-(n-k)]}{1\cdot 2\cdot 3\cdots(n-k)(n-k+1)}$$

$$=\frac{n(n-1)(n-2)\cdots(k+1)k}{1\cdot 2\cdot 3\cdots(n-k)(n-k+1)}$$

$$=\frac{n(n-1)(n-2)\cdots(k+2)(k+1)k\times(k-1)(k-2)\cdots 3\cdot 2\cdot 1}{1\cdot 2\cdot 3\cdots(n-k)(n-k+1)\times(k-1)(k-2)\cdots 3\cdot 2\cdot 1}.$$

Therefore $$\binom{n}{k-1}=\binom{n}{n-k+1}.$$

8. Ex. 1. Find the seventh term in the expansion of $(2x+3y)^{11}$.

In the seventh term the exponent of $3y(=b)$ is 6; the exponent of $2x(=a)$ is $11-6, =5$. The denominator of the coefficient contains six factors beginning with 1, and the numerator contains six factors beginning with 11. Therefore the seventh term is

$$\frac{11\cdot 10\cdot 9\cdot 8\cdot 7\cdot 6}{1\cdot 2\cdot 3\cdot 4\cdot 5\cdot 6}(2x)^5(3y)^6, \ =10777536x^5y^6.$$

Ex. 2. Find the first *five* terms of the expansion of

$$(a^{-\frac{1}{2}} - 2b^{-2})^{11}.$$

We have $(a^{-\frac{1}{2}} - 2b^{-2})^{11}$

$$= (a^{-\frac{1}{2}})^{11} - 11(a^{-\frac{1}{2}})^{10}(2b^{-2}) + \frac{11\cdot 10}{1\cdot 2}(a^{-\frac{1}{2}})^{9}(2b^{-2})^{2}$$

$$- \frac{11\cdot 10\cdot 9}{1\cdot 2\cdot 3}(a^{-\frac{1}{2}})^{8}(2b^{-2})^{3} + \frac{11\cdot 10\cdot 9\cdot 8}{1\cdot 2\cdot 3\cdot 4}(a^{-\frac{1}{2}})^{7}(2b^{-2})^{4} + \cdots$$

$$= a^{-\frac{11}{2}} - 22a^{-5}b^{-2} + 220a^{-\frac{9}{2}}b^{-4} - 1320a^{-4}b^{-6} + 5280a^{-\frac{7}{2}}b^{-8} + \cdots.$$

Ex. 3. Write the term containing x^{10} in the expansion of $\left(2x^2 - \frac{3}{x^{\frac{1}{2}}}\right)^{15}$.

Let k stand for the number of the required term. Then, neglecting the coefficient, we have $(2x^2)^{15-k+1}\left(\frac{3}{x^{\frac{1}{2}}}\right)^{k-1}$.

The power of x obtained from this expansion is

$$(x^2)^{15-k+1}(x^{-\frac{1}{2}})^{k-1}, = x^{32\frac{1}{2}-2\frac{1}{2}k}.$$

In order that this power of x may be equal to x^{10}, we must have

$$32\tfrac{1}{2} - 2\tfrac{1}{2}k = 10;$$

whence $k = 9$, the number of the required term.

We now have

$$\text{ninth term} = \frac{15\cdot 14\cdot 13\cdot \cdots \cdot 9\cdot 8}{1\cdot 2\cdot 3\cdot \cdots \cdot 7\cdot 8}(2x^2)^7\left(\frac{3}{x^{\frac{1}{2}}}\right)^8$$

$$= 13\cdot 11\cdot 5\cdot 9\cdot 2^7\cdot 3^8 x^{10}.$$

EXERCISES.

Write the expansion of each of the following powers:

1. $(x^3 + 3)^4$.
2. $(5ab - 3a^2)^4$.
3. $(1\frac{1}{2}x^{-2} + 1\frac{1}{3}ax)^4$.
4. $(a^{\frac{1}{2}} + 2a^2x^{-1})^4$.
5. $(2a - \sqrt{3}b)^4$.
6. $(x - \sqrt{-3y})^4$.
7. $(a^3 - b^2)^5$.
8. $(ax^{\frac{2}{3}} + 2b)^5$.
9. $(4a^{-2} - 5x)^5$.
10. $(\frac{1}{2}a^{\frac{2}{3}} + \frac{2}{3}x^2)^5$.
11. $(\sqrt{a} + \sqrt{b})^6$.
12. $\left(\frac{2}{\sqrt[3]{a^2}} - \frac{a\sqrt{a}}{2}\right)^5$.

13. $(x\sqrt{-3}+y\sqrt{-2})^5$.
14. $(n-\frac{1}{2})^6$.
15. $(x^2+2a^{-2}x)^6$.

16. $\left(n^2+\frac{2a}{n^{-1}}\right)^6$.
17. $\left(\frac{a^{\frac{3}{2}}x^2}{n}-\frac{2n^{\frac{2}{3}}}{a^2x^{-1}}\right)^6$.
18. $(2\sqrt{x}-\frac{1}{4}a^{-3}x^3)^6$.

19. $[3a^{\frac{1}{2}}-\sqrt{(-2b^3)}]^6$.
20. $(a+x^2)^7$.
21. $(\frac{1}{2}b^{3}+2bx)^7$.

22. $(3n^{\frac{4}{3}}-\frac{2}{3}n^2y^{-2})$.
23. $\left(\frac{a^2b}{2x^{\frac{1}{2}}}+\frac{2x}{a^{-1}b^2}\right)^7$.
24. $(\sqrt{a}-\sqrt{b})^7$

25. $(\sqrt{-2}+2x^{-\frac{2}{3}})^7$.
26. $(a^2+x)^8$.
27. $(a^{-3}+\frac{1}{2})^8$.

28. $(\frac{1}{2}x^{\frac{5}{3}}-2xy)^8$.
29. $(\sqrt[4]{a}+\sqrt[4]{b})^8$.
30. $(a-\sqrt{-a})^8$.

31. $(x^{-1}-x)^9$.
32. $(a^{\frac{2}{3}}+ax)^9$.
33. $(ab^{-2}-b^2x)^9$.

34. $(\sqrt[3]{a}-\sqrt[3]{b})^9$.
35. $(x^2-\sqrt{-x})^9$.
36. $\left(\sqrt{\frac{a}{n}}+\sqrt{\frac{n}{a}}\right)^9$.

37. $(x^2-1)^{10}$.
38. $(a^2b+b^{-3})^{10}$.
39. $\left(x-\frac{a}{x}\right)^{10}$.

40. $[\sqrt{(x+1)}-\sqrt{(x-1)}]^4$.
41. $[\sqrt[3]{(a+b)}+\sqrt[3]{(a-b)}]^6$.

Simplify each of the following expressions:

42. $[x+\sqrt{(x^2-1)}]^6+[x-\sqrt{(x^2-1)}]^6$.

43. $(1+\sqrt{-x})^8+(1-\sqrt{-x})^8$.
44. $(x-\sqrt{-3})^9-(x-\sqrt{-3})^9$.

Write the expansion of each of the following powers:

45. $(1-x+x^2)^3$.
46. $(1+a^{\frac{1}{2}}-a^{-2})^3$.

47. $(2-3x+x^2)^4$.
48. $(1-x\sqrt{2}+x^2\sqrt{3})^4$.

Write the

49. 3d term of $(a+b)^{15}$.
50. 5th term of $(a-b)^{16}$.

51. 3d term of $(x^3+\frac{1}{3})^{13}$.
52. 8th term of $(a^2-b^2)^{12}$.

53. 6th term of $(a^{\frac{1}{10}}+b^{\frac{1}{3}})^{15}$.
54. 7th term of $(a^n-a^{-n})^{14}$.

55. 11th term of $(a^{\frac{2}{3}}-ax^2)^{17}$.
56. 15th term of $\left(a^3+\frac{1}{a}\right)^{20}$.

57. 12th term of $(x-\sqrt{-x})^{20}$.
58. 9th term of $(\sqrt{x}-ax^{\frac{2}{3}})^{16}$.

59. 6th term of $\left(\sqrt[3]{m}-\frac{2x}{\sqrt[3]{m^2}}\right)^{12}$.

60. 10th term of $\left(x^{-\frac{3}{2}}+\frac{\sqrt{2x}}{a^{\frac{1}{2}}}\right)^{15}$.

61. 9th term of $[1 - \sqrt{(1-\sqrt{2})}]^{12}$.

62. Write the middle term of $(x\sqrt{x} - 1)^4$.

63. Write the middle term of $\left(\frac{a}{x} - x^{-\frac{1}{2}}\right)^{16}$.

64. Write the middle terms of $(a^{\frac{1}{3}} + x^{\frac{1}{2}})^9$.

65. Write the middle terms of $(\sqrt{a} - \sqrt[3]{b})^{13}$.

66. Write the term of $\left(5x^3 - \frac{3}{2x^2}\right)^9$ which contains x^{12}.

67. Write the term of $\left(2x^{\frac{5}{2}} - \frac{1}{ax}\right)^{20}$ which contains x^8.

68. Write the term of $(2a^{-3} - 27a^{\frac{3}{5}})^{15}$ which contains $a^{-30\frac{3}{5}}$

Made in United States
Cleveland, OH
05 November 2026

47645048R00416